T0327397

Statistical Intervals

WILEY SERIES IN PROBABILITY AND STATISTICS

The Wiley Series in Probability and Statistics is well established and authoritative. It covers many topics of current research interest in both pure and applied statistics and probability theory. Written by leading statisticians and institutions, the titles span both state-of-the-art developments in the field and classical methods.

Reflecting the wide range of current research in statistics, the series encompasses applied, methodological and theoretical statistics, ranging from applications and new techniques made possible by advances in computerized practice to rigorous treatment of theoretical approaches.

This series provides essential and invaluable reading for all statisticians, whether in academia, industry, government, or research.

A complete list of titles in this series appears at the end of the volume.

Statistical Intervals

A Guide for Practitioners and Researchers

Second Edition

William Q. Meeker
Department of Statistics, Iowa State University

Gerald J. Hahn
General Electric Company, Global Research Center (Retired) Schenectady, NY

Luis A. Escobar
Department of Experimental Statistics, Louisiana State University

Published by John Wiley & Sons, Inc., Hoboken, New Jersey.
Published simultaneously in Canada.

For general information on our other products and services or for technical support, please contact our Customer Care Department within the United States at (800) 762-2974, outside the United States at (317) 572-3993 or fax (317) 572-4002.

Wiley also publishes its books in a variety of electronic formats. Some content that appears in print may not be available in electronic formats. For more information about Wiley products, visit our web site at www.wiley.com.

Library of Congress Cataloging-in-Publication Data:

Names: Meeker, William Q. | Hahn, Gerald J. | Escobar, Luis A.
Title: Statistical intervals : a guide for practitioners and researchers.
Description: Second edition / William Q. Meeker, Gerald J. Hahn, Luis A.
  Escobar. | Hoboken, New Jersey : John Wiley & Sons, Inc., [2017] |
  Includes bibliographical references and index.
Identifiers: LCCN 2016053941 | ISBN 9780471687177 (cloth) | ISBN 9781118595169 (epub)
Subjects: LCSH: Mathematical statistics.
Classification: LCC QA276 .H22 2017 | DDC 519.5/4–dc23 LC record available
at https://lccn.loc.gov/2016053941

MIX
Paper from
responsible sources
FSC® C013604

To Karen, Katherine, Josh, Liam, Ayla, and my parents

W. Q. M.

To Bea, Adrienne and Lou, Susan and John, Judy and Ben, and Zachary, Eli, Sam, Leah and Eliza

G. J. H.

To my grandchildren: Olivia, Lillian, Nathaniel, Gabriel, Samuel, Emmett, and Jackson

L. A. E.

Contents

Preface to Second Edition

OVERVIEW

The first edition of *Statistical Intervals* was published twenty-five years ago. We believe the book successfully met its goal of providing a comprehensive overview of statistical intervals for practitioners and statisticians and we have received much positive feedback. Despite, and perhaps because of this, there were compelling reasons for a second edition. In developing this second edition, Bill Meeker and Gerry Hahn have been most fortunate to have a highly qualified colleague, Luis Escobar, join them.

The new edition aims to:

- Improve or expand on various previously presented statistical intervals, using methods developed since the first edition was published.

- Provide general methods for constructing statistical intervals—some of which have recently been developed or refined—for important situations beyond those previously considered.

- Provide a webpage that gives up-to-date information about available software for calculating statistical intervals, as well as other important up-to-date information.

- Provide, via technical appendices, some of the theory underlying the intervals presented in this book.

In addition to updating the original chapters, this new edition includes new chapters on

- Likelihood-based statistical intervals (Chapter 12).

- Nonparametric bootstrap statistical intervals (Chapter 13).

- Parametric bootstrap and other simulation-based statistical intervals (Chapter 14).

- An introduction to Bayesian statistical intervals (Chapter 15).

- Bayesian statistical intervals for the binomial, Poisson, and normal distributions (Chapter 16).

- Statistical intervals for Bayesian hierarchical models (Chapter 17).

The new edition also includes an additional chapter on advanced case studies (Chapter 18). This chapter further illustrates the use of the newly introduced more advanced general methods for constructing statistical intervals. In totality, well over half of this second edition is new material—an indication of how much has changed over the past twenty-five years.

The first edition tended to focus on simple methods for constructing statistical intervals in commonly encountered situations and relied heavily on tabulations, charts, and simple formulas. The new edition adds methodology that can be readily implemented using easy-to-access software and allows more complicated problems to be addressed.

The purpose and audience for the book, however, remain essentially the same and what we said in the preface to the first edition (see below) still holds. We expect the book to continue to appeal to practitioners and statisticians who need to apply statistical intervals and hope that this appeal will be enhanced by the addition of the new and updated material. In addition, we expect the new edition to have added attraction to those interested in the theory underlying the construction of statistical intervals. With this in mind, we have extended the book title to read *Statistical Intervals: A Guide for Practitioners and Researchers.*

We have added many new applications to illustrate the use of the methods that we present. As in the first edition, all of these applications are based on real data. In some of these, however, we have changed the names of the variables or the scale of the data to protect sensitive information.

Elaboration on New Methods

Chapters 3 and 4 continue to describe (and update) familiar classical statistical methods for confidence intervals, tolerance intervals, and prediction intervals for situations in which one has a simple random sample from an underlying population or process that can be adequately described by a normal distribution. The interval procedures in these chapters have the desirable property of being "exact"—their coverage probabilities (i.e., the probability that the interval constructed using the procedure will include the quantity it was designed to include) are equal to their nominal confidence levels.

For distributions other than the normal, however, we must often resort to the use of approximate procedures for setting statistical intervals. Such procedures have coverage probabilities that usually differ from their (desired or specified) nominal confidence levels. Seven new chapters (Chapters 12–18) describe and illustrate the use of procedures for constructing intervals that are usually approximate. These procedures also have applicability *for constructing statistical intervals in more complicated situations* involving, for example, nonlinear regression models, random-effects models, and censored, truncated, or correlated data, building on the significant recent research in these areas. At the time of the first edition, such advanced methods were not widely used because they were not well known, and tended to be computationally intensive for the then available computing capabilities. Also, their statistical properties had not been studied carefully. Therefore, we provided only a brief overview of such methods in Chapter 12 of the first edition. Today, such methods are considered state of the art and readily achievable computationally. The new methods generally provide coverage probabilities that are closer to the nominal confidence level than the computationally simple Wald-approximation (also known as normal-approximate) methods that are still commonly used today to calculate statistical intervals in some popular statistical computing packages.

Other major changes in the new edition include updates to Chapters 5–7:

- Chapter 5 (on distribution-free statistical intervals) includes recently developed methods for interpolation between order statistics to provide interval coverage probabilities that are closer to the nominal confidence level.

- Chapters 6 and 7 (on statistical intervals for the binomial and Poisson distributions, respectively) now include approximate procedures with improved coverage probability properties for constructing statistical intervals for discrete distributions.

In addition, we have updated the discussion in the original chapters in numerous places. For example, Chapter 1 now includes a section on statistical intervals and big data.

New Technical Appendices

Some readers of the first edition indicated that they would like to see the theory, or at least more technical justification, for the statistical interval procedures. In response, we added a series of technical appendices that provide details of the theory upon which most of the intervals are based and how their statistical properties can be computed. These appendices also provide readers additional knowledge useful in generalizing the methods and adapting them to situations not covered in this book. We maintain, however, our practitioner-oriented focus by placing such technical material into appendices.

The new appendices provide:

- Generic definitions of statistical intervals and development of formulas for computing coverage probabilities (Appendix B).

- Properties of probability distributions that are important in data analysis applications or useful in constructing statistical intervals (Appendix C).

- Some generally applicable results from statistical theory and their use in constructing statistical intervals, including an outline of the general maximum likelihood theory concepts used in Chapter 12 and elsewhere (Appendix D).

- An outline of the theory for constructing statistical intervals for parametric distributions based on pivotal quantities used in Chapters 3, 4, and 14 (Appendix E).

- An outline of the theory for constructing statistical intervals for parametric distributions based on generalized pivotal quantities used in Chapter 14 (Appendix F).

- An outline of the theory for constructing distribution-free intervals based on order statistics, as presented in Chapter 5 (Appendix G).

- Some basic results underlying the construction of the Bayesian intervals used in Chapters 15, 16, and 17 (Appendix H).

- Derivation of formulas to compute the probability of successfully passing a (product) demonstration test based on statistical intervals described in Chapter 9 (Appendix I).

Similar to the first edition, Appendices A and J of the new edition provide, respectively, a summary of notation and acronyms and important tabulations for constructing statistical intervals.

Computer Software

Many commercial statistical software products (e.g., JMP, MINITAB, SAS, and SPSS) compute statistical intervals. New versions of these packages with improved capabilities for constructing statistical intervals, such as those discussed in this book, are released periodically. Therefore, instead of directly discussing current features of popular software packages—which might become rapidly outdated—we provide this information in an Excel spreadsheet accessible from the book's webpage and plan to update this webpage to keep it current.

In many parts of this book we show how to use the open-source R system (http://www.r-project.org/) as a sophisticated calculator to compute statistical intervals. To supplement the capabilities in R, we have developed an R package StatInt that contains some additional functions that are useful for computing statistical intervals. This package, together with its documentation, can be downloaded (for free) from this book's webpage.

More on Book's Webpage

The webpage for this book, created by Wiley, can be found at www.wiley.com/go/meeker/intervals. In addition to the link to the StatInt R package and the Excel spreadsheet on current statistical interval capabilities of popular software, this webpage provides some tables and figures from the first edition that are omitted in the current edition, as well as some additional figures and tables, for finding statistical intervals.

We plan to update this webpage periodically by adding new materials and references, (numerous, we hope) reader comments and experiences, and (few, we hope) corrections.

Summary of Changes from First Edition

Principally for readers of the first edition, we summarize below the changes we have made in the new edition. Chapters 1–10 maintain the general structure of the first edition, but, as we have indicated, include some important updates, and minor changes in the notation, organization, and presentation. Also, new Chapter 11 is an update of old Chapter 13. To complement Chapter 11, we have added the new Chapter 18 which provides advanced case studies that require use of the methods presented in the new chapters. First edition Chapters 11 ("A Review of Other Statistical Intervals") and 12 ("Other Methods for Setting Statistical Intervals") have been omitted in the new edition. The old Chapter 12 is largely superseded and expanded upon by the new Chapters 12–18. Our previous comments in the old Section 11.1 (on simultaneous statistical intervals) now appear, in revised form, in Section 2.9. Some material from the old Sections 11.4 ("Statistical Intervals for Linear Regression Analysis") and 11.5 ("Statistical Intervals for Comparing Populations and Processes") is now covered in the new Sections 4.13 and 4.14, respectively. Most remaining material in the old Chapter 11 has been excluded in the new edition because the situations discussed can generally be better addressed from both a statistical and computational perspective by using the general methods in the new chapters. To make room for the added topics, we have dropped from the earlier edition various tables that are now, for the most part, obsolete, given the readily available computer programs to construct statistical intervals. We do, however, retain those tables and charts that continue to be useful and that make it easy to compute statistical intervals without computer software. In addition, the webpage provides some tabulations that were in the first edition, but not in this edition. We also omit Appendix C of the first edition ("Listing of Computer Subroutines for Distribution-Free Statistical Intervals"). This material has been superseded by the methods described in Chapter 5.

Happy reading!

WILLIAM Q. MEEKER
GERALD J. HAHN
LUIS A. ESCOBAR
June 15, 2016

Preface to First Edition

Engineers, managers, scientists, and others often need to draw conclusions from scanty data. For example, based upon the results of a limited sample, one might need to decide whether a product is ready for release to manufacturing, to determine how reliable a space system really is, or to assess the impact of an alleged environmental hazard. Sample data provide uncertain results about the population or process of interest. Statistical intervals quantify this uncertainty by what is referred to, in public opinion polls, as "the margin of error." In this book, we show how to compute such intervals, demonstrate their practical applications, and clearly state the assumptions that one makes in their use. We go far beyond the discussion in current texts and provide a wide arsenal of tools that we have found useful in practical applications.

We show in the first chapter that an essential initial step is to assure that statistical methods are applicable. This requires the assumption that the data can be regarded as a random sample from the population or process of interest. In evaluating a new product, this might necessitate an evaluation of how and when the sample units were built, the environment in which they were tested, the way they were measured—and how these relate to the product or process of interest. If the desired assurance is not forthcoming, the methods of this book might provide merely a lower bound on the total uncertainty, reflecting only the sampling variability. Sometimes, our formal or informal evaluations lead us to conclude that the best way to proceed is to obtain added or improved data through a carefully planned investigation.

Next, we must define the specific information desired about the population or process of interest. For example, we might wish to determine the percentage of nonconforming product, the mean, or the 10th percentile, of the distribution of mechanical strength for an alloy, or the maximum noise that a customer may expect for a future order of aircraft engines.

We usually do not have unlimited data but need to extract the maximum information from a small sample. A single calculated value, such as the observed percentage of nonconforming units, can then be regarded as a "point estimate" that provides a best guess of the true percentage of nonconforming units for the sampled process or population. However, we need to quantify the uncertainty associated with such a point estimate. This can be accomplished by a statistical interval. For example, in determining whether a product design is adequate, our calculations might show that we can be "reasonably confident" that if we continue to build, use, and measure the product in the same way as in the sample, the long-run percentage of nonconforming units will be between 0.43 and 1.57%. Thus, if our goal is a product with a percentage nonconforming of 0.10% or less, the calculated interval is telling us that additional improvement is needed—since even an optimistic estimate of the nonconforming product for the sampled population or process is 0.43%. On the other hand, should we be willing to accept, at least at first, 2% nonconforming product, then initial product release might be justified (presumably, in parallel with continued product improvement), since this value exceeds our most pessimistic estimate of 1.57%. Finally, if our goal had been to have less than 1% nonconforming product, our results are inconclusive and suggest the need for additional data.

Occasionally, when the available sample is huge (or the variability is small), statistical uncertainty is relatively unimportant. This would be the case, for example, if our calculations show that the proportion nonconforming units for the sampled population or process is between 0.43 and 0.45%. More frequently, we have very limited data and obtain a relatively "huge" statistical interval, e.g., 0.43 to 37.2%. Even in these two extreme situations, the statistical interval is useful. In the first case, it tells us that, if the underlying assumptions are met, the data are sufficient for most practical needs. In the second case, it indicates that unless more precise methods for analysis can be found, the data at hand provide very little meaningful information.

In each of these examples, quantifying the uncertainty due to random sampling is likely to be substantially more informative to decision makers than obtaining a point estimate alone. Thus, statistical intervals, properly calculated from the sample data, are often of paramount interest to practitioners and their management (and are usually a great deal more meaningful than statistical significance or hypothesis tests).

Different practical problems call for different types of intervals. To assure useful conclusions, it is imperative that the statistical interval be appropriate for the problem at hand. Those who have taken one or two statistics courses are aware of confidence intervals to contain, say, the mean and the standard deviation of a sampled population or a population proportion. Some practitioners may also have been exposed to confidence and prediction intervals for regression models. These, however, are only a few of the statistical intervals required in practice. We have found that analysts are apt to use the type of interval that they are most familiar with—irrespective of whether or not it is appropriate. This can result in the right answer to the wrong question. Thus, we differentiate, at an elementary level, among the different kinds of statistical intervals and provide a detailed exposition, with numerous examples, on how to construct such intervals from sample data. In fact, this book is unique in providing a discussion in one single place not only of the "standard" intervals but also of such practically important intervals as confidence intervals to contain a population percentile, confidence intervals on the probability of meeting a specified threshold value, and prediction intervals to include the observations in a future sample.

Many of these important intervals are ignored in standard texts. This, we believe, is partly out of tradition; in part, because the underlying development (as opposed to the actual application) may require advanced theory; and, in part, because the calculations to obtain such intervals can be quite complex. We do not feel restricted by the fact that the methods are based upon advanced mathematical theory. Practitioners should be able to use a method without knowing the theory, as long as they fully understand the assumptions. (After all, one does not need to know what makes a car work to be a good driver.) Finally, we get around the problem of calculational complexity by providing comprehensive tabulations, charts, and computer routines, some of which were specially developed, and all of which are easy to use.

This book is aimed at practitioners in various fields who need to draw conclusions from sample data. The emphasis is on, and many of the examples deal with, situations that we have encountered in industry (although we sometimes disguise the problem to protect the innocent).

Those involved in product development and quality assurance will, thus, find this book to be especially pertinent. However, we believe that workers in numerous other fields, from the health sciences to the social sciences, as well as teachers of courses in introductory statistics, and their students, can also benefit from this book.

We do not try to provide the underlying theory for the intervals presented. However, we give ample references to allow those who are interested to go further. We, obviously, cannot discuss statistical intervals for all possible situations. Instead, we try to cover those intervals, at least for a single population, that we have found most useful. In addition, we provide an introduction, and references to, other intervals that we do not discuss in detail.

It is assumed that the reader has had an introductory course in statistics or has the equivalent knowledge. No further statistical background is necessary. At the same time, we believe that the subject matter is sufficiently novel and important, tieing together work previously scattered throughout the statistical literature, that those with advanced training, including professional statisticians, will also find this book helpful. Since we provide a comprehensive compilation of intervals, tabulations, and charts not found in any single place elsewhere, this book will also serve as a useful reference. Finally, the book may be used to supplement courses on the theory and applications of statistical inference.

Further introductory comments concerning statistical intervals are provided in Chapter 1. As previously indicated, this chapter also includes a detailed discussion of the practical assumptions required in the use of the intervals, and, in general, lays the foundation for the rest of the book. Chapter 2 gives a more detailed general description of different types of confidence intervals, prediction intervals, and tolerance intervals and their applications. Chapters 3 and 4 describe simple tabulations and other methods for calculating statistical intervals. These are based on the assumption of a normal distribution. Chapter 5 deals with distribution-free intervals. Chapters 6 and 7 provide methods for calculating statistical intervals for proportions and percentages, and for occurrence rates, respectively. Chapters 8, 9, and 10 deal with sample size requirements for various statistical intervals.

Statistical intervals for many other distributions and other situations, such as regression analysis and the comparison of populations, are briefly considered in Chapter 11. This chapter also gives references that provide further information, including technical details and examples of more complex intervals. Chapter 12 outlines other general methods for computing statistical intervals. These include methods that use large sample statistical theory and ones based on Bayesian concepts. Chapter 13 presents a series of case studies involving the calculation of statistical intervals; practical considerations receive special emphasis.

Appendix A gives extensive tables for calculating statistical intervals. The notation used in this book is summarized in Appendix B. Listings of some computer routines for calculating statistical intervals are provided in Appendix C.

We present graphs and tables for computing numerous statistical intervals and bounds. The graphs, especially, also provide insight into the effect of sample size on the length of an interval or bound.

Most of the procedures presented in this book can be applied easily by using figures or tables. Some require simple calculations, which can be performed with a hand calculator. When tables covering the desired range are not available (for some procedures, the tabulation of the complete range of practical values is too lengthy to provide here), factors may be available from alternative sources given in our references. However, often a better alternative is to have a computer program to calculate the necessary factors or the interval itself. We provide some such programs in Appendix C. It would, in fact, be desirable to have a computer program that calculates all the intervals presented in this book. One could develop such a program from the formulas given here. This might use available subroutine libraries [such as IMSL (1987) and NAG (1988)] or programs like those given in Appendix C, other algorithms published in the literature [see, e.g., Griffiths and Hill (1985), Kennedy and Gentle (1980), P. Nelson (1985), Posten (1982), and Thisted (1988)], or available from published libraries [e.g., NETLIB, described by Dongarra and Grosse (1985)]. A program of this type, called STINT (for STatistical INTervals), is being developed by W. Meeker; an initial version is available.

GERALD J. HAHN
WILLIAM Q. MEEKER

Acknowledgments

We are highly indebted to various individuals, including many readers of the first edition, who helped make this second edition happen and who contributed to its improvement. Our special appreciation goes to Jennifer Brown, Joel Dobson, Robert Easterling, Michael Hamada, and Shiyao Liu for the long periods of time they spent in reading an early version of the manuscript and in providing us their insights. We also wish to thank Chuck Annis, Michael Beck, David Blough, Frank DeMeo, Necip Doganaksoy, Adrienne Hahn, William Makuch, Katherine Meeker, Wayne Nelson, Robert Rodriguez, and William Wunderlin for their careful review of various drafts, and for providing us numerous important suggestions, and to thank Chris Gotwalt for his advice on the use of the random weighted bootstrap method. H.N. Nagaraja provided helpful comments on Chapter 5 and Appendix G.

Our continued appreciation goes to the cadre of individuals who helped us in writing the first edition: Craig Beam, James Beck, Thomas Boardman, Necip Doganaksoy, Robert Easterling, Marie Gaudard, Russel Hannula, J. Stuart Hunter, Emil Jebe, Jason Jones, Mark Johnson, William Makuch, Del Martin, Robert Mee, Vijayan Nair, Wayne Nelson, Robert Odeh, Donald Olsson, Ernest Scheuer, Jeffrey Smith, William Tucker, Stephen Vardeman, Jack Wood, and William Wunderlin.

We would like to thank Quang Cao (Louisiana State University) and K.P. Poudel (Oregon State University) for providing the tree volume data that we used in Chapter 13.

We would like to express our appreciation to Professors Ken Koehler and Max Morris at Iowa State University (ISU) for their encouragement and support for this project. We would also like to express our appreciation to Professor James Geaghan who provided encouragement, facilities, and support for Bill Meeker and Luis Escobar at Louisiana State University (LSU). In addition, Elaine Miller at LSU and Denise Riker at ISU provided helpful and excellent assistance during the writing of this book.

Finally, we express our sincere appreciation to our wives, Karen Meeker, Bea Hahn, and Lida Escobar, for their understanding and support over the many days (and nights) that we spent in putting this opus together.

William Q. Meeker
Gerald J. Hahn
Luis A. Escobar

About the Companion Website

This book is accompanied by a companion website:

www.wiley.com/go/meeker/intervals

Once registered at the website, the reader will receive access to:

- More extensive figures for
  - ○ Confidence intervals and bounds for a binomial proportion (Figure 6.1)
  - ○ Probability of successful demonstration for tests based on the normal distribution (Figures 9.1)
  - ○ Probability of successful demonstration for tests based on the binomial distribution (Figures 9.2)

- An Excel workbook matrix showing statistical package capabilities of different software packages and a Word document providing some explanation

- Information about the StatInt R package

- Data sets used in the book

- Additional information on various (especially "late-breaking") topics relating to the book

Chapter *1*

Introduction, Basic Concepts, and Assumptions

OBJECTIVES AND OVERVIEW

This chapter provides the foundation for our discussion throughout the book. Its emphasis is on basic concepts and assumptions. The topics discussed in this chapter are:

- The concept of statistical inference (Section 1.1).

- An overview of different types of statistical intervals: confidence intervals, tolerance intervals, and prediction intervals (Section 1.2).

- The assumption of sample data (Section 1.3) and the central role of practical assumptions about the data being "representative" (Section 1.4).

- The need to differentiate between enumerative and analytic studies (Section 1.5).

- Basic assumptions for inferences from enumerative studies, including a brief description of different random sampling schemes (Section 1.6).

- Considerations in conducting analytic studies (Section 1.7).

- Convenience and judgment samples (Section 1.8).

- Sampling people (Section 1.9).

- The assumption of sampling from an infinite population (Section 1.10).

- More on practical assumptions (Sections 1.11 and 1.12).

- Planning the study (Section 1.13).

- The role of statistical distributions (Section 1.14).

Statistical Intervals: A Guide for Practitioners and Researchers, Second Edition.
William Q. Meeker, Gerald J. Hahn and Luis A. Escobar.
© 2017 John Wiley & Sons, Inc. Published 2017 by John Wiley & Sons, Inc.
Companion Website: www.wiley.com/go/meeker/intervals

- The interpretation of a statistical interval (Section 1.15).

- The relevance of statistical intervals in the era of big data (Section 1.16).

- Comment concerning the subsequent discussion in this book (Section 1.17).

1.1 STATISTICAL INFERENCE

Decisions frequently have to be made from limited sample data. For example:

- A television network uses the results of a sample of 1,000 households to determine advertising rates or to decide whether or not to continue a show.

- A company uses data from a sample of five turbines to arrive at a guaranteed efficiency for a further turbine to be delivered to a customer.

- A manufacturer uses tensile strength and other measurements obtained from a laboratory test on ten samples of each of two types of material to select one of the two materials for use in future production.

The sample data are often summarized by statements such as:

- 293 out of the 1,000 sampled households were tuned to the show.

- The mean efficiency for the sample of five turbines was 67.4%.

- The samples using material A had a mean tensile strength 3.2 units larger than those using material B.

The preceding "point estimates" provide a concise summary of the sample results, but they give no information about their precision. Thus, there may be big differences between such point estimates, calculated from the sample data and what one would obtain if unlimited data were available. For example, 67.4% would seem a reasonable estimate (or prediction) of the efficiency of the next turbine. But how "good" is this estimate? By noting the variation in the observed efficiencies of the five turbines, we know that it is unlikely that the turbine to be delivered to the customer will have an efficiency of *exactly* 67.4%. We may, however, expect its efficiency to be "close to" 67.4%. But how close? Can we be reasonably confident that it will be within $\pm 0.1\%$ of the point estimate 67.4%? Or within $\pm 1\%$? Or within $\pm 10\%$? We need to quantify the uncertainty associated with our estimate or prediction. An understanding of this uncertainty is an important input for decision making, for example, in providing a warranty on product performance. Moreover, if our knowledge, as reflected by the width of the uncertainty interval, is too imprecise, we may wish to obtain more data before making an important decision.

The example suggests quantifying uncertainty by constructing statistical intervals around the point estimate. This book shows how to obtain such intervals. We describe frequently needed, but not necessarily well-known, statistical intervals calculated from sample data, differentiate among the various types of intervals, and show their applications. Methods for obtaining each of the intervals are presented and their use is illustrated. We also show how to choose the sample size so as to attain a desired degree of precision, as measured by the width of the resulting statistical interval. Thus, this book provides a comprehensive guide and reference to the use of statistical intervals to quantify the uncertainty in the information about a sampled population or process, based upon a possibly small, but randomly selected sample. The concept of a random sample is discussed further in Section 1.6.3.

1.2 DIFFERENT TYPES OF STATISTICAL INTERVALS: AN OVERVIEW

Various types of statistical intervals may be calculated from sample data. The appropriate interval depends upon the specific application. Frequently used intervals are:

- A *confidence interval* to contain an unknown characteristic of the sampled population or process. The quantity of interest might be a population property or "parameter," such as the mean or standard deviation of the population or process. Alternatively, interest might center on some other property of the sampled population, such as a quantile or a probability. Thus, depending upon the question of interest, one might compute a confidence interval that one can claim, with a specified high degree of confidence, contains (1) the mean tensile strength, (2) the standard deviation of the distribution of tensile strengths, (3) the 0.10 quantile of the tensile strength distribution, or (4) the proportion of specimens that exceed a stated threshold tensile strength value.

- A *statistical tolerance interval* to contain a specified proportion of the units from the sampled population or process. For example, based upon a random sample of tensile strength measurements, we might wish to compute an interval to contain, with a specified degree of confidence, the tensile strengths of at least a proportion 0.90 of the units from the sampled population or process. Hereafter we will generally simply refer to such an interval as a "tolerance interval."

- A *prediction interval* to contain one or more future observations, or some function of such future observations, from a previously sampled population or process. For example, based upon a random sample of tensile strength measurements, we might wish to construct an interval to contain, with a specified degree of confidence, (1) the tensile strength of a randomly selected single future unit from the sampled process (this was of interest in the turbine efficiency example), (2) the tensile strengths for *all* of five future units, or (3) the mean tensile strength of five future units.

Most users of statistical methods are familiar with (the common) confidence intervals for the population mean and for the population standard deviation, but often not for population quantiles or the probability of exceeding a specified threshold value. Some, especially in industry, are also aware of tolerance intervals. Despite their practical importance, however, most practitioners, and even many professional statisticians, know very little about prediction intervals except, perhaps, for their application to regression problems. A frequent mistake is to calculate a confidence interval to contain the population mean when the problem requires a tolerance interval or a prediction interval. At other times, a tolerance interval is used when a prediction interval is needed. Such confusion is understandable, because statistics textbooks typically focus on the common confidence intervals, occasionally make reference to tolerance intervals, and consider prediction intervals only in the context of regression analysis. This is unfortunate because in applications, tolerance intervals, prediction intervals, and confidence intervals on distribution quantiles and on exceedance probabilities are needed almost as frequently as the better-known confidence intervals. Moreover, the calculations for such intervals are generally no more difficult than those for confidence intervals.

1.3 THE ASSUMPTION OF SAMPLE DATA

In this book we are concerned only with situations in which uncertainty is present because the available data are from a random sample (often small) from a population or process. There are, of course, some situations for which there is little or no such statistical uncertainty. This is the case when the relevant information on *every unit* in a finite population has been recorded without measurement error, or when the sample size is so large that the uncertainty in our estimates due to sampling variability is negligible (as we shall see, how large is "large" depends on the specific application). Examples of situations in which one is generally dealing with the entire population are:

- The given data are census information that have been obtained from all residents in a particular city (at least to the extent that the residents could be located and are willing to participate in the study).

- There has been 100% inspection (i.e., all units are measured) of a performance property for a critical component used in a spacecraft.

- A complete inventory of all the parts in a warehouse has been taken.

- A customer has received a one-time order of five parts and has measured each of these parts. Even though the parts are a random sample from a larger population or process, as far as the customer is concerned the five parts make up the entire population of interest.

Even in such situations, intervals to express uncertainty are sometimes needed. Suppose, for example, that based upon extensive data, we *know* that the weight of a product is approximately normally distributed with a mean of 16.10 ounces and a standard deviation of 0.06 ounces. We wish an interval to contain the weight of a single unit randomly selected by a customer, or by a regulatory agency. The calculation of the resulting *probability interval* is described in books on elementary probability and statistics. Such intervals generally assume complete knowledge about the population (e.g., the mean and standard deviation of a normal distribution). In this book, we are concerned with the more complicated problem where, for example, the population mean and standard deviation are *not* known but are estimated, subject to sampling variability. In particular, tolerance intervals and prediction intervals converge to probability intervals as the sample size increases. On the other hand, because there is no statistical uncertainty remaining, the width of confidence intervals converges to zero with increasing sample size.

Statistical uncertainty also exists, even though the entire population has been evaluated, when the readings are subject to measurement error. For example, one might determine that in measuring a particular property, 971 out of 983 parts in a production lot are found to be within specification limits. Due to measurement error, however, the actual number of parts within specifications may not exactly equal 971. Moreover, if something is known about the statistical distribution of measurement error, one can then also quantify the uncertainty associated with the estimated number of parts within specifications (e.g., Hahn, 1982).

Finally, we note that even when there is no quantifiable statistical uncertainty, there may still be other uncertainties of the type suggested in the discussion to follow.

1.4 THE CENTRAL ROLE OF PRACTICAL ASSUMPTIONS CONCERNING REPRESENTATIVE DATA

We have briefly described different statistical intervals that a practitioner might use to express the uncertainty in various estimates or predictions generated from sample data. This book presents the methodology for calculating such intervals. Before proceeding, we need to make clear the major practical assumptions dealing with the "representativeness" of the sample data. We do this in the following sections. Departures from these implicit assumptions are common in practice and can invalidate any statistical analyses. Ignoring such assumptions can lead to a false sense of security, which, in many applications, is the weakest link in the inference process. Thus, for example, product engineers need to question the assumption that the performance observed on prototype units produced in the lab also applies for production units, to be built much later, in the factory. Similarly, a reliability engineer should question the assumption that the results of a laboratory life test will adequately predict field failure rates. In fact, in some studies, the assumptions required for the statistical interval to apply may be so far off the mark that it would be inappropriate, and perhaps even misleading, to use the formal methods presented here.

In the best of situations, one can rely on physical understanding, or information from outside the study, to justify the practical assumptions. Such evaluations, however, are principally the responsibility of the subject-matter expert. Often, the assessment of such assumptions is far from clear-cut. In any case, one should keep in mind that the intervals described in this book reflect only the statistical uncertainty due to limited data. In practice, the actual uncertainty will be larger because the generally unquantifiable deviations of the practical assumptions from reality provide an added *unknown* element of uncertainty beyond that quantified by the statistical interval. If there were formal methods to reflect this further uncertainty (occasionally there are, but often there are not), the resulting interval, expressing the *total* uncertainty, would be wider than the statistical interval alone. This observation suggests a rationale for calculating a statistical interval for situations where the basic assumptions are questionable. If it turns out that the calculated statistical interval is wide, we then know that our estimates have much uncertainty—even *if* the assumptions were all correct. A narrow statistical interval would, on the other hand, imply a small degree of uncertainty *only if* the required assumptions hold.

Because of their importance, we feel it appropriate to review, in some detail, the assumptions and limitations underlying the use and interpretation of statistical intervals before proceeding with the technical details of how to calculate such intervals.

1.5   ENUMERATIVE VERSUS ANALYTIC STUDIES

Deming (1953, 1975, 1986) emphasizes the important differences between "enumerative" and "analytic" studies (a concept that he briefly introduced earlier in Deming, 1950). Despite its central role in making inferences from the sample data, many traditional textbooks in statistics have been slow in giving this distinction the attention that it deserves.

To point out the differences between these two types of studies, and some related considerations, we return to the examples of Section 1.1. The statements there summarize the sample data. In general, however, investigators are concerned with making inferences or predictions *beyond* the sample data. Thus, in these examples, the real interest was, not in the sample data per se, but in:

1. The proportion of households in the *entire country* that were tuned to the show.

2. The efficiency of the, *as yet not manufactured*, turbine to be sent to the customer.

3. A comparison of the mean tensile strengths of the *production units to be built* in the factory *some time in the future* using material A and material B.

In the first example, our interest centers on a finite identifiable unchanging collection of units, or population, from which the sample was drawn. This population, consisting of all the households in the country with access to the show, exists at the time of sampling. Deming uses the term "enumerative study" to describe such situations. More specifically, Deming (1975, page 147), defines an enumerative study as one in which "action will be taken on the material in the frame studied," where he uses the conventional definition of a frame as "an aggregate of identifiable units of some kind, any or all of which may be selected and investigated. The frame may be lists of people, areas, establishments, materials, or other identifiable units that would yield useful results if the whole content were investigated." Thus, the frame provides a finite list, or other identification, of distinct (non-overlapping) and exhaustive sampling units. The frame defines the population to be sampled in an enumerative study.

Some further examples of enumerative studies are:

- Public opinion polls to assess the *current view* on some specified topic(s) of the entire US adult population, or some defined segment thereof, such as all registered voters in a specified locality.

- Sample audits to assess the correctness of last month's bills and to estimate the total error in such bills. In this case, the population of interest consists of all of last month's bills.

- Product acceptance sampling to decide on the disposition of a particular production lot. In this case, the population of interest consists of all units in the production lot being sampled.

In an enumerative study, the correctness of statistical inferences requires a random sample from the frame. Such a sample, is, at least in theory, generally attainable; see Section 1.6.

In contrast, the second two examples of Section 1.1 (dealing, respectively, with the efficiency of a future turbine and the comparison of two materials) illustrate what Deming (1975, page 147) calls "analytic studies." We no longer have an existing, finite, well-defined, unchanging population. Instead, we want to take action to improve or make predictions about the output of a, sometimes hypothetical, future *process* based upon data from an existing (likely different) process.

Specifically, Deming (1975) defines an analytic study as one "in which action will be taken on the process or cause-system . . . the aim being to improve practice in the future . . . . Interest centers in future product, not in the materials studied." He cites as examples "tests of varieties of wheat, comparison of machines, comparisons of ways to advertise a product or service, comparison of drugs, action on an industrial process (change in speed, change in temperature, change in ingredients)." We may wish to use data from an existing process to predict the characteristics of future output from the same or a similar process. Thus, in a prototype study of a new product, interest centers on the process that will manufacture the product in the future.

These examples are representative of many encountered in practice, especially in engineering, medical, and other scientific investigations. In fact, the great majority of applications that we have encountered in practice involve analytical, rather than enumerative studies. It is, moreover, inherently more complicated to draw inferences from analytic studies than from enumerative studies because analytic studies require the critical (and often unverifiable) added assumption that the process about which one wishes to make inferences is statistically identical to that from which the sample was selected.

1.5.1 Differentiating between Enumerative and Analytic Studies

What one wishes to do with the results of the study is often a major differentiator between an enumerative and an analytic study. Thus, if one's interest is limited to describing an existing population, one is dealing with an enumerative study. On the other hand, if one is concerned with a process that is to be improved, or is otherwise subject to change, perhaps as a result of the study, then one is clearly dealing with an analytic study.

Deming (1975) presents a "simple criterion to distinguish between enumerative and analytic studies. A 100% sample of the frame answers the question posed for an enumerative study, subject of course to the limitations of the method of investigation. In contrast, a 100% sample . . . is still inconclusive in an analytic problem." This is because for an analytic study our real interest is in a process that will be operating in the future. Deming's rule can be useful when the differentiation between an analytic and an enumerative study does not seem clear-cut. For example, an "exit poll" to estimate the proportion of voters who have voted (or, at least, assert that they have voted) for a particular candidate, based upon a random sample of individuals leaving the polling booth, is an example of an enumerative study. In this case, a 100% sample provides perfect information (assuming 100% correct responses). In contrast, estimating, before the election, the proportion of voters who will *actually* go to the polls and vote for the candidate involves an analytic study, because it deals with a future process. Thus, between the time of the survey and election day, some voters may change their minds, perhaps as a result of some important external event—or even as a consequence of action taken by one or more of the candidates based upon information obtained in the study. Also, extraneous factors, such as adverse weather conditions on election day (not contemplated on the sunny day on which the poll was conducted), might stop some going to the polls—and the "stay-at-homes" may well differ in their voting preferences from those who do vote. Thus, even if we had sampled every eligible voter prior to the election, we still would not be able to predict the outcome with certainty, because we do not know who will actually vote and who will change their mind in the intervening period. (Special considerations in sampling people are discussed in Section 1.9.)

Taking another example, it is sometimes necessary to sample from inventory to make inferences about a product population or process. If interest focuses merely on characterizing the current inventory, the study is enumerative. If, however, we wish to predict the future performance of the product, perhaps after making design changes, the study is analytic. Finally, drawing conclusions about the performance of a turbine to be manufactured in the future, based upon data on turbines built in the past, involves, as we have indicated, an analytic study. If, however, the measured turbines *and* the turbine to be shipped were all independently and randomly selected from inventory (unlikely to be the case in practice), one would be dealing with an enumerative study.

1.5.2 Statistical Inference for Analytic Studies

We do not agree with the views of some (e.g., Gitlow et al., 1989, page 558) who imply that statistical inference methods, such as statistical intervals, have no place whatsoever in analytic studies. Indeed, such methods have been used successfully for decades in science and industry in studies that have been predominantly analytic. Many statistical methods were, in fact, developed with such studies in mind. Instead, we feel that the decision of whether or not to use statistical intervals in analytic studies needs to be made on a case by case basis. Use of statistical intervals in such studies requires a keen understanding and assessment of the additional assumptions that are being made.

1.5.3 Inferential versus Predictive Analyses

In addition to differentiating between analytic and enumerative studies, it is also useful to differentiate between inferential and predictive analyses. Broadly speaking, the goal of an inferential analysis is to gain an understanding of the mechanism that underlies or resulted in the observed data. A typical example is that of a *manufacturer* wishing to determine how different processing (and possibly environmental) variables impact the performance of a product with the goal of building an improved product in the future.

In contrast, the goal of a predictive analysis is typically to predict future performance—without necessarily understanding the underlying mechanism. In the preceding example, such analyses might be of principal interest to the *purchaser* of the product who wishes to predict future performance.

Both inferential and predictive analyses can involve either enumerative or analytic studies. In our example, the study is analytic when the underlying conditions under which the data were obtained differ from those under which one wishes to draw conclusions—irrespective of whether one is interested in gaining an understanding of the impact of different processing variables or predicting future performance. Thus, the available data might be from in-house testing on early production units, but the inferences or predictions to be made deal with field exposure of high volume production, making the study analytic in both cases. We will now consider, in further detail, the basic assumptions underlying inferences from enumerative and analytic studies.

1.6 BASIC ASSUMPTIONS FOR INFERENCES FROM ENUMERATIVE STUDIES

1.6.1 Definition of the Target Population and Frame

In enumerative studies there is some "target population" about which it is desired to draw inferences. An important first step—though one that is sometimes omitted by analysts—is that of explicitly and precisely defining this target population. For example, the target population may be all the automobile engines of a specified model manufactured on a particular day, or in a specified model year, or over some other defined time period. In addition, one need also make clear the specific characteristic(s) to be evaluated. This may be a measurement or other reading on an engine, or the time to failure of a part on a life test, where "failure" is precisely defined. Also, in many applications, and especially those involving manufactured products, one must clearly state the operating environment in which the defined characteristic is to be evaluated. For a life test, this might be "normal operating conditions," and exactly what constitutes such conditions needs to be clearly stated.

The next step is that of establishing a frame from which the sample is to be taken. Establishing a frame requires obtaining or developing a specific listing, or other enumeration of the population from which the sample will be selected. Examples of frames are the serial numbers of all the automobile engines built over the specified time period, the complete listing of email addresses of the members of an organization, the schedule of incoming commercial flights into an airport on a given day, or a tabulation of all invoices billed during a calendar year. Often, the frame is *not* identical to the target population. For example, a listing of land-line telephone numbers generally corresponds to households, rather than individuals, and omits those who do not have a telephone or who have only a cell phone, people with unlisted phone numbers, new arrivals in the community, etc.—and also may include businesses, which are not always clearly identified as such. If the telephone company wishes to estimate the proportion of listed land-line phones in working order at a given time, a complete listing of such telephones (available to the phone company) will probably coincide with the target population about which inferences are desired. For most other studies, however, there may be an important difference between the frame (i.e., the telephone directory listing) and the target population.

The listing provided by the frame will henceforth be referred to as the "sampled population." Clearly, the inferences from a study, such as those quantified by statistical intervals, will be on the frame and—when the two differ—not on the target population. Thus, our third step—after defining the target population and the frame—is that of evaluating the differences between the two and the possible effect that such differences could have on the conclusions of the study. Moreover, it warrants repeating that these differences introduce uncertainties above and beyond those quantified by the statistical intervals provided in this book. If well understood, these differences can, at least sometimes and to some degree, be dealt with.

1.6.2 The Assumption of a Random Sample

The data are assumed to be a random sample from the frame. Because we deal only with random samples in this book, we will sometimes use the term "sample" to denote a random sample. In enumerative studies, we will be concerned principally with the most common type of random sampling, namely *simple random sampling*. We briefly describe other types of random samples in Section 1.6.3.

Simple random sampling gives every possible sample of n units from the frame the same probability of being selected. A simple random sample of size n can, at least in theory, be obtained from a population of size N by numbering each unit in the population from 1 to N, placing N balls bearing the N numbers into a bin, thoroughly mixing the balls, and then randomly drawing n balls from the bin. The units to be sampled are those with numbers corresponding to the n selected balls. In practice, tables of random numbers generated by computer algorithms (e.g., Rizzo, 2007; Ripley, 2009; Gentle, 2009, 2013) and by statistical computing software (e.g., R Core Team, JMP or Minitab), provide easier ways of obtaining a random sample.

The assumption of random sampling is of critical importance in constructing statistical intervals. This is because such intervals reflect only the randomness due to the sampling process and do not take into consideration biases that might be introduced by not sampling randomly. It is especially important to recognize this limitation because in many studies, and especially ones involving sampling people, one does not have a strict random sample; see Section 1.9.

1.6.3 More Complicated Random Sampling Schemes

There are also other random sampling methods beyond simple random sampling, such as stratified random sampling, cluster random sampling, and systematic random sampling. These are used frequently in such applications as sampling of human populations, auditing, and inventory estimation. For such samples, rather than every possible sample of n units from the sampled population having the *same* probability of being selected, each possible sample has a *known* probability of being selected. Statistical intervals can also be constructed for such more complicated sampling schemes; these intervals are generally more complicated than the ones for simple random samples. The interested reader is referred to books referenced in the Bibliographic Notes section at the end of this chapter.

The more complicated *random* sampling schemes, described briefly below, need to be differentiated from various *nonrandom* sampling schemes, which we describe in Section 1.8 under the general heading of "convenience and judgment sampling."

Stratified random sampling

In some sampling applications, the population is naturally divided into non-overlapping groups or *strata*. For example, a population might be divided according to gender, job rank, age group, geographic region, or manufacturer. It may be important or useful to take account of these strata in the sampling plan because:

- Some important questions may focus on individual strata (e.g., information is needed by geographic region, as well as for the entire country).

- Cost, methods of sampling, access to sample units, or resources may differ among strata (e.g., salary data are generally more readily available for individuals working in the public sector than for those working in the private sector).

- When the response variable has less variability within a stratum than across strata (i.e., across the entire population), stratified random sampling provides more precise estimates than one would obtain from a simple random sample of the same size. The increase in precision results in narrower (but more complicated to construct) statistical intervals concerning the entire population.

In stratified sampling, one takes a simple random sample from each stratum in the population. The methods presented in this book can be used with data from a single stratum to compute statistical intervals for that stratum. However, when stratified sample data are combined across strata (e.g., to obtain a confidence interval for the population mean or total) special methods are needed, as described in textbooks on survey sampling. These books also discuss how to allocate units across strata and how to choose the sample size within each stratum.

Cluster random sampling

In some studies it is less expensive to obtain samples by using "clusters" of "elements" (generic terminology) that are conveniently located together in some manner, instead of taking a simple random sample from the entire population. For example, when items are packed in boxes, it is often easier to take a random sample of boxes and either evaluate all items in each selected box—or take further random samples within each of the selected boxes—rather than to randomly sample individual items irrespective of the box in which they are contained. Also, it may be more natural and convenient to interview some or all adult members of randomly selected families rather than randomly selected individuals from a population of individuals. Finally, it is often easier to find a frame for, and sample groups of, individuals clustered in a random sample of locations rather than taking a simple random sample of individuals spread over, say, an entire city. In other cases, only a listing of clusters, but not of the individuals or items they contain, may be available; clusters are then the natural sampling units. In each of these cases, one needs to define the clusters, obtain a frame that lists all clusters, and then take a random sample of clusters from that frame. Sometimes, as previously indicated, responses are then obtained for all elements (i.e., individuals or items) within each selected cluster, although often subsampling within clusters is conducted.

The value of the information for each additional sample unit within a cluster can be appreciably less than that of an individual unit for a simple random sample, especially if the items in a cluster tend to be similar and many units are chosen from each cluster. On the other hand, if the elements in the clusters are a well-mixed representation of the population, the loss in precision due to the use of cluster sampling may be slight; often, the lower per-element cost of cluster sampling will more than compensate for the loss in statistical efficiency. Thus, given a specified total cost to conduct a study, the net result of cluster sampling can be an improvement in overall statistical efficiency, as evidenced by narrower statistical intervals compared to those for a simple random sample—even though the simple random sample requires a smaller total sample size. Also, when the investigator has a say in choosing the cluster size, the loss of efficiency might be mitigated by taking a larger sample of smaller clusters. (When clusters contain only one element, one is back to simple random sampling.) Books on survey sampling, such as those mentioned in the Bibliographic Notes section at the end of this chapter, provide details of cluster sampling and sample size selection.

Systematic random sampling

It is often much easier to select a sample in a systematic manner than to take a simple random sample. For example, because there is no readily available frame, it might be difficult to obtain a simple random sample of all the customers who come into a store on a particular day. It would, however, be relatively simple to sample every 10th, or other preselected number, person entering the store. Similarly, it might be much easier and more natural to have a clerk examine every 10th item in a file cabinet, instead of choosing a simple random sample of all items. In both examples, the ratio of cost incurred to information gained to conduct the study might be appreciably smaller with systematic sampling than with simple random sampling. In both examples, it would usually be cost effective to use systematic, rather than simple, random sampling. In some situations a systematic sample may, in fact, be the only feasible alternative.

Systematic samples are random samples as long as a random starting point is used. Special methods and formulas, however, are needed to compute statistical intervals. Also, the systematic pattern that is to be used in sampling must be chosen carefully. Serious losses of efficiency or biases may result if there are periodicities in the sampled population and if these are in phase with the systematic sampling scheme. For example, if a motor vehicle bureau measures traffic volume each Wednesday (where Wednesday is a randomly selected weekday), the survey results would likely provide a biased estimate of average weekday traffic volume. On the other hand, if such sampling took place every sixth weekday, the resulting estimate would likely be a lot more reasonable. Books on survey sampling, such as those mentioned in the Bibliographic Notes section at the end of this chapter, provide details of systematic sampling.

1.7 CONSIDERATIONS IN THE CONDUCT OF ANALYTIC STUDIES

1.7.1 Analytic Studies

In an enumerative study, one generally wishes to draw inferences by sampling from a well-defined existing population, the members of which can be enumerated, at least conceptually—even though, as we have seen, difficulties can arise in finding a frame that adequately represents the target population and in obtaining a random sample from that frame. In contrast, in an analytic study one wishes to draw conclusions about a process that may not even exist—or may not be accessible—at the time of the study. As a result, the process that is sampled is likely to differ, in various ways, from the one about which it is desired to draw inferences. As we have indicated, sampling prototype units, made in the lab or on a pilot production line, to draw conclusions about subsequent full-scale production is one common example of an analytic study.

1.7.2 The Concept of Statistical Control

A less evident example of an analytic study arises if, in dealing with a mature production process, one wishes to draw inferences about future production, based upon sample data from current or recent production. Then, if the process is in so-called "statistical control," *and remains so*, the current data may be used to draw inferences about the future performance of the process. The concept of statistical control means, in its simplest form, that the process is stable or unchanging. It implies that the statistical distributions of the characteristics of interest for the current process are identical to these for the process in the future. It also implies that the sequence of data from production is not relevant. Thus, units selected consecutively from production are no more likely to be alike than units selected, say, a day, a week, a month, or even a year, apart. All of this, in turn, means that the only sources of variability are "common cause" within the system, and that variation due to "assignable" or "special" causes, such as differences between raw material lots, operators, and ambient conditions, have been removed.

The concept of statistical control is an ideal state that, in practice, may exist only approximately, although it may often provide a useful working approximation. If a process is in statistical control, then samples from the process are (or can be modeled as) independent and identically distributed. When a process is in statistical control, the statistical intervals provided in this book should yield reasonable inferences about the process. On the other hand, when the process is not in, or near, statistical control, the applicability of the statistical intervals given here for characterizing the process may be undermined by trends, shifts, cycles, and other variations unless they are accounted for in a more comprehensive model.

1.7.3   Other Analytic Studies

Although analytic studies frequently require projecting from the present to a future time period, this is not the only way an analytic study arises. For example, practical constraints, concerns for economy, and a variety of other considerations may lead one to conduct a laboratory scale assessment, rather than perform direct evaluations on a production line, even though production is up and running. In such cases, it is sometimes possible to perform verification studies to compare the results of the sampled process with the process of interest.

1.7.4   How to Proceed

The following operational steps are appropriate for potentially constructing statistical intervals for many analytic studies:

- Have the engineer, scientist or subject-matter expert define the process of interest.

- Determine the possible sources of data that will be useful for making the desired inferences about the process of interest (i.e., define the process to be sampled or evaluated).

- Clearly state the assumptions that are required for the results of the study on the sampled process to be applicable for the process of interest.

- Collect well-targeted data and, to the extent possible, check the assumed model and any other assumptions.

- Jointly decide, in light of the assumptions and the data, and an understanding of the underlying cause mechanism, whether there is value in calculating a statistical interval, or whether this might lead to a false sense of security, and should, therefore, be avoided.

- If it is decided to obtain a statistical interval, ensure that the underlying assumptions are fully recognized and make clear that this interval represents only the uncertainty associated with the random sampling and does not include uncertainties due to differences between the sampled process and the process of interest. Therefore, the actual uncertainty will be greater than that expressed by the width of the interval and in some applications, could be substantially greater.

1.7.5   Planning and Conducting an Analytic Study

In conducting an analytic study, one typically cannot sample directly from the process of interest. This process may, as we have seen, not yet exist. Instead, one needs to define the specific process that is to be sampled and how the sampling should proceed. In so doing, as broad an environment as possible should be considered. For example, in characterizing a production process, one should include the wide spectrum of raw materials and operating conditions that might be encountered in the future. This will require fewer assumptions when using the resulting data to draw inferences about the actual process of interest. It is usually advisable to sample over relatively long time periods because observations taken over a short period are less likely to be representative of the process of interest with regard to both average performance and long-run variability unless the process is in strict statistical control. For example, in studying the properties of a new alloy, specimens produced closely together in time may be more alike than those produced over longer time intervals due to variations in ambient conditions, raw material, operators, machine condition, measuring equipment, etc.

In some analytic studies, one might deliberately make evaluations under extreme conditions. In fact, Deming (1975) asserts that in the early stages of an investigation, "it is nearly always the best advice to start with strata near the extremes of the spectrum of possible disparity in response, as judged by the expert in the subject matter, even if these strata are rare." He cites an example that involves the comparison of the speed of reaching equilibrium for different types of thermometers. He advocates, in this example, an initial study on two groups of people: those with normal temperature and those with high fever. In addition, it is important that information on relevant concomitant variables is recorded, whenever feasible, for inclusion in, possibly graphical, subsequent analyses. For example, in dealing with a production process, data identifying operator, raw material lot, ambient conditions and other factors that might potentially impact the performance of the process should generally be retained for potential future analyses.

1.8 CONVENIENCE AND JUDGMENT SAMPLES

In practice, it is sometimes difficult, or impossible, even in an enumerative study, to obtain a random sample. Often, it is much more convenient to sample without strict randomization. Consider again a product packaged in boxes whose performance is to be characterized. If the product is ball bearings, it might be easy to thoroughly mix the contents of a box and sample randomly. On the other hand, suppose the product is made up of fragile ceramic plates, stacked in large boxes. In this case, it is much easier to sample from the top of the box than to obtain a random sample from among all of the units in the box. Similarly, if the product is produced in rolls of material, it is often simple to cut a sample from either the beginning or the end of the roll, but often impractical to sample from anyplace else. (This is not a systematic random sample, because there is not a random starting point.) Also, when sampling from a production process, it is often more practical to sample periodically, say every 2 hours during an 8-hour shift, than to select samples at four different randomly selected times during each shift. In this, and other applications, a further justification for periodic sampling is the need to consistently monitor the process for changes by the use of control charts, etc.

Selection of product from the top of a box, from either end of a roll, or at prespecified periodic time intervals for a production process, without a random starting point, are examples of what is sometimes referred to as "convenience sampling." Such samples are generally *not* strictly random because some units (e.g., those not at either end of the roll) have no chance of being selected. Because one is not sampling randomly, statistical intervals, strictly speaking, are not applicable for convenience sampling. In practice, however, one uses experience and understanding of the subject matter to decide on the applicability of applying statistical inferences to the results of convenience sampling. Frequently, one might conclude that the convenience sample will provide data that, for all practical purposes, are as "random" as those obtained by a simple random sample. Sampling from an end of a roll might, for example, yield information equivalent to that from simple random sampling *if* production is continuous, the process is in statistical control, and there is no roll end effect. Similar assumptions apply in drawing conclusions about a process based upon selecting samples from production periodically. Thus, treating a convenience sample as if it were a random sample *may sometimes* be reasonable from a practical point of view. However, the fact that this assumption is being made needs to be recognized, and the validity of using statistical intervals as if a random sample had been selected needs to be critically assessed based upon the specific circumstances.

Similar considerations apply in "judgment" or "pseudo-random" sampling. This occurs when personal judgment is used to choose "representative" sample units; a foreman, for example, might by eyeball take what appears to be a "random" selection of production units, without going through the formalities that we have described for selecting a random sample. In many cases, this might yield results that are essentially equivalent to those obtained from a random sample. Sometimes, however, this procedure will result in a higher probability of selecting, for example, conforming or nonconforming units. In fact, studies have shown that what might be called "judgment" can lead to seriously biased samples and, therefore, invalid or misleading results. Thus, the use of judgment, in place of random selection of sample units, invalidates the probabilistic basis for statistical inference and could render statistical intervals meaningless.

Judgment is, of course, important in planning studies, but it needs to be applied carefully in the light of available knowledge and practical considerations. Moreover, where possible, judgment should *not* be used as a substitute for random sampling or other randomization needed to make probabilistic inferential statements, such as constructing statistical intervals. Thus, returning to Deming's example of comparing the speed of reaching equilibrium for different type thermometers, it might well be advantageous to make comparisons for strata of people with normal temperature and with high fever. Within these two strata, however, patients and thermometers should be selected at random, to the degree possible. This will provide the opportunity for valid statistical inferences within strata, even though these inferences may be in a severely limited domain.

1.9 SAMPLING PEOPLE

In many important applications, such as public opinion polls, marketing studies and TV program viewing ratings, the subject of the study is not a product, but people from whom we wish to solicit verbal responses. Such studies typically present added issues, including special considerations to ensure that the frame resembles the target population as closely as possible and the added problem of nonrespondents (e.g., individuals who choose not to participate in a study). If these issues are not handled appropriately, they could make formally constructed statistical intervals meaningless or even misleading. We elaborate below.

Telephone surveys are used extensively in people-response studies. However, such studies can result in the frame seriously failing to represent the target population. This was made evident in the 1936 US Presidential election in which telephone surveys predicted the election of Al Landon over Franklin Roosevelt. An important reason for this erroneous prediction was the fact that in 1936 the characteristics—and, most importantly, the voting preferences—of the frame of the then telephone owners differed appreciably from the target population of the voting electorate. Today, surveys of land-line telephone users can similarly miss the mark by excluding the increasing number of households that rely only on cell phones.

Sampling people at shopping malls or other on-site locations might be used instead of, or in addition to, telephone surveys, especially in marketing studies. If, however, one is interested in a population beyond shopping mall visitors, such studies may again lead to a frame that fails to adequately represent the population of interest by tending to exclude the elderly, poor people, wealthy people, and other demographic groups that visit shopping malls infrequently.

Once defined, it is often possible to secure a random sample from the frame. For example, in telephone surveys, random digit dialing is frequently used. However, unlike sampling a manufactured product, the subjects selected for a people study can choose whether or not to respond—and whether or not to provide a truthful response. Moreover, willingness to participate in a TV viewing survey may be correlated with a respondent's viewing preferences. Also, the fact that the person who answers the phone—or even who happens to be at home at the time of the call—is unlikely to be a random family member, can create additional bias.

Alternatives to telephone surveys tend to have similar or other difficulties. For example, response rates on mail surveys tend to be especially low. On-site studies might yield a higher response rate, but, as already suggested, present other challenges. As already suggested, the problem with nonrespondents is that those who choose to respond might differ in characteristics, views or behavior from those who do not. In particular, people who hold strong views on a subject may be more likely to respond. Special inducements, such as financial compensation, can reduce the number of nonrespondents, but such inducements may also bias the results.

The preceding difficulties are, of course, well known to experts who conduct people-response studies, and various procedures have been developed to mitigate the resulting problems or to compensate for them. Thus, one approach to address the nonresponse problem is to take a follow-up sample of nonrespondents, making a special effort to solicit a response. The results are then compared with those of the initial sample which, if needed, are adjusted accordingly.

Another more general approach that aims to address both nonresponse and inadequacy of the frame is to compare the demographics—especially with regard to variables that are likely to be related to the survey response—of the respondents in the selected sample with those of the population of interest, and then correct for disproportionalities (in a manner similar to the analysis of results from a stratified sample). As a consequence, the results from a particular respondent might be weighted more heavily in the analysis than those of other respondents if the respondent's demographics appear to be underrepresented in the sample.

1.10 INFINITE POPULATION ASSUMPTION

The methods for calculating statistical intervals discussed in this book, except for those in Section 6.3, are based on the assumption that the sampled population or process is infinite, or, at least, very large relative to the sample size. However, in most enumerative studies, the assumption of an infinite population is not met. With a finite population, the sampling itself changes the population available for further sampling by depletion and therefore samples are no longer truly independent. For example, if the population consisted of 1,000 units, selection of the first sampled unit reduces the population available for the second sample to 999 units.

Books on survey sampling show how to use a "finite population correction" to adjust, approximately, for the finite population size. Using such a correction generally results in a narrower statistical interval than one would obtain without such a correction. Thus, ignoring the fact that one is sampling from a finite population usually results in conservative intervals (i.e., intervals that are wider than required for the specified confidence level). In an analytic study, the population or process is conceptually infinite; thus, no finite population correction is needed.

In practice, if the sample size is a small proportion of the population (10% or less is a commonly used figure), ignoring the correction factor will give results that are approximately correct. Thus, in many enumerative studies, it is not unreasonable to assume an infinite population in calculating a statistical interval.

The preceding discussion leads to a closely related, and frequently misunderstood, point concerning sampling from a finite population—namely, that the precision of the results depends principally on the *absolute*, and not the *relative*, sample size. For example, say you want to make a statement about the mean strengths of units produced from two lots from a stable production process. Lot 1 consists of 10,000 units and lot 2 consists of 100 units. Then, a simple random sample of 100 units from lot 1 (a 1% sample) provides more precise information about the mean of lot 1 than a random sample of size 10 from lot 2 (a 10% sample) provides about the mean of lot 2. We will describe methods for sample size selection in Chapters 8, 9, and 10.

1.11 PRACTICAL ASSUMPTIONS: OVERVIEW

In Figure 1.1 we summarize the major points of our discussion in Sections 1.3–1.10 and suggest a possible approach for evaluating the assumptions underlying the calculation of the statistical intervals described in this book. Although it is, of course, not possible to consider all

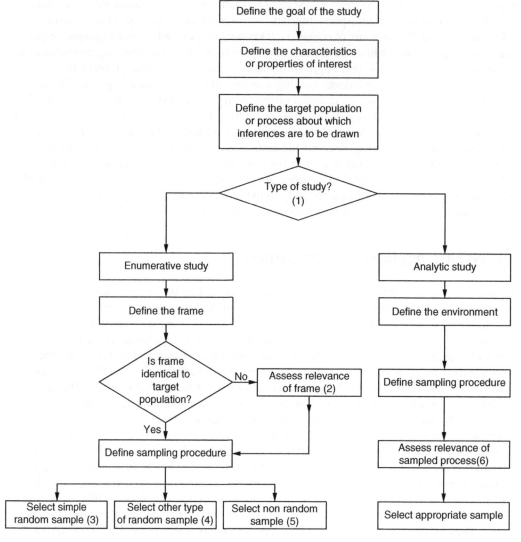

Figure 1.1 Possible approach to evaluating assumptions underlying the calculation of a statistical interval. See the text for explanations of the numbered items.

possible circumstances in such a diagram, we believe that Figure 1.1 provides a useful guide for practitioners on how to proceed for many situations.

The following comments pertain to the numbers shown in parentheses in Figure 1.1:

(1) Is the purpose of the study to draw conclusions about an existing finite population (enumerative study) or is it to act on and/or predict the performance of a (frequently future) process (analytic study)?

(2) Statistical intervals apply to the frame from which the sample is taken. When the frame does not correspond to the target population, inferences about the target population could be biased. A statistical interval quantifies only the sampling uncertainty due to a limited sample size. Actual uncertainty will frequently be larger.

(3) Most statistical intervals in this book assume a simple random sample from the frame.

(4) More complicated statistical intervals than those for simple random samples apply; see Section 1.6.3 and the books on survey sampling referred to in the Bibliographic Notes section at the end of this chapter.

(5) Statistical intervals do not apply. If calculated, they describe uncertainty to the extent that the nonrandom sample provides an approximation to a random sample. Resulting intervals in this case, as in other cases, do not account for uncertainty due to sampling bias.

(6) Statistical intervals apply to the sampled process, and not necessarily to the process of interest. Thus, any statistical interval does not account for uncertainty due to differences between the sampled process and the process of interest.

1.12 PRACTICAL ASSUMPTIONS: FURTHER EXAMPLE

We now cite, taking some liberties, a study (see Semiglazov et al., 1993) conducted for the World Health Organization (WHO) to evaluate the effectiveness of self-examination by women as a means of early detection of breast cancer. The study was conducted on a sample of factory workers in Leningrad (now St. Petersburg) and Moscow. This group of women was presumably selected for such practical reasons as the ready listing of potential participants and the willingness of factory management and workers to cooperate. A major characteristic of interest in this study is the time that self-examination saves in the detection of breast cancer.

Suppose, initially, that the goal was the very limited one of drawing conclusions about breast cancer detection times for female factory workers in Leningrad and Moscow at the time of the study. The frame for this (enumerative) study is the (presumably complete, current, and correct) listing of female factory workers in Leningrad and Moscow. In this case, the frame coincides with the target population, and it may be possible to obtain a simple random sample from this frame. We suppose further that the women selected by the random sample participate in the study and provide correct information and that the sample size is small (i.e., less than 10%), relative to the size of the population. Then the statistical intervals, provided in this book, apply directly for this (very limited) target population. (It is possible in an enumerative study to define the target population so narrowly that it becomes equivalent to the "sample." In that case, one has complete information about the population and, as previously indicated, the confidence intervals presented in this book degenerate to zero width, and the tolerance and prediction intervals become probability intervals.)

Extending our horizons slightly, if we defined the target population to be all women in Moscow and Leningrad at the time of the study, the frame (of female factory workers) is more restrictive than the target population. The statistical uncertainty, as expressed by the appropriate statistical interval, applies only to the sampled population (i.e., the female factory workers), and its relevance to the target population (i.e., all women in Moscow and Leningrad) needs to be assessed.

In actuality, the WHO is likely to be interested in a much wider group of women and a much broader period of time. In fact, the basic purpose of the study likely was that of drawing inferences about the effects of encouraging self-examination for women throughout the world, not only during the period of study, but, say, for the subsequent 25 years. In this case we are, in fact, dealing with an analytic study. In addition to the projection into the future, we need to be concerned with such matters as differences in self-examination learning skills and discipline, alternative ways of detecting breast cancer, the possibility of different manifestations of breast cancer, and many others. The unquantifiable uncertainty involved in translating the results from the sampled population or process (i.e., female factory workers in Moscow and Leningrad at the time of the study) to the (future) population or process of major interest (e.g., all women throughout the world in the subsequent 25 years) may well be much greater than the quantifiable statistical uncertainty.

Our comments are in no way a criticism of the WHO study, the major purpose of which appears to be that of assessing whether, under a particular set of circumstances and over a particular period of time, self-examination can be beneficial. We cite the study only as one example of an analytic study in which statistical intervals, such as those discussed in this book, describe only part of the total uncertainty, and may, in fact, be of very limited relevance.

Fortunately, not all studies are as global in nature and inference as this one. It seems safe to say, however, that in applications, the simple textbook case of an enumerative study in which the frame is in good agreement with the target population, and in which one has a random sample from this frame, is the exception, rather than the rule. Instead it is more common to encounter situations in which:

- One wishes to draw inferences concerning a process (and, thus, is dealing with an analytic, rather than an enumerative, study).

- One is dealing with an enumerative study, but the frame differs from the target population in important respects, and/or sampling from the frame is not (strictly) random.

As indicated, in each of these cases, we need to be concerned with the implications in generalizing our conclusions beyond what is warranted from statistical theory alone—or, as we have repeatedly stated, the calculated statistical interval generally provides an optimistic quantification of the total uncertainty, reflecting only the sampling variability. Thus, in studies like the WHO breast cancer detection evaluation, the prudent analyst needs to decide whether to calculate statistical intervals at all—and, if so, stress their limitations—or to refrain from calculating such intervals in the belief that they may do more harm than good. In any case, such intervals need to be supplemented by, and often are secondary to, the use of statistical graphics to describe the data—as illustrated in the subsequent chapters.

1.13 PLANNING THE STUDY

A logical conclusion from the preceding discussion is that it is of prime importance to properly plan the study to help assure that:

- The target population or process of interest is well defined initially.

- For an enumerative study, the frame matches the target population as closely as practical and the sampling from this frame is random or as close to random as feasible.

- For an analytic study, the investigation is made as broad as possible so as to reduce the almost inevitable gap between the sampled process and the process of interest and randomization is introduced to the degree feasible.

Unfortunately, studies are not always conducted in this way. Often, analysts are handed the results and asked to analyze the data. This requires *retrospectively* defining the target population or process of interest and the frame or process that was actually sampled, and determining how well the critical assumptions for making statistical inferences apply. This is often a frustrating, or even impossible, task because the necessary information is not always available. In fact, one may sometimes conclude that in light of the deficiencies of the investigation or the lack of knowledge about exactly how the study was conducted, it might be misleading to employ any method of statistical inference.

The moral is clear. If one wishes to perform statistical analyses of the data from a study, including calculation of the intervals described here, it is essential to plan the investigation statistically in the first place. One element of planning the study is determining the required sample size; see Chapters 8, 9, and 10. This technical consideration is, however, often secondary to the more fundamental issues described in this chapter. Further details on planning studies are provided in texts on survey sampling (dealing mainly, but not exclusively, with enumerative studies) and books on experimental design (dealing mainly with analytic studies). See the references in the Bibliographic Notes section at the end of this chapter.

1.14 THE ROLE OF STATISTICAL DISTRIBUTIONS

Many of the statistical intervals described in this book assume a distributional model, such as a normal distribution, for the measured variable, possibly after some transformation of the data. Frequently, the assumed model only approximately represents the population or process, although this approximation is often adequate for the problem at hand. With a sufficiently large sample (say, 40 or more observations), it is usually possible to detect important departures from the assumed model and, if the departure is large, to decide whether there is a need to reject or refine the model. For more pronounced departures, fewer observations are needed for such detection. When there is not enough data to detect important departures from the assumed model, the model's correctness must be justified from an understanding of the physical situation and/or past experience. Such understandings, of course, should enter the assessment, irrespective of the sample size.

Some intervals—notably confidence intervals to include the population mean—are relatively insensitive to the assumed distribution; other intervals strongly depend on this assumption. We will indicate the importance of distributional assumptions in our discussion of specific intervals. Hahn (1971) discusses "How Abnormal is Normality?", and numerous books on statistics, including Hahn and Shapiro (1967, Chapter 8), describe graphical methods and statistical tests to evaluate the assumption of normality. We provide a brief introduction to, and example of, this subject in Section 4.11.

Statistical intervals that do not require any distributional assumptions are described in Chapters 5 and 13 and in several of the case studies in Chapters 11 and 18. Such "nonparametric" intervals have the obvious advantage that they do not require the assumption of a specific distribution. Nonparametric intervals are, therefore, especially appropriate for those situations for which the results are sensitive to assumptions about the underlying distribution. A disadvantage of nonparametric intervals is that they tend to be wider (i.e., less precise) than the corresponding interval under an assumed distributional model. Moreover, frequently there is not sufficient data to compute a nonparametric interval at the desired confidence level. Such intervals also still require the other important assumptions discussed in the preceding sections.

1.15 THE INTERPRETATION OF STATISTICAL INTERVALS

Because statistical intervals are based upon limited sample data that are subject to random sampling variation, they will sometimes not contain the quantity of interest that they were calculated to contain, even when all the necessary assumptions hold. Instead, they can be claimed to be correct only a specified percentage (e.g., 90%, 95%, or 99%) of the time they are calculated, that is, they are correct with a specified "degree of confidence." The percentage of such statistical intervals that contain what one claims they contain is known as the confidence level associated with the interval. The selection of a confidence level is discussed in Section 2.6.

The confidence level, at least from a traditional point of view, is a property *of the procedure* for constructing a particular statistical interval, and not a property of the computed interval itself. Thus, the confidence level is the probability that, in any given study, the random sample will result in an interval that contains what it is claimed to contain. Using this (classical) interpretation, a particular confidence interval to, for example, contain the mean of a population *cannot* correctly be described as being an interval that contains the actual population mean with a specified probability. This is because the mean is an unknown fixed characteristic of the population which, in a given situation, is either contained within the interval or is not (i.e., the probability is either one or zero). All we can say is that in calculating many different confidence intervals to contain population means from different (independent) random samples, the calculated confidence interval will actually contain the actual population mean with a specified probability—known as the *coverage probability*—and, due to the vagaries of chance, will fail to do so the other times. We provide further, more specific, elaboration in Sections 2.2.5, 2.3.6, and 2.4.3.

For many of the best-known statistical interval procedures, because of the simplicity of the model assumptions, (e.g., the procedures given in Chapters 3 and 4 for the normal distribution) the coverage probability of the interval procedure is *exactly* equal to the nominal confidence level that is input to the procedure. Outside this relatively narrow set of circumstances, however, statistical interval procedures have coverage probabilities that are only approximately equal to the nominal confidence level. Indeed, it has been a vigorous area of statistical research to find new and better statistical interval procedures that provide better coverage probability approximations. The results of some of this research are used for the interval procedures presented in Chapters 5–7 and 12–18.

Finally, we note that the philosophy of inferences in constructing Bayesian intervals that we present in Chapters 15, 16, 17, and some of the examples in Chapter 18, differs from the preceding non-Bayesian (sometimes referred to as "frequentist") approach to constructing and evaluating statistical intervals. In particular, Bayesian inference methods require the specification of a joint prior distribution to describe our *prior knowledge* about the values of the model parameters (but the use of the probability distribution in this context does *not* imply that the unknown parameter is random). Using conditional probability operations (known as Bayes' theorem), the prior distribution is combined with the data to generate a joint posterior distribution, representing the updated state of knowledge about the parameters. Based on the joint posterior distribution, it is possible, for example, to generate intervals to contain a specific parameter or a particular function of the parameters with a specified *probability*. As a result, such intervals based on Bayesian methods are often referred to as "credible intervals" and not "confidence intervals" (and we will use such terminology in subsequent chapters). One should, however, keep in mind that for given data the probability in the credible interval statement comes directly from the prior distribution. The actual parameter value (fixed, but unknown) is, again, either contained in the interval or not.

Some make a distinction between "subjective Bayesian" and "objective Bayesian" analyses. At the risk of oversimplifying the distinction, in a subjective Bayesian approach, one uses some combination of previous experience, expert opinion, and other subjective information to choose the joint posterior distribution. The objective Bayesian approach, on the other hand, attempts to specify a prior distribution that uses little or no prior information to set the prior distribution (variously referred to as "default," "reference," "noninformative," "vague," or "diffuse" prior distributions) that might, for example, have the objective of producing a Bayesian procedure with good frequentist properties (e.g., that the coverage probability be close to the nominal confidence level). Indeed, the vast majority of Bayesian analyses in practical applications use the objective approach. In such cases one could argue that the use of the term "confidence interval" is still warranted. In many (if not most) practical problems where an informative prior distribution is to be used, the analysis will involve a combination of the subjective and the objective approaches because useful prior information may be available for only one of several parameters.

1.16 STATISTICAL INTERVALS AND BIG DATA

Much has been said about the technological changes that have brought us into the "big data" era. Big data is a consequence of the availability and accessibility of enormously large and complicated data sets. "Big" is a relative term, and how big is "big" is often characterized by the volume, variety, and velocity (know as the "three Vs") of a data set. The arrival of big data, as well as our ability to analyze such data and potentially gain useful information therefrom, were made possible by advances in sensor, communications, data storage, and computational technology.

How do statistical intervals pertain to big data? An immediate answer might be that the confidence intervals discussed in this book have little relevance for big data. This is because confidence intervals are used to quantify the uncertainty in estimating some characteristic(s) of interest due to the (random) sampling of a population or process. This uncertainty is most pronounced when one has limited sample data. In dealing with big data there is little, or often essentially no, sampling uncertainty. In addition, as previously indicated (and as will be discussed in more detail in Chapter 2), tolerance and prediction intervals converge to becoming probability intervals describing a distribution as the sample size increases.

Thus, our major concern in dealing with big data needs to be not with statistical variability, but with the "representativeness" of the data with regard to the population of processes of interest, as discussed throughout this chapter.

So why, in this age of big data, should we be concerned with statistical intervals—much less update a book on the subject? Our answer is simple. Even though the era of big data is, indeed, upon us and raises numerous new challenges, there are still many, and perhaps even most, situations in which circumstances limit us to small data. In assessing a rare disease, one might, for example, have only a relatively few recorded cases. In conducting a sample survey, budget limitations typically restrict us to a small sample. In new product reliability assessments, sample availability and cost concerns often impose severe restrictions on the amount of testing that can be conducted. And these are just a few examples. In fact, in most situations in which we generate new data, such as designed experiments or sample surveys, as opposed to analyzing existing data, one is restricted to relatively small samples and, therefore, it is appropriate to use statistical intervals to quantify the associated statistical uncertainty. In summary, even though this surely is the age of big data, let us not forget the continued need for drawing meaningful inferences from small sets of data.

We also note that the ability to do computer-intensive analyses, which have helped bring about the age of big data, has also led to the development of improved methods for constructing statistical intervals for more complicated inference problems dealing mainly with small to medium-size data, as described in Chapters 12–18, thus making our analyses appreciably more powerful.

1.17 COMMENT CONCERNING SUBSEQUENT DISCUSSION

The assumptions that we have emphasized in this chapter apply throughout this book and warrant restatement each time we present an interval or an example. We have decided, however, to relieve the reader from such repetition. Thus, frequently, we limit ourselves to saying that the resulting interval applies "to the sampled population or process" or more generically, "to the sampled distribution" and often we omit altogether making any such restrictive statement. The reader needs, however, to keep in mind in all applications the underlying assumptions and admonitions stated in this chapter.

BIBLIOGRAPHIC NOTES

General treatment of statistical intervals

Hahn (1970b) describes confidence intervals, tolerance intervals, and prediction intervals for a normal distribution. Both Scheuer (1990) and Vardeman (1992) discuss in detail confidence intervals, prediction intervals, and tolerance intervals that are distribution-free and also ones that depend on the assumption of a normal distribution.

Enumerative and analytic studies

Using different terminology than Deming, the books by Snedecor and Cochran (1967, pages 15–16) and Box et al. (2005, Chapters 1–3) discuss the differences between enumerative and analytical studies. Also, this distinction is explicitly discussed in detail in the book by Gitlow et al. (1989). Some relevant discussion of enumerative and analytical studies appears in Hahn and Meeker (1993), Chatfield (1995, 2002), Hillmer (1996), Wild and Pfannkuch (1999), and MacKay and Oldford (2000). Emphasis on making the distinction between enumerative and analytic studies seems to have waned in recent years—which, in the authors' opinion, is unfortunate.

Books about survey sampling

Statistical intervals can also be constructed for random sampling schemes that are more complicated than simple random sampling. Such intervals generally require special methods for estimating variances and are described in books on survey sampling such as Cochran (1977), Levy and Lemeshow (2008), Groves et al. (2009), Lohr (2010), Lumley (2010), and Little (2014).

Books about experimental design

There are numerous books on the subject of experimental design. These include, for example, Box et al. (2005), Montgomery (2009), Wu and Hamada (2009), Goos and Jones (2011), and Morris (2011).

Chapter *2*

Overview of Different Types of Statistical Intervals

OBJECTIVES AND OVERVIEW

This chapter introduces different kinds of statistical intervals and their applications. A discussion of the procedures for constructing such intervals is postponed to subsequent chapters. We begin with some general comments about the choice of a statistical interval. We end the chapter with discussions of some practical issues concerning the use of statistical intervals.

This chapter explains:

- Reasons for constructing a statistical interval and some examples (Section 2.1).

- Different types of confidence intervals and one-sided confidence bounds (Section 2.2).

- Different types of prediction intervals and one-sided prediction bounds (Section 2.3).

- Tolerance intervals and one-sided tolerance bounds (Section 2.4).

- The selection of an appropriate statistical interval (Section 2.5).

- The selection of a confidence level (Section 2.6).

- The difference between two-sided intervals and one-sided statistical bounds (Section 2.7).

- The advantages of using confidence intervals instead of significance tests (Section 2.8).

- The use of simultaneous statistical intervals (Section 2.9).

2.1 CHOICE OF A STATISTICAL INTERVAL

The appropriate statistical interval for a particular application depends on the application. Thus, the analyst must determine which interval(s) to use, based on the needs of the problem. The

Statistical Intervals: A Guide for Practitioners and Researchers, Second Edition.
William Q. Meeker, Gerald J. Hahn and Luis A. Escobar.
© 2017 John Wiley & Sons, Inc. Published 2017 by John Wiley & Sons, Inc.
Companion Website: www.wiley.com/go/meeker/intervals

Characteristic of interest	General purpose of the statistical interval	
	Description	Prediction
Location	Confidence interval for a distribution mean or median or a specified distribution quantile	Prediction interval for a future sample mean, future sample median, or a particular ordered observation from a future sample
Spread	Confidence interval for a distribution standard deviation	Prediction interval for the standard deviation of a future sample
Enclosure interval	Tolerance interval to contain (or cover) at least a specified proportion of a distribution	Prediction interval to contain all or most of the observations from a future sample
Probability of an event	Confidence interval for the probability of an observation being less than (or greater than) some specified value	Prediction interval to contain the proportion of observations in a future sample that exceed a specified limit

Table 2.1 Examples of some statistical intervals.

following comments give brief guidelines for this selection. Table 2.1 categorizes the statistical intervals discussed in this book and provides some examples. The intervals are classified according to (1) the general purpose of the interval and (2) the characteristic of interest.

2.1.1 Purpose of the Interval

In selecting an interval, one must decide whether the main interest is in *describing the population or process* from which the sample has been selected or in *predicting the results of a future sample* from the same population or process, which we will refer to more generally as a *distribution*. Intervals that describe the sampled distribution (or enclose parameters of a distribution) include confidence intervals for the distribution mean and for the distribution standard deviation and tolerance intervals to contain at least a specified proportion of a distribution. In contrast, prediction intervals for a future sample mean and for a future sample standard deviation and prediction intervals to include all of m future observations deal with predicting the results of a future sample from a previously sampled distribution.

2.1.2 Characteristic of Interest

There are various characteristics of interest that one may wish to enclose with a high degree of confidence. The intervals considered in this book may be roughly classified as dealing with:

- The *location* of a distribution (or future sample) as measured, for example, by its mean or a specified distribution quantile. Specific examples are a confidence interval for the distribution mean and a prediction interval for the sample mean of a future sample.

- The *spread* of a distribution (or future sample) as measured, for example, by its standard deviation. Specific examples are a confidence interval for a distribution standard deviation and a prediction interval to contain the standard deviation of a future sample from the distribution.

- An *enclosure interval*. Specific examples are a tolerance interval to contain at least a specified proportion of a distribution and a prediction interval to contain all, or most, of the observations from a future sample.

- The *probability of an event*. A specific example is a confidence interval for the probability that a measurement will exceed a specified threshold value.

We shall now describe some specific intervals in greater detail.

2.2 CONFIDENCE INTERVALS

Estimates from sample data provide an approximation to some unknown truth. Confidence intervals provide a quantification of the precision of the approximation.

2.2.1 Confidence Interval for a Distribution Parameter

Confidence intervals quantify the precision of our knowledge about a parameter or some other characteristic of a distribution representing a population or process, based upon a random sample. For example, asserting that a 95% confidence interval to contain the mean lifetime of a particular brand of light bulb is 1,000 to 1,200 hours is considerably more informative than simply stating that the mean lifetime is approximately 1,100 hours.

A frequently used type of confidence interval is one to contain the distribution mean. Sometimes, however, one desires a confidence interval to include some other parameter, such as the standard deviation of a normal distribution. For example, our knowledge, based upon past data, about the precision of an instrument that measures air pollution might be expressed by a confidence interval for the standard deviation of the instrument's measurement error. Confidence intervals for parameters of other statistical distributions, such as for the "failure rate" of an exponential distribution or for the shape parameter of a Weibull distribution, are also sometimes desired.

2.2.2 Confidence Interval for a Distribution Quantile

Frequently, our primary interest centers on one or more quantiles (also known as percentage points or percentiles) of a distribution, rather than the distribution's parameters. For example, in evaluating the tensile strength of an alloy, it might be desired to estimate, using data from a random sample of test specimens, the loading that would cause 1% of such specimens to fail. Thus, one would wish to construct a confidence interval for the 0.01 quantile of the tensile strength distribution.

2.2.3 Confidence Interval for the Probability of Meeting Specifications

Many practical problems involve stated performance or specification limits that are required to be met. The sample data are then used to assess the probability of meeting the specification(s). Some examples are:

- A machined part needs to be between 73.151 and 73.157 centimeters in diameter in order to fit correctly with some other part.

- The noise level of an engine must be less than 80 decibels to satisfy a government regulation.

- According to the US Environmental Protection Agency, to be safe the concentration of arsenic in drinking water needs to be less than 10 parts per billion.

- The mean lifetime of a light bulb must be at least 1,000 hours to meet an advertising claim.

- The net contents of a bottled soft drink must be at least 32 fluid ounces to conform with the product labeling.

The preceding situations are characterized by the fact that a particular limit is stated. In such cases, the probability of meeting this stated limit is to be evaluated. If the sampled distribution is *completely specified* (i.e., including its parameters), then elementary methods, described in introductory probability and statistics textbooks, can be used to determine the exact probability of meeting the specification limit. A common example arises in dealing with a normal distribution with a *known* mean and standard deviation. When, however, the available information is limited to a random sample from the distribution, as is frequently the case, the probability of, or the proportion of a distribution, meeting the specification limit cannot be found exactly. One can, however, construct a confidence interval for the unknown probability or proportion. For example, from the available data, one can construct a confidence interval for the proportion of light bulbs from a specified population that will survive 1,000 hours of operation without failure or, equivalently, the probability that any randomly selected bulb will operate for at least 1,000 hours. This probability is often referred to as the product's reliability at 1,000 hours.

2.2.4   One-Sided Confidence Bounds

Much of the discussion so far has implied a two-sided confidence interval with *both* a finite lower endpoint and a finite upper endpoint. Usually these intervals have an equal degree of uncertainty associated with the parameter or other quantity of interest being located outside each of the two interval endpoints. A two-sided interval is generally appropriate, for example, when one is dealing with specifications on dimensions to allow one part to fit with another. In many other problems, the major interest, however, is restricted to the lower endpoint alone or to the upper endpoint alone. This is frequently the case for problems dealing with product quality or reliability. In such situations, one is generally concerned with questions concerning "how bad might things be?" and not "how good might they be?" This may call for the construction of a one-sided lower confidence bound or a one-sided upper confidence bound, depending upon the specific application, rather than a two-sided confidence interval. For example, the results of a reliability demonstration test for a system might be summarized by stating, "The estimated reliability for a mission time of 1,000 hours of operation for the system is 0.987, and a lower 95% confidence bound on the reliability is 0.981." This "pessimistic" bound would be of particular interest when a minimum reliability must be met.

2.2.5   Interpretations of Confidence Intervals and Bounds

Due to random sampling variation, a sample does not provide perfect information about the sampled population. However, using the results of a random sample, a confidence interval for some parameter or other fixed, unknown characteristic of the sampled population provides limits which one can claim, with a specified degree of confidence, contain the actual value of that parameter or characteristic.

A $100(1 - \alpha)\%$ confidence interval for an unknown quantity θ may be formally characterized as follows: "If one repeatedly calculates such intervals from many independent random samples, $100(1 - \alpha)\%$ of the intervals would, in the long run, correctly include the actual value θ.

Equivalently, one would, in the long run, be correct $100(1 - \alpha)\%$ of the time in claiming that the actual value of θ is contained within the confidence interval." More commonly, but less precisely, a two-sided confidence interval is described by a statement such as "we are 95% confident that the interval $[\underset{\sim}{\theta}, \widetilde{\theta}]$ contains the unknown actual parameter value θ." In fact, the observed interval either contains θ or does not. Thus the 95% refers to the *procedure* for constructing a confidence interval, and not to the observed interval itself. One-sided confidence bounds can be similarly interpreted.

2.3   PREDICTION INTERVALS

2.3.1   Prediction Interval to Contain a Single Future Observation

A prediction interval for a single future observation is an interval that will, with a specified degree of confidence, contain a future randomly selected observation from a distribution. Such an interval would interest the purchaser of a single unit of a particular product and is generally more relevant to such an individual than, say, a confidence interval to contain average performance. For example, the purchaser of a new automobile might wish to obtain, from the data on a previous sample of five similar automobiles, an interval that contains, with a high degree of confidence, the gasoline mileage that the new automobile will obtain under specified driving conditions. This interval is calculated from the sample data under the important assumption that the previously sampled automobiles and the future one(s) can be regarded as random samples from the same distribution; this assumes identical production processes and similar driving conditions. In many applications, the population may be conceptual, as per our discussion of analytic studies in Chapter 1.

2.3.2   Prediction Interval to Contain All of m Future Observations

A prediction interval to contain the values of all of m future observations generalizes the concept of a prediction interval to contain a single future observation. For example, a traveler who must plan a specific number of trips may not be interested in the amount of fuel that will be needed on the average for all future trips. Instead, this person would want to determine the amount of fuel that will be needed to complete each of, say, one, or three, or five future trips.

Prediction intervals to contain all of m future observations are often of interest to manufacturers of large equipment who produce only a small number of units of a particular type product. For example, a manufacturer of gas turbines might wish to establish an interval that, with a high degree of confidence, will contain the performance values for all three units in a future shipment of such turbines, based upon the observed performance of similar past units. In this example, the past units and those in the future shipment would conceptually be thought of as random samples from the population of all turbines that the manufacturer might build (see the discussion in Section 1.2).

Prediction intervals are especially pertinent to users of one or a small number of units of a product. Such individuals are generally more concerned with the performance of a specific sample of one or more units, rather than with that of the entire process from which the sample was selected. For example, based upon the data from a life test of 10 systems, one might wish to construct an interval that would have a high probability of including the lives of all of three additional systems that are to be purchased. Prediction intervals to contain all of m future observations are often referred to as *simultaneous* prediction intervals, because one is concerned with simultaneously containing *all* of the m observations within the calculated interval (with the associated level of confidence).

2.3.3   Prediction Interval to Contain at Least k out of m Future Observations

A generalization of a prediction interval to contain all m future observations is one to contain at least k out of m of the future observations. We will refer to this type of interval again in Section 2.4.1.

2.3.4   Prediction Interval to Contain the Sample Mean or Sample Standard Deviation of a Future Sample

Sometimes one desires an interval to contain the *sample mean* (or sample standard deviation or other estimated quantity) of a future sample of m observations, rather than one to contain *all* (or at least k) of the future sample values. Such an interval would be pertinent, for example, if acceptance or rejection of a particular design were to be based upon the sample mean of a future sample from a previously sampled distribution.

Consider the following example: A manufacturer of a high voltage insulating material must provide a potential customer "performance limits" to contain average breakdown strength of the material, estimated from a destructive test on a sample of 10 units. Here "average" is understood to be the sample mean of the readings on the units. The tighter these limits, the better are the chances that the manufacturer will be awarded a forthcoming contract. The manufacturer, however, has to provide the customer five randomly selected units for a test. If the sample mean for these five units does not fall within the performance limits stated by the manufacturer, the product is automatically disqualified. The manufacturer has available a random sample of 15 units from production. Ten of these units will be randomly selected and tested by the manufacturer to establish the desired limits. The remaining five units will be shipped to and tested by the customer. Based on the data from the sample of 10 units, the manufacturer will establish prediction limits for the sample mean of the five future readings so as to be able to assert with 95% confidence that the sample mean of the five units to be tested by the customer will lie in the interval. A 95% prediction interval to contain the future sample mean provides the desired limits.

Alternatively, suppose that in the preceding example the concern is uniformity of performance, as measured by the sample standard deviation, rather than sample mean performance. In this case, one might wish to compute a prediction interval, based upon measurements on a random sample of 10 units, to contain the standard deviation of a future sample of five units from the same process or population.

2.3.5   One-Sided Prediction Bounds

Some applications call for a one-sided prediction bound, instead of a two-sided prediction interval, to contain future sample results. An example is provided by a manufacturer who needs to warranty the efficiency of all units (or of their average) for a future shipment of three motors, based upon the results of a sample of five previously tested motors from the same process. This problem calls for a one-sided lower prediction bound, rather than a two-sided prediction interval.

2.3.6   Interpretation of Prediction Intervals and Bounds

If all the parameters of a probability distribution are known, one can compute a probability interval to contain the values of a future sample. For example, for a normal distribution with mean μ and standard deviation σ, the probability is 0.95 that a single future observation will be contained in the interval $\mu \pm 1.96\sigma$. In the more usual situation where one has only sample data, one can construct a $100(1 - \alpha)\%$ prediction interval to contain the future observation with a specified degree of confidence. Such an interval may be formally characterized as follows: "If from many independent pairs of random samples, a $100(1 - \alpha)\%$ prediction interval is computed from the data of the first sample to contain the value(s) of the second sample, $100(1 - \alpha)\%$ of the intervals would, in the long run, correctly bracket the future value(s)." Equivalently, one would, in the long run, be correct $100(1 - \alpha)\%$ of the time in claiming that the future value(s) will be contained within the prediction interval. The requirement of independence holds both with regard to the different pairs of samples and the observations within each sample.

The probability that a particular prediction interval will contain the future value that it is supposed to contain is unknown because the probability depends on the unknown parameters. As with confidence intervals, the $100(1 - \alpha)\%$ refers to the *procedure* used to construct the prediction interval and not any particular interval that is computed.

2.4 STATISTICAL TOLERANCE INTERVALS

2.4.1 Tolerance Interval to Contain a Proportion of a Distribution

Prediction intervals are in general useful to predict the performance of one, or a small number, of future units. Consider now the case where one wishes to draw conclusions about the performance of a relatively large number of future units, based upon the data from a random sample from the distribution of interest. Conceptually, one can also construct prediction intervals for such situations (e.g., a prediction interval to contain all 100, 1,000, or any number m, future units). Such intervals would, however, often be very wide. Also, the exact number of future units of interest is sometimes not known or may be conceptually infinite. Moreover, rather than requiring that the calculated interval contain *all* of a specified number of units, it is generally sufficient to construct an interval to contain a *large proportion* of such units.

As indicated in Section 2.3.3, there are procedures for calculating prediction intervals to contain at least k out of m future observations, where $k \leq m$. More frequently, however, applications call for the construction of intervals to contain a specified proportion, β, of the entire sampled distribution. This leads to the concept of a tolerance interval.

Specifically, a tolerance interval is an interval that one can claim to contain at least a specified proportion, β, of the distribution with a specified degree of confidence, $100(1 - \alpha)\%$. Such an interval would be of particular interest in setting limits on the process capability for a product manufactured in large quantities. This is in contrast to a prediction interval which, as noted, is of greatest interest in making predictions about a small number of future units.

Suppose, for example, that measurements of the diameter of a machined part have been obtained on a random sample of 25 units from a production process. A tolerance interval calculated from such data provides limits that one can claim, with a specified degree of confidence (e.g., 95%), contains the (measured) diameters of at least a specified proportion (e.g., 0.99) of units from the sampled process.

The two numbers in the preceding statement should not create any confusion when one recognizes that the 0.99 refers to the proportion of the distribution to be contained, and the 95% deals with the degree of confidence associated with the claim.

2.4.2 One-Sided Tolerance Bounds

Practical applications often require the construction of one-sided tolerance bounds. For example, in response to a request by a regulatory agency, a manufacturer has to make a statement concerning the maximum noise that, under specified operating conditions, is met (i.e., is not exceeded) by a high proportion of units, such as 0.99 of a particular model of a jet engine. The statement is to be based upon measurements from a random sample of 10 engines and is to be made with 95% confidence. In this case, the manufacturer desires a one-sided upper 95% tolerance bound that will exceed at least a proportion 0.99 of the population of jet engines, based upon the previous test results. A one-sided upper tolerance bound is appropriate here because the regulatory agency is concerned principally with how noisy, and not how quiet, the engines might be.

A one-sided tolerance bound is equivalent to a one-sided confidence bound on a distribution quantile (see Section 2.2.2). More specifically, a one-sided lower $100(1 - \alpha)\%$ confidence bound on the p quantile of a distribution is equivalent to a one-sided lower tolerance bound that one can claim with $100(1 - \alpha)\%$ confidence is exceeded by at least a proportion $1 - p$ of the distribution. Similarly, a one-sided upper $100(1 - \alpha)\%$ confidence bound on the p quantile of a distribution is equivalent to a one-sided upper tolerance bound that one can claim with $100(1 - \alpha)\%$ confidence exceeds at least a proportion p of the distribution. For this reason, we will, in most parts of this book, focus on the appropriate two-sided confidence interval or one-sided confidence bound for a quantile, instead of discussing one-sided tolerance bounds.

2.4.3 Interpretation of β-Content Tolerance Intervals

As previously noted, if the parameters of a distribution are known, one can use elementary methods to compute a probability interval to contain a specified proportion of the distribution. Typically, however, the distribution parameters are unknown and the available information is limited to a sample. The lack of perfect information about the distribution is taken into consideration by the confidence statement associated with the tolerance interval. Thus, a tolerance interval will bracket at least a certain proportion of the distribution with a specified degree of confidence. A tolerance interval to contain at least a proportion β of the distribution with $100(1 - \alpha)\%$ confidence may be formally characterized as follows: "If one calculates such intervals from many independent random samples, $100(1 - \alpha)\%$ of the intervals would, in the long run, correctly include at least a proportion β of the distribution (or, equivalently, one would, in the long run, be correct $100(1 - \alpha)\%$ of the time in claiming that the actual proportion of the distribution contained in the interval is at least β)." Such tolerance intervals are often referred to as β-content tolerance intervals.

As with confidence and prediction intervals, the $100(1 - \alpha)\%$ refers to the *procedure* used for constructing the tolerance interval and not to any particular computed tolerance interval. The actual proportion of the population contained within the tolerance interval is unknown because this proportion depends on the unknown parameters.

2.4.4 β-Expectation Tolerance Intervals

In the statistical literature, there are references to both "β-content tolerance intervals" and "β-expectation tolerance intervals." A β-content tolerance interval, also sometimes referred to as a $(\beta, 1 - \alpha)$ tolerance interval, is what we call a "tolerance interval" (Section 2.4.1). A β-expectation tolerance interval is what we call a $100\beta\%$ prediction interval for a single future observation (Section 2.3.1). Our terminology is consistent with much of the applications-oriented literature.

2.5 WHICH STATISTICAL INTERVAL DO I USE?

In the previous sections, we differentiated among the various types of statistical intervals. It should be clear from this discussion that the appropriate interval depends upon the problem at hand. That is, the specific questions that need to be answered from the data will determine whether a confidence interval for a distribution mean, a confidence interval for a distribution standard deviation, a confidence interval for a distribution quantile, or some other confidence interval or tolerance interval or prediction interval is needed. Thus, the analyst has to decide, based upon an understanding of the nature of the problem, which specific type of statistical interval is needed in a particular application. Moreover, as detailed as the discussion in this book may seem, we can provide specific information on only a relatively small number of statistical intervals in the forthcoming chapters. General methods for constructing statistical intervals are given in Chapters 12–15 and these general methods can be used in many more situations, as illustrated in these chapters and in Chapters 16–18. Other applications may require a statistical interval that is different from any of those described here (e.g., involving a distribution or random error structure or a correlation structure not described in this book). We hope, however, that our discussion will help the analyst identify when this is the case, and, if necessary, call upon a professional statistician for guidance in developing the needed interval.

2.6 CHOOSING A CONFIDENCE LEVEL

All statistical intervals have an associated confidence level. Loosely speaking, the confidence level is the degree of assurance one desires that the calculated statistical interval contains the value of interest. (See Sections 2.2.5, 2.3.6, and 2.4.3 for a more precise definition.) The analyst must determine the confidence level, based upon what seems to be an acceptable degree of assurance, for each application. Thus, one has to trade off the risk of not including the correct parameter, quantile, future observation, etc., in the interval against the fact that as the degree of assurance is raised, the statistical interval becomes longer.

2.6.1 Further Elaboration

Statistical consultants are frequently requested by their clients to construct 95% confidence intervals to contain, say, the means of different populations from which random samples have been taken. In 95% of such cases, the calculated interval will include the actual population mean—if the various assumptions stated in Chapter 1 hold. Due to chance, however, the client is "misled" in 5% of the cases; that is, 5% of the time the computed interval will not contain the population mean. Clients who are afraid of being among the unlucky 5%, and desire added protection, can request a higher level of confidence, such as 99%. This reduces to one in a hundred the chances of obtaining an interval that does not contain the population mean. The client, however, pays a price for the higher level of confidence. For a fixed sample size, increasing the confidence level results in an increase in the width of the calculated interval. For example, the extreme width of a 99.9% confidence interval may be quite sobering. Besides this, in most cases, the only restriction in selecting a confidence level is that it be less than 100%; to obtain 100% confidence, the entire population must be observed. This, of course, is not possible if our concern is with a future process.

2.6.2 Problem Considerations

The importance of any decision that is to be made from a statistical interval needs to be taken into consideration in selecting a confidence level. For example, at the outset of a research project, one might be willing to take a reasonable chance of drawing incorrect conclusions about the mean of the distribution and use a relatively low confidence level, such as 90%, or even 80%, because the initial conclusions will presumably be corroborated by subsequent analyses. On the other hand, in reporting the final results of a project that may result in the building an expensive new plant or the release of a new drug, one might wish a higher degree of confidence in the correctness of one's claim.

2.6.3 Sample Size Considerations

As we shall see in subsequent chapters, narrower statistical intervals are expected with larger sample sizes for a fixed level of confidence. This is especially the case for confidence intervals to contain a distribution parameter. Increasing the sample size by a factor of k will have the general effect of reducing the width of a confidence interval by a factor of (approximately) the square root of k. For example, increasing the sample size fourfold will have the general effect of (approximately) halving the width of a confidence interval (the result is approximate because the width of a confidence interval is itself random).

The expected width of prediction intervals and tolerance intervals also becomes narrower with increased sample size, but instead of shrinking to zero, converges to that of the probability intervals that one would obtain if the model parameters were known. In Chapters 8–10 we give simple methods for evaluating the effect that sample size has on the expected width of a statistical interval and for finding the sample size that is needed to provide an interval that has a specified expected width. This information can be used to evaluate the trade-off between sample size and choice of confidence level.

Sometimes, one might wish to use relatively high confidence levels (e.g., 99%) with large samples, and lower confidence levels (e.g., 90% or 80%) with small samples. This implies that the more data one has, the surer one would like to be of one's conclusions. Also, such practice recognizes the fact that, for small sample sizes, the calculated statistical interval is often so wide that it has little practical meaning and, therefore, one may wish to obtain a somewhat narrower interval at the cost of reducing one's confidence in its correctness.

2.6.4 A Practical Consideration

As indicated in Chapter 1, statistical intervals take into consideration only the vagaries of random statistical fluctuations due to sampling variation. The sample is assumed to be randomly selected from the distribution of interest. Moreover, unless a nonparametric procedure (i.e., a procedure that does not require a distributional assumption) is used, a statistical model, such as a normal distribution, is assumed. As we have indicated, these assumptions hold only approximately, if at all, in many real-world problems. Errors due to deviations from these assumptions are not included in the confidence level. A high confidence level, thus, may provide a false sense of security. A user of a statistical interval with 99.9% associated confidence might forget that the interval could fail to include the value of interest because of such other factors. A similar point could be made for an interval with 90% associated confidence, but users usually regard 90% intervals with less reverence than 99.9% intervals.

2.6.5 Further Remarks

The 95% confidence level appears to be used more frequently in practice than any other level, perhaps out of custom; 90% and 99% confidence levels seem to be next in popularity. However, confidence levels such as 50%, 75%, 80%, 99.9%, and even 99.99% are also used. The lower confidence levels are, perhaps, used more frequently with prediction and tolerance intervals than with confidence intervals. This is so because, as we shall see, prediction and tolerance intervals tend to be wider than confidence intervals, especially with larger samples.

There is no single, straightforward answer to the question, "What confidence level should I select?" Much depends on the specific situation and the desired trade-off between risk and interval width. One can sidestep the issue by presenting intervals for a number of different confidence levels, and this in fact is often a reasonable approach.

2.7   TWO-SIDED STATISTICAL INTERVALS VERSUS ONE-SIDED STATISTICAL BOUNDS

In Sections 2.2.4, 2.3.5, and 2.4.2, we described one-sided confidence bounds, prediction bounds, and tolerance bounds, respectively. As indicated in these sections, there are many applications for which one is primarily interested in either a lower bound or an upper bound. Even in such situations it is often still convenient to report a two-sided interval. For example, if one wants to predict the number of units that will be returned for warranty repair, major concern will center on the upper prediction bound, indicating how bad things might be. In most cases, however, there will also be at least some interest in the lower prediction bound, giving an indication of how good things might be.

One can combine a one-sided lower and a one-sided upper confidence bound to obtain a two-sided confidence interval. For example, 95% one-sided lower and 95% one-sided upper confidence bounds, taken together, result in a 90% two-sided confidence interval (as shown in Sections B.2.2 and B.6.2). Moreover, most of the commonly used two-sided confidence interval procedures (such as those discussed in Chapters 3 and 4) have equal probabilities of the lower endpoint being larger than the actual value and the upper endpoint being smaller than the actual value. In developing procedures for constructing such intervals, it is generally desirable to have this property. In such situations, the endpoints of, for example, a 90% two-sided confidence interval can be considered to be one-sided 95% confidence bounds. The same properties hold for prediction intervals designed to include a single future observation. We note, however, that the preceding statement generally does not carry over, exactly, to tolerance intervals or simultaneous prediction intervals.

2.8   THE ADVANTAGE OF USING CONFIDENCE INTERVALS INSTEAD OF SIGNIFICANCE TESTS

Statistical significance (or hypothesis) tests and related p-values are frequently used to assess the correctness of a hypothesis about a distribution parameter or characteristic using sample data. For example, a consumer protection agency might wish to test a manufacturer's claim (or hypothesis) that the mean lifetime of a brand of light bulbs is at least 1,000 hours. Statistical significance tests are designed so that the probability of incorrectly rejecting the hypothesis when it is, in fact, true has a specified small value, known as the significance level (or Type I error probability), often denoted by α.

There is a close relationship between confidence intervals and significance tests. Indeed, a confidence interval can generally be used to test a hypothesis. For the light bulb example, suppose the one-sided $100(1 - \alpha)\%$ upper confidence bound for the actual mean lifetime, calculated from the data, exceeds the claimed mean lifetime of 1,000 hours. Then one would conclude that the claim has not been contradicted by the data at the $100\alpha\%$ significance level. On the other hand, if the one-sided $100(1 - \alpha)\%$ upper confidence bound is less than 1,000 hours, there is evidence at the $100\alpha\%$ significance level to reject the claim. Thus, a one-sided $100(1 - \alpha)\%$ confidence bound generally gives the accept or reject information provided by a one-sided significance test at a $100\alpha\%$ level of significance. There is a similar relationship between a two-sided $100(1 - \alpha)\%$ confidence interval and a two-sided significance test at a $100\alpha\%$ significance level.

In addition, confidence intervals generally provide more information and insight than significance tests. This is because confidence intervals give quantitative bounds that express the statistical uncertainty, instead of a mere accept or reject statement. Also, whether or not statistical significance is achieved is highly dependent on sample size. The width of a confidence interval is also highly dependent on sample size. For such an interval, however, the effect of a small sample is evident from noting the width of the interval, while this is not the case for a significance test. Thus, confidence intervals are usually more meaningful than significance tests. One can argue, moreover, that in most practical situations, there is no reason for the statistical hypothesis to hold exactly. For example, two different processes generally would not be expected to have identical means. Thus, whether or not the hypothesis of equal means for the two processes is rejected depends upon the sample sizes and the magnitude of the actual difference between the process means.

For example, a wide confidence interval—often resulting in failure to reject a stated null hypothesis of a significant effect or difference—frequently suggests that only a relatively small sample size was available. It may, of course, also suggest appreciable process variability. Thus, one might conclude that the results are inconclusive and that, if possible, obtaining additional data should be considered. Similarly, a narrow confidence interval—often resulting in the rejection of a null hypothesis—might not necessarily imply that the effect or difference is of practical importance. Instead, it might just be a consequence of a large sample.

In most situations, confidence intervals avoid some of the pitfalls inherent in statistical significance tests. Moreover, statistical intervals are generally easier to explain to management and those with no training in statistics than are significance tests.

Most applied statisticians, and especially those in business and economics and in the physical and engineering sciences, now share the preceding viewpoint. It is, however, not universally accepted and the use of confidence intervals versus significance tests remains a controversial topic, particularly in the social sciences. The Bibliographic Notes section at the end of this chapter gives some references that discuss this topic further.

2.9 SIMULTANEOUS STATISTICAL INTERVALS

Some practical problems require that more than one statistical interval (often many) be computed from the same data and be considered simultaneously. This is the case, for example, with the prediction intervals described in Section 4.8 to contain, with $100(1 - \alpha)\%$ confidence, all of m future observations from a normal distribution. We can view this as the construction of m intervals (one for each future observation), where we want to assert with $100(1 - \alpha)\%$ confidence that *all* of the intervals are correct. Thus, such intervals are called *simultaneous* prediction intervals. Similarly, one might want to compute simultaneous confidence intervals to contain each of the

$$\binom{k}{2} = \frac{k!}{2!(k - 2)!}$$

possible differences between all pairs of means from k different populations.

In practice, constructing simultaneous confidence (or prediction) intervals often make sense because in simultaneously making m interval statements, it can become increasingly likely that at least one of these statements does not hold as m increases, even though each of the intervals individually has a coverage probability of $1 - \alpha$.

Miller (1981) treats, in detail, the theory and methods for making simultaneous inferences, including general methods for computing such intervals, and easy-to-use approximate methods, like the Bonferroni approximation (discussed below), and their application. We review some of the basic practical methods.

Suppose that we want to combine k confidence intervals that, individually, have confidence levels $1 - \alpha_1, \ldots, 1 - \alpha_k$. We want to know the confidence level, $1 - \alpha_J$, of the joint confidence statement (or how to compute the individual intervals so that they have the desired joint confidence level). The following three situations are of interest.

1. The intervals are *functionally dependent* on one another in the sense that the correctness of any one implies the correctness of all of the others (and conversely with regard to incorrectness). For example, a binomial distribution confidence interval for the probability of a future binomial distribution trial is calculated directly from the confidence interval for the binomial distribution parameter, as described in Sections 6.4. Then the coverage probability is the same for both intervals.

2. The intervals are *statistically independent* (i.e., the probability that any one individual confidence interval statement is correct is equal to the conditional probability that it is correct, given the correctness of any combination of the other intervals). The assumption of independence is reasonable if each interval in the set is computed separately from independent data sets (e.g., if various studies had been conducted independently and the data from each analyzed separately). It is, in general, *not* reasonable to assume independence if some or all of the intervals involve computations from the same data set (e.g., in the analysis of variance when confidence intervals are calculated for several group means, using the *same* pooled variance estimate, calculated over all of the groups). If the intervals *are* independent, then the joint confidence level is

$$1 - \alpha_J = (1 - \alpha_1)(1 - \alpha_2) \cdots (1 - \alpha_k).$$

   For example, a set of three independent 95% confidence intervals would have a joint confidence level of $0.95^3 = 0.8574$ or 85.74%.

3. The intervals are *neither functionally dependent nor statistically independent*. In this case, the coverage probability may be either less than or greater than the nominal joint confidence level for the case of independence. For some special cases, it may be possible to compute the exact coverage probability, but, for others, the task is often analytically difficult or intractable or computationally intensive. Elementary probability theory, however, provides a simple, conservative lower bound on the coverage probability for a joint confidence statement—in particular, the joint confidence level

$$1 - \alpha_J \geq 1 - \alpha_1 - \cdots - \alpha_k.$$

This is known as the Bonferroni bound, which is discussed further in Section D.7.1. It provides a useful way for combining confidence statements to give a *conservative* lower bound for the coverage probability. For example, in combining three 95% confidence intervals, the joint confidence level, as calculated from the Bonferroni bound, is at least $1 - 0.05 - 0.05 - 0.05 = 0.85$ or 85%. Similarly, when combining two 99% intervals, we have a joint confidence level of at least $1 - 0.01 - 0.01 = 0.98$ or 98%. In the latter case, the Bonferroni lower bound is close to the joint confidence level that one has under the assumption of independence, $0.99^2 = 0.9801$. The Bonferroni inequality is conservative in the sense that it provides confidence intervals with coverage probabilities that are larger than the nominal confidence levels. The Bonferroni inequality works especially well (i.e., usually gives a close approximation to the nominal confidence level) when k is small (i.e., relatively few individual confidence statements) and the $(1 - \alpha)$ values are close to 1.

BIBLIOGRAPHIC NOTES

Specific areas of application

Katz (1975) provides a discussion involving the use of confidence intervals in courtroom testimony. Altman et al. (2000) argue for the use of confidence intervals in medical research. For some other medical applications, Hall (1989) argues that "statistical models and confidence intervals emphasize parameters rather than distributions." We would suggest that a tolerance interval is often useful to describe the breadth of a distribution.

As described in Section 2.2.2, confidence intervals for reliability are widely used in product life analyses. Nelson (1982) and Meeker and Escobar (1998) describe numerous such applications.

Statistical intervals versus significance testing

The debate between researchers advocating for and against significance tests (versus confidence intervals) continues to be heated. Harlow et al. (1997) present viewpoints on both sides of the controversy. Krantz (1999), in a review of Harlow et al. (1997), defends the use of significance tests in psychological research. Our view is that his arguments do not apply in most other application areas, including the biological, physical, engineering and environmental sciences and in business and economics.

One-sided confidence intervals

Boyles (2008) argues against the use of one-sided confidence intervals because such intervals include extreme values of the quantity of interest that are implausible. We also favor reporting two-sided intervals (see Section 2.2.4), but recognize that in some applications all, or almost all, of the concern for error is on one side or the other of the quantity of interest (e.g., when there is concern for safety). Thus we present methods for constructing one-sided (lower or upper) *confidence bounds*. Agreeing with Boyles's basic concern, we suggest avoiding the term "one-sided confidence interval."

Balancing error probabilities

Fraser (2011, bottom of page 300) supports, for slightly different reasons, our argument (in Section 2.7) for constructing two-sided statistical intervals that balance the error probabilities in the two sides of the interval—allowing one to usefully interpret, with an appropriate adjustment in confidence level, the endpoints of such intervals as one-sided bounds.

Choosing a coverage probability

Landon and Singpurwalla (2008) discuss choosing a coverage probability for prediction intervals by using decision-theoretic considerations.

## Chapter 3

Constructing Statistical Intervals Assuming a Normal Distribution Using Simple Tabulations

OBJECTIVES AND OVERVIEW

This chapter presents, tabulates, and compares factors for calculating the different kinds of intervals, based upon a sample of size n from a normal distribution with unknown mean μ and unknown standard deviation σ. Specifically, we present factors for:

- Two-sided confidence intervals and one-sided confidence bounds to contain the distribution mean μ and the distribution standard deviation σ.

- Two-sided tolerance intervals and one-sided tolerance bounds to contain at least a proportion β of the distribution values, for $\beta = 0.90, 0.95$, and 0.99.

- Two-sided prediction intervals and one-sided prediction bounds to contain the values of all of $m = 1, 2, 5, 10, 20$, and $m = n$ randomly selected future observations, and the sample mean and standard deviation of $m = n$ future observations.

For each kind of interval or bound, the tables provide factors for both 95% and 99% confidence levels. We also present examples illustrating the use of these factors.

The chapter presents:

- The normal distribution, justification for its use, and background information about the factors (Section 3.1).

Statistical Intervals: A Guide for Practitioners and Researchers, Second Edition.
William Q. Meeker, Gerald J. Hahn and Luis A. Escobar.
Companion Website: www.wiley.com/go/meeker/intervals

- The circuit pack voltage application that will be used to illustrate the use of the factors (Section 3.2).

- Use of the factors for two-sided statistical intervals (Section 3.3).

- Use of the factors for one-sided statistical bounds (Section 3.4).

3.1   INTRODUCTION

3.1.1   The Normal Distribution

The normal distribution is the best known and most frequently used statistical model. Its theoretical justification is often based on the "central limit theorem" (CLT). The CLT says that the distribution of the sum of random variables can be approximated by a normal distribution. This result justifies the normal distribution as an appropriate model for phenomena that arise directly as a sum or mean, such as the number of motor vehicle accidents per week in a large city. The CLT also justifies the use of the normal distribution for phenomena that arise as a consequence of the impact of many small factors. Examples of these include the height or weight of men or women. The sensitivity of an interval to the assumption of normality depends on the specific type of interval, the population or process being studied, and the sample size (see Section 4.10).

The normal distribution probability density function (pdf) is

$$f(x; \mu, \sigma) = \frac{1}{\sigma\sqrt{2\pi}} \exp\left[-\frac{1}{2}\left(\frac{x-\mu}{\sigma}\right)^2\right], \quad -\infty < x < \infty,$$

where $-\infty < \mu < \infty$ is the mean (and also the median and mode) and $\sigma > 0$ is the standard deviation of the distribution. The normal pdf is graphed in Figure 3.1 for mean $\mu = 0$ and several values of σ. The corresponding cumulative distribution function (cdf) for the proportion of a distribution below x is

$$F(x; \mu, \sigma) = \Pr(X \le x) = \int_{-\infty}^{x} f(y; \mu, \sigma)dy.$$

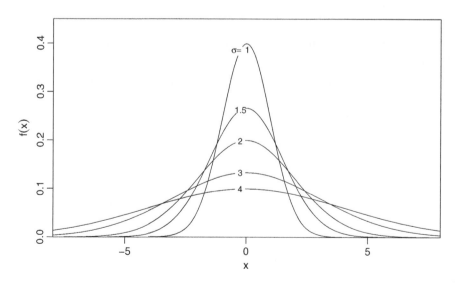

Figure 3.1   Normal distribution probability density function.

This can also be computed from $\Pr(X \le x) = \Phi((x - \mu)/\sigma)$, where $\Phi(z) = F(z; 0, 1)$ is the cdf of the standard normal distribution (i.e., the normal distribution with mean $\mu = 0$ and standard deviation $\sigma = 1$). This function is tabulated in most introductory books on statistics or can be evaluated from statistical calculators in computer packages or on the Web (e.g., `pnorm(0.25)` in R gives $\Pr(Z \le 0.25) = 0.5987$). When the normal distribution is used as a model for a population or process, μ and σ are usually unknown and must be estimated from sample data. Section C.3.2 contains more information about the normal distribution.

3.1.2 Using the Simple Factors

The tables of factors described in this chapter were computed from the more familiar t, χ^2, and F-distribution tables, using standard formulas, which we review in Chapter 4. The simpler factors given here serve several purposes:

- Statistical intervals can be computed with slightly less computation than required with the more general formulas given in Chapter 4.

- Interval widths can be compared directly, providing insight into the differences among the various types of statistical intervals.

- One can easily assess the effect of sample size and confidence level on the width of a particular statistical interval.

In the rest of this chapter we will illustrate the construction, describe the interpretation, and compare the widths of some of the most commonly used statistical intervals for a normal distribution. Chapter 4 discusses, in more detail and generality, the computations for these and some other statistical intervals.

To construct intervals from these factors, it is assumed that a random sample of size n with values x_i, $i = 1, \ldots, n$, has been taken from the population or process of interest, and that the sample mean $\bar{x}$ and the sample standard deviation s have been computed from these observations using the expressions

$$\bar{x} = \frac{\sum_{i=1}^{n} x_i}{n} \quad \text{and} \quad s = \left[\frac{\sum_{i-1}^{n} (x_i - \bar{x})^2}{n - 1} \right]^{1/2}.$$

3.2 CIRCUIT PACK VOLTAGE OUTPUT EXAMPLE

This section introduces an example that we will use to illustrate the computation of statistical intervals based on an assumption of a normal distribution.

Example 3.1 Circuit pack voltage output problem and data. A manufacturer wanted to characterize empirically the voltage outputs for a new electronic circuit pack design. Five prototype units were built and the following measurements were obtained: 50.3, 48.3, 49.6, 50.4, and 51.9 volts. Figure 3.2 is a dot plot of the five voltage measurements.

The sample size for this study was small, as it often is in practice, because of the high cost of manufacturing such prototype units. In this case, the experimenters felt that a sample of five units would be sufficient to answer some initial questions. Other questions might require more units; see Chapters 8–10 for a discussion of the choice of sample size.

From previous experience with the construction and evaluation of similar circuit packs and from engineering knowledge about the components being used in the circuit, the experimenters

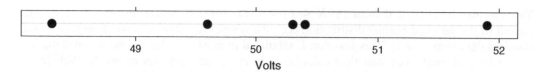

Figure 3.2   Dot plot of the circuit pack output voltages.

felt it would be reasonable to assume that the voltage measurements follow a normal distribution (with an unknown mean and standard deviation).

The sample mean and standard deviation of the voltage output readings are

$$\bar{x} = \frac{50.3 + 48.3 + \cdots + 51.9}{5} = 50.10,$$

$$s = \left[\frac{(50.3 - 50.10)^2 + \cdots + (51.9 - 50.10)^2}{5 - 1}\right]^{1/2} = 1.31.$$ ∎

Example 3.2   Validity of inferences for the circuit pack voltage output application. The experimenters were most interested in making inferences about the circuit packs that would be manufactured in the future. Thus, in the terminology of Chapter 1, this is an analytic study. The statistical intervals assume that the five sample prototype circuit packs *and* the future circuit packs of interest come from the same stable production process and that the unit-to-unit variability in the voltage readings can be adequately described by a normal distribution with a constant mean μ and standard deviation σ. If the process changes (e.g., because of a component substitution) after the sample units are made, inferences about the new state of the process or predictions about future units might not be valid. The assumption of normality can often be assessed from the sample data. This, however, generally requires a sample of at least 20–30 observations to have any reasonable degree of sensitivity. Thus, for the present example, there are not enough data to properly check the assumption of normality, and its justification needs to be based on the experimenter's knowledge of the process and previous experience in similar situations. One can and should, however, evaluate the sensitivity of the conclusions to the assumption of normality and repeat the analysis under alternative assumptions. For example, transforming the response variable, or comparing inferences made with and without distributional assumptions, and with different distributional assumptions can be informative, as illustrated in Section 4.12. ∎

3.3   TWO-SIDED STATISTICAL INTERVALS

3.3.1   Simple Tabulations for Two-Sided Statistical Intervals

Tables J.1a and J.1b provide factors $c_{(1-\alpha;n)}$, where n is the number of observations in the given sample and $100(1 - \alpha)\%$ is the confidence level associated with the calculated interval, so that the two-sided interval

$$\bar{x} \mp c_{(1-\alpha;n)} s$$

is (in turn):

- A two-sided confidence interval for the distribution mean μ.

- A two-sided tolerance interval to contain at least a specified proportion β of the distribution for $\beta = 0.90, 0.95$, and 0.99.

- A two-sided simultaneous prediction interval to contain all of m future observations from the previously sampled normal distribution for $m = 1, 2, 5, 10, 20$, and $m = n$.

- A two-sided prediction interval to contain the sample mean of $m = n$ future observations.

Tables J.2a and J.2b provide factors $c_{L(1-\alpha;n)}$ and $c_{U(1-\alpha;n)}$, where n is the number of observations in the given sample and $100(1 - \alpha)\%$ is the associated confidence level, so that the interval

$$\left[c_{L(1-\alpha;n)} s, \quad c_{U(1-\alpha;n)} s \right]$$

is (in turn):

- A two-sided confidence interval for the distribution standard deviation σ.

- A two-sided prediction interval to contain the standard deviation of $m = n$ future observations.

The tabulated values are for $n = 4(1)10, 12, 15(5)30, 40, 60$, and ∞. Tables J.1a and J.2a are for confidence levels of $100(1 - \alpha)\% = 95\%$ and Tables J.1b and J.2b are for $100(1 - \alpha)\% = 99\%$. Methods for obtaining factors for other values of n and for other confidence levels are given in Chapter 4.

We will use $[\underset{\sim}{\theta}, \ \widetilde{\theta}]$ to denote a two-sided statistical interval that is constructed for θ, some quantity of interest. Thus, $\underset{\sim}{\theta}$ is the lower endpoint of the interval and $\widetilde{\theta}$ is the upper endpoint of the interval. If only a lower or an upper statistical bound is desired, we compute either $\underset{\sim}{\theta}$ or $\widetilde{\theta}$, which we call a one-sided lower or upper bound for θ. For example, $[\underset{\sim}{s_m}, \widetilde{s}_m]$ denotes a two-sided prediction interval to contain the sample standard deviation for a future sample of size m and $\widetilde{\mu}$ denotes an upper confidence bound for the distribution mean μ.

3.3.2 Two-Sided Interval Examples

Using the sample values $\bar{x} = 50.10$ and $s = 1.31$, based on $n = 5$ observed voltage measurements, we can use the factors given in Table J.1a to construct the following statistical intervals:

- A two-sided 95% confidence interval for the mean μ of the distribution of sampled circuit packs is

$$[\underset{\sim}{\mu}, \ \widetilde{\mu}] = 50.10 \mp 1.24 \times 1.31 = [48.5, \quad 51.7].$$

Thus, we are 95% confident that the interval 48.5 to 51.7 volts contains the unknown mean μ of the distribution of circuit pack voltage readings. It is important to remember, as explained in Section 2.2.5, that "95% confident" describes the success rate of the procedure (i.e., the percentage of time that a claim of this type is correct).

- A two-sided 95% tolerance interval to contain at least a proportion $\beta = 0.99$ of the sampled distribution of circuit pack voltages is

$$[\underset{\sim}{T}_{0.99}, \ \widetilde{T}_{0.99}] = 50.10 \mp 6.60 \times 1.31 = [41.5, \quad 58.7].$$

Thus, we are 95% confident that the interval 41.5 to 58.7 volts contains at least a proportion $\beta = 0.99$ of the distribution of circuit pack voltage readings.

- A two-sided (simultaneous) 95% prediction interval to contain the voltage readings of all of 10 additional circuit packs randomly sampled from the same distribution is

$$[\underset{\sim}{Y}_{10}, \ \widetilde{Y}_{10}] = 50.10 \mp 5.23 \times 1.31 = [43.2, \quad 57.0].$$

Thus, we are 95% confident that the voltage readings of all 10 additional circuit packs will be contained within the interval 43.2 to 57.0 volts.

- A two-sided 95% prediction interval to contain the sample mean of the voltage readings of five additional circuit packs randomly sampled from the same distribution is

$$[\underline{\bar{Y}}_5, \ \widetilde{\bar{Y}}_5] = 50.10 \mp 1.76 \times 1.31 = [47.8, \ 52.4].$$

Thus, we are 95% confident that the sample mean of the voltage readings of five additional circuit packs will be in the interval 47.8 to 52.4 volts.

Using the factors in Table J.2a, we can construct the following 95% statistical intervals:

- A two-sided 95% confidence interval for the standard deviation σ of the distribution of sampled circuit pack voltage readings is

$$[\underline{\sigma}, \ \widetilde{\sigma}] = [0.60 \times 1.31, \ 2.87 \times 1.31] = [0.8, \ 3.8].$$

Thus, we are 95% confident that the interval 0.8 to 3.8 volts contains the unknown standard deviation σ of the distribution of circuit pack voltage readings.

- A two-sided 95% prediction interval to contain the standard deviation of the voltage readings of five additional circuit packs randomly sampled from the same distribution is

$$[\underline{S}_5, \ \widetilde{S}_5] = [0.32 \times 1.31, \ 3.10 \times 1.31] = [0.4, \ 4.1].$$

Thus, we are 95% confident that the standard deviation of the voltage readings of five additional circuit packs will be in the interval 0.4 to 4.1 volts.

3.3.3   Comparison of Two-Sided Statistical Intervals

Figure 3.3 provides a comparison of some of the preceding factors for computing two-sided 95% statistical intervals. It illustrates the large differences in interval width between the various interval types. Thus, Figure 3.3 and Tables J.1a and J.1b show that, for a given sample size and confidence level, a confidence interval for the distribution mean is always narrower than a prediction interval to contain all m future observations or a tolerance interval to contain a proportion $\beta = 0.90, 0.95$, or 0.99 of the distribution. On the other hand, whether or not a particular prediction interval is narrower than a particular tolerance interval depends on the number of future observations to be contained in the prediction interval and the proportion of the distribution to be contained in the tolerance interval. Moreover, the relative widths of different prediction intervals depend on the number of future observations to be contained in the prediction interval, and the relative widths of different tolerance intervals depend on the proportion of the distribution to be contained in the tolerance interval, for a given sample size and confidence level. Finally, confidence intervals to contain the distribution mean and the distribution standard deviation are narrower, respectively, than the corresponding prediction intervals to contain the sample mean and standard deviation of a future sample of any size.

3.4   ONE-SIDED STATISTICAL BOUNDS

3.4.1   Simple Tabulations for One-Sided Statistical Bounds

Tables J.3a and J.3b provide factors for calculating one-sided statistical bounds similar to those in Tables J.1a and J.1b (presented in Section 3.3) for calculating two-sided statistical intervals. Thus, Table J.3a (for $100(1 - \alpha)\% = 95\%$) and Table J.3b (for $100(1 - \alpha)\% = 99\%$) provide

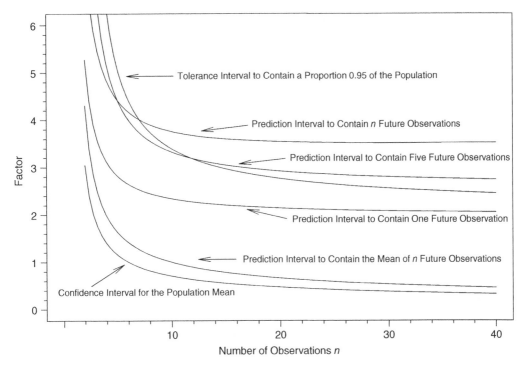

Figure 3.3 Comparison of factors for calculating some two-sided 95% statistical intervals. A similar figure first appeared in Hahn (1970b). Adapted with permission of the American Society for Quality.

factors $c'_{(1-\alpha;n)}$, where n is the number of observations in the given sample and $100(1-\alpha)\%$ is the associated confidence level, so that the one-sided lower bound

$$\bar{x} - c'_{(1-\alpha;n)} s$$

or the one-sided upper bound

$$\bar{x} + c'_{(1-\alpha;n)} s$$

is (in turn):

- A lower (or upper) confidence bound for the distribution mean μ.

- A lower (or upper) tolerance bound to be exceeded by (or exceed) at least a specified proportion β of the distribution for $\beta = 0.90, 0.95$, and 0.99.

- A one-sided lower (or upper) simultaneous prediction bound to be exceeded by (or exceed) all of m future observations from the previously sampled normal distribution for $m = 1, 2, 5, 10, 20$, and $m = n$.

- A one-sided lower (or upper) prediction bound for the sample mean of $m = n$ future observations.

Tables J.4a and J.4b provide factors for calculating one-sided statistical bounds similar to those in Table J.2a and Table J.2b (presented in Section 3.3) for calculating two-sided intervals. Thus, Table J.4a (for $100(1-\alpha)\% = 95\%$) and Table J.4b (for $100(1-\alpha)\% = 99\%$) provide factors $c'_{L(1-\alpha;n)}$ and $c'_{U(1-\alpha;n)}$, where n is the number of observations in the given sample and $100(1-\alpha)\%$ is the associated confidence level, so that the one-sided lower bound

$$c'_{L(1-\alpha;n)}s$$

or the one-sided upper bound

$$c'_{U(1-\alpha;n)}s$$

is (in turn):

- A lower (or upper) confidence bound for the distribution standard deviation σ.

- A lower (or upper) prediction bound for the standard deviation of $m = n$ future observations.

The tabulations are again for $n = 4(1)10$, 12, 15(5)30, 40, 60, and ∞. Methods for obtaining factors other than those tabulated are given in Chapter 4.

3.4.2   One-Sided Statistical Bound Examples

Using the sample values $\bar{x} = 50.10$ and $s = 1.31$, based on $n = 5$ observations, we can use the factors given in Table J.3a to construct the following one-sided statistical bounds:

- A one-sided lower 95% confidence bound for the mean μ of the distribution of sampled circuit pack voltage readings is

$$\underset{\sim}{\mu} = 50.10 - 0.95 \times 1.31 = 48.9.$$

Thus, we are 95% confident that the unknown mean μ of the distribution of circuit pack voltage readings exceeds the lower confidence bound of 48.9 volts.

- A one-sided upper 95% tolerance bound to exceed at least a proportion $\beta = 0.99$ of the sampled distribution of circuit pack voltage readings is

$$\widetilde{T}_{0.99} = 50.10 + 5.74 \times 1.31 = 57.6.$$

Thus, we are 95% confident that at least a proportion $\beta = 0.99$ of the distribution of circuit packs have voltage readings less than 57.6 volts.

- A one-sided lower 95% prediction bound to be exceeded by the voltage readings of all of 10 additional circuit packs randomly sampled from the same distribution is

$$\underset{\sim}{Y}_{10} = 50.10 - 4.42 \times 1.31 = 44.3.$$

Thus, we are 95% confident that the voltage readings of all 10 additional circuit packs will exceed 44.3 volts.

- A one-sided lower 95% prediction bound to be exceeded by the sample mean of the voltage readings of five additional circuit packs from the same distribution is

$$\underset{\sim}{\bar{Y}}_{5} = 50.10 - 1.35 \times 1.31 = 48.3.$$

Thus, we are 95% confident that the sample mean of the voltage readings of five additional circuit packs will exceed 48.3 volts.

ONE-SIDED STATISTICAL BOUNDS   **45**

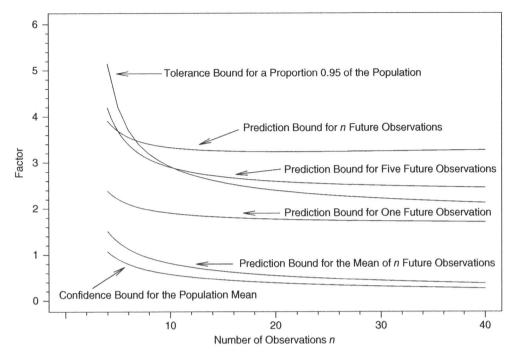

Figure 3.4   Comparison of factors for calculating some one-sided 95% statistical bounds. A similar figure first appeared in Hahn (1970b). Adapted with permission of the American Society for Quality.

Using the factors in Table J.4a, we can construct the following one-sided 95% statistical bounds:

- A one-sided upper 95% confidence bound for the standard deviation σ of the distribution of sampled circuit pack voltage readings is

$$\widetilde{\sigma} = 2.37 \times 1.31 = 3.1.$$

Thus, we are 95% confident that the unknown standard deviation σ of the distribution of circuit pack voltage readings is less than 3.1 volts.

- A one-sided upper 95% prediction bound to exceed the standard deviation of the voltage readings of five additional circuit packs randomly sampled from the same distribution is

$$\widetilde{S}_5 = 2.53 \times 1.31 = 3.3.$$

Thus, we are 95% confident that the standard deviation of the voltage readings of five additional circuit packs will be less than 3.3 volts.

3.4.3   Comparison of One-Sided Statistical Bounds

Figure 3.4 provides a comparison of some of the preceding factors for calculating one-sided 95% statistical bounds. Inspection of this figure and the tabulations leads to conclusions about the relative magnitudes of the factors for calculating different types of one-sided bounds that are similar to those for the two-sided case discussed previously.

Chapter *4*

Methods for Calculating Statistical Intervals for a Normal Distribution

OBJECTIVES AND OVERVIEW

This chapter extends the results of Chapter 3 and gives general methods for calculating various statistical intervals for samples from a population or process that can be approximated by a normal distribution.

The chapter explains:

- How to compute a confidence interval for a normal distribution mean, standard deviation, or quantile (Sections 4.2, 4.3, and 4.4).

- How to compute a confidence interval for the distribution proportion less (greater) than a specified value for a normal distribution (Section 4.5).

- How to compute a tolerance interval to contain a specified proportion of a normal distribution (Section 4.6).

- How to compute a prediction interval to contain a single future observation or the mean of a specified number of future observations from a normal distribution (Section 4.7).

- How to compute a prediction interval to contain at least k of m future observations from a normal distribution (Section 4.8).

- How to compute a prediction interval to contain the standard deviation of a specified number of future observations from a normal distribution (Section 4.9).

- The importance of the assumption of a normal distribution and when the construction of an interval is (and is not) robust to this assumption (Section 4.10).

Statistical Intervals: A Guide for Practitioners and Researchers, Second Edition.
William Q. Meeker, Gerald J. Hahn and Luis A. Escobar.
© 2017 John Wiley & Sons, Inc. Published 2017 by John Wiley & Sons, Inc.
Companion Website: www.wiley.com/go/meeker/intervals

- How to assess the validity of the assumption of a normal distribution and methods for constructing statistical intervals when the data cannot be described by a normal distribution (Section 4.11).

- How to transform the data to achieve approximate normality and draw inferences from transformed data (Section 4.12).

- Statistical intervals for linear regression analysis (Section 4.13).

- Statistical intervals for comparing populations and processes (Section 4.14).

The methods described in this chapter are of interest for situations not covered in the tabulations in Chapter 3 (especially Sections 4.4, 4.5, and 4.8), for developing computer programs to construct statistical intervals, and for gaining general understanding of the methodology. For each case, we describe the construction of both a two-sided interval and one-sided lower and upper bounds. We continue to use the numerical example from Chapter 3 to illustrate the methods.

4.1 NOTATION

The following notation will be used in this chapter:

- $\Phi_{\mathrm{norm}}(z)$ is the standard normal cumulative distribution function, giving $\Pr(Z \leq z)$, where Z is a normally distributed random variable with mean $\mu = 0$ and standard deviation $\sigma = 1$. This cumulative distribution function is tabulated in many elementary statistics textbooks. Computer programs for obtaining $\Phi_{\mathrm{norm}}(z)$ are also available (e.g., pnorm(0.25) in R gives $\Pr(Z \leq 0.25) = 0.5987$).

- $z_{(p)} = \Phi_{\mathrm{norm}}^{-1}(p)$ is the p quantile of the standard normal distribution (i.e., the value below which a normally distributed variate with mean $\mu = 0$ and standard deviation $\sigma = 1$ falls with probability p). For example, $z_{(0.05)} = -1.645$ and $z_{(0.95)} = 1.645$. Standard normal distribution quantiles are tabulated in many elementary statistics textbooks. Computer programs for obtaining $z_{(p)}$ are also available (e.g., qnorm(0.05) in R gives -1.645).

- $t_{(p;r)}$ is the p quantile of Student's t-distribution with r degrees of freedom. For example, $t_{(0.05;5)} = -2.015$ and $t_{(0.95;5)} = 2.015$. Student's t-distribution quantiles are tabulated in many elementary statistics textbooks. Computer programs for obtaining $t_{(p;r)}$ are also available (e.g., qt(0.95,5) in R gives 2.015).

- $t_{(p;r,\delta)}$ is the p quantile of the noncentral t-distribution with r degrees of freedom and noncentrality parameter δ. For example, $t_{(0.05;5,4)} = 2.120$ and $t_{(0.95;5,4)} = 9.025$. Although tables for noncentral t-distribution quantiles exist, they are not commonly available. Computer programs for obtaining $t_{(p,r,\delta)}$ values are available (e.g., qt(0.95,5,4) in R gives 9.025).

- $\chi^2_{(p;r)}$ is the p quantile of the chi-square (χ^2) distribution with r degrees of freedom. For example, $\chi^2_{(0.05;5)} = 1.145$ and $\chi^2_{(0.95;5)} = 11.07$. Chi-square distribution quantiles are tabulated in many elementary statistics textbooks. Computer programs for obtaining $\chi^2_{(p;r)}$ are also available (e.g., qchisq(0.95,5) in R gives 11.07).

- $F_{(p;r_1,r_2)}$ is the p quantile of the Snedecor's F-distribution with r_1 numerator and r_2 denominator degrees of freedom. For example, $F_{(0.95;5,2)} = 19.30$. F-distribution quantiles are tabulated in many elementary statistics textbooks. Computer programs for obtaining $F_{(p;r_1,r_2)}$ are also available (e.g., qf(0.95,5,2) in R gives 19.30).

Web-based applets are also available for computing quantiles and probabilities from these commonly used probability distributions. Technical details about these distributions are given in Appendix C.

In this chapter, we will, as in Chapter 3, assume that the data consist of a random sample of size n from a normal distribution with sample mean $\bar{x}$ and sample standard deviation s.

4.2 CONFIDENCE INTERVAL FOR THE MEAN OF A NORMAL DISTRIBUTION

A two-sided $100(1 - \alpha)\%$ confidence interval for the mean μ of a normal distribution is

$$[\underset{\sim}{\mu}, \widetilde{\mu}] = \bar{x} \mp t_{(1-\alpha/2;n-1)} \frac{s}{\sqrt{n}}. \tag{4.1}$$

One-sided $100(1 - \alpha)\%$ confidence bounds are obtained by replacing $\alpha/2$ by α (and $\mp$ by either $-$ or $+$) in the above expression. The factors $c_{(1-\alpha;n)}$ and $c'_{(1-\alpha;n)}$ for confidence intervals and bounds for μ given in Tables J.1a, J.1b, J.3a, and J.3b were computed from these expressions: $c_{(1-\alpha;n)} = t_{(1-\alpha/2;n-1)}/\sqrt{n}$ and $c'_{(1-\alpha;n)} = t_{(1-\alpha;n-1)}/\sqrt{n}$. When $n - 1$ is large (say greater than 60), $t_{(1-\alpha;n-1)} \approx z_{(1-\alpha)}$. Thus, normal distribution quantiles provide a generally adequate approximation for t-distribution quantiles when n is large and $1 - \alpha/2$ is not too large. For example, $t_{(0.975;60)} = 2.000$ and $z_{(0.975)} = 1.960$.

Most elementary textbooks on statistical methods discuss confidence intervals for the mean of a normal distribution. The underlying theory involves Student's t-distribution and is given in Section E.3.1.

Example 4.1 Confidence Interval for Mean Voltage Output. For the circuit pack output example, presented in Section 3.2, $n = 5$, $\bar{x} = 50.10$ volts, $s = 1.31$ volts, $t_{(0.975;4)} = 2.776$, and $t_{(0.95;4)} = 2.132$. Then $c_{(0.95;5)} = t_{(0.975;4)}/\sqrt{5} = 1.241$ and $c'_{(0.95;5)} = t_{(0.95;4)}/\sqrt{5} = 0.953$. These are the values given in Tables J.1a and J.3a. A two-sided 95% confidence interval for μ is

$$[\underset{\sim}{\mu}, \widetilde{\mu}] = 50.10 \mp 1.241 \times 1.31 = [48.47, \ 51.73].$$

Using R as a calculator gives

```
> 50.10 + c(-qt(0.975,4), qt(0.975,4))*1.31/sqrt(5)
[1] 48.47 51.73
```

A lower 95% confidence bound for μ is

$$\underset{\sim}{\mu} = 50.10 - 0.953 \times 1.31 = 48.85. \qquad \blacksquare$$

4.3 CONFIDENCE INTERVAL FOR THE STANDARD DEVIATION OF A NORMAL DISTRIBUTION

A two-sided $100(1 - \alpha)\%$ confidence interval for the standard deviation σ of a normal distribution is

$$[\underset{\sim}{\sigma}, \widetilde{\sigma}] = \left[s \left(\frac{n-1}{\chi^2_{(1-\alpha/2;n-1)}} \right)^{1/2}, \ s \left(\frac{n-1}{\chi^2_{(\alpha/2;n-1)}} \right)^{1/2} \right].$$

One-sided $100(1 - \alpha)\%$ confidence bounds are obtained by replacing $\alpha/2$ by α in the above expressions. The factors $c_{L(1-\alpha;n)}$, $c_{U(1-\alpha;n)}$, $c'_{L(1-\alpha;n)}$, and $c'_{U(1-\alpha;n)}$ for obtaining confidence intervals and bounds for σ given in Tables J.2a, J.2b, J.4a, and J.4b were computed from these expressions. For example, $c_{L(1-\alpha;n)} = [(n-1)/\chi^2_{(1-\alpha/2;n-1)}]^{1/2}$ and $c'_{L(1-\alpha;n)} = [(n-1)/\chi^2_{(1-\alpha;n-1)}]^{1/2}$.

Many introductory textbooks on statistical methods discuss confidence intervals for the standard deviation (or variance, i.e., the square of the standard deviation) of a normal distribution. The underlying theory involves the chi-square distribution and is given in Section E.3.2.

Example 4.2 Confidence Interval for Standard Deviation of Voltage Output. For the example, $n = 5$, $s = 1.31$ volts, $\chi^2_{(0.975;4)} = 11.14$, $\chi^2_{(0.025;4)} = 0.484$, and $\chi^2_{(0.05;4)} = 0.711$. Then $c_{L(0.95;5)} = \{4/\chi^2_{(0.975;4)}\}^{1/2} = 0.60$, $c_{U(0.95;5)} = \{4/\chi^2_{(0.025;4)}\}^{1/2} = 2.87$, and $c'_{U(0.95;5)} = \{4/\chi^2_{(0.95;4)}\}^{1/2} = 2.37$. These are the values given in Tables J.2a and J.4a. A two-sided 95% confidence interval for σ is

$$[\underline{\sigma}, \; \widetilde{\sigma}] = [1.31 \times 0.60, \; 1.31 \times 2.87] = [0.79, \; 3.76].$$

Using R as a calculator gives

```
> 1.31*sqrt(c(4/qchisq(p=0.975,df=4), 4/qchisq(p=0.025,df=4)))
[1] 0.7848 3.7644
```

An upper 95% confidence bound for σ is $\widetilde{\sigma} = 1.31 \times 2.37 = 3.10$. ∎

4.4 CONFIDENCE INTERVAL FOR A NORMAL DISTRIBUTION QUANTILE

This section presents methods for constructing confidence intervals for a normal distribution quantile. Confidence intervals for distribution quantiles are typically not presented in textbooks on statistical methods. Note that the methods in this section can also be used to obtain one-sided tolerance bounds. This is due to the equivalence of one-sided tolerance bounds and one-sided confidence intervals for a quantile discussed in Sections 2.2.2 and 4.6.

Computational method

A two-sided $100(1 - \alpha)\%$ confidence interval for x_p, the p quantile of the previously sampled normal distribution, is

$$[\underline{x}_p, \; \widetilde{x}_p] = \left[\bar{x} - t_{(1-\alpha/2;n-1,\delta)} \frac{s}{\sqrt{n}}, \; \bar{x} - t_{(\alpha/2;n-1,\delta)} \frac{s}{\sqrt{n}}\right], \tag{4.2}$$

where $t_{(\gamma;n-1,\delta)}$ is the γ quantile of a noncentral t-distribution with $n - 1$ degrees of freedom and noncentrality parameter $\delta = -\sqrt{n}z_{(p)} = \sqrt{n}z_{(1-p)}$. The special case when $p = 0.50$ (so $\delta = 0$) corresponds to a confidence interval for the median (or mean) given in Section 4.2. The underlying theory for this interval is given in Section E.3.3. A one-sided lower or upper confidence bound is obtained by substituting α for $\alpha/2$ in the appropriate endpoint in (4.2).

Example 4.3 Confidence Interval for 0.90 Quantile of Voltage Output. Using the sample data from Example 4.1, suppose that the manufacturer wants a two-sided 95% confidence interval for $x_{0.90}$, the 0.90 quantile of the distribution of circuit pack voltage readings. Using $p = 0.90$ and $\delta = -\sqrt{5}z_{(0.90)} = -2.865$, the required factors in (4.2) can be obtained by using R as a calculator:

```
> -qt(0.975, 4, -2.865636)
[1] 0.868865
> -qt(0.025, 4, -2.865636)
[1] 9.316001
```

Then the desired confidence interval is

$$\left[\underset{\sim}{x}_{0.90}, \ \widetilde{x}_{0.90}\right] = \left[50.10 - 0.8689 \times 1.31/\sqrt{5}, \ 50.10 - 9.316 \times 1.31/\sqrt{5}\right]$$
$$= [50.61, \ 55.56].$$

Putting all of the above together and using R as a calculator gives

```
> 50.10 + c(-qt(0.975, 4, -2.865636), -qt(0.025, 4, -2.865636))
*1.31/sqrt(5)
[1] 50.60902 55.55778
```

■

Example 4.4   One-Sided Upper Confidence Bound for 0.90 Quantile of Voltage Output.
Similar to Example 4.3, suppose that the manufacturer wanted a one-sided upper 95% confidence
bound for $x_{0.90}$, the 0.90 quantile of the distribution of circuit pack voltage readings. Using R
as a calculator gives

```
> 50.10 -qt(0.05, 4, -2.865636)*1.31/sqrt(5)
[1] 54.56269
```

That is, $\widetilde{x}_{0.90} = 54.6$. This bound is equivalent to a 95% upper tolerance bound to exceed at
least a proportion 0.90 of the distribution of voltage outputs.   ■

Tabular method

A two-sided $100(1-\alpha)\%$ confidence interval for x_p, the p quantile of a normal distribution, is

$$\left[\underset{\sim}{x}_p, \ \widetilde{x}_p\right] = \left[\bar{x} - g'_{(1-\alpha/2;p,n)}s, \ \bar{x} - g'_{(\alpha/2;p,n)}s\right] \tag{4.3}$$

for $0.00 < p < 0.50$ and

$$\left[\underset{\sim}{x}_p, \ \widetilde{x}_p\right] = \left[\bar{x} + g'_{(\alpha/2;1-p,n)}s, \ \bar{x} + g'_{(1-\alpha/2;1-p,n)}s\right] \tag{4.4}$$

for $0.50 \leq p < 1.0$, where the factors $g'_{(\gamma;p,n)}$ are given in Tables J.7a–J.7d for values of
p ranging between 0.01 and 0.40. As suggested above, tables for $p > 0.50$ are not needed
because of the relationship $g'_{(\gamma;p,n)} = -g'_{(1-\gamma;1-p,n)}$.
   Related to the computational method given in the first part of this section, the factors $g'_{(\gamma;p,n)}$
can also be computed using the noncentral t-distribution quantile function. In particular, using
notation defined in Section C.3.9, $g'_{(\gamma;p,n)}$ can be obtained for *any* value of $0 < p < 1$ by using

$$g'_{(\gamma;p,n)} = t_{(\gamma;n-1,z_{(1-p)}\sqrt{n})}/\sqrt{n},$$

where $t_{(\gamma;n-1,z_{(1-p)}\sqrt{n})}$ is the γ quantile of the noncentral t-distribution with $n-1$ degrees of
freedom and noncentrality parameter $z_{(1-p)}\sqrt{n}$.

Example 4.5   Confidence Interval for 0.10 Quantile of Voltage Output. Using the sample data from Example 4.1, suppose that the manufacturer wanted a two-sided 95% confidence interval for $x_{0.10}$, the 0.10 quantile of the distribution of circuit pack voltage readings. Thus, $p = 0.10$, and from Table J.7a, $g'_{(0.025;0.10,5)} = 0.389$ and from Table J.7c, $g'_{(0.975;0.10,5)} = 4.166$. Then a two-sided 95% confidence interval for $x_{0.10}$ is

$$[\underset{\sim}{x}_{0.10}, \ \widetilde{x}_{0.10}] = [50.10 - 4.166 \times 1.31, \ \ 50.10 - 0.389 \times 1.31] = [44.64, \ \ 49.59].$$

■

One-sided lower (upper) $100(1 - \alpha)\%$ confidence bounds for x_p are obtained by substituting α for $\alpha/2$ in the appropriate endpoint of either (4.3) or (4.4), depending on whether p is less than or greater than 0.50.

Example 4.6   One-Sided Lower Confidence Bound for 0.10 Quantile of Voltage Output. Similar to Example 4.5, suppose that the manufacturer wanted a one-sided 95% lower confidence bound for $x_{0.10}$, the 0.10 quantile of the distribution of circuit pack voltage readings. Thus, from Table J.7c, $g'_{(0.95;0.10,5)} = 3.407$. Then a lower 95% confidence bound for $x_{0.10}$ is

$$\underset{\sim}{x}_{0.10} = 50.10 - 3.407 \times 1.31 = 45.64.$$

This bound is equivalent to a 95% lower tolerance bound to be exceeded by at least a proportion 0.90 of the distribution of voltage outputs.

■

4.5   CONFIDENCE INTERVAL FOR THE DISTRIBUTION PROPORTION LESS (GREATER) THAN A SPECIFIED VALUE

The probability of an observation from a normal distribution with known mean μ and known standard deviation σ being less than a specified value x can be computed as $p_{LE} = \Pr(X \le x) = \Phi_{\mathrm{norm}}[(x - \mu)/\sigma]$, where $\Phi_{\mathrm{norm}}(z)$ is the standard normal distribution cdf. Similarly, the probability of such an observation being greater than a specified value x is $p_{GT} = \Pr(X > x) = 1 - \Phi_{\mathrm{norm}}[(x - \mu)/\sigma] = \Phi_{\mathrm{norm}}[-(x - \mu)/\sigma]$. Equivalently, p_{LE} and p_{GT} are the proportions of the distribution less than x and greater than x, respectively. When μ and σ are unknown, we obtain point estimates $\widehat{p}_{LE}$ and $\widehat{p}_{GT}$ by substituting $\bar{x}$ for μ and s for σ, respectively, in these formulas.

A two-sided $100(1 - \alpha)\%$ confidence interval for p_{LE} is

$$[\underset{\sim}{p}_{LE}, \ \widetilde{p}_{LE}] = [\texttt{normTailCI}(\alpha/2; k, n), \ \ \texttt{normTailCI}(1 - \alpha/2; k, n)], \quad (4.5)$$

where $k = (x - \bar{x})/s$ and the function $\texttt{normTailCI}$ from the $\texttt{StatInt}$ R package is described in Section E.3.4.

Similarly, a two-sided $100(1 - \alpha)\%$ confidence interval for p_{GT} is

$$
\begin{aligned}
[\underset{\sim}{p}_{GT}, \ \widetilde{p}_{GT}] &= [1 - \widetilde{p}_{LE}, \ \ 1 - \underset{\sim}{p}_{LE}] \\
&= [1 - \texttt{normTailCI}(1 - \alpha/2; k, n), \ \ 1 - \texttt{normTailCI}(\alpha/2; k, n)] \\
&= [\texttt{normTailCI}(\alpha/2; -k, n), \ \ \texttt{normTailCI}(1 - \alpha/2; -k, n)]. \quad (4.6)
\end{aligned}
$$

One-sided lower and upper $100(1 - \alpha)\%$ confidence bounds for p_{LE} are obtained by replacing $\alpha/2$ by α in (4.5). Similarly, replacing $\alpha/2$ by α in (4.6) yields one-sided lower and upper $100(1 - \alpha)\%$ confidence bounds for p_{GT}.

These normal distribution intervals for probabilities are typically not considered in textbooks on statistical methods. The underlying theory involves the noncentral t-distribution and is given in Section E.3.4.

Example 4.7   Confidence Interval for the Proportion of Circuit Packs with Output Greater than 48 Volts. For the example, $n = 5$, $\bar{x} = 50.10$ volts, and $s = 1.31$ volts. The manufacturer wanted to obtain a two-sided 95% confidence interval to contain p_{GT}, the proportion of units in the distribution with voltages greater than $x = 48$ volts (or equivalently, the probability that the voltage of a single randomly selected unit will exceed $x = 48$ volts). A point estimate for this proportion is $\hat{p}_{GT} = \Pr(X > 48) = 1 - \Phi_{\text{norm}}[(48 - \bar{x})/s] = 1 - \Phi_{\text{norm}}(-1.60) = 0.9452$. Then, using $k = (48 - 50.10)/1.31 = -1.60$ and (4.6), a two-sided 95% confidence interval for p_{GT} is

$$[\underset{\sim}{p}_{GT}, \ \widetilde{p}_{GT}] = [\texttt{normTailCI}(0.025; 1.60, 5), \ \texttt{normTailCI}(0.975; 1.60, 5)]$$
$$= [0.57, \ 0.9984],$$

and a one-sided lower 95% confidence bound for p_{GT} is

$$\underset{\sim}{p}_{GT} = \texttt{normTailCI}(0.05; 1.60, 5) = 0.65.$$

A one-sided upper 95% confidence bound for p_{LE}, the proportion of units less than $x = 48$ volts, is

$$[\widetilde{p}_{LE}] = 1 - \underset{\sim}{p}_{GT} = \texttt{normTailCI}(0.95; 5, -1.6) = 0.35.$$

Using R as a calculator with the function $\texttt{normTailCI}$ from the $\texttt{StatInt}$ R package gives

```
> normTailCI(0.025, 1.6, 5)
[1] 0.5745519
> normTailCI(0.975, 1.6, 5)
[1] 0.9984031
> normTailCI(0.05, 1.6, 5)
[1] 0.6492924
> normTailCI(0.95, -1.6, 5)
[1] 0.3507077
```

■

4.6   STATISTICAL TOLERANCE INTERVALS

4.6.1   Two-Sided Tolerance Interval to Control the Center of a Distribution

This section describes methods for computing a normal distribution tolerance interval to control the center of the distribution. Section E.5.1 describes the underlying theory.

Tabular method

A two-sided $100(1 - \alpha)\%$ tolerance interval to contain at least a proportion β of a normal distribution is

$$[\underset{\sim}{T}_{\beta}, \ \widetilde{T}_{\beta}] = \bar{x} \mp g_{(1-\alpha;\beta,n)}s,$$

where the factors $g_{(1-\alpha;\beta,n)}$ are given in Tables J.1a, J.1b, J.5a, and J.5b. Two-sided tolerance intervals to control the center of a distribution are given in some statistical textbooks.

Computational method

For values of $g_{(1-\alpha;\beta,n)}$ that are not in our tables, the function `normCenterTI` from the `StatInt` R package can be used to compute the required factors.

Example 4.8   Tolerance Interval to Contain at Least a Proportion 0.90 of Circuit Pack Output Voltages. For the example, $n = 5$, $\bar{x} = 50.10$ volts, and $s = 1.31$ volts. Suppose now that the manufacturer wanted a two-sided 95% tolerance interval to contain a proportion 0.90 of the distribution.

   We use $g_{(0.95;0.90,5)} = 4.291$ from Table J.5b (also $c_{(0.95;5)} = 4.29$ from Table J.1a). A two-sided 95% tolerance interval to contain at least a proportion 0.90 of the sampled distribution is

$$[\underset{\sim}{T}_{0.90},\ \widetilde{T}_{0.90}] = 50.10 \mp 4.291 \times 1.31 = [44.5,\ 55.7].$$

Using R as a calculator with function `normCenterTI` from the `StatInt` R package gives

```
> 50.10+c(-1,1)*normCenterTI(conf.level=0.95, sample.size=5,
  beta=0.90)*1.31
[1] 44.47931 55.72069
```

∎

4.6.2   Two-Sided Tolerance Interval to Control Both Tails of a Distribution

The two-sided tolerance interval in Section 4.6.1 and Chapter 3 contains, with specified confidence, at least a certain proportion, β, of the distribution between its two endpoints, irrespective of how the proportion of the distribution below the lower endpoint and above the upper endpoint is apportioned. Sometimes, it is more appropriate to use an interval that will separately control the distribution proportions in *each* of the two tails of the distribution. This type of tolerance interval would, for example, be useful if the cost of being above the upper endpoint were different from that of being below the lower endpoint. The two specified distribution proportions need not be equal. A $100(1 - \alpha)\%$ tolerance interval to control the proportion in both tails of the distribution will be denoted by $[\underset{\sim}{T}_{p_{tL}},\ \widetilde{T}_{p_{tU}}]$. Then

$$\Pr(Y < \underset{\sim}{T}_{p_{tL}}) \leq p_{tL} \quad \text{and} \quad \Pr(Y > \widetilde{T}_{p_{tU}}) \leq p_{tU}$$

with $100(1 - \alpha)\%$ confidence, where p_{tL} is the specified maximum proportion of the distribution in the lower tail of the distribution, and p_{tU} is the specified maximum proportion in the upper tail of the distribution. Often, in practice, $p_{tL} = p_{tU} = p$. The tolerance interval is

$$\left[\underset{\sim}{T}_{p_{tL}},\ \widetilde{T}_{p_{tU}}\right] = \left[\bar{x} - g''_{(1-\alpha;p_{tL},n)}s,\ \bar{x} + g''_{(1-\alpha);p_{tU},n}s\right],$$

where the factors $g''_{(1-\alpha,p,n)}$ are given in Tables J.6a and J.6b. Section E.5.2 describes the underlying theory.

Example 4.9   Tolerance Interval to Contain No More than a Proportion 0.05 in Each Tail of the Distribution of Circuit Pack Output Voltages. For the example, we again have $n = 5$, $\bar{x} = 50.10$ volts, and $s = 1.31$ volts. Suppose now that the manufacturer wanted a two-sided 95% tolerance interval to contain no more than a proportion 0.05 in each tail of the distribution.

We use $g''_{(0.95;0.05,5)} = 4.847$ (from Table J.6b). A two-sided 95% tolerance interval to have a proportion less than 0.05 in each distribution tail is

$$\left[\underset{\sim}{T}_{0.05_{tL}}, \ \widetilde{T}_{0.05_{tU}} \right] = 50.10 \mp 4.847 \times 1.31 = [43.8, \ 56.4].$$

This interval is wider than the control-the-center tolerance interval from Example 4.8 that does not control each tail of the distribution. This is because simultaneously controlling both tails is a more stringent requirement than just controlling the total proportion of the distribution outside the interval. ∎

4.6.3 One-Sided Tolerance Bounds

As indicated in Section 2.4.2, a one-sided lower $100(1 - \alpha)\%$ tolerance bound to be exceeded by at least a proportion p of a normal distribution is equivalent to a one-sided lower $100(1 - \alpha)\%$ confidence bound on the $1 - p$ quantile of the distribution. A one-sided upper $100(1 - \alpha)\%$ tolerance bound to exceed at least a proportion p of the distribution is equivalent to a one-sided upper $100(1 - \alpha)\%$ confidence bound on the p quantile of the distribution. A technical demonstration of this result is given in Section B.5. The computation of such one-sided tolerance bounds was illustrated in Examples 4.4 and 4.6.

4.7   PREDICTION INTERVAL TO CONTAIN A SINGLE FUTURE OBSERVATION OR THE MEAN OF m FUTURE OBSERVATIONS

A two-sided $100(1 - \alpha)\%$ prediction interval to contain the mean of m future, independently and randomly selected observations, based upon the results of a previous independent random sample of size n from the same normal distribution, is

$$\left[\underset{\sim}{\bar{Y}}_m, \ \widetilde{Y}_m \right] = \bar{x} \mp t_{(1-\alpha/2;n-1)} \left(\frac{1}{m} + \frac{1}{n} \right)^{1/2} s. \tag{4.7}$$

An important special case arises when $m = 1$; this results in a prediction interval to contain a single future observation. One-sided lower and upper $100(1 - \alpha)\%$ prediction bounds are obtained by replacing $\alpha/2$ by α (and $\mp$ by either $-$ or $+$) in the appropriate part of the preceding expression. The factors for two-sided prediction intervals given in Tables J.1a and J.1b, and the factors for one-sided prediction bounds in Tables J.3a and J.3b for the cases when $m = 1$ and $m = n$ were computed from these formulas.

Few textbooks consider prediction intervals, except in the context of regression analysis. The theory underlying these methods is given in Section E.6.1.

Example 4.10   Prediction Interval to Contain the Voltage Output of a Future New Circuit Pack. Suppose that the manufacturer needs to provide a consumer with a prediction interval that will, with high confidence, contain the voltage of a future circuit pack. From the previous sample, $n = 5$, $\bar{x} = 50.10$ volts, and $s = 1.31$ volts. Also $t_{(0.975;4)} = 2.776$. Thus, a two-sided 95% prediction interval to contain the voltage of single future unit is

$$\left[\underset{\sim}{Y}, \ \widetilde{Y} \right] = 50.10 \mp 2.776 \left(\frac{1}{1} + \frac{1}{5} \right)^{1/2} 1.31 = [46.12, \ 54.08].$$

Using R as a calculator gives

```
> 50.10 + c(-qt(0.975,4), qt(0.975,4))*sqrt(1+1/5)*1.31
[1] 46.12 54.08.
```

An upper 95% prediction bound to exceed the voltage of the future unit is

$$\underset{\sim}{Y} = 50.10 + 2.132 \left(\frac{1}{1} + \frac{1}{5} \right)^{1/2} 1.31 = 53.16.$$

∎

Example 4.11   Prediction Interval to Contain the Mean Voltage Output of Three New Circuit Packs. Suppose now that the manufacturer needs to predict the mean voltage of a future random sample of $m = 3$ circuit packs. From the previous sample, $n = 5$, $\bar{x} = 50.10$ volts, and $s = 1.31$ volts. Also $t_{(0.975;4)} = 2.776$ and $t_{(0.95;4)} = 2.132$. Thus, a two-sided 95% prediction interval to contain the mean voltage of the sample of three future units is

$$\left[\underset{\sim}{\bar{Y}}_3, \ \widetilde{\bar{Y}}_3 \right] = 50.10 \mp 2.776 \left(\frac{1}{3} + \frac{1}{5} \right)^{1/2} 1.31 = [47.44, \ 52.76].$$

Using R as a calculator gives

```
> 50.10 + c(-qt(0.975,4), qt(0.975,4))*sqrt(1/3+1/5)*1.31
[1] 47.44   52.76.
```

A one-sided upper 95% prediction bound to exceed the mean of the three future units is

$$\underset{\sim}{\bar{Y}}_3 = 50.10 + 2.132 \left(\frac{1}{3} + \frac{1}{5} \right)^{1/2} 1.31 = 52.14.$$

∎

4.8   PREDICTION INTERVAL TO CONTAIN AT LEAST k OF m FUTURE OBSERVATIONS

4.8.1   Two-Sided Prediction Interval

A two-sided $100(1 - \alpha)\%$ simultaneous prediction interval to contain the values of at least k of m ($1 \leq k \leq m$) future randomly selected observations from a previously sampled normal distribution is

$$\left[\underset{\sim}{Y}_{k:m}, \ \widetilde{Y}_{k:m} \right] = \bar{x} \mp r_{(1-\alpha;k,m,n)} s,$$

where the $r_{(1-\alpha;k,m,n)}$ has not been tabulated in general, but can be computed by the methods shown in Section E.6.4, which also provides the underlying theory. Section B.7 shows how to use simulation to compute simultaneous prediction interval coverage probabilities for location-scale distributions and Section 14.6 shows how to use such simulations to compute the required factors for these simultaneous prediction intervals.

The factors $r_{(1-\alpha;m,m,n)}$ for prediction intervals to contain *all* m of m future observations are given in Table J.8. The factors for obtaining the intervals given in Tables J.1a and J.1b were also taken from Table J.8. A conservative approximation for $r_{(1-\alpha;m,m,n)}$ is

$$r_{(1-\alpha;m,m,n)} \approx \left(1 + \frac{1}{n} \right)^{1/2} t_{(1-\alpha/(2m);n-1)}. \tag{4.8}$$

This approximation may be used for nontabulated values and for constructing a simple computer program to perform the calculations. It is based on a Bonferroni inequality (see Section 2.9). This approximation is satisfactory for most practical purposes, except for combinations of small n, large m, and small $1 - \alpha$. Also, (4.8) is exact for the special case of $m = 1$ (i.e., a prediction interval to contain a single future observation).

4.8.2 One-Sided Prediction Bounds

One-sided lower and upper $100(1 - \alpha)\%$ (simultaneous) prediction bounds to be exceeded by and to exceed at least k of m future observations from a previously sampled normal distribution are, respectively,

$$\underset{\sim}{Y}_{k:m} = \bar{x} - r'_{(1-\alpha;k,m,n)} s$$

and

$$\widetilde{Y}_{k:m} = \bar{x} + r'_{(1-\alpha;k,m,n)} s,$$

where $r'_{(1-\alpha;k,m,n)}$ has not been tabulated in general, but can be computed by the methods shown in Section E.6.3, which also describes the underlying theory. Section B.7 shows how to use simulation to compute simultaneous one-sided prediction bound coverage probabilities for location-scale distributions and Section 14.6 shows how to use such simulations to compute the required factors for these simultaneous prediction bounds.

The factors $r'_{(1-\alpha;m,m,n)}$ for one-sided prediction bounds to contain all of m future observations are given in Table J.9. The factors given in Tables J.3a and J.3b for obtaining such intervals were taken from Table J.9. A conservative approximation for $r'_{(1-\alpha;m,m,n)}$ is

$$r'_{(1-\alpha;m,m,n)} \approx \left(1 + \frac{1}{n}\right)^{1/2} t_{(1-\alpha/m;n-1)}.$$

This expression provides an adequate approximation in situations similar to the corresponding expression for the two-sided prediction interval. The expression is exact for $m = 1$.

Example 4.12 Prediction Interval to Contain Output Voltages for 10 Future Circuit Packs. For the example, $n = 5, \bar{x} = 50.10$ volts, and $s = 1.31$ volts. The manufacturer wants a two-sided 95% prediction interval to contain the voltages of all $m = 10$ future circuit packs randomly selected from the same process and an upper 95% prediction bound to exceed the voltages of all $m = 10$ future units. From Table J.8, $r_{(0.95;10,10,5)} = 5.229$ and from Table J.9, $r'_{(0.95;10,10,5)} = 4.418$. A two-sided 95% prediction interval to contain all 10 future voltages is

$$\left[\underset{\sim}{Y}_{10:10}, \ \widetilde{Y}_{10:10}\right] = 50.10 \mp 5.229 \times 1.31 = [43.25, \ 56.95].$$

A one-sided upper 95% prediction bound to exceed all 10 future voltages is

$$\widetilde{Y}_{10:10} = 50.10 + 4.418 \times 1.31 = 55.89.$$

■

Example 4.13 Prediction Interval to Contain Output Voltages for at Least 9 of 10 Future Circuit Packs. For the example, $n = 5, \bar{x} = 50.10$ volts, and $s = 1.31$ volts. The manufacturer wants a 95% prediction interval to contain the voltages of at least $k = 9$ of $m = 10$ future circuit packs randomly selected from the same process and an upper 95% prediction bound to exceed the voltages of at least $k = 9$ of $m = 10$ future units. Because only limited tables exist to compute the required factors (described in the Bibliographic Notes section at the end of this chapter), bootstrap-based methods described in Section 14.6 were used to compute $r_{(0.95;9,10,5)} = 3.99$ and $r'_{(0.95;9,10,5)} = 3.16$. A two-sided 95% prediction interval to contain at least 9 of 10 future circuit packs is

$$\left[\underline{Y}_{9:10}, \ \widetilde{Y}_{9:10} \right] = 50.10 \mp 3.99 \times 1.31 = [44.9, \ 55.3].$$

A one-sided upper 95% prediction bound to exceed the voltages of at least 9 of 10 future circuit packs is

$$\widetilde{Y}_{9:10} = 50.10 + 3.16 \times 1.31 = 54.2.$$

$\blacksquare$

4.9   PREDICTION INTERVAL TO CONTAIN THE STANDARD DEVIATION OF m FUTURE OBSERVATIONS

A two-sided $100(1 - \alpha)\%$ prediction interval to contain s_m, the standard deviation of m future observations based upon the results from a previous independent, random sample of size n from the same normal distribution is

$$[\underline{S}_m, \ \widetilde{S}_m] = \left[s \left(\frac{1}{F_{(1-\alpha/2;n-1,m-1)}} \right)^{1/2}, \ s(F_{(1-\alpha/2;m-1,n-1)})^{1/2} \right]. \qquad (4.9)$$

One-sided lower and upper $100(1 - \alpha)\%$ prediction bounds are obtained by replacing $\alpha/2$ by α in the expressions corresponding to $\underline{S}_m$ and $\widetilde{S}_m$, respectively, in (4.9). The factors for obtaining two-sided prediction intervals and one-sided prediction bounds given in Tables J.2a, J.2b, J.4a, and J.4b were computed from these expressions. The theory underlying these methods is given in Section E.6.2.

Example 4.14   Prediction Interval to Contain the Standard Deviation of the Voltages of a Future Sample of Circuit Packs. For the example, the manufacturer asked for a 95% prediction interval to contain the standard deviation of the voltages for a future random sample of $m = 3$ circuit packs from the same process. From the previous sample, $n = 5$, and $s = 1.31$ volts. Also, $F_{(0.975;4,2)} = 39.25$, $F_{(0.975;2,4)} = 10.65$, and $F_{(0.95;2,4)} = 6.944$. Then a two-sided 95% prediction interval to contain s_3, the standard deviation of the voltage of three future units, is

$$[\underline{S}_3, \ \widetilde{S}_3] = \left[1.31 \left(\frac{1}{39.25} \right)^{1/2}, \ 1.31 \times 10.65^{1/2} \right] = [0.209, \ 4.27].$$

Using R as a calculator gives

```
> 1.31*sqrt(c(1/qf(0.975,df1=4,df2=2), qf(p=0.975,df1=2,df2=4)))
[1] 0.2091 4.2749
```

An upper 95% prediction bound to exceed the sample standard deviation of the three future units is

$$\widetilde{S}_3 = 1.31 \times 6.944^{1/2} = 3.45.$$

$\blacksquare$

4.10 THE ASSUMPTION OF A NORMAL DISTRIBUTION

The tabulations and formulas for constructing the statistical intervals in this and the preceding chapter are based on theory for random samples from a normal distribution. Moreover, in the case of a prediction interval, it is also assumed that the future sample is selected randomly and independently from the same normal distribution. Thus, the intervals are strictly valid only under these assumptions.

A confidence interval for the distribution mean, however, tends to be insensitive to deviations from normality in the sampled distribution, unless the sample is very small and the deviation from normality is pronounced. In statistical terminology, a confidence interval to contain the distribution mean is said to be "robust" to deviations from normality. This is the result of the well-known central limit theorem, which states that the distribution of a sample mean is approximately normal if the sample size is not too small and if the underlying distribution is not too skewed. Thus, a confidence interval for the distribution mean, assuming a normal distribution, may be used to construct a confidence interval in many practical situations, even if the assumption of normality is not strictly met. In such cases, the resulting interval, instead of being an exact $100(1 - \alpha)\%$ confidence interval for the distribution mean, is an approximate $100(1 - \alpha)\%$ confidence interval (i.e., the coverage probability for the procedure is not exactly equal to the nominal confidence level $100(1 - \alpha)\%$). The approximation becomes worse as the desired level of confidence $1 - \alpha$ increases.

For similar reasons, prediction intervals to contain the mean of a future sample are also relatively insensitive to deviations from normality, unless either the given sample or the future sample is very small or the deviation from normality is pronounced. This limitation, however, includes the important case of a prediction interval for a single future observation (i.e., $m = 1$). Thus, the prediction interval given here to contain a single future observation may be seriously misleading when sampling from a nonnormal distribution.

Unfortunately, one cannot use the central limit theorem to justify the use of normal distribution based methods for most other types of statistical intervals. For example, the procedure in Section 4.3 for obtaining a confidence interval for the distribution standard deviation is highly sensitive to the assumption of normality, even for large sample sizes. Therefore, serious errors could result in using the methods given there when the sampled distribution is not normal. The same holds for a prediction interval to contain the standard deviation of a future sample.

Tolerance intervals to contain a specified proportion of a distribution and prediction intervals to contain all of m future observations tend to involve inferences or predictions about the tails of a distribution. Deviations from normality are generally most pronounced in the distribution tails. For this reason, such intervals could be seriously misleading when the underlying distribution is not normal, especially for high confidence levels, and for tolerance intervals when the proportion of the distribution to be contained within the interval is close to 1.0. Similar conditions apply to confidence intervals for a distribution tail quantile and confidence intervals to contain the probability of exceeding an extreme specified value. For example, an estimate of the 0.05 quantile under the assumption of a normal distribution, and, even more so, an estimate of the 0.01 quantile, based on only five observations, is strongly dependent on the assumption of normality in the tails of the distribution.

The intuitive reason for these results is that the parametric estimate of the p quantile of a location-scale distribution is the sum of an estimate of a location parameter (the mean in the case of a normal distribution) plus an estimate of a scale parameter (the standard deviation for the normal distribution), multiplied by a factor that depends on p and the assumed distribution; that is, $\widehat{y}_p = \bar{x} + s\Phi_{\mathrm{norm}}^{-1}(p)$. The factor $\Phi^{-1}(p)$ is highly distribution dependent and the dependency becomes stronger for values of p that are in the tails of the distribution. See the references in the Bibliographic Notes section for technical details and further discussion.

4.11   ASSESSING DISTRIBUTION NORMALITY AND DEALING WITH NONNORMALITY

4.11.1   Probability Plots and Q–Q Plots

Before using statistical intervals that depend heavily on the normality assumption (e.g., tolerance intervals and confidence intervals on tail quantiles), one should assess how well a normal distribution actually fits the given data. There are a variety of formal statistical tests, described in numerous textbooks, available for assessing normality. Normal distribution probability plots, however, are a simple and effective, though less formal, tool for doing this, especially if there are 20 or more observations.

Normal distribution probability plots can be constructed by plotting $x_{(i)}$, the ith ordered (from smallest to largest) observation, against the corresponding proportion $p_i = (i - 0.5)/n$ on appropriately scaled axes (i.e., based on the normal distribution assumption), usually employing computer software for this purpose. Normal distribution probabilities $\Pr(X \leq x)$ versus x plot as a *straight* line on such plots. The $(x_{(i)}, p_i)$ pairs are a "nonparametric" estimate (i.e., an estimate that does not depend on an assumption of a particular parametric distribution) of $\Pr(X \leq x)$. Thus, systematic deviations from linearity in the plots p_i versus $x_{(i)}$ are indicative of a departure from normality. Similar probability plots can be constructed to assess other commonly used distributions, such as the lognormal and Weibull. Hahn and Shapiro (1967, Chapter 8), Nelson (1982, Chapter 3) and Meeker and Escobar (1998, Chapter 6) give theory, methods, and examples for probability plots for the normal and other distributions. Probability plots used to be prepared by hand using normal probability paper, but are widely available in statistical software packages today.

Quantile–quantile (Q–Q) plots serve the same purpose as probability plots. To obtain a normal Q–Q plot, one plots, on linear scales, the ith ordered observation against $z_{(p_i)}$, the normal distribution p_i quantile, where $p_i = (i - 0.5)/n$. The only difference between a Q–Q plot and a probability plot is that the probability plot has a probability scale for the p_i while the Q–Q plot has a linear scale for the corresponding normal distribution quantile values $z_{(p_i)}$. The probability scales on probability plots are easier to interpret and explain. Because they use linear scales, Q–Q plots are, however, easier to produce with unsophisticated graphics and statistical computer packages, or with (old-fashioned) ordinary graph paper.

4.11.2   Interpreting Probability Plots and Q–Q Plots

If the plotted points on a normal probability or Q–Q plot deviate appreciably from a straight line, the adequacy of the normal distribution as a model for the population (or process) is in doubt. Judging departures from a straight line requires one to allow for the variability in the sample data; this is highly sample size dependent. Also, for many models, including the normal distribution, one would expect the observations in the tails of the distribution to vary more than those in the center of the distribution. To help judge this variability, Hahn and Shapiro (1967) and Meeker and Escobar (1998) give normal probability plots for simulated data with different sample sizes and with data from both normal and nonnormal distributions. The analyst may wish to do similar simulations, possibly developing a general computer program for routine use.

Example 4.15   Ball Bearing Failure Data. Lieblein and Zelen (1956) describe data from fatigue endurance tests on ball bearings. The purpose of the tests was to study the relationship between fatigue life and stress loading. The data shown in Table 4.1 are a subset of $n = 23$ ball bearing failure times (in millions of cycles to failure) for units tested at one level of stress, previously reported and analyzed by Meeker and Escobar (1998) and Lawless (2003).

17.88	28.92	33.00	41.52	42.12	45.60
48.40	51.84	51.96	54.12	55.56	67.80
68.64	68.64	68.88	84.12	93.12	98.64
105.12	105.84	127.92	128.04	173.40	

Table 4.1 Ball bearing failure data (millions of cycles).

Here we extend the analysis given there and in Meeker and Escobar (1998, Example 11.2). Figure 4.1 gives a histogram and a box plot of these data; Figure 4.2 is a normal probability plot of the data. The plotted points in the normal probability plot tend to scatter around a curve, rather than a straight line. The deviation from a straight line might be more than one would expect due to chance for a sample from a normal distribution—at least, from a visual evaluation. Actually, because of the small sample size in the ball bearing example (23 observations), the statistical evidence against a normal distribution is not terribly strong (based on a graphical goodness of fit test and simulation analysis similar to those described in Meeker and Escobar, 1998, Chapter 6). If similar deviations from linearity were observed with a considerably larger sample size, the evidence against the normal distribution would be considerably stronger. Nevertheless, the curvature in the probability plot and physical considerations (the normal distribution ranges from $-\infty$ to ∞, but failure times must be positive) suggest that the normal distribution might *not* provide an adequate model for the population (or process) from which the data were obtained. ∎

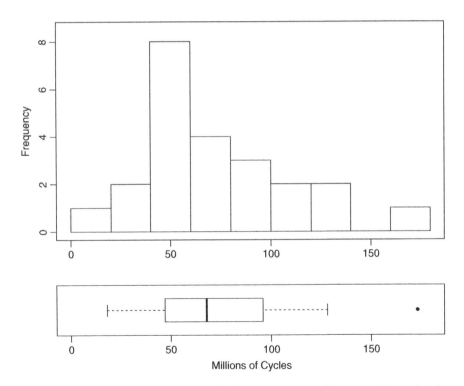

Figure 4.1 A histogram and a box plot of ball bearing cycles to failure in millions of cycles.

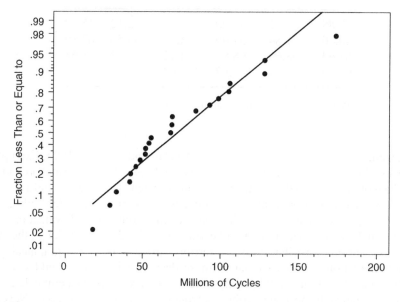

Figure 4.2   Normal distribution probability plot for the ball bearing failure data.

4.11.3   Dealing with Nonnormal Data

If one constructs a probability plot from the given data and determines from this that the normal distribution is not an appropriate model for the problem at hand, there are several ways to proceed:

- Use a smooth curve drawn through the points on a probability plot to obtain a simple and easy-to-understand *nonparametric* graphical estimate of the cumulative distribution function. For example, from Figure 4.2, we estimate that a proportion of approximately $p = 0.30$ of the ball bearings will fail by the end of 50 million cycles.

- Use a "distribution-free method" for constructing the desired statistical interval, as will be described in Chapter 5.

- Seek some other distribution (e.g., the lognormal, gamma, or Weibull) that provides an adequate representation for the data. An alternative model may be justified from physical considerations. In Chapters 12 and 14 we provide further information and references about using statistical distributions other than the normal distribution to model data and to compute confidence intervals. For example, Lieblein and Zelen (1956) used the Weibull distribution as a model for the ball bearing failure data.

- Transform the data in a manner that allows the normal distribution to provide an adequate model for the population or process and a basis for making inferences (using the transformed data). This is equivalent to fitting an alternative distribution. For example, analyzing the logs of a set of data as if they were normally distributed is equivalent to fitting a (two-parameter) lognormal distribution to the original data (see Section 12.4.1 for more information about the lognormal distribution). In the next section we illustrate the use of transformed data to compute statistical intervals.

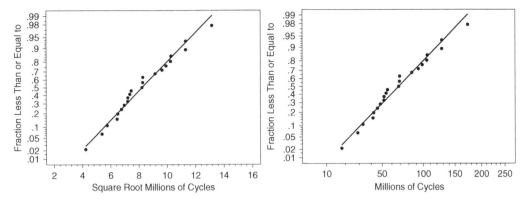

Figure 4.3   Normal distribution probability plot of the square roots of the ball bearing failure data (left) and a normal distribution probability plot for the ball bearing failure data on a square root axis (right).

4.12   DATA TRANSFORMATIONS AND INFERENCES FROM TRANSFORMED DATA

4.12.1   Power Transformations

If the original sample does not appear to have come from a normal distribution, it may still be possible to find a transformation that will allow the data to be adequately represented by a normal distribution. For example, the left-hand plot in Figure 4.3 is a normal distribution probability plot of the square roots of the ball bearing failure data. The right-hand plot is identical, except that the data axis shows the number of cycles on a (nonlinear) square root axis. Although it is easier to make plots like the one on the left with unsophisticated plotting software, the plot on the right is easier to interpret and explain. The two plots in Figure 4.4 are also identical except the plot on the right shows the number of cycles on a logarithmic axis. These plots suggest that either the square root transformation or the log transformation results in transformed data that can be described well by a normal distribution. The fit to the logs is a little better than the fit to the square roots.

The two plots in Figure 4.5 are again similar to the plots in Figures 4.3 and 4.4 but are based on plotting the negative reciprocals of the data using normal probability scales and plotting

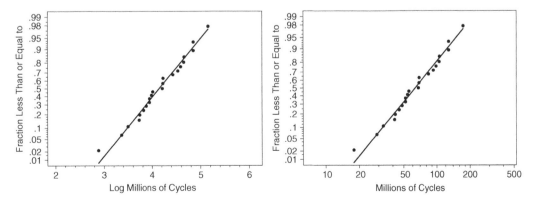

Figure 4.4   Normal distribution probability plot of the natural logs of the ball bearing failure data (left) and a normal distribution probability plot for the ball bearing failure data on a log axis (i.e., a lognormal probability plot) (right).

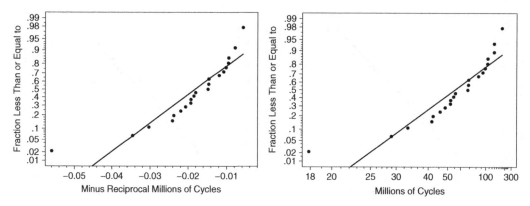

Figure 4.5   Normal distribution probability plot of the reciprocals of the ball bearing failure data (left) and a normal distribution probability plot for the ball bearing failure data on a negative reciprocal data axis (right).

the original data on a negative reciprocal scale axis, respectively. From the curvature in the plotted points around the straight lines on these plots, we conclude that a negative reciprocal transformation of the data fails to lead to a good normal distribution fit.

The preceding transformations are members of the power family of transformations which is defined for $x > 0$ as

$$y^{(\gamma)} = \begin{cases} x^\gamma & \text{if } \gamma > 0, \\ -x^\gamma & \text{if } \gamma < 0, \\ \log(x) & \text{if } \gamma = 0, \end{cases} \qquad (4.10)$$

where γ is a generally unknown parameter that characterizes the transformation. In practice, one tries to find a value (or range of values) for γ that leads to approximate normality. In particular, one may try different values of γ (e.g., $\gamma = 1, 0.5, 0.333, 0$, and -1, corresponding to no transformation, square root, cube root, log, and reciprocal transformations, respectively) to seek a value (or range of values) that results in a probability plot that is nearly linear. Physical considerations or experience may, in some cases, suggest an appropriate value of γ.

4.12.2 Computing Statistical Intervals from Transformed Data

The sample mean and standard deviation of the logs (base e) of the ball bearing data are, respectively, $\bar{y}^{(0)} = 4.150$ and $s^{(0)} = 0.533$ log millions of cycles. These can be used to compute statistical intervals on the log scale that, in turn, can be translated into intervals on the original data scale.

For example, the method in Section 4.4 gives the following lower 95% confidence bound for the 0.10 quantile of the distribution on the log scale:

$$\underset{\sim}{y}^{(0)}_{0.10} = \bar{y}^{(0)} - g'_{(0.95;0.90,23)} s^{(0)} = 4.150 - 1.869 \times 0.533 = 3.15.$$

Taking *antilogs* gives the desired lower 95% confidence bound for the 0.10 quantile on the original data scale,

$$\underset{\sim}{x}_{0.10} = \exp\left(\underset{\sim}{y}^{(0)}_{0.10}\right) = \exp(3.15) = 23.3$$

millions of cycles.

Similar calculations yield confidence intervals for other quantiles, as well as tolerance intervals and prediction intervals on the original data scale. To use the method in Section 4.5 to compute a confidence interval for a normal distribution tail probability, one simply computes $k = [\log(x) - \bar{y}^{(0)}]/s^{(0)}$ (because $\Pr(X \le x) = \Pr[\log(X) \le \log(x)]$) and proceeds as before.

We note, however, that a confidence interval for the mean of the distribution of the transformed data does *not* translate into a confidence interval for the mean on the original scale. It does, however, provide a confidence interval for the median (i.e., 0.50 quantile) of the original distribution. Similarly, a confidence interval for the standard deviation of the distribution of the transformed data does *not* translate into a confidence interval for the standard deviation on the original scale.

4.12.3 Comparison of Inferences Using Different Transformations

Table 4.2 shows the one-sided lower 95% confidence bounds on the 0.01 and 0.10 quantiles and both lower and one-sided upper 95% confidence bounds on the 0.50 quantile for the ball bearing life distribution, using the original untransformed data, three alternative reasonable data transformations, and a distribution-free method. Recall that visual evaluation of the normal probability plots of the original untransformed observations led us to question whether the original untransformed data can be appropriately represented by a normal distribution. The square root and log transformations did, however, provide reasonably good normal distribution fits. The cube root transformation is intermediate between these two transformations and also provides a good fit.

The *distribution-free* intervals are described in Chapter 5. These require no assumptions about the form of the underlying distribution, but are generally wider than the parametric alternatives and sometimes a desired interval does not exist. In the ball bearing example, because there are only 23 observations, it is not possible to find distribution-free lower 95% confidence bounds for the lower quantiles. These call for extrapolation outside the range of the data, and this requires assuming a distribution form. See Chapter 5 for further discussion.

All of the transformations, as well as the untransformed data, give somewhat similar lower and upper confidence bounds for the 0.50 quantile. The lower confidence bounds for the 0.01 quantile, however, differ substantially. In fact, fitting a normal distribution to the original (untransformed) data resulted in a physically impossible negative lower confidence bound for the 0.01 quantile.

Transformation	γ	Sample statistics		One-sided 95% confidence bounds			
		$\bar{y}^{(\gamma)}$	$s^{(\gamma)}$	$\underset{\sim}{x}_{0.01}$	$\underset{\sim}{x}_{0.10}$	$\underset{\sim}{x}_{0.50}$	$\widetilde{x}_{0.50}$
No transformation	1	72.21	37.50	-48.0	2.11	58.8	85.6
Square root	$1/2$	8.234	2.146	1.82	17.8	55.8	81.0
Cube root	$1/3$	4.048	0.707	5.64	20.3	54.9	79.5
Log	0	4.150	0.533	11.5	23.3	52.5	76.7
Distribution-free	—	—	—	—	—	51.8	84.1

Table 4.2 Comparison of one-sided 95% confidence bounds for various quantiles using different transformations for the ball bearing life data. γ is the parameter of the power distribution transformation.

The preceding comparison leads to some general conclusions about fitting parametric models to data. First, it suggests that it is useful to analyze the data under alternative reasonable models. We could have gone further and plotted the lower 95% confidence bounds for various quantiles versus γ for the range of reasonable values of γ based on probability plots of the data. Second, it shows that alternative models that fit the data well and distribution-free methods are likely to give similar results for inferences *within* the range of the data (i.e., for inferences that only involve interpolation among data points). Inferences that call for extrapolation outside the range of the data are, in general, much more sensitive to the assumed model. Moreover, although a poorly fitting model may provide credible inferences within the range of the data, it will often give unreasonable answers (e.g., the negative value for $\underset{\sim}{x}_{0.01}$) for extrapolations outside the data range.

4.12.4   Box–Cox Transformations

The Box–Cox family of transformations is closely related to the power transformation family. This transformation is defined for $x > 0$ as

$$y^{(\gamma)} = \begin{cases} \dfrac{x^\gamma - 1}{\gamma} & \text{if } \gamma \neq 0, \\ \log(x) & \text{if } \gamma = 0, \end{cases}$$

where γ is a (generally unknown) parameter that characterizes the transformation and x^γ is a power transformation. When $\gamma \neq 0$, $y^{(\gamma)}$ is a standardized form of the power transformation x^γ in (4.10). In this sense, the Box–Cox transformation includes all power transformations, and when γ approaches 0 the transformation $y^{(\gamma)}$ approaches the limit $\log(x)$. Thus, when γ is close to 0, the Box–Cox transformation is close to a log transformation. Probability plots can be used in the same way as for the power transformation to guide the choice of γ. As with the power transformations, physical considerations or experience may suggest how to choose γ. Also analytical methods have been proposed to help determine the "best" value for γ (see references in the Bibliographic Notes section at the end of this chapter).

4.13   STATISTICAL INTERVALS FOR LINEAR REGRESSION ANALYSIS

Regression analysis is a widely used statistical technique, usually used to relate the mean of a response variable (also sometimes known as the "dependent variable") to one or more explanatory variables (sometimes called "independent variables" or "covariates"). Given sample data and an assumed relationship between the response and the explanatory variables, the method of least squares is often used to estimate unknown parameters in the relationship. There are many applications for regression analysis and the resulting statistical intervals. Because—unlike many of the other topics considered in this book—regression analysis is discussed in much detail in numerous elementary and advanced textbooks (see the Bibliographic Notes section), our comments here will be brief.

As in other situations, the validity of statistical intervals generally depends on the correctness of the assumed model. The commonly used linear regression model assumes that the observed response variable y follows a normal distribution with mean

$$\mu = \beta_0 + \beta_1 x_1 + \cdots + \beta_p x_p.$$

Here the x_j values are known values of the explanatory variables and the β_j coefficients are unknown coefficients that are to be estimated from the data using the method of least squares.

The standard model also assumes that the observed y values are statistically independent with a variance σ^2 that does not depend on the x_j values. In some applications, these x_j values might be known functions of some explanatory variables (e.g., 1/temperature, log(voltage), (time)^2).

Users of regression analysis should watch for departures from the model assumptions. Methods for doing this can involve graphical analysis of the residuals (i.e., the differences between the observed and the predicted response values) from the fitted model, as described in textbooks on applied regression analysis. Some ways of handling possible departures from the assumed model are:

1. If the mean of y cannot be expressed as a linear function of the parameters, special (usually iterative) nonlinear least squares methods for estimating the parameters may be required; see, for example, Bates and Watts (1988), Seber and Wild (1989), or Ritz and Streibig (2008).

2. If σ^2 is not the same for all observations, a transformation—for example, a Box and Cox (1964) transformation, given in Section 4.12.4—of the response variable might be appropriate. Sometimes the method of weighted least squares is used. Both approaches are described in Carroll and Ruppert (1988). In other cases, it might be desirable to model both the mean and the standard deviation as separate functions of the explanatory variable (see Nelson, 1984, for an example).

3. If the observed response values (y) are not statistically independent, either generalized least squares or time series analysis methods—see, for example, Wei (2005), Cryer and Chan (2008), Bisgaard and Kulahci (2011), or Box et al. (2015)—might be appropriate.

4. If some of the values of the response variable are censored (i.e., the actual response is unknown other than being less than a known left-censoring value, greater than a known right-censoring value, or to lie between known lower and upper censoring values), or if they do not follow a normal distribution, the method of maximum likelihood, rather than the method of least squares, should be used for estimating the parameters. See Nelson (1990), Meeker and Escobar (1998), or Lawless (2003) for details.

5. If the observed values of the explanatory variables contain significant measurement error, the methods given by Fuller (1987), Carroll et al. (2006), or Buonaccorsi (2010) might be appropriate.

The rest of this section provides references to methods for computing statistical intervals for the standard linear regression model, when all of the assumptions hold. For other situations, some of the references given above provide similar methods. Some of the procedures require factors that are too numerous to tabulate and thus require specialized computer software (e.g., Eberhardt et al., 1989). In Chapter 12 we will describe some other more general methods that can be applied to regression analysis, but these also generally require special computer software.

4.13.1 Confidence Intervals for Linear Regression Analysis

Many textbooks on statistical methods and specialized textbooks on regression analysis give details on the construction of confidence intervals for

- The parameters $(\beta_0, \beta_1, \ldots, \beta_p)$ of the regression model.
- The expected value or mean $\mu = \beta_0 + \beta_1 x_1 + \cdots + \beta_p x_p$ of the response variable for a specified set of conditions for the explanatory variable(s).

- Quantiles of the distribution of the response variable for a specified set of conditions for the explanatory variable(s).

- The variance σ^2 (or standard deviation σ) of the observations (which may or may not depend on a given set of x_j values).

4.13.2   Tolerance Intervals for Linear Regression Analysis

Tolerance intervals for the response variable for one or more conditions of the explanatory variable are *not* provided in most of the standard textbook chapters and textbooks on regression analysis (one exception is Graybill, 1976). This may be because special factors are required to compute these intervals.

4.13.3   Prediction Intervals for Regression Analysis

Many textbooks on statistical methods and specialized textbooks on regression analysis show how to compute a prediction interval for a single future observation on a response variable for a specified set of conditions for the explanatory variable(s). In fact, in introductory textbooks, regression analysis is the only situation in which prediction intervals are generally discussed. The methods for a single future observation easily extend to a prediction interval for the mean of m future response variable observations. It is also possible to construct prediction intervals to contain at least k of m future observations, but special methods would be required.

4.14   STATISTICAL INTERVALS FOR COMPARING POPULATIONS AND PROCESSES

Designed experiments are often used to compare two or more competing products, designs, treatments, packaging methods, etc. Statistical intervals are useful for presenting the results of such experiments. Most textbooks on elementary statistical methods show how to compute confidence intervals for the difference between the means of two randomly sampled populations or processes, assuming normal distributions and using various assumptions concerning the variances of the two distributions, based upon either paired or unpaired observations. References dealing with the use of statistical intervals to compare more than two normal distributions are also provided. The statistical methods for (fixed effects) analysis of variance and analysis of covariance models frequently used in the analysis of such data are special cases of the more general regression analysis methods that were outlined in the previous section. See the Bibliographic Notes section at the end of this chapter for references to some books that cover these and related topics.

BIBLIOGRAPHIC NOTES

General theory

Odeh and Owen (1980) describe the distribution theory methods behind the computation of many of the intervals in this chapter.

Tolerance intervals

Guttman (1970) describes the general theory behind tolerance intervals and tolerance regions (i.e., multivariate tolerance intervals), including Bayesian tolerance intervals (covered in our

Chapter 15). Krishnamoorthy and Mathew (2009) provide theory, methods, and references for non-Bayesian tolerance intervals and tolerance regions.

Simultaneous prediction intervals

The simultaneous prediction interval conservative approximation for $r_{(1-\alpha;m,m,n)}$ given by (4.8) in Section 4.8.1 is based on a Bonferroni inequality (see Section 2.9 or Miller, 1981) and was suggested by Chew (1968), who also describes a second approximation. Both approximations were evaluated by Hahn (1969). The one given in (4.8) was found to be satisfactory for most practical purposes, except for combinations of small n, large m, and small $1 - \alpha$. The theory for prediction intervals to contain all of m future observations was originally given by Hahn (1969, 1970a). Theory and tables for more general prediction intervals to contain at least k of m future observations are given in Fertig and Mann (1977) for the one-sided case and Odeh (1990) for the two-sided case. General concepts of prediction intervals and additional references are given by Hahn and Nelson (1973). The factors for two-sided prediction intervals and one-sided prediction bounds to contain the standard deviation of a future sample were described in Hahn (1972b).

Normality and nonnormality

Detailed discussion and specific evaluations of the effect of nonnormality are given by Scheffé (1959, Sections 10.2 and 10.3). See Hahn (1971) and Canavos and Kautauelis (1984) for further discussion of the strong sensitivity of estimates of small (or large) quantiles to departures from the assumption of normality.

The Box–Cox family of transformations was first described in Box and Cox (1964). Analytical methods to help determine the "best" value for the Box–Cox transformation parameter γ are described, for example, in Draper and Smith (1981, page 225) and Atkinson (1985).

Regression analysis

Most textbooks on statistical methods devote one or more chapters to regression analysis, and there are a number of textbooks that deal exclusively with this subject. See, for example, Draper and Smith (1981), Kutner et al. (2005), Gelman and Hill (2006), Montgomery et al. (2015), and Seber and Lee (2012).

Thomas and Thomas (1986) give confidence bands for the quantiles of the assumed normal distribution of the response variable for a specified set of conditions for the explanatory variable(s).

Lieberman and Miller (1963) give tabulations for, and examples of, simultaneous tolerance intervals for the distribution of a response variable for a range of conditions for the explanatory variable(s). Miller (1981, Chapter 3), Wallis (1951), Bowden (1968), Turner and Bowden (1977, 1979), and Limam and Thomas (1988a) describe other methods of computing simultaneous tolerance bands for a regression model. Mee et al. (1991) describe a procedure for computing simultaneous tolerance intervals for a regression model and describe applications to calibration problems.

Approximate simultaneous prediction intervals to contain all of m future observations for a regression model are described in Lieberman (1961). Exact intervals and improved approximations are given by Hahn (1972a), and a related application is discussed by Nelson (1972b). Miller (1981, Chapter 3) provides a comprehensive discussion of these and related methods.

Comparisons

Lawless (2003, Chapters 3–6) and Nelson (1982, Chapters 10–12) discuss methods for comparing various nonnormal distributions and for making comparisons when the data are censored.

When a normal distribution cannot be assumed, distribution-free procedures may be used. For example, Gibbons (1997) shows how to obtain a distribution-free confidence interval for the difference between the location parameters of two distributions and on the ratio of their scale parameters.

Hahn (1977) shows how to obtain a prediction interval for the difference between future sample means from two previously sampled populations, assuming that normal distributions can be used to describe the populations. Meeker and Hahn (1980) present similar prediction intervals for the ratio of exponential distribution means and normal distribution standard deviations. Both papers describe applications where one desires to make a statement about a future comparison of two populations, based upon the results of past samples from the same populations. Such an interval might, for example, be desired by a manufacturer who wants to predict the results of a future comparison, to be conducted by a regulating agency, of two previously tested products.

Tolerance intervals for a subpopulation corresponding to any particular group in a "fixed effects" analysis of variance model can be computed by using the methods referenced in Section 4.13, because, as indicated in Section 4.14, fixed effects analysis of variance models are special cases of linear regression models. Mee (1989) gives a method for computing a tolerance interval for a population, based on a stratified sampling scheme (i.e., when samples are taken from each subpopulation).

Random-effects models

In "random effects" analysis of variance models (see Mendenhall, 1968, Chapter 12), one is dealing with two or more sources of random variability from a single population. These might, for example, consist of a random sample of batches, as well as a random sample of units within batches. Lemon (1977) gives an approximate procedure for constructing one-sided tolerance bounds for such situations, assuming the batch means follow a normal distribution, and Mee and Owen (1983) provide improved factors. Mee and Owen (1983) provide tolerance bounds that are exact if the ratio of the within batch variance to the between batch variance is known; otherwise, their factors provide approximate bounds. Mee (1984) gives similar factors for two-sided tolerance intervals (referred to as β-content tolerance intervals) and for two-sided prediction intervals for a single observation from the entire population (referred to as β-expectation tolerance intervals). Limam and Thomas (1988b) provide methods for constructing similar tolerance intervals for a one-way random-effects model with explanatory variables.

The advanced methods of interval construction given in Chapters 12–17 can also be used to construct statistical intervals for models with random effects, and we illustrate the construction of such intervals in Sections 18.2 and 18.3.

Other Tables

In modern practice, the use of computer algorithms has largely replaced the use of extensive tables that once were important for computing certain kinds of statistical intervals. For some purposes, however, the tables might still be useful.

Table 1 of Odeh and Owen (1980) gives factors $g'_{(\gamma;p,n)}$ used in constructing confidence intervals for normal distribution quantiles and one-sided tolerance bounds. Factors are provided for all combinations of $p = 0.75, 0.90, 0.95, 0.975, 0.99, 0.999$, $n = 2(1)100(2)180(5)$

$300(10)400(25)650(50)1,000, 1,500, 2,000,$ and $\gamma = 0.995, 0.99, 0.975, 0.95, 0.9, 0.75,$ $0.50, 0.25, 0.10, 0.05, 0.025, 0.01, 0.005.$

Table 3 of Odeh and Owen (1980) gives factors $g_{(\gamma;\beta,n)}$ used to construct two-sided tolerance intervals to control the center of a distribution. Factors are provided for all combinations of $\beta = 0.75, 0.90, 0.95, 0.975, 0.99, 0.995,$ $n = 2(1)100(2)180(5)300(10)400(25)650(50)1,000,$ $1,500, 2,000, 3,000, 5,000, 10,000, \infty,$ and $1 - \alpha = 0.50, 0.75, 0.90, 0.95, 0.975, 0.99,$ $0.995.$ Note that Odeh and Owen (1980) use p (instead of β in our notation) for the content probability and γ (instead of $1 - \alpha$) for the confidence level.

Table 4 of Odeh and Owen (1980) gives values of the factors $g''_{(\gamma,p,n)}$ used to construct two-sided tolerance intervals to control both tails of a distribution. Factors are provided for all combinations of $\gamma = 0.995, 0.99, 0.975, 0.95, 0.90, 0.75, 0.50,$ $p = 0.125, 0.10, 0.05,$ $0.025, 0.01, 0.005, 0.0005,$ and $n = 2(1)100(2)180(5)300(10)400(25)650(50)1,000,$ $1,500, 2,000, 3,000, 5,000, 10,000, \infty.$

Table 7 of Odeh and Owen (1980) provides values similar to those provided by the R function normTailCI (but parameterized in a slightly different manner) for computing confidence intervals for normal distribution tail probabilities. Values are provided for all combinations of $k = -3.0(0.20)3.0,$ $n = 2(1)18(2)30, 40(20)120, 240, 600, 1000, 1200,$ and $\gamma = 0.50, 0.75, 0.90, 0.95, 0.975, 0.99, 0.995.$

Distribution-Free Statistical Intervals

OBJECTIVES AND OVERVIEW

This chapter shows how to calculate "distribution-free" two-sided statistical intervals and one-sided statistical bounds. Such intervals and bounds, which are based on order statistics, do not require the assumption of a particular underlying distribution, such as the normal distribution used in Chapters 3 and 4, for their construction. Moreover, as implied by the term distribution-free, the statistical properties of these procedures, such as their coverage probabilities, do not depend on the underlying distribution. This is in contrast to nonparametric methods, such as those presented in some subsequent chapters, whose construction does not depend on the underlying distribution, but whose statistical properties do. The subtle difference between distribution-free and nonparametric procedures is discussed further at the beginning of Chapter 11. In either case, the important assumption that sampling is random from the population (or process) of interest, and, more generally, the assumptions discussed in Chapter 1, still pertain.

The topics discussed in this chapter are distribution-free:

- Confidence intervals for a distribution quantile, such as the median (Section 5.2).

- Tolerance intervals to contain at least a specified proportion of a distribution (Section 5.3).

- Prediction intervals to contain a specified ordered observation in a future sample (Section 5.4).

- Prediction intervals to contain at least k of m future observations (Section 5.5).

Corresponding one-sided bounds are also considered.

Statistical Intervals: A Guide for Practitioners and Researchers, Second Edition.
William Q. Meeker, Gerald J. Hahn and Luis A. Escobar.
© 2017 John Wiley & Sons, Inc. Published 2017 by John Wiley & Sons, Inc.
Companion Website: www.wiley.com/go/meeker/intervals

5.1 INTRODUCTION

5.1.1 Motivation

One might ask "When should I use distribution-free statistical methods?" The answer, we assert, is "Whenever possible." If one can do a study with minimal distributional assumptions, then the resulting conclusions are based on a more solid foundation. Moreover, it is often appropriate to have the data analysis begin with a nonparametric approach as, for example, provided by the probability plots discussed in Section 4.11. Then, if needed, one can proceed to a more structured analysis involving a particular parametric distribution, such as the normal distribution, perhaps following a transformation chosen so that the distribution better fits the data.

It is interesting and useful to compare conclusions drawn from analyses that use and do not use an assumed distribution, as well as ones that use different assumed distributions. If an assumed distribution fits the data well, the point estimates obtained from the analyses (e.g., quantile estimates) within the range of the data often do not differ much from those obtained from a distribution-free approach. If an assumed distribution does not fit the data well, such point estimates could differ substantially. In this situation it is generally inappropriate to calculate a statistical interval assuming such a distribution. A distribution-free interval (if one exists) will generally be wider than the corresponding interval based on a particular distribution.

In particular, let $x_1, x_2, \ldots, x_n$ represent n independent observations from any continuous distribution and let $x_{(1)}, x_{(2)}, \ldots, x_{(n)}$ denote the same observations, ordered from smallest to largest. These ordered observations are commonly called the "order statistics" of the sample. The distribution-free confidence intervals, tolerance intervals, and prediction intervals discussed in this chapter use selected order statistics as interval endpoints. A distribution-free two-sided interval requires the use of two order statistics from the sample, such as the smallest and largest observations. The distribution-free one-sided bounds discussed in this chapter use a single order statistic as the bound.

Because the distribution-free statistical intervals and bounds are restricted to particular observed order statistics, it is generally not possible to obtain an interval with precisely the desired confidence level. Therefore, a frequent practice is to accept a somewhat larger (i.e., conservative) confidence level than that originally specified. An alternative, presented here, is to interpolate between the conservative interval (or bound) and a nonconservative interval (or bound) to obtain an approximately distribution-free interval.

Also, we need to emphasize that, as we will see throughout this chapter, it may not be possible, especially for small samples, to obtain a distribution-free interval with a confidence level as large as desired—even if one uses the extreme observations of the sample (i.e., the smallest and/or largest order statistics for an interval centered at the median of a sample) as the interval endpoints. Then one must settle for an interval with the largest achievable confidence level, even though this might be less than the desired level. Sometimes it will be impossible to obtain a meaningful distribution-free interval. For example, with a sample of size 50, it is impossible to calculate a lower confidence bound for the 0.01 quantile of the distribution at any reasonable confidence level without making an assumption about the form of the underlying distribution.

Another shortcoming of distribution-free intervals is that they can be much wider than distribution-dependent intervals. Thus, relatively large samples are often needed to attain an acceptable interval width or even to provide an interval at the desired confidence level, even if the extreme order statistic(s) are used to construct the interval. This is part of the price that one pays for not making a distributional assumption, and may limit the value of the distribution-free approach.

Although distribution-free intervals have limitations, such intervals still warrant serious consideration in many applications. This is because the alternative of using methods that require the assumption of a particular distribution (such as the normal distribution) can lead to seriously incorrect intervals if the assumption is incorrect.

To obtain a distribution-free two-sided statistical interval from a random sample from a specified distribution one generally proceeds as follows:

1. Specify the desired confidence level for the interval.

2. Determine (from tabulations or calculations) the order statistics of the sample that— if they exist—provide the statistical interval with at least the desired confidence level for the given sample size. We recommend that these order statistics be chosen with appropriate *symmetry* (the nature of which depends on the type of interval being computed, as explained in detail subsequently). If no such order statistics exist, use the interval endpoints that come closest to providing the desired confidence level. Sometimes, the resulting confidence interval is so wide that the interval is of little practical value.

3. Determine the coverage probability, or confidence level, associated with the preceding interval. This coverage probability will be greater than or equal to the specified (nominal) confidence level.

4. In those cases for which the coverage probability meets or exceeds the desired confidence level,

    (a) Use the selected order statistics as the endpoints of a distribution-free conservative interval or,

    (b) Use interpolation between the conservative interval (or bound) and a nonconservative interval (or bound), providing an approximate distribution-free interval that has a confidence level that more closely approximates the desired confidence level.

Distribution-free one-sided bounds are obtained in a similar manner, except that only one order statistic is used as the desired lower or upper bound. Indeed, we recommend in Sections 5.2 and 5.4 that, operationally, a two-sided conservative (or interpolated approximate) $100(1 - \alpha)\%$ confidence interval be obtained by combining lower and upper one-sided conservative (or interpolated approximate) $100(1 - \alpha/2)\%$ confidence bounds, thereby ensuring the symmetry mentioned above.

The particular order statistics defining a distribution-free statistical interval or bound proce-dure (which is equivalent to specifying the interval procedure to be used) should be determined *before* noting the values of the observations. Thus, for example, one might decide, initially, based upon the sample size and the desired confidence level, to use the extreme order statistics (i.e., the smallest and largest observations) as endpoints for a tolerance interval, irrespective of what their observed values turn out to be. Finally, we note that for samples from a continuous distribution (generally the case for physical measurements made with good precision), the cov-erage probabilities for the order-statistic based statistical intervals in this chapter are exactly as computed and distribution-free (although, as indicated, generally different than the originally specified desired confidence level due to the discrete number of order statistic choices). For samples from a discrete distribution (i.e., for which the reported data might include ties), the resulting coverage probability is again distribution-free but somewhat larger than the coverage probability computed under the continuous-distribution assumption.

5.1.2 Notation

This section outlines the basic notation that will be used in calculating the distribution-free intervals and bounds provided in this chapter. The notation pertains to probability distributions that relate to dichotomous outcomes—that is, populations that contain units that belong to one category or another. Following quality control terminology, we generically refer to these units as "conforming" and "nonconforming," respectively. Technical details about the discrete probability distributions used in this chapter (to characterize the distribution of order statistics from unspecified continuous distributions) are given in Appendix C. Appendix G gives most of the underlying theory and derivations of results used in this chapter.

The following probability distribution and quantile functions are used in this chapter:

- `pbinom`$(x; n, \pi)$ and `qbinom`$(p; n, \pi)$ are, respectively, the binomial cdf and quantile functions (Section C.4.1).

- `pnhyper`$(x; k, D, N)$ and `qnhyper`$(p; k, D, N)$ are, respectively, the negative hypergeometric "waiting time" cdf and quantile functions (Section C.4.6).

Detailed information, including the interpretation of the parameters and R functions for computing probabilities and quantiles for these distributions, is given in the referenced section from Appendix C.

The following application will be used to illustrate the intervals presented in this chapter.

Example 5.1   Amount of a Compound Present in Composite Samples from 100 Randomly Selected Batches from a Chemical Process. A production engineer wants to evaluate the capability of a chemical process to produce a particular compound. Measurements are available, in parts per million (ppm), of the amount of compound present in composite samples taken from each of $n = 100$ randomly selected batches from the process. Each batch was thoroughly mixed before sampling and can, therefore, be regarded as homogeneous. Measurement error was small enough to be ignored. If measurement error had been a problem, it could have been reduced by taking a sufficient number of independent measurements from each batch and averaging these measurements within each batch. Table 5.1 gives the resulting readings, ordered from smallest to largest; ordering these observations will facilitate the application of the methods given in this chapter.

Whenever data are collected over time, it is important to check for possible time-related dependencies. For example, measuring instruments may drift with time or use, or readings may depend on ambient temperature which might change over the time that the readings are taken. A time-order plot of the readings is given in Figure 5.1. The data do not exhibit any trend, cycle,

Order index	1	2	3	4	5	6	7	8	9	10
1–10	1.49	1.66	2.05	2.24	2.29	2.69	2.77	2.77	3.10	3.23
11–20	3.28	3.29	3.31	3.36	3.84	4.04	4.09	4.13	4.14	4.16
21–30	4.57	4.63	4.83	5.06	5.17	5.19	5.89	5.97	6.28	6.38
31–40	6.51	6.53	6.54	6.55	6.83	7.08	7.28	7.53	7.54	7.68
41–50	7.81	7.87	7.94	8.43	8.70	8.97	8.98	9.13	9.14	9.22
51–60	9.24	9.30	9.44	9.69	9.86	9.99	11.28	11.37	12.03	12.32
61–70	12.93	13.03	13.09	13.43	13.58	13.70	14.17	14.36	14.96	15.89
71–80	16.57	16.60	16.85	17.18	17.46	17.74	18.40	18.78	19.84	20.45
81–90	20.89	22.28	22.48	23.66	24.33	24.72	25.46	25.67	25.77	26.64
91–100	28.28	28.28	29.07	29.16	31.14	31.83	33.24	37.32	53.43	58.11

Table 5.1 Ordered measurements of the amount of a compound (in ppm) for 100 batches from a chemical process.

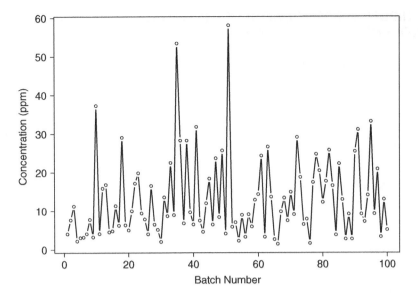

Figure 5.1   Time-order plot of the chemical process data.

or other departure from randomness. There are several large values, but these were known to be caused by random, uncontrollable shocks to the process. Otherwise, the process appears to be in statistical control.

Figure 5.2 is a histogram of the data showing that the distribution of the amount of compound in the batches is sharply skewed to the right, with values close to the lower limit of 0 occurring most frequently. The normal distribution clearly is not a good model for these data. Applying a Box–Cox transformation (see Section 4.12) to the measurements might lead to an improved normal distribution approximation. In this chapter, however, we will use distribution-free methods instead to characterize the population or process from which the 100 batches are a random sample.   ∎

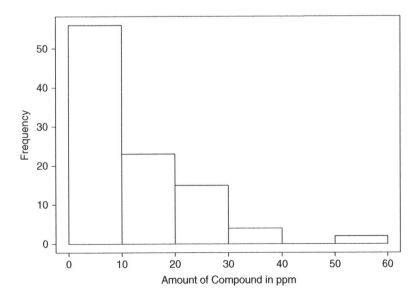

Figure 5.2   Histogram of the chemical process data.

5.2 DISTRIBUTION-FREE CONFIDENCE INTERVALS AND ONE-SIDED CONFIDENCE BOUNDS FOR A QUANTILE

This section shows how to obtain distribution-free two-sided confidence intervals and one-sided confidence bounds for a distribution quantile. After providing some initial comments about coverage probabilities, we present methods and examples showing how to calculate such intervals and bounds. We note that when taking a distribution-free approach, one is frequently interested in quantiles of the distribution, rather than parameters like the distribution mean or standard deviation.

5.2.1 Coverage Probabilities for Distribution-Free Confidence Intervals or One-Sided Confidence Bounds for a Quantile

A two-sided distribution-free confidence interval for x_p, the p quantile of the distribution (where $0 < p < 1$), based on a random sample of size n, is obtained from $[\underset{\sim}{x}_p, \; \widetilde{x}_p] = [x_{(\ell)}, \; x_{(u)}]$ where, as before, $x_{(\ell)}$ and $x_{(u)}$ are, respectively, the ℓth and the uth ordered observations from the given sample (see subsequent discussion). As shown in Section G.2.1, for given ℓ and u, the coverage probability that the interval $[\underset{\sim}{x}_p, \; \widetilde{x}_p]$ will contain x_p is

$$\mathrm{CPXP}(n, \ell, u, p) = \texttt{pbinom}(u - 1; n, p) - \texttt{pbinom}(\ell - 1; n, p), \quad 1 \leq \ell < u \leq n. \quad (5.1)$$

A distribution-free one-sided *upper* confidence bound for x_p is given by $\widetilde{x}_p = x_{(u)}$. The coverage probability for given u is

$$\mathrm{CPXP}(n, 0, u, p) = \texttt{pbinom}(u - 1; n, p), \quad 1 \leq u \leq n. \quad (5.2)$$

A distribution-free one-sided *lower* confidence bound is given by $\underset{\sim}{x}_p = x_{(\ell)}$. The associated coverage probability for given ℓ is

$$\mathrm{CPXP}(n, \ell, n + 1, p) = 1 - \texttt{pbinom}(\ell - 1; n, p), \quad 1 \leq \ell \leq n. \quad (5.3)$$

An equivalent expression for $\mathrm{CPXP}(n, \ell, n + 1, p)$, which can be inverted to provide a direct computation of the value of ℓ giving a desired coverage probability, is

$$\mathrm{CPXP}(n, \ell, n + 1, p) = \texttt{pbinom}(n - \ell; n, 1 - p). \quad (5.4)$$

Derivations of these coverage probabilities are given in Section G.2.

We refer to CPXP as providing the "coverage probability" in describing a *procedure* to compute a confidence interval or bound before the order statistics are observed from a particular data set. After the data are observed, we use the common terminology "confidence level" to describe the associated level of confidence for the calculated confidence interval.

5.2.2 Using Interpolation to Obtain Approximate Distribution-Free Confidence Bounds or Confidence Intervals for a Quantile

The choice of order statistic indices ℓ and u defines the distribution-free confidence interval procedure. Because of the limited number of possibilities for choosing ℓ and u, there is a limited number of possibilities for the coverage probability for the distribution-free interval. Suppose, for example, that the conservative confidence interval with the lowest coverage probability that still exceeds a desired 95% confidence level has a coverage probability of 0.982. At the same time, say that there is another pair of order statistics that provides a coverage probability of 0.944. This suggests a third alternative: interpolate between the preceding two intervals. The resulting modified procedure, though not exactly distribution-free and not necessarily conservative, is approximately distribution-free and generally closer to the desired confidence level than the conservative procedure.

This type of interpolation can be similarly applied to obtain one-sided lower or upper confidence bounds. The following subsections show how to choose ℓ and/or u and do the interpolation. In contrast to the presentations in previous chapters, we treat one-sided confidence bounds first. We do this because the procedure that we recommend for obtaining a distribution-free two-sided $100(1 - \alpha)\%$ confidence interval for x_p is to combine one-sided lower and upper $100(1 - \alpha/2)\%$ confidence bounds.

5.2.3 Distribution-Free One-Sided Upper Confidence Bounds for a Quantile

Tabular/graphical method

One can use Table J.11 to determine the appropriate order statistic to provide a distribution-free one-sided *upper* conservative $100(1 - \alpha)\%$ confidence bound for x_p with $p = 0.75, 0.90, 0.95$, and 0.99, and $1 - \alpha = 0.90, 0.95$, and 0.99. Alternatively, one can use Figure 5.3a or 5.3b to determine such an order statistic for values of p ranging from 0.5 to 0.998 for $1 - \alpha = 0.90$ and 0.95. (The table and figures use β in place of p.) This table and these figures were originally developed to aid in the construction of distribution-free tolerance intervals and bounds, but are also useful for finding two-sided confidence intervals and one-sided confidence bounds for quantiles. The table is more precise, but the figures give more insight. This table and these figures allow one to determine the value of ν (explained below) for n ranging from 10 to 1,000.

The integer $\nu - 1$ is the number of extreme observations to be removed from the upper end of the sample of size n to obtain the order statistic that provides the desired one-sided upper confidence bound for x_p. Thus, $u = n - (\nu - 1) = n - \nu + 1$. The coverage probability

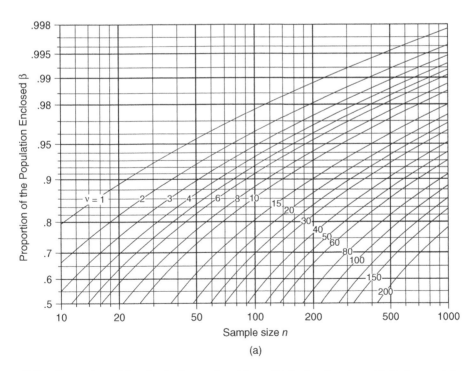

(a)

Figure 5.3a Proportion of the distribution enclosed by a distribution-free two-sided tolerance interval (or one-sided tolerance bound) with 90% confidence when $\nu - 2$ (or $\nu - 1$) extreme observations are excluded from the ends (end) of an ordered sample of size n. The figure is also used to obtain distribution-free two-sided confidence intervals and one-sided confidence bounds for distribution quantiles. A similar figure first appeared in Murphy (1948). Adapted with permission of the Institute of Mathematical Statistics. See also Table J.11.

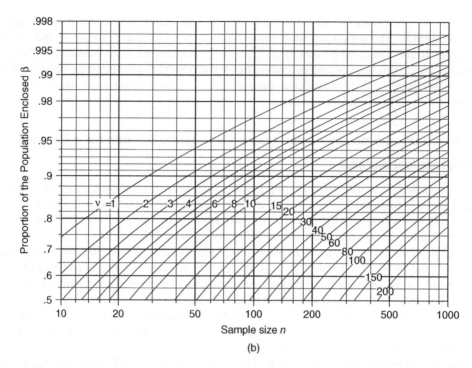

Figure 5.3b   Proportion of the distribution enclosed by a distribution-free two-sided tolerance interval (or one-sided tolerance bound) with 95% confidence when $\nu - 2$ (or $\nu - 1$) extreme observations are excluded from the ends (end) of an ordered sample of size n. The figure is also used to obtain distribution-free two-sided confidence intervals and one-sided confidence bounds for distribution quantiles. A similar figure first appeared in Murphy (1948). Adapted with permission of the Institute of Mathematical Statistics. See also Table J.11.

is also shown for all entries in Table J.11. When this coverage probability is less than the desired confidence level, it is marked with a *. In using Figure 5.3a or 5.3b, one enters the appropriate figure (determined by the selected value of $1 - \alpha$) with the specified values n and p (interpolating, if necessary) to find ν. When the result, using the figures, is between two values of ν, as is usually the case, to be conservative one would select the line with the smaller value of ν. The resulting upper conservative $100(1 - \alpha)\%$ confidence bound is $\widetilde{x}_p = x_{(n-\nu+1)}$.

Computational method

The following method can be used for situations that are not covered in the figures or tabulations. A distribution-free one-sided upper conservative $100(1 - \alpha)\%$ confidence bound for x_p can be obtained as $\widetilde{x}_p = x_{(u)}$, where u is chosen as the smallest integer such that $\mathrm{CPXP}(n, 0, u, p)$ in (5.2) is greater than or equal to $1 - \alpha$. More directly, one can obtain u as *one plus* the $1 - \alpha$ quantile of the binomial distribution with sample size n and parameter $\pi = p$ (i.e., $u = \mathrm{qbinom}(1 - \alpha; n, p) + 1$). If $u = n + 1$, there is no distribution-free one-sided upper confidence bound for x_p having coverage probability greater than or equal to $1 - \alpha$. The coverage probability is given by (5.2).

Approximate interpolation method

An interpolated one-sided upper confidence bound for x_p is found as follows. Suppose that $x_{(u_c)}$ is the one-sided upper confidence bound having the smallest conservative coverage probability $\text{CPXP}(n, 0, u_c, p)$ and that $x_{(u_n)}$ is the one-sided upper confidence bound with the largest nonconservative coverage probability $\text{CPXP}(n, 0, u_n, p)$. We can then obtain an interpolated confidence bound between $x_{(u_c)}$ and $x_{(u_n)}$ that has a coverage probability that will generally be closer to the nominal $1 - \alpha$ confidence level than either of the alternatives by finding

$$\omega = \frac{(1 - \alpha) - \text{CPXP}(n, 0, u_n, p)}{\text{CPXP}(n, 0, u_c, p) - \text{CPXP}(n, 0, u_n, p)}. \tag{5.5}$$

Then the interpolated one-sided upper approximate $100(1 - \alpha)\%$ confidence bound for x_p is

$$\omega x_{(u_c)} + (1 - \omega) x_{(u_n)}. \tag{5.6}$$

Example 5.2   One-Sided Upper Confidence Bound for the 0.90 Quantile of the Compound Amount Distribution. A distribution-free one-sided upper conservative 95% confidence bound for the 0.90 quantile, $x_{0.90}$, for the output of the chemical process described in Example 5.1 is obtained as follows. Entering Table J.11 (with $p = 0.9$, $1 - \alpha = 0.95$ and $n = 100$) or Figure 5.3b (for $1 - \alpha = 0.95$) and reading up to the next higher curve with $n = 100$ and $p = 0.9$, gives $\nu = 5$. Then $u = n - \nu + 1 = 96$ and $\widetilde{x}_{0.90} = x_{(96)} = 31.83$ (from Table 5.1) provides the desired upper conservative confidence bound. The coverage probability for this bound is seen from Table J.11 to be 0.9763—which is appreciably larger than the desired 0.95. Thus, we turn to the interpolation method to obtain an improved bound. Using $u = 95$ gives the "neighboring" one-sided nonconservative bound $\widetilde{x}_{0.90} = x_{(95)} = 31.14$, with a coverage probability (from Table J.11) of 0.9424. Interpolating between the two bounds, using (5.5) and (5.6), gives $\omega = (0.95 - 0.9424)/(0.9763 - 0.9424) = 0.2242$ and so $\widetilde{x}_{0.90} = 0.2242 \times 31.83 + (1 - 0.2242) \times 31.14 = 31.29$ as an approximate distribution-free one-sided upper 95% confidence bound for $x_{0.90}$. Alternatively, applying the computational method with interpolation and using R as a calculator for the binomial quantile function and cdf gives

```
> qbinom(p=0.95, size=100, prob=0.90)+1
[1] 96
> pbinom(q=96-1, size=100, prob=0.90)
[1] 0.9763
> pbinom(q=95-1, size=100, prob=0.90)
[1] 0.9424
> omega <- (0.950 - 0.9424)/(0.9763 - 0.9424)
> omega*31.83 + (1 - omega)*31.14
[1] 31.29
```

The figures and tables in this book do not provide a direct means to obtain upper confidence bounds for quantiles less than 0.50, and one needs to use the computational method instead.

Example 5.3   One-Sided Upper Confidence Bound for the 0.10 Quantile of the Compound Amount Distribution. Following the approach at the end of Example 5.2, we find the one-sided upper conservative and nonconservative 95% confidence bounds and then interpolate to obtain the approximate upper 95% confidence bound for $x_{0.10}$. Using R as a calculator for the binomial distribution quantile function and cdf gives

```
> qbinom(p=0.95, size=100, prob=0.10)+1
[1] 16
> pbinom(q=16-1, size=100, prob=0.10)
[1] 0.9601
```

```
> pbinom(q=15-1, size=100, prob=0.10)
[1] 0.9274
> omega <- (0.950 - 0.9274)/(0.9601 - 0.9274)
> omega*4.04 + (1 - omega)*3.84
[1] 3.978
```

Thus the one-sided upper approximate 95% confidence bound for $x_{0.10}$, based on interpolation, is $\widetilde{x}_{0.10} = 3.98$. ∎

5.2.4   Distribution-Free One-Sided Lower Confidence Bounds for a Quantile

Tabular/graphical method

The procedure for finding distribution-free one-sided *lower* confidence bounds for a quantile is similar to that for upper confidence bounds presented in Section 5.2.3 and Examples 5.2 and 5.3. One can again use Table J.11 to determine the appropriate order statistic to provide a distribution-free one-sided lower conservative $100(1 - \alpha)\%$ confidence bound for x_p for $p = 0.25, 0.10, 0.05$, and 0.01 (using $1 - p$ in place of p), and $1 - \alpha = 0.90, 0.95$, and 0.99. The coverage probability is shown for all entries in Table J.11. When this coverage probability is less than the desired confidence level, it is marked with a * in the table. Alternatively, one can use Figure 5.3a or 5.3b to determine the appropriate order statistic for values of p ranging from 0.002 to 0.5 (again, using $1 - p$ in place of p) for $1 - \alpha = 0.90$ and 0.95. This table and these figures provide ν for values of n ranging from 10 to 1,000. In using Figure 5.3a or 5.3b, one enters the appropriate figure (determined by the selected value of $1 - \alpha$) with the specified values n and p (interpolating, if necessary) to find ν. When the result, using the figures, is between two values of ν, to be conservative, one would select the line with the smaller value of ν. Then the resulting lower conservative $100(1 - \alpha)\%$ confidence bound is $x_{(\nu)}$.

Computational method

The following method can be used for situations that are not covered in the figures or tabulations. A distribution-free one-sided lower conservative $100(1 - \alpha)\%$ confidence bound for x_p can be obtained as $x_p = x_{(\ell)}$, where ℓ is chosen as the largest integer such that $\mathrm{CPXP}(n, \ell, n + 1, p) = \mathrm{pbinom}(n - \ell; n, 1 - p)$ is greater than or equal to $1 - \alpha$. Thus $n - \ell = \mathrm{qbinom}(1 - \alpha; n, 1 - p)$ which implies $\ell = n - \mathrm{qbinom}(1 - \alpha; n, 1 - p)$. If $\ell = 0$, there is no distribution-free one-sided lower confidence bound for x_p having coverage probability greater than or equal to $1 - \alpha$. The coverage probability is given by (5.3).

Approximate interpolation method

An interpolated one-sided lower confidence bound for x_p is found as follows. Suppose that $x_{(\ell_c)}$ is the one-sided lower confidence bound having the smallest conservative coverage probability $\mathrm{CPXP}(n, \ell_c, n + 1, p)$ and that $x_{(\ell_n)}$ is the one-sided lower confidence bound with the largest nonconservative coverage probability $\mathrm{CPXP}(n, \ell_n, n + 1, p)$. We can then obtain an interpolated confidence bound between $x_{(\ell_c)}$ and $x_{(\ell_a)}$ that has a coverage probability that will generally be closer to the nominal $1 - \alpha$ confidence level than either of the alternatives by finding

$$\omega = \frac{(1 - \alpha) - \mathrm{CPXP}(n, \ell_n, n + 1, p)}{\mathrm{CPXP}(n, \ell_c, n + 1, p) - \mathrm{CPXP}(n, \ell_n, n + 1, p)}. \tag{5.7}$$

Then the interpolated one-sided lower approximate $100(1 - \alpha)\%$ confidence bound for x_p is

$$\omega x_{(\ell_c)} + (1 - \omega)x_{(\ell_n)}. \tag{5.8}$$

Example 5.4   One-Sided Lower Confidence Bound for the 0.10 Quantile of the Compound Amount Distribution. A distribution-free one-sided lower conservative 95% confidence bound for $x_{0.10}$ for Example 5.1 can be obtained as follows. Entering Table J.11 or Figure 5.3b with $n = 100$, $1 - p = 0.9$, and $1 - \alpha = 0.95$ (and reading up to the next higher curve) gives $\ell = \nu = 5$. Then $\underset{\sim}{x}_{0.10} = x_{(5)} = 2.29$ provides the desired lower conservative confidence bound. The coverage probability for the lower bound is seen from Table J.11 to be 0.9763. Using instead $\ell = 6$ gives the neighboring nonconservative confidence bound $\underset{\sim}{x}_{0.10} = x_{(6)} = 2.69$ which has a coverage probability (again from Table J.11) of 0.9424. Interpolating between these two bounds using (5.7) and (5.8) gives $\omega = (0.95 - 0.9424)/(0.97631 - 0.9424) = 0.2241$; thus $\underset{\sim}{x}_{0.10} = 0.2241 \times 2.29 + (1 - 0.2241) \times 2.69 = 2.600$ is an approximate distribution-free 95% lower confidence bound for $x_{0.10}$.

Alternatively, applying the computational method with interpolation and using R as a calculator for the binomial quantile function and cdf gives

```
> 100-qbinom(p=0.95, size=100, prob=1-0.10)
[1] 5
> 1 - pbinom(q=5-1, size=100, prob=0.10)
[1] 0.9763
> 1 - pbinom(q=6-1, size=100, prob=0.10)
[1] 0.9424
> omega <- (0.95 - 0.9424)/(0.9763 - 0.9424)
> omega* 2.29 + (1 - omega)*2.69
[1] 2.6
```

The figures and tables available in this book do not provide a direct means to obtain lower confidence bounds for quantiles greater than 0.50. As we have seen, however, the computational method is easy to implement with a sophisticated calculator, such as R.

Example 5.5   One-Sided Lower Confidence Bound for the 0.90 Quantile of the Compound Amount Distribution. Following the approach at the end of Example 5.4, we find the one-sided lower conservative and nonconservative 95% confidence bounds and then interpolate to obtain the one-sided lower approximate 95% confidence bound for $x_{0.90}$. Using R as a calculator for the binomial quantile function and cdf gives

```
> 100-qbinom(p=0.95, size=100, prob=1-0.90)
[1] 85
> 1-pbinom(q=85-1, size=100, prob=0.90)
[1] 0.9601
> 1-pbinom(q=86-1, size=100, prob=0.90)
[1] 0.9274
> omega <- (0.950 - 0.9274)/(0.9601 - 0.9274)
> omega*24.33 + (1 - omega)*24.72
[1] 24.45
```

Thus the approximate lower 95% confidence bound for $x_{0.90}$ is $\underset{\sim}{x}_{0.90} = 24.45$. ∎

5.2.5   Distribution-Free Two-Sided Confidence Interval for a Quantile

Tabular method

A distribution-free two-sided conservative $100(1 - \alpha)\%$ confidence interval for x_p, the p quantile of the sampled distribution, is obtained from a sample of size n as $[\underline{x}_p, \ \widetilde{x}_p] = [x_{(\ell)}, \ x_{(u)}]$, where ℓ and u are given for $p = 0.10, 0.20$, and 0.50 in Tables J.10a–J.10c, respectively, for various values of n and $1 - \alpha$. These tabulations provide integer values of ℓ and u that are symmetric or nearly symmetric, around $[np] + 1$, where $[r]$ is the integral part of r. This approach is used because the distribution-free point estimate of the p quantile lies between $x_{([np])}$ and $x_{([np]+1)}$. One can also use Tables J.10a and J.10b to find distribution-free confidence intervals for $p = 0.80$ and 0.90, respectively. In this case, one enters the table corresponding to $1 - p$ and uses

$$[\underline{x}_p, \ \widetilde{x}_p] = [x_{(n-u+1)}, \ x_{(n-\ell+1)}]$$

as the desired interval.

As indicated in Section 5.1.1 and in the description of the computational methods, with a limited sample size, it is sometimes impossible to construct a distribution-free statistical interval that has at least the desired confidence level. This problem is particularly acute when estimating quantiles in the tail of a distribution from a small sample. Situations for which a desired confidence level cannot be met are indicated by a * in Tables J.10a–J.10c.

Computational method

For situations that are not covered in the tabulations, we recommend combining a one-sided upper $100(1 - \alpha/2)\%$ confidence bound (Section 5.2.3) and a one-sided lower $100(1 - \alpha/2)\%$ confidence bound (Section 5.2.4) to give a two-sided $100(1 - \alpha)\%$ confidence interval. This approach also facilitates the use of the interpolation method, given there, to provide a procedure with a coverage probability that is generally closer to the desired confidence level (and provides a somewhat narrower interval).

Example 5.6   Confidence Interval for the Median of the Compound Amount Distribution. Assume that a distribution-free 95% confidence interval is needed for $x_{0.50}$, the median of the compound amount distribution in Example 5.1. For $n = 100$ and $1 - \alpha = 0.95$, Table J.10c gives $\ell = 41$ and $u = 61$. Thus, the interval enclosed by $x_{(41)}$ and $x_{(61)}$ provides the following distribution-free conservative 95% confidence interval for $x_{0.50}$:

$$[\underline{x}_{0.50}, \ \widetilde{x}_{0.50}] = [x_{(41)}, \ x_{(61)}] = [7.81, 12.93].$$

The coverage probability for the procedure used to compute this confidence interval is given in Table J.10c, or can be calculated as $\text{CPXP}(100, 41, 61, 0.50) = \text{pbinom}(60; 100, 0.50) - \text{pbinom}(40; 100, 0.50) = 0.9540$ (i.e., the confidence level is 95.40%, instead of the desired 95%). The preceding interval is not exactly symmetric around the center of the data.

The *symmetric conservative* procedure uses $\ell = 40$ and $u = 61$ and has a coverage probability of 0.9648. The *symmetric nonconservative* procedure uses $\ell = 41$ and $u = 60$ and has a coverage probability of 0.9431. Interpolating between these two confidence bounds, in a manner similar to that in Examples 5.2 and 5.5, provides a procedure that can be expected to have a coverage probability closer to the desired 0.95. Using R as a calculator for the binomial quantile function and cdf gives

```
> # lower endpoint first
> 100-qbinom(p=0.975, size=100, prob=0.50)
[1]  40
```

```
> 1-pbinom(q=40-1, size=100, prob=0.50)
[1] 0.9824
> 1-pbinom(q=41-1, size=100, prob=0.50)
[1] 0.9716
> omega <- (0.975 - 0.9716)/(0.9824 - 0.9716)
> omega*7.68 + (1 - omega)*7.81
[1] 7.769
> # upper endpoint
> qbinom(p=0.975, size=100, prob=0.50)+1
[1] 61
> pbinom(q=61-1, size=100, prob=0.50)
[1] 0.9824
> pbinom(q=60-1, size=100, prob=0.50)
[1] 0.9716
> omega <- (0.975 - 0.9716)/(0.9824 - 0.9716)
> omega*12.93 + (1 - omega)*12.32
[1] 12.51
```

Thus the approximate distribution-free 95% confidence interval for $x_{0.50}$ based on interpolation is $[\underset{\sim}{x}_{0.50}, \; \widetilde{x}_{0.50}] = [7.77, \; 12.51]$. ∎

Confidence intervals for other quantiles are found similarly.

Example 5.7 Confidence Interval for the 0.10 Quantile of the Compound Amount Distribution. A 90% confidence interval for $x_{0.10}$ can be obtained by combining the one-sided upper approximate 95% confidence bound from Example 5.3 and the one-sided lower approximate 95% confidence bound from Example 5.4, giving $[\underset{\sim}{x}_{0.10}, \; \widetilde{x}_{0.10}] = [2.60, 3.98]$.

Using the tabular method, we can find a 95% conservative confidence interval for $x_{0.10}$ by entering Table J.10a with $n = 100$ and $1 - \alpha = 0.95$, giving $\ell = 4$ and $u = 16$. Thus,

$$[\underset{\sim}{x}_{0.10}, \; \widetilde{x}_{0.10}] = [x_{(4)}, \; x_{(16)}] = [2.24, \; 4.04]$$

is a distribution-free two-sided conservative 95% confidence interval for $x_{0.10}$. The coverage probability for the procedure that generated this interval is shown in Table J.10a or calculated as $\mathrm{CPXP}(100, 4, 16, 0.10) = \mathtt{pbinom}(15; 100, 0.10) - \mathtt{pbinom}(3; 100, 0.10) = 0.9523$. ∎

Example 5.8 Confidence Interval for the 0.90 Quantile of the Compound Amount Distribution. A 90% confidence interval for $x_{0.90}$ can be obtained by combining the one-sided upper approximate 95% confidence bound from Example 5.2 and the one-sided lower approximate 95% confidence bound from Example 5.5, giving $[\underset{\sim}{x}_{0.90}, \; \widetilde{x}_{0.90}] = [24.45, \; 31.29]$.

To find a 95% conservative confidence interval for $x_{0.90}$, we again use Table J.10a with $n = 100$ and $1 - p = 1 - 0.90 = 0.10$. We obtain $\ell = 4$ and $u = 16$, giving $n - u + 1 = 100 - 16 + 1 = 85$ and $n - \ell + 1 = 100 - 4 + 1 = 97$ for the desired lower and upper order statistic indices, respectively. Thus

$$[\underset{\sim}{x}_{0.90}, \; \widetilde{x}_{0.90}] = [x_{(85)}, \; x_{(97)}] = [24.33, \; 33.24]$$

is a two-sided distribution-free conservative 95% confidence interval for $x_{0.90}$. The coverage probability for the procedure that generated this interval is $\mathrm{CPXP}(100, 85, 97, 0.90) = \mathtt{pbinom}(96; 100, 0.90) - \mathtt{pbinom}(84; 100, 0.90) = 0.9523$. ∎

Note from Table J.10a that for $n = 50$ or less, a two-sided distribution-free 99% confidence interval for $x_{0.10}$ cannot be obtained with symmetrically chosen order statistics. Therefore, if one wishes to choose ℓ and u symmetrically or approximately symmetrically about $[np] + 1 = 5 + 1 = 6$, one must settle for the largest achievable confidence level from the first and 11th order statistics. In particular, for $n = 50$, the largest achievable confidence level for a near-symmetric interval is 98.55%, corresponding to $\ell = 1$ and $u = 11$. Alternatively, one may increase the upper end of the interval and use the first and 12th ordered observations. This would give a nonsymmetric distribution-free confidence interval of $[1.49, \ 3.29]$ with a confidence level of 99.16% (i.e., $\text{CPXP}(50, 1, 12, 0.1) = \texttt{pbinom}(11; 50, 0.1) - \texttt{pbinom}(0; 50, 0.1) = 0.9916$).

5.3   DISTRIBUTION-FREE TOLERANCE INTERVALS AND BOUNDS TO CONTAIN A SPECIFIED PROPORTION OF A DISTRIBUTION

5.3.1   Distribution-Free Two-Sided Tolerance Intervals

A distribution-free two-sided conservative $100(1 - \alpha)\%$ tolerance interval to contain at least a proportion β of the sampled distribution from a sample of size n is obtained as

$$[T_\beta, \ \widetilde{T}_\beta] = [x_{(\ell)}, \ x_{(u)}].$$

To determine the specific order statistics that will provide the desired confidence level, one generally chooses ℓ and u symmetrically or nearly symmetrically within the ordered sample data to provide, if possible, the desired coverage probability. The coverage probability for this distribution-free tolerance interval procedure is

$$\text{CPTI}(n, \ell, u, \beta) = \texttt{pbinom}(u - \ell - 1; n, \beta), \quad 0 \leq \ell < u \leq n + 1, \qquad (5.9)$$

where $0 < \beta < 1$. Derivations of this coverage probability are given in Section G.3.

Similar to our previous discussion in Section 5.2.1, CPTI refers to the tolerance interval coverage probability prior to observing the data. After the data have been observed, the numbers represent a confidence level.

Tabular/graphical method

Table J.11 or Figure 5.3a or 5.3b can be used to find appropriate values of ℓ and u for various values of β and $1 - \alpha$. Specifically, from either the table or the figure, choose the value of ν that gives the desired level of confidence for enclosing at least a proportion β of the distribution (reading down to the next smallest integer when using the figures). The integer $\nu - 2$ is the total number of observations to be removed from the extremes of the ordered sample, giving the order statistics that define the desired tolerance interval. To find the particular order statistics to be used for the tolerance interval, we need to divide ν into two parts as follows:

- Let $\nu_1 = \nu_2 = \nu/2$ if ν is even.
- Let $\nu_1 = \nu/2 - 1/2$ and $\nu_2 = \nu_1 + 1$ or $\nu_1 = \nu/2 + 1/2$ and $\nu_2 = \nu_1 - 1$ if ν is odd (either will give a tolerance interval with the same level of confidence). Then use $\ell = \nu_1$ and $u = n - \nu_2 + 1$. The desired tolerance interval is then formed by the values of $x_{(\ell)}$ and $x_{(u)}$.

When ν is an odd integer, the tolerance interval will not be exactly symmetric about the center of the distribution.

Computational method

For situations that are not covered in the figures or tabulations, one can obtain ν directly as $\nu = n - \texttt{qbinom}(1 - \alpha; n, \beta)$ (see (G.6) for the derivation of this result) and then proceed as described for the tabular and graphical methods. If $\nu \leq 1$, then there is not a two-sided distribution-free tolerance interval to contain at least a proportion β with coverage probability greater than or equal to $1 - \alpha$. The coverage probability is given by (5.9).

Example 5.9 Two-Sided Conservative Tolerance Interval to Contain at Least a Proportion 0.90 of the Compound Amount Distribution. The manufacturer needs a two-sided tolerance interval that will, with 95% confidence, contain the compound amounts for at least a proportion 0.90 of the batches from the sampled production process, based on the sample data from 100 batches. From Table J.11, using $n = 100$, $\beta = 0.90$ and $1 - \alpha = 0.95$, we find $\nu = 5$; we choose $\nu_1 = 2$ and $\nu_2 = 3$, yielding $\ell = 2$ and $u = 98$. Thus,

$$[\underset{\sim}{T}_{0.90}, \; \widetilde{T}_{0.90}] = [x_{(2)}, \; x_{(98)}] = [1.66, \; 37.32]$$

is a two-sided conservative 95% tolerance interval to contain at least a proportion 0.90 of the distribution values with (at least) 95% confidence. Alternatively, $\nu_1 = 3$ and $\nu_2 = 2$ might have been chosen, yielding $\ell = 3$ and $u = 99$, resulting in the interval

$$[\underset{\sim}{T}_{0.90}, \; \widetilde{T}_{0.90}] = [x_{(3)}, \; x_{(99)}] = [2.05, \; 53.43].$$

The coverage probability for this procedure (for either choice of ℓ and u) is $\mathrm{CPTI}(100, 2, 98, 0.90) = \mathtt{pbinom}(98 - 2 - 1; 100, 0.90) = 0.9763$. ■

The ambiguity in the previous example is due to the odd value of ν and the resulting asymmetry. The approximate method described next avoids this ambiguity.

Approximate interpolation method

As with the other conservative statistical interval methods that are based on order statistics, the conservative coverage probability of the procedure may appreciably exceed the desired $100(1 - \alpha)\%$ confidence level. In this case, interpolation (between the closest conservative and the closest nonconservative interval), similar to that described in Section 5.2.2, can be used to obtain a narrower interval with a confidence level that is generally approximately equal to $1 - \alpha$. The resulting interval will be approximately distribution-free and will no longer necessarily be conservative.

Using this approach, we can again find ν directly as described in the above computational method in this section. Note that ν must be at least 2, and if it is an odd number, we subtract 1 so as to maintain the conservativeness of the procedure and so that the resulting values of $\ell = \nu/2$ and $u = n - \nu/2 + 1$ are symmetric about the center of the data set.

Let $\ell_c = \nu/2$, $\ell_n = \ell_c + 1$, and $u_i = n - \ell_i + 1$, $(i = c, n)$, where c and n indicate "conservative" and "nonconservative" order statistic indices, respectively. If we now set $u_n = u_c - 1$ and $\ell_n = \ell_c + 1$, then $[x_{(\ell_c)}, \; x_{(n - \ell_c + 1)}]$ and $[x_{(\ell_n)}, \; x_{(n - \ell_n + 1)}]$ are, respectively, conservative and nonconservative tolerance intervals with coverage probabilities $\mathrm{CPTI}(n, \ell_c, u_c, \beta)$ and $\mathrm{CPTI}(n, \ell_n, u_n, \beta)$ given by (5.9). Thus $\mathrm{CPTI}(n, \ell_n, u_n, \beta) < 1 - \alpha < \mathrm{CPTI}(n, \ell_c, u_c, \beta)$. Now if we let

$$\omega = \frac{(1 - \alpha) - \mathrm{CPTI}(n, \ell_n, u_n, \beta)}{\mathrm{CPTI}(n, \ell_c, u_c, \beta) - \mathrm{CPTI}(n, \ell_n, u_n, \beta)},$$

the approximate tolerance interval based on the interpolation method is

$$[\underset{\sim}{T}_{\beta}, \; \widetilde{T}_{\beta}] = [\omega x_{(\ell_c)} + (1 - \omega) x_{(\ell_n)}, \; \omega x_{(u_c)} + (1 - \omega) x_{(u_n)}].$$

Because of symmetry of the tolerance interval procedure, the same interpolation weight is used for both endpoints of the interval.

Example 5.10   Symmetric Two-Sided Approximate Tolerance Interval to Contain at Least a Proportion 0.90 of the Compound Amounts. This example is similar to Example 5.9 except that it uses the interpolation method to find a symmetric approximate tolerance interval. Using R as a calculator for the binomial quantile function and cdf gives

```
> # find v
> 100-qbinom(p=0.95, size=100, prob=0.90)
[1] 5      # is odd, so decrease by 1 to v=4
> #  conservative indices are 4/2=2 and 100-4/2+1=99
> pbinom(q=(99-2)-1, size=100, prob=0.90)   #conservative coverage
[1] 0.9922
> #  nonconservative indices are 6/2=3 and 100-6/2+1=98
> pbinom(q=(98-3)-1, size=100, prob=0.90)   #nonconservative coverage
[1] 0.9424
> omega <- (0.95 - 0.9424)/(0.9922 - 0.9424)
> omega*1.66 + (1 - omega)*2.05
[1] 1.99
> omega*53.43 + (1 - omega)*37.32
[1] 39.78
```

Thus the resulting approximate distribution-free 95% tolerance interval to contain at least a proportion 0.90 of the distribution is $[T_{0.90}, \ \widetilde{T}_{0.90}] = [1.99, \ 39.78]$. ∎

5.3.2   Distribution-Free One-Sided Tolerance Bounds

Coverage probabilities for distribution-free one-sided tolerance bounds can be computed by using (5.9). Specifically, using $u = n + 1$ gives the coverage probability when $x_{(\ell)}$ represents a one-sided lower tolerance bound and using $\ell = 0$ gives the coverage probability when $x_{(u)}$ represents a one-sided upper tolerance bound.

As described in Section 2.4.2, a distribution-free one-sided lower tolerance bound that one can claim with $100(1 - \alpha)\%$ confidence is exceeded by at least a proportion $1 - p$ of the distribution is equivalent to a distribution-free one-sided lower $100(1 - \alpha)\%$ confidence bound on the p quantile of a distribution. Similarly, a distribution-free one-sided upper tolerance bound that one can claim with $100(1 - \alpha)\%$ confidence exceeds at least a proportion p of the distribution is equivalent to a distribution-free one-sided upper $100(1 - \alpha)\%$ confidence bound on the p quantile of a distribution. Such confidence bounds were considered in Sections 5.2.3 and 5.2.4 and, therefore, will not be discussed further here.

5.3.3   Minimum Sample Size Required for Constructing a Distribution-Free Two-Sided Tolerance Interval

As described in Section 5.1.1, a given sample size might be inadequate to construct a distribution-free tolerance interval with a desired level of confidence, even if one uses the extreme observations of the sample as the interval endpoints. Note from Table J.11 that a sample of size $n = 100$ is not sufficient to obtain a distribution-free two-sided 95% tolerance interval to contain at least a proportion 0.99 of a distribution. In fact, using the extreme values of a given sample of this size gives only a coverage probability of 0.2642 for containing at least a proportion 0.99 of the sampled distribution (i.e., $\text{CPTI}(100, 1, 100, 0.99) =$ pbinom$(100 - 1 - 1; 100, 0.99) = 0.2643$).

Table J.12 gives the smallest sample size n needed to provide $100(1 - \alpha)\%$ confidence that the tolerance interval defined by the extreme values of the sample, $[x_{(1)}, \ x_{(n)}]$, will contain a proportion p of the sampled distribution for selected values of $1 - \alpha$ and β. This table was obtained by substituting $u = n$ and $\ell = 1$ into (5.9), yielding the expression

$$1 - \alpha = 1 - n\beta^{n-1} + (n-1)\beta^n,$$

which is solved for n using standard numerical techniques. To obtain a 95% confidence interval to contain at least a proportion 0.99 of the distribution, we see from Table J.12 that a minimum of $n = 473$ observations are needed. The associated coverage probability is $\mathrm{CPTI}(473, 1, 473, 0.99) = \texttt{pbinom}(473 - 1 - 1; 473, 0.99) = 0.9502$.

Table J.13 gives the smallest sample size needed to have $100(1 - \alpha)\%$ confidence that the largest (smallest) observation in the sample will exceed (be exceeded by) at least $100\beta\%$ of the distribution for selected values of $1 - \alpha$ and β. This table was obtained by substituting $\ell = 0$ and $u = n$ (or equivalently $\ell = 1$ and $u = n + 1$) into (5.9), yielding the expression

$$1 - \alpha = 1 - \beta^n.$$

Solving for n yields $n = \log(\alpha)/\log(\beta)$. If the computed value of n is not an integer, n should be increased to the next highest integer.

5.4 PREDICTION INTERVALS AND BOUNDS TO CONTAIN A SPECIFIED ORDERED OBSERVATION IN A FUTURE SAMPLE

This section shows how to obtain distribution-free two-sided prediction intervals and one-sided prediction bounds for $Y_{(j)}$, the jth ordered observation from a future sample of size m. First we provide some general information about coverage probabilities for these prediction intervals, followed by detailed methods and examples showing how to obtain the intervals.

General tables for prediction intervals to contain a specified ordered observation in a future sample would be too voluminous to include in this book. Instead, we present a computational method that can be used with a sophisticated calculator (or a simple computer program) to obtain the needed intervals.

As in Section 5.2, we treat one-sided prediction bounds first because the procedure that we recommend for obtaining a distribution-free two-sided $100(1 - \alpha)\%$ prediction interval for $Y_{(j)}$ is to combine one-sided lower and upper $100(1 - \alpha/2)\%$ prediction bounds for $Y_{(j)}$.

5.4.1 Coverage Probabilities for Distribution-Free Prediction Intervals and One-Sided Prediction Bounds for a Particular Ordered Observation

A two-sided distribution-free prediction interval for $Y_{(j)}$, the jth ordered observation from a future sample of size m, is obtained as $[\underset{\sim}{Y}_{(j)}, \; \widetilde{Y}_{(j)}] = [x_{(\ell)}, \; x_{(u)}]$. The coverage probability for given ℓ and u is

$$\mathrm{CPYJ}(n, \ell, u, m, j) = \texttt{pnhyper}(u - 1; j, m, m + n) - \texttt{pnhyper}(\ell - 1; j, m, m + n),$$
(5.10)

where $1 \le \ell < u \le n$.

A distribution-free one-sided *upper* prediction bound for $Y_{(j)}$ is given by $\widetilde{Y}_{(j)} = x_{(u)}$. The coverage probability for a given u is

$$\mathrm{CPYJ}(n, 0, u, m, j) = \texttt{pnhyper}(u - 1; j, m, m + n), \quad 1 \le u \le n. \qquad (5.11)$$

A distribution-free one-sided *lower* prediction bound for $Y_{(j)}$ is given by $\underset{\sim}{Y}_{(j)} = x_{(\ell)}$. The coverage probability for a given ℓ is

$$\mathrm{CPYJ}(n, \ell, n + 1, m, j) = 1 - \texttt{pnhyper}(\ell - 1; j, m, m + n), \quad 1 \le \ell \le n,$$

where $1 \leq j \leq m$ in all of the above. An equivalent expression for $\mathrm{CPYJ}(n, \ell, n+1, m, j)$, which can be inverted to provide a direct computation of the value of ℓ giving a desired coverage probability, is

$$\mathrm{CPYJ}(n, \ell, n+1, m, j) = \mathtt{pnhyper}(n - \ell; m - j + 1, m, m + n). \tag{5.12}$$

Derivations of these coverage probabilities are given in Section G.4.

Similar to our previous discussion in Section 5.2.1, CPYJ refers to the prediction interval coverage probability prior to observing the data. After the data have been observed, the number represents a confidence level.

5.4.2   Distribution-Free One-Sided Upper Prediction Bound for $Y_{(j)}$

Computational method

A distribution-free one-sided upper conservative $100(1 - \alpha)\%$ prediction bound for $Y_{(j)}$ from a future sample of size m is obtained as $\widetilde{Y}_{(j)} = x_{(u)}$, where u is chosen as the smallest integer such that the coverage probability $\mathrm{CPYJ}(n, 0, u, m, j)$ in (5.11) is greater than or equal to $1 - \alpha$. More directly, one can obtain u as *one plus* the $(1 - \alpha)$ quantile of the negative hypergeometric distribution with parameters $k = j$, $D = m$, and $N = m + n$ (i.e., $u = \mathtt{qnhyper}(1 - \alpha; j, m, m + n) + 1$). If $u = n + 1$, there is no distribution-free one-sided upper prediction bound for $Y_{(j)}$ having coverage probability greater than or equal to $1 - \alpha$. The coverage probability is given by (5.11).

Approximate interpolation method

Suppose that $x_{(u_c)}$ is the one-sided upper prediction bound having the smallest conservative coverage probability $\mathrm{CPYJ}(n, 0, u_c, m, j)$ and that $x_{(u_n)}$ is the one-sided upper prediction bound with the largest nonconservative coverage probability $\mathrm{CPYJ}(n, 0, u_n, m, j)$. We can then find an interpolated prediction bound between $x_{(u_c)}$ and $x_{(u_n)}$ that has a coverage probability that will generally be closer to the nominal $1 - \alpha$ confidence level than either of the alternatives by letting

$$\omega = \frac{(1 - \alpha) - \mathrm{CPYJ}(n, 0, u_n, m, j)}{\mathrm{CPYJ}(n, 0, u_c, m, j) - \mathrm{CPYJ}(n, 0, u_n, m, j)}.$$

Then the interpolated one-sided upper approximate $100(1 - \alpha)\%$ prediction bound for $Y_{(j)}$ is

$$\omega x_{(u_c)} + (1 - \omega) x_{(u_n)}.$$

Example 5.11   One-Sided Upper Prediction Bound for a Future Sample Median Compound Amount. A distribution-free one-sided upper conservative 95% prediction bound for the sample median of a future sample of size 59 for the output of the chemical process described in Example 5.1 is desired. Thus an upper prediction bound is needed for $Y_{(30)}$. Using R as a calculator for the negative hypergeometric quantile function $\mathtt{qnhyper}$ and cdf function $\mathtt{pnhyper}$ from the $\mathtt{StatInt}$ R package gives

```
> qnhyper(p=0.95, k=30, D=59, N=100+59)+1
[1] 64
> pnhyper(q=64-1, k=30, D=59, N=100+59)
[1] 0.9522
```

showing that the desired upper 95% prediction bound is $\widetilde{\widetilde{Y}}_{0.50} = \widetilde{Y}_{(30)} = x_{(64)} = 13.43$ with a confidence level of 0.9522. Although 0.9522 is close to the nominal 0.95 confidence level, using interpolation in (5.13) and (5.14) will provide a slightly tighter bound and a confidence level that is closer to the nominal confidence level. In particular, the nonconservative prediction bound $\widehat{Y}_{0.50} = \widetilde{Y}_{(30)} = x_{(63)} = 13.09$ is combined with the previously computed conservative bound. Using R as a calculator gives

```
> pnhyper(q=63-1, k=30, D=59, N=100+59)
[1] 0.9383
> omega <- (0.950 - 0.9383)/(0.9522 - 0.9383)
> omega*13.43 + (1 - omega)*13.09
[1] 13.38
```

so that $\widetilde{\widehat{Y}}_{0.50} = 13.38$.   ∎

5.4.3   Distribution-Free One-Sided Lower Prediction Bound for $Y_{(j)}$

Computational method

A distribution-free one-sided lower conservative $100(1 - \alpha)\%$ prediction bound for $Y_{(j)}$ from a future sample of size m is obtained as $Y_{(j)} = x_{(\ell)}$, where ℓ is chosen as the largest integer such that the coverage probability $\mathrm{CPYJ}(n, \ell, n + 1, m, j) = \mathrm{pnhyper}(n - \ell; m - j + 1, m, m + n)$ is greater than or equal to $1 - \alpha$. This gives $\ell = n - \mathrm{qnhyper}(1 - \alpha; m - j + 1, m, m + n)$. If $\ell = 0$, there is no distribution-free one-sided lower prediction bound for $Y_{(j)}$ having coverage probability greater than or equal to $1 - \alpha$. The coverage probability is given by (5.12).

Approximate interpolation method

The interpolation for a one-sided lower prediction bound for $Y_{(j)}$ is done in the following manner. Suppose that $x_{(\ell_c)}$ is the one-sided lower prediction bound having the smallest conservative coverage probability $\mathrm{CPYJ}(n, \ell_c, n + 1, m, j)$ and that $x_{(\ell_n)}$ is the one-sided lower prediction bound with the largest nonconservative coverage probability $\mathrm{CPYJ}(n, \ell_n, n + 1, m, j)$. We can then find an interpolated prediction bound between $x_{(\ell_c)}$ and $x_{(\ell_n)}$ that has a coverage probability that will generally be closer to the nominal $1 - \alpha$ confidence level than either of the alternatives as follows. Let

$$\omega = \frac{(1 - \alpha) - \mathrm{CPYJ}(n, \ell_n, n + 1, m, j)}{\mathrm{CPYJ}(n, \ell_c, n + 1, m, j) - \mathrm{CPYJ}(n, \ell_n, n + 1, m, j)}. \tag{5.13}$$

Then the interpolated one-sided lower approximate $100(1 - \alpha)\%$ prediction bound for $Y_{(j)}$ is

$$\omega x_{(\ell_c)} + (1 - \omega) x_{(\ell_n)}. \tag{5.14}$$

Example 5.12   One-Sided Lower Prediction Bound for the Future Sample Median Compound Amount. A distribution-free one-sided lower conservative 95% prediction bound for the sample median of a future sample of size 59 for the output of the chemical process described in Example 5.1 is desired. Thus a lower prediction bound is needed for $Y_{(30)}$. Using R as a calculator for the negative hypergeometric quantile function and cdf gives

```
> 100-qnhyper(p=0.95, k=59-30+1, D=59, N=100+59)
[1] 37
> 1-pnhyper(q=37-1, k=30, D=59, N=100+59)
[1] 0.9522
```

showing that the desired lower 95% prediction bound is $\widehat{Y}_{0.50} = \underset{\sim}{Y}_{(30)} = x_{(37)} = 7.28$ with conservative coverage probability 0.9522. Using interpolation in (5.13) and (5.14) will provide a coverage probability that is closer to the nominal confidence level. In particular, the non-conservative bound $\widehat{Y}_{0.50} = \underset{\sim}{Y}_{(30)} = x_{(38)} = 7.53$ is combined with the previously computed conservative bound. Using R gives

```
> 1-pnhyper(q=38-1, k=30, D=59, N=100+59)
[1] 0.9383
> omega <- (0.950 - 0.9383)/(0.9522 - 0.9383)
> omega*7.28 + (1 - omega)*7.53
[1] 7.32
```

so that $\widehat{Y}_{0.50} = 7.32$. ∎

The similarity in the results for Examples 5.11 and 5.12 is due to symmetry in dealing with the median.

5.4.4 Distribution-Free Two-Sided Prediction Interval for $Y_{(j)}$

A distribution-free two-sided conservative $100(1 - \alpha)\%$ prediction bound for $Y_{(j)}$ from a future sample of size m can be obtained by combining a one-sided upper conservative $100(1 - \alpha/2)\%$ prediction bound (Section 5.4.2) and a one-sided lower conservative $100(1 - \alpha/2)\%$ prediction bound (Section 5.4.3). This approach also provides the first step for the interpolation method to give a procedure with a coverage probability that is closer to the desired confidence level and that results in a somewhat narrower interval.

Example 5.13 Prediction Interval for Future Sample Median Compound Amount. A distribution-free two-sided prediction interval is needed for $\widehat{Y}_{0.50}$, the median of a future sample of 59 batches from the chemical process described in Example 5.1. Combining the upper and lower one-sided conservative 95% prediction bounds from Examples 5.11 and 5.12 gives the two-sided conservative 90% prediction interval $[\widehat{Y}_{0.50}, \ \widehat{Y}_{0.50}] = [\underset{\sim}{Y}_{(30)}, \ \widetilde{Y}_{(30)}] = [x_{(37)}, \ x_{(64)}] = [7.28, \ 13.43]$. Also, combining the two one-sided approximate 95% prediction bounds based on the interpolation scheme gives a two-sided approximate 90% prediction interval $[\widehat{Y}_{0.50}, \ \widehat{Y}_{0.50}] = [7.32, \ 13.38]$.

In order to construct two-sided 95% prediction intervals we can retrace the steps in Examples 5.11 and 5.12 using one-sided nominal confidence levels of 0.975. The symmetric conservative procedure uses $\ell = 34$ and $u = 67$ and has a coverage probability of 0.9597. The symmetric nonconservative procedure uses $\ell = 35$ and $u = 66$ and has a coverage probability of 0.9453. Interpolating in a manner similar to that done in Examples 5.11 and 5.12 provides a procedure that will have a coverage probability closer to the desired 0.95 confidence level. Using R as a calculator for the negative hypergeometric quantile function and cdf gives

```
> # lower endpoint first
> 100-qnhyper(p=0.975, k=59-30+1, D=59, N=100+59)
[1] 34
> 1-pnhyper(q=34-1, k=30, D=59, N=100+59)
[1] 0.9798
> 1-pnhyper(q=35-1, k=30, D=59, N=100+59)
[1] 0.9727
```

```
> omega <- (0.975 - 0.9727)/(0.9798 - 0.9727)
> omega*6.55 + (1 - omega)*6.83
[1] 6.739
> # upper endpoint
> qnhyper(p=0.975, k=30, D=59, N=100+59)+1
[1] 67
> pnhyper(q=67-1, k=30, D=59, N=100+59)
[1] 0.9798
> pnhyper(q=66-1, k=30, D=59, N=100+59)
[1] 0.9727
> omega <- (0.975 - 0.9727)/(0.9798 - 0.9727)
> omega*14.17 + (1 - omega)*12.32
[1] 12.92
```

Thus the distribution-free two-sided approximate 95% prediction interval for $\widehat{Y}_{0.50}$ based on interpolation is $[\widetilde{Y}_{0.50}, \ \widehat{Y}_{0.50}] = [6.74, \ 12.92]$. The coverage probabilities for the lower and upper bounds in this example are the same because the interval endpoints for the median are symmetric around the center of the sample order statistics.   ∎

5.5 DISTRIBUTION-FREE PREDICTION INTERVALS AND BOUNDS TO CONTAIN AT LEAST k OF m FUTURE OBSERVATIONS

5.5.1 Distribution-Free Two-Sided Prediction Intervals to Contain at Least k of m Future Observations

A distribution-free two-sided conservative $100(1 - \alpha)\%$ prediction interval to contain at least k of m future observations from a previously sampled distribution is obtained as

$$[\underset{\sim}{Y}_{k;m}, \ \widetilde{Y}_{k;m}] = [x_{(\ell)}, \ x_{(u)}].$$

To determine the particular order statistics that will provide a prediction interval with the desired confidence level, one generally chooses ℓ and u symmetrically or nearly symmetrically within the ordered sample data to provide, if possible, the desired coverage probability. The coverage probability for a distribution-free prediction interval, to be based on a previous sample of n observations, is

$$\mathrm{CPKM}(n, \ell, u, k, m) = \mathrm{pnhyper}(u - \ell - 1; k, m, m + n), \quad 0 \leq \ell < u \leq n + 1. \quad (5.15)$$

Note that the case for which $\ell = 0$ corresponds to an upper prediction bound and the case for which $u = n + 1$ corresponds to a lower prediction bound. Derivation of this coverage probability is given in Section G.5.

When the prediction interval is formed by the extreme values of the given sample (i.e., $\ell = 1$ and $u = n$), the coverage probability associated with the inclusion of all $k = m$ future observations is

$$1 - \alpha = \frac{n(n - 1)}{(n + m)(n + m - 1)}. \quad (5.16)$$

As in Section 5.2.1, CPKM refers to the prediction interval coverage probability prior to observing the data. After the data have been observed, we use the standard terminology "confidence level."

Tabular method

One searches Tables J.14a–J.14c, with given n, m, and the desired confidence level $1 - \alpha$, for the largest value of ν that provides the desired value of k shown in the body of the table. Then one chooses ℓ and u symmetrically in a manner similar to that for constructing distribution-free two-sided tolerance intervals, as described in Section 5.3.1 and illustrated in Example 5.9. Tables J.14a–J.14c contain entries for selected values of $1 - \alpha$, n, m, and ν. Note that, as with the tolerance intervals in Section 5.3.1, when ν is an odd integer, the tolerance interval will not be exactly symmetric.

Example 5.14   Two-Sided Conservative Prediction Interval to Contain the Compound Amount for All 5 of 5 Future Batches. Based on the previous sample of $n = 100$ batches, the manufacturer wants a distribution-free 90% prediction interval to contain *all* the observations from a future sample of five batches from the same distribution (i.e., $k = 5$ and $m = 5$). Searching Table J.14a with $n = 100$, $m = 5$, and $1 - \alpha = 0.90$ to find the largest ν giving $k = 5$ in the body of the table, we obtain $\nu = 2$. Proceeding as in Section 5.3.1, we divide ν into $\nu_1 = 1$ and $\nu_2 = 1$ to obtain $\ell = \nu_1 = 1$ and $u = n - \nu_2 + 1 = 100$. Thus, the desired 90% two-sided prediction interval to contain all $k = 5$ of the $m = 5$ future observations is the range formed by the smallest and largest observations of the given sample:

$$[\underset{\sim}{Y}_{5;5}, \; \widetilde{Y}_{5;5}] = [x_{(1)}, \; x_{(100)}] = [1.49, \; 58.11].$$

Because this interval uses the two extreme observations in the previous sample as the endpoints of the prediction interval, we determine from (5.16) that the confidence level is $100(100 - 1)/[(100 + 5)(100 + 5 - 1)] = 0.9066$. ∎

Example 5.15   Two-Sided Conservative Prediction Interval to Contain the Compound Amount for 4 out of 5 Future Batches. Now suppose we require a 95% prediction interval to contain at least $k = 4$ of $m = 5$ future observations. Searching Tables J.14a and J.14b with $n = 100$, $m = 5$, and $1 - \alpha = 0.95$, we find the largest ν such that $k = 4$ to be $\nu = 7$. Partitioning $\nu = 7$ into $\nu_1 = 3$ and $\nu_2 = 4$ gives $\ell = \nu_1 = 3$ and $u = n - \nu_2 + 1 = 97$, following the procedure in Section 5.3.1. Thus, the desired 95% prediction interval to contain at least four of five future observations is

$$[\underset{\sim}{Y}_{4;5}, \; \widetilde{Y}_{4;5}] = [x_{(3)}, \; x_{(97)}] = [2.05, \; 33.24].$$

The confidence level for this interval, computed using the CPKM function in (5.15), is found to be 0.9545. If we had partitioned $\nu = 7$ into $\nu_1 = 4$ and $\nu_2 = 3$ instead, we would have obtained $\ell = 4$ and $u = 98$, and

$$[\underset{\sim}{Y}_{4;5}, \; \widetilde{Y}_{4;5}] = [x_{(4)}, \; x_{(98)}] = [2.24, \; 37.32].$$

The confidence level associated with this interval remains the same (i.e., 0.9545). ∎

The ambiguity in the previous example (with regard to which of the preceding two intervals to use) is due to the odd value of ν and the resulting asymmetry. The approximate method based on interpolation described below avoids this ambiguity.

Computational method

Alternatively, ν can be obtained directly as $\nu = n - \text{qnhyper}(1 - \alpha; k, m, n + m)$ where $\text{qnhyper}(\cdot)$ is the quantile of the negative hypergeometric distribution (see Section G.5 for the derivation of this result). This method can be used for combinations of m, n, k, and $1 - \alpha$ not covered by the tabulations. If $\nu \leq 1$, then there is not a two-sided distribution-free prediction interval to contain at least k out of m future observations with coverage probability greater than or equal to $1 - \alpha$. The coverage probability is given by (5.15).

Approximate interpolation method

As with other conservative statistical interval methods based on order statistics, the coverage probability of the preceding prediction interval may appreciably exceed the desired $100(1 - \alpha)\%$ confidence level. In this case, interpolation, similar to that described in Section 5.2.2, can be used to obtain a narrower approximate distribution-free interval with coverage probability that is generally closer to $1 - \alpha$. First, we find ν directly as in the computational method description, given above. Note that ν must be at least 2, and if it is an odd number, we need to subtract 1 to remain conservative and to ensure that the resulting values of $\ell = \nu/2$ and $u = n - \nu/2 + 1$ are symmetric about the center of the ordered data.

Let $\ell_c = \nu/2$, $\ell_n = \ell_c + 1$, and $u_i = n - \ell_i + 1$ $(i = c, n)$, where c and n indicate "conservative" and "nonconservative" order statistic indices, respectively. If $u_n = u_c - 1$ and $\ell_n = \ell_c + 1$, then $[x_{(\ell_c)}, \; x_{(n-\ell_c+1)}]$ and $[x_{(\ell_n)}, \; x_{(n-\ell_n+1)}]$ are, respectively, conservative and nonconservative prediction intervals and their individual coverage probabilities $\mathrm{CPKM}(n, \ell_c, u_c, k, m)$ and $\mathrm{CPKM}(n, \ell_n, u_n, k, m)$ are given by (5.15). Thus, $\mathrm{CPKM}(n, \ell_n, u_n, k, m) < (1 - \alpha) < \mathrm{CPKM}(n, \ell_c, u_c, k, m)$. Also, let

$$\omega = \frac{(1 - \alpha) - \mathrm{CPKM}(n, \ell_n, u_n, k, m)}{\mathrm{CPKM}(n, \ell_c, u_c, k, m) - \mathrm{CPKM}(n, \ell_n, u_n, k, m)}.$$

Then the approximate prediction interval based on the interpolation method is

$$[\underset{\sim}{Y}_{k;m}, \; \widetilde{Y}_{k;m}] = [\omega x_{(\ell_c)} + (1 - \omega)x_{(\ell_n)}, \; \omega x_{(u_c)} + (1 - \omega)x_{(u_n)}].$$

Because of symmetry, the same interpolation weight is used for both endpoints of the interval.

Example 5.16   Application of the Interpolation Method to Example 5.14. We now apply the interpolation method to Example 5.14 to find an improved symmetric approximate distribution-free prediction interval. First, to obtain a 90% prediction interval to contain *all* of the observations from a future sample of five batches from the same population (i.e., $k = 5$ and $m = 5$), we interpolate between the conservative and nonconservative intervals. To do this we use the R negative hypergeometric quantile function and cdf to find the intervals and interpolate, as follows:

```
# find v
> 100-qnhyper(p=1-0.10, k=5,   D=5, N=100+5)
[1] 2  #   is an even number, so is OK
#   conservative indices are 2/2=1 and 100-2/2+1=100
> pnhyper(q=(100-1)-1, k=5,   D=5, N=100+5)  #conservative coverage
[1] 0.9066
#   nonconservative indices are 1+1=2 and 100-1=99
> pnhyper(q=(99-2)-1, k=5,   D=5, N=100+5)  #nonconservative coverage
[1] 0.8203
> omega <- (0.90 - 0.8203)/(0.9066 - 0.8203)
> omega*1.49  + (1 - omega)*1.66
[1] 1.503
> omega*58.11 + (1 - omega)*53.43
[1] 57.75
```

yielding the 95% prediction interval $[\underset{\sim}{Y}_{4;5}, \; \widetilde{Y}_{4;5}] = [1.50, \; 57.75]$.

To find a 95% prediction interval to contain at least 4 of 5 observations from a future sample of batches from the same process (i.e., $k = 4$ and $m = 5$), we similarly use R as a calculator, proceeding as follows:

```
# find v
> 100-qnhyper(p=1-0.05, k=4,  D=5, N=100+5)
[1]  7   # is an odd number, so need to subtract one and use v=6
#  conservative indices are 6/2=3 and 100-6/2+1=98
> pnhyper(q=(98-3)-1, k=4,  D=5, N=100+5)   #conservative coverage
[1] 0.9652
#  nonconservative indices are 3+1=4 and 98-1=97
> pnhyper(q=(97-4)-1, k=4,  D=5, N=100+5)   #nonconservative coverage
[1] 0.9426
> omega <- (0.95 - 0.9426)/(0.9652 - 0.9426)
> omega*2.05  + (1 - omega)*2.24
[1] 2.178
> omega*37.32 + (1 - omega)*33.24
[1] 34.58
```

yielding the 95% prediction interval $[\underset{\sim}{Y}_{4;5}, \ \widetilde{Y}_{4;5}] = [2.18, \ 34.58]$. ∎

5.5.2 Distribution-Free One-Sided Prediction Bounds to Exceed or Be Exceeded by at Least k of m Future Observations

In (5.15), using $u = n + 1$ corresponds to having $x_{(\ell)}$ provide a one-sided lower prediction bound and using $\ell = 0$ corresponds to having $x_{(u)}$ provide a one-sided upper prediction bound.

A one-sided lower prediction bound to be exceeded by at least $m - k + 1$ of m future observations is equivalent to a one-sided lower prediction bound to be exceeded by the k largest of m future observations, described in Section 5.4.3. Similarly, a one-sided upper prediction bound to exceed at least k of m future observations is equivalent to a one-sided upper prediction bound to exceed the kth largest of m future observations, described in Section 5.4.2.

Also, a one-sided prediction bound that uses the smallest (or largest) of n observations as the lower (or upper) limit, to be exceeded by (exceed) all of m future observations has associated coverage probability

$$1 - \alpha = \frac{n}{n + m}. \tag{5.17}$$

Tables J.16a–J.16c give the necessary sample size n so that a one-sided lower (upper) distribution-free prediction bound defined by the smallest (largest) observation from the previous sample will be exceeded by (will exceed)

- all m,
- at least $m - 1$, and
- at least $m - 2$

observations in a future sample of size m from the previously sampled distribution using a $100(1 - \alpha)\%$ confidence level for selected values $1 - \alpha$ and m.

BIBLIOGRAPHIC NOTES

The theory underlying the use of order statistics and some of the distribution-free statistical intervals presented in this chapter is outlined in Appendix G. A substantial treatment of the theory of order statistics can be found, for example, in David and Nagaraja (2003) who also discuss distribution-free statistical intervals.

Distribution-free confidence intervals for a quantile

Murphy (1948) first presented charts like those in Figures 5.3a and 5.3b to construct distribution-free two-sided tolerance intervals and one-sided tolerance bounds. Somerville (1958) gave tabulations that are similar to Table J.11, but without the coverage probabilities.

Distribution-free confidence intervals for a quantile

Because of the equivalence of one-sided tolerance bounds and one-sided confidence bounds on a distribution quantile (described in Section 2.4.2), the charts given by Murphy (1948) and the tables given by Somerville (1958) can also be used to obtain distribution-free one-sided confidence bounds and two-sided confidence intervals for a quantile.

Owen (1988) noted that the method of empirical likelihood, when applied to construct confidence intervals for a distribution quantile, gives the classical result, described in Section 5.2, that the interval endpoints are order statistics and that, due to their discreteness, the coverage probability may differ appreciably from the desired nominal confidence level. Chen and Hall (1993) give a smoothed empirical likelihood procedure, based on a kernel smoother applied to the empirical distribution function, that has coverage probabilities that are closer to the nominal confidence level. Adimari (1998) uses a simpler continuous approximation to the empirical distribution function and shows via simulation that its finite sample properties are as good as those of the Chen–Hall procedure. Beran and Hall (1993) show that simple interpolation procedures, like those presented in this chapter, provide an effective method for improving the classical intervals based on order statistics for both confidence intervals and prediction intervals and that deviations of coverage probabilities from the nominal confidence level tend to be conservative.

Gibbons (1997) shows that if the underlying (unspecified) distribution can be assumed to be symmetric, somewhat narrower distribution-free confidence intervals for the median (i.e., $y_{0.50}$) can be obtained. Gilat and Hill (1996) provide distribution-free confidence intervals for quantiles of an arbitrary unknown distribution, including discontinuous distributions.

Distribution-free prediction intervals

The properties of the negative hypergeometric distribution (also known as the "inverse hypergeometric distribution"), used in computing distribution-free prediction intervals, are discussed by Guenther (1975).

Danziger and Davis (1964) provide tables for prediction intervals to contain at least k of m future observations (which they refer to as "tolerance intervals"). Hall et al. (1975) provide tables for the coverage probability associated with distribution-free one-sided prediction bounds and two-sided prediction intervals for the special case for which the extreme observation(s) from the given sample is (are) used as the interval endpoint(s).

Fligner and Wolfe (1976) show how to use the probability integral transform to derive expressions for coverage probabilities for certain distribution-free two-sample statistical interval procedures, including the prediction-interval procedures in this chapter. They also show that when the sampled distribution is discrete, the coverage probability is larger than the probability computed by formulas (5.1), (5.9), (5.10), and (5.15). Fligner and Wolfe (1979b) describe applications for prediction intervals for the sample median and provide tables for obtaining such intervals. Fligner and Wolfe (1979a) provide theory and large-sample approximations for a distribution-free prediction interval for a sample median.

Guilbaud (1983) gives more general theory for the particular problem of setting prediction intervals for sample medians. Davis and McNichols (1999) show how to use nonparametric prediction intervals to conduct multiple-comparisons-with-control hypothesis tests.

Chapter 6

Statistical Intervals for a Binomial Distribution

OBJECTIVES AND OVERVIEW

This chapter describes statistical intervals for proportions or percentages. Such intervals are used, for example, when each observation is either a "conforming" or a "nonconforming" unit and the data consist of the number, or equivalently, the proportion or percentage, of nonconforming units, in a random sample of n units from a population or process. Two examples are:

- An integrated circuit passes an operational test only if it successfully completes a specified set of operations after a 48-hour "burn-in" at 85°C and 85% relative humidity. Thus, the given data consist of the proportion of the n units that failed or passed the test. The goal is to estimate the proportion of potentially failing (nonconforming) units in the sampled manufacturing process.

- Federal regulations require that the level of a certain pollutant in the exhaust from an internal combustion engine be less than 10 parts per million (ppm). If an engine fails to meet this standard, it must undergo expensive rework. Management wants to estimate the proportion of units from a specified process that will require such rework. The available data consist of the number of units that needed rework in a random sample of n engines from the manufacturing process.

Our discussion will be mainly in terms of "nonconforming" units to suggest the common quality control application. The applicability of the intervals is, however, much more general. For example, a nonconforming unit could be an individual indicating a preference for a particular candidate in a forthcoming election, the survival of an animal in a biological experiment, and so on.

Statistical Intervals: A Guide for Practitioners and Researchers, Second Edition.
William Q. Meeker, Gerald J. Hahn and Luis A. Escobar.
© 2017 John Wiley & Sons, Inc. Published 2017 by John Wiley & Sons, Inc.
Companion Website: www.wiley.com/go/meeker/intervals

The following topics are discussed in this chapter:

- Confidence intervals for π, the actual proportion nonconforming in the sampled distribution (Section 6.2).

- Confidence interval for the number of nonconforming units in a finite population (Section 6.3).

- Confidence intervals for the probability that the number of nonconforming units in a sample of size m will be less than or equal to (or greater than) a specified number (Section 6.4).

- Confidence intervals for the quantile of the distribution of the number of nonconforming units in a future sample of size m (Section 6.5).

- Tolerance intervals and bounds for the distribution of the number of nonconforming units in a future sample of size m (Section 6.6).

- Prediction intervals for the number of nonconforming units in a future sample of size m (Section 6.7).

6.1 INTRODUCTION

6.1.1 The Binomial Distribution

Problems involving the proportion of nonconforming units in a random sample of size n from a large population (or a stable process) can often be modeled with the binomial distribution. In particular, the probability of observing x nonconforming units in a random sample of size n independent observations, assuming a constant proportion π of nonconforming units, is given by the binomial probability function

$$\Pr(X = x) = \texttt{dbinom}(x; n, \pi) = \frac{n!}{x!(n-x)!}\pi^x(1-\pi)^{n-x}, \quad x = 0, 1, \ldots, n.$$

The probability of observing x *or fewer* nonconforming units in a random sample of size n is

$$\Pr(X \leq x) = \texttt{pbinom}(x; n, \pi) = \sum_{i=1}^{x} \texttt{dbinom}(i; n, \pi), \quad x = 0, 1, \ldots, n.$$

More technical details about the binomial distribution are given in Section C.4.1.

6.1.2 Other Distributions and Related Notation

In addition to the binomial distribution, this chapter also uses the normal and F-distributions described in Section 4.1, the beta distribution, described in Section C.3.3, and the beta-binomial distribution, described in Section C.4.2. Sections 6.3 and 6.7 use the hypergeometric distribution, which is closely related to the binomial distribution and is described in Section C.4.5. Notationally, $\texttt{phyper}(x; n, D, N)$ is the hypergeometric distribution cdf giving the probability of x or fewer nonconforming units from a sample of n units when sampling without replacement from a finite population of size N that initially contains D nonconforming units.

6.1.3   Notation for Data and Inference

In this chapter we will assume that the data consist of x observed nonconforming units (a realization of the random variable X) out of a sample of size n from a binomial distribution with an actual proportion nonconforming π. In Sections 6.4–6.7, inferences or predictions will be made for the distribution of Y, the number of nonconforming units from a future sample or samples of size m from the same distribution with the same proportion nonconforming π.

6.1.4   Binomial Distribution Statistical Interval Properties

For the binomial distribution problems considered here, the parameter π is unknown. In Section 6.2 we seek a confidence interval for π based on the observed x nonconforming units from the past random sample of size n. In subsequent sections, we show how to construct other statistical intervals of interest from the given binomial distribution sample.

   The binomial is a discrete distribution. A binomial random variable can take on only the integer values $X = 0, 1, 2, \ldots, n$, and cannot take on values between these integers. Because of this, statistical interval methods given in this chapter generally do not have exactly the desired nominal confidence level. Instead, the coverage probabilities depend on the unknown value of π (as illustrated in Section 6.2.6). Thus, the statistical intervals given here are either *approximate* or *conservative* (i.e., the coverage probability is larger than the nominal confidence level), depending on the method used. There are numerous methods for computing binomial distribution statistical intervals. In Section 6.2, we will present and illustrate four important methods for constructing confidence intervals on π. References given in the Bibliographic Notes section at the end of this chapter provide further description and evaluation of these and other methods.

6.1.5   Two Examples, Motivation, and a Caution

The binomial distribution is used in a wide range of applications in science, engineering, and business. In this section we present two examples that will be used throughout this chapter.

Example 6.1   Proportion of Defective Integrated Circuits. A random sample of $n = 1,000$ integrated circuits had been selected from production and $x = 20$ of these units failed a post-production test. From the data, an estimate of π, the proportion of defective units generated by the sampled production process, is $\widehat{\pi} = x/n = 20/1,000 = 0.02$. The resulting data can be used to make inferences or predictions about the production process, assuming that it is in statistical control.   ∎

Example 6.2   Proportion of Nonconforming Engines. The amount of a pollutant was measured for each engine in a random sample of $n = 10$ engines from some defined population. If one can assume that the pollution measurements can be described by a normal distribution, desired statistical intervals can be obtained by using the methods in Chapter 4, based upon the sample mean and sample standard deviation of the pollutant measurements. These were $\bar{x} = 8.05$ ppm and $s = 1.09$ ppm. The assumption of a normal distribution, however, might be questionable, and one might want to make inferences concerning the population proportion outside specification limits without making such an assumption. An alternative nonparametric approach would be to simply count the number of observations outside the specified limits, define this as the number of "nonconforming units," and then use the techniques described in this chapter to construct the desired statistical intervals. In the example, a pollutant level above 10 ppm was deemed to be nonconforming, and one of the 10 engine exhaust measurements exceeded this value. Thus, there was $x = 1$ nonconforming unit in a sample of size $n = 10$.

Such dichotomizing of measurement data might be especially useful when there are multiple (possibly correlated) measurements, each of which must meet some stated specification limit. In that case, a nonconforming unit would simply be one that fails to meet the specification limit for at least one of the measurements. ■

As discussed in Chapter 5, the effect of abandoning the assumption of a specific distribution, such as the normal distribution, is that information is sacrificed by not using the exact measurements; thus the resulting intervals will generally be wider. As a result, in the second example, a confidence interval to contain the proportion of nonconforming engines in the sampled population will tend to be wider if the interval is based only on a count of the number of such engines in the sample, rather than on the pollution measurements, assuming a particular distribution. Such loss of precision, in some situations, is a serious concern, especially in evaluating a high-reliability product. Thus, one should avoid dichotomizing measurement data, if possible. In some other examples, the data may be inherently of a binary nature. This would be the case, for example, for the operation of a switch, the cure of a patient, and the integrated circuit test in Example 6.1. Then one has no choice other than to use the methods of this chapter.

As in previous chapters, to assure valid inferences, we reiterate the importance of the random selection of sample units from the population or process of interest. The analyst must consider the practical considerations for this to be the case; see Chapter 1.

6.2   CONFIDENCE INTERVALS FOR THE ACTUAL PROPORTION NONCONFORMING IN THE SAMPLED DISTRIBUTION

6.2.1   Preliminaries

The sample proportion $\widehat{\pi} = x/n$ is a point estimate for π, the actual distribution proportion. The estimate $\widehat{\pi}$, however, differs from π due to sampling fluctuations. Thus, one frequently desires to compute, from the sample data, a two-sided confidence interval or a one-sided confidence bound for π. This section presents and motivates the use of several of the most commonly used methods for computing confidence intervals for π. The underlying theory for the methods is given in Sections D.6.2, D.5.6, and H.3.1. The Bibliographic Notes section at the end of this chapter gives references to articles that describe and give technical details about the methods for computing these and other confidence interval methods for π. As in previous chapters, we mainly present two-sided confidence intervals; one-sided lower and upper $100(1-\alpha)\%$ confidence bounds are obtained by replacing $\alpha/2$ with α in the appropriate formula for obtaining a two-sided confidence interval.

Each of the four methods for obtaining a confidence interval for a binomial proportion presented in this section will be illustrated by either the nonconforming engine example or the defective integrated circuit example. Tables 6.1 and 6.2 compare the confidence intervals from all four methods for both examples. Table 6.2 also contains other statistical intervals described in subsequent sections of this chapter. Section 6.2.6 compares the coverage probability properties of the confidence interval methods and provides recommendations.

The methods discussed in this section (and Sections 6.4–6.7) assume that sampling is from a process that is in statistical control (see Section 1.7.2) or from a finite population for which the sample size n is small relative to the population size N or sampling *with* replacement. When sampling *without* replacement, the guideline for "small" is typically given as $n/N < 0.10$. The situation where $n/N \geq 0.10$ will be discussed in Section 6.3.

6.2.2   The Conservative Method

For x observed nonconforming units in a sample of size n, a two-sided conservative $100(1 - \alpha)\%$ confidence interval for π is

$$[\underset{\sim}{\pi}, \; \widetilde{\pi}] = [\texttt{qbeta}(\alpha/2; x, n - x + 1), \; \texttt{qbeta}(1 - \alpha/2; x + 1, n - x)] \tag{6.1}$$

$$= \left[\left(1 + \frac{(n - x + 1)F_{(1-\alpha/2; 2n-2x+2, 2x)}}{x} \right)^{-1}, \right.$$

$$\left. \left(1 + \frac{n - x}{(x + 1)F_{(1-\alpha/2; 2x+2, 2n-2x)}} \right)^{-1} \right], \tag{6.2}$$

where $\texttt{qbeta}(p; a, b)$ is the p quantile of the beta distribution with shape parameters a and b (described in Section C.3.3) and $F_{(p; r_1, r_2)}$ is the p quantile of Snedecor's F-distribution with r_1 numerator and r_2 denominator degrees of freedom (see Sections 4.1 and C.3.11). The lower endpoint is defined to be $\underset{\sim}{\pi} = 0$ if $x = 0$, and the upper endpoint is $\widetilde{\pi} = 1$ if $x = n$. The derivation of this interval (see Section D.6.2) leads naturally to the use of the beta distribution quantile for performing the computation. The equivalent formulas that use F-distribution quantiles are derived from the relationship between the beta distribution and the F-distribution in (C.11). Tables of Snedecor's F-distribution quantiles are more readily available, but the beta quantile method is more convenient when a sophisticated calculator, such as R, is available.

This method is conservative in the sense that the coverage probability is guaranteed to be greater than or equal to the nominal confidence level (as illustrated in Section 6.2.6). The method is referred to in some places in the literature as the "exact" method. We avoid using this terminology because it is misleading and conflicts with the usual meaning of an exact (as opposed to approximate or conservative) statistical interval; see Section B.2.1.

The preceding method has been used to construct the plots in Figure 6.1. These plots provide a simple method to obtain two-sided conservative 90% or 95% confidence intervals (or one-sided 95% or 97.5% confidence bounds) for π. To use the charts, one computes $\widehat{\pi} = x/n$ from the data, locates this value on the horizontal axis, draws a vertical line from this point to the curves corresponding to the sample size n, and then draws a horizontal line to the vertical axis to obtain the endpoint(s) of the desired interval (bound).

Example 6.3   Conservative Confidence Interval for the Proportion of Nonconforming Engines. For the engine exhaust pollutant application, management wants a confidence interval for π, the proportion of engines in the sampled population with pollutant levels greater than 10 ppm. Using the conservative method given by (6.2) with $n = 10$ and $x = 1$, we obtain $F_{(0.975; 20, 2)} = 39.45$, $F_{(0.975; 4, 18)} = 3.608$, and $F_{(0.95; 4, 18)} = 2.928$ from tables in various statistics textbooks or from R. Then, substituting into (6.2), a conservative 95% confidence interval for π is

$$[\underset{\sim}{\pi}, \; \widetilde{\pi}] = \left[\left(1 + \frac{10 \times 39.45}{1} \right)^{-1}, \; \left(1 + \frac{9}{2 \times 3.608} \right)^{-1} \right] = [0.0025, \; 0.44].$$

A one-sided upper conservative 95% confidence bound for π is

$$\widetilde{\pi} = \left(1 + \frac{9}{2 \times 2.928} \right)^{-1} = 0.39.$$

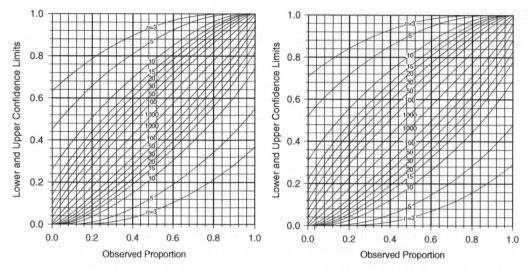

Figure 6.1   Two-sided conservative 90% confidence intervals (one-sided conservative 95% confidence bounds) for a binomial proportion (left) and two-sided conservative 95% confidence intervals (one-sided conservative 97.5% confidence bounds) for a binomial proportion (right). Similar figures first appeared in Clopper and Pearson (1934). Adapted with permission of the Biometrika Trustees.

Alternatively, using R as a calculator for the beta quantile function method in (6.1) gives

```
> c(qbeta(p=0.025, shape1=1, shape2=10), qbeta(p=0.975, shape1=2,
  shape2=9))
[1] 0.0025 0.4450
> qbeta(0.95, 2, 9)
[1] 0.3942
```

for the conservative 95% confidence interval and one-sided 95% confidence bound. These results could also have been read (approximately) from the right plot in Figure 6.1 (the left plot for the one-sided bound).   ■

Type of confidence interval/bound for π	Lower	Upper
Conservative two-sided	0.0025	0.4450
Wald two-sided	−0.0859	0.2859
Agresti–Coull two-sided	−0.0039	0.4260
Jeffreys two-sided	0.0110	0.3813
Conservative one-sided	0.0051	0.3942
Wald one-sided	−0.0560	0.2560
Agresti–Coull one-sided	0.0059	0.3644
Jeffreys one-sided	0.0179	0.3306

Table 6.1   Two-sided 95% confidence intervals and one-sided 95% confidence bounds for π, the proportion of nonconforming engines.

6.2.3   The Wald (Normal Theory) Approximate Method

A two-sided approximate $100(1 - \alpha)\%$ confidence interval for π is

$$[\underline{\pi}, \ \widetilde{\pi}] = \widehat{\pi} \mp z_{(1-\alpha/2)} \left[\frac{\widehat{\pi}(1 - \widehat{\pi})}{n} \right]^{1/2}, \tag{6.3}$$

where $z_{(1-\alpha/2)}$ is the $1 - \alpha/2$ quantile of the standard normal distribution. This simple method was especially relevant when computational capabilities were limited and still appears frequently in elementary statistics textbooks. It is often incorrectly stated that the method provides adequate accuracy when both $n\widehat{\pi}$ and $n(1 - \widehat{\pi})$ exceed 10. Unfortunately, as has been widely reported in the more recent statistical literature (see the Bibliographic Notes section at the end of this chapter), even when these conditions are met, the properties of this method can be extremely poor, and we do not recommend it for general use (see Section 6.2.6). We describe the method because of its popularity (and historical significance) and because it is still used in some situations (e.g., when no appropriate software or tables are readily available, such as in an informal conversation) and when a simple, crude approximation will suffice.

Example 6.4   Wald-Approximation Confidence Interval for the Proportion of Defective Integrated Circuits. In Example 6.1, $x = 20$ nonconforming units were found in the sample of $n = 1,000$ integrated circuits. Because $n\widehat{\pi} = 1,000 \times 0.02 = 20$ and $n(1 - \widehat{\pi}) = 1,000 \times 0.98 = 980$, the commonly proposed guideline suggests that the approximate method given by (6.3) will be adequate. An approximate 95% confidence interval for π, using $\widehat{\pi} = 0.02$ and $z_{(0.975)} = 1.96$, is

$$[\underline{\pi}, \ \widetilde{\pi}] = 0.02 \mp 1.96 \left[\frac{0.02 \times 0.98}{1,000} \right]^{1/2} = [0.0113, \ 0.0287].$$

An upper approximate 95% confidence bound for π, using $z_{(0.95)} = 1.645$, is

$$\widetilde{\pi} = 0.02 + 1.645 \left[\frac{0.02 \times 0.98}{1,000} \right]^{1/2} = 0.027.$$

Using R as a calculator gives

```
> pihat <- 20/1000
> pihat  + c(-1, 1)*qnorm(0.975)*sqrt(pihat*(1-pihat)/1000)
[1]  0.0113 0.0287
> pihat  + qnorm(0.95)*sqrt(pihat*(1-pihat)/1000)
[1]  0.02728
```

for the approximate 95% confidence interval and upper 95% confidence bound.   ■

6.2.4   The Agresti–Coull Adjusted Wald-Approximation Method

Due to the poor performance of the Wald method in many situations, numerous alternative methods have been suggested with the goal of having a method that is almost as simple as the Wald method, but with better properties. One such method adds a constant to the observed number of nonconforming units and another constant to the sample size, where the constants depend only on the nominal confidence level. This adjustment to the simple Wald method provides a surprising improvement in performance. In particular, the Agresti–Coull method uses

$$x^\dagger = x + z^2_{(1-\alpha/2)}/2 \quad \text{and} \quad n^\dagger = n + z^2_{(1-\alpha/2)}$$

in place of x and n, giving the adjusted point estimate $\widehat{\pi}^\dagger = x^\dagger/n^\dagger$. Then $\widehat{\pi}^\dagger$ and $n^\dagger$ are simply substituted for $\widehat{\pi}$ and n, respectively, in (6.3) to compute the Agresti–Coull adjusted Wald-approximation confidence interval for π. For a 95% confidence interval, $z_{(0.975)} = 1.96 \approx 2$ so that $x^\dagger = x + 2$ and $n^\dagger = n + 4$, providing an easy-to-remember adjustment known as the "add two" method (add two nonconforming and two conforming units to the data). This method outperforms the simple Wald method (see Section 6.2.6) and should be used instead, when possible.

Example 6.5 Agresti–Coull Adjusted Approximate Confidence Interval for the Proportion of Defective Integrated Circuits. From Example 6.4, $x^\dagger = 20 + 1.96^2/2 = 21.921$, $n^\dagger = 1,000 + 1.96^2 = 1,003.84$, and $\widehat{\pi}^\dagger = 21.921/1,003.84 = 0.02184$. Then

$$[\underaccent{\tilde}{\pi}, \quad \widetilde{\pi}] = 0.02184 \mp 1.96 \left[\frac{0.02184(1 - 0.02184)}{1,003.84} \right]^{1/2} = [0.013, \quad 0.031]. \qquad (6.4)$$

Interval/bound type	for	Lower	Upper
Conservative two-sided CI	π	0.0123	0.0307
Wald two-sided CI	π	0.0113	0.0287
Agresti–Coull two-sided CI	π	0.0128	0.0309
Jeffreys two-sided CI	π	0.0127	0.0301
Conservative one-sided CB	π	0.0133	0.0289
Wald one-sided CB	π	0.0127	0.0273
Agresti–Coull one-sided CB	π	0.0138	0.0288
Jeffreys one-sided CB	π	0.0137	0.0283
Conservative two-sided CI	$\Pr(Y \leq 2), m = 50$	0.8016	0.9765
Jeffreys two-sided CI	$\Pr(Y \leq 2), m = 50$	0.8093	0.9745
Conservative two-sided CI	$y_{0.90}, m = 1,000$	17	38
Jeffreys two-sided CI	$y_{0.90}, m = 1,000$	17	37
Conservative one-sided CB	$y_{0.10}, m = 1,000$	9	22
Jeffreys one-sided CB	$y_{0.10}, m = 1,000$	9	22
Conservative one-sided CB	$y_{0.90}, m = 1,000$	19	36
Jeffreys one-sided CB	$y_{0.90}, m = 1,000$	19	35
Conservative two-sided TI	Control center 0.80, $m = 1,000$	9	36
Jeffreys two-sided TI	Control center 0.80, $m = 1,000$	9	35
Conservative two-sided PI	$Y, m = 1,000$	9	35
Normal two-sided PI	$Y, m = 1,000$	7	33
Joint-sample two-sided PI	$Y, m = 1,000$	10	34
Jeffreys two-sided PI	$Y, m = 1,000$	10	34

Table 6.2 Two-sided 95% confidence intervals and one-sided 95% confidence bounds for π, the proportion of defective integrated circuits, and other related intervals. CI, TI, and PI indicate confidence, tolerance, and prediction interval, respectively. CB indicates confidence bound.

A one-sided upper approximate 95% confidence bound for π, using $z_{(0.95)} = 1.645$, is

$$\widetilde{\pi} = 0.0213 + 1.645 \left[\frac{0.0213(1 - 0.0213)}{1,003.84} \right]^{1/2} = 0.029.$$

Equivalently, using R as a calculator gives

```
> xadj <- 20 + qnorm(0.975)^2/2
> nadj <- 1000 + qnorm(0.975)^2
> pihatadj <- xadj/nadj
> pihatadj  + c(-1, 1)*qnorm(0.975)*sqrt(pihatadj*(1-pihatadj)/nadj)
[1] 0.01280 0.03088
> xadj <- 20 + qnorm(0.95)^2/2
> nadj <- 1000 + qnorm(0.95)^2
> pihatadj <- xadj/nadj
> pihatadj  + qnorm(0.95)*sqrt(pihatadj*(1-pihatadj)/nadj)
[1] 0.02879
```

for the approximate 95% confidence interval and one-sided confidence bound. If one used the simpler "add 2" rule instead, the confidence interval would be $[\underset{\sim}{\pi}, \ \widetilde{\pi}] = [0.01286, \ 0.03097]$ which, after rounding to three significant digits, gives the same result as (6.4).   ∎

6.2.5 The Jeffreys Approximate Method

Using the Jeffreys method (see the Bibliographic Notes section at the end of this chapter and Section H.4.1 for a description of the origin of this method), a two-sided approximate $100(1 - \alpha)\%$ confidence interval for π, based on x observed nonconforming units in a sample of size n, is

$$[\underset{\sim}{\pi}, \ \widetilde{\pi}] = [\text{qbeta}(\alpha/2; x + 0.5, n - x + 0.5), \ \text{qbeta}(1 - \alpha/2; x + 0.5, n - x + 0.5)]$$

$$= \left[\left(1 + \frac{(n - x + 0.5)F_{(1-\alpha/2;2n-2x+1,2x+1)}}{x + 0.5} \right)^{-1}, \right. \tag{6.5}$$

$$\left. \left(1 + \frac{n - x + 0.5}{(x + 0.5)F_{(1-\alpha/2;2x+1,2n-2x+1)}} \right)^{-1} \right].$$

The structure of the formulas for this method is the same as that for the conservative method given by (6.1) and (6.2), but with different degrees of freedom for the F quantiles (and their corresponding multipliers) or different parameters for the beta distribution quantiles. Unlike the conservative method, the Jeffreys method does not guarantee a coverage probability that is greater than or equal to the nominal confidence level for all values of π. Instead, the Jeffreys method has excellent *mean* coverage properties, as we will see in Section 6.2.6.

Example 6.6   Jeffreys Confidence Interval for the Proportion of Nonconforming Engines. In the engine exhaust pollutant application in Example 6.3, using the Jeffreys method given by (6.5), with $n = 10$ and $x = 1$, we obtain $F_{(0.975;19,3)} = 14.181$, $F_{(0.975;3,19)} = 3.903$, and $F_{(0.95;3,19)} = 3.127$ from tables in statistics textbooks or by using R. Then a 95% confidence interval for π is

$$[\pi, \ \widetilde{\pi}] = \left[\left(1 + \frac{9.5 \times 14.181}{1.5}\right)^{-1}, \ \left(1 + \frac{9.5}{1.5 \times 3.903}\right)^{-1} \right] = [0.011, \ 0.38].$$

An upper 95% confidence bound for π is

$$\widetilde{\pi} = \left(1 + \frac{9.5}{1.5 \times 3.127}\right)^{-1} = 0.33.$$

Using R as a calculator for the beta quantile function gives

```
> c(qbeta(p=0.025, shape1=1.5, shape2=9.5), qbeta(p=0.975, shape1=1.5,
shape2=9.5))
[1] 0.0110 0.3813
> qbeta(p=0.95, shape1=1.5, shape2=9.59.5)
[1] 0.3306
```

for the approximate 95% confidence interval and one-sided confidence bound.   ∎

6.2.6 Comparisons and Recommendations

If it is desirable to avoid arguments about the use of approximations (e.g., in court proceedings), one should use the conservative method to assure a coverage probability that is always greater than or equal to the nominal confidence level. When, however, $n\pi$ or $n(1 - \pi)$ is small, the conservative method can tend to result in intervals that will be wider than the competing methods. Otherwise we recommend the Jeffreys method for general use for constructing both two-sided confidence intervals and one-sided confidence bounds for π, the Agresti–Coull method when an especially simple computational method is needed, and the Wald method for situations for which an immediate, even though crude, ballpark assessment is desired.

In the remainder of this section we compare the performance of the different methods to provide some insight into these recommendations. Generally, one should compare the properties of different methods and make a choice on the method to use *before* looking at the data, thus avoiding any appearance of lack of objectiveness. The coverage probability for a binomial confidence interval method depends on the actual π and can be computed by using (B.6) in Section B.2.4.

Figures 6.2 and 6.3 show the coverage probabilities versus the actual binomial proportion π for the 95% confidence intervals constructed using each of the four methods presented in this section. These plots are for small sample sizes ($n = 5$ and 20) and larger sample sizes ($n = 100$ and 1,000) in the two figures, respectively. Although these plots deal only with 95% confidence intervals, similar results hold for other confidence levels.

Note the symmetry of these plots around $\pi = 0.5$. Also shown on the plots are the mean coverage and the minimum coverage computed at 200 equally spaced values of π between 0.001 and 0.999. The plots show that the coverage probability properties of the methods are similar and close to the nominal 95% value when $n = 1,000$ (except for extreme values of π) but can differ appreciably when sample sizes are more moderate.

The top rows of Figures 6.2 and 6.3 show that the conservative method has, as predicted by theory (see Section D.6.2), a coverage probability that is greater than the nominal 95% confidence level for all values of π. For small sample sizes the coverage probability is appreciably larger than the nominal. In addition, as previously stated, it can be shown that intervals

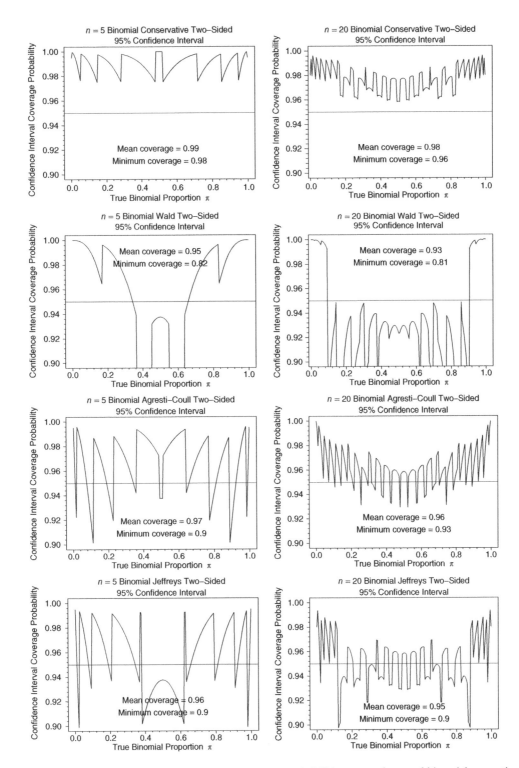

Figure 6.2　Plots of confidence interval (for π) coverage probabilities versus the actual binomial proportion π for the conservative (top row), Wald (second row), Agresti–Coull (third row), and Jeffreys (bottom row) nominal 95% confidence interval methods with $n = 5$ (left) and $n = 20$ (right).

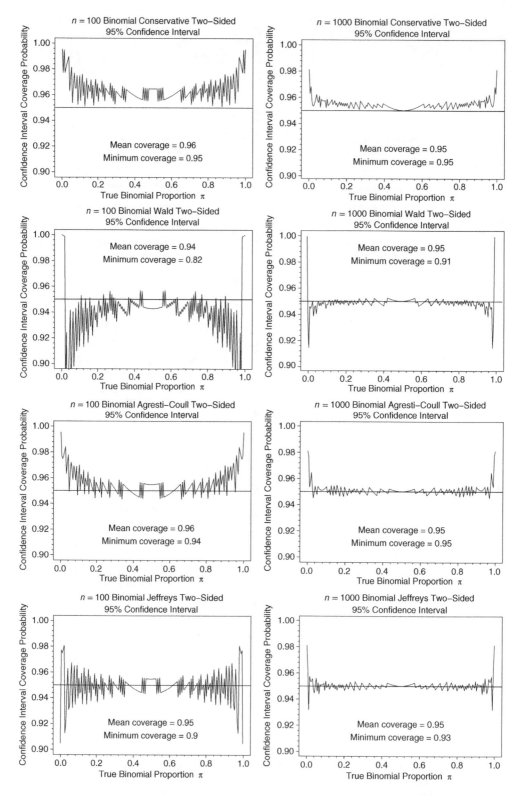

Figure 6.3 Plots of confidence interval (for π) coverage probabilities versus the actual binomial proportion π for the conservative (top row), Wald (second row), Agresti–Coull (third row), and Jeffreys (bottom row) nominal 95% confidence interval methods with $n = 100$ (left) and $n = 1,000$ (right).

constructed using the conservative method tend to be wider than those calculated using other methods.

The plots in the second row of Figures 6.2 and 6.3 show that the Wald-approximation method usually provides a coverage probability that is less, and sometimes substantially less, than the nominal confidence level (i.e., the method is usually nonconservative). For this reason, we recommend against use of the Wald method.

The Agresti–Coull and Jeffreys methods, shown in the bottom two rows of Figures 6.2 and 6.3, have reasonably good approximate coverage properties with mean coverage close to (or slightly above in the case of Agresti–Coull) the nominal confidence level, even for small samples. Because it tends to be more conservative, the Agresti–Coull method typically results in slightly wider intervals than the Jeffreys method. In addition—as we will see in the discussion to follow dealing with coverage probabilities for one-sided confidence bounds—the Jeffreys method provides good balance between the error probabilities of being outside of the upper and lower endpoints for a two-sided confidence interval, while, in contrast, the Agresti–Coull method *does not*.

In contrast to Figures 6.2 and 6.3, which show two-sided confidence interval coverage probabilities, Figure 6.4 gives one-sided confidence bound coverage probabilities (for the lower bound on the left and the upper bound on the right), comparing the Agresti–Coull method (top) with the Jeffreys method (bottom) for $n = 20$. Note that the plots for the lower confidence

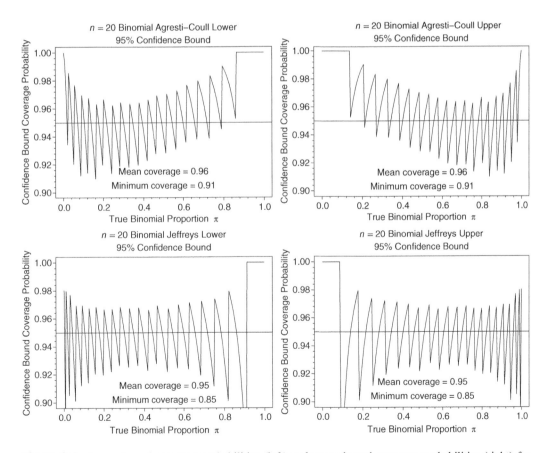

Figure 6.4 Lower bound coverage probabilities (left) and upper bound coverage probabilities (right) for nominal one-sided 95% confidence bounds for the binomial distribution parameter π constructed using the Agresti–Coull (top) and Jeffreys (bottom) confidence bound methods with $n = 20$.

bound on the left are mirror images of the plots for the upper confidence bound on the right. This comparison shows that the Jeffreys method has an important advantage over the Agresti–Coull interval of providing better balance between the lower and upper coverage probabilities. In particular, the Agresti–Coull method one-sided lower (upper) confidence bounds tend to be nonconservative for small (large) values of π. As explained in Section 2.7, it is usually desirable to have the kind of balance that the Jeffreys method provides (i.e., no trend with respect to the actual parameter value).

The Jeffreys upper confidence bound coverage has a minimum of 0.85 when π is in the neighborhood of 0.10. This is because, with a sample size of $n = 20$, the expected number nonconforming is close to 2 and thus there is little information in the data and desirable large-sample properties are not present. In such situations, it may be preferable to use the conservative method, but the upper bound will then be considerably larger. The behavior is similar for the lower bound when π is in the neighborhood of 0.90.

The results of applying each of the four methods to the nonconforming engines and defective integrated circuits were summarized in Tables 6.1 and 6.2, respectively. For the integrated circuit example, the four methods each yielded similar confidence intervals and bounds. This is as expected, in light of the relatively large number of nonconforming units in the sample data (i.e., $x = 20$ from among $n = 1,000$ engines).

The preceding result is in sharp contrast to that for the engine example. In this application there was only one defective unit in the sample of 10 engines (i.e., $x = 1$, $n = 10$) and the four methods yielded appreciably different 95% confidence intervals and bounds. Moreover, the Wald and Agresti–Coull methods both resulted in negative lower bounds. Such nonsensical values are not possible using the conservative and Jeffreys methods, both of which resulted in lower (positive) confidence bounds close to 0. In addition, the Wald method gave appreciably lower values for the upper confidence bound than the other three methods. Further evaluations suggest a consistently similar "shift" to larger lower confidence bounds when π is estimated to exceed 0.5. This shift is corrected by the Agresti–Coull method, resulting in improved coverage probabilities.

The purpose of this discussion has been to make readers further aware of the performance of four commonly used and/or theoretically useful methods for constructing confidence intervals and bounds on the binomial distribution parameter π—a frequently encountered problem. In practice, we would follow our own advice for both examples—that is, decide on the method for constructing confidence intervals or bounds before seeing the data. This would lead us to use either the conservative method or the Jeffreys method, unless there is some special reason, per our discussion, to do otherwise.

Finally, we note that each of the four methods gave appreciably wider 95% confidence intervals and larger upper 95% confidence bounds than the interval $[\underset{\sim}{\pi}, \ \widetilde{\pi}] = [0.0026, \ 0.227]$ and upper bound of $\widetilde{\pi} = 0.18$ that would be obtained using the original measurements ($n = 10$, $\bar{x} = 8.05$, and $s = 1.09$), assuming a normal distribution and the methods presented in Section 4.5. This is the price we pay for dichotomizing the data.

6.3 CONFIDENCE INTERVAL FOR THE PROPORTION OF NONCONFORMING UNITS IN A FINITE POPULATION

In some applications it is desired to obtain a confidence interval for the proportion noncon-forming when sampling from a small population, such as a batch of a manufactured product, and estimates are needed for particular batches. To do this, we obtain a $100(1 - \alpha)\%$ confidence interval for D, the number of nonconforming units in a batch of N units, based upon having x nonconforming units in a random sample of n units taken from the batch without replacement. The resulting interval can then be translated to a confidence interval for the proportion $\pi = D/N$ of nonconforming units in the batch.

The methods presented in this section apply for any values of sample size n and population size N. When the sample proportion n/N is less than 0.10, the simpler methods presented in Section 6.2 generally provide a good approximation to the methods presented in this section. Thus, the methods presented in this section are principally needed for situations in which n/N is greater than 0.10.

6.3.1 The Conservative Method

The distribution of X, the number of nonconforming units, is hypergeometric (i.e., $X \sim \text{HYPER}(n, D, N)$), as described in Section C.4.5, and the method to obtain a confidence interval for D is based on this distribution. That is, a conservative $100(1 - \alpha)\%$ confidence interval $[\underset{\sim}{D}, \widetilde{D}]$ for D can be obtained by finding the smallest value $\underset{\sim}{D}$ and the largest value $\widetilde{D}$ such that

$$1 - \text{phyper}(x - 1; n, \underset{\sim}{D}, N) > \frac{\alpha}{2} \quad \text{and} \quad \text{phyper}(x; n, \widetilde{D}, N) > \frac{\alpha}{2}, \qquad (6.6)$$

where $\underset{\sim}{D} = 0$ if $x = 0$, $\widetilde{D} = N - n$ if $x = n$, and $\text{phyper}(\cdot)$ is the hypergeometric cumulative distribution function, defined in Section C.4.5. One-sided lower and upper $100(1 - \alpha)\%$ confidence bounds for Y are obtained by replacing $\alpha/2$ with α in the lower and upper endpoints of (6.6), respectively. The conservative method can be derived by inverting a significance test for the number of nonconforming units in a finite population or by using the equivalent method outlined in Section D.6.2.

Example 6.7   Confidence Interval for the Number of Nonconforming Units in a Batch. A product is shipped in batches of size $N = 1,000$ units. A customer desires a confidence interval for D, the number of nonconforming units in a specific batch. A random sample, without replacement, of size $n = 200$ from the batch was inspected and $x = 4$ defects were found. The conservative method using (6.6) yields $[\underset{\sim}{D}, \widetilde{D}] = [7, \ 47]$ for a 95% confidence interval for D and $\widetilde{Y} = 42$ for an upper 95% confidence bound.

Using (6.6) and R as a calculator with function `phyper2` (from R package `StatInt`), trying values of D between 5 and 9 gives

```
> D.try <-  5:9
> 1-phyper2(q=4-1, size=200, D=D.try, N=1000)
[1] 0.00658 0.01665 0.03283 0.05555 0.08471
```

to be compared with 0.025. So the smallest value $\underset{\sim}{D}$ meeting the left-hand restriction in (6.6) is 7. Similarly, trying values of D between 45 and 49 gives

```
> D.try <- 45:49
> phyper2(q=4, size=200, D=D.try, N=1000)
[1] 0.0351 0.0303 0.0260 0.0224 0.0192
```

again to be compared with 0.025. So the largest value $\widetilde{D}$ meeting the right-hand restriction (6.6) is 47. The interval for D can be translated into an interval for π:

$$[\underset{\sim}{\pi}, \ \widetilde{\pi}] = \left[\frac{\underset{\sim}{D}}{N}, \frac{\widetilde{D}}{N}\right] = \left[\frac{7}{1,000}, \frac{47}{1,000}\right] = [0.007, \ 0.047].$$

■

6.3.2   Large-Population Approximate Method

If the population size N is large enough (relative to n), the methods for proportions based on the binomial distribution (described in Section 6.2) apply. It is commonly suggested that the approximation will be adequate if $n/N < 0.10$; see also the discussion on infinite population assumptions in Section 1.10.

Example 6.8   Approximate Confidence Interval for the Number of Nonconforming Units in a Batch. In Example 6.7, $n/N = 200/1{,}000 = 0.20$ and, therefore applying any one of the four methods presented in Section 6.2 would be questionable. We do so only for illustrative purposes. It requires first obtaining a confidence interval for $\pi = D/N$ by one of the four methods and then converting this to a confidence interval for D. In particular, using the conservative method in Section 6.2.2 gives $[\underaccent{\tilde}{\pi},\ \widetilde{\pi}] = [0.0055,\ 0.0504]$. Then $[\underaccent{\tilde}{D},\ \widetilde{D}] = 1{,}000 \times [\underaccent{\tilde}{\pi},\ \widetilde{\pi}] = [5,\ 50]$. Using R as a calculator gives

```
>  1000*c(qbeta(p=0.025, shape1=4, shape2=197), qbeta(p=0.975,
   shape1=5, shape2=196))
[1]   5.48 50.41
```

for the approximate 95% confidence interval. Because the sample size in this application is 20% of the population (more than the 10% commonly suggested for using the large-population approximation), we note that a 95% confidence interval for D is $[5,\ 50]$ is somewhat wider than the interval $[7,\ 47]$ obtained by the more appropriate, already conservative, approach used in Example 6.7.   ■

6.4   CONFIDENCE INTERVALS FOR THE PROBABILITY THAT THE NUMBER OF NONCONFORMING UNITS IN A SAMPLE IS LESS THAN OR EQUAL TO (OR GREATER THAN) A SPECIFIED NUMBER

Some applications require inferences concerning the probability that Y, the number of nonconforming units in a sample of size m, is less than or equal to (or greater than) some specified number y. For example, units may be selected at random from a production process and put in packages of size m. Based on the information in a previous random sample of size n, the manufacturer wants to find a confidence interval for the probability p_{LE} that the number of nonconforming units in a sample of size m is less than or equal to y, where y is a prespecified number (which could be any integer from 0 to m).

   If the proportion of nonconforming units from the distribution were *known* to be π, the probability p_{LE} that Y, the number of nonconforming units in a sample of size m, will be less than or equal to a prespecified number y is computed from the binomial cumulative distribution function as

$$p_{LE} = \Pr(Y \le y) = \texttt{pbinom}(y; m, \pi), \tag{6.7}$$

where $\texttt{pbinom}(y; m, \pi)$ is the binomial cdf defined in Sections 6.1.1 and C.4.1. Usually, π is unknown and only sample data on the number of nonconforming units x in the previous sample of size n are available. Because p_{LE} is a decreasing function of π (see (C.20) in Section C.4.1), the following two-step procedure is used to find an approximate two-sided confidence interval for p_{LE}:

1. Obtain a two-sided confidence interval for π, based on the data, using one of the methods given in Section 6.2.

2. Substitute these values for π into (6.7) to obtain the desired two-sided confidence interval for p_{LE}.

If $[\underset{\sim}{\pi}, \ \widetilde{\pi}]$ is a two-sided $100(1 - \alpha)\%$ confidence interval for π, a two-sided $100(1 - \alpha)\%$ confidence interval for p_{LE} is

$$[\underset{\sim}{p_{LE}}, \ \widetilde{p}_{LE}] = [\texttt{pbinom}(y; m, \widetilde{\pi}), \ \texttt{pbinom}(y; m, \underset{\sim}{\pi})].$$

Similarly, if the proportion of nonconforming units from the distribution is known to equal π, the probability p_{GT} that Y, the number of nonconforming units in a sample of size m, will be greater than y is the complement of the binomial cumulative distribution function,

$$p_{GT} = 1 - p_{LE} = \Pr(Y > y) = 1 - \texttt{pbinom}(y; m, \pi).$$

When π is unknown, because p_{GT} is an increasing function of π, a $100(1 - \alpha)\%$ confidence interval for p_{GT} is

$$[\underset{\sim}{p_{GT}}, \ \widetilde{p}_{GT}] = [1 - \widetilde{p}_{LE}, \ 1 - \underset{\sim}{p_{LE}}] = [1 - \texttt{pbinom}(y; m, \underset{\sim}{\pi}), \ 1 - \texttt{pbinom}(y; m, \widetilde{\pi})].$$

Because the function $\texttt{pbinom}(y; m, \pi)$ is a continuous monotone function of the parameter π, the coverage properties of each of the preceding methods is exactly the same as that for the corresponding confidence interval method (from Section 6.2) used to obtain a confidence interval for π.

Example 6.9 Confidence Interval for the Probability of Two or Fewer Defective Integrated Circuits in a Package of Size 50. Suppose that the manufacturer of integrated circuits in Example 6.1 ships packages of 50 units. A 95% confidence interval is desired for the probability that a package will have two or fewer nonconforming units, or, equivalently, for the proportion of packages that will have two or fewer nonconforming units. Using, for example, the conservative 95% confidence interval $[\underset{\sim}{\pi}, \ \widetilde{\pi}] = [0.0123, \ 0.0307]$ from Table 6.2, the desired conservative 95% confidence interval for $p_{LE} = \Pr(Y \leq 2)$ is

$$\begin{aligned}[\underset{\sim}{p_{LE}}, \ \widetilde{p}_{LE}] &= [\texttt{pbinom}(2; 50, 0.0307), \ \texttt{pbinom}(2; 50, 0.0123)] \\ &= [0.80, \ 0.98].\end{aligned}$$

Thus, we are (at least) 95% confident that between 80% and 98% of such packages will have no more than two defective units.

A one-sided lower conservative 95% confidence bound for p_{LE}, using the previously obtained one-sided upper conservative 95% confidence bound $\widetilde{\pi} = 0.0289$, is

$$\underset{\sim}{p_{LE}} = \texttt{pbinom}(2; 50, 0.0289) = 0.82.$$

Thus, we are 95% confident that at least 82% of the packages of 50 units will have no more than two defective units.

Using R as a calculator and the conservative confidence interval for π from Section 6.2.2 gives

```
> pbinom(q=2, size=50, c(qbeta(p=0.975, shape1=21, shape2=980),
        qbeta(p=0.025, shape1=20, shape2=981)))
[1]   0.8016 0.9765
> pbinom(2, size=50, qbeta(0.95, 21, 980))
[1]   0.8242
```

for the conservative 95% confidence interval and the one-sided lower conservative 95% confidence bound. Similarly, using the Jeffreys confidence interval from Section 6.2.5 gives

```
> pbinom(2, size=50, qbeta(p=c(0.975, 0.025), shape1=20.5,
shape2=980.5))
[1]    0.8093 0.9745
> pbinom(2,size=50, qbeta(p=0.95, shape1=20.5, shape2=980.5))
[1] 0.8314
```

for the approximate 95% confidence interval and one-sided bound. The two-sided results are summarized in Table 6.2.

This example depends heavily on the assumption that π, the proportion nonconforming, is constant over time (i.e., that the production process is "in control"). If the process produces varying proportions of nonconforming units during differing periods of time this assumption is not met, and the preceding confidence interval or bound (as well as the other statistical intervals presented in this chapter) could be misleading. ■

6.5   CONFIDENCE INTERVALS FOR THE QUANTILE OF THE DISTRIBUTION OF THE NUMBER OF NONCONFORMING UNITS

Some applications require, based on the available data, a confidence interval (or a one-sided confidence bound), for a quantile, y_p, of the distribution of Y, the number of nonconforming units in samples of m units. For example, suppose that units from a production process are put in packages of size m and it is desired to make a statement about a value y_p that exceeds a certain proportion p of these packages, based on a past random sample of size n. This problem is the inverse of the one in Section 6.4; there y was specified and it was desired to obtain a confidence interval for the probability $p_{LE} = \Pr(Y \leq y)$ of having y or fewer nonconforming units in samples (e.g., packages) of size m.

The p quantile y_p of a binomial distribution is defined as the smallest value of y such that $\Pr(Y \leq y) = \texttt{pbinom}(y; m, \pi) \geq p$ and is denoted by $\texttt{qbinom}(p; m, \pi)$. If the proportion of nonconforming units from the process were *known* to be π, then the desired quantile could be computed directly from $\texttt{qbinom}(p; m, \pi)$. Usually, however, π is unknown and only sample data on the number of nonconforming units x in the previous sample of size n are available.

6.5.1   Two-Sided Confidence Interval for y_p

In a manner similar to that used in Section 6.4, because $\texttt{qbinom}(p; m, \pi)$ is a nondecreasing function of π, the following two-step procedure is used to find a two-sided confidence interval for y_p:

1. Obtain a two-sided confidence interval for π, based on the data, using one of the methods given in Section 6.2.

2. Substitute these values for π into $\texttt{qbinom}(p; m, \pi)$ to obtain the desired two-sided confidence interval for y_p.

Thus, if $[\underset{\sim}{\pi}, \ \widetilde{\pi}]$ is a two-sided approximate (or conservative) $100(1 - \alpha)\%$ confidence interval for π, a two-sided approximate (or conservative) $100(1 - \alpha)\%$ confidence interval for y_p is

$$[\underset{\sim}{y_p}, \ \widetilde{y}_p] = [\texttt{qbinom}(p; m, \underset{\sim}{\pi}), \ \texttt{qbinom}(p; m, \widetilde{\pi})]. \tag{6.8}$$

Unlike the confidence interval procedure for binomial probabilities in Section 6.4, the quantile function is *not* a continuous function of π (it is an integer-valued step function). Therefore, the coverage probability as a function of the actual value of π will not be exactly the same as that for the confidence interval method for π. The general formula (B.5) in Section B.2.4 can, however, be used to do coverage probability evaluations and allow calibration (see Sections 6.6.4 and B.8).

Example 6.10   Confidence Interval for the 0.90 Quantile of the Distribution of the Number of Defective Integrated Circuits in a Package of Size $m = 1,000$. Suppose that the manufacturer of integrated circuits in Example 6.1 ships packages of $m = 1,000$ units. A 95% confidence interval is desired for $y_{0.90}$, the 0.90 quantile of the distribution of the number of defects for such packages. Using, for example, the conservative 95% confidence interval $[\underset{\sim}{\pi}, \ \widetilde{\pi}] = [0.0123, \ 0.0307]$ from Table 6.2, the resulting conservative 95% confidence interval for $y_{0.90}$ is

$$[\underset{\sim}{y}_{0.90}, \ \widetilde{y}_{0.90}] = [\mathtt{qbinom}(0.90; 1000, 0.0123), \ \mathtt{qbinom}(0.90; 1000, 0.0307)]$$

$$= [17, \ 38].$$

Thus, we are at least 95% confident that the 0.90 quantile of the distribution of the number of defects in such packages is between 17 and 38.

Using R as a calculator, the conservative confidence interval for π from Section 6.2.2 gives

```
> qbinom(p=0.90, size=1000, c(qbeta(p=0.025, shape1=20, shape2=981),
          qbeta(p=0.975, shape1=21, shape2=980)))
[1] 17 38
```

as a conservative 95% confidence interval for $y_{0.90}$. Similarly the Jeffreys confidence interval from Section 6.2.5 gives

```
> qbinom(p=0.90, size=1000, qbeta(p=c(0.025, 0.975), shape1=20.5,
shape2=980.5))
[1] 17 37
```

as an approximate 95% confidence interval for $y_{.90}$.

Because of the discreteness of the quantile function and because there is only a small difference between the conservative and Jeffreys methods for obtaining a confidence interval for π, the confidence intervals for the two methods in this example are nearly the same. The preceding results are summarized in Table 6.2. ∎

6.5.2   One-Sided Confidence Bounds for y_p

A one-sided lower (upper) $100(1 - \alpha)\%$ confidence bound for y_p is found by substituting a one-sided lower (upper) $100(1 - \alpha)\%$ confidence bound for π into the appropriate endpoint of the two-sided confidence interval in (6.8). These one-sided confidence bounds are of particular interest because of their relationship to one-sided tolerance bounds and two-sided tolerance intervals, to be described in Section 6.6.

Example 6.11   One-Sided 95% Confidence Bounds for the 0.10 and 0.90 Quantiles of the Distribution of Defects in a Package of Size $m = 1,000$. Using the data from Example 6.1 and the one-sided lower conservative 95% confidence bound $\underset{\sim}{\pi} = 0.0133$ from Table 6.2 and proceeding as in Example 6.10, a one-sided lower conservative 95% confidence bound for $y_{0.10}$ is

$$\underset{\sim}{y}_{0.10} = \mathtt{qbinom}(0.10; 1000, 0.0133) = 9.$$

Similarly, using the one-sided upper conservative 95% confidence bound $\widetilde{\pi} = 0.0289$, a 95% one-sided upper confidence bound for $y_{0.90}$ is

$$\widetilde{y}_{0.90} = \texttt{qbinom}(0.90; 1000, 0.0289) = 36.$$

Using R as a calculator, with the conservative confidence interval method for π from Section 6.2.2 gives the one-sided conservative 95% confidence bounds

```
> qbinom(p=0.10, size=1000, qbeta(p=0.05, shape1=20, shape2=981))
[1] 9
> qbinom(p=0.90, size=1000, qbeta(p=0.95, shape1=21, shape2=980))
[1] 36
```

for $y_{0.10}$. Similarly, using the Jeffreys confidence interval from Section 6.2.5 gives the one-sided approximate 95% confidence bounds

```
> qbinom(p=0.10, size=1000, qbeta(p=0.05, shape1=20.5, shape2=980.5))
[1] 9
> qbinom(p=0.90, size=1000, qbeta(p=0.95, shape1=20.5, shape2=980.5))
[1] 35
```

for $y_{0.10}$. These results are summarized in Table 6.2 and will be used in the next section to construct two-sided tolerance intervals and one-sided tolerance bounds for a binomial distribution. ∎

6.6   TOLERANCE INTERVALS AND ONE-SIDED TOLERANCE BOUNDS FOR THE DISTRIBUTION OF THE NUMBER OF NONCONFORMING UNITS

Some applications require two-sided tolerance intervals or one-sided tolerance bounds for the distribution of Y, the number of nonconforming units in samples of m units. For example, units from a production process are packaged in groups of size m and it is desired find a value of y so that one can state, with a specified degree of confidence, that at least a proportion 0.90 of such packages contain y or fewer nonconforming units. This statement calls for a one-sided lower tolerance bound. Other problems may require a one-sided upper tolerance bound or a two-sided tolerance interval. As indicated in Section 2.4.2, a one-sided tolerance bound is equivalent to a one-sided confidence bound on a quantile of the distribution. Also, approximate two-sided tolerance intervals can be obtained by appropriately combining one-sided lower and upper tolerance bounds.

6.6.1   One-Sided Lower Tolerance Bound for a Binomial Distribution

A one-sided lower $100(1 - \alpha)\%$ tolerance bound $\underset{\sim}{T'_{\beta}}$ to be exceeded by at least a proportion β of the distribution is the same as a lower $100(1 - \alpha)\%$ confidence bound on $y_{(1-\beta)}$, the $1 - \beta$ quantile of the distribution. Thus $\underset{\sim}{T'_{\beta}} = \underset{\sim}{y}_{(1-\beta)}$, which can be computed as described in Section 6.5.2.

Example 6.12   One-Sided Lower Tolerance Bound for the Number of Defective Integrated Circuits. Due to the equivalence described above, the desired one-sided lower conservative 95% tolerance bound to be exceeded by a proportion 0.90 of the distribution of the number of defective integrated circuits is, from Example 6.11, $\underset{\sim}{T'_{0.90}} = \underset{\sim}{y}_{0.10} = 9$. ∎

6.6.2 One-Sided Upper Tolerance Bound for a Binomial Distribution

A one-sided upper $100(1-\alpha)\%$ tolerance bound $\widetilde{T}'_\beta$ to exceed at least a proportion β of the distribution is the same as an upper $100(1-\alpha)\%$ confidence bound on y_β. Thus $\widetilde{T}'_\beta = \widetilde{y}_\beta$, which can be computed as described in Section 6.5.2.

Example 6.13 One-Sided Upper Tolerance Bound for the Number of Defective Integrated Circuits. Due to the equivalence described above, the desired one-sided upper conservative 95% tolerance bound to exceed a proportion 0.90 of the distribution of the number of defective integrated circuits is, from Example 6.11, $\widetilde{T}'_{0.90} = \widetilde{y}_{0.90} = 36$. ∎

6.6.3 Two-Sided Tolerance Interval for a Binomial Distribution

A method of constructing a two-sided approximate tolerance interval $[\underset{\sim}{T}_\beta, \ \widetilde{T}_\beta]$ for a binomial distribution with sample size m is to combine two one-sided tolerance bounds (or one-sided confidence bounds on appropriate quantiles) for the distribution of interest. In particular, as explained in Section D.7.4, to obtain an approximate $100(1-\alpha)\%$ tolerance interval to contain at least a proportion β of the distribution, one can use a one-sided lower $100(1-\alpha)\%$ confidence bound on the $(1-\beta)/2$ quantile for the lower endpoint and a one-sided upper $100(1-\alpha)\%$ confidence bound on the $(1+\beta)/2$ quantile for the upper endpoint. That is, $[\underset{\sim}{T}_\beta, \ \widetilde{T}_\beta] = [\underset{\sim}{y}_{(1-\beta)/2}, \ \widetilde{y}_{(1+\beta)/2}]$.

Example 6.14 Two-Sided Tolerance Interval for the Number of Defective Integrated Circuits. A two-sided approximate 95% tolerance interval to contain at least a proportion 0.80 of the distribution of the number of defective integrated circuits in Example 6.1 is obtained by using a one-sided lower 95% confidence bound on the 0.10 quantile for the lower endpoint and a one-sided upper 95% confidence bound on the 0.90 quantile for the upper endpoint. Taking the results based on the one-sided conservative confidence bounds $\underset{\sim}{y}_{(0.10)}$ and $\widetilde{y}_{(0.90)}$ from Example 6.11 gives $[\underset{\sim}{T}_{0.80}, \ \widetilde{T}_{0.80}] = [\underset{\sim}{y}_{(0.10)}, \ \widetilde{y}_{(0.90)}] = [9, \ 36]$. ∎

6.6.4 Calibrating Tolerance Intervals

Figure 6.5 gives plots of coverage probabilities versus the actual binomial proportion non-conforming π for the conservative and Jeffreys two-sided tolerance interval procedures with different nominal confidence levels. As a specific example, we show the coverage probabilities for these two procedures for the particular combination of $n = 100$, $m = 100$, and $\beta = 0.80$. These plots show the procedures to be consistently, and frequently appreciably, conservative (i.e., the coverage probability exceeds the nominal confidence level) not only for the conservative method, but also (although generally less so) for the Jeffreys method. Similar results, not shown here, hold for other combinations of n, m, and β and for one-sided confidence bounds.

The method of *statistical interval calibration*, discussed in Section B.8, can be used to control the coverage properties for situations in which the coverage probability consistently exceeds or consistently falls below the nominal confidence level. In particular, if an interval procedure tends to be too conservative, one can adjust the procedure by using a smaller nominal confidence level as input. For example, row 3 in Figure 6.5 shows that the tolerance interval procedures based on both the conservative method and the Jeffreys method result in mean coverage probabilities (0.98 and 0.97, respectively) greater than the nominal confidence level 95%. If one desires a procedure that has mean coverage that approximately equals 0.95, one could use the Jeffreys procedure with a nominal confidence level between 90% and 92.5%.

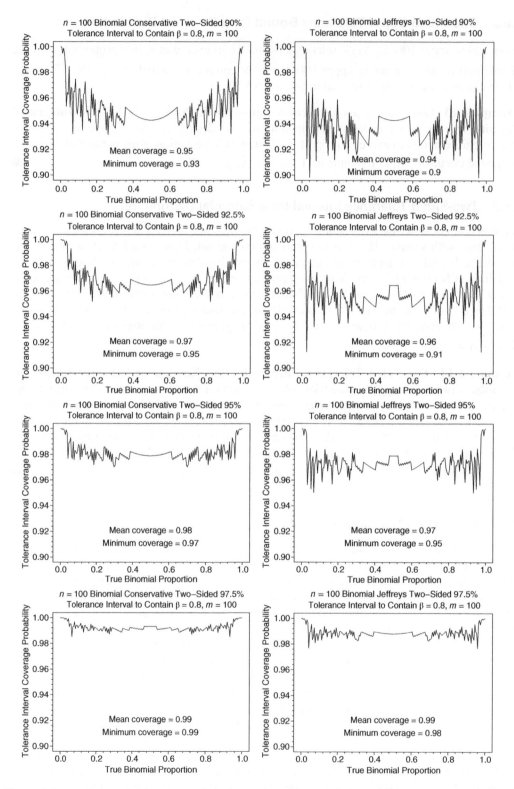

Figure 6.5 Coverage probabilities versus the actual binomial proportion nonconforming π for the conservative (left) and Jeffreys (right) two-sided tolerance interval methods to contain at least a proportion $\beta = 0.80$ for $n = 100$ and $m = 100$ with nominal confidence levels 90% (top row), 92.5% (second row), 95% (third row), and 97.5% (bottom row).

6.7   PREDICTION INTERVALS FOR THE NUMBER NONCONFORMING IN A FUTURE SAMPLE

Suppose, as before, that x nonconforming units have been observed in a random sample of size n from a distribution. From these data, it is desired to find a prediction interval that will, with some specified degree of confidence, contain Y, the number of nonconforming units in a future random sample of size m from the same distribution. We assume that the two samples are independent and that the number of nonconforming units in each sample can be described by a binomial distribution with parameter π.

6.7.1   The Conservative Method

Consider the combined samples of size n and m and let $N = n + m$. Let X and Y denote the nonconforming units in the two samples. Before observing X, the conditional distribution of X, given $D = X + Y$, is HYPER(D, n, N), which (see (C.26)) is equivalent to HYPER(n, D, N). After observing $X = x$, then $D = x + Y$ and the hypergeometric cdf $\mathrm{phyper}(x; n, D, N)$ (defined in Section C.4.5) is nonincreasing with respect to D (as shown in Section C.4.5). Then, using the result in Section 6.3.1, the following procedure gives a conservative $100(1 - \alpha)\%$ confidence interval for D; this, in turn, is converted into a prediction interval for Y, the number of nonconforming units in the future sample of m units.

A conservative $100(1 - \alpha)\%$ prediction interval $[\underset{\sim}{Y}, \ \widetilde{Y}]$ for Y can be obtained by finding the smallest value $\underset{\sim}{Y}$ and the largest value $\widetilde{Y}$ such that

$$1 - \mathrm{phyper}(x - 1; n, x + \underset{\sim}{Y}, N) > \frac{\alpha}{2} \quad \text{and} \quad \mathrm{phyper}(x; n, x + \widetilde{Y}, N) > \frac{\alpha}{2}, \quad (6.9)$$

where $\underset{\sim}{Y} = 0$ if $x = 0$ and $\widetilde{Y} = m$ if $x = n$. The prediction interval is conservative because it is based on a conservative confidence interval method for D. One-sided lower and upper conservative $100(1 - \alpha)\%$ prediction bounds for Y are obtained by replacing $\alpha/2$ with α in the lower and upper endpoints of (6.9), respectively.

Example 6.15   Conservative Prediction Interval for the Number of Defective Integrated Circuits. Suppose that, based on the $x = 20$ nonconforming integrated circuits from $n = 1,000$ randomly selected units, the manufacturer desires a 95% prediction interval to contain the number of nonconforming units in a future sample of $m = 1,000$ randomly sampled units from the same production process. The conservative method given by (6.9) gives $[\underset{\sim}{Y}, \ \widetilde{Y}] = [9, \ 35]$ for a conservative 95% prediction interval and $\widetilde{Y} = 32$ for an upper conservative 95% prediction bound for Y. Thus, based on $x = 20$ nonconforming units from the $n = 1,000$ sample units, one can, for example, assert, with (at least) 95% confidence, that the number of nonconforming units in the future sample of $m = 1,000$ will not exceed 32 units.

The hypergeometric probabilities needed for the preceding prediction interval or bound can be computed using R as a calculator with function `phyper2` from R package `StatInt`. Following (6.9), we try values of Y between 7 and 11:

```
> Y.try <- 7:11
> 1-phyper2(q=20-1, size=1000, D=20+Y.try, N=1000+1000)
[1] 0.0092 0.0172 0.0298 0.0481 0.0732
```

to be compared with 0.025. So the smallest value $\underset{\sim}{Y}$ meeting the left-hand restriction in (6.9) is 9. Similarly, trying values of Y between 33 and 37 gives

```
> Y.try <- 33:37
> phyper2(q=20, size=1000, D=20+Y.try, N=1000+1000)
[1]  0.0470 0.0360 0.0273 0.0206 0.0153
```

also to be compared with 0.025. So the largest value $\widetilde{Y}$ meeting the right-hand restriction (6.9) is 35.   ∎

6.7.2   The Normal Distribution Approximation Method

A large-sample approximate $100(1 - \alpha)\%$ prediction interval for Y, based on the assumption that

$$\frac{m\widehat{\pi} - Y}{\sqrt{\widehat{\mathrm{Var}}(m\widehat{\pi} - Y)}} = \frac{m\widehat{\pi} - Y}{\sqrt{m\widehat{\pi}(1 - \widehat{\pi})(1 + m/n)}} \tag{6.10}$$

can be adequately approximated by a $\mathrm{NORM}(0, 1)$ distribution, is

$$[\underset{\sim}{Y}, \ \widetilde{Y}] = m\widehat{\pi} \mp z_{(1-\alpha/2)}\left[m\widehat{\pi}(1 - \widehat{\pi})\frac{m + n}{n}\right]^{1/2}. \tag{6.11}$$

Because the coverage probabilities of this method can be nonconservative (depending on π, n, and m), we recommend rounding in a conservative manner (i.e., round down for the lower endpoint and round up for the upper endpoint). This approximate interval is easier to compute than the conservative method given by (6.9). Again, one-sided lower and upper $100(1 - \alpha)\%$ prediction bounds for Y are obtained by replacing $\alpha/2$ with α in the lower and upper endpoints of (6.11), respectively. The interval in (6.11) is not defined if $x = 0$ (or $x = n$). In such cases one can still use (6.11) by taking $x = 0.5$ (or $x = n - 0.5$), although in either of these cases the assumption that the expressions in (6.10) are approximately $\mathrm{NORM}(0, 1)$ would not be valid.

Example 6.16   Normal Distribution Approximation Prediction Interval for the Number of Defective Integrated Circuits. For the problem considered in Example 6.15, the approximate 95% prediction interval for Y, using the normal distribution approximation (6.11), is

$$[\underset{\sim}{Y}, \ \widetilde{Y}] = 1{,}000 \times 0.02 \mp 1.96 \times \left[1{,}000 \times 0.02 \times 0.98 \times \frac{2{,}000}{1{,}000}\right]^{1/2} = [7.7, \ \ 32.3],$$

which, when rounded in the conservative manner, gives $[7, \ \ 33]$. This differs somewhat, but not appreciably, from the previously computed prediction interval using the conservative method. Using R as a calculator gives

```
> 1000*0.02 + c(-1, 1)*qnorm(0.975)*sqrt(1000*0.02*0.98*2000/1000)
[1]  7.7 32.3
```

for the approximate prediction interval.   ∎

One-sided prediction bounds can be obtained similarly.

6.7.3 The Joint-Sample Approximate Method

Let $\widehat{\pi}_{xy} = (X + Y)/(n + m)$. A large-sample approximate $100(1 - \alpha)\%$ prediction interval for Y can be based on the assumption that

$$\frac{m\widehat{\pi}_{xy} - Y}{\sqrt{\widehat{\text{Var}}(m\widehat{\pi}_{xy} - Y)}} = \frac{mX - nY}{\sqrt{mn\widehat{\pi}_{xy}(1 - \widehat{\pi}_{xy})(n + m)}} \tag{6.12}$$

can be adequately approximated by a $\text{NORM}(0, 1)$ distribution. Squaring the right-hand side of (6.12), setting it equal to $z^2_{(1-\alpha/2)}$ and solving the resulting quadratic equation for the two roots in Y gives the interval

$$[\underset{\sim}{Y},\ \widetilde{Y}] = \frac{\left[\widehat{Y}\left(1 - \frac{z^2_{(1-\alpha/2)}}{m+n}\right) + \frac{m z^2_{(1-\alpha/2)}}{2n}\right] \mp z_{(1-\alpha/2)}\sqrt{\widehat{Y}(m - \widehat{Y})\left(\frac{1}{m} + \frac{1}{n}\right) + \left(\frac{m z_{(1-\alpha/2)}}{2n}\right)^2}}{1 + \frac{m z^2_{(1-\alpha/2)}}{n(m+n)}},$$

$$\tag{6.13}$$

where $\widehat{Y} = mx/n$ and it is best (in terms of coverage probability properties) to round the lower endpoint upward and to round the upper endpoint downward (which we call the *nonconservative* manner of rounding). In such cases one can still use (6.11) by taking $x = 0.5$ (or $x = n - 0.5$), although in either of these cases the assumption that the expressions in (6.13) are approximately $\text{NORM}(0, 1)$ would not be valid.

This approximate interval is more complicated than the normal-approximation method in (6.11), but the coverage probability properties are much better, as described in Section 6.7.5. Again, one-sided lower and upper $100(1 - \alpha)\%$ prediction bounds for Y are obtained by replacing $\alpha/2$ with α in the lower and upper endpoints of (6.13), respectively.

Example 6.17 Joint-Sample Approximate Prediction Interval for the Number of Defective Integrated Circuits. For the application considered in Example 6.15, the 95% prediction interval for Y using the joint-sample approximation (6.13) is

$$[\underset{\sim}{Y},\ \widetilde{Y}] = \frac{\left[20\left(1 - \frac{1.96^2}{2,000}\right) + \frac{1,000\times1.96^2}{2,000}\right] \mp 1.96\sqrt{20(1,000 - 20)\frac{2}{1,000} + \left(\frac{1,000\times1.96}{2,000}\right)^2}}{1 + \frac{1,000\times1.96^2}{1,000\times2,000}}$$

$$= [9.44,\ 34.2],$$

which when rounded in the nonconservative manner gives $[10,\ 34]$. This compares reasonably well with the intervals based on the other methods.

Using R as a calculator gives

```
> Yhat <- 20
> zvalue <- qnorm(0.975)
> YhatFactor <- (1-zvalue^2/(2000))
> YhatAdd <- 1000*zvalue^2/(2000)
> varAdd <- ((1000*zvalue)/(2000))^2
> denom <- 1+1000*zvalue^2/(1000*2000)
> YhatCenter <- Yhat*YhatFactor+YhatAdd
> YhatSE <- sqrt(Yhat*(1000-Yhat)*(2/1000)+varAdd)
```

```
> (YhatCenter+c(-1,1)*zvalue*YhatSE)/denom
[1]   9.44 34.24
```

for the approximate prediction interval.   ∎

One-sided lower or upper prediction bounds can be obtained similarly by replacing $1 - \alpha/2$ with $1 - \alpha$ and using the appropriate interval endpoint.

6.7.4   The Jeffreys Method

The Jeffreys method for obtaining a binomial distribution prediction interval is an extension of the Jeffreys method for obtaining a confidence interval for π, given in Section 6.2.5, and is based on quantiles of the beta-binomial distribution, which can be viewed as a Bayesian binomial predictive distribution (see Section H.6.1 for technical details). Thus, given x observed nonconforming units in a sample of size n, a two-sided approximate $100(1 - \alpha)\%$ Jeffreys prediction interval for Y is

$$[\underset{\sim}{Y}, \ \widetilde{Y}] = [\text{qbetabinom}(\alpha/2; m, x + 0.5, n - x + 0.5),$$
$$\text{qbetabinom}(1 - \alpha/2; m, x + 0.5, n - x + 0.5)], \qquad (6.14)$$

where $\text{qbetabinom}(p; n, a, b)$ is the p quantile of the beta-binomial distribution (see Section C.4.2) with sample-size parameter n and shape parameters a and b.

Example 6.18   Jeffreys Prediction Interval for the Number of Defective Integrated Circuits. For the problem in Examples 6.15 and 6.16 a Jeffreys approximate 95% prediction interval for Y using (6.14) is

$$[\underset{\sim}{Y}, \ \widetilde{Y}] = [\text{qbetabinom}(0.025; 1000, 20.5, 980.5),$$
$$\text{qbetabinom}(0.975; 1000, 20.5, 980.5)] = [10, \quad 34].$$

Using R as a calculator, the beta-binomial quantile function qbetabinom (in R package StatInt) gives

```
> qbetabinom(p=c(0.025, 0.975), size=1000, shape1=20+0.50,
shape2=1000-20+0.50)
[1] 10 34
```

for the approximate prediction interval. This interval again differs, but not appreciably, from the intervals obtained using the two previous methods.   ∎

6.7.5   Comparisons and Recommendations

The recommendations in this section are based on evaluations of coverage probabilities that are not shown here. Equation (B.24) in Section B.6.4 provides an expression for computing the coverage probabilities.

If having at least the nominal level of confidence is important (e.g., to avoid arguments about the use of approximations), the conservative method should be used. The conservative method, however, tends to be overly conservative, resulting in prediction intervals that are excessively wide, especially when any of x, $n - x$, $m\widehat{\pi}$, or $m(1 - \widehat{\pi})$ is small (e.g., less than 10).

Computations have shown that the joint-sample interval has a coverage probability close to the nominal confidence level as long as x, $n - x$, $m\widehat{\pi}$, and $m(1 - \widehat{\pi})$ are all larger than 5. The Jeffreys method tends to be a little more conservative than the joint-sample method. Thus we recommend the joint-sample method when an approximate prediction interval method (as opposed to the conservative method) is acceptable. If an easy-to-compute method is desired (e.g., because appropriate software is not readily available), the normal-approximation method can be used as a crude approximation, but should otherwise be avoided. More generally, we would recommend comparing the coverage properties of different intervals for particular situations (before looking at the data) to help decide which method to use.

BIBLIOGRAPHIC NOTES

Basic textbooks on probability, such as Ross (2012) and Ross (2014), provide detailed treatments of the properties and applications of the binomial distribution.

Below we present references that give original sources or further details about the various methods that we have presented in this chapter. We also provide references that present yet additional methods for calculating binomial distribution confidence intervals. We do not discuss these further methods here for space considerations and because we feel that the intervals that we do present provide readers ample useful choices.

Confidence intervals

Numerous methods have been suggested to compute confidence intervals for a binomial distribution proportion. The conservative method is due to Clopper and Pearson (1934) who presented charts similar to those in Figure 6.1. Odeh and Owen (1983) provide extensive tabulations for these binomial confidence intervals with values of n ranging from 20 to 1,000.

Fujino (1980) recognized the poor performance of the Wald (normal distribution approximation) method and suggested a simple correction, depending on the desired confidence level, that is similar to the "add two" method later suggested by Agresti and Coull (1998). Blyth and Still (1983) and Blyth (1986) did further early work comparing different methods and suggesting improved alternative methods.

Leemis and Trivedi (1996) review the literature on binomial confidence interval methods and compare the accuracy of two approximate methods (the Wald method and a method based on the Poisson distribution). Newcombe (1998) compared the properties of seven different methods of computing binomial confidence intervals. Agresti and Coull (1998) also compared several methods and showed that the most commonly used methods (at the time) had poor coverage performance. In particular, the Wald method has erratic coverage properties that can be far from the nominal value and the conservative method can be extremely conservative (resulting in overly wide confidence intervals). They then recommended an alternative method, based on the inversion of the score test, first proposed by Wilson (1927). The formula for this method is, however, considerably more complicated than the Wald approximation. So they suggested, as a compromise, the "add two" modification for 95% confidence intervals (add two successes and two failures to the data) of the Wald method. Brown et al. (2001) defined the closely related method that we give in Section 6.2.4 and called it the Agresti–Coull method. This method is motivated by having the same center point as the Wilson score method.

Brown et al. (2001) provide a detailed overview and comparison of different methods for constructing binomial confidence intervals. They confirm the poor performance of the Wald method earlier reported by Agresti and Coull (1998) and others and recommend the Wilson score method and the Jeffreys method. The Jeffreys method given in Section 6.2.5 is derived from a Bayesian method based on what is known as a Jeffreys prior distribution (see Sections 16.1.2 and H.4.1, respectively, for more applications of Bayesian methods for the binomial distribution and technical details). Brown et al. (2002) use asymptotic expansions for coverage probabilities to compare the performance of five different binomial confidence interval methods. Their approach confirms the conclusions of Brown et al. (2001). Brown et al. (2003) extend the results of Brown et al. (2002) to other members of the exponential family of distributions (Poisson and negative binomial).

Cai (2005) focuses on the development of methods to evaluate and construct confidence interval methods that have good one-sided properties. This is in line with our philosophy of recommending two-sided confidence interval methods that have nearly equal error probabilities for each side of the interval (see Section 2.7). The results show that the Jeffreys method is far superior to the Wilson score method in this important regard and thus the Jeffreys method is especially recommended when computing one-sided confidence bounds.

Confidence intervals when sampling from a finite population

Chung and De Lury (1950) give charts for obtaining conservative confidence intervals for a proportion when sampling from a finite population for N larger than 500. Katz (1953) provided alternative approximate methods that were used at the time (when tables or computer routines for hypergeometric probabilities did not exist). The conservative method is also presented by Konijn (1973) and Tomsky et al. (1979) who provide tables for population sizes N from 2 to 100, for various values of the observed number of nonconforming units (x in our notation) and sample sizes (n in our notation). Odeh and Owen (1983) provide tables for $N = 400(200)2,000$. Buonaccorsi (1987) investigates and compares the conservative method for constructing such confidence intervals with an alternative method that he shows to be inferior.

Tolerance intervals

Tolerance intervals and one-sided tolerance bounds for the binomial (and Poisson) distribution were first given by Hahn and Chandra (1981) and are also presented in Krishnamoorthy and Mathew (2009). Wang and Tsung (2009) describe methods for evaluating the coverage probabilities of binomial (and Poisson) distribution tolerance intervals and present algorithms for developing methods with improved coverage probability properties. In particular, evaluation of the coverage properties of the Hahn and Chandra (1981) method shows that the method is highly conservative. Cai and Wang (2009) present methods for constructing one-sided approximate tolerance bounds (equivalent to one-sided confidence bounds on a quantile) for discrete distributions based on "probability matching" that use high-order approximations for the coverage probability and match these to the desired confidence level. Then they show how to combine two one-sided tolerance bounds to provide two-sided approximate tolerance intervals. Their method tends to have coverage probabilities that are closer to the nominal confidence level than those of other proposed methods. Finally, Krishnamoorthy et al. (2011) provide a simple approximate method for constructing binomial (and Poisson) distribution tolerance intervals.

Prediction intervals

Thatcher (1964) suggested the conservative method based on the hypergeometric distribution in (6.9). Hahn and Nelson (1973) suggested the method in (6.11) based on a normal distribution approximation. The Jeffreys prediction interval is derived from a Bayesian method, based on the posterior predictive distribution of the number of nonconforming units in a future sample of size m, assuming a Jeffreys prior distribution for π, as described in Sections 16.1.5 and H.6.1. Wang (2008) shows how to compute the coverage probabilities of prediction interval procedures for discrete distributions. Krishnamoorthy and Peng (2011) describe earlier methods, suggest other simple closed-form methods for prediction intervals (especially the joint-sample method) for the binomial (and Poisson) distributions, and compare these methods with respect to their coverage probability properties.

Chapter 7

Statistical Intervals for a Poisson Distribution

OBJECTIVES AND OVERVIEW

This chapter describes statistical intervals for the number of events over some interval of time or region of space, assuming independent events and a constant event-occurrence rate. Such situations can often be modeled by the Poisson distribution. For example, the Poisson distribution might provide an adequate description of the number of flaws on the surface of a product. This would require (among other technical conditions) that the product units are all of the same size and that flaws occur at random and independently of each other at a constant rate λ. Similarly, the number of unscheduled shutdowns of a computer system over some specified period of time might be described by a Poisson distribution. This would require that unscheduled shutdowns occur independently of one another and that the event-occurrence rate be constant over time and from one system to another. This assumption would not be correct if, for example, environmental factors that cause failure (e.g., lightning) simultaneously affect more than one system or if the failure rate changes with time (this might be the case if some system components are subject to wearout). In this chapter, our discussion will frequently be in terms of x, the number of events (e.g., unscheduled shutdowns of a computer system) in a given time interval of length n. We could similarly have discussed variables that describe events over constant length, area, or volume, such as the number of defects per foot of wire or the number of flaws per square meter of a finished surface.

The following topics are discussed in this chapter:

- Confidence intervals for λ, the (actual) event-occurrence rate of the sampled population or process (Section 7.2).

- Confidence intervals for the probability that the number of events in a specified amount of exposure will be less than or equal to (or greater than) a specified number (Section 7.3).

Statistical Intervals: A Guide for Practitioners and Researchers, Second Edition.
William Q. Meeker, Gerald J. Hahn and Luis A. Escobar.
© 2017 John Wiley & Sons, Inc. Published 2017 by John Wiley & Sons, Inc.
Companion Website: www.wiley.com/go/meeker/intervals

- Confidence intervals for a quantile of the distribution of the number of events in a specified amount of exposure (Section 7.4).

- Tolerance intervals and one-sided tolerance bounds for the distribution of the number of events in a specified amount of exposure (Section 7.5).

- Prediction intervals for the number of events in a future amount of exposure (Section 7.6).

The preceding are analogous to most of the topics that were considered in Chapter 6 for statistical intervals for proportions and percentages (binomial distribution) and the discussion that follows resembles that of the previous chapter.

7.1   INTRODUCTION

7.1.1   The Poisson Distribution

As indicated, problems involving the number of occurrences of independent, randomly occurring events per unit of space or time can often be modeled with the Poisson distribution. The probability function for the Poisson distribution (i.e., the probability of exactly x events), is

$$\Pr(X = x) = \mathtt{dpois}(x; \lambda) = \frac{\exp(-\lambda)\,\lambda^x}{x!},$$

where the parameter λ is the mean event-occurrence rate (i.e., $\mathrm{E}(X)$). Also, if the number of events per unit exposure has a Poisson distribution with occurrence rate λ, the number of events in n units of exposure (where n is not necessarily an integer) has a Poisson distribution with expectation $n\lambda$.

The function

$$\Pr(X \leq x) = \mathtt{ppois}(x; \lambda) = \sum_{i=0}^{x} \frac{\exp(-\lambda)\,\lambda^i}{i!}$$

denotes the cumulative Poisson probability of observing x or fewer events in a unit of exposure, where λ is the constant event-occurrence rate per unit of exposure. See Section C.4.4 for more technical details about the Poisson distribution.

7.1.2   Poisson Distribution Statistical Interval Properties

In the problems considered here, λ, the Poisson distribution event-occurrence rate per unit of exposure is unknown. Instead, all that is known is that over a past n units of exposure, there were x events of interest. One of our problems, then, is to obtain a confidence interval for λ.

The Poisson, like the binomial, is a discrete distribution. A Poisson random variable can take on the integer values $x = 0, 1, 2, \ldots$ (with no theoretical upper limit). Because the distribution of x is discrete, statistical intervals do not generally have exactly the desired confidence level. Instead, the coverage probabilities of the statistical interval methods given in this chapter depend on the unknown value of the mean number of events $n\lambda$ (as illustrated in Section 7.2.6). Thus, the statistical intervals given here are either *conservative* (i.e., the coverage probability confidence level is larger than the nominal confidence level) or *approximate*, depending on the method that is used. There are numerous methods for computing Poisson distribution statistical intervals. In Section 7.2.1 we will present and illustrate four important methods for constructing confidence intervals on λ. References given in the Bibliographic Notes section at the end of this chapter provide further description and evaluation of these and other Poisson statistical interval methods.

Example 7.1   Number of Unscheduled Shutdowns for a Group of Computing Systems. In $n = 5.0$ system-years of operation of a group of computing systems there have been $x = 24$ unscheduled shutdowns. We assume that unscheduled shutdowns occur independently of one another at a constant rate from one system to the next and from one year to the next, according to a Poisson distribution. (Such assumptions can be assessed from the observed unscheduled shutdown times; see Ascher and Feingold, 1984). The mean number of unscheduled shutdowns per year (or unscheduled shutdown rate) is then estimated from the given data as $\widehat{\lambda} = x/n = 24/5.0 = 4.8$. We note that in this example a system-year of operation was taken to be the unit of exposure.   ■

7.2   CONFIDENCE INTERVALS FOR THE EVENT-OCCURRENCE RATE OF A POISSON DISTRIBUTION

7.2.1   Preliminaries

The observed event-occurrence rate $\widehat{\lambda} = x/n$ is a point estimate of the actual event-occurrence rate λ. The estimate $\widehat{\lambda}$, however, differs from λ due to random sampling fluctuations. Thus, one frequently desires to compute, from the sample data, a two-sided confidence interval or a one-sided confidence bound for λ.

This section presents and motivates the use of four of the most commonly used methods for computing confidence intervals for the Poisson distribution parameter λ. The underlying theory for the methods is given in Sections D.6.2, D.5.6, and H.3.2. The Bibliographic Notes section at the end of this chapter gives references to articles that describe and give more technical details for these and other confidence interval procedures for λ. As in previous chapters, we mainly present two-sided confidence intervals; one-sided lower and upper $100(1 - \alpha)\%$ confidence bounds are obtained by replacing $\alpha/2$ with α in the appropriate formula for obtaining a two-sided confidence interval.

Each of the four methods for obtaining a confidence interval for λ presented in this section will be illustrated by the computing system unscheduled shutdown example. Table 7.1 compares and Section 7.2.6 discusses the confidence intervals obtained by applying each of the four methods to this example. Table 7.1 also contains results for other intervals presented later in this chapter.

7.2.2   The Conservative Method

For x observed events in n units of exposure, a two-sided conservative $100(1 - \alpha)\%$ confidence interval for λ is

$$[\underset{\sim}{\lambda}, \ \widetilde{\lambda}] = [\text{qgamma}(\alpha/2; x, n), \ \text{qgamma}(1 - \alpha/2; x + 1, n)] \tag{7.1}$$

$$= \left[\frac{0.5\,\chi^2_{(\alpha/2;2x)}}{n}, \ \frac{0.5\,\chi^2_{(1-\alpha/2;2x+2)}}{n} \right], \tag{7.2}$$

where $\text{qgamma}(p; \alpha, n)$ is the p quantile of the gamma distribution with shape parameter α and rate parameter n (as mentioned in Section C.3.5, the gamma distribution can be parameterized with either a rate parameter or a scale parameter) and $\chi^2_{(p;r)}$ is the p quantile of a chi-square distribution with r degrees of freedom. The lower limit is defined to be $\underset{\sim}{\lambda} = 0$ if $x = 0$. The derivation of this interval (see Section D.6.2) leads directly to the use of the gamma distribution quantile for performing the computation. The equivalent formulas that use chi-square distribution quantiles follow from the close relationship between the gamma distribution and the chi-square distribution (see Section C.3.6). Tables of chi-square quantiles are more readily available, but the gamma quantile method is more convenient when a sophisticated calculator, such as R, is used.

This method is conservative in the sense that the coverage probability is guaranteed to be greater than or equal to the nominal confidence level (see Section 7.2.6). It has been referred to in many places in the literature as the "exact" method. We avoid this terminology because it is misleading and conflicts with the usual meaning of an exact (as opposed to an approximate or conservative) statistical interval; see Section B.2.1.

Example 7.2   Conservative Confidence Interval for the Computer Systems Unscheduled Shutdown Rate. For the computer systems unscheduled shutdown application introduced in Example 7.1, management wants a confidence interval for λ, the actual or long-term rate of unscheduled shutdowns for the computer systems. Using the conservative method given by (7.2) with $x = 24$ unscheduled shutdowns in $n = 5$ years of operation, we obtain $\chi^2_{(0.025;2\times24)} = 30.75$, $\chi^2_{(0.975;2\times24+2)} = 71.42$, and $\chi^2_{(0.95;2\times24+2)} = 67.50$ from tables in numerous statistics textbooks or from R. Then a conservative 95% confidence interval for λ is obtained by substituting into (7.2), giving

$$[\underset{\sim}{\lambda}, \ \widetilde{\lambda}] = \left[\frac{0.5 \times 30.75}{5}, \ \frac{0.5 \times 71.42}{5} \right] = [3.075, \ 7.142].$$

Interval/bound type	For	Lower	Upper
Conservative two-sided CI	λ	3.075	7.142
Wald two-sided CI	λ	2.880	6.720
Score two-sided CI	λ	3.226	7.143
Jeffreys two-sided CI	λ	3.155	7.022
Conservative one-sided CB	λ	3.310	6.750
Wald one-sided CB	λ	3.188	6.412
Score one-sided CB	λ	3.550	6.818
Jeffreys one-sided CB	λ	3.393	6.634
Conservative two-sided CI	$\Pr(Y \le 5), m = 0.50$	0.848	0.995
Jeffreys two-sided CI	$\Pr(Y \le 5), m = 0.50$	0.856	0.994
Conservative two-sided CI	$y_{0.90}, m = 2$	9	19
Jeffreys two-sided CI	$y_{0.90}, m = 2$	10	19
Conservative one-sided CB	$y_{0.10}, m = 2$	3	9
Jeffreys one-sided CB	$y_{0.10}, m = 2$	4	9
Conservative one-sided CB	$y_{0.90}, m = 2$	10	18
Jeffreys one-sided CB	$y_{0.90}, m = 2$	10	18
Conservative two-sided TB	Control center 0.80, $m = 2$	3	18
Jeffreys two-sided TB	Control center 0.80, $m = 2$	4	18
Conservative two-sided PI	$Y, m = 4$	9	33
Normal two-sided PI	$Y, m = 4$	8	31
Joint-sample two-sided PI	$Y, m = 4$	10	32
Jeffreys two-sided PI	$Y, m = 4$	9	32

Table 7.1   Two-sided 95% confidence intervals and one-sided 95% confidence bounds for λ, the Poisson distribution computer systems unscheduled shutdown rate, and other related intervals. CI and PI denote confidence and prediction interval; CB and TB denote confidence bound and tolerance bound.

A one-sided upper conservative 95% confidence bound for λ is $\widetilde{\lambda} = (0.5 \times 67.50)/5 = 6.750$. Alternatively, using R as a calculator for the gamma quantile function, (7.1) gives

```
> qgamma(p=c(0.025, 0.975), shape=c(24, 24+1), rate=5)
[1]   3.075 7.142
> qgamma(p=0.95, shape=24+1, rate=5)
[1]   6.75
```

for the conservative 95% confidence interval and one-sided upper 95% confidence bound. ∎

7.2.3 The Wald (Normal Theory) Approximate Method

A simple two-sided approximate $100(1 - \alpha)\%$ confidence interval for λ, based on the assumption that the so-called "Wald quantity"

$$\frac{\lambda - \widehat{\lambda}}{\sqrt{\widehat{\lambda}/n}} \tag{7.3}$$

can be approximated by a $\mathrm{NORM}(0, 1)$ distribution, is obtained by inverting a Wald significance test (i.e., setting (7.3) equal to $z_{(1-\alpha)}$ and solving for λ), giving

$$[\underset{\sim}{\lambda}, \ \widetilde{\lambda}] = \widehat{\lambda} \mp z_{(1-\alpha/2)} \left(\frac{\widehat{\lambda}}{n}\right)^{1/2}, \tag{7.4}$$

where $z_{(1-\alpha/2)}$ is the $1 - \alpha/2$ quantile of the standard normal distribution.

This simple method was especially relevant when computational capabilities were limited and still appears frequently in elementary statistics textbooks. It is often incorrectly stated that the method provides adequate accuracy when x exceeds 20. Unfortunately, even when x is as large as 40 or 50, the performance of this procedure is poor, and we do not recommended it for general use (see Section 7.2.6 for more details). We describe the Wald method because of its popularity (as well as historical significance) and because it is still used in some situations (e.g., when no appropriate software or tables are readily available, such as in an informal conversation) and when a simple, crude approximation will suffice.

Example 7.3 Wald-Approximation Confidence Interval for the Computer Systems Unscheduled Shutdown Rate. For the computer systems shutdown application (Example 7.1), management wants a confidence interval for λ, the rate of unscheduled shutdowns for the computer systems. Using the Wald method given by (7.4) with $x = 24$ unscheduled shutdowns in $n = 5$ years of operation, we obtain an approximate 95% confidence interval for λ, using $z_{(0.975)} = 1.960$, as

$$[\underset{\sim}{\lambda}, \ \widetilde{\lambda}] = 4.8 \mp 1.96 \left(\frac{4.8}{5.0}\right)^{1/2} = [2.88, \ 6.72].$$

A one-sided upper approximate 95% confidence bound for λ is obtained by using $z_{(0.95)} = 1.645$ in place of $z_{(0.975)} = 1.960$, giving 6.41. Using R as a calculator gives

```
> lambdahat <- 24/5
> lambdahat + c(-1,1)*qnorm(0.975)*sqrt(lambdahat/5)
[1] 2.88 6.72
> lambdahat + qnorm(0.95)*sqrt(lambdahat/5)
[1] 6.41
```

for the approximate 95% confidence interval and one-sided 95% confidence bound. ∎

7.2.4 The Score Approximate Method

An improved (relative to the Wald method) two-sided approximate $100(1 - \alpha)\%$ confidence interval for λ is based on the assumption that the so-called "score quantity"

$$\frac{\lambda - \widehat{\lambda}}{\sqrt{\lambda/n}} \tag{7.5}$$

can be approximated by a $\mathrm{NORM}(0, 1)$ distribution and is obtained by inverting a score significance test (i.e., setting (7.5) equal to $z_{(1-\alpha)}$ and solving for λ), giving

$$[\underset{\sim}{\lambda}, \ \widetilde{\lambda}] = \widehat{\widehat{\lambda}} \mp z_{(1-\alpha/2)} \frac{1}{\sqrt{n}} \left(\widehat{\lambda} + \frac{z^2_{(1-\alpha/2)}}{4n} \right)^{1/2}, \tag{7.6}$$

where $\widehat{\widehat{\lambda}} = (x + z^2_{(1-\alpha/2)}/2)/n$ is an alternative center of the interval and $z_{(1-\alpha/2)}$ is again the $1 - \alpha/2$ quantile of the standard normal distribution. This method is slightly more complicated than the Wald method, but has much improved properties (see Section 7.2.6). A derivation of the score method is given in Example D.10.

Example 7.4 Score Approximate Confidence Interval for the Computer Systems Unscheduled Shutdown Rate. For the computer systems unscheduled shutdown application (Example 7.1), an approximate 95% confidence interval for λ is obtained using the score method by substituting $x = 24$, $n = 5$, and $z_{(0.975)} = 1.960$ into (7.6) giving

$$\widehat{\widehat{\lambda}} = [24 + 1.96^2/2]/5 = 5.184$$

$$[\underset{\sim}{\lambda}, \ \widetilde{\lambda}] = 5.184 \mp 1.960 \frac{1}{\sqrt{5}} \left[4.8 + \frac{1.960^2}{4 \times 5} \right]^{1/2} = [3.23, \ 7.14].$$

An upper approximate 95% confidence bound for λ, using $z_{(0.95)} = 1.645$ in place of 1.960, is 6.82. Using R as a calculator gives

```
> lambdahat <- 24/5
> lambdahathat <- (24+qnorm(0.975)^2/2)/5
> lambdahathat + c(-1,1)*qnorm(0.975)*(1/sqrt(5))*
    sqrt(lambdahat+qnorm(0.975)^2/(4*5))
[1] 3.226 7.143
> lambdahathat + qnorm(0.95)*(1/sqrt(5))*
    sqrt(lambdahat+qnorm(0.95)^2/(4*5))
[1] 6.818
```

■

7.2.5 The Jeffreys Approximate Method

Using the Jeffreys method (see the Bibliographic Notes section at the end of this chapter and Section H.4.2 for a description of the origin of this method), a two-sided approximate

$100(1 - \alpha)\%$ confidence interval for λ, based on x observed events in n units of exposure, is

$$[\underset{\sim}{\lambda}, \ \widetilde{\lambda}] = [\text{qgamma}(\alpha/2; x + 0.50, n), \quad \text{qgamma}(1 - \alpha/2; x + 0.50, n)] \tag{7.7}$$

$$= \left[\frac{0.5\, \chi^2_{(\alpha/2; 2x+1)}}{n}, \ \frac{0.5\, \chi^2_{(1-\alpha/2; 2x+1)}}{n} \right]. \tag{7.8}$$

The structure of the formulas for this method is the same as that for the conservative method given by (7.1) and (7.2), but with different parameters for the gamma distribution quantiles or different degrees of freedom for the chi-square quantiles. Unlike the conservative method, the Jeffreys method does not guarantee a coverage probability that is greater than or equal to the nominal confidence level for all values of λ. Instead, the Jeffreys method has excellent *mean* coverage properties, as we will see in Section 7.2.6.

Example 7.5 Jeffreys Confidence Interval for the Computer Systems Unscheduled Shutdown Rate. For the computer systems unscheduled shutdown application introduced in Example 7.1, to calculate a confidence interval for λ based on the Jeffreys method given by (7.8), with $x = 24$ unscheduled shutdowns in $n = 5$ system-years of operation, we first obtain $\chi^2_{(0.025; 2 \times 24+1)} = 31.55$, $\chi^2_{(0.925; 2 \times 24+1)} = 70.22$, and $\chi^2_{(0.95; 2 \times 24+1)} = 66.34$ from tables in statistics textbooks or by using R. Then a 95% confidence interval for λ is

$$[\underset{\sim}{\lambda}, \ \widetilde{\lambda}] = \left[\frac{0.5 \times 31.55}{5}, \ \frac{0.5 \times 70.22}{5} \right] = [3.16, \ 7.02].$$

An upper 95% confidence bound for λ is

$$\widetilde{\lambda} = \frac{0.5 \times 66.34}{5} = 6.63.$$

Using R as a calculator for the gamma quantile function in (7.7) gives

```
> qgamma(p=c(0.025,0.975), shape=24+0.50, rate=5)
[1]   3.155 7.022
> qgamma(p=0.95, shape=24+0.50, rate=5)
[1]   6.634
```

7.2.6 Comparisons and Recommendations

We have presented four different approximate methods for constructing a confidence interval or confidence bound for the Poisson distribution parameter λ. Which method do we recommend for a specific application?

If it is necessary to avoid any arguments about the use of approximations (e.g., in court proceedings), one should use the conservative method to assure a coverage probability that is always greater than or equal to the nominal confidence level. When $n\lambda$ is small, however, the conservative method typically results in intervals that will be wider than the competing methods. The score method is useful when a simple computational procedure is needed, and the Wald method could be used in situations for which an immediate, even though crude, ballpark assessment is desired. Otherwise we recommend the Jeffreys method for general use for constructing both two-sided confidence intervals and one-sided confidence bounds for λ.

In the remainder of this section we compare the performance of the different methods to provide some insights into the basis for these recommendations. Because the different methods

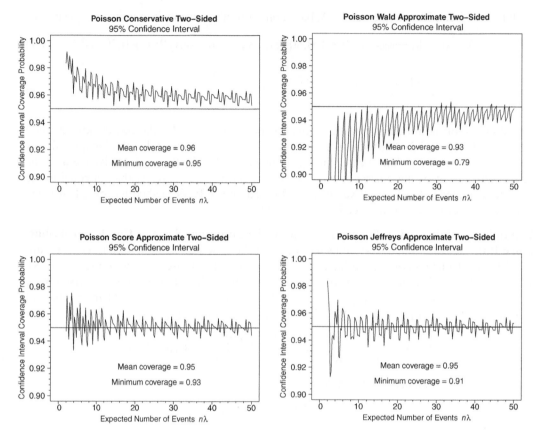

Figure 7.1 Plots of confidence interval (for λ) coverage probabilities versus the expected number of events $n\lambda$ for the conservative and Wald (top row) and score and Jeffreys (bottom row) methods for a nominal 95% confidence level.

for constructing a confidence interval on the Poisson occurrence rate λ have different coverage-probability behaviors, one should compare the properties of these procedures before deciding which one to use (this should be done before looking at the data to avoid any appearance of lack of objectiveness). The coverage probability depends on $n\lambda$, the expected number of events in n units of exposure, and can be computed by using (B.7) in Section B.2.4. The following results deal with 95% confidence intervals and bounds, but similar results are obtained for other confidence levels.

Figure 7.1 shows the coverage probabilities versus $n\lambda$, the expected number of events in n units of exposure, for the 95% confidence intervals for each of the four confidence interval methods presented in this section. The plots also show the mean coverage and the minimum coverage probabilities computed at 200 equally spaced values of $n\lambda$ between 2 and 50.

The top left-hand plot in Figure 7.1 shows that the conservative method has (as expected from theory; see Section D.6.2) a coverage probability that is, for all values of $n\lambda$, greater (and for small $n\lambda$, appreciably greater) than the nominal 95% confidence level. The top right-hand plot shows that the Wald-approximation method almost always provides a coverage probability that is less than (and for small values of $n\lambda$, substantially less than) the nominal confidence level (i.e., the method is nonconservative). This method, in fact, tends to remain (at least slightly)

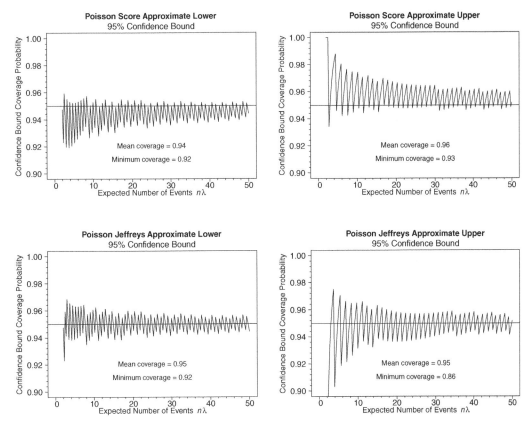

Figure 7.2   Plots of one-sided lower (left) and upper (right) confidence bound (for λ) coverage probabilities versus the expected number of events $n\lambda$ for the score (top) and Jeffreys (bottom) methods for a nominal 95% confidence level.

nonconservative for values of $n\lambda$ between 30 and 50 (and even above 50). For this reason, we recommend against use of the Wald method except for situations in which a hand-calculated interval is needed and a ballpark approximation will suffice.

The score method in the bottom left-hand plot and the Jeffreys method in the bottom right-hand plot of Figure 7.1 have reasonably good approximate coverage properties (i.e., coverage probabilities close to the nominal 95% confidence level), even for fairly small values of $n\lambda$. In addition—as we will see in the discussion to follow dealing with coverage probabilities for one-sided confidence bounds—the Jeffreys method provides good balance between the error probabilities of being outside of the upper and lower endpoints for a two-sided confidence interval, while the score method *does not*.

Figure 7.2 gives *one-sided* confidence bound coverage probabilities (lower bound on the left and upper bound on the right), comparing the score method (top) with the Jeffreys method (bottom), for values of $n\lambda$ between 2 and 50. These plots show that the score method one-sided lower (upper) confidence bounds are generally nonconservative (conservative), and especially so for small values of $n\lambda$. This comparison shows that the Jeffreys method has the advantage over the score method because it provides better balance between errors in the lower and upper endpoints of a two-sided interval. As explained in Section 2.7, it is preferable to have the kind of balance that the Jeffreys method provides.

In summary, the preceding comparisons show that the Jeffreys method possesses the best statistical properties among the three approximate methods presented for constructing both two-sided confidence intervals and one-sided confidence bounds. This is especially so for small values of $n\lambda$.

7.2.7   Comparison of Results from Applying Different Methods

The two-sided 95% confidence intervals and one-sided 95% confidence bounds obtained for λ using each of the four methods are compared in the top half of Table 7.1. This tabulation also shows the one-sided lower 95% confidence bounds. We note that for this example the results obtained using these different methods are quite similar.

7.3   CONFIDENCE INTERVALS FOR THE PROBABILITY THAT THE NUMBER OF EVENTS IN A SPECIFIED AMOUNT OF EXPOSURE IS LESS THAN OR EQUAL TO (OR GREATER THAN) A SPECIFIED NUMBER

Some applications require inferences concerning the probability that Y, the number of events in m units of exposure, will be less than or equal to (or greater than) some prespecified nonnegative integer y. In particular, based on the information in a previous exposure amount n, an analyst might want to find a confidence interval for the probability p_{LE} that the number of events in a specified exposure amount m is less than or equal to y, where y is a prespecified nonnegative integer.

If the rate of occurrence of events from the distribution were *known* to be λ, the probability p_{LE} that Y, the number of events in a sample of m units of exposure, will be less than or equal to a prespecified number y is computed from the Poisson cumulative distribution function as

$$p_{LE} = \Pr(Y \leq y) = \texttt{ppois}(y; m\lambda), \tag{7.9}$$

where $\texttt{ppois}$ is the Poisson cdf defined in Section 7.1.1 and Section C.4.4. Usually, λ is unknown and only sample data on the number of events x in n units of exposure are available. Because p_{LE} is a decreasing function of λ (see (C.24) in Section C.4.4), the following two-step procedure is used to find an approximate two-sided confidence interval for p_{LE}:

1. Obtain a two-sided confidence interval for λ, based on the data, using one of the methods given in Section 7.2.

2. Substitute these values for λ into (7.9) to obtain the desired two-sided confidence interval for p_{LE}.

Thus, if $[\underset{\sim}{\lambda}, \ \widetilde{\lambda}]$ is a two-sided $100(1 - \alpha)\%$ confidence interval for λ, a two-sided $100(1 - \alpha)\%$ confidence interval for p_{LE} is

$$[\underset{\sim}{p}_{LE}, \ \widetilde{p}_{LE}] = [\texttt{ppois}(y; m\widetilde{\lambda}), \ \texttt{ppois}(y; m\underset{\sim}{\lambda})].$$

Similarly, if the rate of occurrence of events from the distribution is known to equal λ, the probability p_{GT} that Y, the number of events in a specified amount of exposure m, will be greater than y is the complement of the Poisson cumulative distribution function

$$p_{GT} = 1 - p_{LE} = \Pr(Y > y) = 1 - \texttt{ppois}(y; m\lambda).$$

When λ is unknown, because p_{GT} is an increasing function of λ, a $100(1 - \alpha)\%$ confidence interval for p_{GT} is

$$[\underset{\sim}{p_{GT}}, \ \widetilde{p}_{GT}] = [1 - \widetilde{p}_{LE}, \ 1 - \underset{\sim}{p_{LE}}] = [1 - \text{ppois}(y; m\underset{\sim}{\lambda}), \ 1 - \text{ppois}(y; m\widetilde{\lambda})].$$

Because the function $\text{ppois}(y; m)$ is a continuous monotone function of λ, the coverage properties of each of the preceding procedures is exactly the same as that for the corresponding confidence interval procedure (from Section 7.2) used to obtain a confidence interval for λ.

Example 7.6 Confidence Interval for the Probability of Five or Fewer Unscheduled Computer Shutdowns in 6 Months. Suppose that in Example 7.1 we would like to compute a 95% confidence interval for the probability that $y = 5$ or fewer unscheduled shutdowns will occur in the next $m = 1/2$ year of system operation. In Example 7.2, a conservative 95% confidence interval for λ was found to be $[\underset{\sim}{\lambda}, \ \widetilde{\lambda}] = [3.075, \ 7.142]$. Thus, a conservative 95% confidence interval for $p_{LE} = \Pr(Y \leq 5)$ is

$$[\underset{\sim}{p_{LE}}, \ \widetilde{p}_{LE}] = [\text{ppois}(5; 0.50 \times 7.142), \ \text{ppois}(5; 0.50 \times 3.075)] = [0.848, \ 0.995].$$

Thus, we are 95% confident that the probability of five or fewer unscheduled shutdowns in the next half year of operation is between 0.848 and 0.995. Similarly, a one-sided lower conservative 95% confidence bound on p_{LE} is

$$\underset{\sim}{p_{LE}} = \text{ppois}(5; 0.50 \times 6.750) = 0.874.$$

This example depends heavily on the assumption that the actual Poisson occurrence rate λ does not change in the next half year from what it was during the previous 5 years.

Using R as a calculator, based on the conservative confidence interval for λ from Section 7.2.2, gives

```
>ppois(q=5, 0.50*qgamma(p=c(0.975, 0.025), shape=c(24+1, 24), rate=5))
[1]   0.8481 0.9950
> ppois(q=5, 0.50*qgamma(p=0.95, shape=c(24+1), rate=5))
[1]  0.8737
```

Similarly, using the Jeffreys confidence interval from Section 7.2.5 gives

```
> ppois(q=5, 0.50*qgamma(p=c(0.975, 0.025), shape=24+0.50, rate=5))
[1]    0.8561 0.9943
> ppois(q=5, 0.50*qgamma(p=0.95, shape=24+0.50, rate=5))
[1]  0.8808
```

The preceding two-sided confidence intervals are displayed in Table 7.1. ∎

7.4 CONFIDENCE INTERVALS FOR THE QUANTILE OF THE DISTRIBUTION OF THE NUMBER OF EVENTS IN A SPECIFIED AMOUNT OF EXPOSURE

Some applications require a confidence interval (or a one-sided confidence bound) for a quantile y_p of the distribution of Y, the number of events in m units of exposure. For example, suppose that a characterization of the distribution of the number of computer system unscheduled shutdowns for exposure periods of length m is needed, and it is desired to make a statement about a value y_p that exceeds a certain proportion p of such *exposure periods*. This problem is the inverse of the one in Section 7.3; there y was specified and it was desired to obtain a confidence interval for the probability $p_{LE} = \Pr(Y \leq y)$ of having y or fewer events in m units of exposure.

The p quantile y_p of a Poisson distribution is defined as the smallest value of y such that $\Pr(Y \leq y) = \mathtt{ppois}(y; m\lambda) \geq p$ and is denoted by $\mathtt{qpois}(p; m\lambda)$. If the event-occurrence rate were *known* to be λ, then the desired quantile could be computed directly from $\mathtt{qpois}(p; m\lambda)$. Usually, however, λ is unknown and only sample data on the number of events x in a previous exposure of size n are available.

7.4.1 Two-Sided Confidence Interval for y_p

In a manner similar to that used in Section 7.3, because $\mathtt{qpois}(p; m\lambda)$ is an increasing function of λ, the following two-step procedure is used to find an approximate two-sided confidence interval for y_p:

1. Obtain a two-sided confidence interval for λ, based on the data, using one of the methods given in Section 7.2.

2. Substitute these values for λ into $\mathtt{qpois}(p; m\lambda)$ to obtain the desired two-sided confidence interval for y_p.

Thus, if $[\underset{\sim}{\lambda}, \ \widetilde{\lambda}]$ is a two-sided approximate (or conservative) $100(1 - \alpha)\%$ confidence interval for λ, a two-sided approximate (or conservative) $100(1 - \alpha)\%$ confidence interval for y_p is

$$[\underset{\sim}{y_p}, \ \widetilde{y}_p] = [\mathtt{qpois}(p; m\underset{\sim}{\lambda}), \ \mathtt{qpois}(p; m\widetilde{\lambda})]. \tag{7.10}$$

Unlike the confidence interval procedure for Poisson probabilities in Section 7.3, the quantile function is *not* a continuous function of λ (it is an integer-valued step function). Therefore, the coverage probability as a function of the actual value of λ will not be exactly the same as that for the confidence interval procedure for λ. The general formula (B.5) in Section B.2.4 can, however, be used to conduct such coverage probability evaluations and perform calibration, discussed in Section B.8.

Example 7.7 Confidence Interval for the 0.90 Quantile of the Distribution of the Number of Unscheduled Computer Shutdowns in 2 Years of Operation. The operators of the computer systems in Example 7.1 desire a 95% confidence interval for $y_{0.90}$, the 0.90 quantile of the distribution of the number of unscheduled shutdowns in $m = 2$ years of operation. Using the conservative 95% confidence interval $[\underset{\sim}{\lambda}, \ \widetilde{\lambda}] = [3.155, \ 7.022]$ from Table 7.1, the resulting conservative 95% confidence interval for $y_{0.90}$ is

$$[\underset{\sim}{y}_{0.90}, \ \widetilde{y}_{0.90}] = [\mathtt{qpois}(0.90; 2 \times 3.075), \ \mathtt{qpois}(0.90; 2 \times 7.142)]$$
$$= [9, \ 19].$$

Thus, we are at least 95% confident that the 0.90 quantile of the distribution of the number of unscheduled shutdowns is between 9 and 19.

Using R as a calculator, based on the conservative confidence interval for λ from Section 7.2.2, gives

```
>qpois(p=0.90, 2*qgamma(p=c(0.025, 0.975), shape=c(24, 24+1), rate=5))
[1]   9 19
```

as a conservative 95% confidence interval for $y_{0.90}$. Similarly, using the Jeffreys confidence interval from Section 7.2.5 gives

```
> qpois(p=0.90, 2*qgamma(p=c(0.025,0.975), shape=24+0.50, rate=5))
[1] 10 19
```

as an approximate 95% confidence interval for $y_{0.90}$.

The preceding results are summarized in Table 7.1. Because of the discreteness in the quantile function and because there is only a small difference between the conservative and Jeffreys methods for obtaining a confidence interval for λ, the confidence intervals for the two methods in this example are almost identical.   ∎

7.4.2 One-Sided Confidence Bounds for y_p

A one-sided lower (upper) $100(1-\alpha)\%$ confidence bound for y_p is found by substituting a one-sided lower (upper) $100(1-\alpha)\%$ confidence bound for λ into the appropriate endpoint of the two-sided confidence interval in (7.10). These one-sided confidence bounds are of particular interest because of their relationship to one-sided tolerance bounds and two-sided tolerance intervals, to be described in Section 7.5.

Example 7.8 One-Sided 95% Confidence Bounds for the 0.10 and 0.90 Quantiles of the Distribution of the Number of Unscheduled Computer Shutdowns in 2 Years of Operation. For the computer systems unscheduled shutdown application introduced in Example 7.1, using the one-sided lower conservative 95% confidence bound $\utilde{\lambda} = 3.310$ from Table 7.1 and proceeding as in Example 7.7, a conservative 95% lower confidence bound for $y_{0.10}$ for $m = 2$ years of operation is

$$\utilde{y}_{0.10} = \texttt{qpois}(0.10; 2 \times 3.310) = 3.$$

Similarly, using the one-sided upper conservative 95% confidence bound $\tilde{\lambda} = 6.750$, a one-sided upper conservative 95% confidence bound for $y_{0.90}$ for $m = 2$ years of operation is

$$\tilde{y}_{0.90} = \texttt{qpois}(0.90; 2 \times 6.750) = 18.$$

Using R as a calculator, based on the conservative confidence interval method for λ from Section 7.2.2, gives the conservative 95% one-sided confidence bounds

```
> qpois(p=0.10, 2*qgamma(p=0.05, shape=24, rate=5))
[1] 3
> qpois(p=0.90, 2*qgamma(p=0.95, shape=24+1, rate=5))
[1] 18
```

for $y_{0.10}$ and $y_{0.90}$, respectively. Similarly, using the Jeffreys confidence interval for λ from Section 7.2.5 gives the one-sided approximate 95% confidence bounds

```
> qpois(p=0.10, 2*qgamma(p=0.05, shape=24.5, rate=5))
[1] 4
> qpois(p=0.90, 2*qgamma(p=0.95, shape=24.5, rate=5))
[1] 18
```

for $y_{0.10}$ and $y_{0.90}$, respectively. The preceding results are summarized in Table 7.1 and will be used in the next section to construct two-sided tolerance intervals and one-sided tolerance bounds.   ∎

7.5   TOLERANCE INTERVALS AND ONE-SIDED TOLERANCE BOUNDS FOR THE DISTRIBUTION OF THE NUMBER OF EVENTS IN A SPECIFIED AMOUNT OF EXPOSURE

Some applications require two-sided tolerance intervals or one-sided tolerance bounds for the distribution of Y, the number of events in m units of exposure. Thus, in Example 7.1 one may want to find a value y so one can state, with a specified degree of confidence, that at least a proportion 0.90 of computing systems experience y or fewer unscheduled shutdowns in m units of operation. This statement calls for a one-sided lower tolerance bound. Other problems may require a one-sided upper tolerance bound or a two-sided tolerance interval. As indicated in Section 2.4.2, however, a one-sided tolerance bound is equivalent to a one-sided confidence bound on a quantile of the distribution.

Also, approximate two-sided tolerance intervals can be obtained by appropriately combining one-sided lower and upper tolerance bounds. We provide further details below.

7.5.1   One-Sided Lower Tolerance Bound for a Poisson Distribution

A one-sided lower $100(1 - \alpha)\%$ tolerance bound $\underset{\sim}{T}'_\beta$ to be exceeded by at least a proportion β of the distribution is the same as a lower $100(1 - \alpha)\%$ confidence bound on $y_{(1-\beta)}$, the $1 - \beta$ quantile of the distribution. Thus $\underset{\sim}{T}'_\beta = \underset{\sim}{y}_{(1-\beta)}$, which can be computed as described in Section 7.4.2.

Example 7.9   One-Sided Lower Tolerance Bound for the Number of Unscheduled Computer Shutdowns in 2 Years of Operation. Due to the equivalence described above, a one-sided lower conservative 95% tolerance bound to be exceeded by a proportion 0.90 of the distribution of the number of computer unscheduled shutdowns in $m = 2$ years of operation in Example 7.1 is, from Example 7.8, $\underset{\sim}{T}'_{0.90} = \underset{\sim}{y}_{0.10} = 3$.   ∎

7.5.2   One-Sided Upper Tolerance Bound for a Poisson Distribution

A one-sided upper $100(1 - \alpha)\%$ tolerance bound $\widetilde{T}'_\beta$ to exceed at least a proportion β of the distribution is the same as an upper $100(1 - \alpha)\%$ confidence bound on y_β, the β quantile of the distribution. Thus $\widetilde{T}'_\beta = \widetilde{y}_\beta$, which can be computed as described in Section 7.4.2.

Example 7.10   One-Sided Upper Tolerance Bound for the Number of Unscheduled Computer Shutdowns in 2 Years of Operation. A one-sided upper conservative 95% tolerance bound to exceed a proportion 0.90 of the distribution of the number of computer unscheduled shutdowns in $m = 2$ years of operation in Example 7.1 is, from Example 7.8, $\widetilde{T}'_{0.90} = \widetilde{y}_{0.90} = 18$.   ∎

7.5.3   Two-Sided Tolerance Interval for a Poisson Distribution

A method of constructing a two-sided approximate tolerance interval $[\underset{\sim}{T}_\beta, \ \widetilde{T}_\beta]$ for a Poisson distribution is to combine two one-sided tolerance bounds (or one-sided confidence bounds on appropriate quantiles) for the distribution of interest. In particular, as explained in Section D.7.4, to obtain an approximate two-sided $100(1 - \alpha)\%$ tolerance interval to contain at least a proportion β of the distribution, one can use a one-sided lower $100(1 - \alpha)\%$ confidence bound on the $(1 - \beta)/2$ quantile for the lower endpoint and a one-sided upper $100(1 - \alpha)\%$ confidence bound on the $(1 + \beta)/2$ quantile for the upper endpoint. That is, $[\underset{\sim}{T}_\beta, \ \widetilde{T}_\beta] = [\underset{\sim}{y}_{(1-\beta)/2}, \ \widetilde{y}_{(1+\beta)/2}]$.

Example 7.11   Two-Sided Tolerance Interval for the Number of Unscheduled Computer Shutdowns in 2 Years of Operation. For Example 7.1, a 95% two-sided tolerance interval to contain at least a proportion 0.80 of the distribution of the number of computer unscheduled shutdowns in $m = 2$ years of operation is obtained by using a one-sided lower 95% confidence bound on the 0.10 quantile for the lower endpoint and a one-sided upper 95% confidence bound on the 0.90 quantile for the upper endpoint. Using the results from Example 7.8 gives $[\underset{\sim}{T}_{0.80}, \ \widetilde{T}_{0.80}] = [\underset{\sim}{y}_{(0.10)}, \ \widetilde{y}_{(0.90)}] = [3, \ 18]$. ∎

7.5.4   Calibrating Tolerance Intervals

Figure 7.3 gives plots of coverage probabilities versus $n\lambda$, the expected number of events in n units of exposure, for the conservative and Jeffreys two-sided tolerance interval methods for four nominal confidence levels, ranging from 90% to 97.5%, for $n = m$, and $\beta = 0.90$.

Figure 7.3 shows that both methods tend to give highly conservative results. The plots show that these methods are conservative for not only small values of $n\lambda$ but up to $n\lambda = 50$. Although these results are for one special case, they also hold for other combinations of n, m, and β.

As a consequence of the preceding consistent conservativeness of both the conservative and Jeffreys methods for constructing Poisson tolerance intervals, these methods are especially appropriate candidates for adjustment by applying the *statistical interval calibration method*, introduced in Section 6.6.4 and described further in Section B.8. In particular, if an interval procedure tends to be consistently too conservative, one can adjust the procedure by using a smaller nominal confidence level as input. For example, the third row in Figure 7.3 shows that the procedures for calculating tolerance intervals based on both the conservative method and the Jeffreys method are both consistently conservative with mean coverage probabilities (0.98 and 0.97, respectively) appreciably greater than the nominal confidence level of 95%. If one desires a procedure that has mean coverage that approximately equals 0.95, one could use the Jeffreys procedure with a nominal confidence level between 90% and 92.5%.

7.6   PREDICTION INTERVALS FOR THE NUMBER OF EVENTS IN A FUTURE AMOUNT OF EXPOSURE

Suppose, as before, that x events have been observed in n units of exposure. From this information, it is desired to find a prediction interval that will, with some specified degree of confidence, contain Y, the number of events in m future units of exposure. Also, as before, we assume that the two exposures are independent and that the number of events in each can be described by a Poisson distribution with the same rate parameter λ.

7.6.1   The Conservative Method

Let X and Y denote the number events during exposures of n and m units, respectively, from a Poisson distribution with event-occurrence rate λ. Prior to observing X, the conditional distribution of X, given $X + Y$, is BINOM$(X + Y, \pi)$, where $\pi = n/(n + m)$. After observing $X = x$, `pbinom`$(x; x + Y, \pi)$ (the binomial cumulative distribution function, defined in Section C.4.1) is nonincreasing in $x + Y$. Then the following procedure gives a conservative $100(1 - \alpha)\%$ confidence interval for $x + Y$ that can then be converted (because x is known) into a prediction interval for Y, the number of events in the future exposure of m units.

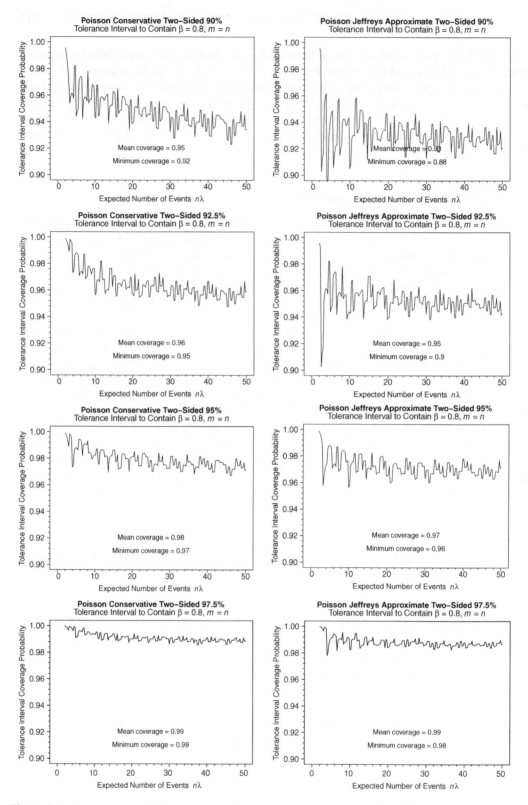

Figure 7.3 Coverage probabilities versus $n\lambda$, the expected number of events in n units of exposure for the conservative (left) and Jeffreys (right) two-sided tolerance interval methods to contain at least a proportion $\beta = 0.80$ for $n = m$ with nominal confidence levels 90% (top), 92.5% (second row), 95% (third row), and 97.5% (bottom).

A conservative $100(1 - \alpha)\%$ prediction interval $[\underset{\sim}{Y}, \widetilde{Y}]$ for Y is obtained by finding the smallest value $\underset{\sim}{Y}$ and the largest value $\widetilde{Y}$ such that

$$1 - \texttt{pbinom}(x - 1; x + \underset{\sim}{Y}, \pi) > \frac{\alpha}{2} \quad \text{and} \quad \texttt{pbinom}(x; x + \widetilde{Y}, \pi) > \frac{\alpha}{2}, \quad (7.11)$$

where $\underset{\sim}{Y} = 0$ if $x = 0$. The prediction interval is conservative because it is based on a conservative confidence interval procedure for $x + Y$. One-sided lower and upper $100(1 - \alpha)\%$ prediction bounds for Y are obtained by replacing $\alpha/2$ with α in the lower and upper endpoints of (7.11).

From the relationships between the binomial cdf and the beta cdf (given in Section C.4.1) and that between the binomial cdf and Snedecor's F-distribution (given in Section C.3.11), the inequalities in (7.11) are equivalent to the following inequalities:

$$\texttt{pbeta}(\pi; x, \underset{\sim}{Y} + 1) > \frac{\alpha}{2} \quad \text{and} \quad 1 - \texttt{pbeta}(\pi; x + 1, \widetilde{Y}) > \frac{\alpha}{2},$$

$$\texttt{pf}\left(\frac{\underset{\sim}{Y} + 1}{x} \times \frac{n}{m}; 2x, 2\underset{\sim}{Y} + 2\right) > \frac{\alpha}{2} \quad \text{and} \quad \texttt{pf}\left(\frac{\widetilde{Y}}{x + 1} \times \frac{n}{m}; 2x + 2, 2\widetilde{Y}\right) > \frac{\alpha}{2} \quad (7.12)$$

where again $\pi = n/(n + m)$.

By inverting (7.12), the same conservative prediction interval given by (7.11) can be obtained by finding the smallest value $\underset{\sim}{Y}$ and the largest value $\widetilde{Y}$ such that

$$\frac{m}{\underset{\sim}{Y} + 1} < \frac{n}{x} F_{(1 - \alpha/2; 2\underset{\sim}{Y} + 2, 2x)} \quad \text{and} \quad \frac{\widetilde{Y}}{m} > \left(\frac{x + 1}{n}\right) F_{(1 - \alpha/2; 2x + 2, 2\widetilde{Y})}. \quad (7.13)$$

The preceding approach, as opposed to the use of (7.11), is useful when tables of Snedecor's F-distribution quantiles are available and a sophisticated calculator like R is not.

Example 7.12 Conservative Prediction Interval and One-Sided Prediction Bound for the Number of Unscheduled Computer Shutdowns in 4 System-Years of Operation. In
Example 7.1, a 95% prediction interval to contain the number of unscheduled shutdowns in $m = 4$ future years of system operation is desired.

The conservative method given by (7.11) yields $[\underset{\sim}{Y}, \widetilde{Y}] = [9, 33]$ as a 95% prediction interval and $\widetilde{Y} = 31$ as an upper 95% prediction bound for Y. Thus, based on $x = 24$ past unscheduled shutdowns in $n = 5$ system-years of operation, one can, for example, assert with (at least) 95% confidence that the number of unscheduled shutdowns in $m = 4$ future system-years of operation will not exceed 31.

The binomial probabilities needed for the preceding prediction interval or bound can be computed with the R function pbinom. Thus, using R as a calculator with function pbinom and trying values of Y between 7 and 11, gives

```
> Y.try <- 7:11
> 1-pbinom(q=24-1, size=24+Y.try, prob=5/(5+4))
[1]  0.00996 0.01893 0.03311 0.05391 0.08249
```

to be compared with $\alpha/2 = 0.025$. So the smallest value $\underset{\sim}{Y}$ meeting the left-hand restriction in (7.11) is 9. Similarly, trying values of Y between 31 and 35 gives

```
> Y.try <- 31:35
> pbinom(q=24, size=24+Y.try, prob=5/(5+4))
[1]  0.0506 0.0379 0.0281 0.0206 0.0150
```

also to be compared with $\alpha/2 = 0.025$. So the largest value $\widetilde{Y}$ meeting the right-hand restriction (7.11) is 33. One-sided prediction bounds are computed in a similar manner.   ∎

Example 7.13 Conservative One-Sided Upper Prediction Bound for the Number of Unscheduled Computer Shutdowns in One Half Year of Operation. Suppose that a consumer who is installing one of the computer systems in Example 7.1 wants to use the manufacturer's data to obtain an upper 95% prediction bound for Y, the number of unscheduled shutdowns during a future half year of operation (i.e., $m = 0.5$). Using R as a calculator with function pbinom and trying values of Y between 4 and 8 gives

```
> Y.try <- 4:8
> pbinom(q=24, size=24+Y.try, prob=5/(5+0.5))
[1]  0.247208598 0.118135409 0.050078636 0.019143740 0.006689431
```

to be compared with $\alpha = 0.05$. So the largest value $\widetilde{Y}$ meeting the right-hand restriction (7.11) is 6, giving the desired upper prediction bound.   ∎

7.6.2   The Normal Distribution Approximation Method

A large-sample approximate $100(1 - \alpha)\%$ prediction interval for Y, based on the assumption that

$$\frac{m\widehat{\lambda} - Y}{\sqrt{\widehat{\mathrm{Var}}(m\widehat{\lambda} - Y)}}$$

can be adequately approximated by a $\mathrm{NORM}(0, 1)$ distribution, is

$$[\underset{\sim}{Y}, \ \widetilde{Y}] = m\widehat{\lambda} \mp z_{(1-\alpha/2)}m\left[\widehat{\lambda}\left(\frac{1}{n} + \frac{1}{m}\right)\right]^{1/2}. \tag{7.14}$$

This approximate interval is easier to compute than that using the conservative method given by (7.11) or (7.13). Again, one-sided lower and upper $100(1 - \alpha)\%$ prediction bounds for Y are obtained by replacing $\alpha/2$ with α in the lower and upper endpoints of (7.14).

Example 7.14   Normal Distribution Approximation Prediction Interval for the Number of Unscheduled Computer Shutdowns in 4 System-Years of Operation. For Example 7.1, a 95% prediction interval for Y in $m = 4$ future system-years of operation, using the normal distribution approximation (7.14), is

$$[\underset{\sim}{Y}, \ \widetilde{Y}] = 4 \times 4.8 \mp 1.96 \times 4.0\left[4.8\left(\frac{1}{5} + \frac{1}{4}\right)\right]^{1/2} = [7.7, \ 30.7],$$

which, after rounding to the nearest integer gives [8, 31]. This interval compares reasonably well with the interval calculated using the conservative method in Example 7.12.

Using R as a calculator gives

```
> lambdahat <- 24/5
> Yhat <- 4*lambdahat
> Yhat + c(-1,1)*qnorm(0.975)*4*sqrt(lambdahat*(1/5+1/4))
[1]   7.7 30.7
```

∎

One-sided prediction bounds can be obtained similarly.

7.6.3 The Joint-Sample Approximate Method

Let $\widehat{\lambda}_{xy} = (X + Y)/(n + m)$ and suppose that

$$\frac{m\widehat{\lambda}_{xy} - Y}{\sqrt{\widehat{\text{Var}}\left(m\widehat{\lambda}_{xy} - Y\right)}} = \frac{mX - nY}{\sqrt{mn(X + Y)}} \tag{7.15}$$

can be adequately approximated by a $\text{NORM}(0, 1)$ distribution. Then squaring the right-hand side of (7.15), setting it equal to $z_{(1-\alpha/2)}^2$, and solving the resulting quadratic equation for the two roots in Y gives the following large-sample $100(1 - \alpha)\%$ prediction interval for Y:

$$[\underset{\sim}{Y}, \ \widetilde{Y}] = m\widehat{\lambda} + \frac{mz_{(1-\alpha/2)}^2}{2n} \mp z_{(1-\alpha/2)}\left[m\widehat{\lambda}\left(\frac{1}{n} + \frac{1}{m}\right) + \left(\frac{mz_{(1-\alpha/2)}}{2n}\right)^2\right]^{1/2}. \tag{7.16}$$

This method is known as the "joint-sample (approximate) method." It has been shown that in using this method, the best agreement between coverage probabilities and the nominal confidence level $1 - \alpha$ is obtained by rounding in a nonconservative manner (i.e., rounding the noninteger lower endpoint upward to the next integer and rounding the noninteger upper endpoint downward to the next integer).

This approximate interval is a more complicated than the normal-approximation method in (7.14), but the coverage probability properties are much better (as described in Section 7.6.5). Again, one-sided lower and upper $100(1 - \alpha)\%$ prediction bounds for Y are obtained by replacing $\alpha/2$ with α in the lower and upper endpoints of (7.16), respectively.

Example 7.15 Joint-Sample Approximate Prediction Interval for the Number of Unscheduled Computer Shutdowns in 4 System-Years of Operation. In Example 7.1, the 95% prediction interval for Y using the joint-sample approximation (7.16) is

$$[\underset{\sim}{Y}, \ \widetilde{Y}] = 4 \times 4.8 + \frac{4 \times 1.96^2}{2 \times 5} \mp 1.96\left[4.0 \times \frac{24}{5}\left(\frac{1}{5} + \frac{1}{4}\right) + \left(\frac{4 \times 1.96}{2 \times 5}\right)^2\right]^{1/2}$$

$$= [9.1, \ 32.4],$$

which when rounded in the nonconservative manner gives $[10, \ 32]$. This interval compares reasonably well with the intervals calculated using the conservative method in Example 7.12 and the normal-approximation method in Example 7.14. Using R as a calculator gives

```
> lambdahat <- 24/5
> Yhat <- 4*lambdahat
> Yhat + (4*1.96^2)/(2*5) +
    c(-1,1)*qnorm(0.975)*sqrt(4*Yhat*(1/5+1/4) +
    ((4*1.96)/(2*5))^2)
[1] 9.1 32.4
```

One-sided prediction bounds can be obtained similarly.

7.6.4 The Jeffreys Method

The Jeffreys method for obtaining a Poisson distribution prediction interval is an extension of the Jeffreys method for obtaining a confidence interval for λ, given in Section 7.2.5, and is based on quantiles of the negative binomial distribution, which can be viewed as a Bayesian Poisson predictive distribution (see Section H.6.2 for technical details). Thus, given x observed events in n units of exposure, a two-sided approximate $100(1 - \alpha)\%$ Jeffreys prediction interval for Y in m future units of exposure is

$$[\underset{\sim}{Y},\ \widetilde{Y}] = [\texttt{qnbinom}(\alpha/2; x + 0.5, n/(n + m)),$$
$$\texttt{qnbinom}(1 - \alpha/2; x + 0.5, n/(n + m))], \qquad (7.17)$$

where $\texttt{qnbinom}(p; k, \pi)$ is the p quantile of the negative binomial distribution with "stopping parameter" k and "proportion parameter" π.

Example 7.16   Jeffreys Prediction Interval for the Number of Unscheduled Computer Shutdowns in 4 System-Years of Operation. For Example 7.1, a Jeffreys approximate 95% prediction interval for the number of unscheduled shutdowns Y in $m = 4$ system-years of future operation, using (7.17), is

$$[\underset{\sim}{Y},\ \widetilde{Y}] = [\texttt{qnbinom}(0.025; 24 + 0.5, 5/(5 + 4)),\ \ \texttt{qnbinom}(0.975; 24 + 0.5, 5/(5 + 4))]$$
$$= [9,\ 32].$$

Using R as a calculator, the negative binomial quantile function $\texttt{qnbinom}$ gives

```
> qnbinom(p=c(0.025, 0.975), size=24+0.5, prob=5/(5+4))
[1]   9 32
```

This interval again differs, but not appreciably, from the intervals obtained using the three previous methods.   ■

7.6.5   Comparisons and Recommendations

The recommendations in this section are based on evaluations of coverage probabilities that are not shown here. Equation (B.25) in Section B.6.4 provides an expression for computing coverage probabilities.

If having at least the nominal level of confidence is important (e.g., to avoid arguments about the use of approximations), the conservative method should be used. The conservative method, however, can be overly conservative, resulting in prediction intervals that are excessively wide, especially when either $n\lambda$ or $m\lambda$ is small (e.g., less than 10).

The joint-sample method was found to have coverage probabilities close to the nominal confidence level when $n\lambda$ and $m\lambda$ were both 10 or larger. Thus we recommend the joint-sample method when an approximate prediction interval method (as opposed to the conservative method) is acceptable. The Jeffreys method tends to be somewhat more conservative than the joint-sample method, especially when the ratio m/n is small (say, less than 1), but not as conservative as the conservative method. The normal-approximate method requires that $n\lambda$ and $m\lambda$ both be 30 or larger for the coverage probabilities to be reliably close to the nominal confidence level. More generally, we recommend, if possible, comparing the coverage properties of different intervals for particular situations to help decide which method to use.

BIBLIOGRAPHIC NOTES

Basic textbooks on probability, such as Ross (2012) and Ross (2014), provide detailed treatments of the properties and applications of the Poisson distribution.

We present below references that provide original sources or further details about the various methods presented in this chapter. We also provide references that present yet additional methods for calculating Poisson distribution confidence intervals. We do not discuss these further methods here for space considerations and because we feel that the intervals that we do present provide readers ample choices.

Confidence intervals

Garwood (1936) initially proposed the conservative confidence interval method for the Poisson distribution. Nelson (1972a) gives charts to find upper confidence bounds for the Poisson distribution parameter λ. The Jeffreys confidence interval method given in Section 7.2.5 derives from Bayesian estimation with a Jeffreys prior distribution, as illustrated in Sections 16.2.2 and H.4.2. Byrne and Kabaila (2005) compare different methods for obtaining confidence intervals for λ. Brown et al. (2003) present conservative, Wald, score, and Jeffreys confidence intervals for the distribution parameter of distributions in the exponential family—which includes the Poisson distribution as a special case—and evaluate the coverage probabilities of the various procedures.

Tolerance intervals

One-sided tolerance bounds for the Poisson distribution were first given by Hahn and Chandra (1981) and are also presented in Krishnamoorthy and Mathew (2009). Wang and Tsung (2009) describe methods for evaluating the coverage probability of Poisson (and binomial) distribution tolerance intervals and present algorithms for developing methods with improved coverage probability properties. Cai and Wang (2009) present methods for constructing one-sided approximate tolerance bounds (equivalent to one-sided confidence bounds on a quantile) for discrete distributions that are based on "probability matching" that use high-order approximations for the coverage probability and match these to the desired confidence level. Then they show how to combine the two one-sided tolerance bounds to provide approximate two-sided tolerance intervals. Their methods, when compared to other methods, tend to have coverage probabilities that are closer to the nominal confidence level. Krishnamoorthy et al. (2011) provide a further simple approximate method for constructing Poisson (and binomial) tolerance intervals.

Prediction intervals

Nelson (1970) describes the equivalent of the conservative Poisson distribution prediction interval method given in Section 7.6.1. Nelson (1982) also gives the normal distribution approximate method described in Section 7.6.2. The Jeffreys prediction interval method given in Section 7.6.4 derives from Bayesian prediction with a Jeffreys prior distribution, as illustrated in Sections 16.2.5 and H.6.2. Wang (2008) gives expressions for the coverage probability for prediction intervals for discrete distributions and suggests an alternative method for constructing prediction intervals. Krishnamoorthy and Peng (2011) review various existing methods and suggest other simple closed-form methods for prediction intervals for both the Poisson and binomial distributions, including the joint-sample method given in Section 7.6.3. They also compare coverage probabilities and expected length properties.

Chapter 8

Sample Size Requirements for Confidence Intervals on Distribution Parameters

OBJECTIVES AND OVERVIEW

This chapter addresses the frequently asked question "How large a sample do I need to obtain a confidence interval?" To determine sample size requirements, one generally starts with a statement of the needed precision (e.g., in terms of interval width) and then uses the procedures for constructing statistical intervals described in the previous chapters "in reverse."

This and the following two chapters are concerned with data *quantity* (sample size). We need, however, to reiterate that the issue of data quantity is often secondary to that of the *quality* of the data. In particular, in making a statistical estimate or constructing a statistical interval, one assumes that the available data were obtained by using a random sample from a defined population or process of interest. As stated previously, when this is not the case, all bets are off. Just increasing the sample size—without broadening the scope of the investigation—does not compensate for lack of randomness; all it does is allow one to obtain a possibly biased estimate with greater precision. Putting it another way, increasing the sample size per se usually improves the precision of an estimate, but not necessarily its accuracy.

Section 8.1 describes basic requirements for sample size determination. Subsequent sections of this chapter deal with sample size determination methods to estimate a:

- Normal distribution mean (Section 8.2).

- Normal distribution standard deviation (Section 8.3).

- Normal distribution quantile (Section 8.4).

Statistical Intervals: A Guide for Practitioners and Researchers, Second Edition.
William Q. Meeker, Gerald J. Hahn and Luis A. Escobar.
© 2017 John Wiley & Sons, Inc. Published 2017 by John Wiley & Sons, Inc.
Companion Website: www.wiley.com/go/meeker/intervals

- Binomial proportion (Section 8.5).

- Poisson occurrence rate (Section 8.6).

In this chapter and the following two chapters we present sample-size-determination methods for a variety of situations described in this book. Sometimes, these methods will lead to the finding that to attain the needed degree of precision, one requires a larger sample than is practical. Discouraging as this may be, it is better to know before starting an investigation than at its conclusion.

8.1 BASIC REQUIREMENTS FOR SAMPLE SIZE DETERMINATION

To determine the required sample size, one generally requires:

- A specification of the objectives of the investigation.

- A statement of the needed precision.

- A decision of what statistical distribution, if any, is to be assumed, and, frequently initial guesses or estimates, to be referred to as "planning values," for one or more parameters of that distribution.

8.1.1 The Objectives of the Investigation

Before one can determine how large a sample is needed, one must specify what is to be computed from the resulting data. This could, for example, be a confidence interval for a specified parameter, such as the mean or standard deviation of a normal distribution, a binomial proportion, a Poisson occurrence rate, a tolerance interval to contain a specified proportion of the distribution, or a prediction interval to contain one or more future observations. One also needs to decide whether a two-sided interval or a one-sided bound is to be constructed and the desired confidence level.

8.1.2 Statement of Needed Precision

Statisticians are often asked how many observations are needed to estimate some quantity (e.g., the mean of a sampled distribution) with "95% confidence." Such a question, however, is generally insufficient per se. In fact, a literal answer is often one or two observations, depending upon the specific quantity being estimated. Unfortunately, the resulting precision is often so poor (reflected by a very wide interval) that the resulting interval has little value.

For example, suppose a vendor wants to provide a customer an interval that contains, with 95% confidence, the proportion of conforming units in a large manufacturing lot. Such an interval can be computed even if one has only a single randomly selected unit. If the unit is in conformance, the 95% confidence interval for the proportion of conforming units in the lot is [0.025, 1.00] (see Section 6.2). Similarly, if two units were randomly selected and both are found to be conforming, the calculated 95% confidence interval is [0.16, 1.00]. In these cases, a statistical confidence interval has been found, but it is of little practical value.

As a second example, suppose that the mean of a normal distribution is to be estimated from a sample of two observations. A random sample of two units has resulted in readings of 15.13 and 15.25. A 95% confidence interval to include the distribution mean is [14.4, 15.95] (see Section 4.2). The 99% confidence interval is much wider still: [11.37, 19.01]!

These examples illustrate that it is often possible to compute confidence intervals from a sample of two and, in some cases, even a single observation. The resulting intervals, however, are, because of their immense width, generally of little value (except to demonstrate how little the data tell us about the characteristic of interest).

Thus, to determine the sample size required to obtain a *useful* interval, one must specify not only the desired confidence level, but also the needed precision. Such precision is measured, for example, by the allowable error in the resulting estimate or the half-width of the statistical interval that is to be constructed from the data. For example, for a product packaged in jars labeled to contain "one pound net weight," one might desire a sufficiently large sample to be able to estimate the actual mean content weight within 0.1 ounce with 95% confidence.

8.1.3 Assumed Statistical Distribution and Parameter Planning Values

Chapters 4, 6, and 7 describe statistical intervals for a normal distribution, a binomial distribution, and a Poisson distribution, respectively. The problem context should make clear which of these models apply. There are also many other distributions that might be appropriate in a particular situation, as illustrated in Chapters 12–18. Also, Chapter 5 describes distribution-free intervals as an alternative to those for the normal, or some other specific, distribution.

The formulas for computing statistical intervals depend on the assumed distribution. Thus, the assumed distribution, if any, must be specified before the required size of the sample can be determined. Moreover, if a particular distribution is assumed, the sample size determination often requires a "planning value" for an unknown distribution parameter. For example, to determine the sample size:

- To estimate the mean of a normal distribution will require a "planning value" of the distribution standard deviation.

- To estimate the proportion of nonconforming units in a binomial distribution requires a planning value of the proportion to be estimated.

In general, such information is unknown before the investigation. If it were known, the investigation would be unnecessary in many cases, such as the second example. One can, however, usually provide conservative planning values. These, in turn, will usually result in conservative (i.e., larger than needed) sample sizes. We will use the superscript □ to indicate a planning value.

8.2 SAMPLE SIZE FOR A CONFIDENCE INTERVAL FOR A NORMAL DISTRIBUTION MEAN

8.2.1 Introduction

This section shows how to choose a sample size large enough to estimate, with a specified precision, the mean of a normal distribution. The width (or half-width) of the resulting confidence interval (described in Section 4.2) is a convenient way to specify the needed precision. We will show how to choose the sample size n such that the resulting confidence interval for μ has the form $\bar{x} \pm d$, where d is the desired confidence interval half-width.

The first few methods described here require a planning value for the distribution standard deviation. Alternative methods, briefly described at the end of this section, use two-stage sampling to avoid having the width of the confidence interval depend on a planning value.

8.2.2   Tabulations and a Simple Formula for the Case when σ is Assumed to Be Known

Table J.17a gives the sample size needed to estimate μ within $\pm k\sigma$ with confidence level $1 - \alpha$, when $\sigma^\square = \sigma$ is assumed to be *known*. Specifically, the table provides the sample size needed to obtain a confidence interval that has a half-width of $d = k\sigma^\square$, as a function of k and $1 - \alpha$. The table provides entries for values of $1 - \alpha$ from 0.50 to 0.999 and for k from 0.01 to 2.00.

The quantities in Table J.17a were computed from the simple approximate formula

$$n = \left[\frac{z_{(1-\alpha/2)}\sigma^\square}{d} \right]^2 \tag{8.1}$$

and then rounding n to the next largest integer. One can also use (8.1) directly in place of the tabulations or for nontabulated values of k and α.

Example 8.1   Sample Size to Estimate Mean Alloy Tensile Strength. An experiment to estimate the mean tensile strength of a new alloy is to be conducted. The experimenters must decide how many specimens to test so that the 95% confidence interval for μ will have a half-width of 1,500 kilograms (i.e., $d = 1,500$). Experience with similar alloys suggests that variability in specimen strength can be modeled by a normal distribution. A conservative (high) guess for the standard deviation of the distribution of tensile strength is $\sigma^\square = 2,500$ kilograms.

From the preceding, we calculate $d = k\sigma^\square = 1,500$. Then $k = d/\sigma^\square = 1,500/2,500 = 0.60$, and we find from Table J.17a that the necessary sample size is 11. As a check, (8.1) gives

$$n = \left[\frac{1.96 \times 2,500}{1,500} \right]^2 = 10.67$$

which, when rounded up to 11, agrees with the value obtained from Table J.17a.   ∎

8.2.3   Tabulations for the Case when σ is Unknown

When σ is unknown, the situation is more complicated because the confidence interval half-width is now itself a random variable, and therefore cannot be determined exactly ahead of time. Many elementary textbooks give (8.1) as a simple way to determine the approximate sample size needed to obtain a confidence interval that will be close to the specified half-width. In this case, a planning value $\sigma^\square$ is used in place of σ as a *prediction* of s, the estimate of σ that will be obtained from the sample. At the time the study is planned, neither σ nor s is known. To be on the safe side, however, one can use a conservatively large planning value $\sigma^\square$ for σ.

Even if the planning value $\sigma^\square$ is exactly the same as the actual value σ, there is a substantial probability (generally greater than 0.50) that using (8.1) will result in a confidence interval with half-width that is *larger* than the needed value d. The main reason for this is that the sample standard deviation s obtained from the sample, and used in (4.1), is likely to be larger than σ.

As we have indicated, the width of the confidence interval is a random variable. Thus it is appropriate to select the sample size so that, with a prespecified probability, the resulting interval half-width is not more than d, assuming that the planning value $\sigma^\square$ is, indeed, equal to σ. Table J.17b provides a means for doing this by making an upward adjustment to the sigma-known sample size provided by (8.1). The table is used as follows. Suppose that we want the future sample to yield a $100(1 - \alpha)\%$ confidence interval that has a half-width that is no larger than d with $100(1 - \gamma)\%$ probability if $\sigma = \sigma^\square$. First one uses Table J.17a or (8.1) to get an initial value n, based again upon a planning value $\sigma^\square$ for σ. Then one enters Table J.17b with this value and the values of $1 - \alpha$ and $1 - \gamma$ and reads the adjusted sample size from the body of the table. The γ' column in Table J.17b gives the probability that the confidence interval half-width is less than d when the initial sample size from the simple formula in (8.1) is used if indeed $\sigma^\square = \sigma$, but the sample estimate s, subsequently obtained from the data, is used for $\sigma^\square$ in (8.1).

Example 8.2   Tabular Method to Find the Sample Size to Estimate Mean Alloy Tensile Strength with Unknown Sigma. Here we continue with Example 8.1, where (8.1) gave $n \approx 11$. Entering Table J.17b with $1 - \alpha = 0.95$ and $1 - \gamma = 0.90$, interpolating between $n = 10$ and 15 for $n = 11$, gives a final sample size of about 19. Thus, the price one pays to be 90% sure that the half-width of the confidence interval does not exceed $d = 1,500$, as compared to using the simpler, less conservative method (that does not provide such assurance), is an increase in the sample size from 11 to 19. We note, moreover, from Table J.17b that if σ were indeed equal to $\sigma^\square$, then the probability is only 0.34 that the half-width of the confidence interval will be less than d if a sample of size $n = 11$ were used. ∎

8.2.4   Iterative Formula for the Case when σ is Unknown

For situations not covered in Table J.17b, and to explain how the table works, we give the following iterative method of finding the sample size for unknown σ.

We want to find the smallest sample size that allows us to be $100(1 - \gamma)\%$ sure that the resulting $100(1 - \alpha)\%$ confidence interval will have a half-width less than d (assuming that the planning value $\sigma^\square > \sigma$). A suitably modified version of (8.1) can be obtained by substituting $t_{(1-\alpha/2;n-1)}$ for $z_{(1-\alpha/2)}$ and $\widetilde{S}$ for $\sigma^\square$ in this expression, where $\widetilde{S}$ is the upper $100(1 - \gamma)\%$ prediction bound for S, the sample standard deviation of the future sample of size n that we wish to determine. Thus, we start with the expression

$$ n \geq \left[\frac{t_{(1-\alpha/2;n-1)}\widetilde{S}}{d} \right]^2 . \tag{8.2} $$

The upper $100(1 - \gamma)\%$ prediction bound for S (assuming that $\sigma^\square$ is the *actual* value of σ) is

$$ \widetilde{S} = \sigma^\square \left[\frac{\chi^2_{(1-\gamma;n-1)}}{n - 1} \right]^{1/2} . $$

Substituting this for $\widetilde{S}$ in (8.2) gives

$$ n \geq \left[\frac{t_{(1-\alpha/2;n-1)}\sigma^\square}{d} \right]^2 \left[\frac{\chi^2_{(1-\gamma;n-1)}}{n - 1} \right] . \tag{8.3} $$

Because n appears on both sides of (8.3), iteration is required to find a solution. In particular, we need to find the smallest value of n such that the left-hand side is greater than the right-hand side. It is easy to write a computer program or a simple R function to do this using standard numerical methods. Working manually, one can use the following steps:

1. Use (8.1) to get a starting value n_1 for n, based upon the planning value $\sigma^\square$.

2. Use the formula

$$ n_2 = \left[\frac{t_{(1-\alpha/2;n_1-1)}\sigma^\square}{d} \right]^2 \left[\frac{\chi^2_{(1-\gamma;n_1-1)}}{n_1 - 1} \right] $$

and round to the next higher integer to obtain an adjusted sample size value n_2.

3. Obtain successive additional values for n by using the recursion formula

$$n_i = \left[\frac{t_{(1-\alpha/2;n_{i-1}-1)}\sigma^\square}{d} \right]^2 \left[\frac{\chi^2_{(1-\gamma;n_{i-1}-1)}}{n_{i-1}-1} \right]$$

and round up at each stage until $n_i = n_{i-1}$ or $n_i = n_{i-1} - 1$.

Generally no more than about five to seven iterations are required.

Example 8.3 Iterative Method to Find the Sample Size to Estimate Mean Alloy Tensile Strength with Unknown Sigma. As in Example 8.2, suppose that we want to be 90% sure that the half-width of the 95% confidence interval to contain μ does not exceed 1,500, and our planning value for σ is 2,500. Thus, as before, $\sigma^\square = 2,500$, $d = 1,500$, $1 - \alpha = 0.95$, and $1 - \gamma = 0.90$. Our first guess is $n_1 = 11$—the solution for the case for which σ is known. Then, using $t_{(0.975;10)} = 2.228$ and $\chi^2_{(0.90,10)} = 15.99$ in (8.3), we obtain

$$n_2 = \left(\frac{2.228 \times 2,500}{1,500} \right)^2 \frac{15.99}{10} = 22.047.$$

Rounding up to $n_2 = 23$ and using $t_{(0.975;22)} = 2.074$ and $\chi^2_{(0.90;22)} = 30.81$ gives

$$n_3 = \left(\frac{2.074 \times 2,500}{1,500} \right)^2 \frac{30.81}{22} = 16.73.$$

The next three iterations give, after rounding up, $n_4 = 19$, and $n_5 = 18$, and $n_6 = 19$, allowing us to terminate the iterations. Thus $n = 19$ is the smallest integer such that the left-hand side of (8.3) is greater than the right-hand side. This also agrees with the value obtained from Table J.17b in Example 8.2. ∎

8.2.5 Using an Upper Prediction Bound from a Previous Sample when σ is Unknown

If one has an estimate of σ from a previous random sample from the same distribution, one can obtain an alternative sample size formula by using the upper prediction bound for σ (described in Section 4.9) instead of the planning value $\sigma^\square$.

In particular, if $\sigma_\ell^\square$ is a planning value (in this case, replaced by the sample standard deviation) based on a previous sample with $\ell - 1$ degrees of freedom, one needs to solve

$$n \geq \left[\frac{t_{(1-\alpha/2;n-1)}\sigma_\ell^\square}{d} \right]^2 F_{(1-\gamma;n-1,\ell-1)}$$

for the smallest value of n such that this inequality holds. This can be done by using a root-finding algorithm (or simple trial and error). A simple (large-sample) approximation for the required sample size is

$$n \approx n_1 F_{(1-\gamma;n_1-1,\ell-1)},$$

where n_1 is obtained from (8.1) or (8.2), using $\sigma_\ell^\square$ for $\sigma^\square$.

8.2.6 A Two-Stage Sampling Method

When an investigation can be conducted in two stages, it is possible to obtain a confidence interval with exactly the needed half-width d, even if σ is unknown. This is done by using the sample standard deviation of the first stage to compute a confidence interval, and ignoring the sample standard deviation of the second stage in constructing the confidence interval. The sample size n_1 for the first stage should be chosen as large as possible, but smaller than the anticipated total sample size. Specifically, obtain n_1, say, from (8.1) using a planning value for $\sigma^{\square}$, which now need not be conservative. The sample size for the second stage of the investigation is then

$$n_2 = \left[\frac{t_{(1-\alpha/2;n_1-1)} s_1}{d} \right]^2 - n_1,$$

where s_1 is the sample standard deviation from the first stage. The resulting $100(1 - \alpha)\%$ confidence interval for μ is

$$\bar{x} \pm d,$$

where $\bar{x}$ is the mean of the $n_1 + n_2$ observations from both stages. If n_2 turns out to be negative, then the data from the first stage will give an interval with a half-width that is less than d. In this case the second sample is not needed.

This two-stage sampling procedure has been rightfully subject to criticism because, though exact, it does not, as previously indicated, use the information from the second stage to estimate σ. This, however, will not be a serious practical concern if the sample size in the first stage is sufficiently large (e.g., $n_1 > 30$).

Example 8.4   Use of Two-Stage Sampling to Estimate Mean Alloy Tensile Strength. We continue with Example 8.1, but now use $d = 500$. Suppose that an initial sample of 20 units yielded $s_1 = 2{,}500$. Then, using $t_{(0.975;19)} = 2.093$, the investigation would require a total of

$$n_1 + n_2 = \left(\frac{2.093 \times 2{,}500}{500} \right)^2 \approx 110$$

observations; that is, $n_2 = 110 - 20 = 90$ observations in the second stage.

If an initial sample of size 10, instead of size 20, had been taken, and had again given $s_1 = 2{,}500$, the total sample size requirements would have been estimated, using $t_{(0.975;9)} = 2.262$, to be

$$n_1 + n_2 = \left(\frac{2.262 \times 2{,}500}{500} \right)^2 \approx 128$$

instead of 110. ∎

8.3   SAMPLE SIZE TO ESTIMATE A NORMAL DISTRIBUTION STANDARD DEVIATION

8.3.1   Introduction

This section gives easy-to-use methods for choosing the sample size needed to estimate a normal distribution standard deviation σ.

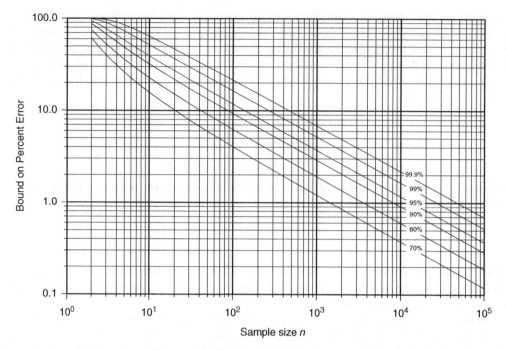

Figure 8.1   Sample size needed to estimate a normal distribution standard deviation for various probability levels. This figure is based on methodology described by Greenwood and Sandomire (1950).

8.3.2 Computational Method

The exact upper probability bound giving the percent error for estimating σ from a sample of size n is computed as

$$100p = 100\left[1 - \left(\frac{\chi^2_{(\alpha;n-1)}}{n-1}\right)^{1/2}\right]. \tag{8.4}$$

Solving 8.4 for n gives the sample size needed so that, with a specified probability $1 - \alpha$, σ will be underestimated by no more than $100p\%$. One can use a root-finding algorithm (or simple trial and error) to find the needed value of n for specified p and α.

8.3.3 Graphical Method

Figure 8.1 gives solutions of (8.4) for n for $1 - \alpha = 0.70, 0.80, 0.90, 0.95, 0.99$, and 0.999 for values of percent error $100p$ between 0.10 and 100 and n up to 100,000. These curves can be used in reverse to find n for a given $100p$. In particular, to find n, enter Figure 8.1 on the "bound on percent error" axis with needed percent error $100p$ to the curve with the specified confidence level. Then read down from that curve to obtain the needed sample size n.

8.3.4 Tabular Method

Table J.18 gives solutions of (8.4) for n for $1 - \alpha = 0.80, 0.85, 0.90, 0.95, 0.99$, and 0.999 for 22 values of percent error $(100p)$ from 0.70 to 100.

Example 8.5 Sample Size Needed to Estimate the Measurement Error of a Chemical Assay Procedure. An experiment is to be conducted to estimate σ, the standard deviation of the measurement error of a chemical assay procedure. From previous experience with similar procedures, one can assume that the measurement error is normally distributed. The experiment will be conducted with specimens that are known to contain exactly the same amount of a chemical and that are destroyed during the assay. The experimenters need to know how many specimens to prepare so that, with 95% probability, the sample standard deviation s will underestimate the actual measurement error standard deviation σ by no more than 20%.

To find the needed sample size, enter Figure 8.1 on the "bound on percent error" axis with $100p\% = 20\%$ and read down from the 95% curve, giving $n \approx 36$. As a check, using (8.4) with $n - 1 = 35$ and $\chi^2_{(0.05;35)} = 22.46$ gives the 95% upper bound on the percent error for estimating σ as

$$100\left[1 - \left(\frac{22.46}{35}\right)^{1/2}\right] = 19.8\%,$$

or approximately 20%. For another check, the tabular method, entering Table J.18 with $100p\% = 20\%$ and $1 - \alpha = 0.95$, gives $n \approx 36$. Using R as a calculator yields a similar result:

```
> 100*(1-sqrt(qchisq(p=0.05,df=36-1)/(36-1)))
[1] 19.88
```

∎

8.4 SAMPLE SIZE TO ESTIMATE A NORMAL DISTRIBUTION QUANTILE

As described in Section 2.4.2, a one-sided confidence bound on a distribution quantile is equivalent to a one-sided tolerance bound. Methods for sample size determination for a one-sided tolerance bound are given in Section 9.1 and will not be repeated here.

8.5 SAMPLE SIZE TO ESTIMATE A BINOMIAL PROPORTION

8.5.1 Introduction

This section shows how to determine the approximate sample size needed to estimate a population proportion, with specified precision. More specifically, we want the resulting confidence interval to be no larger than $\widehat{\pi} \pm d$, where d is the confidence interval half-width. To make this determination, one must provide a "planning value," to be denoted by $\pi^{\square}$, for the sample estimate $\widehat{\pi}$ of π that will be obtained from the data. A conservative planning value for $\pi^{\square}$ (in the sense that, if incorrect, it will tend to overestimate the required sample size) is the value closest to 0.50 that still appears plausible.

8.5.2 Graphical Method

As explained in Section 6.2, Figure 6.1 can be used to obtain confidence intervals or bounds for a population proportion π. The figures can be used in reverse to determine the required sample size for estimating π. To do this, one must first choose a planning value $\pi^{\square}$. One then uses Figure 6.1 and sets the observed proportion $\widehat{\pi} = \pi^{\square}$ on the horizontal axis. The resulting confidence intervals, for various sample sizes, are read on the (vertical) π axis. Thus, at a glance, one can readily assess the effect on the confidence interval half-width of using different sample sizes.

8.5.3   A Simple Computational Procedure

The expression

$$n = \left[\pi^{\square} \left(1 - \pi^{\square} \right) \right] \left[\frac{z_{(1-\alpha/2)}}{d} \right]^2 \tag{8.5}$$

rounded to the next largest integer gives the approximate sample size needed to obtain a $100(1 - \alpha)\%$ confidence interval for π with half-width d. This normal distribution large sample size approximation is generally satisfactory if both $n\pi$ and $n(1 - \pi)$ exceed 10. Like the graphical procedure, use of this expression requires a planning value $\pi^{\square}$. The most conservative (largest) n is again obtained by choosing $\pi^{\square} = 0.50$.

Example 8.6   Sample Size Needed to Estimate the Proportion of Incorrectly Assembled Devices. Some unknown proportion π of a large number of field-installed devices were assembled incorrectly and need to be repaired. To assess the magnitude of the problem, the manufacturer of the devices needs to estimate π, the proportion of incorrectly assembled devices in the field. In particular, the manufacturer needs to know how many units to sample at random so that $\widehat{\pi}$ will be within ± 0.08 of π with 90% confidence (i.e., so that the 90% confidence interval around the estimate $\widehat{\pi}$ should be no larger than ± 0.08). It is possible that π is close to 0.50, and, therefore, $\pi^{\square} = 0.50$ will be taken as the planning value for π.

In Figure 6.1, using $\pi^{\square} = 0.50$, we see that a sample of size 10 gives a 90% confidence interval of $[0.22, \quad 0.78]$, a sample of size 30 gives a 90% confidence interval of $[0.34, \quad 0.66]$, a sample size of 100 gives a 90% confidence interval of $[0.41, \quad 0.59]$, and, using interpolation, a sample size of 200 gives a confidence interval of $[0.44, \quad 0.56]$. Thus, to estimate π within ± 0.08 with 90% confidence requires a sample size somewhat above 100 if $\pi^{\square}$ is taken to be 0.50. After the sample has been taken, one generally uses the resulting sample estimate $\widehat{\pi}$, in place of $\pi^{\square}$ in obtaining the needed 90% confidence interval to contain π, and this interval would have a smaller half-width than 0.08 unless $\widehat{\pi} = 0.50$. Note also that if $\pi^{\square}$ is taken either smaller or larger than 0.50, the required sample size to achieve an interval of the same half-width would be somewhat smaller.

Using (8.5) with $d = 0.08$, $1 - \alpha = 0.90$ (or $1 - \alpha/2 = 0.95$), $\pi^{\square} = 0.50$, and $z_{(0.95)} = 1.645$, the required sample size is

$$n = [0.50(1 - 0.50)] \left(\frac{1.645}{0.08} \right)^2 \approx 106.$$

This approximation is reasonable because both $n\pi^{\square} = n(1 - \pi^{\square}) = 53$ appreciably exceed 10.

Suppose, on the other hand, one expects $\widehat{\pi}$ to be less than 0.20 (or to be greater than 0.80). Then using the planning value $\pi^{\square} = 0.20$, the required sample size to obtain a 90% confidence interval for π with half-width $d = 0.08$ is only

$$n = [0.20(1 - 0.20)] \left(\frac{1.645}{0.08} \right)^2 \approx 68.$$

8.6 SAMPLE SIZE TO ESTIMATE A POISSON OCCURRENCE RATE

8.6.1 Introduction

In this section, we will be dealing with observed events, where the number of such events is assumed to follow a Poisson distribution with parameter λ (constant rate per unit of exposure), which we wish to estimate. This section shows how to choose the approximate sample size to estimate a Poisson occurrence rate λ with a specified precision. That is, we require the sample to be sufficiently large so that the upper or lower confidence bound does not differ from the sample estimate by more than a prespecified percentage. It is necessary to provide a "planning value" $\lambda^{\square}$ for $\widehat{\lambda}$, the sample estimate of λ. A conservative approach (tending to result in a larger sample size than actually needed) is to take $\lambda^{\square}$ to be the largest value expected for $\widehat{\lambda}$.

8.6.2 Graphical Method

For a specified confidence level, $100(1 - \alpha)\%$, Figure 8.2a shows the percentage by which the one-sided lower confidence bound for a Poisson occurrence rate is less than the sample estimate as a function of the number of occurrences in the sample. In particular, curves are given for $1 - \alpha = 0.70, 0.80, 0.90, 0.95, 0.99$, and 0.999. Similarly, Figure 8.2b shows the percentage by which the one-sided upper confidence bound for a Poisson occurrence rate exceeds the sample estimate. In Figure 8.2a, the percent error was computed as $100[1 - (\underset{\sim}{\lambda}/\widehat{\lambda})]$ and for Figure 8.2b, the percent error was computed as $100[(\widetilde{\lambda}/\widehat{\lambda}) - 1]$.

To use these figures to determine the needed sample size, enter the graph on the vertical axis at the point corresponding to the needed bound on the percent error in the estimate of the occurrence rate. Then draw a line horizontally to intersect the line corresponding to the needed degree of confidence. Now move down from the point of intersection to read x, the "needed" number of occurrences, from the horizontal axis. Using a planning value $\lambda^{\square}$ for the occurrence rate λ, the approximate required sample size is $n \approx x/\lambda^{\square}$.

Example 8.7 Sample Size to Compute an Upper Confidence Bound on the Flaw Occurrence Rate. Flaws on the painted surface of an appliance occur independently of one another at a constant rate λ per appliance. This implies a Poisson distribution for the number of flaws per constant surface area. A new improved process that is believed to have a flaw rate λ of not more than 0.10 flaws per appliance has been developed. A study is to be conducted to estimate this mean flaw rate precisely enough so that the one-sided upper 95% confidence bound on λ will exceed the sample estimate $\widehat{\lambda}$ by not more than 20%. It is needed to determine how many appliances must be selected at random from the process to meet this criterion. This example calls for the use of Figure 8.2b because we wish to construct an upper confidence bound on λ.

From Figure 8.2b, we note that the one-sided upper 95% confidence bound $\widetilde{\lambda}$ will exceed $\widehat{\lambda}$ by about 20% if we observe 82 flaws. Using the planning value $\lambda^{\square} = 0.10$, the study will require approximately $n = x/\lambda^{\square} = 82/0.10 = 820$ test units. If this problem had instead called for a lower 95% confidence bound $\underset{\sim}{\lambda}$, Figure 8.2a tells us that $\underset{\sim}{\lambda}$ would be about 20% less than $\widehat{\lambda}$ if we had observed 63 flaws, and thus the study would have required approximately 630 appliances. The large sample size required to obtain the desired level of precision in this example is daunting—but, as previously observed, it is important to have this information before starting the study so that we can act appropriately. ∎

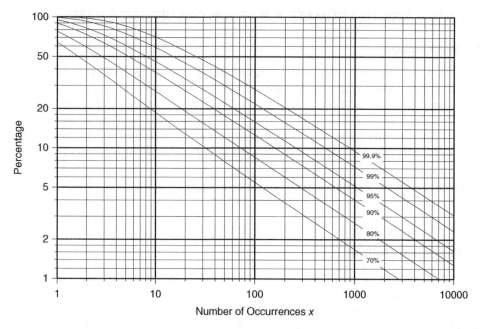

Figure 8.2a Percentage by which the lower confidence bound for the Poisson parameter λ is less than $\widehat{\lambda}$ for various confidence levels.

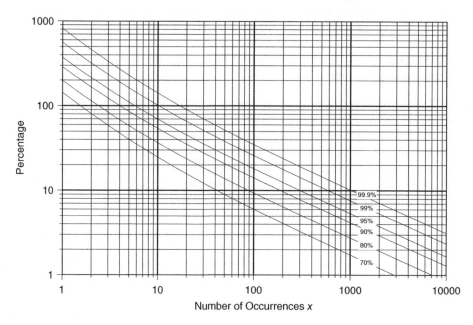

Figure 8.2b Percentage by which the upper confidence bound for the Poisson parameter λ exceeds $\widehat{\lambda}$ for various confidence levels.

8.6.3 A Simple Approximate Computational Procedure

The expression

$$n = \lambda^\square \left[\frac{z_{(1-\alpha/2)}}{d} \right]^2 \tag{8.6}$$

rounded to the next larger integer gives the approximate sample size needed to obtain a $100(1 - \alpha)\%$ confidence interval with a half-width d for the Poisson occurrence rate. For a one-sided bound, replace $1 - \alpha/2$ with $1 - \alpha$. This approximation, based on the fact that $\widehat{\lambda}$ follows approximately a normal distribution in large samples, is often adequate for practical purposes when the sample size n is large enough to yield at least 10 occurrences.

Example 8.8 Sample Size to Compute an Upper Confidence Bound on the Flaw Occurrence Rate. Continuing with Example 8.7, $\lambda^\square = 0.10$, $z_{(0.95)} = 1.645$, and $d = 0.02$ (20% of $\lambda^\square$). Thus, to construct the desired upper 95% confidence bound on λ, the study will require

$$n = 0.10 \left(\frac{1.645}{0.02} \right)^2 \approx 677$$

appliances. The difference between this result and the previous approximate sample size of 630, even though $n\lambda^\square = 677 \times 0.10 \approx 68$, is due to the lack of symmetry in the sampling distribution of $\widehat{\lambda}$ and the fact that Figure 8.2b is based on a different confidence interval procedure than (8.6). ∎

BIBLIOGRAPHIC NOTES

Much has been written about how to choose a sample size in a statistical study. A considerable amount of this work focuses on the power of hypothesis tests, but some of it is concerned with the width of confidence intervals. The many books devoted to this topic include Mace (1964), Brush (1988), Mathews (2010), Desu and Raghavarao (2012), and Ryan (2013). The books Odeh and Fox (1975), Odeh et al. (1977), and Odeh and Owen (1980, 1983) provide tables giving confidence and tolerance intervals, one-sided bounds, and sampling plans for a range of different statistical applications, and many of these tables and charts can be used to help design statistical studies (and choose a sample size). In addition, the more general books Natrella (1963), Beyer (1968), and Dixon and Massey (1969) also contain tables, charts, and figures that can be used to determine the necessary sample size for various problems.

Adcock (1997) provides a review of methods for determining a sample size and compares Bayesian and non-Bayesian approaches. Lindley (1997) and Sahu and Smith (2006) also outline Bayesian methods for determining a sample size. Lenth (2001) discusses some practical matters and describes a computer package interface for such problems.

Kupper and Hafner (1989) show that even if the planning value $\sigma^\square$ were exactly the same as the actual value σ the chances are (i.e., with a probability between 0.53 and 0.87 for the cases considered) that in using (8.1) the half-width of the resulting confidence interval will be larger than the desired value d. Table J.17b was patterned after a similar figure in Kupper and Hafner (1989). The two-stage sampling procedure in Section 8.2.6 was first suggested by Stein (1945).

Meeker and Escobar (1998) give a general approach for determining the sample size needed to estimate a function of parameters (e.g., a distribution quantile) with a given degree of precision.The approach is based on maximum likelihood estimation theory and illustrated with (log-)location-scale distributions.

A figure similar to Figure 8.1 was given in Greenwood and Sandomire (1950).

Chapter *9*

Sample Size Requirements for Tolerance Intervals, Tolerance Bounds, and Related Demonstration Tests

OBJECTIVES AND OVERVIEW

This chapter shows how to determine sample size requirements for tolerance intervals and for related demonstration tests concerning the proportion of product that exceeds (or is exceeded by) a specified value. This chapter explains sample size determination methods for:

- Normal distribution tolerance intervals and bounds (Section 9.1).

- A one-sided demonstration test based on normally distributed measurements to give a desired probability of successful demonstration (Section 9.2).

- Minimum sample size for distribution-free two-sided tolerance intervals and one-sided tolerance bounds (Section 9.3).

- Distribution-free two-sided tolerance intervals and one-sided distribution-free tolerance bounds with a specified amount of precision (Section 9.4).

- A one-sided demonstration test based on binomial data to give a desired probability of successful demonstration (Section 9.5).

Statistical Intervals: A Guide for Practitioners and Researchers, Second Edition.
William Q. Meeker, Gerald J. Hahn and Luis A. Escobar.
© 2017 John Wiley & Sons, Inc. Published 2017 by John Wiley & Sons, Inc.
Companion Website: www.wiley.com/go/meeker/intervals

9.1   SAMPLE SIZE FOR NORMAL DISTRIBUTION TOLERANCE INTERVALS AND ONE-SIDED TOLERANCE BOUNDS

This section provides simple methods for finding the sample size needed to achieve a specified level of precision when the data are to be used to compute a two-sided tolerance interval or a one-sided tolerance bound to contain at least a specified proportion of a sampled normal distribution. As the sample size increases, the computed tolerance interval will approach the probability interval that *actually* contains the specified distribution proportion. Small sample sizes can, however, result in a tolerance interval that is much wider than this limiting probability interval.

9.1.1   Criterion for the Precision of a Tolerance Interval

We will use the following criterion for finding the sample size to control the size of a tolerance interval. Choose the sample size to be large enough such that both the following hold:

1. The probability is $1 - \alpha$ (large) that at least a proportion β of the distribution will be included within the tolerance interval.

2. The probability is δ (small) that more than a proportion β^* of the distribution will be included, where β and β^* are specified proportions and β^* is greater than or equal to β.

The idea is that, with fixed $\beta^* > \beta$, and probability

$$1 - \alpha = \Pr(\text{interval will contain at least a proportion } \beta \text{ of the distribution}),$$

the probability

$$\delta = \Pr(\text{interval will contain at least a proportion } \beta^* \text{ of the distribution})$$

is a decreasing function of the sample size n. That is, δ, the probability that the interval is so wide that it will contain a proportion β^* of the distribution, will decrease to zero as n increases. This criterion can be used for both two-sided tolerance intervals and one-sided tolerance bounds. As we will see in subsequent sections in this chapter, when a one-sided tolerance bound (or a one-sided confidence bound on a quantile) is used as a criterion for a demonstration test, the probability of successful demonstration will be $1 - \delta$.

9.1.2   Tabulations for Tolerance Interval/Bound Sample Sizes

Tables J.19 and J.20 give, respectively, the necessary sample sizes for two-sided tolerance intervals and one-sided tolerance bounds for a normal distribution for $\beta = 0.50, 0.75, 0.90, 0.95, 0.99, 1 - \alpha = 0.80, 0.90, 0.95, 0.99, \delta = 0.20, 0.10, 0.05, 0.01$, and several values of $\beta^* > \beta$, depending on β.

Example 9.1   Sample Size for a Normal Distribution Two-Sided Tolerance Interval for a Part Dimension. The engineers responsible for a machined part want to establish limits for a critical dimension so that, for marketing purposes, they can claim, with 95% confidence, that the interval contains the dimension for a large proportion of the parts. Based on experience with similar processes, the engineers feel that the dimensions can be adequately modeled by a normal distribution. The measurements on the dimensions for a random sample of the parts will be used to compute a two-sided tolerance interval to contain a proportion 0.90 of the distribution of parts produced from the process with 95% confidence, and this will provide the desired limits.

If the sample size is too small, the tolerance interval providing the desired coverage with the specified level of confidence may be so wide that it will appreciably overestimate the scatter in the distribution of the dimensions. Thus, in addition to the requirement that the tolerance interval contain at least a proportion $\beta = 0.90$ of the distribution with 95% confidence (i.e., $1 - \alpha = 0.95$), the manufacturer wants to choose the sample size n sufficiently large so that the probability is only $\delta = 0.10$ that the interval will actually contain a proportion $\beta^* = 0.96$ or more of the dimensions of the manufactured parts. From Table J.19, the sample size needed to accomplish this is $n = 91$ units. After the sample has been obtained, the desired tolerance interval is calculated using the methods given in Section 4.6.1.   ∎

Example 9.2   Sample Size for a Normal Distribution One-Sided Lower Tolerance Bound for Component Strength. The designers of a system want a one-sided lower 95% tolerance bound for the strength of a critical component. Again, the normal distribution model is felt to adequately describe the distribution of strengths. An experiment is to be conducted to obtain data to compute a lower tolerance bound that will, with 95% confidence, be exceeded by the strengths of at least 99% of the components in the sampled product population. As described in Section 4.6.3, this lower tolerance bound is equivalent to a one-sided lower 95% confidence bound on the 0.01 quantile of the strength distribution.

If the chosen sample size is too small, the resulting lower tolerance bound for strength will be unduly conservative. Thus, in addition to the requirement that the lower tolerance bound be exceeded by at least a proportion $\beta = 0.99$ of the components in the population, with 95% confidence (i.e., $1 - \alpha = 0.95$), the engineers want to choose a sample large enough so that the probability is only $\delta = 0.01$ that the resulting lower tolerance bound will be exceeded by the strengths of a proportion $\beta^* = 0.997$ or more of the components in the population. From Table J.20, the necessary sample size is $n = 370$. After the sample has been obtained the desired tolerance bound is found using the methods of Section 4.6.3.   ∎

9.2   SAMPLE SIZE TO PASS A ONE-SIDED DEMONSTRATION TEST BASED ON NORMALLY DISTRIBUTED MEASUREMENTS

9.2.1   Introduction

A one-sided tolerance bound (or equivalently a one-sided confidence bound on a quantile) is often used to conduct a demonstration test. This section shows how to find the sample size needed to conduct a test to demonstrate, with $100(1 - \alpha)\%$ confidence, that the $p^\dagger$ quantile of a normal distribution, denoted by $x_{p^\dagger}$, is less than or equal to a specified value $x^\dagger$ (i.e., $x_{p^\dagger} \leq x^\dagger$). The value $x^\dagger$ is often an upper specification limit that most of the product values (denoted by x) should not exceed. The demonstration will be successful if the one-sided upper confidence bound for the quantile is less than $x^\dagger$ (i.e., if $\widetilde{x}_{p^\dagger} \leq x^\dagger$).

This problem is equivalent to finding the sample size needed to demonstrate that $p = \Pr(X \leq x^\dagger)$ is greater than or equal to $p^\dagger$ for specified $x^\dagger$ and $p^\dagger$. The demonstration will be successful if the lower one-sided confidence bound for p is greater than or equal to $p^\dagger$ (i.e., if $p \geq p^\dagger$). We want the demonstration to be successful with high probability p_{dem} when the actual probability p is a specified value that exceeds $p^\dagger$. An expression for computing p_{dem} is derived in Section I.1.1.

In reliability applications, it is often necessary to demonstrate that $x_{p^\dagger} \geq x^\dagger$, where $p^\dagger$ is typically small, corresponding to a small quantile of the life distribution. In this case, $x^\dagger$ is a lower, rather than an upper, specification limit (e.g., minimum life). This situation can be handled by the same methods as those to be described for an upper specification limit. This is because the sample size needed to demonstrate $1 - p = \Pr(X \geq x^\dagger) \leq 1 - p^\dagger$ is the same as that needed to demonstrate $p = \Pr(X \leq x^\dagger) \geq p^\dagger$.

Example 9.3 Sample Size for a Normal Distribution One-Sided Tolerance Bound to Demonstrate Compliance with a Noise-Level Standard. A federal standard requires that the measured noise level of a particular type of machinery not exceed 60 decibels at a distance of 20 meters from the source for at least a proportion 0.95 of the units. A manufacturer of such machinery needs to test a random sample of the many units in the field to show compliance. This requires a demonstration test with $x^\dagger = 60$ and $p^\dagger = 0.95$ (i.e., it is required to demonstrate that $x_{0.95} = x_{p^\dagger} < x^\dagger = 60$).

The resulting data will be used to find an upper 90% confidence bound $(1 - \alpha = 0.90)$ on $x_{0.95}$, the 0.95 quantile of the distribution of noise emitted from the population of units in the field. The demonstration will be successful with 90% confidence if the upper 90% confidence bound on the 0.95 quantile of the population (i.e., $\widetilde{x}_{0.95}$) does not exceed $x^\dagger = 60$ decibels. Equivalently, the demonstration requires $p \geq p^\dagger$, where $\underset{\sim}{p}$ is a lower 90% confidence bound for $p = \Pr(X \leq x^\dagger)$, the proportion of units with noise levels less than or equal to $x^\dagger$ decibels. In addition, the manufacturer wishes the sample size to be sufficiently large so that the probability of a successful demonstration is $p_{\text{dem}} = \Pr(\underset{\sim}{p} \geq p^\dagger) \geq 0.95$ when the actual proportion nonconforming is $p = 0.98$. ∎

9.2.2 Graphical Method

Figures 9.1a–9.1d show p_{dem}, the probability of successfully demonstrating that the proportion nonconforming in the distribution is greater than $p^\dagger$ at a $100(1 - \alpha)\%$ confidence level. This probability is a function of the actual proportion nonconforming p, the sample size n, confidence level $100(1 - \alpha)\%$, and the probability to be demonstrated $p^\dagger$, assuming a normal distribution. These charts were developed from theory based on the noncentral t-distribution, described in Section I.1, and cover all combinations of $1 - \alpha = 0.90$ and 0.95, and $p^\dagger = 0.95$ and 0.99.

Example 9.4 Graphical Method to Find the Sample Size Needed to Demonstrate Compliance with a Noise-Level Standard. For Example 9.3, we use Figure 9.1a, designed for determining the required sample size for demonstrating with $100(1 - \alpha)\% = 90\%$ confidence that $p > p^\dagger = 0.95$. In particular, we enter the horizontal scale at $p = 0.98$ and move up, and simultaneously enter the vertical scale at the desired $p_{\text{dem}} = 0.95$ and move to the right, to find the point of intersection. We then find the closest curves and interpolate between them to find the needed value of n. In this case, we interpolate between 100 and 150 to get $n \approx 140$. ∎

9.2.3 Tabular Method

The tabular method to determine the required sample size for such problems uses Table J.20. Table J.20 is in terms of the previously discussed equivalent problem of setting a one-sided confidence bound on a quantile (equivalent to a one-sided tolerance bound) and uses the equivalent terminology where $\beta = p^\dagger$ corresponds to the probability to be demonstrated, β^* is the unknown actual probability, and $\delta = 1 - p_{\text{dem}}$ is the complement of the probability of successful demonstration. This table provides the required sample size for all combinations of $1 - \alpha = 0.80, 0.90, 0.95,$ and $0.99, \beta = 0.50, 0.75, 0.90, 0.95,$ and $0.99, \delta = 1 - p_{\text{dem}} = 0.01,$ $0.05, 0.10,$ and $0.20,$ and various values of β^*, depending on β.

Example 9.5 Tabular Method to Find the Sample Size Needed to Demonstrate Compliance with a Noise-Level Standard. For the application described in Example 9.3, we want to show, with 90% confidence $(1 - \alpha = 0.90)$, that the conforming proportion is $p > p^\dagger = 0.95$, while having a probability of successful demonstration $p_{\text{dem}} = 0.95$ $(\delta = 0.05)$ when the actual conforming proportion is $p = \beta^* = 0.98$. From Table J.20, we find the necessary sample size to be $n = 138$. The demonstration will be successful if $\widetilde{y}_{0.95} = \bar{x} + 1.826s$, the resulting upper

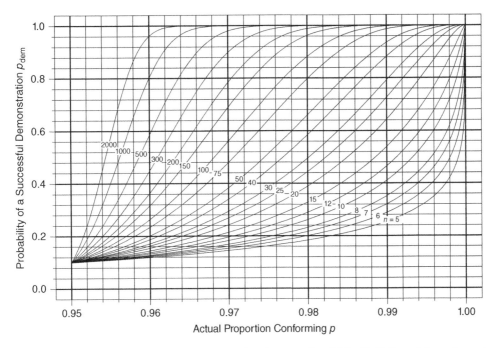

Figure 9.1a Probability of successfully demonstrating that $p > p^{\dagger} = 0.95$ with 90% confidence for various sample sizes (normal distribution).

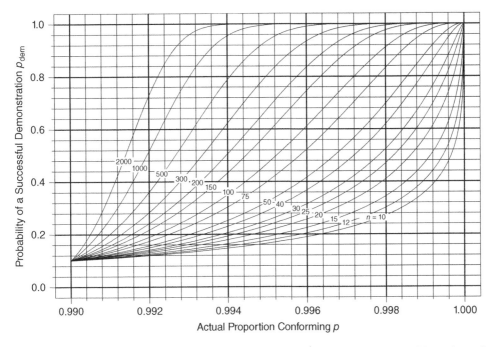

Figure 9.1b Probability of successfully demonstrating that $p > p^{\dagger} = 0.99$ with 90% confidence for various sample sizes (normal distribution).

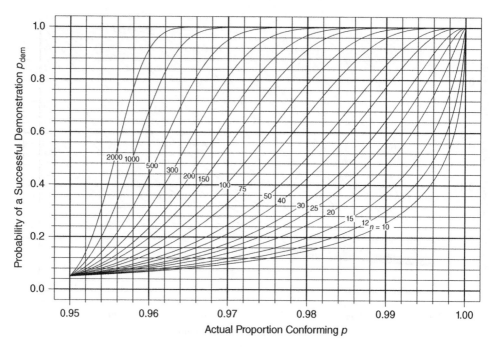

Figure 9.1c   Probability of successfully demonstrating that $p > p^\dagger = 0.95$ with 95% confidence for various sample sizes (normal distribution).

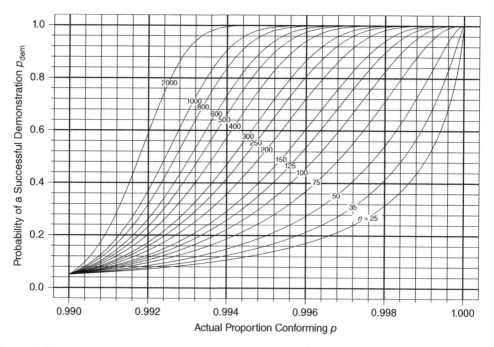

Figure 9.1d   Probability of successfully demonstrating that $p > p^\dagger = 0.99$ with 95% confidence for various sample sizes (normal distribution).

confidence bound for $y_{0.95}$, is less than 60. Here $\widetilde{y}_{0.95}$ is calculated as shown in Section 4.4, using interpolation in Table J.7c to obtain $g'_{(0.90;0.05,138)} = 1.83$. Alternatively, using R as a calculator gives

```
> qt(p=0.90,df=138-1,ncp=qnorm(0.95)*sqrt(138))/sqrt(138)
[1] 1.826222
```

■

9.2.4 Computational Method

The probability of successful demonstration for given n, $1 - \alpha$, $p^{\dagger}$, and p is derived in Section I.1.1 to be

$$p_{\text{dem}} = \text{pt}[\text{qt}(\alpha; n - 1, \delta_{p^{\dagger}}); n - 1, \delta_p],$$

where pt and qt are, respectively, the cdf and quantile functions of the noncentral t-distribution given in Section C.3.9. Section I.1.3 gives a procedure to find the smallest sample size n to provide the desired p_{dem}.

Example 9.6 Computational Method to Find the Sample Size Needed to Demonstrate Compliance with a Noise-Level Standard. For the application described in Example 9.3, we want to show, with 90% confidence ($1 - \alpha = 0.90$), that the conforming proportion is $p > p^{\dagger} = 0.95$, while having a probability of successful demonstration $p_{\text{dem}} = 0.95$ ($\delta = 0.05$) when the actual conforming proportion is $p = \beta^* = 0.98$. One can use the above graphical or tabular methods or trial and error to get a starting range of possible values for n. Using R as a calculator and trying values of n between 135 and 140 gives

```
> ntry <- 135:140
> pt(qt(0.10, ntry-1, -qnorm(0.95)*sqrt(ntry)), ntry-1,
  -qnorm(0.98)*sqrt(ntry))
[1]  0.9468584 0.9480746 0.9492642 0.9504277 0.9515657 0.9526787
```

suggesting that a sample size of $n = 138$ will meet the requirement.   ■

9.3 MINIMUM SAMPLE SIZE FOR DISTRIBUTION-FREE TWO-SIDED TOLERANCE INTERVALS AND ONE-SIDED TOLERANCE BOUNDS

9.3.1 Minimum Sample Size for Two-Sided Tolerance Intervals

Section 5.3 describes how to compute two-sided distribution-free tolerance intervals. As indicated there, Table J.12 gives the smallest sample size needed to provide $100(1 - \alpha)\%$ confidence that the interval defined by the range of sample observations will contain at least a proportion β of the sampled distribution for $1 - \alpha = 0.50, 0.75, 0.90, 0.95, 0.98, 0.99, 0.999$ and $\beta = 0.50(0.05)0.95(0.01)0.99, 0.995, 0.999$.

Example 9.7 Minimum Sample Size for a Distribution-Free Two-Sided Tolerance Interval for a Part Dimension. Suppose that for the application in Example 9.1, the manufacturer now wants a tolerance interval that does not require the assumption that the dimensions follow a normal distribution. We still, however, must assume that we are dealing with a random sample from the population of interest. Now the manufacturer wants to find the smallest sample size that will use the minimum and the maximum observations for a two-sided distribution-free tolerance interval to contain, with 95% confidence, the critical dimension for at least a proportion 0.90 of the units in the sampled population. From Table J.12 we see that a minimum random sample of size $n = 46$ units is needed.

One might be surprised that the required sample size for a distribution-free tolerance interval ($n = 46$) is smaller than that when normality is assumed ($n = 91$ from Example 9.1). The two intervals, however, are not comparable. The normal distribution-based interval is more demanding because, in addition to specifying that we wished to construct a tolerance interval to include at least a proportion 0.90 of the population with 95% confidence, we also required the probability to be no more than $\delta = 0.10$ that the interval will cover a proportion 0.96 or more of the population values. On the other hand, the sample size determination for the distribution-free case did *not* have this second requirement and the distribution-free interval can be expected to be much wider than the interval that is based on the assumption of normality. In fact, it called for the *smallest* possible sample size to include at least a proportion 0.90 of the population with 95% confidence, irrespective of the interval's precision (see Section 9.4 for sample size determination that controls the precision of a distribution-free tolerance interval).   ∎

9.3.2   Minimum Sample Size for One-Sided Tolerance Bounds

A distribution-free one-sided tolerance bound is equivalent to a one-sided distribution-free confidence bound for a quantile of that distribution (see Section 2.4.2 for details). Methods for constructing such distribution-free bounds are given in Sections 5.2.3, 5.2.4, and 5.3.2. As indicated in Section 5.3.3, Table J.13 gives the smallest sample size needed to obtain a one-sided tolerance bound that will, with $100(1 - \alpha)\%$ confidence, be exceeded by (exceed) at least a proportion β of the distribution. This table is based on the expression $n = \log(\alpha)/\log(\beta)$, which can be used directly for nontabulated values. Then the lower (upper) bound will be the smallest (largest) observed value in the sample.

Example 9.8   Minimum Sample Size for a Distribution-Free One-Sided Tolerance Bound for Bearing Life. A group of reliability engineers is planning a test to estimate the life of a newly designed engine bearing. They wish to use the test results to compute a lower tolerance bound that they can claim, with 95% confidence (i.e., $1 - \alpha = 0.95$), will be exceeded by the lifetimes of at least a proportion $\beta = 0.99$ of the population of bearings. This tolerance bound is equivalent to a one-sided lower 95% confidence bound for the 0.01 quantile of the bearing lifetime distribution.

The engineers want to know the minimum required sample size to obtain the desired lower bound with no distributional assumptions. This requires using the first bearing failure time as the bound. In this case, if the bearings are placed on test simultaneously, the test can be terminated after the first failure. The practical usefulness of the results will depend on the magnitude of the first failure. In particular, if the first failure occurs very early, all we know is how little we know. From Table J.13, we note that 299 bearings need to be tested. Equivalently, we find that $n = \log(0.05)/\log(0.99) \approx 299$. One then uses the first bearing failure time as the desired one-sided lower tolerance bound. Moreover, assuming that all bearings are put on test at the same time, one could terminate testing after the first failure. However, it is often good practice to continue the test to get added results.   ∎

9.4   SAMPLE SIZE FOR CONTROLLING THE PRECISION OF TWO-SIDED DISTRIBUTION-FREE TOLERANCE INTERVALS AND ONE-SIDED DISTRIBUTION-FREE TOLERANCE BOUNDS

Distribution-free tolerance intervals based on the smallest possible sample size, as described in the preceding section, are often too wide for the intended application. Thus, in Example 9.1, a 95% tolerance interval to contain at least a proportion $\beta = 0.90$ of the distribution, based on the smallest possible sample size ($n = 46$), results in a *distribution-free* interval that is so wide that the probability is 0.554 (computed using (5.9)) that the tolerance interval will, in fact, contain more than a proportion 0.96 of the distribution.

Table J.21 is similar to Tables J.19 and J.20 for the normal distribution tolerance intervals and bounds. It provides the sample size needed to control the precision of distribution-free two-sided tolerance intervals and one-sided tolerance bounds, using the same criterion described in Section 9.1.1. For the distribution-free case, the criterion results in the same sample size for both two-sided tolerance intervals and one-sided tolerance bounds. This is because the level of confidence for the distribution-free tolerance intervals and bounds depends on the total number of observations that are removed from the end(s) of the ordered sample values to make the interval and *not* the end(s) from which they are removed.

Example 9.9   Sample Size for a Distribution-Free Two-Sided Tolerance Interval for a Part Dimension. Consider the application in Examples 9.1 and 9.7. Suppose that the distribution-free tolerance interval to contain a proportion $\beta = 0.90$ of the population with 95% confidence (i.e., $1 - \alpha = 0.95$) should, in addition, be sufficiently narrow so that the probability is only $\delta = 0.10$ that the interval will contain more than a proportion $\beta^* = 0.96$ of the population. From Table J.21, the necessary sample size is $n = 154$. Section 5.3.1 shows how to obtain the interval. In particular, we note that increasing the sample size from $n = 46$ (Example 9.7) to $n = 154$ (i.e., a 235% increase) reduces the probability that the distribution-free tolerance interval (formed by the range from the smallest to the largest observed value) will contain more than a proportion 0.96 of the population from $\delta = 0.554$ to $\delta = 0.091$ (computed exactly using (5.9)). We also note that the sample size of $n = 154$ is about 70% larger than the sample of $n = 91$ which was required to achieve the same coverage probabilities under the assumption that the measured dimension is normally distributed (see Example 9.1).   ∎

Example 9.10   Sample Size for a Distribution-Free One-Sided Tolerance Bound for Bearing Life. Continuing from Example 9.8, suppose that, in addition to requiring that the tolerance bound be exceeded by the lifetimes of at least a proportion $\beta = 0.99$ of the population of bearings with 95% confidence (i.e., a lower 95% confidence bound on the 0.01 quantile of the bearing lifetime distribution so $1 - \alpha = 0.95$), we now require that the sample size also should be large enough so that the probability is only $\delta = 0.10$ that the bound is exceeded by more than a proportion $\beta^* = 0.997$ of the population. From Table J.21, we note that the required sample size is $n = 1,050$. Section 5.2.4 shows how to obtain the resulting interval. Moreover, using (5.9), we find that increasing the sample size from $n = 299$ (Example 9.8) to $n = 1,050$ reduces the probability that the bound will be exceeded by a proportion $\beta^* = 0.997$ of the population from $\delta = 0.593$ to $\delta = 0.0995$.   ∎

Example 9.11   Sample Size for a Distribution-Free One-Sided Lower Tolerance Bound for Component Strength. For Example 9.2—which assumed a normal distribution—we note that a sample size of $n = 370$ was required to obtain a tolerance interval with the desired probabilities. If no distributional assumption is made, we enter Table J.21 instead, with $\beta = 0.99$, $1 - \alpha = 0.95$, $\delta = 0.01$, and $\beta^* = 0.997$, and we find that the required sample size is $n = 1,941$. The substantial increase in the required sample is the price paid for dropping the normality assumption and is due to the high degree of precision required in the tail of the distribution (which, unfortunately, is the part of the distribution where the normality assumption is most likely to be in doubt).   ∎

9.5   SAMPLE SIZE TO DEMONSTRATE THAT A BINOMIAL PROPORTION EXCEEDS (IS EXCEEDED BY) A SPECIFIED VALUE

This section shows how to find the sample size needed to demonstrate, with a specified confidence $100(1 - \alpha)\%$, that the population proportion conforming (or nonconforming) to a specified requirement, denoted by π, is greater than or equal to (less than or equal to) a specified value, denoted by $\pi^{\dagger}$. Assuming that conforming units occur independently of each other and with a constant probability implies that the binomial distribution is an appropriate model for the number of conforming (nonconforming) units in a random sample of units. The demonstration requires that a one-sided lower (upper) confidence bound for π be greater than or equal to (less than or equal to) $\pi^{\dagger}$. We want to choose the sample size so that there is a specified high probability p_{dem} of a successful demonstration when $\pi \geq \pi^{\dagger}$. Note that demonstrating that $1 - \pi \leq 1 - \pi^{\dagger}$ is equivalent to demonstrating that $\pi \geq \pi^{\dagger}$. Thus the methods in this section can be used for both the conformance and the nonconformance problems. Expressions for p_{dem} are derived in Sections I.2.1 and I.2.2.

In a manner similar to the normal distribution demonstration tests described in Section 9.2, the demonstration test described in the previous paragraph is basically equivalent to a demonstration test based on a distribution-free one-sided tolerance bound (or a distribution-free one-sided confidence interval on a quantile). Therefore, Table J.21 can be used to find the needed sample size for both problems.

Example 9.12 Distribution-Free Demonstration Test for Integrated Circuit Conformance. A new electronic integrated circuit chip must pass a battery of diagnostic tests to conform to specifications. Suppose that the producer of the chips must demonstrate, with 90% confidence, that at least a proportion 0.99 of the manufactured units conform to specifications (i.e., show that $\pi \geq \pi^{\dagger} = 0.99$). This is done by obtaining $\underset{\sim}{\pi}$, a lower 90% confidence bound for π, based upon the results of a random sample of chips (using methods in Section 6.2). The demonstration will be successful if $\underset{\sim}{\pi} \geq \pi^{\dagger}$. Assuming that a proportion 0.999 of the units in the sampled population are actually in conformance, how large does the sample size n have to be so that the probability of passing the demonstration is 0.95? In our notation, $\pi^{\dagger} = 0.99$, $1 - \alpha = 0.90$, and we want the probability of a successful demonstration to be $p_{\text{dem}} = 0.95$ when $\pi = 0.999$. ∎

9.5.1 Graphical Method

Figures 9.2a–9.2d show p_{dem}, the probability of a successful demonstration that $\pi \geq \pi^{\dagger}$ at the $100(1 - \alpha)\%$ confidence level as a function of the actual proportion nonconforming π, and the sample size n for all combinations of $1 - \alpha = 0.90$ and 0.95, and $\pi^{\dagger} = 0.95$ and 0.99. These can be used to determine n for specified values of $\pi^{\dagger}$, π, p_{dem}, and $1 - \alpha$. The figure also shows c, the maximum number of nonconforming units in the resulting random sample that are allowable for the demonstration to be successful.

Note that each line corresponds to the discrete maximum number of nonconforming units in the sample. When entering the figure, one generally will not find a line at exactly the desired point. Rather than interpolating, one will generally go to the next higher line, in order to be conservative.

Example 9.13 Graphical Method to Find the Sample Size for a Distribution-Free Demonstration Test for Integrated Circuit Conformance. For the application in Example 9.12, we use Figure 9.2b which was designed specifically for demonstrating with $100(1 - \alpha)\% = 90\%$ confidence that $\pi \geq \pi^{\dagger} = 0.99$. Enter the horizontal scale at the value $\pi = 0.999$ and move up from that point. Simultaneously enter the vertical scale at $p_{\text{dem}} = 0.95$ and move to the right. After finding the point of intersection, read the values $n = 531$ and $c = 2$ from the line immediately above this point. Thus, the required sample size is $n = 531$, and the demonstration will be successful if there are $c = 2$ or fewer nonconforming units in the sample. ∎

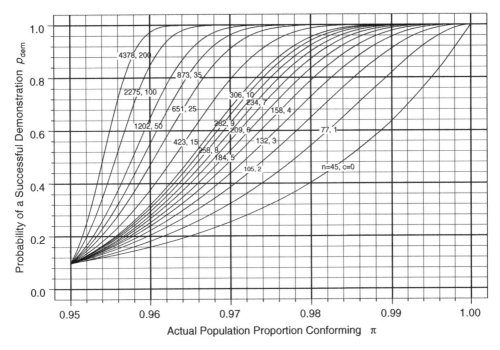

Figure 9.2a Probability of successfully demonstrating that $\pi > \pi^\dagger = 0.95$ with 90% confidence for various sample sizes (binomial distribution).

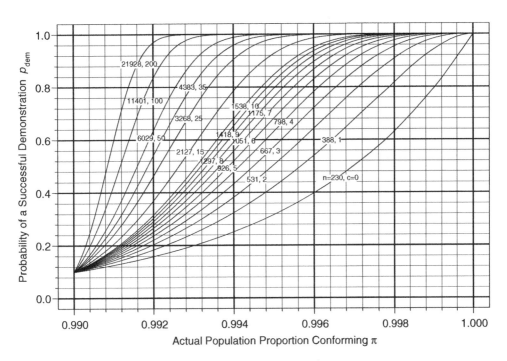

Figure 9.2b Probability of successfully demonstrating that $\pi > \pi^\dagger = 0.99$ with 90% confidence for various sample sizes (binomial distribution).

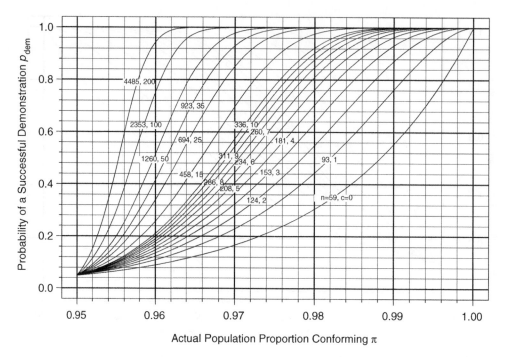

Figure 9.2c   Probability of successfully demonstrating that $\pi > \pi^\dagger = 0.95$ with 95% confidence for various sample sizes (binomial distribution).

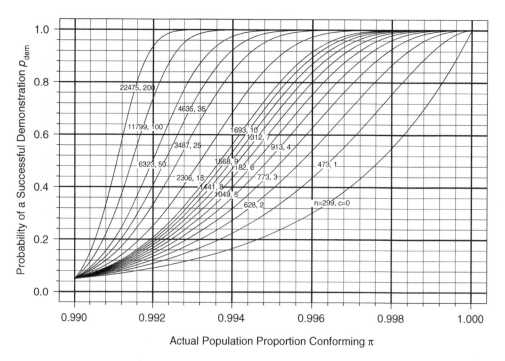

Figure 9.2d   Probability of successfully demonstrating that $\pi > \pi^\dagger = 0.99$ with 95% confidence for various sample sizes (binomial distribution).

9.5.2 Tabular Method

The tabular method to determine the required sample size for such problems uses Table J.21, as illustrated in the following example.

Example 9.14   Tabular Method to Find the Sample Size for a Distribution-Free Demonstration Test for Integrated Circuit Conformance. For the application in Example 9.12, enter Table J.21 with $1 - \alpha = 0.90$, $\beta = \pi^\dagger = 0.99$, $\beta^* = \pi = 0.999$, and $\delta = 1 - p_{\text{dem}} = 1 - 0.95 = 0.05$, and read the sample size to be $n = 531$. ∎

9.5.3 Computational Method

The probability of successful demonstration that $\pi \geq \pi^\dagger$ in a binomial-distribution-based demonstration test for given n, $1 - \alpha$, $\pi^\dagger$, and π is derived in Sections I.2.1 and I.2.2 to be

$$p_{\text{dem}} = \text{pbinom}(c; n, 1 - \pi),$$

where the demonstration is successful if the number of nonconforming units is less than or equal to

$$c = n - \text{qbinom}(1 - \alpha; n, \pi^\dagger) - 1.$$

Section I.2.3 gives a procedure to find the smallest sample size n to provide the desired p_{dem}.

Example 9.15   Computational Method to Find the Sample Size for a Distribution-Free Demonstration Test for Integrated Circuit Conformance. For the application in Example 9.12, it is desired to demonstrate with 90% confidence (so $(1 - \alpha) = 0.90$) that the proportion of conforming integrated circuits is at least $\pi^\dagger = 0.99$. Moreover, there is a requirement that the probability of successful demonstration p_{dem} be at least 0.95 when the true conformance probability is $\pi = 0.999$. One can use the above graphical or tabular methods or trial and error to get a starting range of possible values for n. Using R as a calculator and trying values of n between 528 and 535 gives

```
> ntry <- 528:533
> pbinom(ntry-qbinom(1-0.10, ntry, 0.99)-1, ntry, 1-0.999)
[1] 0.9012626 0.9009509 0.9006390 0.9832133 0.9831304 0.9830473
>
> 531-qbinom(1-0.10, 531, 0.99)-1
[1] 2
```

indicating that the needed sample size is $n = 531$ and that the demonstration will be successful if the number of nonconforming integrated circuits is fewer than $c = 2$. ∎

BIBLIOGRAPHIC NOTES

Faulkenberry and Weeks (1968) suggested the criterion for finding the sample size to control the size of a tolerance interval given in Section 9.1.1. Tables like Table J.19 and J.20 were first presented by Faulkenberry and Daly (1970). Wilks (1941) suggested an alternative criterion based on the "stability" of the repeated-sampling probability within a tolerance interval. Jílek (1982) suggests some other criteria.

McKane et al. (2005) provide methods for computing the probability of successful demonstration for any (log-)location-scale distribution, allowing for censoring that often arises in life tests.

Chapter *10*

Sample Size Requirements for Prediction Intervals

OBJECTIVES AND OVERVIEW

This chapter provides guidelines for choosing the sample size required to obtain a prediction interval to contain a future single observation, a specified number of future observations, or some other quantity to be calculated from a future sample from a previously sampled distribution. The topics discussed in this chapter are:

- The factors that determine the width of a prediction interval (Section 10.1).

- Sample size determination for a normal distribution prediction interval (Section 10.2).

- Sample size determination for a distribution-free prediction interval for at least k of m future observations (Section 10.3).

10.1 PREDICTION INTERVAL WIDTH: THE BASIC IDEA

There are two sources of imprecision in statistical prediction: First, there is the random variation in the future sample. Second, because the given data are limited, there is uncertainty with respect to the characteristics (e.g., parameters) of the previously sampled distribution. Say, for example, that the results of an initial sample of size n from a normal distribution with unknown mean μ and unknown standard deviation σ are to be used to predict the value of a single future randomly selected observation from the same distribution. The sample mean $\bar{X}$ of the initial sample will be used to predict the future observation. First, $\bar{X} = \mu + \varepsilon_1$, where ε_1, the random variation associated with the mean of the given sample, is itself normally distributed with mean 0 and variance σ^2/n. The future observation to be predicted is $Y = \mu + \varepsilon_2$, where ε_2 is the random variation associated with the future observation, and is normally distributed with mean 0 and variance σ^2, independently of ε_1. Thus, the prediction error is $Y - \bar{X} = \varepsilon_2 - \varepsilon_1$, and

Statistical Intervals: A Guide for Practitioners and Researchers, Second Edition.
William Q. Meeker, Gerald J. Hahn and Luis A. Escobar.
© 2017 John Wiley & Sons, Inc. Published 2017 by John Wiley & Sons, Inc.
Companion Website: www.wiley.com/go/meeker/intervals

has variance $\sigma^2 + \sigma^2/n$. The width of a normal-theory prediction interval to contain Y will be proportional to the square root of the estimate of this quantity (see Section 4.7). Increasing the size of the initial sample will reduce the uncertainty associated with the sample mean $\bar{X}$ (i.e., σ^2/n), but it will reduce only the sampling error in the estimate of the variation (σ^2) associated with the future sample. Thus, an increase in the size of the initial sample beyond the point where the inherent variation in the future sample tends to dominate will not materially reduce the width of the prediction interval.

10.2 SAMPLE SIZE FOR A NORMAL DISTRIBUTION PREDICTION INTERVAL

10.2.1 Introduction

This section deals with selecting the size of the initial sample that will be used to construct a prediction interval to contain the mean of a *future* sample of size m from the previously sampled normal distribution. Figures for the frequently encountered special case where the future sample size is $m = 1$, and also for $m = 10$, are provided, followed by a numerical example. Finally, it is shown how these ideas can be applied to assessing sample size requirements for some other prediction intervals.

Unlike a confidence interval to contain a distribution parameter, which converges to a point (generally the actual parameter value) as the sample size increases, a prediction interval converges to an interval. This limiting interval is, as previously indicated, the probability interval (introduced in Section 2.3.6) that one would obtain from a past sample of very large size (i.e., essentially from knowledge of the normal distribution parameters μ and σ). It is thus not possible to obtain a prediction interval consistently narrower than this limiting interval, irrespective of how large an initial sample is taken. Thus, we suggest that the criterion for assessing the effect of sample size on prediction interval width be expressed in terms of this limiting interval. More specifically, because the width of the calculated prediction interval is an observed value of a random variable, we propose that one decide on the initial sample size based on either of the following two relative widths:

- The expectation of the ratio between the prediction interval width and the width of the limiting interval or

- The ratio of an appropriate upper prediction bound on the prediction interval width, relative to the width of the limiting interval.

10.2.2 Relative Width of the Prediction Interval

The ratio of the width of the two-sided $100(1 - \alpha)\%$ prediction interval for the mean of a future sample of size m from an initial sample of size n, given by (4.7). to that of the limiting (or probability) interval assuming an infinite initial sample size (i.e., $\mu \pm z_{(1-\alpha/2)}\sigma/\sqrt{m}$) is given by

$$W = \left[\frac{t_{(1-\alpha/2;n-1)}S_n}{z_{(1-\alpha/2)}\sigma}\right]\left(1 + \frac{m}{n}\right)^{1/2}, \tag{10.1}$$

where $t_{(1-\alpha/2;r)}$ is the $1 - \alpha/2$ quantile of the Student's t-distribution with r degrees of freedom.

The relative width W involves the random variable S_n—the estimated standard deviation of the initial sample—and, therefore, is itself a random variable. The expected value of W is obtained by substituting the expected value of S_n into (10.1) to obtain

$$
\mathrm{E}(W) = \left[\frac{t_{(1-\alpha/2;n-1)}}{z_{(1-\alpha/2)}}\right]\left(\frac{2}{n-1}\right)^{1/2}\left[\frac{\Gamma\left(\frac{n}{2}\right)}{\Gamma\left(\frac{n-1}{2}\right)}\right]\left(1+\frac{m}{n}\right)^{1/2}, \tag{10.2}
$$

where $\Gamma(\cdot)$ is the gamma function (defined in Appendix A). We propose $\mathrm{E}(W)$, the expected value of the relative width of the prediction interval to the limiting interval, as one criterion for selecting the initial sample size n.

One can also obtain a one-sided upper bound on the ratio of the prediction interval width to the limiting interval width. In particular, a one-sided upper $100\gamma\%$ prediction bound on W is obtained by substituting an upper $100\gamma\%$ prediction bound for S_n/σ into (10.1), as described in Section 4.9. Thus

$$
\widetilde{W}_U = \left[\frac{t_{(1-\alpha/2;n-1)}}{z_{(1-\alpha/2)}}\right]\left(\frac{\chi^2_{(\gamma;n-1)}}{n-1}\right)^{1/2}\left(1+\frac{m}{n}\right)^{1/2}, \tag{10.3}
$$

where $\chi^2_{(\gamma;r)}$ is the γ quantile of the chi-square distribution with r degrees of freedom. The interpretation of $\widetilde{W}_U$ is that, in repeated constructions of a $100(1-\alpha)\%$ prediction interval for $\bar{y}_m$ using (10.1), from independent samples of size n, the relative width of the interval W will exceed this bound only $100(1-\gamma)\%$ of the time. We propose $\widetilde{W}_U$ as an alternative criterion for choosing the initial sample size n.

10.2.3   Figures for Two-Sided Prediction Intervals

Figure 10.1 gives the expected relative width of a two-sided prediction interval for a single future observation (i.e., $m = 1$) from a normal distribution, as a function of the initial sample size for the confidence levels $1 - \alpha = 0.5, 0.8, 0.9, 0.95,$ and 0.99 associated with the prediction interval. Figure 10.2 gives upper 95% prediction bounds on the relative widths for the same values of $1 - \alpha$. Figures 10.3 and 10.4 provide information similar to that in Figures 10.1 and 10.2 for two-sided prediction intervals to contain the mean of $m = 10$ future observations. These figures were calculated from (10.2) and (10.3), respectively. They can be used to assess the effect of the initial sample size on (1) the expected relative width and (2) the upper prediction bound for the relative width of prediction intervals. They can be used in reverse to determine the required size of the initial sample.

Example 10.1   Sample Size for a Prediction Interval to Contain the Thrust-Delivery Time for a Future Engine. A rocket engine is to be used in a critical, self-destructive operation. An experiment is being planned on a random sample of engines to determine how long they can deliver a certain amount of thrust for a specified amount of fuel. The delivery time is assumed to follow a normal distribution with unknown mean and standard deviation. It is desired to determine the effect of initial sample size on a two-sided 90% prediction interval to contain the thrust-delivery time for a subsequently randomly selected single future engine, using as criteria (a) the expected relative width and (b) the upper 95% prediction bound on the relative width of the two-sided prediction interval.

- From Figure 10.1, to obtain, for a single future observation, a two-sided 90% prediction interval whose expected width is 20% (i.e., expected relative width 1.2) larger than that of the smallest achievable interval would require an initial sample size close to $n = 7$. An exact computation using (10.2) gives $\mathrm{E}(W) = 1.21$ (i.e., 21% larger) for $n = 7$ and $\mathrm{E}(W) = 1.18$ (i.e., 18% larger) for $n = 8$. Note, from examining the curves in

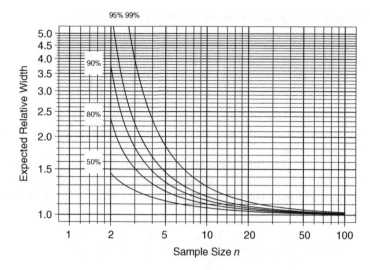

Figure 10.1   Prediction interval expected width relative to the limiting interval for $m = 1$ future observation. A similar figure first appeared in Meeker and Hahn (1982). Adapted with permission of the American Society for Quality.

Figure 10.1, that precision improves only slightly beyond $n = 20$. This is because beyond this point, the major part of the variability is not in the uncertainty in the initial sample but in that for the single future observation.

- From Figure 10.2, to obtain, for a single future observation, a two-sided 90% prediction interval whose width we can expect with 95% confidence to be no more than twice as wide as the smallest achievable interval width would require an initial sample size of $n = 6$. An exact computation using (10.3) gives $\widetilde{W}_U = 1.97$ for $n = 6$.

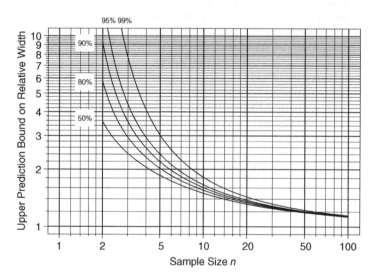

Figure 10.2   Upper 95% prediction bound on prediction interval width relative to the limiting interval for $m = 1$ future observation. A similar figure first appeared in Meeker and Hahn (1982). Adapted with permission of the American Society for Quality.

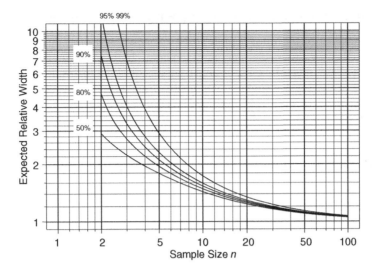

Figure 10.3 Prediction interval expected width relative to the limiting interval for the mean of $m = 10$ future observations. A similar figure first appeared in Meeker and Hahn (1982). Adapted with permission of the American Society for Quality.

Example 10.1 dealt with sample size determination for a prediction interval for a single future observation ($m = 1$)—the case most frequently encountered in practice. The results also apply, however, for sample size determination for a prediction interval for the mean of $m > 1$ future observations. As previously indicated, Figure 10.3 and 10.4 can be used for sample size determination for a prediction interval for the mean of $m = 10$ future observations. One can use (10.2) and (10.3) to construct similar curves for other values of m—as well as for other situations not covered in the tabulations, such as other values of $1 - \alpha$—or to use directly in a particular application.

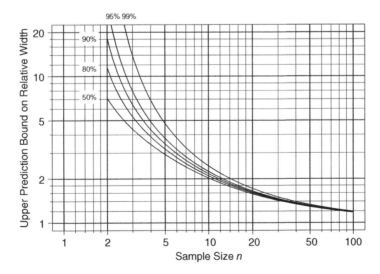

Figure 10.4 Upper 95% prediction bound on prediction interval width relative to the limiting interval for the mean of $m = 10$ future observations. A similar figure first appeared in Meeker and Hahn (1982). Adapted with permission of the American Society for Quality.

10.2.4 One-Sided Prediction Bounds for a Future Sample Mean

The procedure for evaluating the initial sample size for a one-sided normal distribution prediction bound is similar to that for a two-sided prediction interval. In this case, however, relative width deals with that part of the interval (below or above the mean) of interest, and one uses $t_{(1-\alpha;n-1)}$ and $z_{(1-\alpha)}$ in place of $t_{(1-\alpha/2;n-1)}$ and $z_{(1-\alpha/2)}$ in (10.2) and (10.3). As a result, in Figures 10.1 and 10.2 the confidence levels 0.5, 0.8, 0.9, 0.95, and 0.99 are replaced by confidence levels of 0.75, 0.90, 0.95, 0.975, and 0.995, respectively.

10.2.5 Other Prediction Intervals

The concepts presented in the previous sections can be readily applied to determine the initial sample size for other types of prediction intervals. Some specific cases are discussed below.

Simultaneous prediction intervals to contain all of m future observations

Using results from Section 4.8.1, the limiting width of a simultaneous two-sided prediction interval to contain all m future observations from a normal distribution, as the initial sample size n becomes large (i.e., known μ and σ), is

$$\mu \pm z_{(\delta)}\sigma,$$

where $\delta = (1 - \alpha/2)^{1/m}$. Thus, the ratio of the prediction interval width to the limiting interval width is

$$\frac{r_{(1-\alpha;m,m,n)}S_n}{z_{(\delta)}\sigma},$$

where $r_{(1-\alpha;m,m,n)}$ is the conservative approximate factor defined in (4.8) for obtaining a simultaneous two-sided prediction interval to contain all of m future observations from a normal distribution. As before, the expected relative width and upper prediction bound on the relative width can be obtained, respectively, by replacing S_n/σ by its expected value and by its appropriate upper prediction bound. The resulting expressions can then be used to assess the effect of the sample size on the relative width of the desired two-sided prediction interval, and to guide sample size determination (iteratively). A one-sided prediction bound is handled similarly, but now one uses $\delta = (1 - \alpha)^{1/m}$.

Prediction interval to contain the standard deviation of a future sample

Using results from Section 4.9, the limiting two-sided probability interval to contain the standard deviation of a future sample from a normal distribution as the initial sample size n becomes large (i.e., resulting in known σ) is

$$\left[\sigma\left(\frac{\chi^2_{(\alpha/2;m-1)}}{m-1} \right)^{1/2}, \ \sigma\left(\frac{\chi^2_{(1-\alpha/2;m-1)}}{m-1} \right)^{1/2} \right].$$

Thus, the ratio of the width of the prediction interval to its limiting width is

$$\frac{S_n[(F_{(1-\alpha/2;m-1,n-1)})^{1/2} - (F_{(\alpha/2;m-1,n-1)})^{1/2}]}{\sigma[(\chi^2_{(1-\alpha/2;m-1)})^{1/2} - (\chi^2_{(\alpha/2;m-1)})^{1/2}]/(m-1)^{1/2}},$$

where $F_{(\gamma; r_1, r_2)}$ is the γ quantile of the F-distribution with r_1 numerator and r_2 denominator degrees of freedom. Expressions for the expected relative width and for an upper prediction bound on this width can be obtained by replacing S_n/σ by its expectation and by its upper prediction bound, respectively, as above. These results can then be used for sample size determination.

10.3 SAMPLE SIZE FOR DISTRIBUTION-FREE PREDICTION INTERVALS FOR AT LEAST k OF m FUTURE OBSERVATIONS

10.3.1 Tabular Method for Two-Sided Prediction Intervals

As indicated in Chapter 5, Table J.15 gives the sample size n so that a two-sided prediction interval, that has as its endpoints the largest and the smallest observations of this initial sample, will enclose all m observations in a future sample of size m with $100(1-\alpha)\%$ confidence, for $1-\alpha = 0.50, 0.75, 0.90, 0.95, 0.98, 0.99, 0.999$ and $m = 1(1)25(5)50(10)100$. These tables can be used directly to help choose the size of an initial sample that is to be used to set a prediction interval to contain all, or almost all, observations in a future sample from the same distribution.

These distribution-free sample size criteria are inherently different from those discussed in the preceding two sections, dealing with normal distribution prediction intervals, in that they provide information about the *minimum* size sample that is needed to construct any such interval with the desired level of confidence. As noted in Chapter 5, if the initial sample is too small it is not possible to construct a distribution-free interval at the desired confidence level. Moreover, because the interval uses the endpoints of the initial sample, it may be unsatisfactorily large. In contrast, in our earlier discussion of normal distribution prediction intervals, we were concerned with taking a sufficiently large sample to satisfy specified requirements on precision relative to the ideal situation of having so large a sample that the distribution parameters are essentially known.

Example 10.2 Sample Size for a Prediction Interval to Contain the Concentration Amount for Five Future Batches. Based on the measured values of the sample of $n = 100$ units given in Table 5.1, the manufacturer wants to find a distribution-free prediction interval to contain all of the measured values of a future sample of $m = 5$ units from the same distribution, without making any assumptions about the form of the distribution. From Table J.15, we note for $m = 5$ that, even if one uses the extreme observations of the past sample, a minimum sample of size 193 would be required to obtain a 95% prediction interval. Thus, the past sample of size 100 would be inadequate. However, Table J.15 indicates that a 90% prediction interval can be obtained. In that case an initial sample of 93 observations would suffice. ∎

10.3.2 Tabular Method for One-Sided Prediction Bounds

As indicated in Section 5.5.2, Tables J.16a–J.16c give the sample size n so that a one-sided lower (upper) prediction bound defined by the smallest (largest) observation of this initial sample will be exceeded by (will exceed)

- all m,
- at least $m - 1$, and
- at least $m - 2$

observations in a future sample of size m with $100(1-\alpha)\%$ confidence, for selected values of $1-\alpha$ and m. These tables can be used directly to help choose the size of an initial sample for setting a one-sided upper (lower) distribution-free prediction bound to exceed (to be exceeded by) all, or almost all, observations in a future sample from the same distribution.

Example 10.3   Sample Size for a Lower Prediction Bound for Future Battery Lifetimes. A satellite will contain 12 rechargeable batteries of which 10 must survive for a time that is to be determined. The manufacturer needs a lower prediction bound that will, with 99% confidence, be exceeded by at least 10 of 12 failure times for the batteries that will be installed in a future single satellite. Because the batteries have at least two causes of failure and little is known about the life distribution, a distribution-free bound is to be used. Also, the batteries are very expensive, thus, only a limited number can be procured for the life test. The time of the first failure will be used for the lower prediction bound. The manufacturer needs to know how many randomly selected batteries to test to be 99% confident that at least 10 of the batteries in the future shipment of 12 will not fail prior to the time of the first failure in the initial sample. From Table J.16c, we obtain $n = 40$; thus, the first failure in a sample of $n = 40$ batteries will provide the desired prediction bound. We note that, in this example, testing can be terminated after the first failure. However, the results may be of limited value if the first failure occurs too early—other than telling us how little we know.   ∎

BIBLIOGRAPHIC NOTES

The criteria used in this chapter for choosing the sample size needed for a prediction interval when sampling from a normal distribution were initially given in Meeker and Hahn (1982). Straightforward extensions include determining sample size requirements for constructing:

- A prediction interval to contain the mean of a future sample from an exponential distribution (Hahn, 1975).

- A prediction interval to contain the difference between the means of two future samples (Hahn, 1977).

- A prediction interval to contain the ratio of two future sample standard deviations for a normal distribution, or the ratio of two future sample means from an exponential distribution (Meeker and Hahn, 1980).

Chapter *11*

Basic Case Studies

OBJECTIVES AND OVERVIEW

This chapter presents a series of case studies that illustrate the methods in the first 10 chapters of this book. They are a representative sample of frequently occurring problems that we have encountered recently. We present these problems as they were presented to us, rather than in a "clean" textbook style. Then we describe our proposed solution. We stress the basic underlying assumptions and the practical aspects of using and interpreting statistical intervals.

We illustrate some of the most important topics covered in the earlier chapters. Thus, there is some repetition of techniques, but each example has some new feature. In some of the case studies we compare different approaches for answering a question. In one example we use methods from Section 4.5 to estimate the probability that an observation will exceed a threshold, assuming that the data came from a normal distribution. Then, without making the normal distribution assumption, we show how to estimate the same probability by using as data only the number of observations that exceed the threshold (a nonparametric method using binomial distribution methods from Section 6.2).

The following applications are discussed in this chapter:

- Demonstration that the operating temperature of most manufactured devices will not exceed a specified value (Section 11.1).

- Forecasting future demand for spare parts (Section 11.2).

- Estimating the probability of passing an environmental emissions test (Section 11.3).

- Planning a demonstration test to verify that a radar system has a satisfactory probability of detection (Section 11.4).

- Estimating the probability of exceeding a regulatory limit (Section 11.5).

- Estimating the reliability of a circuit board (Section 11.6).

Statistical Intervals: A Guide for Practitioners and Researchers, Second Edition.
William Q. Meeker, Gerald J. Hahn and Luis A. Escobar.
© 2017 John Wiley & Sons, Inc. Published 2017 by John Wiley & Sons, Inc.
Companion Website: www.wiley.com/go/meeker/intervals

- Using sample results to estimate the probability that a demonstration test will be successful (Section 11.7).

- Estimating the proportion within specifications for a two-variable problem (Section 11.8).

- Determining the minimum sample size for a demonstration test (Section 11.9).

Appendix A outlines and defines notation used in this book. Bringing together parametric and nonparametric methods for some of the examples in this chapter exposes a minor, but unavoidable, conflict in our choice of notation. We consistently use p to denote a particular quantile of a distribution or the tail probability of a distribution. In Chapter 6 (and in some other chapters) we use π to denote the Bernoulli or binomial distribution probability of a single randomly selected unit being nonconforming (or having some other characteristic of interest) and p to denote a particular quantile of a binomial distribution or the tail probability of a binomial distribution. In some of the applications in this chapter we use notation such as p_{GT} to denote the probability of a particular event (e.g., the probability of being greater than a specification limit) that could be described by either a normal (or some other parametric) distribution or a binomial distribution.

Relatedly, as described in the introduction to Chapter 5, nonparametric methods for constructing statistical intervals do not require specification of a particular parametric distribution. Additionally, some nonparametric statistical interval methods (e.g., the ones presented in Chapter 5) are distribution-free because their coverage probabilities do not depend on the form of the actual underlying distribution. Distribution-free procedures are nonparametric, but not all nonparametric methods are distribution-free. When we use the methods in Chapter 5 to compute a confidence interval for a quantile, the method is nonparametric and distribution-free. When we use methods in Chapter 6 to estimate a tail probability of a distribution, the method is nonparametric, but not distribution-free. In general, if the endpoints of a statistical interval are defined by order statistics (as in Chapter 5), the method is distribution-free. Otherwise (e.g., confidence intervals for tail probabilities based on a binomial distribution), the method is *not* distribution-free.

11.1 DEMONSTRATION THAT THE OPERATING TEMPERATURE OF MOST MANUFACTURED DEVICES WILL NOT EXCEED A SPECIFIED VALUE

11.1.1 Problem Statement

The designers of a solid-state electronic device wanted to "demonstrate that the surface temperature of most devices will not exceed 180°C in operation," based on measurements on a sample of such devices.

11.1.2 Some Basic Assumptions

To make the demonstration, it is necessary that the selected devices be a random sample from the production process. This assumption deserves careful scrutiny. If, for example, early prototype devices are used in the demonstration test, inferences from the test might not apply to subsequent production. In this example, special care was taken to assure that the test units were randomly selected from those made in a pilot production process that closely simulated actual manufacturing conditions. For example, raw materials were obtained from the same sources as those used in production. Also, the operating conditions and performance measurements for the demonstration test must be the same as those to be encountered in operation. Thus, additional efforts were made to have the test condition simulate as closely as possible the actual field environment, and to use comparable measuring instruments. We also assume that the production process is stable, now and in the future. One should keep in mind that, as indicated in Chapter 1, if these assumptions are not met, the resulting statistical intervals (in this analytic study) express only one part of the total uncertainty and are likely to be too narrow.

11.1.3 Statistical Problem

After discussion with the device designers, the following more statistically precise problem statement was agreed upon. It is desired to show, with 90% confidence, that p_C, the proportion of units produced by the process with surface temperatures less than or equal to $L = 180°C$, is at least $p_C^\dagger = 0.99$ (or some other high proportion); that is, to show with high confidence that $p_C = \Pr(\text{Temp} \le 180) \ge 0.99$.

There are two general methods of making such a demonstration:

1. Make no assumption about the form of the statistical distribution of the surface temperatures. In this case, one dichotomizes the data by classifying each sampled device as either conforming (temperature less than or equal to $180°C$) or nonconforming (temperature greater than $180°C$). Then one can use the procedures for proportions (i.e., those based on the binomial distribution) given in Chapter 6 to compute a one-sided lower confidence bound on p_C.

2. Assume that the operating temperatures (or some transformation of the operating temperatures) have a particular probability distribution (such as the normal distribution) and use the temperature readings to find a one-sided lower confidence bound for p_C. Thus, for a normal distribution, one would use the procedures given in Section 4.5.

In either case, if the one-sided lower 90% confidence bound for p_C exceeds 0.99, the needed demonstration is achieved.

The second method provides a more efficient use of the data—especially if all, or the great majority, of the values are well within bounds—if one can assume an appropriate distribution for the operating temperatures. Of course, when one can only observe that a unit is conforming or nonconforming, the second approach is not applicable. This occurs, for example, with a detonator which, when tested, either works or does not. We will use both approaches and compare the findings.

11.1.4 Results from a Preliminary Experiment

The available data were limited to a random sample of only six devices from the pilot production process. These yielded the following surface temperature readings (in °C):

$$170.5, 172.5, 169.5, 174.0, 176.0, 168.0.$$

A normal probability plot of these data is shown in Figure 11.1. There is no obvious deviation from normality (i.e., the points in the plot tend to scatter around a straight line). The sample, however, is clearly too small to draw any definitive conclusions about the underlying distribution. We observe that none of the six observations was above the $180°C$ threshold. From the data, we calculate the sample mean and standard deviation to be

$$\bar{x} = \frac{170.5 + 172.5 + \cdots + 168.0}{6} = 171.75$$

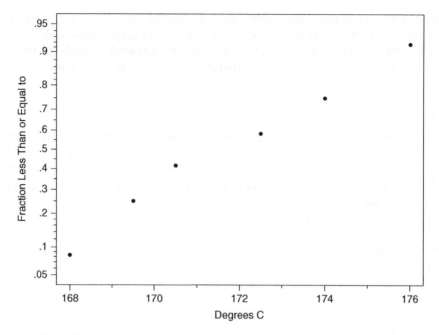

Figure 11.1   Normal probability plot of device surface temperature readings.

and

$$s = \left[\frac{(170.5 - 171.75)^2 + \cdots + (168.0 - 171.75)^2}{6 - 1} \right]^{1/2} = 2.98.$$

We will use a nonparametric method based on the binomial distribution as well as a method that assumes an underlying normal distribution to obtain one-sided lower confidence bounds for p_C, the proportion of conforming devices (i.e., those with operating temperatures less than 180°C) for the sampled process.

11.1.5   Nonparametric One-Sided Lower Confidence Bound on the Proportion Conforming

Because there were $x = 6$ conforming devices in a sample of size $n = 6$, the point estimate of the proportion conforming is $\widehat{p}_C = x/n = 6/6 = 1.0$. Using (6.1) from Section 6.2, a one-sided lower $100(1 - \alpha)\%$ confidence bound for p_C, based on the binomial distribution, is computed as

$$\underset{\sim}{p_C} = \texttt{qbeta}(\alpha; x, n - x + 1)$$

where $\texttt{qbeta}(p; a, b)$ is the p quantile of the beta distribution with shape parameters a and b (see Section C.3.3). With $n = x = 6$, a one-sided lower 90% confidence bound for p_C is

$$\underset{\sim}{p_C} = \texttt{qbeta}(0.10; 6, 1) = 0.68.$$

This value can also be obtained from (6.2). Thus, all we can say with 90% confidence, using a nonparametric approach, is that at least a proportion 0.68 of the devices for the sampled process have temperatures less than or equal to 180°C. More generally, the second column of Table 11.1 compares nonparametric one-sided lower confidence bounds for various confidence

	One-sided lower confidence bound on p_C	
Confidence level	Nonparametric (binomial)	Normal distribution
75%	0.79	0.98
90%	0.68	0.93
95%	0.61	0.88
99%	0.46	0.74

Table 11.1   One-sided lower confidence bounds on the proportion of conforming devices for various confidence levels.

levels. Therefore, even though all of the test units were in conformance, we cannot claim that the actual p_C exceeds the specified proportion $p_C^\dagger = 0.99$ with 90% (or even with 75%) confidence. This, of course, is not surprising due to the small sample size (i.e., $n = 6$). Indeed, these results are the best that one can obtain with a sample of six devices, using nonparametric methods. This fact, however, was known before taking the sample. (See Sections 9.5 and 11.1.8.)

11.1.6   One-Sided Lower Confidence Bound for the Proportion Conforming Assuming a Normal Distribution

We now assume that a normal distribution with an unknown mean and standard deviation adequately describes the distribution of the surface temperatures of the manufactured devices. We can use the sample estimates $\bar{x} = 171.75$ and $s = 2.98$ to compute a point estimate for $p_C = \Pr(X \leq 180)$. Substituting $\bar{x}$ for the mean and s for the standard deviation of the normal distribution, we get the point estimate $\widehat{p}_C = \Phi_{\text{norm}}[(L - \bar{x})/s] = \Phi_{\text{norm}}[(180 - 171.75)/2.98] = \Phi_{\text{norm}}(2.77) = 0.9972$. Also, under the normal distribution assumption, one can use the methods given in Section 4.5 to compute a one-sided lower 90% confidence bound for p_C, the process proportion less than 180°C. This would also be a one-sided lower 90% confidence bound on the probability that a single randomly selected device will have a surface temperature less than or equal 180°C. To proceed, we first compute

$$k = \frac{L - \bar{x}}{s} = \frac{180 - 171.75}{2.98} = 2.77.$$

Then a one-sided lower 90% confidence bound is obtained from $p_C = \texttt{normTailCI}(0.10; 2.77, 6) = 0.93$. Thus we are 90% confident that the proportion of devices less than or equal to 180°C is at least 0.93. The third column of Table 11.1 shows one-sided lower confidence bounds based on the normal distribution assumption for various confidence levels.

Even though these bounds are more favorable than those computed without making any distributional assumptions, they are still not good enough to achieve the desired demonstration. We need to emphasize that the limited data do *not* contradict the claim that 99% of the devices from the process are in conformance, because, after all, our point estimate is 0.9972. Rather, in this example, in which the burden of proof was placed on the designers, the limited sample of size $n = 6$ was just not big enough to achieve the desired demonstration with 90% confidence. (Unlike the nonparametric case, this was not known prior to obtaining the data.) Moreover, under the normal distribution assumption the data *do* allow one to claim with 75% confidence that at least a proportion 0.98 of the devices in the sampled population have temperatures less than or equal to 180°C.

11.1.7 An Alternative Approach: Confidence Interval for a Normal Distribution Quantile for Device Temperatures

It was not possible to demonstrate with 90% confidence that at least a proportion 0.99 of the devices from the sampled process meet the 180°C requirement, based on the sample of six devices, even under normal distribution assumptions. Thus, the designers asked: What surface temperature value can be demonstrated with 90% confidence to be met by at least a proportion 0.99 of the devices? This new question calls for a one-sided upper 90% confidence bound on the 0.99 quantile. One now uses the methods described in Section 4.4, with $n = 6, \bar{x} = 171.75, s = 2.98, p = 0.99, 1 - \alpha = 0.90$ and $g'_{(0.90;0.01,6)} = 4.243$ from Table J.7c. Then a one-sided upper 90% confidence interval for the 0.99 quantile for the distribution of surface temperatures (equivalent to a one-sided upper 90% tolerance bound to exceed the surface temperatures for at least a proportion 0.99) of the devices from the sampled process is

$$\widetilde{x}_{0.99} = \widetilde{T}_{0.99} = \bar{x} + g'_{(1-\alpha;1-p,n)}s = 171.75 + 4.243 \times 2.98 = 184.4.$$

Thus we can claim, with 90% confidence, that at least a proportion 0.99 of the devices from the sampled process have surface temperatures that are less than 184.4°C.

11.1.8 Sample Size Requirements

The analyses failed to demonstrate that 99% of the devices from the sampled process have surface temperatures less than or equal to 180°C. Thus it is clear that a larger sample will be required to achieve the desired demonstration. The designers now need to know how large a random sample from the process is required for this purpose.

Assuming that p_C is really greater than the specified value $p_C^\dagger = 0.99$ to be demonstrated, we need a sample that is large enough to demonstrate this fact with some specified probability (i.e., to have $\Pr(p_C \geq p_C^\dagger) \geq p_{\text{dem}}$). A very large sample is required if the actual (unknown) p_C is close to (but greater than) $p_C^\dagger = 0.99$. On the other hand, the sample size could be smaller if p_C is very close to 1 (e.g., $p_C = 0.9999$).

For analysis method 1 (the nonparametric approach, based on the binomial distribution), Figure 9.2b shows p_{dem}, the probability of a successful demonstration at the 90% confidence level that p_C is at least $p_C^\dagger = 0.99$, as a function of $\pi = p_C$, the actual process proportion conforming, and the sample size n. For example, if the actual proportion of units less than or equal to 180°C in the sampled process is really $\pi = p_C = 0.996$, a sample size of $n = 1{,}538$ has a probability $p_{\text{dem}} = 0.95$ of resulting in a successful demonstration at the 90% confidence level. Moreover, for the demonstration to be successful, it is necessary that no more than $c = 10$ of the 1,538 sampled devices have measured temperatures greater than 180°C. That is, using (I.4), $\Pr(X \leq 10) = \texttt{pbinom}(10; 1538, 1 - 0.996) = 0.9511$.

Graphs like Figure 9.2b are easy to construct with a computer program. We provide such graphs for several combinations of problem parameters (i.e., confidence level and $p_C^\dagger$, the specified proportion conforming to be demonstrated) in Chapter 9. As described in Section 9.5.2, the needed sample size can also be found from Table J.21. In particular, we enter Table J.21 to determine the necessary sample size so as to demonstrate with confidence level $1 - \alpha = 0.90$ that the conformance probability is greater than $\beta = p_C^\dagger = 0.99$, subject to the requirement that $\delta = 1 - p_{\text{dem}} \leq 0.05$ when the actual conformance probability is $\beta^* = 0.996$. The table gives the necessary sample size as 1,538.

For analysis method 2 (i.e., assuming that the temperatures have a normal distribution with unknown mean μ and standard deviation σ), the probability of a successful demonstration, p_{dem}, again depends on the confidence level to be associated with the demonstration test, on the sample size n, and on p_C, the unknown actual proportion conforming. Figure 9.1b shows the probability of a successful demonstration at the 90% confidence level as a function of the actual proportion conforming, for various sample sizes. In this case, the demonstration would be successful if $\tilde{x}_{0.99}$, the one-sided upper 90% confidence bound for $x_{0.99}$ (computed as in Sections 4.4 or 11.1.7) is less than 180°C or, equivalently, if p_C (computed as in Sections 4.5 or 11.1.6) is greater than 0.99. For example, we can see from Figure 9.1b that a sample of $n \approx 325$ units gives $p_{\mathrm{dem}} \approx 0.95$ when $p_C = 0.996$.

The required sample size using method 2 is considerably smaller than that required for method 1 ($n = 1{,}538$) which, however, did not require the assumption of a normal distribution. More generally, comparison of Figures 9.2b and 9.1b shows the potential gain from using the actual measurements and assuming a normal distribution when this assumption is warranted, as compared to the nonparametric approach based on the binomial distribution.

As described in Section 9.2.3, the needed sample size under a normal distribution assumption can also be found from Table J.20. We enter Table J.20 with confidence level $1 - \alpha = 0.90$, to demonstrate that the conformance proportion is greater than $\beta = p_C^\dagger = 0.99$ so that $\delta = 1 - p_{\mathrm{dem}} \leq 0.05$ when the actual conformance proportion is $\beta^* = 0.996$. The table gives the necessary sample size as 329. The slight difference between this and the sample size 325 obtained from Figure 9.1b is due to our inability to interpolate much more than two significant digits from the graph.

Using results from Section 4.4, we now obtain an expression for $\tilde{x}_{0.99}$, the one-sided upper 90% confidence bound for the 0.99 quantile of the distribution of surface temperatures that will be used in the demonstration test. To compute the needed factor $g'_{(0.90;0.99,329)}$ we can use R to obtain

```
> qt(p=0.90, df=329-1, ncp=qnorm(0.99)*sqrt(329))/sqrt(329)
[1] 2.470618
```

or use interpolation in Table J.7c. The demonstration will be successful if the one-sided upper confidence bound, calculated from the future sample of 329 randomly selected devices,

$$\tilde{x}_{0.99} = \bar{x} + 2.471s,$$

is less than the specified limit of 180°C.

11.2 FORECASTING FUTURE DEMAND FOR SPARE PARTS

11.2.1 Background and Available Data

A company manufactures replacement bearings for an electric motor. Demand has been stable over recent years. The company, however, is planning to discontinue production of this product. Before so doing, they want to produce and stockpile enough bearings so they can claim, with 95% confidence, that demand can be met for at least 7 years. The numbers of bearings sold in each of the past 5 years (in thousands of units) were

$$27.7, 37.1, 35.7, 30.8, \ 32.7.$$

These data are graphed against time in Figure 11.2.

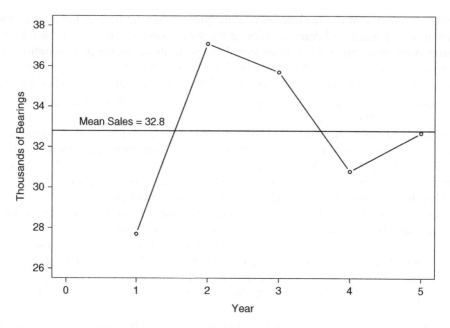

Figure 11.2   Yearly bearing sales in units of thousands.

11.2.2 Assumptions

Based on previous history with similar products and from physical considerations, it was deemed reasonable to assume that:

- The number of units sold per year has a normal distribution with a mean and standard deviation that are constant from year to year, and will continue to be so.

- The number of units sold each year is statistically independent of the number sold in any other year.

Under these assumptions, the number of units sold in each of the past 5 years and each of the next 7 years can be regarded as independent observations from the same normal distribution.

Often, a time series (i.e., a sequence of observations taken over time) will exhibit some trend or correlation among consecutive observations. In such cases, the preceding assumptions would not be true. One might then account for trend by using regression analysis (if extrapolation of the trend appears to be justified) or, more generally, by using special techniques for the statistical analysis of time series; see, for example, Box et al. (2015). Generally, however, one needs much more data than available here to obtain meaningful estimates of the nature of the trend or general correlation structure.) Although there are statistical tests and informal graphical methods to check for departures from the important stated assumptions (e.g., Kutner et al., 2005, Chapter 4), it is possible to detect only very extreme departures with just 5 observations. No such gross departures are evident in Figure 11.2. We need to emphasize, however, that these assumptions are critical for our analysis, and are often not satisfied in practice. Also, even if such assumptions were reasonable in the past, they might not hold in the future. For example, the fact that the producer is stopping production might itself impact future demand. In the problem at hand, however, sales of the product containing the bearing had been without a significant trend for many years and the bearings in motors that are less than 7 years old (which would be less heavily represented in the field in the future) rarely failed. Thus, demand for replacement bearings could be expected to remain fairly constant over the next 7 years.

11.2.3 Prediction and a One-Sided Upper Prediction Bound for Future Total Demand

From given data for the past 5 years, the sample mean is

$$\bar{x} = \frac{27.7 + 37.1 + 35.7 + 30.8 + 32.7}{5} = 32.80,$$

and the sample standard deviation is

$$s = \left[\frac{(27.7 - 32.80)^2 + \cdots + (32.7 - 32.80)^2}{5 - 1}\right]^{1/2} = 3.77.$$

Under the stated assumptions, $\bar{x}$ provides a prediction for the average yearly demand. Thus, $7 \times 32.8 = 229.6$ provides a prediction for the total 7-year demand (in thousands of units) for the replacement bearings. However, because of statistical variability in both the past and the future yearly demands, the actual total demand would be expected to differ from this prediction. Using the method outlined in Section 4.7 and $t_{(0.95;4)} = 2.132$, a one-sided upper 95% prediction bound for the mean of the yearly sales for the next $m = 7$ years is

$$\tilde{Y} = \bar{x} + t_{(1-\alpha;n-1)} \left(\frac{1}{m} + \frac{1}{n}\right)^{1/2} s,$$

$$= 32.80 + 2.132 \left(\frac{1}{7} + \frac{1}{5}\right)^{1/2} 3.77 = 37.5.$$

Thus, a one-sided upper 95% prediction bound for $7 \times Y$, the total 7-year demand, is $7 \times \tilde{Y} = 7 \times 37.5$, or 262,500 bearings. That is, although our point prediction for the 7-year demand is 229,600 bearings, we can claim, with 95% confidence, that the total demand for the next 7 years will not exceed 262,500 bearings. At the same time, if the producer actually built 262,500 bearings, we would predict that the inventory would, most likely, last for 262.5/32.80 or approximately 8 years. In passing, we note that a one-sided *lower* 95% prediction bound for the total demand for the next 7 years is 196,700 bearings.

11.2.4 An Alternative One-Sided Upper Prediction Bound Assuming a Poisson Distribution for Demand

An alternative one-sided upper prediction bound for the total demand in the next 7 years can be obtained by assuming that yearly demand can be modeled with a Poisson distribution with a constant rate (at least for the past 5 years) and that this will continue to be so (at least for the next 7 years). In this case, $\widehat{\lambda} = \bar{x} = 32.80$ is an estimate for the yearly demand rate. Thus, a point prediction for the demand (in thousands of units) in the next 7 years is again $7 \times 32.80 = 229.6$. Using the normal distribution approximation method given in Section 7.6.2, with $n = 5$, $m = 7$, and $z_{(0.95)} = 1.645$, an approximate one-sided upper 95% prediction bound for total demand for the next 7 years is

$$\tilde{Y} = m\widehat{\lambda} + z_{(1-\alpha)} m \left[\widehat{\lambda}\left(\frac{1}{m} + \frac{1}{n}\right)\right]^{1/2}$$

$$= 7 \times 32.80 + 1.645 \times 7 \left[32.80 \left(\frac{1}{7} + \frac{1}{5}\right)\right]^{1/2} = 229.6 + 38.6 = 268.2.$$

73.2	67.8	68.5	73.8	69.3	70.9	65.4	71.2	72.4	69.6
67.1	69.2	66.5	72.9	75.4	74.2	69.1	64.0	68.9	70.2

Table 11.2   Engine emissions measurements. Measurements are in manufacturing sequence (reading across).

These results agree reasonably well with those obtained under the assumption that yearly demand has a normal distribution. One would, of course, not expect full agreement because somewhat different models have been assumed. As frequently happens, it is not clear from the underlying situation, or the limited data, which model is more appropriate. Therefore, in this case, as in many others, it is useful to compute intervals under various plausible models and compare the results. We should note, however, that for this problem, the Poisson distribution approach is subject to restrictive assumptions similar to those described for the normal distribution in Section 11.2.2.

11.3 ESTIMATING THE PROBABILITY OF PASSING AN ENVIRONMENTAL EMISSIONS TEST

11.3.1 Background

A manufacturer must submit three engines, randomly selected from production, for an environmental emissions test. To pass the test, the measurement for a particular pollutant on each of three engines to be tested must be less than $L = 75$ ppm. Based on the measurements in Table 11.2 from a previous test, involving 20 (presumably randomly selected) engines, the manufacturer wants to construct a one-sided lower (i.e., worst case) 95% confidence bound for the probability of passing the test.

11.3.2 Basic Assumptions

We make the important assumption that the 20 past engines and the three future engines are all randomly selected from the same "in statistical control" production process. The reasonableness of this assumption needs to be carefully evaluated based on an understanding of the problem and, possibly, a plot of the available data against manufacturing order. In this case, such a plot, shown in Figure 11.3, indicates no obvious trends or other nonrandom behavior.

11.3.3 Nonparametric Approach

The following approach requires no assumptions about the form of the underlying distribution of emissions measurements. Because the values for 19 out of 20 of the previously tested engines did not exceed the 75-ppm threshold, an estimate of the proportion of conforming engines is $\widehat{p}_C = 19/20 = 0.95$. Thus, a point estimate of the probability that all three future engines will meet the specified limit of 75 ppm is $\widehat{p}_{\text{dem}} = (\widehat{p}_C)^3 = 0.95^3 = 0.86$.

A one-sided lower 95% confidence bound on p_C can be obtained by using one of the confidence interval methods for proportions described in Section 6.2. For $x = 19$ and $n = 20$ and confidence level $1 - \alpha = 0.95$, using R as a calculator for the beta quantile formula (6.1) in Section 6.2.2 gives

```
> qbeta(p=0.05, shape1=19, shape2=20-19+1)
[1] 0.7839
```

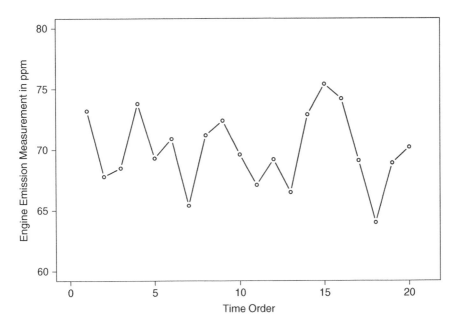

Figure 11.3   Time-ordered plot of engine emission measurements.

so $p_C = 0.784$ is a one-sided lower conservative 95% confidence bound. Thus, under the stated assumptions, we can claim with 95% confidence that the proportion of conforming engines is at least 0.784. The one-sided lower 95% confidence bound on the probability that *all three* future engines will conform is then

$$p_{\text{dem}} = (p_C)^3 = 0.784^3 = 0.48.$$

Thus, based on the limited available data, we are 95% confident that the probability of passing the test is at least 0.48—unfortunately, not a very high number.

11.3.4   Normal Distribution Approach

Figure 11.4 is a normal probability plot of the data in Table 11.2. This plot suggests that a normal distribution adequately describes such emission measurements. From the data, $\bar{x} = 69.98$ and $s = 3.04$. A point estimate for $p_C = \Pr(X \leq 75)$, assuming that the emission measurements have a normal distribution, is $\widehat{p}_C = \Phi_{\text{norm}}[(L - \bar{x})/s] = \Phi_{\text{norm}}[(75 - 69.98)/3.04] = \Phi_{\text{norm}}(1.65) \approx 0.95$. Under the same normal distribution assumption, one can use the methods of Section 4.5 to compute a one-sided lower 95% confidence bound for the proportion of conforming units. This would also be a one-sided lower 95% confidence bound on the probability that a single randomly selected unit conforms to the specified limit. First, compute

$$k = \frac{L - \bar{x}}{s} = \frac{75 - 69.98}{3.04} = 1.65.$$

Based on the $n = 20$ observations, a one-sided lower 95% confidence bound on $\Pr(X \leq 75)$, the probability that a single engine will meet the specified limit, is

$$p_C = \texttt{normTailCI}(0.05; 1.65, 20) = 0.856.$$

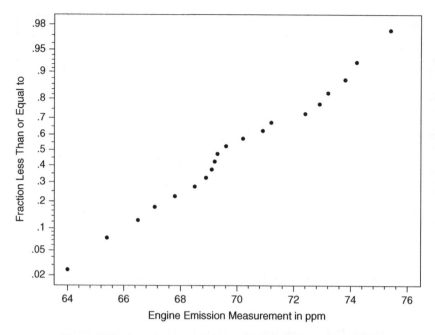

Figure 11.4   Normal probability plot of engine emission readings.

Then a one-sided lower 95% confidence bound on the probability of all three units meeting specifications is

$$\underset{\sim}{p}_{\text{dem}} = (\underset{\sim}{p}_C)^3 = 0.856^3 \approx 0.63.$$

Thus, we can say, with 95% confidence, that the probability that all three engines will conform to the specification is at least 0.63—still not a very high value.

This method gives a somewhat more precise bound on the desired probability than does the nonparametric method. The price paid for this gain is the need to assume that the measurements have a normal distribution. Needless to say, neither confidence bound was very comforting to the experimenters. Greater assurance might be gained by increasing the sample size, or possibly by decreasing the measurement error associated with the pollution readings (i.e., reducing σ).

11.3.5   Finding a One-Sided Upper Prediction Bound on the Emission Level

Because it was not clear that all three submitted engines would pass the emissions test at a threshold of 75 ppm, it was desired to determine a new threshold value for which we could be highly confident of passing the test for all three future engines. This calls for finding a one-sided upper prediction bound that will, with a specified degree of confidence, not be exceeded by the future measurement on each of the three engines.

For the distribution-free approach given in Section 5.5.2, one can use (5.17) to compute the confidence level associated with using the largest of the $n = 20$ previous observations as a one-sided upper prediction bound to exceed the $m = 3$ future observations. The confidence level is

$$1 - \alpha = \frac{n}{n + m} = \frac{20}{20 + 3} = 0.87.$$

Thus we can be 87% confident that the emission levels of all three future observations will be less than $x_{(20)} = 75.4$ ppm.

We note that the preceding distribution-free approach does not allow us to construct a one-sided upper 95% prediction bound to exceed the next three observations, unless we take a larger sample. If, however, we can assume that the emission readings can be adequately described by a normal distribution, the methods outlined in Section 4.8.2 can be used to compute such a bound. For this example, using $n = 20$, $k = m = 3$, and $1 - \alpha = 0.95$, Table J.9 gives $r'_{(0.95;3,3,20)} = 2.331$. Then

$$\widetilde{Y}_{3:3} = \bar{x} + r'_{(0.95;3,3,20)} s = 69.98 + 2.331 \times 3.04 = 77.1.$$

That is, we are 95% confident that all three future engines will have emissions less than 77.1 ppm.

11.4 PLANNING A DEMONSTRATION TEST TO VERIFY THAT A RADAR SYSTEM HAS A SATISFACTORY PROBABILITY OF DETECTION

11.4.1 Background and Assumptions

The manufacturer of an aircraft radar system needs to demonstrate with a high degree of confidence that the system can, under specified conditions, detect a target at a distance of 35 miles with a probability of detection p_D that exceeds the specified value $p_D^{\dagger} = 0.95$. The verification test consists of a sequence of passes under the specified conditions between the radar equipped airplane and an approaching target plane, both traveling approximately the same path for each pass. The pass is deemed to be a success if detection occurs before the planes are 35 miles apart, and a failure otherwise. The demonstration involves calculating a one-sided lower confidence bound ($\underset{\sim}{p_D}$) for p_D, the probability of detection at a distance exceeding 35 miles, and will be deemed successful if $\underset{\sim}{p_D} \geq p_D^{\dagger} = 0.95$. The problem is to determine the required number of passes (i.e., the sample size). It is assumed that each pass will give an independent observation of miles to detection from the defined process consisting of all similar passes (i.e., approximately the same direction, weather, system configuration, etc.) As always, properly defining this process to assure that it really represents the situation of interest, and planning the passes accordingly, is critical. After the passes have taken place, one might consider two possible methods for computing $\underset{\sim}{p_D}$:

1. Make no assumption about the form of the distribution of miles to detection and use the observed proportion of successes (i.e., the proportion of passes exceeding 35 miles). In this case, one uses the binomial distribution as a basis to compute a one-sided lower confidence bound on p_D, the probability of detection, using one of the procedures outlined in Section 6.2.

2. Assume that miles to detection (or some transformation, such as log miles) has a normal distribution, and use one of the methods described in Section 4.5 to compute $\underset{\sim}{p_D}$.

It was decided to base the sample size on the first method because (a) there was little prior information about the form of the distribution of miles to detection and (b) it leads to a more conservative procedure (i.e., fewer assumptions, but a larger sample size). If the resulting data supported the assumption of a normal distribution, then the second method might be used in the subsequent data analysis (thus giving more precise assessments).

11.4.2 Choosing the Sample Size

Suppose that it is desired to demonstrate, with 90% confidence, that the probability of detection p_D is greater than $p_D^\dagger = 0.95$. One would clearly want to have a successful demonstration if p_D "appreciably exceeds" 0.95, where the term "appreciably exceeds" still requires elaboration.

Using the methods outlined in Section 9.5, based on the nonparametric binomial distribution model, the probability of a successful demonstration is graphed in Figure 9.2a as a function of n, the actual probability of detection for various number of passes. Thus we see, for example, if $n = 77$ passes are used, the demonstration that $p_D^\dagger \geq 0.95$ will be successful if there are no more than $c = 1$ passes that *fail* to detect the target at a distance of 35 miles or more. Further, if the actual probability of detection is $\pi = p_D = 0.98$ (and, thus, the probability of an unsuccessful pass is 0.02), then the probability of a successful demonstration (i.e., $c = 0$ or 1) for $n = 77$ is

$$p_{\text{dem}} = \Pr(\underset{\sim}{p_D} \geq 0.95) = \Pr(X \leq 1) = \texttt{pbinom}(1; 77, 0.02) = 0.543,$$

where X is the number of unsuccessful passes. A probability of 0.543 of successful demonstration would generally not be regarded as adequate. Similarly, we determine from Figure 9.2a that if we felt that the actual probability of detection is really as high as 0.995, the probability of a successful demonstration with $n = 77$ is 0.94, which would likely be regarded as satisfactory. Thus, the required sample size to achieve a high probability of successful demonstration when $\pi = p_D$ exceeds 0.95 depends heavily on the actual (but unknown) probability of detection $\pi = p_D$.

We also note from Figure 9.2a that for the probability of successful demonstration to be about 0.90 when the actual probability of detection is $\pi = p_D = 0.98$, $n = 234$ passes are required. In this case, the demonstration will require that there be 7 or fewer unsuccessful passes. The probability of successful demonstration when $\pi = p_D = 0.98$ (i.e., the probability of obtaining seven or fewer passes without a detection in the 234 passes) is

$$p_{\text{dem}} = \Pr(\underset{\sim}{p_D} \geq 0.95) = \Pr(X \leq 7) = \texttt{pbinom}(7; 234, 0.02) = 0.89998.$$

Alternatively, we can use Table J.21 to determine the required sample size. Entering this table, (a) to demonstrate with confidence level $1 - \alpha = 0.90$ that the probability of detection p_D is greater than $\beta = p_D^\dagger = 0.95$, while (b) assuring that the probability of failing to achieve demonstration is $\delta = 1 - p_{\text{dem}} < 0.10$ when the actual probability of detection is $\beta^* = 0.98$, gives the required sample size as $n = 258$. This is greater than the sample size indicated by Figure 9.2a because $n = 234$ gives only $p_{\text{dem}} = 0.89998$ (slightly less than the desired p_{dem} of 0.90), while $n = 258$ is the smallest sample size to give $p_{\text{dem}} \geq 0.90$, and gives an actual p_{dem} of

$$p_{\text{dem}} = \Pr(\underset{\sim}{p_D} \geq 0.95) = \Pr(X \leq 8) = \texttt{pbinom}(8; 258, 0.02) = 0.923.$$

Also, a sample of $n = 258$ will permit demonstration as long as there are eight or fewer, rather than seven or fewer, unsuccessful passes.

11.5 ESTIMATING THE PROBABILITY OF EXCEEDING A REGULATORY LIMIT

11.5.1 Background and Assumptions

A company has taken readings on the concentration level of a chemical compound at a particular point in a river. One reading (each reading is actually an average of five measurements) was taken during the first week of the quarter in each of the past 27 quarters. The data are given in Table 11.3 (which also shows the day of the week when the measurement was taken) and are plotted against time in the top row of Figure 11.5. The company was asked to use the past

Observation number	Quarter	Day (during the first week of the quarter)	Chemical concentration (in ppm)
1	Q2	Monday	48
2	Q3	Wednesday	94
3	Q4	Monday	112
4	Q1	Friday	44
5	Q2	Wednesday	93
6	Q3	Thursday	198
7	Q4	Tuesday	43
8	Q1	Monday	52
9	Q2	Wednesday	35
10	Q3	Friday	170
11	Q4	Monday	25
12	Q1	Wednesday	22
13	Q2	Tuesday	44
14	Q3	Thursday	16
15	Q4	Friday	139
16	Q1	Tuesday	92
17	Q2	Friday	26
18	Q3	Monday	116
19	Q4	Thursday	91
20	Q1	Thursday	113
21	Q2	Friday	14
22	Q3	Monday	50
23	Q4	Wednesday	75
24	Q1	Friday	66
25	Q2	Monday	43
26	Q3	Tuesday	10
27	Q4	Friday	83

Table 11.3 Chemical concentration readings.

data to estimate the probability that a future reading will exceed the regulatory limit of 300 ppm, even though all of the 27 past readings have been appreciably below this value. In order to respond, it will be necessary to use a statistical model to describe the relationship between the past and future readings. The simplest such model assumes that all past and future readings are random observations from the same process. However, because we are dealing with a time series, generated by a process that might change over time (due to changes in production level, pollutant processing methods, etc.), this model may not be appropriate. The physical process must be reviewed and the data carefully checked to assess the existence of a trend, a cyclical or a seasonal pattern, or other correlations, among the observations. A trend might be present if the mean of the process is changing with time, due, for example, to changes in production level or pollution abatement measures. Seasonal effects might occur because of differences in concentration due to seasonal variations in production or the impact of changes in weather conditions. Differences might arise due to varying levels of production on different days of the week. Fortunately, there were no physical reasons to expect the simple model not to apply in this

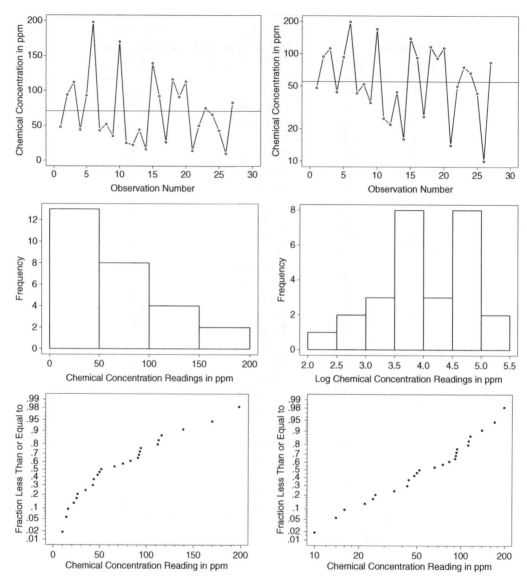

Figure 11.5 Time series plots (top row, with log axis on the right), histograms (middle row, plotting logs on the right), and probability plots (bottom row, with normal on the left and lognormal on the right) for the concentration readings.

example. It is hoped, however, that there might be a reduction in pollution levels in the future. In this case, the results obtained under the assumed model would tend to be overly conservative in the sense that they are likely to overpredict future levels of pollution. In any case, an empirical assessment of the validity of the assumptions, based on the past data, is in order.

11.5.2 Preliminary Graphical Analysis

The top row of Figure 11.5 shows time series plots of the concentration readings on a linear axis (on the left) and on a log axis (on the right). The plots show that there is appreciable variability in the readings (when compared to past experience with similar studies), there are several large extreme observations (checks at the original source of the data did not suggest any recording errors), the possibility of a downward trend, but no clear indication of cyclical or periodic behavior. The middle row of Figure 11.5 provides histograms of the concentration readings (on the left) and the logs of the concentration readings (on the right).

The bottom row of Figure 11.5 shows a normal probability plot (on the left) and a lognormal probability plot (on the right). The plots on the left indicate that the distribution of concentration readings is appreciably skewed to the right and, especially because the points on the normal probability plot deviate importantly from a straight line, there is evidence that such readings are not normally distributed. In particular, there are some large extreme observations. The plots on the right, based on a log transformation, indicate approximate symmetry of the distribution, suggesting that the readings may be better approximated by a *lognormal*, than by a normal, distribution. (A lognormal distribution is frequently found to be an appropriate model in pollution assessment problems.) A log transformation, moreover, also tended to accommodate the extreme observations. Thus, henceforth, we will consider the logs of the concentration readings.

11.5.3 Formal Tests for Periodicity and Autocorrelation

Checks for periodicity were first performed by using an analysis of variance to test for differences among the four quarters of the year and among the different days of the week. No statistically significant differences were found.

For a time series with a constant mean and standard deviation (often referred to as "stationary"), let r_k denote an estimate of the correlation between observations that are k time periods apart. The set of values r_k, $k = 1, 2, 3, \ldots$, is known as the sample autocorrelation function (ACF) and is an important tool for modeling time series data. Figure 11.6 is a plot of the sample ACF for the log chemical concentration readings. The value of r_0 is, by definition, equal to 1. The ACF can be computed easily with many of the popular data analysis computer programs (e.g., R). Although exact sampling theory for these statistics is complicated, in large samples of independent observations (say, $n > 60$), the r_k can be assumed to be approximately normally distributed with a standard error approximated by Bartlett's formula (e.g., Chapter 2 of Box et al., 2015). Although crude for the current sample of size 27, this approximation can be used to roughly assess the statistical significance of the correlations or to construct approximate confidence intervals to contain the actual correlations. This provides an approximate formal check for the assumption that the readings are uncorrelated. The dashed lines shown in Figure 11.6 indicate approximate limits outside of which a sample autocorrelation would be statistically significant at a 5% significance level. We note that the estimated autocorrelations are all contained within their bracketed bounds (equivalently, confidence intervals for the actual correlations all enclose 0). Thus there is no evidence of autocorrelation within the quarterly data.

11.5.4 Formal Test for Trend

Formal statistical procedures can be used to supplement the graphical analyses. As mentioned earlier, the top row of Figure 11.5 might suggest to some that there is a linear (downward) trend over time in the plotted values. A formal check for such a time trend can be done by fitting a simple linear regression of the readings (in this case, the logs of the readings) versus time. If a trend is present and can be assumed to continue into the future, the regression model—or some further generalization, such as those discussed in books on time series analysis such as Wei (2005), Bisgaard and Kulahci (2011), and Box et al. (2015)—might be used for forecasting.

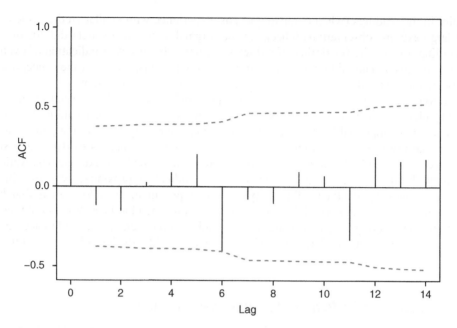

Figure 11.6 Sample autocorrelation function of the log chemical concentration readings.

Figure 11.7 gives a summary of the simple regression analysis for the log of concentration versus time. The assumed model for this analysis is

$$\log(\text{concentration}) = \beta_0 + \beta_1(\text{time}) + \varepsilon,$$

where time $= 1, 2, \ldots, 27$ (i.e., the 27 quarters for which data were available) and ε is a random noise (or error) term which is assumed to have a normal distribution with a mean of 0 and a standard deviation that is constant over time. We also assume that the ε values are independent of each other. A two-sided 95% confidence interval for β_1, the slope of the regression line, is

$$\widehat{\beta}_1 \mp t_{(1-\alpha/2;n-2)} s_{\widehat{\beta}_1} = -0.0214 \pm 2.060 \times 0.0190 = [-0.060, \ 0.018],$$

where $\widehat{\beta}_1$ is the least squares estimate of β_1, $s_{\widehat{\beta}_1}$ the standard error of this estimate, and $t_{(1-\alpha/2;n-2)}$ the $1 - \alpha/2$ quantile of Student's t-distribution with $n - 2$ degrees of freedom (e.g., qt(0.975,25) in R gives 2.060); see the brief discussion in Section 4.13.1. Because this confidence interval for the slope coefficient contains the value 0, there is no statistical evidence of a trend in the data.

```
Coefficients:
                Estimate  Std. Error  t ratio  Pr(>|t|)
(Intercept)       4.3087      0.3042    14.16    <0.001
time             -0.0214      0.0190    -1.13     0.27

Residual standard error: 0.769 on 25 degrees of freedom
Multiple R-squared: 0.0482
```

Figure 11.7 Summary of regression analysis of the log chemical concentration readings versus time.

If any of the above tests had given positive results, it would be an indication that more sophisticated time series analyses might be required. In this case, all tests came out negative, as expected from physical considerations, and so it seems satisfactory to proceed using simple methods. At the same time, we need to note that:

- The power of a statistical test to establish significance is highly dependent upon sample size ($n = 27$ is quite modest).

- The fact that certain patterns were not exhibited in the past does not guarantee that they will not happen in the future.

11.5.5 Nonparametric Binomial Model

Without making any assumptions about the form of the distribution of the readings, it is possible to estimate p_{GT}, the probability of exceeding the regulatory limit of $L = 300$ ppm. We assume only that both the past and the future readings are independently and randomly chosen from the same stationary process. Because none of the 27 past quarterly readings exceeded 300 ppm, an estimate of this probability is $\widehat{p}_{GT} = 0/27 = 0$. The methods given in Section 6.2, based on the binomial distribution, can be used to compute a one-sided upper confidence bound on this probability. In particular, using the conservative method outlined in Section 6.2.2, a one-sided upper 95% confidence bound on the probability of exceeding the regulatory limit on a randomly selected day (at least in the first week of a forthcoming quarter) is

$$\widetilde{p}_{GT} = \texttt{qbeta}(0.95; 1, 27) = 0.105.$$

Thus, we are 95% confident that p_{GT} is less than $\widetilde{p}_{GT} = 0.105$. Alternatively, we could have obtained this value from Figure 6.1.

The result might seem a little disappointing, in light of the data. Even though there were no readings above, or even close to, 300 ppm for the sampled day in each of the past 27 quarters, the data have not established with 95% confidence that the probability is satisfactorily small, even if we assume that there will be no change in the process. This analysis, however, ignores the actual values of the readings (other than whether or not they exceed 300 ppm). An alternative analysis (see below) that assumes a distributional model would be expected to be more informative (i.e., provide a tighter bound for p_{GT}).

11.5.6 Lognormal Distribution Model

Our graphical analyses of the 27 readings indicated that the logs of the readings might be modeled adequately by a normal distribution. The sample mean and standard deviation of the log readings are $\bar{x} = 4.01$ and $s = 0.773$, respectively. A point estimate for the proportion of days that the limit will be exceeded, assuming that the logs of the chemical concentration readings have a normal distribution, is $\widehat{p}_{GT} = \widehat{\Pr}[X \geq \log(300)] = 1 - \Phi_{\text{norm}}[(\log(L) - \bar{x})/s] = 1 - \Phi_{\text{norm}}[(5.70 - 4.01)/0.773] = 1 - \Phi_{\text{norm}}(2.19) = 0.0143$. Under the same normal distribution assumption, one can use the methods given in Section 4.5 to compute a one-sided upper 95% confidence bound for the proportion of days that the limit will be exceeded. This would also be a one-sided upper 95% confidence bound on the probability that the limit will be exceeded on a single randomly selected day. Specifically, first compute

$$k = \frac{\log(L) - \bar{x}}{s} = \frac{5.70 - 4.01}{0.773} = 2.19.$$

Then, using (4.6), the one-sided upper 95% confidence bound for p_{GT} is $\widetilde{p}_{GT} =$ normTailCI$(0.95; -2.19, 27) = 0.05620$. Thus, we are 95% confident that the probability of a reading exceeding 300 ppm is less than 0.056. This value is smaller and, thus, more satisfactory than the one-sided upper 95% confidence bound of 0.105, which was obtained with the nonparametric binomial distribution model. This improvement was obtained in return for the (not unreasonable) assumption that the readings have a lognormal distribution.

In passing, we note that if we had incorrectly assumed a normal (rather than a lognormal) distribution for the readings, we would have obtained the one-sided upper 95% confidence bound for p_{GT} to be $\widetilde{p}_{GT} = 0.00015$ or 0.015%. The contrast between 0.00015 and 0.05620 (a ratio of about 375) illustrates our statement in Section 4.10, that confidence bounds on probabilities in the tail of a distribution are not robust to an incorrect distributional assumption.

11.6 ESTIMATING THE RELIABILITY OF A CIRCUIT BOARD

11.6.1 Background and Assumptions

A company manufactures a circuit board that contains 110 similar integrated circuit chips that must operate in a field environment for 30,000 hours. Successful operation requires that all chips in a board operate without failure in service. It is reasonable to assume that chips on the same board fail independently of one another. This assumption would be incorrect if, for example, failures are caused by shocks that affect more than one chip, or if failure of one chip increases the stress on the others. The assumption appears reasonable, however, because failure from internal defects in the chips is the dominant failure mode. Thus, one can assess life in service from tests on chips, rather than requiring tests on boards. In fact, an accelerated test on individual chips has been developed for this purpose. This test simulates the 30,000 hours of operation in a normal service environment by a 1,000-hour dynamic high temperature–high humidity exposure at 85°C and 85% relative humidity.

Due to a (hopefully) one-time manufacturing problem, some of the chips in a special shipment of 50,000 chips may be prone to failure during field service (such chips will, henceforth, be referred to as being "defective"). To estimate p_C, the proportion of defective chips in the shipment, a random sample of 1,000 chips was selected from the inventory of 50,000 chips, and these chips were subjected to the 1,000-hour accelerated test. Two chips failed the test.

We note that in this example, unlike most of the others in this chapter, the sample is from a well-defined population concerning which we wish to draw inferences—namely, the 50,000 chips in inventory from this shipment. Thus, in this sense, using the terminology of Chapter 1, this is an enumerative study. This would not be the case if we had wished to draw inferences about future chips from the process from which this shipment came. However, because (a) the chips are being tested in an accelerated test environment that is meant to simulate operational conditions, and (b) testing on chips is to be used to draw conclusions about results on boards under the assumption of independent failures, one might argue that, in totality, this is an analytic, rather than an enumerative, study. We will not get hung up in this discussion on terminology but simply emphasize the importance of these fundamental assumptions to drawing conclusions from this evaluation.

We also note that we are dealing here with a finite population of 50,000 chips. Our sample of 1,000 chips is, however, a small percentage of the population (appreciably less than 10%), and, therefore, the finiteness of the population can be ignored for practical purposes, as indicated in Chapter 1. If this had not been the case (e.g., there had been only 5,000 chips in the shipment) then the methods described in Section 6.3 could be applied to draw inferences about the unsampled chips, instead of those to be described below. The "finite population correction factor" method mentioned in Section 1.10, if it were applied using (6.3), would not provide an adequate approximate confidence interval, in this case. This is because (6.3) is based upon a normal distribution approximation to the sampling distribution of $\widehat{p}$, and the approximate confidence interval method using (6.3) is inappropriate for a situation with only two nonconforming units; see Section 6.2.3.

The following information is desired for the in-service operating conditions, together with appropriate statistical bounds:

- An estimate of the proportion of defective chips in the shipment.

- An estimate of the proportion of boards (each containing 110 chips from the shipment) that will contain one or more defective chips.

- An estimate of the probability that at least 9 of 10 boards that use chips from the shipment, and are to be installed in a system, will operate successfully in service.

11.6.2 Estimate of the Proportion of Defective Chips

An estimate of the actual proportion (p_C) of defective chips for the shipment is $\widehat{p}_C = 2/1{,}000 = 0.002$, or 0.2%. Using the conservative method given in Section 6.2.2, a two-sided 90% confidence interval for p_C is

$$[\underset{\sim}{p}_C, \ \widetilde{p}_C] = [0.0003555, \ 0.006282].$$

Thus, we are 90% confident that the proportion of defective chips in inventory is between 0.00036 and 0.0063.

11.6.3 Estimating the Probability that an Assembled Circuit Board Will Be Defective

A board is defective if it contains one or more defective chips. The number of defective chips on a board has a binomial distribution with parameters p_C and $n = 110$. Because the circuit board contains 110 chips, the probability that a board is not defective is the probability that none of the 110 chips is defective, assuming that the only reason for board failure is the independent failure of the chips. Under the previously stated assumption of independence, this probability is

$$(1 - p_C)^{110}.$$

Thus, the probability that the board is defective (i.e., has one or more defective chips) is

$$p_B = 1 - (1 - p_C)^{110}.$$

An estimate of this probability is, therefore,

$$\widehat{p}_B = 1 - (1 - \widehat{p}_C)^{110} = 1 - (1 - 0.002)^{110} = 0.20.$$

Because p_B is a monotonic increasing function of p_C, confidence bounds for p_B can be obtained directly from those for p_C (see Section 6.4). In particular, the endpoints of a 90% confidence interval for the probability that a circuit board will be defective are

$$\underset{\sim}{p}_B = 1 - (1 - \underset{\sim}{p}_C)^{110} = 1 - (1 - 0.0003555)^{110} = 0.038,$$

$$\widetilde{p}_B = 1 - (1 - \widetilde{p}_C)^{110} = 1 - (1 - 0.006282)^{110} = 0.50.$$

Thus, we are 90% confident that the proportion of defective boards constructed from chips in inventory is between 0.038 and 0.50. This wide interval is not surprising when we recognize that the available data are on the equivalent of only about nine boards.

11.6.4 Estimating System Reliability

A system that contains 10 circuit boards requires at least 9 such boards to operate successfully. We assume that the number of boards that fail in service has a binomial distribution with parameters $n = 10$ and p_B. This model would apply under assumptions similar to those stated in Section 11.6.1.

Following the approach outlined in Section 6.4, the probability of successful system operation is

$$p_D = \Pr(0\,\text{boards fail}) + \Pr(1\,\text{board fails}) \qquad (11.1)$$
$$= (1 - p_B)^{10} + 10(p_B)^1(1 - p_B)^9.$$

Because p_D is a monotonically decreasing function of p_B, a one-sided lower (upper) confidence bound on p_D is obtained by substituting the corresponding one-sided upper (lower) confidence bound for p_B into (11.1). That is,

$$\underset{\sim}{p_D} = (1 - \widetilde{p}_B)^{10} + 10\widetilde{p}_B(1 - \widetilde{p}_B)^9$$
$$= (1 - 0.50008)^{10} + 10(0.50008)(1 - 0.50008)^9 = 0.011,$$
$$\widetilde{p}_D = (1 - \underset{\sim}{p_B})^{10} + 10\underset{\sim}{p_B}(1 - \underset{\sim}{p_B})^9$$
$$= (1 - 0.038357)^{10} + 10(0.038357)(1 - 0.038357)^9 = 0.946.$$

Thus, based on the fact that the random sample of 1,000 chips contained 2 defective units, we can say, with 90% confidence, that the probability of successful system operation is between 0.011 and 0.946! Thus, our evaluation has been highly uninformative. This is not very surprising because the available data on 1,000 chips are used to draw conclusions, with a high degree of confidence, about a system involving 1,100 chips.

11.7 USING SAMPLE RESULTS TO ESTIMATE THE PROBABILITY THAT A DEMONSTRATION TEST WILL BE SUCCESSFUL

Audio quality performance scores have been obtained on a random sample from production of 16 high fidelity speakers. The data are shown in Table 11.4. A future demonstration test will be successful if the score for *each* of 32 additional randomly selected speakers exceeds $L = 450$. We need to estimate p_{dem}, the probability of passing the demonstration test. The speaker manufacturing process has been established to be stable (i.e., in statistical control) and our analysis is based on the important assumption that this will continue to be the case.

11.7.1 Nonparametric Binomial Model Approach

Because all of the initial 16 units had performance scores greater than 450, an estimate of p_{GT}, the proportion of units from production with scores above 450, is $\widehat{p}_{GT} = 16/16 = 1.0$. Using the method discussed in Section 6.2.2, a one-sided lower conservative 90% confidence bound for p_{GT} is

$$\underset{\sim}{p_{GT}} = \texttt{qbeta}(0.10; 16, 1) = 0.866.$$

That is, we are 90% confident that at least a proportion 0.866 of the units from production would, if measured, score above 450.

552	586	702	722	742	790	800	838
838	921	960	981	994	1,035	1,110	1,405

Table 11.4 Audio quality performance measurements in increasing order.

The probability that all 32 future units will score higher than 450 is $p_{\text{dem}} = (p_{GT})^{32}$, and, because $\widehat{p}_{GT} = 1.0$, the resulting point estimate is also $\widehat{p}_{\text{dem}} = 1.0$. Because p_{dem} is an increasing function of p_{GT}, a one-sided lower 90% confidence bound for p_{dem} can be obtained by using $\underset{\sim}{p}_{GT}$ in this formula in place of p_{GT}. That is,

$$\underset{\sim}{p}_{\text{dem}} = (\underset{\sim}{p}_{GT})^{32} = 0.866^{32} = 0.010.$$

Thus, we are 90% confident that the probability of passing the demonstration test is at least 0.01! This result is not very useful except to tell us that we cannot be assured of a successful demonstration about 32 future units from successful go/no-go data on only 16 past units.

11.7.2 Normal Distribution Approach

Figure 11.8 is a normal probability plot of the 16 scores for the initially tested units. Except for the largest observation (1,405), the data seem to be well represented by a normal distribution. Thus, this distribution may provide a reasonable model, at least for the lower tail of the distribution of audio performance scores—and it is this lower tail that is of concern to us. Sample statistics for these data are $\bar{x} = 873.5$ and $s = 211.5$. Note, however, that both of these values may be inflated estimates if the largest observation is erroneous—a topic that we will investigate further in Section 11.7.3. A point estimate for the probability that the performance of a single randomly selected unit will exceed 450, assuming that the performance scores have a normal distribution, is $\widehat{p}_{GT} = \widehat{\Pr}(X \geq 450) = 1 - \Phi_{\text{norm}}[(450 - \bar{x})/s] = 1 - \Phi_{\text{norm}}[(450 - 873.5)/211.5] = 1 - \Phi_{\text{norm}}(-2.002) \approx 0.9773$. Under the normal distribution assumption, one can use the method given in Section 4.5 to compute a one-sided lower 90% confidence bound for $p_{GT} = \Pr(X \geq 450)$. In particular, with

$$k = \frac{L - \bar{x}}{s} = \frac{450 - 873.5}{211.5} = -2.002,$$

and $n = 16$, $\underset{\sim}{p}_{GT} = \texttt{normTailCI}(0.10; 2.002, 16) = 0.9207$. That is, assuming that the data are a random sample from a normal distribution, we can be 90% confident that the proportion of units with scores above $L = 450$ is at least 0.9207.

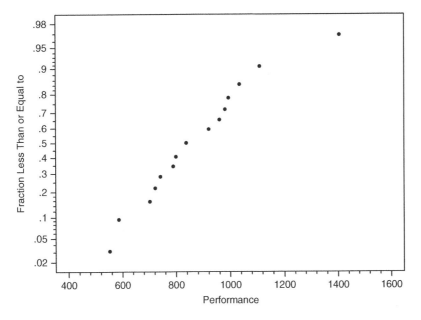

Figure 11.8 Normal probability plot of audio quality performance measurements.

As in Section 11.7.1, we compute a one-sided lower 90% confidence bound for p_{dem} as

$$\underset{\sim}{p}_{\text{dem}} = (\underset{\sim}{p}_{GT})^{32} = 0.9207^{32} = 0.07.$$

We can similarly calculate the upper 90% confidence bound on p_{dem} to be $\widetilde{p}_{\text{dem}} = 0.84$, resulting in a two-sided 80% confidence interval of 0.07 to 0.84. This tells us that the available data are insufficient for our purposes, and suggests that additional sampling is needed to be able to draw any definitive conclusions about passing the demonstration test.

11.7.3 Testing Sensitivity of the Conclusions to Changes in the Extreme Observation(s)

Because the largest observation (i.e., 1,405) was suspiciously large for the performance scores, it is worthwhile to investigate the sensitivity of our conclusions to this observation. Thus, the following alternative analyses were performed:

1. The largest observation was ignored altogether and the analysis was repeated. This would be a reasonable approach if the extreme observation were totally incorrect (e.g., due to a data recording mistake that is completely independent of the actual value) and if the correct observation could not be recovered. It would not be correct otherwise.

2. The largest observation was treated as a "right-censored" observation with a value equal to or larger than the second largest observation (a value of 1,110). In this case, we assume that we do not know the exact value of the largest observation, but believe that its value was no less than that of the second largest observation. This (like the next method) may be a reasonable approach if we do not want one, or a few, large extreme observations in the upper tail of the distribution to unduly affect the inferences concerning the lower tail of the distribution (where the normal distribution seems to provide a good model).

3. The largest observations were assumed to have been right censored at several other scores: 800, 900, 1,000, and 1,100. Such censoring might provide reasonable estimates in the lower tail of the distribution (as desired here) if one felt that the normal distribution provides a good representation of that part of the distribution, but not of the distribution as a whole. See Oppenlander et al. (1988) for further exposition of this approach.

When analyzing censored data, maximum likelihood (ML) is generally used to estimate model parameters and functions of model parameters; see Nelson (1982), Meeker and Escobar (1998), or Lawless (2003) for a description of the theory and methods for analyzing censored data. ML methods will be discussed and illustrated in Chapter 12. The analyses for methods 2 and 3 were conducted using ML.

The results of these alternative analyses are compared with those of the original analyses in Table 11.5. This tabulation, in addition to the estimates and approximate 80% confidence intervals for p_{GT} and p_{dem}, also provides ML estimates and 80% confidence intervals for the normal distribution parameters μ and σ. The estimates for p_{GT} and p_{dem} in Table 11.5 differ slightly from those computed in Section 11.7.2. This is because the ML estimates of σ (dividing by n instead of $n - 1$) used in Table 11.5 differ slightly from those computed in Section 11.7.2 so as to make the results directly comparable with those for the other analyses, which require the ML method.

The results, at first glance, suggest a fairly large difference among the estimates and confidence intervals for p_{dem}. This is, however, overshadowed by the continued large statistical uncertainty, reflected by the widths of the confidence intervals.

Estimates using	μ	σ	p_{GT}	p_{dem}
Original data (16 observations)	874 [808, 939]	205 [163, 257]	0.981 [0.933, 0.996]	0.536 [0.110, 0.874]
Largest observation ignored	838 [786, 890]	157 [124, 198]	0.993 [0.965, 0.999]	0.806 [0.315, 0.973]
Censored at 1,110 (1 censored observation)	860 [804, 916]	175 [138, 222]	0.990 [0.956, 0.999]	0.736 [0.239, 0.955]
Censored at 1,000 (3 censored observations)	857 [801, 914]	171 [132, 223]	0.991 [0.956, 0.999]	0.755 [0.236, 0.964]
Censored at 900 (7 censored observations)	861 [795, 927]	178 [127, 249]	0.990 [0.943, 0.999]	0.713 [0.153, 0.962]
Censored at 800 (9 censored observations)	826 [765, 887]	144 [98, 211]	0.996 [0.960, 1.000]	0.866 [0.272, 0.992]

Table 11.5 Maximum likelihood estimates and approximate two-sided 80% confidence intervals (in brackets) for μ, σ, p_{GT} and p_{dem} for audio quality performance measurements.

One disadvantage of the censoring procedure used here is that some subjectivity is needed to choose a censoring point. A good rule of thumb for doing this, when interest centers on the lower (upper) tail of the distribution, is to let a normal probability plot serve as a guide, and to censor those observations that cause departure from linearity in the upper (lower) tail in the plot (note that it would be totally inappropriate to extrapolate into the region where data were censored). Of course, whenever possible, physical considerations should enter into the choice of the censoring point.

In practice, we frequently do not know which specific model is appropriate and which analysis is best. Thus, performing a variety of analyses, as we have done here, provides useful insights into the "robustness" of the results under varying assumptions. Examination of the ML estimates and 80% confidence intervals for p_{dem} in Table 11.5 leads us to the same conclusion irrespective of the approach used (i.e., the available data are insufficient to allow us to draw any definitive conclusions about the outcome of the demonstration test).

11.8 ESTIMATING THE PROPORTION WITHIN SPECIFICATIONS FOR A TWO-VARIABLE PROBLEM

11.8.1 Problem Description

Specifications for an electronic device require:

- Forward voltage must exceed 0.50 volts,

- Reverse breakdown voltage must exceed 95 volts,

- A device may not have both forward voltage below 0.55 volts *and* reverse breakdown voltage below 100 volts.

Device number	Forward voltage	Reverse breakdown voltage	Meet all criteria
1	0.52	101	Yes
2	0.65	110	Yes
3	0.57	97	Yes
4	0.53	98	No
5	0.59	105	Yes
6	0.64	107	Yes
7	0.60	100	Yes
8	0.48	93	No
9	0.60	105	Yes
10	0.54	102	Yes

Table 11.6 Electronic device forward and reverse breakdown voltage measurements.

The available data consist of a sample of ten devices randomly selected from the process that builds the device. This has resulted in the measurements given in Table 11.6. Based upon this information, it is desired to obtain an upper 95% confidence bound on the proportion of devices outside of specifications (i.e., the proportion nonconforming) from the sampled process. Assumptions concerning the "representativeness" of the sample, similar to those described for the previous examples, apply here also.

11.8.2 Nonparametric Approach

If one can represent the measurements by a bivariate normal distribution, one can then use statistical methods for this distribution to get a point estimate of the process proportion nonconforming (using the sample estimates in place of the unknown distribution parameters). However, obtaining a confidence interval or bound on this proportion, based on bivariate normal distribution assumptions, is a complex problem, especially because the forward voltage and reverse breakdown voltage measurements are clearly correlated. Thus, we propose instead a much simpler, first-cut, nonparametric solution that, though less efficient statistically, also does not require any assumptions about the form of the statistical distribution.

We note that two of the ten sampled devices are nonconforming (in particular, device number 4 fails to meet the third requirement, and device number 8 fails on all three requirements). Thus, the observed proportion of nonconforming units is 0.2. From this information, we use Figure 6.1 or methods in Section 6.2 to obtain the desired upper 95% confidence bound on the process proportion nonconforming to be 0.51.

As in previous examples, the information lost by ignoring the actual measurements, and using only the information on whether or not a device is nonconforming, depends on the specific situation. Thus, if all devices had been well within the acceptance region, this first-cut simple approach would have resulted in a greater loss of information than was the case with the data at hand.

11.9 DETERMINING THE MINIMUM SAMPLE SIZE FOR A DEMONSTRATION TEST

11.9.1 Problem Description

A manufacturer feels that a production process provides essentially zero nonconforming units, with regard to a long list of specifications, some of which require a destructive test to evaluate.

An unconvinced customer, however, before accepting the product, requires the manufacturer to demonstrate "with 95% confidence" that the process results in no more than a proportion 0.05 nonconforming units.

Thus, each unit in a random sample from the process is to be evaluated and classified as conforming or nonconforming. Because this involves an expensive series of tests, the manufacturer wants to minimize the required random sample size to achieve the desired demonstration.

11.9.2 Solution

Because the manufacturer is confident that the process yields essentially no nonconforming units, a random sample would also be free of such units. Then one can use the methods given in Chapter 6 to obtain an upper confidence bound on the process proportion nonconforming, based upon the selected sample size. Thus, one can use the results of Chapter 6 in reverse to find how large a sample is needed.

In particular, from Figure 6.1 one notes that a one-sided upper 95% confidence bound of 0.05 is achieved with a sample of about size 60—if that sample, indeed, has no nonconforming units. Thus, the minimum required sample size is approximately 60. Actually, more precise methods lead to a required sample size of 59 units (as can also be seen from Table J.13).

11.9.3 Further Comments

The desired demonstration will be achieved only if the sample really results in zero nonconforming units. For this to be likely, the actual process proportion nonconforming must, indeed, be quite small. In fact, from the lowest curve in Figure 9.2a (which is expressed in terms of the process proportion conforming), we note that even if the process proportion conforming is as large as 0.99, there is only a probability 0.64 of successful demonstration with a sample of 45 conforming units. In fact, we see from this curve that a process conformance rate of close to 0.998 is required for there to be a 0.90 probability of successful demonstration. This is why in discussing sample size requirements in Section 9.5, we did not use a "minimum sample size" approach. Instead, we required specification not only of the process proportion conforming that is to be demonstrated, but also of the proportion conforming for which we desire a high probability that the demonstration test be successful.

Chapter *12*

Likelihood-Based Statistical Intervals

OBJECTIVES AND OVERVIEW

Previous chapters dealt with statistical intervals for complete samples (i.e., no censoring or truncation) from common statistical distributions, focusing on the normal distribution (Chapters 3 and 4), the binomial distribution (Chapter 6), and the Poisson distribution (Chapter 7). In addition, Chapter 5 provided methods for constructing distribution-free intervals. This chapter and subsequent chapters describe and illustrate more general methods for constructing statistical intervals that can be applied to many other distributions and to more complicated models and types of data.

The following topics are discussed in this chapter:

- The motivation for likelihood-based inference and model selection (Sections 12.1).

- The construction of a likelihood function and maximum likelihood (ML) estimation for a parametric model for different types of data (Section 12.2).

- Likelihood-based confidence intervals for a single-parameter distribution, illustrated by the exponential distribution (Section 12.3).

- Likelihood and ML estimators for location-scale and log-location-scale distributions, illustrated by the lognormal and Weibull distributions (Section 12.4).

- Likelihood-based confidence intervals for location-scale and log-location-scale distributions, illustrated by the lognormal and Weibull distributions (Section 12.5).

- Confidence intervals based on computationally simpler Wald approximations of the likelihood-based intervals (Section 12.6).

- Brief comments on the likelihood-based and Wald confidence intervals for other models and brief introductions to likelihood-based tolerance and prediction intervals (Section 12.7).

This chapter emphasizes concepts, methods, examples, and interpretation of data. Due to the nature of the material, the discussion in this and subsequent chapters is somewhat more technical than that in earlier chapters. In addition, Section D.5 outlines the general underlying theory of likelihood and Wald methods for constructing confidence intervals. The Bibliographic Notes section at the end of this chapter identifies sources of more detailed technical information.

12.1 INTRODUCTION TO LIKELIHOOD-BASED INFERENCE

12.1.1 Motivation for Likelihood-Based Inference

There are numerous situations in practice that involve continuous distributions other than the normal distribution (such as the lognormal and Weibull distributions). In addition, the available data are frequently incomplete (i.e., the exact value of an observation is not known). This is the case, for example, in dealing with censored data (i.e., data for which one knows only that the observed value is below some lower observation limit or above some upper observation limit) or with binned data (i.e., grouped or interval censored). For example, in dealing with environmental data, one frequently encounters left-censored observations for which one knows only that an observation is below some threshold detection limit. Similarly, for life data, observations on unfailed units are right censored because all that is known about such units is that their failure times exceed their current survival times. In this chapter, we describe general approaches, based on the likelihood function, for constructing intervals for such situations.

ML is a highly versatile method for fitting statistical models to data. Roughly speaking, the ML method provides that model fit to the data—from among all possible model fits—that makes the data most probable. In most applications, the ML method is applied to a parametric statistical model (as opposed to the distribution-free nonparametric methods in Chapter 5) to describe a set of data or a process or population that generated the data. The appeal of ML methods stems from the fact that they can be applied to a wide variety of statistical models and types of data (e.g., continuous, discrete, categorical, censored, and truncated) for which other popular methods, such as least squares, are not, in general, applicable. In particular, ML methods are used extensively in life data analysis; see Meeker and Escobar (1998, Chapter 8). Moreover, statistical theory shows that, under standard regularity conditions, ML estimators are "optimal" in large samples. That is, ML estimators are consistent and asymptotically efficient as the sample size (number of failures in the case of right-censored failure-time data) increases. Thus, among consistent competitors to ML estimators, none has a smaller large-sample approximate variance. Software that use ML methods has over recent years become increasingly accessible through commercially available products, thereby tremendously expanding the feasible areas of application.

Example 12.1 Time between α-Particle Emissions of Americium-241. Berkson (1966) investigated α-particle emissions of americium-241 (which has a half-life of about 458 years). Physical theory suggests that, over a short period of time, the observed interarrival times of particles from a specimen are independent of one another and come from an exponential distribution with cumulative distribution function (cdf)

$$F(t; \theta) = 1 - \exp\left(-\frac{t}{\theta}\right), \tag{12.1}$$

	Time		Interarrival times Frequency of occurrence	
	Interval endpoints		All times	Sample of frequencies
i	Lower t_{i-1}	Upper t_i	$n = 10{,}220$ d_i	$n = 200$ d_i
1	0	100	1,609	41
2	100	300	2,424	44
3	300	500	1,770	24
4	500	700	1,306	32
5	700	1,000	1,213	29
6	1,000	2,000	1,528	21
7	2,000	4,000	354	9
8	4,000	∞	16	0

Table 12.1 Binned α-particle interarrival time data.

where θ is the mean time between arrivals. The corresponding homogeneous Poisson process model that describes the number of emissions over time has an arrival rate with intensity $\lambda = 1/\theta$. For the interarrival times of α-particles, λ is proportional to the americium-241 decay rate. The value of λ depends on the size of the specimen, the size and efficiency of the detector/counter, and various other factors. See Section C.3.7 or Meeker and Escobar (1998, Section 4.1) for more information about the exponential distribution. See Ross (2012, Chapter 4) for more information about the homogeneous Poisson process.

The original data consisted of 10,220 interarrival times of α-particles (the time unit is equal to 1/5,000 second throughout this example). The interarrival times were placed into intervals (or bins) running from 0 to 4,000 time units with interval lengths ranging from 100 to 2,000 time units, and with one additional interval for observed times exceeding 4,000 time units. For purposes of illustration, we will suppose that the 10,220 times represent a population and we will consider samples from this population.

Initially, consider a random sample of $n = 200$ interarrival times, binned as described previously. This reduced sample size is more typical of what one encounters in common applications. The $n = 200$ sample interarrival times were obtained by sampling from a multinomial distribution with probabilities equal to the proportion of interarrival times in each of the bins. The counts in the bins for the 10,220 interarrival times and $n = 200$ sample interarrival times are shown in Table 12.1. We focus on estimating θ, the mean time between arrivals, and the rate of arrivals $\lambda = 1/\theta$, using the $n = 200$ sample observations. ∎

12.1.2 Model Selection

Applications of ML methods typically involve a tentatively assumed statistical model for the data, often aided by a graphical analysis. In practice, the search is for a physically reasonable model that adequately describes the population or process of interest, without being unnecessarily complicated. Usually the search involves iteratively assessing alternative models. Tentative models may be suggested by physical theory, previous experience with similar data, and other expert knowledge. Meeker and Escobar (1998, Chapter 6) explain and illustrate the use of probability plots to help identify a suitable distribution, providing more detail than our introduction to the subject in Section 4.11.

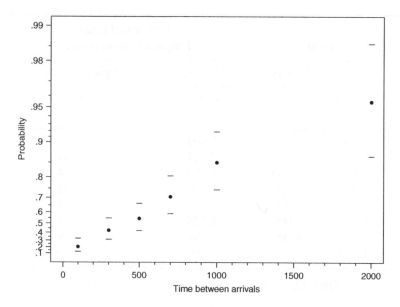

Figure 12.1   Exponential probability plot of the $n = 200$ sample observations for the α-particle interarrival time data with simultaneous nonparametric approximate 95% confidence bands.

The selection of the scales for a probability plot results in assuming a particular probability distribution. The adequacy of an assumed distribution can be assessed subjectively by the degree to which the plotted points scatter around a straight line. More formally, Nair (1981) shows that if one can draw an arbitrary straight line with a positive slope all the way *within* the simultaneous confidence band that has been constructed from the data (i.e., a straight line that falls within the simultaneous confidence bands), then one can conclude that the data are consistent with the distribution assumed by the probability scale of the plot. If, on the other hand, one cannot draw a straight line within the calculated simultaneous confidence band, then there is statistical evidence that the data did not come from the distribution used to construct the probability plot. Meeker and Escobar (1998, Chapters 3 and 6) show how the preceding simultaneous confidence bands are calculated and then used on probability plots for purposes of distributional assessment.

Example 12.2   Exponential Distribution Probability Plot for the α-Particle Data. Figure 12.1 shows an exponential probability plot (i.e., a plot using special probability scales for which an exponential cdf is a straight line, similar to other probability plots in Section 4.11.1). The plot shows a nonparametric estimate of the interarrival time cdf, together with simultaneous nonparametric approximate 95% confidence bands. This plot does not include the interval from 2,000 to 4,000 because an exponential distribution probability plot cannot accommodate a probability estimate of 1. The approximate linearity of this plot indicates that the exponential distribution provides a good fit to the data. This is reinforced by noting that one *is* able to draw a straight line within the area between the simultaneous nonparametric confidence bands in Figure 12.1.   ∎

12.2   LIKELIHOOD FUNCTION AND MAXIMUM LIKELIHOOD ESTIMATION

In this section we formally define the likelihood function and the method of maximum likelihood for estimating model parameters (and functions of model parameters). The following section applies such concepts to the single-parameter exponential distribution. This will be followed, in Section 12.4, by a discussion and examples involving probability distributions with two parameters.

12.2.1 Probability of the Data

The likelihood is a function of the data and a model's (unknown) parameter(s), which we will henceforth denote as $\boldsymbol{\theta}$. The likelihood function must be equal to or proportional to the *probability of the observed data*. For simple parametric models (i.e., models without explanatory variables), the number of parameters is usually small (e.g., less than 3). The exponential distribution has only one parameter. The normal, lognormal and Weibull distributions have two parameters.

For a set of n independent observations, the (total) likelihood function can be written as the joint probability

$$L(\boldsymbol{\theta}) = L(\boldsymbol{\theta}; \mathrm{DATA}) = \mathcal{C} \prod_{i=1}^{n} L_i(\boldsymbol{\theta}; \mathrm{data}_i), \tag{12.2}$$

where $L_i(\boldsymbol{\theta}; \mathrm{data}_i)$ is the likelihood of observation i. The factor $\mathcal{C}$ in (12.2) is a constant that may depend on the data but generally does not depend on $\boldsymbol{\theta}$. Thus, for computational purposes, we can let $\mathcal{C} = 1$.

Let T denote a random variable from the probability distribution of interest. If an observation is known to have occurred between t_{i-1} and t_i (*interval-censored* or *binned data*), the probability of the observed event (i.e., its likelihood) is

$$L_i(\boldsymbol{\theta}; \mathrm{data}_i) = L_i(\boldsymbol{\theta}) = \Pr(t_{i-1} < T \le t_i) = \int_{t_{i-1}}^{t_i} f(t; \boldsymbol{\theta}) \, dt = F(t_i; \boldsymbol{\theta}) - F(t_{i-1}; \boldsymbol{\theta}),$$

$$\tag{12.3}$$

where $f(t; \boldsymbol{\theta})$ is the pdf and $F(t; \boldsymbol{\theta})$ is the cdf of T. Such interval-censored data arise, for example, when (typically a large number of) observations are "binned" into a (usually small) number of intervals (or bins), generally to compress data. If an interval is small or the observation is reported as t_i an (*exact observation*), then

$$L_i(\boldsymbol{\theta}; \mathrm{data}_i) = L_i(\boldsymbol{\theta}) = f(t_i; \boldsymbol{\theta})$$

is approximately proportional to the probability of the observed event and can be used instead of (12.3).

If an observation is known only to be greater than t_i (*right censored*), the probability of the observed event (i.e., its likelihood) is

$$L_i(\boldsymbol{\theta}; \mathrm{data}_i) = L_i(\boldsymbol{\theta}) = \Pr(T > t_i) = \int_{t_i}^{\infty} f(t; \boldsymbol{\theta}) \, dt = 1 - F(t_i; \boldsymbol{\theta}).$$

Right-censored observations occur, for example, when the value of a measurand is greater than the upper limit of a measuring instrument or in a lifetime study when a unit has not failed by the end of the study.

If an observation is known only to be less than or equal to t_i (*left censored*), its probability (i.e., its likelihood) is

$$L_i(\boldsymbol{\theta}; \mathrm{data}_i) = L_i(\boldsymbol{\theta}) = \Pr(T \le t_i) = \int_{-\infty}^{t_i} f(t; \boldsymbol{\theta}) \, dt = F(t_i; \boldsymbol{\theta}).$$

Left-censored observations occur, for example, when the value of a measurand is less than the lower limit of a measuring instrument (resulting in a non-detect observation) or in a lifetime study when a unit has failed before the unit's first inspection time.

For a given set of data, $L(\boldsymbol{\theta})$ is viewed as a function of $\boldsymbol{\theta}$. The dependence of $L(\boldsymbol{\theta})$ on the data is understood and is suppressed in the notation. Values of $\boldsymbol{\theta}$ for which $L(\boldsymbol{\theta})$ is relatively large are more plausible than values of $\boldsymbol{\theta}$ for which the probability of the data is relatively small.

The method of ML provides an estimate $\widehat{\boldsymbol{\theta}}$ of $\boldsymbol{\theta}$ by finding the value of $\boldsymbol{\theta}$ that maximizes $L(\boldsymbol{\theta})$. Values of $\boldsymbol{\theta}$ with relatively large $L(\boldsymbol{\theta})$ can be used to define confidence regions for $\boldsymbol{\theta}$, as described in subsequent sections. ML is also used to estimate *functions* of $\boldsymbol{\theta}$ such as distribution quantiles or probabilities associated with the model.

12.2.2 The Likelihood Function and its Maximum

For a sample of n independent observations, denoted generically by data_i, $i = 1, \ldots, n$, and a specified model, the total likelihood $L(\boldsymbol{\theta})$ for the sample is given by (12.2). For some purposes, it is convenient to use the log-likelihood $\mathcal{L}(\boldsymbol{\theta}) = \log[L(\boldsymbol{\theta})]$. For example, some theory for ML is developed more naturally in terms of sums like

$$\mathcal{L}(\boldsymbol{\theta}) = \log[L(\boldsymbol{\theta})] = \sum_{i=1}^{n} \log[L_i(\boldsymbol{\theta})] = \sum_{i=1}^{n} \mathcal{L}_i(\boldsymbol{\theta})$$

instead of the product in (12.2). Because $\mathcal{L}(\boldsymbol{\theta})$, for any fixed value of $\boldsymbol{\theta}$, is a monotone increasing function of $L(\boldsymbol{\theta})$, the maximum of $\mathcal{L}(\boldsymbol{\theta})$, if one exists, occurs at the same value of $\boldsymbol{\theta}$ as the maximum of $L(\boldsymbol{\theta})$. Also, for practical problems, $\mathcal{L}(\boldsymbol{\theta})$ can be represented in computer memory without special scaling. This may not be the case for $L(\boldsymbol{\theta})$ because of possible extreme exponent values. For example, the values of some likelihoods may be less than 10^{-400}.

Example 12.3   Likelihood for the α-Particle Data. The α-particle example, because it involves only a single unknown parameter θ, provides a simple illustration of the basic concepts. Substituting (12.1) into (12.3) and (12.3) into (12.2), and letting $\mathcal{C} = 1$ results in the following exponential distribution likelihood function for the interval-censored data in Table 12.1:

$$L(\theta) = \prod_{i=1}^{n} L_i(\theta) = \prod_{i=1}^{n} [F(t_i; \theta) - F(t_{i-1}; \theta)]$$

$$= \prod_{i=1}^{8} [F(t_i; \theta) - F(t_{i-1}; \theta)]^{d_i} = \prod_{i=1}^{8} \left[\exp\left(-\frac{t_{i-1}}{\theta}\right) - \exp\left(-\frac{t_i}{\theta}\right) \right]^{d_i}, \quad (12.4)$$

were d_i is the number of interarrival times in interval i. Note that in the first line of (12.4), the product is over the $n = 200$ observed times. In the second line, the product is over the eight bins into which the data were grouped.   ∎

The ML estimate of $\boldsymbol{\theta}$ is found by maximizing $L(\boldsymbol{\theta})$. When there is a unique global maximum, $\widehat{\boldsymbol{\theta}}$ denotes the value of $\boldsymbol{\theta}$ that maximizes $L(\boldsymbol{\theta})$. In some applications, the maximum is not unique. The function $L(\boldsymbol{\theta})$ may have multiple local maxima or can have relatively flat spots along which $L(\boldsymbol{\theta})$ changes slowly, if at all. Such flat spots may or may not be at the maximum value of $L(\boldsymbol{\theta})$. The shape and magnitude of $L(\boldsymbol{\theta})$ relative to $L(\widehat{\boldsymbol{\theta}})$ over all possible values of $\boldsymbol{\theta}$ describe the information about $\boldsymbol{\theta}$ that is contained in data_i, $i = 1, \ldots, n$. The *relative likelihood function* is defined as $R(\theta) = L(\theta)/L(\widehat{\theta})$. $R(\theta)$ allows one to assess the probability of the data for values of θ, *relative* to the probability at the ML estimate. For example, $R(\theta) = 0.1$ implies that the probability of the data is 10 times larger at $\widehat{\theta}$ than at θ.

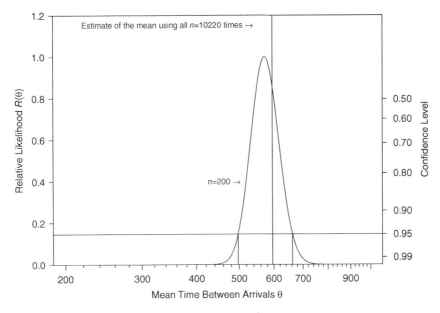

Figure 12.2   Relative likelihood function $R(\theta) = L(\theta)/L(\widehat{\theta})$ for the $n = 200$ α-particle interarrival times. The tall vertical line indicates the ML estimate of θ based on all 10,220 interarrival times (considered to be the population mean). The short vertical lines show the likelihood-based approximate 95% confidence interval for θ from the $n = 200$ sample data.

Example 12.4   Relative Likelihood for the α-Particle Data. Figure 12.2 shows the relative likelihood function for the $n = 200$ sample data. The maximum of $R(\theta)$ in Figure 12.2 is at the ML estimate $\widehat{\theta} = 572.3$. The tall vertical line at 596.34 shows the ML estimate of θ based on all of the 10,220 arrival times. Section 12.3.1 explains the right-hand-side vertical axis and shows how to use $R(\theta)$ to compute confidence intervals for θ.   ∎

After computing the ML estimate of a distribution $F(t; \theta)$, it is good practice to plot this estimate along with parametric confidence intervals for $F(t; \theta)$ on a probability plot. This provides a visual comparison with the nonparametric estimate of the cdf (i.e., the plotted points). Such a plot is also useful for presenting the results of the analysis and providing an assessment of distributional goodness of fit.

Example 12.5   Summary of Estimates for the α-Particle Data. Figure 12.3 is another exponential probability plot for the $n = 200$ sample observations. The solid line is the ML estimate, $F(t; \widehat{\theta})$, of the exponential cdf $F(t; \theta)$. The dotted lines are drawn through a set of pointwise likelihood-based approximate 95% confidence intervals for $F(t; \theta)$; these intervals will be explained in Section 12.3.3. These intervals can be compared with the nonparametric simultaneous confidence bands shown in Figure 12.1.

Table 12.2 summarizes the results from fitting an exponential distribution to the 200 sample arrival times, showing ML estimates, standard errors, and approximate 95% confidence intervals for θ, the mean of the exponential interarrival time distribution, and for $\lambda = 1/\theta$, the arrival intensity rate, using three different methods described in Sections 12.3 and 12.6. Due to the relatively large sample size in this example ($n = 200$), there is little difference in the results among the three methods.   ∎

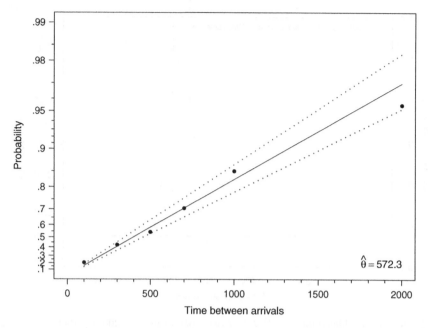

Figure 12.3 Exponential probability plot of the $n = 200$ sample observations for the α-particle interarrival time data. The solid line is the ML estimate of the exponential distribution $F(t; \theta)$ and the dotted lines are pointwise approximate 95% confidence intervals for $F(t; \theta)$.

12.3 LIKELIHOOD-BASED CONFIDENCE INTERVALS FOR SINGLE-PARAMETER DISTRIBUTIONS

The likelihood function is a versatile tool for assessing the information that the data contains on parameters or on functions of the parameters that are of practical interest, such as distribution quantiles and values of the cdf. Most importantly, it provides a useful method for finding approximate confidence intervals for parameters and functions of parameters.

	Mean time between arrivals θ		Arrival rate $\lambda \times 10^5$	
ML estimate $\widehat{\theta}$		572.3	ML Estimate $\widehat{\lambda} \times 10^5$	175
Standard error $\widehat{\text{se}}_{\widehat{\theta}}$		41.72	Standard error $\widehat{\text{se}}_{\widehat{\lambda} \times 10^5}$	13
Approximate 95% confidence intervals			Approximate 95% confidence intervals	
Likelihood-based		[498, 662]	Likelihood-based	[151, 201]
Wald $Z_{\log(\widehat{\theta})} \overset{.}{\sim} \text{NORM}(0, 1)$		[496, 660]	Wald $Z_{\log(\widehat{\lambda})} \overset{.}{\sim} \text{NORM}(0, 1)$	[152, 202]
Wald $Z_{\widehat{\theta}} \overset{.}{\sim} \text{NORM}(0, 1)$		[490, 653]	Wald $Z_{\widehat{\lambda}} \overset{.}{\sim} \text{NORM}(0, 1)$	[149, 200]

Table 12.2 Results of fitting an exponential distribution to the $n = 200$ observations for the α-particle interarrival time data.

12.3.1 Confidence Intervals for the Exponential Distribution Mean

An approximate $100(1 - \alpha)\%$ likelihood-based confidence interval $[\underset{\sim}{\theta},\ \widetilde{\theta}]$ for the exponential distribution mean θ is the set of all values of θ such that

$$-2\log[R(\theta)] \leq \chi^2_{(1-\alpha;1)}$$

or, equivalently, the set defined by

$$R(\theta) \geq \exp\left[-\chi^2_{(1-\alpha;1)}/2\right], \tag{12.5}$$

where $\chi^2_{(1-\alpha;1)}$ is the $1 - \alpha$ quantile of the chi-square distribution with 1 degree of freedom. The theoretical justification for this interval is given in Section D.5.

Example 12.6 Likelihood-Based Confidence Interval for the Mean Time between Arrivals of α-Particles. Figure 12.2 shows the likelihood-based approximate 95% confidence interval for θ, based on the $n = 200$ sample observations. In particular, in this figure the interval determined by the intersections of the relative likelihood function $R(\theta)$ with the horizontal line at $\exp[-\chi^2_{(0.95;1)}/2] = 0.147$ (corresponding to 0.95 on the confidence level axis), provides an approximate 95% confidence interval for θ. Thus, the two short vertical lines drawn down from this intersection give the endpoints of this confidence interval as [498, 662], as shown in Table 12.2.   ∎

An approximate one-sided confidence bound can be obtained by using the appropriate endpoint of a two-sided confidence interval and appropriately adjusting the confidence level. In particular, a one-sided $100(1 - \alpha)\%$ lower or upper confidence bound is the corresponding endpoint of a two-sided $100(1 - 2\alpha)\%$ confidence interval.

Example 12.7 Likelihood-Based One-Sided Confidence Bounds for the Mean Time between Arrivals of α-Particles. In Figure 12.2, the intersections of the relative likelihood function $R(\theta)$ with the horizontal line at $\exp[-\chi^2_{(0.95;1)}/2] = 0.147$ provide approximate one-sided 97.5% confidence bounds for θ. For approximate one-sided 95% confidence bounds, the horizontal line would be drawn at $\exp[-\chi^2_{(0.90;1)}/2] = 0.259$ (corresponding to 0.90 on the right-hand scale of Figure 12.2).   ∎

12.3.2 Confidence Intervals for a Monotone Function of the Exponential Distribution Mean

The arrival rate $\lambda = 1/\theta$ is a *monotone decreasing* function of θ. As a consequence, the confidence interval $[1/\widetilde{\theta},\ 1/\underset{\sim}{\theta}]$ for λ will contain λ if and only if the corresponding confidence interval for θ contains θ. Thus this confidence interval for λ has the same confidence level as the corresponding confidence interval for θ. Confidence intervals for other monotone functions of θ can be obtained in a similar manner.

Example 12.8 Likelihood-Based Confidence Interval for the Arrival Rate of α-Particles. The ML estimate of the arrival rate λ is obtained from the $n = 200$ sample observations as $\widehat{\lambda} = 1/\widehat{\theta} = 0.00175$. The likelihood-based approximate 95% confidence interval for λ is

$$[\underset{\sim}{\lambda},\ \widetilde{\lambda}] = \left[\frac{1}{\widetilde{\theta}},\ \frac{1}{\underset{\sim}{\theta}}\right] = [0.00151,\ 0.00201],$$

as given in Table 12.2.   ∎

12.3.3   Confidence Intervals for $F(t; \theta)$

Because the exponential cdf $F(t; \theta)$ is a decreasing function of θ, a confidence interval for $F(t_e; \theta)$ at time t_e is

$$[\underset{\sim}{F}(t_e), \; \widetilde{F}(t_e)] = [F(t_e; \widetilde{\theta}), \; F(t_e; \underset{\sim}{\theta})]. \qquad (12.6)$$

Example 12.9   Confidence Intervals for the cdf of the α-Particle Time between Arrivals.
The dotted lines in Figure 12.3 show pointwise likelihood-based approximate 95% confidence intervals for $F(t; \theta)$, based on the $n = 200$ sample observations, computed using (12.6). ∎

12.3.4   Effect of Sample Size on the Likelihood and Confidence Interval Width

Statistical theory shows that, for most statistical models, the variance (not the standard error) of an estimator is inversely proportional to the sample size. Thus, increasing the sample size by a factor k approximately reduces the width of a confidence interval by a factor $1/\sqrt{k}$. For example, we need to increase the sample size by a factor of 4 to cut the expected width of a confidence interval approximately in half.

Example 12.10   The Effect of Sample Size on the Width of Confidence Intervals for the α-Particle Mean Time between Arrivals. To illustrate the effect of sample size on the likelihood function and the resulting likelihood-based confidence intervals, we have constructed, and shown in Table 12.3, binned pseudo-samples of size $n = 20$, $n = 2,000$ and $n = 20,000$, in addition to the previous sample of size $n = 200$, from the complete α-particle data, which consisted of $n = 10,220$ observations. These pseudo-samples were constructed to have a constant proportion of observations within each bin so that the ML estimate of θ is the same for each of the four samples (this is why we call them "pseudo"-samples).

Figure 12.4 is a plot of $R(\theta)$ for the $n = 20$, 200 and 2,000 samples. $R(\theta)$ for the $n = 20,000$ sample is not shown; it is too narrow. The vertical line at the center of the figure shows the estimate of the mean of the $n = 20,000$ sample. For each sample, the short vertical lines drawn from the intersections of the horizontal line with the corresponding relative likelihood give the likelihood-based approximate 95% confidence interval for θ. We note from Figure 12.4 that the

	Time Interval endpoints		Interarrival times Frequency of occurrence Samples of frequencies			
Interval	Lower t_{i-1}	Upper t_i	$n = 20,000$ d_i	$n = 2,000$ d_i	$n = 200$ d_i	$n = 20$ d_i
1	0	100	3,000	300	30	3
2	100	300	5,000	500	50	5
3	300	500	3,000	300	30	3
4	500	700	3,000	300	30	3
5	700	1,000	2,000	200	20	2
6	1,000	2,000	3,000	300	30	3
7	2,000	4,000	1,000	100	10	1
8	4,000	∞	0	0	0	0

Table 12.3   Alpha-particle pseudo-samples constructed to have a constant proportion within each bin.

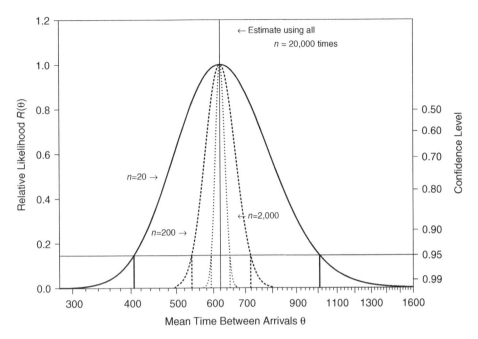

Figure 12.4 Relative likelihoods $R(\theta) = L(\theta)/L(\widehat{\theta})$ for the $n = 20$, 200, and 2,000 pseudo-samples from the α-particle data, showing the likelihood-based 95% confidence intervals for θ (short vertical lines) and the mean of the $n = 20,000$ sample (long vertical line).

spread of the likelihood function for a sample size of 2,000 is much tighter than that for the smaller samples, indicating that the larger samples contain much more information about θ. ∎

12.4 LIKELIHOOD-BASED ESTIMATION METHODS FOR LOCATION-SCALE AND LOG-LOCATION-SCALE DISTRIBUTIONS

This section describes methods for computing likelihood-based confidence intervals for common distributions with more than one parameter. The methods presented in this section and Section 12.5 will be illustrated by the two-parameter Weibull and lognormal distributions, but they apply to many other distributions.

12.4.1 Background on Location-Scale and Log-Location-Scale Distributions

The location-scale and log-location-scale families of distributions include the continuous probability distributions used most frequently in practical applications. They include the normal (location-scale), lognormal (log-location-scale), and Weibull (log-location-scale) distributions. The general cdf and pdf for the location-scale distribution family are

$$F(x) = \Pr(X \le x) = \Phi\left(\frac{x - \mu}{\sigma}\right) \quad \text{and} \quad f(x) = \frac{dF(x)}{dx} = \frac{1}{\sigma}\phi\left(\frac{x - \mu}{\sigma}\right), \quad -\infty < x < \infty,$$

where μ is the location parameter and σ is the scale parameter of the distribution of X. If, for example, $\Phi(z)$ and $\phi(z)$ are replaced by the standard normal cdf and pdf, $\Phi_{\text{norm}}(z)$ and $\phi_{\text{norm}}(z)$, respectively (defined in Section 3.1.1), then the distribution of X is normal with mean (location parameter) μ and standard deviation (scale parameter) σ.

The general cdf and pdf for the log-location-scale distribution family are, respectively,

$$F(t) = \Phi\left[\frac{\log(t) - \mu}{\sigma}\right] \quad \text{and} \quad f(t) = \frac{dF(t)}{dt} = \frac{1}{\sigma t}\phi\left[\frac{\log(t) - \mu}{\sigma}\right], \ t > 0.$$

If, for example, $\Phi(z)$ and $\phi(z)$ are replaced by the standard normal cdf and pdf, $\Phi_{\text{norm}}(z)$ and $\phi_{\text{norm}}(z)$, respectively, then the distribution of T is lognormal. In this case, $\exp(\mu)$ is the lognormal distribution median (also a scale parameter) and σ is the lognormal distribution shape parameter. If $\Phi(z)$ and $\phi(z)$ are replaced by the standard smallest extreme value distribution (also known as the Gumbel distribution of minima), $\Phi_{\text{sev}}(z) = 1 - \exp[-\exp(z)]$ and $\phi_{\text{sev}}(z) = \exp[z - \exp(z)]$, respectively, then the distribution of T is Weibull. Thus the Weibull cdf is

$$\Pr(T \le t; \mu, \sigma) = F(t; \mu, \sigma) = \Phi_{\text{sev}}\left[\frac{\log(t) - \mu}{\sigma}\right] = 1 - \exp\left[-\left(\frac{t}{\eta}\right)^\beta\right], \quad t > 0,$$

where $\beta = 1/\sigma$ is the Weibull distribution shape parameter and $\eta = \exp(\mu)$ is the Weibull distribution scale parameter. Section C.3.1 provides further definitions and properties of these and other location-scale and log-location-scale distributions.

12.4.2 Likelihood Function for Location-Scale Distributions

The likelihood function for a sample $x_1, \ldots, x_n$ from a location-scale distribution, $F(x; \mu, \sigma) = \Phi[(x - \mu)/\sigma]$, consisting of a combination of left-censored, exact (i.e., uncensored), and right-censored observations, is

$$L(\mu, \sigma) = \prod_{i=1}^n L_i(\mu, \sigma; \text{data}_i)$$

$$= \prod_{i=1}^n [F(x_i; \mu, \sigma)]^{\kappa_i} [f(x_i; \mu, \sigma)]^{\delta_i(1-\kappa_i)} [1 - F(x_i; \mu, \sigma)]^{(1-\delta_i)(1-\kappa_i)}$$

$$= \prod_{i=1}^n \left[\Phi\left(\frac{x_i - \mu}{\sigma}\right)\right]^{\kappa_i} \times \left[\frac{1}{\sigma}\phi\left(\frac{x_i - \mu}{\sigma}\right)\right]^{\delta_i(1-\kappa_i)} \times \left[1 - \Phi\left(\frac{x_i - \mu}{\sigma}\right)\right]^{(1-\delta_i)(1-\kappa_i)},$$

$$(12.7)$$

where

$$\delta_i = \begin{cases} 1 & \text{if } x_i \text{ is an exact observation} \\ 0 & \text{if } x_i \text{ is a left- or right-censored observation} \end{cases}$$

and

$$\kappa_i = \begin{cases} 1 & \text{if } x_i \text{ is a left-censored observation} \\ 0 & \text{if } x_i \text{ is an exact or a right-censored observation.} \end{cases}$$

When there is no censoring, the normal distribution likelihood function simplifies and the values of μ and σ that maximize it are $\widehat{\mu} = \bar{x}$ and $\widehat{\sigma} = \left[\sum_{i=1}^n (x_i - \bar{x})^2/n\right]^{1/2}$, where $\bar{x}$ is the sample mean of the observations $x_1, \ldots, x_n$.

12.4.3 Likelihood Function for the Lognormal, Weibull, and Other Log-Location-Scale Distributions

Because the logarithms of lognormal, Weibull, and other log-location-scale random variables are location-scale random variables, the likelihood functions for these distributions can also be written in terms of the standardized location-scale distributions. In particular, for a sample consisting of a combination of left-censored, exact, and right-censored observations, the likelihood function is

$$L(\mu, \sigma) = \prod_{i=1}^{n} \left\{ \Phi \left[\frac{\log(t_i) - \mu}{\sigma} \right] \right\}^{\kappa_i} \times \left\{ \frac{1}{\sigma t_i} \phi \left[\frac{\log(t_i) - \mu}{\sigma} \right] \right\}^{\delta_i (1 - \kappa_i)}$$
$$\times \left\{ 1 - \Phi \left[\frac{\log(t_i) - \mu}{\sigma} \right] \right\}^{(1 - \delta_i)(1 - \kappa_i)}, \tag{12.8}$$

where δ_i and κ_i are as defined in Section 12.4.2.

Some computer programs omit the $1/t_i$ term in (12.8). Because this term does not depend on the unknown parameters, this has no effect on the ML estimates or the likelihood ratio. It does, however, affect the reported value of the likelihood (or more commonly the log-likelihood) at the maximum (and elsewhere). Thus readers need to be cautious when comparing values of maximum (log-)likelihoods from different software.

12.4.4 Maximum Likelihood Estimation and Relative Likelihood for Log-Location-Scale Distributions

As discussed in Section 12.2.1, ML methods provide estimates from an assumed distribution, even with censored data. Similar to the distribution fitting in Chapter 4, fitting a lognormal (Weibull) distribution is equivalent to fitting a straight line through the nonparametric estimate of the cdf on a lognormal (Weibull) probability plot, using ML to fit the line. Comparison of the nonparametric estimate of the cdf with the fitted line provides an assessment of the distributional fit. For initial exploratory evaluations of data and models, it is useful to plot the relative likelihood $R(\mu, \sigma) = L(\mu, \sigma)/L(\hat{\mu}, \hat{\sigma})$ or a similarly defined $R[\exp(\mu), \sigma]$. With two parameters, a contour plot of the relative likelihood (see Example 12.11) provides a helpful initial assessment of the plausible region of parameter values.

Example 12.11 Atrazine Concentration Data Weibull and Lognormal ML Estimates.
Atrazine is a herbicide widely used in the United States and other parts of the world. Experiments with animals have suggested adverse health effects of exposure to atrazine (e.g., Tillitt et al., 2010). Use of atrazine has been banned in the European Union since 2004.

Junk et al. (1980) compared June 1978 readings (taken before the growing season) from a sample of 24 Nebraska wells with similar readings obtained in September (after the growing season). The data were also analyzed by Helsel (2005). We consider the June 1978 data. Our analyses to characterize atrazine concentration at that time will be based on the assumption that the 24 Nebraska wells are a random sample from a defined larger population of Nebraska wells. The data are shown in Table 12.4. For nine of the wells, the concentration of atrazine was below

0.38	0.04	< 0.01	0.03	0.03	0.05
0.02	< 0.01	< 0.01	< 0.01	0.11	0.09
< 0.01	< 0.01	< 0.01	< 0.01	0.02	0.03
0.02	0.02	0.05	0.03	0.05	< 0.01

Table 12.4 Atrazine concentration data. Concentration is in units of μg/L. The nine observations identified by < 0.01 were below the detection limit of 0.01 μg/L.

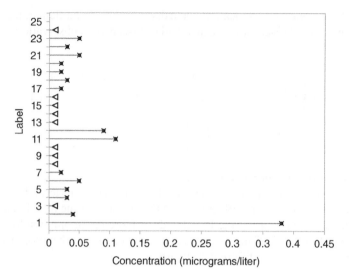

Figure 12.5   Event plot of the atrazine data with left-censoring (denoted by left-pointing triangle) for observations below the detection limit.

the detection limit of 0.01 μg/L. Thus these observations are left censored. Figure 12.5 is an event plot that displays both the observed values and the left-censored data.

Figure 12.6 is a lognormal distribution contour plot of the relative likelihood function $R[\exp(\mu), \sigma] = L[\exp(\mu), \sigma]/L[\exp(\widehat{\mu}), \widehat{\sigma}]$ for these data. The plot indicates plausible ranges of values for $\exp(\mu)$ and σ. The relative likelihood surface exhibits a unique maximum at the ML estimates $[\exp(\widehat{\mu}) = 0.01747$ and $\widehat{\sigma} = 1.3710]$.

Figures 12.7 and 12.8 are lognormal and Weibull probability plots, respectively, of the data, with ML estimates of $F(t)$ shown by straight lines. The small-dashed curves in Figures 12.7

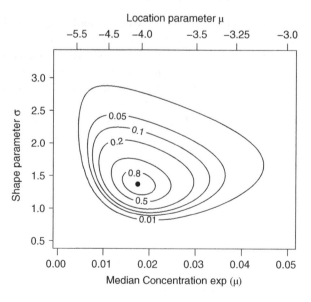

Figure 12.6   Lognormal distribution relative likelihood function contour plot for $\exp(\mu)$ and σ for the atrazine data. The ML estimates are indicated by the dot.

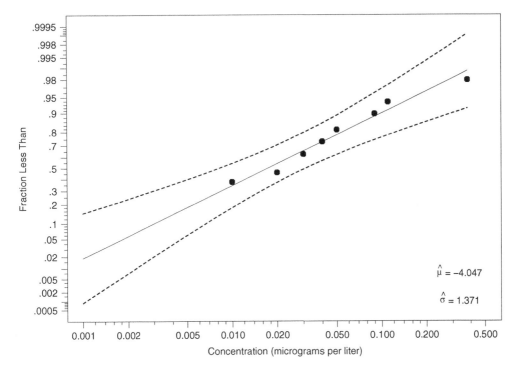

Figure 12.7 Lognormal probability plot of the atrazine data with the ML estimate and pointwise likelihood-based approximate 95% confidence intervals for $F(t)$.

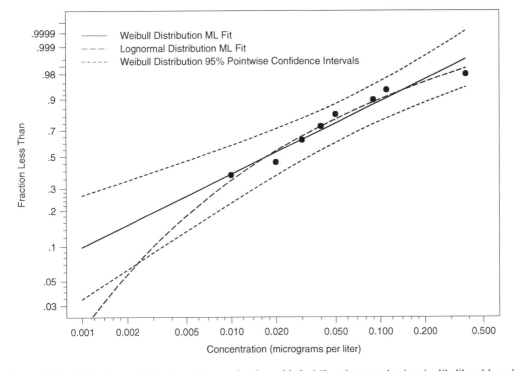

Figure 12.8 Weibull probability plot of the atrazine data with the ML estimate and pointwise likelihood-based approximate 95% confidence intervals for $F(t)$. The long-dashed curved line shows the lognormal distribution ML estimate of $F(t)$.

and 12.8 are drawn through a set of pointwise likelihood-based approximate 95% confidence intervals for $F(t)$ (computed as described in Section 12.5.4). The long-dashed curve on the Weibull probability plot in Figure 12.8 is the ML estimate of the lognormal distribution $F(t)$.

Table 12.5 shows ML estimates for μ, $\exp(\mu)$, σ, $t_{0.90}$ (the 0.90 distribution quantile), and $F(0.30)$, for the fitted lognormal and Weibull distributions. The parameters μ and σ for the lognormal and Weibull distributions are not directly comparable because these parameters have different interpretations for these two distributions. The tabulation also shows standard errors and likelihood-based and Wald-approximation 95% confidence intervals that will be discussed subsequently.

Figure 12.8 and Table 12.5 show that inferences with the lognormal and Weibull distributions are in good agreement *within the range of the data*, as seen, for example, by the estimates of $t_{0.90}$ and $F(0.30)$. Figure 12.8, however, shows increasing differences between the two distributions when extrapolating outside the range of the data (especially in the lower tail of the distribution). This illustrates the importance of exercising extreme caution in making inferences about the tails of an assumed distribution outside the data range. Such extrapolations tend to be highly dependent on the assumed distribution and may differ appreciably even between models that seem to fit well within the range of the data, as in this example. We also note that the confidence intervals shown in Figures 12.7 and 12.8 are based on the assumption of a lognormal and a Weibull distribution, respectively, and do not reflect possible departures from these distributional assumptions. Subsequent discussion of the atrazine data will focus on the lognormal fit to the data. ∎

12.5   LIKELIHOOD-BASED CONFIDENCE INTERVALS FOR PARAMETERS AND SCALAR FUNCTIONS OF PARAMETERS

This section describes methods for computing likelihood-based confidence intervals for distributions or models with more than one parameter. Our examples will deal with two-parameter location-scale and log-location-scale distributions with parameters μ and σ (per our discussion in Section 12.4) in general and the Weibull and lognormal distributions in particular. The methods, however, also apply to other distributions and models with a relatively small number of parameters (e.g., two or three).

12.5.1   Relative Likelihood Contour Plots and Likelihood-Based Joint Confidence Regions for μ and σ

Using the large-sample chi-square approximation for the distribution of the likelihood-ratio statistic (see Section D.5.1), each of the constant-likelihood contour lines in a relative likelihood function contour plot, such as that shown for the atrazine data in Figure 12.6, defines an approximate joint confidence region for μ and σ with coverage probability close to its nominal value $100(1 - \alpha)\%$, even in moderately small samples (e.g., 15–20 uncensored observations, as established in simulation studies). In particular, for a two-dimensional relative likelihood, the region $R(\theta_i, \theta_j) > \exp[-\chi^2_{(1-\alpha;2)}/2] = \alpha$ provides an approximate likelihood-based $100(1 - \alpha)\%$ joint confidence region for θ_i and θ_j, where θ_i and θ_j are the parameters of a two-parameter distribution.

Example 12.12   Joint Confidence Region for the Atrazine Data Lognormal Distribution Parameters (μ, σ). The region $R(\mu, \sigma) > \exp(-\chi^2_{(0.90;2)}/2) = 0.05$ in Figure 12.6 provides a joint likelihood-based approximate 95% confidence region for μ and σ for the atrazine data. Figure 12.9, similar to Figure 12.6, plots contours of constant values of

	Distribution	
	Lognormal	Weibull
ML estimate $\widehat{\mu}$	-4.047	-3.488
Standard error $\widehat{\text{se}}_{\widehat{\mu}}$	0.3096	0.3383
Approximate 95% confidence intervals for μ		
Likelihood-based	$[-4.785, \ -3.467]$	$[-4.984, \ -3.351]$
Wald $Z_{\widehat{\mu}} \overset{.}{\sim} \text{NORM}(0,1)$	$[-4.654, \ -3.441]$	$[-4.151, \ -2.825]$
ML estimate $\exp(\widehat{\mu})$	0.01747	0.01759
Standard error $\widehat{\text{se}}_{\exp(\widehat{\mu})}$	0.005407	0.006897
Approximate 95% confidence intervals for $\exp(\mu)$		
Likelihood-based	$[0.00836, \ 0.0312]$	$[0.006847, \ 0.03507]$
Wald $Z_{\exp(\widehat{\mu})} \overset{.}{\sim} \text{NORM}(0,1)$	$[0.006870, \ 0.02807]$	$[0.004074, \ 0.03111]$
Wald $Z_{\widehat{\mu}} \overset{.}{\sim} \text{NORM}(0,1)$	$[0.009522, \ 0.03204]$	$[0.008158, \ 0.0379]$
ML estimate $\widehat{\sigma}$	1.3710	1.5078
Standard error $\widehat{\text{se}}_{\widehat{\sigma}}$	0.2752	0.2820
Approximate 95% confidence intervals for σ		
Likelihood-based	$[0.965, \ 2.141]$	$[1.089, \ 2.294]$
Wald $Z_{\log(\widehat{\sigma})} \overset{.}{\sim} \text{NORM}(0,1)$	$[0.925, \ 2.032]$	$[1.045, \ 2.175]$
Wald $Z_{\widehat{\sigma}} \overset{.}{\sim} \text{NORM}(0,1)$	$[0.832, \ 1.910]$	$[0.955, \ 2.061]$
ML estimate $\widehat{t}_{0.90}$	0.1012	0.1075
Standard error $\widehat{\text{se}}_{\widehat{t}_{0.90}}$	0.04004	0.03481
Approximate 95% confidence intervals for $t_{0.90}$		
Likelihood-based	$[0.05332, \ 0.2849]$	$[0.0609, \ 0.2423]$
Wald $Z_{\log(\widehat{t}_{0.90})} \overset{.}{\sim} \text{NORM}(0,1)$	$[0.04662, \ 0.2198]$	$[0.0570, \ 0.2028]$
Wald $Z_{\widehat{t}_{0.90}} \overset{.}{\sim} \text{NORM}(0,1)$	$[0.02275, \ 0.1797]$	$[0.03929, \ 0.1757]$
ML estimate $\widehat{F}(0.30)$	0.9810	0.9894
Standard error $\widehat{\text{se}}_{\widehat{F}(0.30)}$	0.01911	0.01351
Approximate 95% confidence intervals for $F(0.30)$		
Likelihood-based	$[0.9046, \ 0.9982]$	$[0.9235, \ 0.9995]$
Wald $\widehat{Z}_e \overset{.}{\sim} \text{NORM}(0,1)$	$[0.8975, \ 0.9980]$	$[0.9275, \ 0.9996]$
Wald $Z_{\widehat{F}} \overset{.}{\sim} \text{NORM}(0,1)$	$[0.9435, \ 1.0185]$	$[0.9629, \ 1.0159]$

Table 12.5 Atrazine concentration data ML estimates, standard errors and likelihood-based and Wald-approximation 95% confidence intervals.

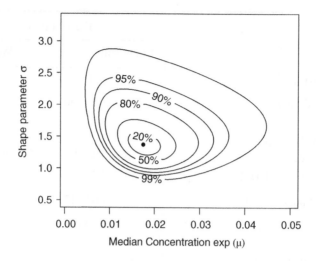

Figure 12.9   Contour plot of the lognormal joint likelihood-based confidence regions for $\exp(\mu)$ and σ for the atrazine data. The ML estimates are indicated by the dot.

$100 \Pr\{X^2_{(2)} \leq -2 \log[R(\mu, \sigma)]\}$, giving approximate confidence levels for joint likelihood-based confidence regions for μ and σ. Figure 12.9 is identical to Figure 12.6, except that the contours are labeled with nominal confidence levels and each represents an approximate joint confidence region for $\exp(\mu)$ and σ.   ∎

12.5.2   The Profile Likelihood and Likelihood-Based Confidence Intervals for μ and $\exp(\mu)$

In practice, for models that have more than one parameter, we frequently focus on just a single parameter (or a single scalar function of multiple parameters, to be discussed shortly). Then we use a *profile likelihood* to provide information about—and construct a confidence interval for—the focus parameter. Profile likelihoods are based on the theory of likelihood-ratio tests (see Section D.5.5). The profile likelihood gets its name from the maximization operation, described below, that defines the profile function. If there are two parameters, the likelihood can be visualized as a mountain and the profile likelihood as the projection of the mountain against the background from the direction of the parameter of interest. The profile likelihood is then used to construct a confidence interval for the focus parameter in a manner similar to the relative likelihood for a one-parameter distribution (e.g., Figure 12.2).

In particular, for a two-parameter location-scale or log-location-scale distribution with parameters μ and σ, the profile likelihood for μ is

$$R(\mu) = \max_{\sigma} \left[\frac{L(\mu, \sigma)}{L(\widehat{\mu}, \widehat{\sigma})} \right]. \tag{12.9}$$

The interval over which $R(\mu) > \exp[-\chi^2_{(1-\alpha;1)}/2]$ is an approximate $100(1 - \alpha)\%$ confidence interval for μ (see Section D.5.5 for the underlying theory). A confidence interval for μ includes all values of μ that have a relatively high profile likelihood. Using (12.9), for every fixed value of μ, we find the point of *highest* relative likelihood by maximizing $R(\mu, \sigma)$ with respect to σ. This gives the profile likelihood value for that value of μ. Values of μ with relatively high profile likelihood are more plausible than those with low profile likelihood.

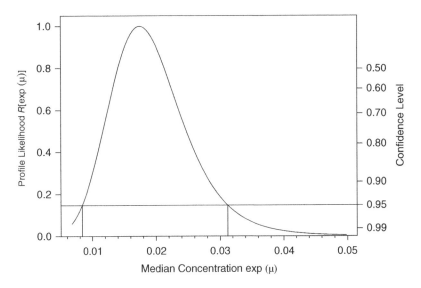

Figure 12.10   Lognormal distribution profile likelihood $R[\exp(\mu)]$ for the median atrazine concentration.

Due to the *invariance* property of ML estimators (e.g., if $\widehat{\mu}$ is the ML estimate of μ, then $\exp(\widehat{\mu})$ is the ML estimate of $\exp(\mu)$), a profile likelihood for μ can be translated directly into a profile likelihood for $\exp(\mu)$, corresponding to the lognormal median or Weibull scale parameter. Likelihood-based confidence intervals translate in a similar manner. For example, if $[\mu, \ \widetilde{\mu}]$ is a likelihood-based confidence interval for μ, then $[\exp(\mu), \ \exp(\widetilde{\mu})]$ is the corresponding likelihood-based confidence interval for $\exp(\mu)$.

Example 12.13   Profile Likelihood and Confidence Interval for the Lognormal Median Atrazine Concentration and the Lognormal Distribution Scale Parameter. Figure 12.10 shows the profile likelihood $R[\exp(\mu)]$ for the atrazine data, assuming a lognormal distribution, and also the likelihood-based 95% confidence interval for the distribution median $\exp(\mu)$. The right-hand scale of Figure 12.10 gives the confidence level for the likelihood-based confidence interval (the relationship between the profile relative likelihood scale and the confidence level scale is based on (12.5)). To obtain a two-sided confidence interval for $\exp(\mu)$ one draws a horizontal line at the desired confidence level and then, at the two points at which this line intersects the profile likelihood curve, one draws vertical lines down to the median concentration axis to determine the lower and upper endpoints of the resulting confidence interval. Thus, one obtains a two-sided 95% confidence interval for $\exp(\mu)$ given by $[0.00836, \ 0.0312]$, as shown in Table 12.5. The likelihood-based 95% confidence interval for μ is $[-4.785, \ -3.467]$; these values are the natural logs of the endpoints of the likelihood-based confidence interval for $\exp(\mu)$, and are also shown in Table 12.5. These interval endpoints can also be obtained from the profile likelihood in a manner similar to that described for Figures 12.2 and 12.4.   ■

12.5.3   Likelihood-Based Confidence Intervals for σ

For two-parameter location-scale and log-location-scale distributions with parameters μ and σ, the profile likelihood for σ is

$$R(\sigma) = \max_{\mu} \left[\frac{L(\mu, \sigma)}{L(\widehat{\mu}, \widehat{\sigma})} \right].$$

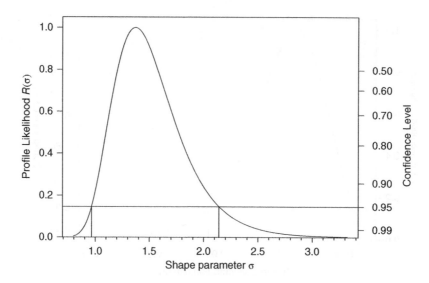

Figure 12.11   Lognormal distribution profile likelihood $R(\sigma)$ for the atrazine data.

The interval over which $R(\sigma) > \exp[-\chi^2_{(1-\alpha;1)}/2]$ is a likelihood-based approximate $100(1 - \alpha)\%$ confidence interval for σ. A 95% confidence interval for the Weibull shape parameter $\beta = 1/\sigma$ is $[\underset{\sim}{\beta}, \;\; \widetilde{\beta}] = [1/\widetilde{\sigma}, \;\; 1/\underset{\sim}{\sigma}]$.

Example 12.14   Profile Likelihood $R(\sigma)$ and Confidence Interval for σ for the Atrazine Concentration Data. Figure 12.11 shows the profile likelihood $R(\sigma)$ for σ for the atrazine data, assuming a lognormal distribution. This plot also shows the construction of the likelihood-based 95% confidence interval for σ, which was found to be $[0.965, \;\; 2.141]$, as shown in Table 12.5. ∎

12.5.4   Likelihood-Based Confidence Intervals for Scalar Functions of μ and σ

The parameterization of a statistical model is typically chosen for some combination of tradition, scientific meaning, and numerical/computational convenience. The location parameter μ and the scale parameter σ are commonly used to describe location-scale distributions and the corresponding log-location-scale distributions such as the lognormal and Weibull distributions (although the Weibull scale parameter $\eta = \exp(\mu)$ and shape parameter $\beta = 1/\sigma$ are commonly used for that distribution). Prime interest, however, often centers on functions of these distribution parameters such as cdf probabilities $p = F(t) = \Phi[(\log(t) - \mu)/\sigma]$ and distribution quantiles $t_p = F^{-1}(p) = \exp[\mu + \Phi^{-1}(p)\sigma]$. Such quantities could also be considered to be alternative "parameters" of the distribution.

In general, the ML estimator of a function $g(\mu,\sigma)$ of (μ,σ) is $\widehat{g} = g(\widehat{\mu},\widehat{\sigma})$. Due to the *invariance* property of ML estimators, likelihood-based methods can, in principle, be readily applied, to make inferences about such functions. In particular, for a scalar function, say $g_1(\mu,\sigma)$, this can be done by defining a one-to-one transformation (or reparameterization), $g(\mu,\sigma) = (\omega_1,\omega_2)$, where $\omega_1 = g_1(\mu,\sigma)$ and $\omega_2 = g_2(\mu,\sigma)$ are functions of μ and σ, defined or chosen such that $g(\mu,\sigma)$ is a one-to-one transformation. In the case that ω_1 is a newly defined parameter, it sometimes suffices to take $\omega_2 = \sigma$ or $\omega_2 = \mu$. The likelihood in terms of the new parameters (ω_1,ω_2) is $L^*(\omega_1,\omega_2) = L(\mu,\sigma)$, where $(\mu,\sigma) = g^{-1}(\omega_1,\omega_2)$. This approach can be used to compute confidence intervals for the elements of $g(\mu,\sigma)$ if the first partial derivatives of $g(\mu, \sigma)$ with respect to μ and σ are continuous. Then ML fitting can be conducted and profile plots obtained for the reparameterization using the approach previously described for μ and for σ. This, in turn, leads to a procedure for obtaining likelihood-based confidence intervals for any scalar or vector function of μ and σ. The method is simple to implement if one can readily compute $g(\mu, \sigma)$ and its inverse, as is the case for lognormal and Weibull distribution quantiles or cdf values. Otherwise, iterative numerical methods are needed for obtaining the required inverse function.

Confidence interval for t_p

The profile likelihood for the p quantile of a location-scale distribution $t_p = \exp[\mu + \Phi^{-1}(p)\sigma]$ is

$$R(t_p) = \max_{\sigma} \left\{ \frac{L^*(t_p, \sigma)}{L^*(\widehat{t_p}, \widehat{\sigma})} \right\} = \max_{\sigma} \left\{ \frac{L[\log(t_p) - \sigma\Phi^{-1}(p), \sigma]}{L(\widehat{\mu}, \widehat{\sigma})} \right\}.$$

This, in turn, allows calculation of a confidence interval for t_p, in a manner that is similar to that used for the parameters μ and σ, as described in Sections 12.5.2 and 12.5.3.

Example 12.15 Profile Likelihood and Confidence Interval for the 0.90 Quantile for the Lognormal Distribution Fit to the Atrazine Concentration Data. Figure 12.12 shows a contour plot for the relative likelihood $R(t_{0.90}, \sigma)$, where $t_{0.90}$ is the 0.90 quantile of the atrazine concentration distribution. This plot provides a sense of the plausible ranges of values for $t_{0.90}$ and σ. Figure 12.13 shows the corresponding profile likelihood for $t_{0.90}$. The resulting likelihood-based 95% confidence interval for $t_{0.90}$, calculated in a manner similar to that in Figure 12.10, is $[0.05332, \ 0.2849]$. This interval is shown in Figure 12.13 and in Table 12.5. ∎

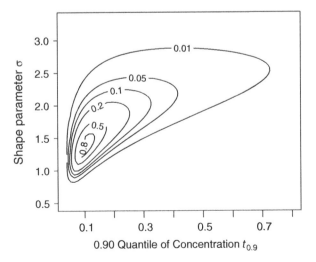

Figure 12.12 Contour plot of lognormal distribution relative likelihood $R(t_{0.90}, \sigma)$ for the atrazine data.

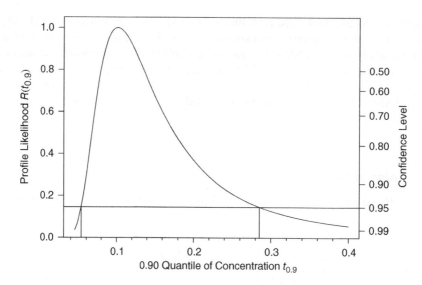

Figure 12.13   Lognormal distribution profile likelihood $R(t_{0.90})$ for the atrazine data.

Confidence interval for $F(t_e)$

The profile likelihood for $F(t_e) = \Phi\{[\log(t_e) - \mu]/\sigma\}$, the cdf of a location-scale distribution at a specified value t_e, is

$$R[F(t_e)] = \max_{\sigma}\left\{ \frac{L^*[F(t_e), \sigma]}{L^*\left[\widehat{F}(t_e), \widehat{\sigma}\right]} \right\} = \max_{\sigma}\left\{ \frac{L[\log(t_e) - \Phi^{-1}[F(t_e)]\sigma, \sigma]}{L(\widehat{\mu}, \widehat{\sigma})} \right\}.$$

This allows calculation of a confidence interval for $F(t_e)$ like that for μ and σ, as described in Sections 12.5.2 and 12.5.3.

Example 12.16   Profile Likelihood and Confidence Interval for $F(0.30)$ for the Lognormal Distribution Fit to the Atrazine Concentration Data. Figure 12.14 gives the profile likelihood for $F(0.30)$, the population fraction of wells for which the atrazine concentration is less than $t_e = 0.30\,\mu\text{g/L}$. The likelihood-based 95% confidence interval for $F(0.30)$ is [0.9046, 0.9982]. This interval is shown in Figure 12.14 and Table 12.5. The width of this interval is explained by the fact that an atrazine concentration of 0.30 is above all but one of the 24 observations in Table 12.4.   ∎

12.6   WALD-APPROXIMATION CONFIDENCE INTERVALS

This section describes Wald-approximation confidence intervals both with and without transformations to improve the approximation and illustrates their application to the atrazine data. These intervals can be viewed as large-sample approximations of likelihood-based confidence intervals.

Wald confidence intervals—also known as *normal-approximation confidence intervals*—are generally easy to compute from ML estimates for the model parameters and the corresponding estimates of their standard errors (information typically provided by computer programs). Wald

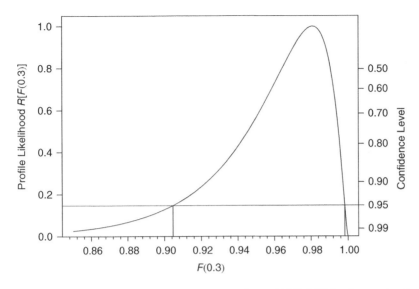

Figure 12.14   Lognormal distribution profile likelihood $R[F(0.30)]$ for the atrazine data.

confidence intervals are provided by many statistical computer packages. As a result, these intervals have been in common use much longer than the likelihood-based intervals discussed so far in this chapter. Therefore, they are better known, especially among those who received their formal statistical training some time ago. With moderate to large samples, Wald intervals can provide useful approximations that can be computed easily.

Wald confidence intervals, however, can have major shortcomings. Simulation studies have shown that in some applications, especially with small sample sizes, heavy censoring, and nonnormal distributions, Wald interval procedures can have coverage probabilities that are appreciably different from their nominal confidence levels. Thus, although the likelihood-based intervals and the Wald intervals are both based on large-sample approximations, the likelihood-based procedures generally give an appreciably better approximation.

Wald confidence regions and intervals are based on a *quadratic approximation* to the profile log-likelihood. This approximation tends to be adequate when the profile log-likelihood is approximately quadratic over the confidence region. With large samples, and the usual regularity conditions (see Section D.3.2), a profile log-likelihood is approximately quadratic. Then the Wald and the likelihood-based intervals will be in close agreement. In other situations, however, the Wald approximation can be seriously inadequate. Moreover, the sample size required to have an adequate Wald approximation is difficult to determine because it depends on the model and on the particular quantity that is to be estimated. Thus, when the quadratic approximation to the log-likelihood is poor, likelihood-based confidence intervals, simulation-based confidence intervals (Chapter 14), or Bayesian credible intervals (Chapters 15 and 16) should be used instead of Wald-approximation intervals.

For censored data or data from truncated distributions, (see Meeker and Escobar, 1998, Section 11.6, for a discussion differentiating these two types of situations), the adequacy of the Wald approximation (and other "large-sample" approximations) also depends principally on the amount of censoring and/or truncation. In particular, for censored data, the adequacy of a large-sample approximation is more a function of the expected number of uncensored observations, rather than the actual sample size. In fact, in some extreme examples, with heavy censoring and estimation in the distribution tail, sample sizes in the thousands may still be insufficient for a good approximation. This also holds for large-sample approximations, including likelihood-based intervals, in general. But the approximation provided by a likelihood-based procedure is usually better, and sometimes much better, than the Wald approximation.

An additional difficulty with Wald confidence intervals is that, unlike likelihood-based confidence intervals, they are *not* transformation invariant. Thus, the choice of a transformation can have a substantial effect on the resulting interval. A good choice for a transformation is one that results in a profile likelihood that is approximately symmetric.

12.6.1 Parameter Variance-Covariance Matrix

Section D.5.6 describes the general theory for computing Wald confidence intervals. This requires an estimate of the variance-covariance matrix for the ML estimates of the model parameters. For a location-scale distribution (with location and scale parameters μ and σ, respectively), one computes the *local* estimate $\widehat{\Sigma}_{\widehat{\theta}}$ of $\Sigma_{\widehat{\theta}}$ as the inverse of the *observed* information matrix

$$\widehat{\Sigma}_{\widehat{\mu},\widehat{\sigma}} = \begin{bmatrix} \widehat{\mathrm{Var}}(\widehat{\mu}) & \widehat{\mathrm{Cov}}(\widehat{\mu},\widehat{\sigma}) \\ \widehat{\mathrm{Cov}}(\widehat{\mu},\widehat{\sigma}) & \widehat{\mathrm{Var}}(\widehat{\sigma}) \end{bmatrix} = \begin{bmatrix} -\dfrac{\partial^2 \mathcal{L}(\mu,\sigma)}{\partial \mu^2} & -\dfrac{\partial^2 \mathcal{L}(\mu,\sigma)}{\partial \mu \partial \sigma} \\ -\dfrac{\partial^2 \mathcal{L}(\mu,\sigma)}{\partial \sigma \partial \mu} & -\dfrac{\partial^2 \mathcal{L}(\mu,\sigma)}{\partial \sigma^2} \end{bmatrix}^{-1},$$

where the partial derivatives are evaluated at $\mu = \widehat{\mu}$ and $\sigma = \widehat{\sigma}$. The intuitive motivation for this estimator is that the partial second derivatives describe the curvature of the log-likelihood, evaluated at the ML estimate. A higher degree of curvature in the log-likelihood surface near the maximum implies a concentrated likelihood near $(\widehat{\mu},\widehat{\sigma})$, implying better precision (i.e., smaller variance) in the estimates.

Example 12.17 Estimate of the Variance-Covariance Matrix for the Atrazine Concentration Data Lognormal ML Estimates. For the atrazine data and the lognormal distribution model,

$$\widehat{\Sigma}_{\widehat{\mu},\widehat{\sigma}} = \begin{bmatrix} 0.095828 & -0.024867 \\ -0.024867 & 0.075726 \end{bmatrix}.$$

An estimate of the correlation between $\widehat{\mu}$ and $\widehat{\sigma}$ is

$$\widehat{\rho}_{\widehat{\mu},\widehat{\sigma}} = \widehat{\mathrm{Cov}}(\widehat{\mu},\widehat{\sigma})/\left[\widehat{\mathrm{Var}}(\widehat{\mu})\widehat{\mathrm{Var}}(\widehat{\sigma})\right]^{1/2} = -0.024867/(0.095828 \times 0.0757)^{1/2} = -0.292.$$

This negative correlation is reflected by the orientation of the likelihood contours in Figure 12.6. ∎

12.6.2 Wald-Approximation Confidence Intervals for μ and $\exp(\mu)$

Approximating the distribution of $Z_{\widehat{\mu}} = (\widehat{\mu} - \mu)/\widehat{\mathrm{se}}_{\widehat{\mu}}$ by a $\mathrm{NORM}(0,1)$ distribution yields a Wald $100(1-\alpha)\%$ confidence interval for the location parameter μ of a location-scale distribution as

$$[\underset{\sim}{\mu}, \ \widetilde{\mu}] = \widehat{\mu} \mp z_{(1-\alpha/2)}\widehat{\mathrm{se}}_{\widehat{\mu}}, \tag{12.10}$$

where $\widehat{\mathrm{se}}_{\widehat{\mu}} = \sqrt{\widehat{\mathrm{Var}}(\widehat{\mu})}$. As noted previously, a one-sided (lower or upper) approximate $100(1-\alpha)\%$ lower (or upper) confidence bound for μ is obtained by replacing $z_{(1-\alpha/2)}$ with $z_{(1-\alpha)}$ and using the appropriate endpoint of the two-sided confidence interval.

One can readily construct a confidence interval for a monotone function (or transformation) of a parameter by applying the function to the endpoints of the confidence interval calculated for the parameter. For example, an approximate $100(1 - \alpha)\%$ confidence interval for the median $t_{0.50} = \exp(\mu)$ of a lognormal distribution (still based on the $Z_{\widehat{\mu}} \sim \text{NORM}(0, 1)$ approximation) is $[\underset{\sim}{t}_{0.50}, \ \widetilde{t}_{0.50}] = [\exp(\underset{\sim}{\mu}), \ \exp(\widetilde{\mu})]$. Similarly, for the Weibull distribution, a confidence interval for the scale parameter η (approximate 0.63 quantile) is $[\underset{\sim}{\eta}, \ \widetilde{\eta}] = [\exp(\underset{\sim}{\mu}), \ \exp(\widetilde{\mu})]$.

Example 12.18 Wald-Approximation Confidence Interval for the Median Atrazine Concentration and Lognormal Distribution Scale Parameter. Again using a lognormal distribution for the atrazine data, substituting $\widehat{\mu} = -4.047$ and $\widehat{\text{se}}_{\widehat{\mu}} = \sqrt{0.095828} = 0.30956$ into (12.10) gives

$$[\underset{\sim}{\mu}, \ \widetilde{\mu}] = -4.047 \mp 1.960 \times 0.30956 = [-4.654, \ -3.441],$$

which is an approximate 95% confidence interval for μ (the mean of the logs of the atrazine concentrations). From this, the corresponding approximate 95% confidence interval for the lognormal distribution median (which is also the lognormal distribution scale parameter $t_{0.50} = \exp(\mu)$) is

$$[\underset{\sim}{t}_{0.50}, \ \widetilde{t}_{0.50}] = [\exp(\underset{\sim}{\mu}), \ \exp(\widetilde{\mu})] = [\exp(-4.6537), \ \exp(-3.4403)]$$
$$= [0.009522, \ 0.03204].$$

This interval indicates that we are approximately 95% confident that the interval [0.009522, 0.03204] μg/L contains the actual $\exp(\mu)$. Recall that $\exp(\mu)$ is the median of the lognormal distribution, interpreted in this application as the concentration level exceeded by 50% of the wells in the sampled population. ∎

12.6.3 Wald-Approximation Confidence Intervals for σ

Because σ must be positive, we follow the common practice of using the log transformation to obtain a confidence interval for this parameter. Approximating the sampling distribution of $Z_{\log(\widehat{\sigma})} = [\log(\widehat{\sigma}) - \log(\sigma)]/\widehat{\text{se}}_{\log(\widehat{\sigma})}$ by a $\text{NORM}(0, 1)$ distribution, an approximate $100(1 - \alpha)\%$ confidence interval for σ is

$$[\underset{\sim}{\sigma}, \ \widetilde{\sigma}] = [\widehat{\sigma}/w, \ \widehat{\sigma} \times w],$$

where $w = \exp[z_{(1-\alpha/2)}\widehat{\text{se}}_{\widehat{\sigma}}/\widehat{\sigma}]$ and $\widehat{\text{se}}_{\widehat{\sigma}} = \sqrt{\widehat{\text{Var}}(\widehat{\sigma})}$.

Example 12.19 Wald-Approximation Confidence Interval for σ for the Atrazine Concentration Data. For the lognormal distribution fitted to the atrazine data, the shape parameter σ is estimated to be $\widehat{\sigma} = 1.3710$, an estimate of its standard error is $\widehat{\text{se}}_{\widehat{\sigma}} = \sqrt{0.075726} = 0.2752$, and $w = \exp[1.960 \times 0.2752/1.3710] = 1.4821$. Thus, an approximate 95% confidence interval for σ based on the approximation $Z_{\log(\widehat{\sigma})} \sim \text{NORM}(0, 1)$, is

$$[\underset{\sim}{\sigma}, \ \widetilde{\sigma}] = [1.3710/1.4821, \ 1.3710 \times 1.4821] = [0.9251, \ 2.0319].$$

The comparison, given in Table 12.5, shows that the preceding confidence interval for σ based on the Wald approximation $Z_{\log(\widehat{\sigma})} \sim \text{NORM}(0, 1)$ agrees reasonably well with the likelihood-based confidence interval. In contrast, the untransformed Wald confidence interval based on the approximation $Z_{\widehat{\sigma}} \sim \text{NORM}(0, 1)$ (computed as $\widehat{\sigma} \mp 1.960 \times \widehat{\text{se}}_{\widehat{\sigma}}$), and also shown in Table 12.5, differs considerably from both of these intervals. This provides further support for the common practice of using the log transformation in computing confidence intervals for distribution parameters that need to be positive. ∎

12.6.4   Wald-Approximation Confidence Intervals for Functions of μ and σ

Following the general theory in Section D.5.6, a Wald confidence interval for a function of μ and σ, say $g_1 = g_1(\mu, \sigma)$, can be based on the large-sample approximate $\text{NORM}(0, 1)$ distribution of $Z_{\widehat{g}_1} = (\widehat{g}_1 - g_1)/\widehat{\text{se}}_{\widehat{g}_1}$. Then an approximate $100(1 - \alpha)\%$ confidence interval for g_1 is

$$[\underset{\sim}{g_1}, \ \widetilde{g}_1] = \widehat{g}_1 \mp z_{(1-\alpha/2)}\widehat{\text{se}}_{\widehat{g}_1},$$

where, using a special case of (D.28) in Section D.5.6 (and also an implementation of the delta method, described in Section D.2),

$$\widehat{\text{se}}_{\widehat{g}_1} = \sqrt{\widehat{\text{Var}}(\widehat{g}_1)}$$

$$= \left[\left(\frac{\partial g_1}{\partial \mu}\right)^2 \widehat{\text{Var}}(\widehat{\mu}) + 2\left(\frac{\partial g_1}{\partial \mu}\right)\left(\frac{\partial g_1}{\partial \sigma}\right)\widehat{\text{Cov}}(\widehat{\mu}, \widehat{\sigma}) + \left(\frac{\partial g_1}{\partial \sigma}\right)^2 \widehat{\text{Var}}(\widehat{\sigma})\right]^{1/2}. \quad (12.11)$$

The partial derivatives in (12.11) are evaluated at $\mu = \widehat{\mu}$ and $\sigma = \widehat{\sigma}$.

Confidence interval for t_p

An approximate $100(1 - \alpha)\%$ confidence interval for the distribution quantile $t_p = \exp[\mu + \Phi^{-1}(p)\sigma]$ based on the large-sample approximate $\text{NORM}(0, 1)$ distribution of $Z_{\log(\widehat{t}_p)} = [\log(\widehat{t}_p) - \log(t_p)]/\widehat{\text{se}}_{\log(\widehat{t}_p)}$ is

$$[\underset{\sim}{t_p}, \ \widetilde{t}_p] = [\widehat{t}_p/w, \ \widehat{t}_p \times w], \quad (12.12)$$

where $w = \exp[z_{(1-\alpha/2)}\widehat{\text{se}}_{\widehat{t}_p}/\widehat{t}_p]$. Applying (12.11) gives

$$\widehat{\text{se}}_{\widehat{t}_p} = \left[\widehat{\text{Var}}(\widehat{t}_p)\right]^{1/2} = \left\{\widehat{t}_p^2 \widehat{\text{Var}}[\log(\widehat{t}_p)]\right\}^{1/2}$$

$$= \widehat{t}_p \left\{\widehat{\text{Var}}(\widehat{\mu}) + 2\Phi^{-1}(p)\widehat{\text{Cov}}(\widehat{\mu}, \widehat{\sigma}) + [\Phi^{-1}(p)]^2 \widehat{\text{Var}}(\widehat{\sigma})\right\}^{1/2}. \quad (12.13)$$

Example 12.20   Wald-Approximation Confidence Intervals for the 0.90 Quantile for the Lognormal Distribution Fit to the Atrazine Concentration Data. The lognormal distribution ML estimate of $t_{0.90}$ is

$$\widehat{t}_{0.90} = \exp\left[\widehat{\mu} + \Phi_{\text{norm}}^{-1}(0.9)\widehat{\sigma}\right] = \exp[-4.0474 + 1.2815 \times 1.3710] = 0.101$$

and substituting into (12.13) gives

$$\widehat{\text{se}}_{\widehat{t}_{0.90}} = 0.1012\left[0.095828 + 2 \times 1.2815 \times (-0.024867) + 1.2815^2 \times 0.075726\right]^{1/2}$$

$$= 0.040042.$$

An approximate 95% confidence interval for $t_{0.90}$ based on $Z_{\log(\widehat{t}_{0.90})} \stackrel{.}{\sim} \text{NORM}(0, 1)$ is obtained by substituting into (12.12), giving

$$[\underset{\sim}{t_{0.90}}, \ \widetilde{t}_{0.90}] = [0.1012/2.1712, \ 0.1012 \times 2.1712 = [0.0466, \ 0.220],$$

where $w = \exp[1.960 \times 0.04004/0.1012] = 2.1712$.

An approximate 95% confidence interval for $t_{0.90}$ based on $Z_{\widehat{t}_{0.90}} \stackrel{.}{\sim} \text{NORM}(0, 1)$ (no transformation) is

$$
\begin{aligned}
[\underset{\sim}{t}_{0.90}, \ \widetilde{t}_{0.90}] &= [\widehat{t}_{0.90} \mp z_{(0.975)}\widehat{\text{se}}_{\widehat{t}_{0.90}}] \\
&= [0.10123 - 1.960 \times 0.040042, \ \ 0.10123 + 1.960 \times 0.040042] \\
&= [0.0227, \ \ 0.180].
\end{aligned}
$$

The preceding results are compared in Table 12.5. The two Wald intervals for $t_{0.90}$ deviate moderately and appreciably, respectively, from the likelihood-based interval. Also, both deviate in the same direction. This is related to the left-skewed shape of $R(t_{0.90})$, as seen in Figure 12.13. The log transformation on $t_{0.90}$ improves the symmetry of the profile likelihood for $\widehat{t}_{0.90}$. Thus, the Wald interval based on a log transformation again does a better job approximating the likelihood-based interval than does the Wald interval that does not employ this transformation. ∎

Confidence interval for $F(t_e)$

Let t_e be a specified value for which an estimate of the cdf $F(t)$ is desired. The ML estimate for $F(t_e)$ is $\widehat{F}(t_e) = F(t_e; \widehat{\mu}, \widehat{\sigma}) = \Phi(\widehat{z}_e)$ where $\widehat{z}_e = [\log(t_e) - \widehat{\mu}]/\widehat{\sigma}$. An approximate confidence interval for $F(t_e)$ can be obtained from

$$
[\underset{\sim}{F}(t_e), \ \widetilde{F}(t_e)] = \widehat{F}(t_e) \mp z_{(1-\alpha/2)}\widehat{\text{se}}_{\widehat{F}}, \tag{12.14}
$$

where applying the delta method in (12.11) gives

$$
\widehat{\text{se}}_{\widehat{F}} = \frac{\phi(\widehat{z}_e)}{\widehat{\sigma}}\left[\widehat{\text{Var}}(\widehat{\mu}) + 2\widehat{z}_e\widehat{\text{Cov}}(\widehat{\mu}, \widehat{\sigma}) + \widehat{z}_e^2\widehat{\text{Var}}(\widehat{\sigma})\right]^{1/2}. \tag{12.15}
$$

The interval in (12.14) is based on the $\text{NORM}(0, 1)$ approximation for $Z_{\widehat{F}} = [\widehat{F}(t_e) - F(t_e)]/\widehat{\text{se}}_{\widehat{F}}$. This approximation may, however, be poor with a small to moderate number of (uncensored) observations; the endpoints of the interval might even fall outside the range $0 \leq F(t_e) \leq 1$.

As with other Wald approximations, an appropriate transformation $g_1 = g_1(F)$ will generally result in a procedure having a coverage probability that is closer to its nominal value if $Z_{\widehat{g}_1} = (g_1 - \widehat{g}_1)/\widehat{\text{se}}_{\widehat{g}_1}$ has a distribution that is closer than $Z_{\widehat{F}}$ to $\text{NORM}(0, 1)$. An interval of this type that usually performs much better than the interval (12.14) is based on the assumption that $\widehat{Z}_e = [\log(t_e) - \widehat{\mu}]/\widehat{\sigma}$ approximately follows a $\text{NORM}(0, 1)$ distribution; this is known as the $\widehat{z}$ procedure. We first construct a confidence interval for $z_e = [\log(t_e) - \mu]/\sigma$ as

$$
[\underset{\sim}{z}_e, \ \widetilde{z}_e] = \widehat{z}_e \mp z_{(1-\alpha/2)}\widehat{\text{se}}_{\widehat{z}_e}, \tag{12.16}
$$

where $\widehat{z}_e$ is the observed value of $\widehat{Z}_e$ (i.e., computed from data). Using the delta method (see Section D.2), we obtain

$$
\widehat{\text{se}}_{\widehat{z}_e} = \frac{1}{\widehat{\sigma}}\left[\widehat{\text{Var}}(\widehat{\mu}) + 2\widehat{z}_e\widehat{\text{Cov}}(\widehat{\mu}, \widehat{\sigma}) + \widehat{z}_e^2\widehat{\text{Var}}(\widehat{\sigma})\right]^{1/2}. \tag{12.17}
$$

Then the confidence interval for $F(t_e) = \Phi(z_e)$, found by applying the monotone transformation $\Phi(z)$ to the endpoints of the $[\underset{\sim}{z}_e, \ \widetilde{z}_e]$ interval from (12.16), is

$$
[\underset{\sim}{F}(t_e), \ \widetilde{F}(t_e)] = [\Phi(\underset{\sim}{z}_e), \ \Phi(\widetilde{z}_e)]. \tag{12.18}
$$

The endpoints of this interval will always be between 0 and 1.

Example 12.21   Wald-Approximation Confidence Intervals for $F(0.30)$ **for the Lognormal Distribution Fit to the Atrazine Concentration Data.** Table 12.5 gives two Wald confidence intervals for $F(t_e)$ for $t_e = 0.30 \, \mu$g/L (i.e., the probability that the atrazine concentration will be less than $0.30 \, \mu$g/L) based, respectively, on assuming that $\widehat{Z}_e$ and $Z_{\widehat{F}}$ are approximately NORM$(0, 1)$ distributed. For the no-transformation method, we first compute $\widehat{z}_e = [\log(0.30) + 4.0474]/1.3710 = 2.074$ from which $\widehat{F}(0.30) = \Phi(2.074) = 0.9810$. Then, from (12.15),

$$\widehat{\text{se}}_{\widehat{F}} = \frac{\phi(2.074)}{1.3710} \left[0.095828 + 2 \times 2.074 \times (-0.024867) + 2.074^2 \times 0.075726 \right]^{1/2}$$
$$= 0.019112;$$

and from (12.14),

$$[\underset{\sim}{F}(0.30), \ \widetilde{F}(0.30)] = 0.9810 \mp 1.960 \times 0.019112 = [0.944, \ 1.018]. \qquad (12.19)$$

For the $\widehat{z}$ procedure, using (12.15) and (12.17) gives

$$\widehat{\text{se}}_{\widehat{z}_e} = \widehat{\text{se}}_{\widehat{F}}/\phi(\widehat{z}_e) = 0.019112/\phi(2.074) = 0.41158.$$

Then, using (12.16),

$$[\underset{\sim}{z}_e, \ \widetilde{z}_e] = 2.074 \mp 1.960 \times 0.41158 = [1.2673, \ 2.8807].$$

Finally, using (12.18) gives

$$[\underset{\sim}{F}(0.030), \ \widetilde{F}(0.030)] = [\Phi(1.2673), \ \Phi(2.8807)] = [0.897, \ 0.998]. \qquad (12.20)$$

The comparison in Table 12.5 shows that the $\widehat{z}$ Wald interval from (12.20) agrees well with the likelihood-based interval, but the Wald interval from (12.19) has an endpoint exceeding 1.0, a clear indication of an inadequate large-sample approximation.   ∎

12.6.5   Using the Wald Approximation to Compute a Confidence Interval for a Correlation Coefficient

The correlation coefficient is frequently used to describe the linear association between two continuous random variables. Formally, the correlation coefficient between two random variables X and Y is defined as

$$\rho = \frac{\text{E}[(X - \mu_X)(Y - \mu_Y)]}{\sigma_X \sigma_Y},$$

where (μ_X, μ_Y) and (σ_X^2, σ_Y^2) are, respectively, the means and variances of X and Y. The correlation coefficient is also a parameter of the well-known bivariate normal distribution (e.g., Johnson and Wichern, 2002, Chapter 4). For a random sample (x_i, y_i), $i = 1, \ldots, n$, from jointly normally distributed variables (X, Y), the ML estimator of the correlation coefficient is

$$\widehat{\rho} = \frac{\sum_{i=1}^{n}(x_i - \bar{x})(y_i - \bar{y})}{\left[\sum_{i=1}^{n}(x_i - \bar{x})^2 \right]^{1/2} \left[\sum_{i=1}^{n}(y_i - \bar{y})^2 \right]^{1/2}}, \qquad (12.21)$$

where $\bar{x}$ and $\bar{y}$ are the sample means of the observed x_i and y_i values, $i = 1, \ldots, n$, respectively.

A commonly used method to construct a confidence interval for ρ is to use a Wald approximation on the so-called Fisher's z-transformation scale; that is,

$$\widehat{z} = \frac{1}{2} \log \left(\frac{1 + \widehat{\rho}}{1 - \widehat{\rho}} \right). \tag{12.22}$$

This transformation maps the ∓ 1 range of $\widehat{\rho}$ into a range of $\mp \infty$ for $\widehat{z}$. A commonly used estimate of the standard error of $\widehat{z}$ is $\widehat{se}_{\widehat{z}} = \sqrt{1/(n-3)}$ and an approximate 95% confidence interval for $z = \frac{1}{2} \log[(1 + \rho)/(1 - \rho)]$ is

$$[\underset{\sim}{z}, \ \widetilde{z}] = \widehat{z} \mp z_{(1-\alpha/2)} \widehat{se}_{\widehat{z}}. \tag{12.23}$$

Then a Wald confidence interval for ρ is obtained by inverting the z-transformation using

$$[\underset{\sim}{\rho}, \ \widehat{\rho}] = [\tanh(\underset{\sim}{z}), \ \tanh(\widetilde{z})],$$

where $\tanh(z) = [\exp(2z) - 1]/[\exp(2z) + 1]$ is the hyperbolic tangent function.

Example 12.22 Wald-Approximation Confidence Interval for the Correlation between Body Weight and Pulse Rate. Table 12.6 gives pulse rate and body weight measurements for a, presumably random, sample of 20 middle-aged male members of a health fitness club. The data set (which includes several further response variables) is from Jackson (1991, page 267) and is attributed to Professor A.C. Linnerud from North Carolina State University. Figure 12.15 is a scatter plot of the observations. The investigators desired a confidence interval for the coefficient of correlation between body weight and pulse measurement.

Body weight (pounds)	Pulse rate (beats/minute)
191	50
189	52
193	58
162	62
189	46
182	56
211	56
167	60
176	74
154	56
169	50
166	52
154	64
247	50
193	46
202	62
176	54
157	52
156	54
138	68

Table 12.6 Body weight and pulse rate measurements for a sample of 20 middle-aged men.

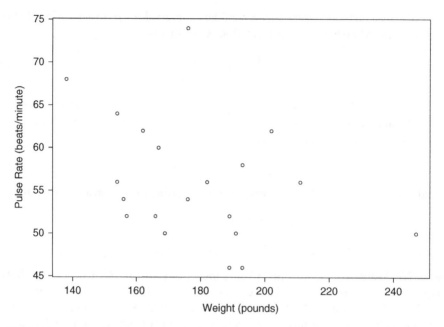

Figure 12.15   Scatter plot of pulse rate versus body weight measurements for a sample of 20 middle-aged men.

Using (12.21) yields $\widehat{\rho} = -0.365762$. Substituting into (12.22) gives

$$\widehat{z} = \frac{1}{2} \log\left(\frac{1 - 0.365762}{1 + 0.365762}\right) = -0.3835218,$$

and because $n = 20$, $\widehat{\mathrm{se}}_{\widehat{z}} = \sqrt{1/(20 - 3)} = 0.2425356$. Then substituting into (12.23) gives

$$[\underset{\sim}{z}, \ \widetilde{z}] = -0.3835218 \mp 1.959964 \times 0.2425356 = [-0.8588828, \ \ 0.09183933].$$

Thus, the Wald 95% confidence interval for ρ is

$$[\underset{\sim}{\rho}, \ \widetilde{\rho}] = [\tanh(\underset{\sim}{z}), \ \tanh(\widetilde{z})] = [-0.696, \ \ 0.092].$$

The resulting rather wide interval can be attributed to the relatively small sample size. Also, we note that, because the 95% confidence interval includes the value 0 (although barely), the correlation is not statistically significantly different from 0 at the 5% significance level. The preceding interval will be compared in Chapter 13 to similar confidence intervals computed using nonparametric bootstrap methods.

The preceding confidence interval applies to the population from which the sample was assumed to be randomly selected (in this case the health club members). The investigators' interest, however, is likely to be not in the middle-aged men from *this particular* health club, but in the general population of middle-aged men that belong to health clubs, or even middle-aged men in general. It warrants repeating that statistical inferences about a population or process of interest apply only to the degree that the observations can be regarded as a random sample from the population or process of interest. In this case, this requires assuming that the health club members are "representative" of the larger population of interest. As a consequence—and as we have tried to emphasize in Chapter 1—the calculated statistical interval describes only the statistical uncertainty due to taking a (presumably) random sample from the population or process of interest. The actual uncertainty is larger due to likely departures from assumptions that were (implicitly or explicitly) made in computing the interval. ∎

12.7 SOME OTHER LIKELIHOOD-BASED STATISTICAL INTERVALS

12.7.1 Likelihood-Based and Wald-Approximation Confidence Intervals for Other Distributions

The discussion in much of this chapter has focused on the construction of approximate confidence intervals for location-scale and, especially, log-location-scale distributions in general, and for lognormal and Weibull distributions in particular. The concepts presented, however, apply for a wide range of statistical models; see Meeker and Escobar (1998, Chapter 11) for examples.

12.7.2 Likelihood-Based One-Sided Tolerance Bounds and Two-Sided Tolerance Intervals

We describe in Section 2.4.2 how a one-sided lower (upper) tolerance bound to be exceeded by (to exceed) at least a proportion p of a population is equivalent to a one-sided confidence bound for the $1 - p$ (for the p) distribution quantile. We can, therefore, use likelihood-based or Wald-approximation one-sided confidence bounds on quantiles (see Sections 12.5.4 and 12.6.4) to construct one-sided tolerance bounds.

Furthermore, as explained in Section D.7.3, an approximate two-sided tolerance interval to contain p can be obtained by combining a lower $100(1 - \alpha/2)\%$ confidence bound for the $(1 - p)/2$ distribution quantile with an upper $100(1 - \alpha/2)\%$ confidence bound for the $(1 + p)/2$ distribution quantile. For example, to construct an approximate 95% tolerance interval to contain at least a proportion 0.90 of the population, one would combine a lower 97.5% confidence bound for the distribution quantile $t_{0.05}$ with an upper 97.5% confidence bound for the $t_{0.95}$ quantile. The coverage probability for this approximate tolerance interval will be conservative (i.e., greater than the nominal confidence level) if the confidence intervals for the quantiles are exact. The justification for this approximation (based on a Bonferroni inequality) is described in Section D.7.3.

12.7.3 Likelihood-Based Prediction Intervals

Likelihood-based prediction intervals are constructed by using a likelihood function that is proportional to the joint probability distribution of the unknown parameters *and* the quantity to be predicted (future random variable), treating the quantity to be predicted as an unknown parameter. Then methods analogous to those presented earlier in this chapter can be applied to construct a profile likelihood for the future value of the random variable and from this obtain the desired likelihood-based prediction interval. Theory and technical details for likelihood-based prediction intervals are provided in Butler (1986, 1989) and Bjørnstad (1990). Some simple applications are described by Nelson (2000) and Nordman and Meeker (2002).

BIBLIOGRAPHIC NOTES

Most textbooks on mathematical statistics—for example, Casella and Berger (2002) and Boos and Stefanski (2013)—give a simple introduction to likelihood and maximum likelihood estimation. Several books, including Edwards (1985), Kalbfleisch (1985), Severini (2000), and Pawitan (2001), place special emphasis on the subject of likelihood. Meeker and Escobar (1995) describe some of the important concepts underlying profile likelihood plots and how Wald confidence intervals can be used as approximations to likelihood-based confidence intervals. Hong et al. (2008a) demonstrate the potential high degree of sensitivity that Wald intervals can have to the chosen transformation and provide a method to construct such intervals for distribution probabilities. Hong et al. (2008b) prove that a set of pointwise confidence intervals for $F(t)$ for a range of t values is equivalent to a set of pointwise confidence intervals for t_p for any corresponding range of p, if the intervals are based on likelihood (and this is *not* generally true for other confidence interval methods). Jeng and Meeker (2000) report simulation results comparing the coverage probabilities of likelihood, Wald, and parametric bootstrap confidence interval procedures (to be described in Chapter 14) for estimating quantiles of Weibull and lognormal distributions.

Nair (1981, 1984) and Meeker and Escobar (1998, Chapters 3) show how to compute simultaneous confidence bands for a cdf, such as those used in probability plots.

Chapter *13*

Nonparametric Bootstrap Statistical Intervals

OBJECTIVES AND OVERVIEW

This chapter describes and illustrates computationally intensive *nonparametric bootstrap* methods to compute statistical intervals, primarily for continuous distributions. These methods require obtaining a sequence of simulated *bootstrap samples*, based on the given data. Then these bootstrap samples are used to generate corresponding *bootstrap estimates*.

As mentioned in the introduction to Chapter 5, *nonparametric* implies that no particular parametric distribution needs to be specified when applying the statistical method. The distribution-free methods introduced in Chapter 5 were also nonparametric. The nonparametric methods presented in this chapter, however, are not distribution-free because the statistical properties (e.g., coverage probabilities) of the procedures depend on the unspecified underlying probability distribution.

Although nonparametric bootstrap procedures do not require specification of a particular parametric distribution for the underlying data, they generally do not work well with small samples (e.g., fewer than ten observations for some applications) in the sense that the procedures have coverage probabilities that might be far from the specified nominal confidence level.

The alternative parametric bootstrap methods presented in Chapter 14 require one to specify the form of a parametric distribution for the given data. Such methods can lead to excellent approximate, or sometimes exact, procedures for computing statistical intervals, even for small samples, when the chosen distribution is correct. Parametric bootstrap methods may, however, result in misleading answers if the chosen distribution is seriously in error.

The topics discussed in this chapter are:

- The basic concept of using nonparametric bootstrap methods to obtain confidence intervals (Section 13.1).

Statistical Intervals: A Guide for Practitioners and Researchers, Second Edition.
William Q. Meeker, Gerald J. Hahn and Luis A. Escobar.
© 2017 John Wiley & Sons, Inc. Published 2017 by John Wiley & Sons, Inc.
Companion Website: www.wiley.com/go/meeker/intervals

- Nonparametric methods for generating bootstrap samples and obtaining bootstrap estimates (Section 13.2).

- Choice of the number of bootstrap samples and how to manage such samples (Section 13.3).

- How to obtain nonparametric bootstrap confidence intervals (Section 13.4).

13.1 INTRODUCTION

13.1.1 Basic Concepts

In Chapter 4, we presented exact methods for computing confidence intervals for a *normal distribution*. The computer-intensive nonparametric bootstrap methods described in this chapter provide alternatives for constructing approximate confidence intervals for distribution characteristics such as the mean and standard deviation, without having to make an assumption about the underlying distribution.

The general idea behind the nonparametric bootstrap procedures presented in this chapter (as well as the parametric simulation/bootstrap procedures presented in Chapter 14) is to replace mathematical approximations or intractable distribution theory with Monte Carlo simulation, taking advantage of the power of modern computer hardware and relatively recent developments in statistical theory. This approach, like other methods, needs to be implemented with much care and should be guided by the theoretical statistical principles that we outline later in this chapter. Improper application can result in poor or seriously incorrect results.

13.1.2 Motivating Example

We use the following example to illustrate the nonparametric bootstrap methods for computing statistical intervals.

Example 13.1 Estimating Total Tree Volume. Table 13.1 gives calculated tree volume measurements for a, presumed random, sample of $n = 29$ trees from a much larger population of 25-year-old loblolly pines (*Pinus taeda* L.). The data were obtained from the Southwide Seed Source Study described in Poudel and Cao (2013) (and given to us by Professor Quang Cao from Louisiana State University). The original observations provided tree diameters at breast height (dbh, 4.5 feet above ground) in centimeters (cm) and the total tree height in meters (m). The tree volumes, V, in cubic meters (m^3) were then estimated using the equation $V = 0.00017 + 0.0000281 \times (\text{dbh})^2 h$ proposed by Matney and Sullivan (1982).

Figure 13.1 is a histogram of the data. The volume measurement of 0.307 m^3 seems to be an outlier which, however, was confirmed to be a correct measurement. It is desired to estimate the total volume of the trees in the sampled population. To do so, it is necessary to estimate mean tree volume; this was found to be $\bar{x} = 0.1091$. To quantify the statistical uncertainty of this estimate, a confidence interval is required. Figure 13.1 suggests that a normal distribution

0.149	0.086	0.149	0.194	0.044	0.104
0.156	0.122	0.117	0.079	0.179	0.307
0.049	0.165	0.043	0.079	0.109	0.102
0.195	0.063	0.068	0.029	0.079	0.124
0.151	0.115	0.023	0.016	0.067	

Table 13.1   Tree volumes in cubic meters (m^3).

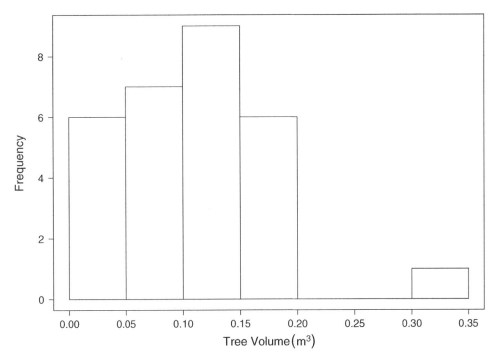

Figure 13.1   Histogram of the tree volume data.

does not provide a good representation of the tree volume distribution and, in fact, the form of this distribution is uncertain. Therefore, we construct a confidence interval that does not require any distributional assumptions.   ■

13.2   NONPARAMETRIC METHODS FOR GENERATING BOOTSTRAP SAMPLES AND OBTAINING BOOTSTRAP ESTIMATES

Statistical intervals are computed as a function of the available data, consisting of n observations denoted by DATA. Bootstrap interval procedures employ, in addition, a set of B bootstrap samples, DATA_j^*, $j = 1, \ldots, B$, generated by Monte Carlo simulation that, in some sense (depending on the type of bootstrap procedure) mimics the original sampling procedure. For each of the B bootstrap samples, one or more bootstrap statistics are computed. There are several different methods for generating nonparametric bootstrap samples. This section describes two methods and reasons for using each of them. Section 13.4 shows how the resulting bootstrap samples and statistics are used to compute nonparametric confidence intervals.

13.2.1   Nonparametric Bootstrap Resampling

Figure 13.2 illustrates the nonparametric bootstrap *resampling* method. In this method a point estimate, $\widehat{\theta}$, of the scalar quantity of interest θ (or a particular function of interest computed from θ) is obtained initially directly from the data. Then B bootstrap samples (also called *resamples*), each of size n, are obtained by sampling, *with replacement*, from the n cases in the given data set. Specifically, to obtain the jth bootstrap sample DATA_j^*, we select with replacement a sample of size n from the n original observations in DATA. Each observation in DATA has an equal probability of being chosen on each draw. Because each draw from the n

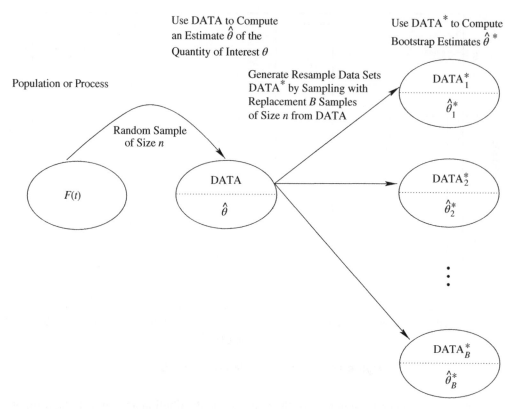

Figure 13.2 Illustration of nonparametric bootstrap resampling for obtaining bootstrap samples DATA* and bootstrap estimates $\widehat{\theta}^*$.

original observations is replaced before the next draw, some observations in the original DATA may be selected more than once and others not at all in a single bootstrap sample (as will be illustrated in columns 2, 3, and 4 of Table 13.2).

For each bootstrap resample, an estimate of the desired distribution characteristic of interest θ (or a function of interest related to θ) is computed from the n resample values, giving $\widehat{\theta}_j^*$, $j = 1, \ldots, B$. The resulting B values of $\widehat{\theta}^*$ can then be used to compute the desired statistical interval or intervals as described in Section 13.4.

Example 13.2 Nonparametric Bootstrap Samples for Mean Tree Volume. We illustrate nonparametric bootstrap resampling with the mean tree volume estimation problem (Example 13.1). First, $B = 200,000$ samples, each of size $n = 29$, were resampled with replacement from the 29 tree volumes given in Table 13.1. The mean (the quantity of interest here) of each of these B samples was then calculated. Figure 13.3 is a histogram of the resulting B nonparametric bootstrap sample means. The solid vertical line in the center of the figure shows the median of the 200,000 bootstrap sample means, while the tall dotted line is the mean of the original data. The closeness of these two lines indicates that there is almost no median bias in the bootstrap estimates of the mean. The short vertical lines in the tails of the figure indicate the 0.025 and 0.975 quantiles of the empirical distribution of bootstrap sample means (i.e., the sample means computed for each of the B bootstrap samples). We will see in Section 13.4.2 that these two quantiles provide a crude but simple approximate 95% nonparametric bootstrap confidence interval for the population mean tree volume. ∎

Tree volume	Uniform Multinomial distribution Integer weights			Uniform Dirichlet distribution Continuous weights		
	$j = 1$	$j = 2$	$j = 3$	$j = 1$	$j = 2$	$j = 3$
0.149	1	1	0	1.839	0.829	1.626
0.086	1	1	1	0.344	0.920	0.135
0.149	0	1	0	0.230	4.115	1.708
0.194	0	1	1	0.400	0.199	0.483
0.044	0	2	2	0.340	0.681	0.908
0.104	4	3	0	0.099	0.132	1.714
0.156	1	0	1	0.546	0.467	0.344
0.122	0	0	1	1.208	1.057	4.715
0.117	2	1	1	3.495	0.268	0.542
0.079	2	0	1	1.245	0.139	0.168
0.179	0	3	2	1.176	0.796	0.244
0.307	2	1	0	1.638	3.473	0.265
0.049	2	3	1	0.132	0.603	1.201
0.165	1	1	0	1.961	0.601	1.043
0.043	1	0	0	0.917	1.072	3.113
0.079	0	0	1	0.608	0.198	0.084
0.109	1	2	1	1.045	2.040	0.323
0.102	2	0	0	0.207	0.463	0.160
0.195	0	0	2	1.219	0.697	2.600
0.063	1	1	2	2.316	0.501	0.181
0.068	2	0	1	1.289	0.302	3.126
0.029	2	1	3	1.042	0.172	0.666
0.079	0	2	0	2.118	1.807	0.073
0.124	1	1	2	0.386	0.227	0.136
0.151	1	2	1	0.177	3.453	0.662
0.115	0	1	3	0.188	1.170	0.592
0.023	0	0	1	0.505	0.686	1.038
0.016	1	0	1	0.782	1.289	0.417
0.067	1	1	0	1.549	0.643	0.734
Mean $\widehat{\mu} = 0.1091$	$\widehat{\mu}_j^* = 0.105$	0.115	0.100	0.115	0.135	0.107
Standard deviation $\widehat{\sigma} = 0.0624$	$\widehat{\sigma}_j^* = 0.067$	0.059	0.056	0.067	0.078	0.056
Standard error $\widehat{se} = 0.0116$	$\widehat{se}_j^* = 0.012$	0.011	0.010	0.012	0.014	0.010
Bootstrap-t ratio $t_j^* = (\widehat{\mu}_j^* - \widehat{\mu})/\widehat{se}_j^* = -0.322$	0.513	-0.909	0.491	1.778	-0.228	

Table 13.2   Examples of integer and continuous random-weight bootstrap sampling from the tree volume data.

13.2.2 Nonparametric Random-Weight Bootstrap Sampling

The resampling method described in Section 13.2.1 can also be viewed as a random-weight method of sampling where n integer weights $(\omega_1, \ldots, \omega_n)$, one for each observation in the data set, are a sample from an n-cell multinomial distribution with equal probability $1/n$ for each of the n cells. Some of the original observations will be resampled more than once (and thus have integer weights greater than 1) and others will be not be sampled at all (and thus will

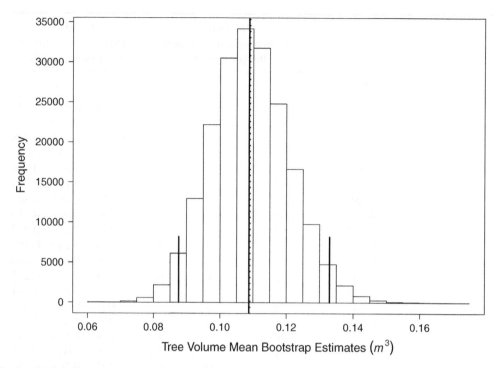

Figure 13.3 Histogram of the sample means for $B = 200{,}000$ bootstrap resamples from the tree volume data, showing the mean of the original data (tall dotted vertical line), the median or 0.50 quantile (tall solid vertical line), and the 0.025 and 0.975 quantiles (short vertical lines) of the empirical distribution of the bootstrap sample means.

have weight 0). A potential problem with the resampling method described in Section 13.2.1 is that, in some applications, certain resamples of n observations may not be able to estimate the quantity of interest (in the case of nonparametric bootstrap) or all of the model parameters (in the case of parametric bootstrap), even if the original data are able to do so. Situations for which this is likely to happen include:

- When data are censored and there are only a limited number of noncensored observations. In such cases, it is possible to obtain resamples with all observations censored.

- Even when data are not censored, we have encountered applications with small to moderate sample sizes where the resampling method resulted in noticeable instability in estimating parameters from resampled data. See Section 14.7.3 for an example.

- Logistic regression applications for which the probability of a response tends to have either a high or a low probability of a "success." For example, in an experiment to estimate the probability of a success as a function of dose and if successes and only successes occur above a given dose level, resamples will not be able to estimate the model parameters.

- The analysis of data from designed experiments for which the number of model parameters is close to the number of observational units. In this case resamples will often contain repeats at certain conditions (and no observations at others), making it impossible to estimate all model parameters.

Random-weight bootstrap sampling is an appealing alternative to the method described in Section 13.2.1 that can be applied when the estimation method allows noninteger weights (e.g., maximum likelihood or least squares). In particular, nonnegative weights can be generated from a *continuous* distribution of a positive random variable that has the same mean and standard deviation (usually taken to be equal to 1). Weights generated independently from an exponential distribution with mean 1 is a common choice. Another alternative is to generate the weights from a uniform Dirichlet distribution, which can be achieved by standardizing the independent exponential weights to sum to n. In either of these cases bootstrap estimates are obtained by applying an appropriate weighted estimation method, using the B sets of random weights. Then if the original data were capable of doing the desired estimation, the estimation will usually be successful for each set of random weights.

Generating the random weights does not require any assumptions about the underlying distribution of the data, and thus this method is a nonparametric method of generating bootstrap samples. The random-weight bootstrap method can, however, be used in both nonparametric and parametric (see Chapter 14) bootstrap applications. In situations for which there is little or no risk of estimation problems with resampling, the resampling (integer weight) method and the random continuous weight method will give similar bootstrap results.

Standard weighted-observation estimation formulas can be used to compute the random-weight bootstrap estimates for some simple statistics. For example, let $x_1, \ldots, x_n$ denote a random sample from a distribution of interest. Random-weight bootstrap estimates of the mean and standard deviation would be

$$\widehat{\mu}^* = \frac{1}{\sum_{i=1}^{n} \omega_i} \sum_{i=1}^{n} \omega_i x_i \quad \text{and} \quad \widehat{\sigma}^* = \left[\frac{1}{\sum_{i=1}^{n} \omega_i} \sum_{i=1}^{n} \omega_i (x_i - \widehat{\mu}^*)^2 \right]^{1/2}. \tag{13.1}$$

Note that the formula for $\widehat{\sigma}^*$ corresponds to the formula for unweighted data that divides the sum of squares by the sample size n (which is the maximum likelihood estimator for a normal distribution), and we will use such estimates of standard deviations throughout this chapter.

To get the complete set of bootstrap estimates, the generation of the n random weights and the computation of the estimates from (13.1) would be repeated B times. Note that the data are constant and the random weights induce randomness in the computed bootstrap statistics.

Table 13.2 contrasts the generation of bootstrap estimates from simple resampling discussed in Section 13.2.1 (equivalent to multinomial generation of integer weights) and Dirichlet distribution continuous weights, showing three bootstrap samples and bootstrap statistics from each method.

The important difference between the integer and continuous weight methods of generating bootstrap samples illustrated in Table 13.2 is that some of the integer weights are 0, indicating that the associated observations are completely ignored in computing the bootstrap statistics using this method. In contrast, when the continuous weights are used, each of the original observations has a contribution to the computation of the bootstrap statistics.

In addition to the situations listed above, the random-weight bootstrap method has been applied most commonly in situations with complicated data and/or a complicated parametric model and when maximum likelihood estimation is used. For maximum likelihood estimation, the weighted likelihood is

$$L(\boldsymbol{\theta}) = L(\boldsymbol{\theta}; \text{DATA}) = C \prod_{i=1}^{n} [L_i(\boldsymbol{\theta}; \text{data}_i)]^{\omega_i}$$

where, as in (12.2), $L_i(\boldsymbol{\theta}; \text{data}_i)$ is the likelihood contribution for observation i. This weighted likelihood is maximized just like a regular likelihood to obtain bootstrap estimates $\widehat{\boldsymbol{\theta}}^*$ of the model parameters in $\boldsymbol{\theta}$. Use of the random-weight bootstrap will be illustrated in Chapter 14.

13.3   BOOTSTRAP OPERATIONAL CONSIDERATIONS

13.3.1   Choosing the Number of Bootstrap Samples

As we have seen, bootstrap methods are based on Monte Carlo simulation. Thus, if a procedure is repeated, one should expect to obtain different answers due to Monte Carlo sampling error (i.e., variability in the generation of bootstrap samples). This variability may be small or large in magnitude, depending on B, the number of bootstrap samples, and the sampling variability in the estimates. One can, however, make the Monte Carlo sampling error arbitrarily small by choosing B to be sufficiently large. Therefore, if it is easy to generate samples, analysts might decide to play it safe and set B to be large so as not to have to be concerned with this source of variability. There has, in fact, been some inflation in suggested bootstrap sample sizes with the growth of computing power over the years. We have seen recommendations to use values of $B = 1,000,000$, or even larger, to obtain intervals that, for all practical purposes, eliminate Monte Carlo sampling error altogether in most applications. In our examples, we have generally used $B = 200,000$ so as to obtain statistical intervals with small Monte Carlo sampling error (resulting in some variability in the third significant digit of an interval endpoint in most cases).

If the application requires an estimation method that is computationally intensive, smaller values of B will be favored. The specific number will depend on the goal of the computation. Historically, bootstrap sample sizes as small as $B = 200$ have been suggested for estimating the bias or standard error of an estimate and $B = 2,000$ or $B = 4,000$ samples for constructing confidence intervals (and a larger number for higher confidence levels). For some applications, such as constructing confidence intervals on a distribution tail quantile, values of B need to be larger than those for, say, constructing confidence intervals for the median. Today, with more computer power than ever available, the recommendations have generally increased to values like $B = 10,000$ to $B = 50,000$. One argument for the smaller values of B being acceptable is that even with small B, Monte Carlo error is generally small relative to the width of the interval, and thus may be of little practical importance. One should, however, strive to avoid reporting digits that are mostly or completely in error. Another possible approach is to automatically review the results every b bootstrap samples (where b might be 1,000 or 10,000) and stop sampling when the interval endpoints have stabilized.

13.3.2   Saving Bootstrap Results

It is often useful, especially for parametric and computationally intensive bootstrap procedures, to retain the individual simulated sample results for possible future use and assessment. This will, for example, allow one to perform graphical assessment of the Monte Carlo simulation results (which we highly recommend for new and unfamiliar situations) and to quickly compute alternative intervals for different quantities or intervals with different confidence levels. Thus, when using a parametric bootstrap approach to estimate functions of a parameter vector $\boldsymbol{\theta}$, one might retain in a computer file, the parameter estimates $\widehat{\boldsymbol{\theta}}_j^*$, $j = 1, \ldots, B$, and, in some cases, the lower-triangular elements of the corresponding variance-covariance matrices of the parameter estimates.

13.3.3   Calculation of Quantiles of a Bootstrap Distribution

Bootstrap inference generally requires one to calculate quantiles of the empirical distribution of bootstrap estimates of the quantity of interest (or of a particular function related to this quantity). We define the p quantile of an empirical distribution as the kth order statistic, where $k = pB$ when pB is an integer and k is equal to pB rounded to the next largest integer when pB is not an integer. There are, however, alternative definitions for the p quantile (e.g., rounding to the nearest integer). Hyndman and Fan (1996) describe nine such methods used in various statistical packages. All nine of these methods are available, by option, in the `quantile` function in R. Most of these methods use more sophisticated interpolation than the simple method employed in this book. When B is large (as in our examples), however, the differences in the results obtained among the alternative methods tend to be small.

13.4 NONPARAMETRIC BOOTSTRAP CONFIDENCE INTERVAL METHODS

As in other chapters, our discussion in this and subsequent sections in this chapter deals primarily with two-sided confidence intervals. As described in Section 2.7, however, one can obtain a one-sided lower (or upper) confidence bound from the corresponding two-sided interval by substituting α for $\alpha/2$ in the expression for the lower (or upper) endpoint of the two-sided interval.

13.4.1 Methods for Computing Nonparametric Bootstrap Confidence Intervals from the Bootstrap Samples

One classic approach for constructing a nonparametric bootstrap confidence interval for a quantity of interest is to use "appropriate" quantiles of the empirical bootstrap distribution of that quantity. There are a number of ways to select such quantiles. We present three of these: the simple percentile method, the bias-corrected and accelerated (BCa) percentile method, and the bias-corrected (BC) percentile method. We also present the nonparametric "bootstrap-t" method, based on the idea of an approximate pivotal quantity (see Appendix E) which is an extension of the nonparametric "basic bootstrap" method, which we also discuss.

The simple percentile method is attractive due to its ease of application and intuitive appeal. Bootstrap large-sample theory shows, however, that the BCa and bootstrap-t methods tend to have a coverage probability that is closer to the nominal confidence level. We, therefore, advise practitioners to use one of these two methods, when possible, rather than the simple percentile method. The BC method also gives better results than the simple percentile method; it is useful when it is difficult or impossible to use the BCa or the bootstrap-t method. More details are given in the next five subsections.

13.4.2 The Simple Percentile Method

Description of the method

The simple percentile method is a straightforward approach for using the generated empirical bootstrap distribution (described in Sections 13.2.1 and 13.2.2) to obtain nonparametric bootstrap confidence intervals. It uses the $\alpha/2$ and $1 - \alpha/2$ quantiles of the empirical bootstrap distribution of the estimates of the quantity of interest as the estimated endpoints of the desired confidence interval. That is,

$$\left[\underset{\sim}{\theta}, \ \widetilde{\theta} \right] = \left[\widehat{\theta}^*_{(\alpha/2)}, \ \widehat{\theta}^*_{(1-\alpha/2)} \right], \tag{13.2}$$

where $\widehat{\theta}^*_{(p)}$ is the p quantile of the empirical distribution of bootstrap estimates for θ, the quantity of interest. If, for example, a confidence interval for the mean of a distribution is desired, then one uses the $\alpha/2$ and $1 - \alpha/2$ quantiles, respectively, of the empirical distribution of bootstrap sample means as the lower and upper bounds of the desired confidence interval.

Method	Section	Interval
Parametric traditional normal distribution	4.2	[0.085, 0.133]
Nonparametric simple percentile bootstrap	13.4.2	[0.087, 0.133]
Nonparametric BCa percentile bootstrap	13.4.3	[0.089, 0.135]
Nonparametric BC percentile bootstrap	13.4.4	[0.088, 0.134]
Nonparametric basic	13.4.5	[0.085, 0.131]
Nonparametric bootstrap-t	13.4.6	[0.082, 0.130]
Parametric Wald using an assumed Weibull distribution	12.6.4	[0.089, 0.134]
Parametric GPQ bootstrap using an assumed Weibull distribution	14.4.3	[0.088, 0.138]

Table 13.3   Approximate 95% confidence intervals, using different methods, for mean tree volume.

Example 13.3   The Simple Percentile Bootstrap Confidence Interval for Mean Tree Volume. Consider again the data in Example 13.1. The histogram of the 200,000 bootstrap sample mean tree volumes displayed in Figure 13.3 is approximately symmetric and centered near the mean of the simulated data (indicated by the tall vertical line). This leads us to expect that in this application, the simple percentile method will provide an acceptable confidence interval for the mean. Applying this method to construct a 95% confidence interval for mean tree volume based on the 200,000 bootstrap means displayed in the histogram, one takes the 0.025 and 0.975 quantiles of this empirical bootstrap distribution as the desired lower and upper confidence bounds, respectively. Using the method in Section 13.3.3 leads to the 5,000th (i.e., $0.025 \times 200,000$) and 195,000th (i.e., $0.975 \times 200,000$) ordered observations, resulting in the 95% confidence interval of $[\widehat{\mu}^{*}_{(0.025)}, \ \widehat{\mu}^{*}_{(0.975)}] = [0.087, \ 0.133]$ for the mean tree volume. It might be argued that for this example one does not need bootstrap methods to construct a confidence interval for mean tree volume because such an interval can be easily obtained using the well-known traditional method, assuming an underlying normal distribution (see Section 4.2). Moreover, due to the central limit theorem and the sample size ($n = 29$), the assumption of normality of the sampled population is not expected to be critical. However, because there is evidence of skewness in the original data (Figure 13.1), the nonparametric bootstrap method is more accurate and requires little or no extra cost of computation. Thus, both 95% confidence intervals were calculated for this application and are compared in Table 13.3, along with other confidence intervals that will be discussed shortly. The differences among the intervals in this example might, however, be judged to be sufficiently small so as to be of little practical importance. ∎

Adequacy of simple percentile method

Statistical theory suggests that the simple bootstrap percentile method works well when there exists a monotone transformation of the point estimator of the quantity of interest (although the specific form of this transformation does not need to be known) and the distribution of this transformed estimator has a normal distribution with a median equal to the transformed median of the bootstrap distribution of the quantity of interest. Furthermore, it is required that the variance of this transformed estimator does not depend on the value of the quantity of interest. When such a transformation does not exist, a bootstrap-based confidence interval using the simple percentile method could be incorrect in one or both of two ways:

- The coverage probability differs from the specified nominal confidence level. For example, a nominal 95% simple percentile bootstrap confidence interval procedure may have a coverage probability that differs appreciably from 0.95.

- The two tail probabilities differ appreciably. Then, even though a 95% simple percentile bootstrap confidence interval may have a coverage probability close to 0.95, one calculated tail probability is appreciably less than 0.025, while the other is correspondingly greater than 0.025. This problem is of particular concern when using the simple percentile method to construct one-sided confidence bounds.

Much research has addressed the preceding concerns (see the Bibliographic Notes section at the end of this chapter). This research has yielded more sophisticated approaches for constructing confidence intervals from the empirical bootstrap distribution, such as the methods described next.

13.4.3   The BCa Percentile Method

Motivation for the method

The theory for the simple percentile method described briefly in the preceding subsection requires the existence of a normalizing, unbiasing, and variance-stabilizing transformation. The bias-corrected and accelerated (BCa) method was developed for situations for which the assumptions for use of the simple percentile method are not, at least approximately, satisfied. The BCa method is more complicated to implement than the simple percentile method, but it adjusts for both bias and variance dependency in the distribution of the estimator. Thus, the BCa method provides an improved way to construct nonparametric bootstrap confidence intervals, as compared to the simple percentile method. Technical details are in references given in the Bibliographic Notes section at the end of this chapter.

Description of the method

The basic idea of the BCa method is to replace the quantiles $\alpha/2$ and $1 - \alpha/2$ used in the simple percentile method by the adjusted quantiles α_1 and α_2. In particular, the confidence interval using the BCa percentile method is given by

$$\left[\underset{\sim}{\theta}, \ \widetilde{\theta}\right] = \left[\widehat{\theta}^*_{(\alpha_1)}, \ \widehat{\theta}^*_{(\alpha_2)}\right],$$

where the adjusted quantiles are

$$\alpha_1 = \Phi_{\mathrm{norm}}\left[z_{(\widehat{b})} + \frac{z_{(\widehat{b})} - z_{(1-\alpha/2)}}{1 - \widehat{a}[z_{(\widehat{b})} - z_{(1-\alpha/2)}]}\right], \tag{13.3}$$

$$\alpha_2 = \Phi_{\mathrm{norm}}\left[z_{(\widehat{b})} + \frac{z_{(\widehat{b})} + z_{(1-\alpha/2)}}{1 - \widehat{a}[z_{(\widehat{b})} + z_{(1-\alpha/2)}]}\right]. \tag{13.4}$$

Here $z_{(\alpha)}$ is the α quantile of the standard normal distribution, and $\widehat{b}$ is the fraction of the B values of $\widehat{\theta}^*$ that are less than $\widehat{\theta}$. In the preceding expressions, $z_{(\widehat{b})}$ is the bias-correction value that corrects for median bias in the distribution of $\widehat{\theta}^*$ (on the standard normal scale) and $\widehat{a}$ is the acceleration constant that corrects for the dependence on θ of the variance of the transformed value of $\widehat{\theta}^*$.

   Different methods have been suggested for computing $\widehat{a}$. It can be shown that the acceleration constant is related to the skewness of the sampling distribution of the estimator of the quantity of interest. One frequently used and relatively simple method employs the delete-one-observation-at-a-time "jackknife method" of estimating the skewness coefficient (i.e., the standardized third central moment) of the distribution of the estimator of the quantity of interest. This estimate of the acceleration constant is described, for example, in Efron and Tibshirani (1993, Chapter 11) and is given by

$$\widehat{a} = \frac{\sum_{i=1}^{n}(\widehat{\theta}_{[\cdot]} - \widehat{\theta}_{[i]})^3}{6[\sum_{i=1}^{n}(\widehat{\theta}_{[\cdot]} - \widehat{\theta}_{[i]})^2]^{3/2}}, \tag{13.5}$$

where $\widehat{\theta}_{[i]}$ is the estimate of θ, calculated from the original sample with the ith data point deleted and $\widehat{\theta}_{[\cdot]} = \sum_{i=1}^{n}\widehat{\theta}_{[i]}/n$ is the jackknife sample mean. When the jackknife method is not appropriate for computing the acceleration constant (e.g., when there is censoring), a more complicated method, such as Monte Carlo simulation, is required to estimate its value.

Example 13.4 The BCa Percentile Bootstrap Confidence Interval for Mean Tree Volume. For the tree volume example, $\widehat{b} = 0.5115926$. Thus, $z_{(0.5115926)} = 0.02906232$ and $\widehat{a} = 0.0296337$. These two estimates suggest, respectively, only a small amount of positive bias and right-skewness in the empirical bootstrap distribution of $\widehat{\theta}^*$. Then using $1 - \alpha = 0.95$ and $z_{(0.975)} = 1.959964$ and substituting into (13.3) and (13.4) gives

$$\alpha_1 = \Phi_{\text{norm}}\left[0.02906232 + \frac{0.02906232 - 1.959964}{1 - 0.0296337(0.02906232 - 1.959964)}\right] = 0.03614,$$

$$\alpha_2 = \Phi_{\text{norm}}\left[0.02906232 + \frac{0.02906232 + 1.959964}{1 - 0.0296337(0.02906232 + 1.959964)}\right] = 0.98393.$$

Thus, the BCa percentile bootstrap confidence interval for the mean tree volume is given by the 0.03614 and 0.98393 quantiles of the empirical distribution of $\widehat{\theta}^*$, namely the 7,228th and 196,786th ordered values of the empirical bootstrap distribution, resulting in a 95% confidence interval for mean tree volume of $[0.089, \ 0.135]$. As expected from the relatively small corrections needed, the confidence interval endpoints for the BCa method are, in this example, close to those from the simple percentile method (see Table 13.3). ∎

13.4.4   The BC Percentile Method

Motivation for the method

In some applications (e.g., when some observations are censored) it might be difficult to compute the value $\widehat{a}$ required for the BCa method. In such cases, using the BCa method with $\widehat{a} = 0$ can still provide an important improvement over the simple percentile method with little additional effort. This approach is called the bias-corrected percentile (BC) method.

Description of the method

Setting $\widehat{a} = 0$ in (13.3) and (13.4) gives

$$\alpha_1 = \Phi_{\text{norm}}\left[2z_{(\widehat{b})} - z_{(1-\alpha/2)}\right], \tag{13.6}$$

$$\alpha_2 = \Phi_{\text{norm}}\left[2z_{(\widehat{b})} + z_{(1-\alpha/2)}\right]. \tag{13.7}$$

Example 13.5   The BC Percentile Bootstrap Confidence Interval for Mean Tree Volume.
Using the same inputs as in Example 13.4 and substituting into (13.6) and (13.7) gives

$$\alpha_1 = \Phi_{\text{norm}}[2 \times 0.02906232 - 1.959964] = 0.02860,$$

$$\alpha_2 = \Phi_{\text{norm}}[2 \times 0.02906232 + 1.959964] = 0.97821.$$

Thus, the BC percentile confidence interval for the mean tree volume is given by the 0.02860 and 0.97821 quantiles of the empirical distribution of $\widehat{\theta}^*$, namely the 5,719th and 195,641th ordered values of the empirical bootstrap distribution, resulting in a 95% confidence interval for mean tree volume of $[0.088, \ 0.134]$. As expected from the relatively small correction needed, the confidence interval endpoints for the BC method are again close to those from the simple percentile method. We also note that, even though the resulting confidence intervals are very similar, the correction in α values is more moderate using the BC method, as compared to the BCa method (see Table 13.3). ∎

13.4.5   The Nonparametric Basic Bootstrap Method

Motivation for the method

The bootstrap method described here is sometimes called the "basic bootstrap method" and is a method as easy to apply as the simple percentile bootstrap method. Suppose that the quantity of interest is θ and that the estimator of θ is $\widehat{\theta}$. This basic method assumes that the distributions of $\widehat{\theta} - \theta$ and $\widehat{\theta}^* - \widehat{\theta}$ are approximately the same.

Description of the method

The confidence interval using the basic bootstrap method is

$$\left[\underset{\sim}{\theta}, \ \widetilde{\theta}\right] = \left[2\widehat{\theta} - \widehat{\theta}^*_{(1-\alpha/2)}, \ 2\widehat{\theta} - \widehat{\theta}^*_{(\alpha/2)}\right],$$

where the quantiles $\widehat{\theta}^*_{(\alpha/2)}$ and $\widehat{\theta}^*_{(1-\alpha/2)}$ are the same quantiles of the empirical distribution of bootstrap estimates used in (13.2) for the simple percentile method.

When the sampling distribution of $\widehat{\theta}$ is skewed, the distribution of $\widehat{\theta} - \theta$ will depend strongly on the value of θ and thus the motivating assumption given above will not hold. For this reason, this bootstrap method is not recommended for general use.

Example 13.6   The Basic Bootstrap Confidence Interval for Mean Tree Volume. Using $\widehat{\mu} = 0.1091$ (from Table 13.2) and the same quantiles from Example 13.3, the basic bootstrap method gives

$$\left[\underset{\sim}{\mu}, \ \widetilde{\mu}\right] = \left[2\widehat{\mu} - \widehat{\mu}^*_{(0.975)}, \ 2\widehat{\mu} - \widehat{\mu}^*_{(0.025)}\right]$$

$$= [2 \times 0.1091 - 0.1329, \ 2 \times 0.1091 - 0.0873] = [0.085, \ 0.131],$$

again comparing well with the confidence intervals obtained using the other methods (see Table 13.3). ∎

13.4.6   The Nonparametric Bootstrap-t Method

Motivation for the method

The bootstrap-t method is an extension of the basic bootstrap method described in Section 13.4.5 that corrects for skewness of the sampling distribution of $\widehat{\theta}$. The bootstrap-t method does not require specification of an acceleration constant and is thus conceptually simpler and more broadly applicable than the BCa method. Also, theoretical assessments have shown the bootstrap-t method to be a strong competitor to the BCa method in providing a coverage probability that is close to the nominal confidence level. In order to use the bootstrap-t method it is, however, necessary to be able to estimate the standard error of the estimate of the quantity of interest from each bootstrap sample.

Description of method

Suppose that the quantity of interest is θ and that from the given data one can compute the estimate $\widehat{\theta}$ and $\widehat{se}_{\widehat{\theta}}$, a corresponding estimate of the standard error of $\widehat{\theta}$. Then the bootstrap estimates $\widehat{\theta}_j^*$ and their corresponding estimated standard errors $\widehat{se}_{\widehat{\theta}^*,j}$ are computed from each bootstrap sample $j = 1, \ldots, B$. From these, the bootstrap-t (studentized) statistics

$$z_{\widehat{\theta},j}^* = \frac{\widehat{\theta} - \widehat{\theta}_j^*}{\widehat{se}_{\widehat{\theta}^*,j}}, \quad j = 1, \ldots, B, \tag{13.8}$$

are computed. The bootstrap-t confidence interval is

$$\left[\underset{\sim}{\theta}, \ \widetilde{\theta} \right] = \left[\widehat{\theta} + z_{\widehat{\theta}_{(\alpha/2)}}^* \ \widehat{se}_{\widehat{\theta}}, \ \widehat{\theta} + z_{\widehat{\theta}_{(1-\alpha/2)}}^* \ \widehat{se}_{\widehat{\theta}} \right],$$

where $z_{\widehat{\theta}_{(\alpha)}}^*$ is the α quantile of the empirical distribution of the $z_{\widehat{\theta},j}^*$, $j = 1, \ldots, B$, values.

Intuitively, the bootstrap-t method can be thought of as a refinement of the Wald method described in Section 12.6. Instead of using a normal distribution approximation to the t-like statistic, the distribution is approximated by a bootstrap simulation. Like the Wald method, the confidence intervals generated by the bootstrap-t method depend on the transformation scale used in constructing the intervals. Thus, unlike the simple percentile and the BCa and BC percentile methods, the bootstrap-t method does *not* have the important property of being transformation invariant.

Example 13.7   The Bootstrap-t Confidence Interval for Mean Tree Volume. The sample mean and standard deviation of the tree volume data were previously calculated as $\widehat{\mu} = 0.1091$ and $\widehat{\sigma} = 0.0624$. Then $\widehat{se}_{\widehat{\mu}} = 0.0624/\sqrt{20} = 0.0116$. Using (13.8), the bootstrap-t statistics

$$z_{\widehat{\mu},j}^* = \frac{\widehat{\mu} - \widehat{\mu}_j^*}{\widehat{se}_{\widehat{\mu}^*,j}}$$

were computed for $j = 1, \ldots, 200{,}000$. The 0.025 and 0.975 quantiles of the empirical distribution of the $z_{\widehat{\mu}}^*$ values are $z_{\widehat{\mu}_{(0.025)}}^* = -2.3677$ and $z_{\widehat{\mu}_{(0.975)}}^* = 1.8266$. Note that these quantiles differ appreciably from the ∓ 1.96 quantiles of the standard normal distribution used in the traditional normal distribution and Wald methods. Then the 95% bootstrap-t confidence interval for the mean tree volume is

$$\left[\underset{\sim}{\mu}, \ \widetilde{\mu} \right] = \left[\widehat{\mu} + z_{\widehat{\mu}_{(0.025)}}^* \ \widehat{se}_{\widehat{\mu}}, \ \widehat{\mu} + z_{\widehat{\mu}_{(0.975)}}^* \ \widehat{se}_{\widehat{\mu}} \right]$$
$$= [0.1091 - 2.3677 \times 0.0116, \ 0.1091 + 1.8266 \times 0.0116] = [0.082, \ 0.130]. \quad \blacksquare$$

As previously observed, the various nonparametric bootstrap confidence intervals for estimating the mean in the tree volume example turned out to be similar to each other (and were

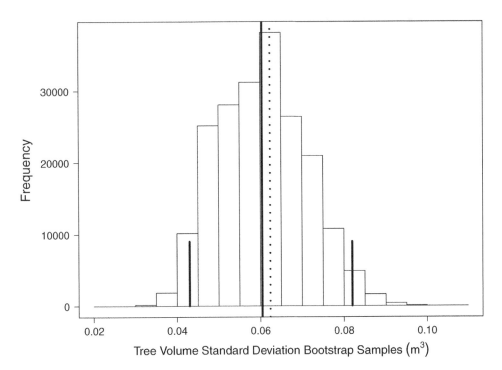

Figure 13.4 Histogram of the sample standard deviations from $B = 200,000$ bootstrap resamples for the tree volume data, showing the sample standard deviation of the original data (tall dotted vertical line), the median or 0.50 quantile (tall solid vertical line), and the 0.025 and 0.975 quantiles (short vertical lines) of the empirical distribution of the bootstrap sample standard deviations.

also close to the traditional normal distribution confidence interval and to the parametric bootstrap confidence intervals still to be discussed). This similarity is despite the differences in the manner in which the bootstrap samples were used to compute the confidence intervals, as seen from Table 13.3. This is not always the case, as demonstrated by the following example.

Example 13.8   Bootstrap Methods for Computing a Confidence Interval for the Standard Deviation of Tree Volume. To calculate the bootstrap confidence interval for σ, we again draw $B = 200,000$ bootstrap resamples of size $n = 29$ (with replacement) from the original data. We then compute $\widehat{\sigma}^*$ for each resample. Figure 13.4 is a histogram of the resulting $\widehat{\sigma}^*_j$, $j = 1, \ldots, 200,000$. The solid and dotted vertical lines in the middle of Figure 13.4 are the sample median of the $\widehat{\sigma}^*$ values and $\widehat{\sigma}$, the sample standard deviation of the original data, respectively. The short vertical lines indicate the 0.025 and 0.975 quantiles of the empirical distribution of the $\widehat{\sigma}^*$ values, and show the simple percentile bootstrap 95% confidence interval for σ. This interval is $[0.042, \ 0.082]$.

The distance between the two vertical lines in the center of Figure 13.4 suggests some median bias in the bootstrap sample estimates and the likely need for a bias correction. Applying the BCa method for determining the adjusted quantiles of the empirical bootstrap distribution, we obtain $\widehat{b} = 0.579733$ and $z_{(0.579733)} = 0.20121$, indicating some positive median bias in the estimate. From (13.5), the jackknife estimate of the acceleration factor is $\widehat{a} = 0.131735$, indicating some right-skewness in the sampling distribution of $\widehat{\sigma}^*$. Then using $z_{(0.975)} = 1.959964$ for

Method	Section	Interval
Parametric traditional normal distribution	4.2	[0.050, 0.086]
Nonparametric simple percentile bootstrap	13.4.2	[0.042, 0.082]
Nonparametric BCa percentile bootstrap	13.4.3	[0.047, 0.094]
Nonparametric BC percentile bootstrap	13.4.4	[0.045, 0.086]
Nonparametric basic bootstrap	13.4.5	[0.043, 0.083]

Table 13.4 Approximate 95% confidence intervals, using different methods, for σ, the population standard deviation of tree volume.

$1 - \alpha = 0.95$ and substituting into (13.3) and (13.4) gives

$$\alpha_1 = \Phi_{\text{norm}} \left[0.20121 + \frac{0.20121 - 1.959964}{1 + 0.131735(0.20121 - 1.959964)} \right] = 0.109966,$$

$$\alpha_2 = \Phi_{\text{norm}} \left[0.20121 + \frac{0.20121 + 1.959964}{1 + 0.131735(0.20121 + 1.959964)} \right] = 0.999365.$$

Thus the BCa percentile bootstrap confidence interval is given by the 0.109966 and 0.999365 quantiles, or the 21,993th and 199,872th ordered observations, of the empirical distribution of $\widehat{\sigma}^*$. This yields a 95% confidence interval for σ of [0.047, 0.094]. Similarly, the BC percentile bootstrap confidence interval is given by the 0.0596707 and 0.9909211 quantiles, or the 11,934th and 198,184th ordered observations, of the empirical distribution of $\widehat{\sigma}^*$, yielding a 95% confidence interval for σ of [0.045, 0.086]. The basic method, using simple computations similar to those used in Example 13.6, with $\widehat{\sigma} = 0.06242$ gives a 95% confidence interval for σ of [0.043, 0.083]. We did not use the bootstrap-t method in this example because of the difficulty of obtaining a nonparametric estimate for the standard error of $\widehat{\sigma}$.

The preceding results, together with the traditional normal distribution confidence interval (see Section 4.3) are shown in Table 13.4. We note that the various bootstrap confidence intervals vary considerably from each other and also from the traditional normal distribution interval. Following our discussion in Section 13.4.1 (and lacking an interval calculated using the bootstrap-t method), we suggest that the practitioner seeking a single interval use the interval obtained by the BCa percentile bootstrap method. ∎

13.4.7 Nonparametric Bootstrap Confidence Intervals for a Correlation Coefficient

Section 12.6.5 showed, with the body weight and pulse rate data in Example 12.22, how to construct a confidence interval for a correlation coefficient using the Wald-approximation method. This approach assumed a joint normal distribution for body weight and pulse rate—an assumption that seems reasonable from physical considerations, but could not be verified with much confidence from the given sample of only 20 observations. We also note that the Wald method for computing a confidence interval for the correlation coefficient requires the knowledge that the Fisher z-transformation is an appropriate normalizing and variance-stabilizing transformation, thus allowing this Wald approximation to be reasonably accurate. The BCa method for obtaining a confidence interval for a correlation coefficient method works automatically without such assumptions, knowledge, or other inputs and also has a higher degree of theoretical accuracy. We, therefore, now construct nonparametric bootstrap confidence intervals for the correlation coefficient for this example.

Example 13.9   Nonparametric Bootstrap Methods for Computing a Confidence Interval for the Correlation between Body Weight and Pulse Rate. To apply the bootstrap method to obtain a confidence interval for the correlation between body weight and pulse rate for the population sampled in Example 12.22, we again begin by drawing $B = 200{,}000$ resamples of size $n = 20$ (with replacement) from the original data and compute $\widehat{\rho}^*$ for each such resample. A histogram of the resulting bootstrap sample values of $\widehat{\rho}^*$ (not shown here) was again reasonably symmetric and centered near $\widehat{\rho}$, suggesting that the simple percentile method should be adequate for selecting the appropriate bootstrap distribution quantile. Thus, applying this method to obtain a 95% confidence interval, we use the 0.025 and 0.975 quantiles (the 5,000th and 195,000th ordered observations) of the empirical distribution of the $\widehat{\rho}^*$ values to obtain a 95% confidence interval for ρ of $[-0.660, \ -0.0001]$.

Applying the BCa method for determining the appropriate quantile of the empirical bootstrap distribution, we obtain $\widehat{b} = 0.521788$ and $z_{(0.519576)} = 0.05464061$, indicating a small amount of positive bias, and $\widehat{a} = -0.03051945$, indicating a small amount of left-skewness in the distribution of $\widehat{\rho}^*$. Then using $z_{(0.975)} = 1.959964$ for $1 - \alpha = 0.95$ and substituting into (13.3) and (13.4) gives

$$\alpha_1 = \Phi_{\text{norm}}\left[0.05464061 + \frac{0.05464061 - 1.959964}{1 + 0.03051945(0.05464061 - 1.959964)}\right] = 0.02452,$$

$$\alpha_2 = \Phi_{\text{norm}}\left[0.05464061 + \frac{0.05464061 + 1.959964}{1 + 0.03051945(0.05464061 + 1.959964)}\right] = 0.97456.$$

Thus the BCa percentile bootstrap confidence interval is given by the 0.02452 and 0.97456 quantiles, or the 4,903th and 194,912th ordered observations, of the distribution of $\widehat{\rho}^*$, yielding a 95% confidence interval for ρ of $[-0.661, \ 0.0019]$. Similarly, the BC percentile bootstrap confidence interval is obtained from the 0.03211 and 0.98074 quantiles, or the 6,421th and 196,147th ordered observations, of the distribution of $\widehat{\rho}^*$, yielding a 95% confidence interval for ρ of $[-0.645, \ 0.0260]$. Finally, the bootstrap-t method when based on Fisher's z-transformation (computational details not shown here), results in a confidence interval of $[-0.660, \ -0.0001]$. Interestingly, this interval is identical to the interval obtained by the simple percentile method. This equivalence arises because for Fisher's z-transformation, $\widehat{se}_{\widehat{z}} = [1/(n-3)]^{1/2}$ does not depend on $\widehat{z}$ (see Section 12.6.5).

The preceding results, together with the Wald interval, are summarized in Table 13.5. Note that the Wald method used in Section 12.6.5 required the knowledge that Fisher's z-transformation was an appropriate normalizing and variance-stabilizing transformation, allowing the Wald method to provide a reasonably accurate procedure for computing a confidence interval for the correlation coefficient. The BCa method works automatically without such knowledge or other inputs and also has a higher degree of theoretical accuracy. Again, as expected from the small corrections needed, the 95% confidence intervals obtained by the simple percentile, BCa, and bootstrap-t methods are all similar to one another. The 95% confidence interval obtained using the BC method is somewhat different from the other nonparametric bootstrap confidence intervals because, in this application, the bias correction and the acceleration correction in the BCa method correct in opposite directions. However, all of the nonparametric bootstrap 95% confidence intervals for ρ differ somewhat from the interval obtained using the Wald method.

Method	Section	Interval
Wald	12.6.5	$[-0.696, \quad 0.0916]$
Nonparametric simple percentile bootstrap	13.4.2	$[-0.660, \; -0.0001]$
Nonparametric BCa percentile bootstrap	13.4.3	$[-0.661, \; -0.0019]$
Nonparametric BC percentile bootstrap	13.4.4	$[-0.645, \quad 0.0260]$
Nonparametric basic method	13.4.5	$[-0.731, \; -0.0720]$
Nonparametric bootstrap-t	13.4.6	$[-0.660, \; -0.0001]$

Table 13.5 Approximate 95% confidence intervals for the correlation between body weight and pulse rate.

Finally, we observe that the 95% confidence intervals on ρ using the Wald and the BC bootstrap methods include the value $\rho = 0$ and those obtained by the other methods do not. Thus ρ would be deemed statistically significantly different from 0 at the 5% significance level using the simple percentile, BCa, basic, and bootstrap-t methods, but not using the Wald or BC bootstrap methods. ∎

13.4.8 Cautions on the Use of Nonparametric Bootstrap Confidence Interval Methods

A key advantage of using nonparametric bootstrap methods to construct statistical intervals, as compared, for example, to parametric bootstrap methods (to be discussed in Chapter 14) is that they do *not* require one to assume a statistical distribution underlying the observed data. The statistical theory that justifies the use of nonparametric bootstrap methods does, however, depend on several other conditions. These include that:

- The observations come from a distribution with a finite variance.

- The observations are statistically independent.

- The number of observations in the given data set is sufficiently large.

- The statistic being bootstrapped is a smooth function of the observations.

- An appropriate scale, depending on the possible use of a transformation, has been chosen for computing the interval.

The parametric bootstrap methods in Chapter 14, as we shall see, have fewer restrictions and thus are more flexible than the nonparametric bootstrap methods, at the cost of requiring one to assume a particular parametric distribution.

Finite variance requirement

The requirement that the observations come from a distribution with finite variance is met in most practical applications.

Independence assumption

Data from a simple random sample from a population generally result in statistically independent observations. Unfortunately, in many applications one does not have a random sample from the population or process of interest. Also, times series data are often autocorrelated and, therefore, do not provide statistically independent observations. Similarly, spatial data generally have spatial correlations. Special bootstrap methods have been developed for applications involving correlated data. These use sample blocks that retain the approximate correlation structure of the data; see the references given in the Bibliographic Notes section at the end of this chapter for further details.

Sample size requirements

In general, to obtain comparable precision, nonparametric statistical methods require larger samples than do parametric methods. To effectively use bootstrap resampling methods, moreover, requires that the sample size be large enough so that the number of possible unique resamples is sufficiently large. Thus, it is generally suggested that one have a sample size of at least $n = 10$ observations. For a detailed discussion of this issue, see Hall (1992, Appendix I). Another reason for needing a sufficiently large sample is that the theoretical basis for the bootstrap procedures depends on large-sample theory (as do the methods in Chapter 12). This theory states that, under certain conditions, a bootstrap procedure has a coverage probability that approaches the nominal confidence level as the sample size increases. The theory, depending on specified conditions, can also give the rate of convergence.

Smoothness

A further requirement for nonparametric bootstrap methods to provide accurate results for moderate size samples is that the statistic being estimated be a smooth function of the observations. This is the case for moment statistics such as sample means, variances, correlations and ordinary least squares regression estimators.

Unfortunately, many other statistics that are often desired to be calculated from the data, such as sample quantiles, are *not* smooth functions of the observations. Figure 13.5 is a histogram of sample medians computed from $B = 200,000$ bootstrap resamples generated from the 29 tree volumes in Table 13.1. This plot is in sharp contrast to the smooth empirical bootstrap distribution of the sample means in Figure 13.3. The fact that over 80,000 resample medians are between 0.10 and 0.11 and over 20,000 are between 0.08 and 0.09, and yet none are between 0.09 and 0.10, seems, at first glance, hard to believe. Yet there is a simple explanation. Because $n = 29$ is odd, the sample median will be one of the observations in the given data. But examining the original 29 observations in Table 13.1, we note that there are no observations between 0.09 and 0.10. Therefore, it is *impossible* for a resample median to be between 0.09 and 0.10. We note that some small changes in the observed data (e.g., the observed value of 0.102 being changed to 0.098) could result in an appreciable change in the form of the empirical bootstrap distribution shown in Figure 13.5.

The nonsmooth or discrete-like nature of the nonparametric bootstrap distribution for the sample median suggests that application of bootstrap methods may yield erroneous results in constructing a confidence interval for the distribution median, even with $n = 29$ observations, and undoubtedly would be worse for smaller n or if we were estimating a tail quantile.

With larger samples, the problem is less severe, but convergence to yield well-behaved results may be slow and, for some situations (e.g., estimating distribution tail quantiles), sample sizes that only moderately exceed 29 may be insufficient to provide a procedure that has a coverage probability close to the nominal confidence level. Methods to improve nonparametric bootstrap methods by introducing smoothing into the bootstrap procedure have been developed. Again, these are discussed in some of the references given in the Bibliographic Notes section at the end of this chapter. As an alternative to bootstrap methods in obtaining nonparametric confidence

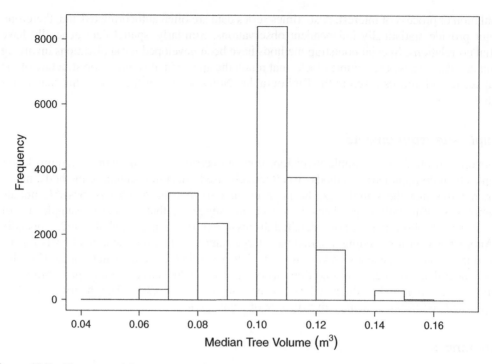

Figure 13.5   Histogram of the sample medians from $B = 20,000$ nonparametric bootstrap resamples from the tree volume data.

intervals for a distribution tail quantile, we recommend the simple order statistics method with interpolation described in Section 5.2.

Transformation choice

The simple percentile, BCa, and BC methods are *transformation invariant* (also known as *transformation preserving*). This implies that if you construct a confidence interval for a transformed scale (such as the log of concentration) and then convert the endpoints back to the original scale, you obtain the same interval that you would get by applying the method without the transformation. The basic bootstrap and percentile-t methods, however, are not transformation preserving. Thus, as with Wald intervals, one must decide on which scale to use. Generally a transformation that transforms a bounded parameter range to an unbounded range is a good choice (e.g., log of a positive quantity or Fisher's z-transformation for a quantity bounded between -1 and 1).

BIBLIOGRAPHIC NOTES

History

This chapter provides only an introduction to the bootstrap methods for constructing statistical intervals. Since the first paper on bootstrap methods by Efron (1979), numerous, frequently more advanced, bootstrap and related methods have been proposed. Also, entire books about bootstrap methods in general, and computing statistical intervals using such methods in particular, have been written. One recent book (Chernick, 2008) devotes 86 and 54 pages, respectively, of its 329 pages (excluding the index) to references prior to 1999 and from 1999 to 2007, respectively; although not all of these references deal with bootstrap methods, many do.

General introductions

Chernick (2008) provides an introduction to bootstrap methods aimed at practitioners. This book begins with a discussion of bootstrap history and of the variety of applications, including some relatively advanced methods, such as nonlinear regression, time series analysis, and spatial data analysis. Each chapter ends with a section containing historical notes. As already mentioned, this book also includes an extensive bibliography. Hesterberg (2015) provides an overview and description of nonparametric bootstrap methods, providing intuitive and graphical explanations of situations for which bootstrap methods do not work well.

Theory

Efron and Tibshirani (1986) provide a reasonably accessible overview of bootstrap methods. Efron and Tibshirani (1993) is a more detailed book-length treatment. Also Davison and Hinkley (1997) give a comprehensive presentation of bootstrap and related methods, along with a large number of applications and special topics, such as applying bootstrap methods to correlated data. Hall (1992) and Shao and Tu (1995) present the theory of bootstrap methods and related topics. DasGupta (2008, Chapter 18) provides a concise summary of many of the important theoretical results related to bootstrap methods. Lahiri (2003) describes theory and methods of applying bootstrap methods to problems involving correlated data. Beran (2003) and Efron (2003) are somewhat more recent review articles on bootstrap methods. Efron (2003) provides a retrospective view of bootstrap methods, 25 years after the initial publications on the subject.

A few classic theoretical papers deserve special note. Efron (1987) describes the theory behind the BCa method and shows that the method has second-order correctness over a range of different situations. Hall (1988) presents a theoretical framework for comparing the properties of different bootstrap procedures. His results show that both the BCa percentile and the bootstrap-t methods are second-order correct as parametric bootstrap methods for the multivariate exponential family of distributions and as nonparametric bootstrap methods in situations for which the estimators can be written as a function of a multivariate vector of means. This paper also shows that simpler nonparametric bootstrap methods, such as the simple percentile method, have inferior asymptotic properties, compared to more advanced methods, such as the BCa percentile method.

Hall and Martin (1989) show why the standard nonparametric bootstrap methods do not work well for calculating confidence intervals for distribution quantiles. Ho and Lee (2005) provide theoretical arguments demonstrating that refinements such as smoothing and iterated bootstrap can be used to obtain nonparametric bootstrap confidence interval procedures for distribution quantiles that have improved performance. The refinements and amount of improvement, however, depend on the specification of appropriate bandwidth values for the required smoothing operations. The need for specifying these values make practical implementation difficult.

Random-weight bootstrap

Rubin (1981) presents a Bayesian analog to nonparametric bootstrap sampling. In his method, log-likelihood terms are randomly weighted to obtain simulated posterior distributions. Weights are generated from a uniform Dirichlet distribution. Newton and Raftery (1994) apply similar ideas to a sequence of parametric inference examples. Even though these random-weight bootstrap methods were developed within a nonparametric Bayesian framework, they also apply to non-Bayesian and parametric inference problems, as will be illustrated in Chapter 14. Jin et al. (2001) show that random-weight bootstrap estimators have good properties if positive independent and identically distributed weights are generated from a continuous distribution that has the same mean and standard deviation (e.g., an exponential distribution with mean 1). Chatterjee and Bose (2005) present a generalized bootstrap for which the traditional resampling and various weighted likelihood and other weighted estimating equation methods are special cases. Barbe and Bertail (1995) provide a highly technical presentation of the asymptotic theory of various random-weight methods for generating bootstrap estimates. They show how to choose the distribution of the random weights by using Edgeworth expansions. Chiang et al. (2005) apply random-weight bootstrap methods to a recurrent events application with informative censoring in a semi-parametric model. Hong et al. (2009) apply random-weight bootstrap methods to a prediction interval application involving complicated censoring and truncation.

Chapter *14*

Parametric Bootstrap and Other Simulation-Based Statistical Intervals

OBJECTIVES AND OVERVIEW

This chapter describes and illustrates computationally intensive *parametric bootstrap* and other simulation-based methods to compute statistical intervals, primarily for continuous distributions. These methods, like the nonparametric bootstrap methods in Chapter 13, require obtaining a sequence of simulated *bootstrap samples* based on the given data that are then used to generate corresponding *bootstrap estimates*. The parametric bootstrap procedures presented in this chapter (like all of the other chapters in this book other than Chapters 5 and 13) require one to specify a parametric distribution for the given data. These methods can lead to excellent approximate, or sometimes exact, procedures for computing statistical intervals, even for small samples, when the chosen distribution is correct. Parametric bootstrap methods may, however, result in misleading answers if the chosen distribution is seriously in error.

The topics discussed in this chapter are:

- The basic concept of using simulation and parametric bootstrap methods to obtain confidence intervals (Section 14.1).

- Methods for generating parametric bootstrap samples and obtaining bootstrap estimates (Section 14.2).

- How to obtain parametric confidence intervals by using the simulated distribution of a pivotal quantity (Section 14.3).

- How to obtain parametric confidence intervals by using the simulated distribution of a generalized pivotal quantity (Section 14.4).

Statistical Intervals: A Guide for Practitioners and Researchers, Second Edition.
William Q. Meeker, Gerald J. Hahn and Luis A. Escobar.
© 2017 John Wiley & Sons, Inc. Published 2017 by John Wiley & Sons, Inc.
Companion Website: www.wiley.com/go/meeker/intervals

- Methods for using simulation to compute *parametric* tolerance intervals (Section 14.5).

- Methods for using simulation to compute *parametric* prediction intervals (Section 14.6).

- Discussion of other simulation and parametric bootstrap methods and applications for other than the (log-)location-scale distributions discussed in the preceding sections (Section 14.7).

14.1   INTRODUCTION

In Chapter 12 we presented approximate likelihood-based and commonly used normal distribution approximation (Wald) statistical intervals for situations for which exact procedures may be unavailable or difficult to obtain. The computer-intensive parametric bootstrap and simulation-based methods described in this chapter provide alternatives for constructing approximate (and, for some special cases, exact) confidence intervals for distribution parameters, as well as for other characteristics that are functions of distribution parameters (such as the mean, quantiles, and probabilities). Parametric bootstrap procedures can also be used to compute tolerance and prediction intervals.

Specifically, when exact methods for obtaining statistical intervals are not readily available, one generally employs an approximate method, resulting in a procedure that will have a coverage probability that is approximately equal to the desired nominal confidence level (see Appendix B for details). The popular Wald confidence interval procedures described in Section 12.6 may be adequate for initial casual or informal analyses, particularly when the sample size is large. But, as discussed in Chapter 12, likelihood-based methods for constructing confidence intervals, in general, outperform the Wald methods. Bootstrap methods provide useful alternatives to both Wald and likelihood-based methods and may yield more accurate approximate confidence interval procedures. We can, in addition, sometimes apply bootstrap methods for situations for which other reasonable alternatives do not exist (e.g., when likelihood-based confidence interval methods are too demanding computationally).

As in other chapters, our discussion in this section (and subsequent sections in this chapter) deals primarily with two-sided confidence intervals. As described in Section 2.7, however, one can obtain a one-sided lower (or upper) confidence bound from the corresponding two-sided interval by substituting α for $\alpha/2$ in the expression for the lower (or upper) endpoint of the two-sided interval.

14.1.1   Basic Concepts

The general idea behind the parametric bootstrap and simulation-based procedures presented in this chapter is to replace mathematical approximations or intractable distribution theory with Monte Carlo simulation, taking advantage of the power of modern computers and relatively recent developments in statistical theory.

As explained in Section 2.2, a key criterion for judging an approximate procedure for constructing a statistical interval is how well the procedure would perform if it were repeated over and over. The coverage probability (i.e., the probability that the procedure provides an interval that contains the quantity of interest) should be equal or close to the chosen nominal confidence level $1 - \alpha$. Moreover, we generally favor two-sided intervals for which the error probability α is split equally or approximately equally between the upper and lower interval bound; that is the probability of being incorrect is close to $\alpha/2$ for each side of the interval.

In practice, we usually cannot actually repeat the sampling process over and over. We can, however, simulate the sampling process to create bootstrap samples. Then the empirical sampling distribution of the appropriate statistics from the resulting bootstrap samples is used to

| 0.289 | 0.281 | 0.315 | 0.319 | 0.311 | 0.323 | 0.296 | 0.323 | 0.311 | 0.266 |
| 0.259 | 0.345 | 0.330 | 0.304 | 0.293 | 0.248 | 0.304 | | | |

Table 14.1   Pipeline thickness measurements (inches).

compute the desired statistical interval, reducing the reliance on sometimes crude large-sample approximations.

For a simple example, consider again the discussion in Section 12.5.2. There the Wald confidence interval for μ was based on the fact that in large samples the distribution of $Z_{\widehat{\mu}} = (\widehat{\mu} - \mu)/\widehat{se}_{\widehat{\mu}}$ can be approximated by a $\mathrm{NORM}(0, 1)$ distribution. An alternative to this approximation is to use a bootstrap approach. Bootstrap methods using a sufficiently large number of bootstrap samples can provide improved approximations to the *actual* distribution of $Z_{\widehat{\mu}}$ and, therefore, a better confidence interval procedure for μ—especially when dealing with small data sets.

Some important operational considerations for using bootstrap methods are given in the nonparametric bootstrap chapter (Section 13.3); these considerations also apply to the simulation and parametric bootstrap methods given in this chapter.

14.1.2   Motivating Examples

Example 14.1   Estimating Pipeline Thickness. Table 14.1 shows pipeline thickness measurements taken at $n = 17$ randomly selected one-foot segments along a multi-mile-long pipeline. The recorded value for each segment is the minimum thickness found for that segment. To assess the risk of a leak, it was desired to estimate the probability that the minimum thickness of a randomly selected one-foot segment taken from the entire pipeline will be less than 0.15 inches, as well as the 0.0001 quantile of the one-foot segment minimum thickness distribution, together with confidence intervals on these estimates. Based on previous experience and extreme value theory (e.g., Coles, 2001), the Weibull distribution was believed to provide an adequate description of the distribution of minimum thickness for such one-foot segments. This assumption is critical in light of the extreme extrapolations from the data required to obtain the desired estimates. Figure 14.1 shows a Weibull probability plot for the $n = 17$ pipeline thickness measurements. The plot and maximum likelihood (ML) line fitted to the plotted points suggest that the Weibull distribution provides an excellent fit to the distribution of minimum pipeline thickness for one-foot segments, at least within the range of the data. Equivalently, the smallest extreme value (SEV) distribution (see Section C.3) provides an excellent fit to the logs of the measurements. The ML estimates of the parameters of the SEV distribution fit to the logs of the thickness values are $\widehat{\mu} = -1.16406$ and $\widehat{\sigma} = 0.07003$. The corresponding Weibull distribution ML estimates are $\widehat{\eta} = \exp(\widehat{\mu}) = 0.3122$ and $\widehat{\beta} = 1/\widehat{\sigma} = 14.28$. For simplicity we henceforth use the term "pipeline thickness" to denote the minimum thickness within each of the one-foot segments that make up the entire pipeline.   ■

Example 14.2   Fracture Strengths of a Carbon-Epoxy Composite Material. Table 14.2 gives data described in Dirikolu et al. (2002) on the fracture strengths of a random sample of 19 specimens of a carbon-epoxy composite material from a specified population. The Weibull probability plot of the data, shown in Figure 14.2, suggests that the Weibull distribution provides an adequate fit to fracture strength within the range of the data. A Weibull distribution is also suggested on physical grounds because the composite material is brittle and failures tend to occur at the weakest part of the structure. The ML estimates of the Weibull distribution parameters are $\widehat{\eta} = 510.2$ and $\widehat{\beta} = 18.86$.   ■

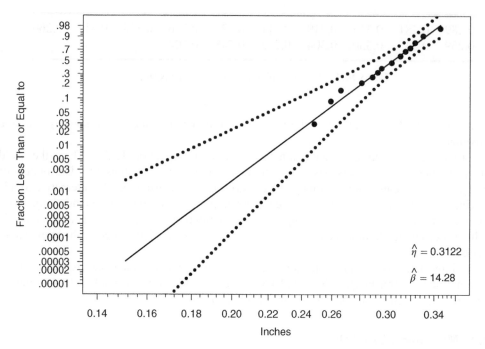

Figure 14.1 Weibull probability plot of the pipeline thickness data described in Example 14.1 and corresponding pointwise 95% parametric bootstrap confidence intervals for $F(t)$, as described in Example 14.7.

14.2 PARAMETRIC BOOTSTRAP SAMPLES AND BOOTSTRAP ESTIMATES

Statistical intervals are computed as a function of the available data, consisting of n observations denoted by DATA. Similar to the nonparametric bootstrap methods in Chapter 13, parametric bootstrap and other simulation-based interval procedures employ, in addition, a set of B samples, DATA_j^*, $j = 1, \ldots, B$, generated by Monte Carlo simulation, based on the given data, that, in some sense (depending on the type of bootstrap/simulation procedure) mimics the original sampling procedure.

Sections 13.2.1 and 13.2.2 show how to generate nonparametric bootstrap samples. Nonparametric bootstrap samples can also be used with parametric bootstrap methods. As explained and illustrated in Section 14.7, there are often important advantages to using one of the nonparametric bootstrap sampling methods when data are complicated (e.g., involving censoring or truncation).

This section describes a parametric simulation-based method of generating bootstrap samples that can be used in situations for which there is no censoring or when censoring is easy to simulate. Using this method will, in some cases, lead to statistical interval procedures that are exact.

532.7	502.5	442.0	47.03	519.0	502.7	477.0	510.0	522.0	552.0
522.0	439.0	513.6	497.5	521.6	450.9	476.5	507.3	463.5	

Table 14.2 Fracture strength of carbon-epoxy composite material specimens (megapascals).

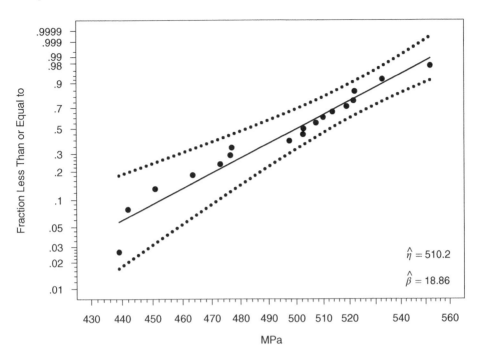

Figure 14.2 Weibull probability plot of carbon-epoxy composite material fracture strengths and pointwise 95% parametric bootstrap confidence intervals for $F(t)$.

Figure 14.3 illustrates the parametric bootstrap sampling method of obtaining bootstrap samples and bootstrap estimates. With parametric bootstrap sampling, one initially uses the n data cases to compute the ML estimate $\widehat{\boldsymbol{\theta}}$ of the unknown parameter vector $\boldsymbol{\theta}$. An estimate of the assumed underlying parametric distribution $F(t; \boldsymbol{\theta})$ is $F(t; \widehat{\boldsymbol{\theta}})$. Then B bootstrap samples of size n are simulated from $F(t; \widehat{\boldsymbol{\theta}})$ and these are denoted by DATA_j^*, $j = 1, \ldots, B$. For each of these B samples, the ML bootstrap estimate of the parameter vector, denoted by $\widehat{\boldsymbol{\theta}}_j^*$, is computed.

Let $v(\boldsymbol{\theta})$ denote a scalar quantity of interest to be estimated (e.g., a mean, standard deviation, probability, or distribution quantile). Bootstrap estimates of $v(\boldsymbol{\theta})$ are denoted by $v(\widehat{\boldsymbol{\theta}}_j^*)$, $j = 1, \ldots, B$. The values of $v(\widehat{\boldsymbol{\theta}}_j^*)$ and/or $\widehat{\boldsymbol{\theta}}_j^*$ can be used, in a variety of ways, to construct parametric bootstrap statistical intervals. Some of the most important methods for doing this are described and illustrated in the remaining sections of this chapter.

The adequacy of the statistical interval procedure obtained using the parametric bootstrap approach could be questionable if the distributional assumption is inadequate. Also, when dealing with data involving censoring or truncation (or some other special feature), it is necessary to completely specify (in a probabilistic model sense) how the censored and/or truncated data were generated so that the bootstrap data can be properly simulated. The advantage of the parametric bootstrap approach is that it can lead to excellent approximate or exact statistical intervals, even for small sample sizes, if the assumed distribution is correct. Also the method is easy to implement when the sampling is simple (e.g., statistically independent observations with no censoring or truncation).

Example 14.3 Parametric Bootstrap Samples for Estimating the Distribution of Pipeline Thickness. Figure 14.4 is a scatter plot of the first 1,000 (out of 200,000) generated pairs of ML

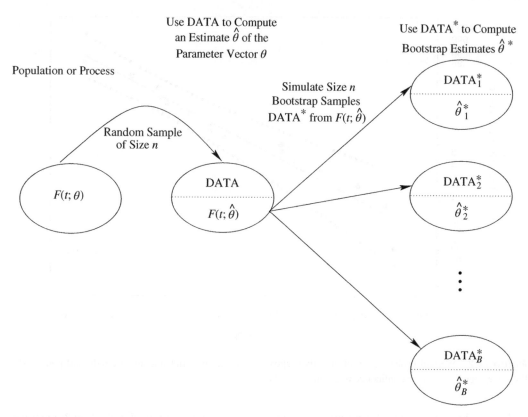

Figure 14.3   Illustration of parametric bootstrap sampling for obtaining bootstrap samples DATA* and bootstrap estimates $\widehat{\theta}^*$.

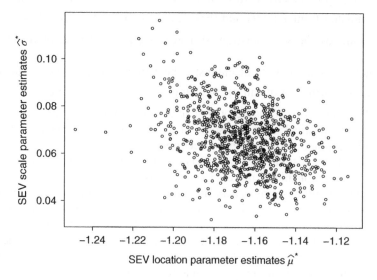

Figure 14.4   Scatter plot of 1,000 pairs of ML estimates $\widehat{\mu}^*$ and $\widehat{\sigma}^*$ from simulated samples of size $n = 17$ from the SEV$(\widehat{\mu}, \widehat{\sigma})$ distribution with $\widehat{\mu} = -1.16406$ and $\widehat{\sigma} = 0.07003$.

estimates $\widehat{\mu}_j^*$ and $\widehat{\sigma}_j^*$, $j = 1, \ldots, B$, from samples of size $n = 17$ (the sample size for the pipeline thickness data) simulated from an $\mathrm{SEV}(\widehat{\mu}, \widehat{\sigma})$ distribution with $\widehat{\mu} = -1.16406$ and $\widehat{\sigma} = 0.07003$ (i.e., the ML estimates obtained in Example 14.1). These simulated values will be used to construct statistical intervals in a series of examples in Sections 14.3–14.6. ∎

14.3 BOOTSTRAP CONFIDENCE INTERVALS BASED ON PIVOTAL QUANTITIES

14.3.1 Introduction

Simulation-based methods for constructing statistical intervals based on pivotal quantities (PQs) pre-date bootstrap methods. With modern computing capabilities (both hardware and software), however, these methods are much more accessible. Because of their similarity to the more general parametric bootstrap methods, these simulation-based methods are often also referred to as *parametric bootstrap* methods.

Chapter 4 presented different types of statistical intervals for the normal distribution. The well-known intervals there are based on PQs, as described in Appendix E. For example, if $\bar{X}$ is the sample mean and S is the sample standard deviation computed from a sample of size n from a normal distribution with population mean μ and standard deviation σ, then $(\mu - \bar{X})/(S/\sqrt{n})$ has a t-distribution with $n - 1$ degrees of freedom. Then, using the definition of quantiles of the t-distribution,

$$\Pr\left[t_{(\alpha/2;n-1)} \leq \frac{\mu - \bar{X}}{S/\sqrt{n}} \leq t_{(1-\alpha/2;n-1)}\right] = 1 - \alpha. \tag{14.1}$$

Because the distribution of $(\mu - \bar{X})/(S/\sqrt{n})$ does not depend on any unknown parameters, we say that it is a PQ. This terminology arises because using simple algebra to pivot the inequality in (14.1) (i.e., arrange to have μ by itself in the center of the inequality) we obtain

$$\Pr\left[\bar{X} + t_{(\alpha/2;n-1)}S/\sqrt{n} \leq \mu \leq \bar{X} + t_{(1-\alpha/2;n-1)}S/\sqrt{n}\right] = 1 - \alpha.$$

This probability statement is a special case of the more general statement (B.1) from Section B.2.1 and implies that

$$\left[\underset{\sim}{\mu}, \; \widetilde{\mu}\right] = \left[\bar{x} + t_{(\alpha/2;n-1)}s/\sqrt{n}, \; \bar{x} + t_{(1-\alpha/2;n-1)}s/\sqrt{n}\right]$$

is a $100(1 - \alpha)\%$ confidence interval for μ, where $\bar{x}$ is the observed value of $\bar{X}$ and s is the observed value of S.

As shown in Appendix E and illustrated in subsequent examples in this chapter, a PQ can also be used to construct other kinds of statistical intervals (e.g., tolerance intervals and prediction intervals) and for other distributions. For distributions other than the normal (and lognormal), however, or when censoring is involved, tables or computer functions for the needed quantiles of the distributions of PQs are generally not available. In these cases one can instead use parametric bootstrap methods, as described in this section, to obtain the desired distribution quantiles. Moreover, in contrast to the likelihood or Wald methods in Chapter 12, the parametric bootstrap simulation methods used in this chapter are exact when they are based directly on PQs and the parametric bootstrap samples described in Section 14.2. Moreover, when the methods are not exact (e.g., because they are based on a generalized PQ, as presented briefly in Section 14.4, or an otherwise approximate PQ), the coverage probability will generally be close to the nominal confidence level, with the approximation improving with larger sample sizes.

14.3.2   Confidence Intervals for the Location Parameter of a Location-Scale Distribution or the Scale Parameter of a Log-Location-Scale Distribution

As described in Section 12.4.1, μ is a location parameter for a location-scale distribution and $\exp(\mu)$ is a scale parameter for a log-location-scale distribution. In a manner similar to the discussion in Section 14.3.1 (see also Section E.7.2), $Z_{\widehat{\mu}} = (\mu - \widehat{\mu})/\widehat{\sigma}$ is a PQ, where $\widehat{\mu}$ and $\widehat{\sigma}$ are ML estimators of the location-scale distribution parameters. Then an exact $100(1 - \alpha)\%$ confidence interval for μ can be computed as

$$\left[\underset{\sim}{\mu},\ \widetilde{\mu}\right] = \left[\widehat{\mu} + z_{\widehat{\mu}(\alpha/2;n)}\,\widehat{\sigma},\ \ \widehat{\mu} + z_{\widehat{\mu}(1-\alpha/2;n)}\,\widehat{\sigma}\right], \tag{14.2}$$

where $z_{\widehat{\mu}(\gamma;n)}$ is the γ quantile of the distribution of $Z_{\widehat{\mu}}$ for a sample of size n. The corresponding $100(1 - \alpha)\%$ confidence interval for the log-location-scale distribution scale parameter $\eta = \exp(\mu)$ is

$$\left[\underset{\sim}{\eta},\ \widetilde{\eta}\right] = \left[\exp(\underset{\sim}{\mu}),\ \exp(\widetilde{\mu})\right].$$

The distribution of $Z_{\widehat{\mu}}$ (and thus quantiles of the distribution) can be readily obtained by using parametric bootstrap methods. In particular, following the approach described in Section 13.2, ML estimates $\widehat{\mu}$ and $\widehat{\sigma}$ of the assumed location-scale distribution parameters μ and σ are initially obtained using the n observations in the sample data. Then B simulated samples of size n are generated from the resulting fitted distribution (i.e., from the assumed location-scale distribution with location parameter $\widehat{\mu}$ and scale parameter $\widehat{\sigma}$). From each of these B samples, we calculate bootstrap ML estimates $\widehat{\mu}_j^*$ and $\widehat{\sigma}_j^*$, $j = 1, \ldots, B$, from which we compute $z_{\widehat{\mu},j}^* = (\widehat{\mu} - \widehat{\mu}_j^*)/\widehat{\sigma}_j^*$, $j = 1, \ldots, B$. The desired quantiles of $Z_{\widehat{\mu}}$ are then obtained from the ordered $z_{\widehat{\mu},j}^*$ values, as described in Section 13.3.3.

Example 14.4   Bootstrap Confidence Interval for the Pipeline Thickness Weibull Distribution Scale Parameter. In Example 14.1, the ML estimates for the Weibull distribution scale and shape parameters for the pipeline thickness data were found, based on the sample of $n = 17$ observations, to be $\widehat{\eta} = 0.31222$ and $\widehat{\beta} = 14.28013$. The corresponding estimated SEV distribution location and scale parameters were found to be $\widehat{\mu} = \log(\widehat{\eta}) = -1.16406$ and $\widehat{\sigma} = 1/\widehat{\beta} = 0.07003$ (which could also have been obtained by fitting an SEV distribution to the logs of the data). In Example 14.3, the parametric bootstrap ML estimates $\widehat{\mu}_j^*$ and $\widehat{\sigma}_j^*$, $j = 1, \ldots, 200,000$, were simulated from the $\text{SEV}(\mu, \sigma)$ distribution with $\mu = \widehat{\mu} = -1.16406$ and $\sigma = \widehat{\sigma} = 0.07003$. These simulated values were then used to compute the values $z_{\widehat{\mu},j}^* = (\widehat{\mu} - \widehat{\mu}_j^*)/\widehat{\sigma}_j^*$, $j = 1, \ldots, 200,000$.

To obtain an exact parametric bootstrap 95% confidence interval for μ, one first determines the 0.025 and 0.975 quantiles of the distribution of $Z_{\widehat{\mu}}$ as the 5,000th and 195,000th ordered $z_{\widehat{\mu}}^*$ observations. These quantiles are $z_{\widehat{\mu}(0.025;17)} = -0.55841$ and $z_{\widehat{\mu}(0.975;17)} = 0.57035$. The 95% confidence interval for μ is then obtained by substituting into (14.2):

$$\left[\underset{\sim}{\mu},\ \widetilde{\mu}\right] = [-1.16406 - 0.55841 \times 0.07003,\ \ -1.16406 + 0.57035 \times 0.07003]$$

$$= [-1.203,\ \ -1.125].$$

Finally, the 95% confidence interval for the Weibull distribution scale parameter $\eta = \exp(\mu)$ is

$$\left[\underset{\sim}{\eta},\ \widetilde{\eta}\right] = \left[\exp(\underset{\sim}{\mu}),\ \exp(\widetilde{\mu})\right] = [0.3002,\ \ 0.3249].$$

∎

14.3.3 Confidence Intervals for the Scale Parameter of a Location-Scale Distribution or the Shape Parameter of a Log-Location-Scale Distribution

As previously noted, σ is a scale parameter for a location-scale distribution and a shape parameter for a log-location-scale distribution. For the Weibull distribution, $\beta = 1/\sigma$ is more commonly used to represent the distribution shape parameter. Here, as described in Section E.7.3, $Z_{\widehat{\sigma}} = \sigma/\widehat{\sigma}$ is a PQ, where $\widehat{\sigma}$ is the ML estimator of σ. Then an exact $100(1 - \alpha)\%$ confidence interval for σ can be computed as

$$\left[\underset{\sim}{\sigma}, \, \widetilde{\sigma}\right] = \left[z_{\widehat{\sigma}(\alpha/2;n)}\widehat{\sigma}, \, z_{\widehat{\sigma}(1-\alpha/2;n)}\widehat{\sigma}\right] \tag{14.3}$$

where $z_{\widehat{\sigma}(\gamma;n)}$ is the γ quantile of the distribution of $Z_{\widehat{\sigma}}$, based on a sample of size n. The quantiles of $Z_{\widehat{\sigma}}$ can be readily approximated with parametric bootstrap methods, in a manner similar to that described in Section 14.3.2. Then an exact $100(1 - \alpha)\%$ confidence interval for β is

$$\left[\underset{\sim}{\beta}, \, \widetilde{\beta}\right] = \left[1/\widetilde{\sigma}, \, 1/\underset{\sim}{\sigma}\right].$$

Example 14.5 Confidence Interval for the Pipeline Thickness Weibull Distribution Shape Parameter. The ML estimates for the pipeline thickness data in Example 14.1 and the bootstrap ML estimates $\widehat{\mu}_j^*$ and $\widehat{\sigma}_j^*$ from Example 14.3 are used in a manner similar to Example 14.4 to compute $z_{\widehat{\sigma},j}^* = \widehat{\sigma}/\widehat{\sigma}_j^*$, $j = 1, \ldots, 200{,}000$. To obtain a 95% confidence interval for σ, one then uses the 0.025 and 0.975 quantiles of the empirical distribution of $Z_{\widehat{\sigma}}$. These quantiles are the 5,000th and 195,000th ordered $z_{\widehat{\sigma}}^*$ observations, $z_{\widehat{\sigma}(0.025;17)} = 0.74582$ and $z_{\widehat{\sigma}(0.975;17)} = 1.63655$. The exact 95% confidence interval for σ using this parametric bootstrap procedure is obtained by substituting into (14.3), giving

$$\left[\underset{\sim}{\sigma}, \, \widetilde{\sigma}\right] = [0.74582 \times 0.07003, \, 1.63655 \times 0.07003] = [0.0522, \, 0.1146].$$

The corresponding 95% confidence interval for the Weibull distribution shape parameter β is

$$\left[\underset{\sim}{\beta}, \, \widetilde{\beta}\right] = \left[1/\widetilde{\sigma}, \, 1/\underset{\sim}{\sigma}\right] = [1/0.1146, \, 1/0.0522] = [8.726, \, 19.157]. \qquad \blacksquare$$

14.3.4 Confidence Intervals for the p Quantile of a Location-Scale or a Log Location-Scale Distribution

The p quantile of a location-scale distribution $F(x; \mu, \sigma) = \Phi[(x - \mu)/\sigma]$ is $x_p = \mu + \Phi^{-1}(p)\sigma$, and its ML estimator is $\widehat{x}_p = \widehat{\mu} + \Phi^{-1}(p)\widehat{\sigma}$. The p quantile for a log-location-scale distribution is $t_p = \exp[\mu + \Phi^{-1}(p)\sigma]$. As explained in Section E.7.4,

$$Z_{\widehat{x}_p} = \frac{x_p - \widehat{x}_p}{\widehat{\sigma}} = \left[\frac{\mu - \widehat{\mu}}{\widehat{\sigma}} + \left(\frac{\sigma}{\widehat{\sigma}} - 1\right)\Phi^{-1}(p)\right]$$

is a PQ. This implies that

$$\left[\underset{\sim}{x}_p, \, \widetilde{x}_p\right] = \left[\widehat{x}_p + z_{\widehat{x}_p(\alpha/2;n)}\widehat{\sigma}, \, \widehat{x}_p + z_{\widehat{x}_p(1-\alpha/2;n)}\widehat{\sigma}\right]$$

is an exact $100(1 - \alpha)\%$ confidence interval for x_p, where $z_{\widehat{x}_p(\gamma;n)}$ is the γ quantile of the distribution of $Z_{\widehat{x}_p}$ for samples of size n. The corresponding exact $100(1 - \alpha)\%$ confidence interval for $t_p = \exp(x_p)$ is

$$\left[\underline{t}_p, \; \tilde{t}_p\right] = \left[\exp(\underline{x}_p), \; \exp(\tilde{x}_p)\right]$$

$$= \left[\hat{t}_p \exp\left(z_{\hat{x}_p(\alpha/2;n)}\hat{\sigma}\right), \; \hat{t}_p \exp\left(z_{\hat{x}_p(1-\alpha/2;n)}\hat{\sigma}\right)\right]. \tag{14.4}$$

As with the other PQs in this chapter, the quantiles of $Z_{\hat{x}_p}$ can be readily calculated using parametric bootstrap methods in a manner similar to that described in Section 14.3.2. In particular, the ML bootstrap estimates $\hat{\mu}_j^*$ and $\hat{\sigma}_j^*$ for $j = 1, \ldots, B$ are used to compute

$$z_{\hat{x}_p,j}^* = \left[\frac{\hat{\mu} - \hat{\mu}_j^*}{\hat{\sigma}_j^*} + \left(\frac{\hat{\sigma}}{\hat{\sigma}_j^*} - 1\right)\Phi^{-1}(p)\right], \quad j = 1, \ldots, B. \tag{14.5}$$

The desired quantiles and confidence interval are then obtained from the ordered values of $z_{\hat{x}_p,j}^*$, $j = 1, \ldots, B$, as described in Section 13.3.3.

Example 14.6   Confidence Interval for the Weibull 0.0001 Quantile for Pipeline Thickness.

The ML estimate of the 0.0001 quantile $x_{0.0001}$ for the SEV distribution of the logs of pipeline thickness in Example 14.1 is

$$\hat{x}_{0.0001} = \hat{\mu} + \Phi_{\text{sev}}^{-1}(0.0001)\hat{\sigma} = -1.16406 - 9.21029 \times 0.07003 = -1.80906.$$

Thus $\hat{t}_{0.0001} = \exp(-1.80906) = 0.164$ inches. This estimate involves much extrapolation from a sample of only 17 observations, as can be seen by the extrapolation into the lower left-hand corner of Figure 14.1. Those involved with the application believed, however, that the Weibull distribution was justified, even in the extreme lower tail of the distribution. This was, at least in part, because each observation was the minimum thickness over the entire area of a one-foot-long sampled segment. To compute a confidence interval for $t_{0.0001}$, the 0.0001 quantile of the distribution of thickness, we proceed as in Examples 14.4 and 14.5. The bootstrap ML estimates $\hat{\mu}_j^*$ and $\hat{\sigma}_j^*$ from the SEV$(\hat{\mu},\hat{\sigma})$ distribution from Example 14.3 are used to compute $z_{\hat{x}_{0.0001},j}^*$, $j = 1, \ldots, 200{,}000$, by substituting into (14.5). Figure 14.5 is a histogram of the resulting 200,000 simulated values of $z_{\hat{x}_{0.0001}}^*$.

To obtain a 95% confidence interval for $x_{0.0001}$, we use the 0.025 and 0.975 quantiles of the empirical distribution of $Z_{\hat{x}_{0.0001}}$, namely the 5,000th and 195,000th ordered $z_{\hat{x}_{0.0001}}^*$ values. These quantiles are $z_{\hat{x}_{0.0001}}(0.025;17) = -5.972$ and $z_{\hat{x}_{0.0001}}(0.975;17) = 2.551$. The 95% confidence interval for $t_{0.0001} = \exp(x_{0.0001})$ using this exact parametric bootstrap procedure is then obtained by substituting into (14.4), giving

$$\left[\underline{t}_{0.0001}, \; \tilde{t}_{0.0001}\right] = [0.1638\exp(-5.972 \times 0.07003), \; 0.1638\exp(2.551 \times 0.07003)]$$

$$= [0.108, \; 0.196].$$

Finally, we note from Figure 14.5 that the distribution of $z_{\hat{x}_{0.0001}}^*$ is somewhat skewed to the left. This indicates that Wald confidence intervals would, unlike in Example 14.4, result in an inadequate approximation here. Thus the parametric bootstrap methods provide an important improvement over the Wald approximation for constructing confidence intervals on distribution quantiles.   ∎

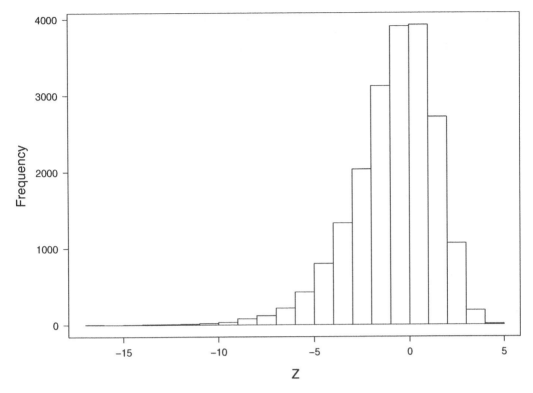

Figure 14.5 Histogram of the 200,000 simulated pivotal quantity values $z^*_{\widehat{x}_{0.0001}}$ for the Weibull distribution 0.0001 quantile for the pipeline thickness data.

14.4 GENERALIZED PIVOTAL QUANTITIES

Pivotal quantities are not available for all inferences of interest. In some cases for which a PQ does not exist, there may be a generalized pivotal quantity (GPQ) that can be used to construct a confidence interval for a distribution parameter or a function of parameters. A GPQ is similar to a PQ in that it is a scalar function of the random parameter estimator or estimators (e.g., $\widehat{\mu}$ and $\widehat{\sigma}$) and the parameters to be estimated (e.g., μ and σ). A GPQ differs from a PQ in that the *unconditional* sampling distribution of the GPQ (i.e., the distribution that includes variability from repeated sampling) may depend on the unknown parameters (e.g., μ and σ). To be a GPQ, a function must have the following two properties:

1. Conditional on the data (or on the observed value(s) of the parameter estimates calculated from the data, such as $\widehat{\mu}$ and $\widehat{\sigma}$ for a location-scale distribution), the distribution of a GPQ does not depend on any unknown parameters.

2. If the random bootstrap parameter estimators (e.g., $\widehat{\mu}^*$ and $\widehat{\sigma}^*$) in a GPQ are replaced by the corresponding observed values of the parameter estimates (e.g., $\widehat{\mu}$ and $\widehat{\sigma}$), the GPQ must be equal to the actual value of the function of the parameters that is being estimated.

Use of a GPQ-based procedure will, in general, lead to only an approximate confidence interval. Research has shown, however, that GPQ methods tend to provide procedures with a coverage probability that is very close to the nominal confidence level. Section F.4 gives conditions under which a GPQ-based confidence interval procedure is exact. Other technical details about GPQs are given in Appendix F and in references given in the Bibliographical Notes section at the end of this chapter.

As illustrated in examples in this section, the computation of a GPQ-based confidence interval is similar to but simpler than the computation of a PQ interval. In particular, to obtain a (usually approximate) $100(1 - \alpha)\%$ GPQ confidence interval, one simulates a large number B of realizations of the GPQ. The confidence interval, as with the simple percentile method in Section 13.4.2, is obtained from the $\alpha/2$ and $1 - \alpha/2$ quantiles of the empirical GPQ distribution.

14.4.1 Generalized Pivotal Quantities for μ and σ of a Location-Scale Distribution and for Functions of μ and σ

This section gives expressions for GPQs for the location parameter μ and the scale parameter σ of a location-scale distribution. These GPQs could be used to compute confidence intervals for μ and σ, respectively. Such intervals, however, would agree with those obtained from the simpler pivotal quantity methods discussed in Sections 14.3.2 and 14.3.3, respectively. The main reason for providing these GPQs here is that they are used as building blocks for obtaining GPQs and corresponding confidence intervals for functions of μ and σ for which no PQ exists.

Generalized pivotal quantities for the location parameter μ

As shown in Section F.3.1,

$$Z_{\widehat{\mu}} = Z_{\widehat{\mu}}(\mu, \widehat{\mu}, \widehat{\sigma}, \widehat{\mu}^*, \widehat{\sigma}^*) = \widehat{\mu} + \left(\frac{\mu - \widehat{\mu}^*}{\widehat{\sigma}^*} \right) \widehat{\sigma} \qquad (14.6)$$

is a GPQ for μ.

Generalized pivotal quantities for the scale parameter σ

As shown in Section F.3.2,

$$Z_{\widehat{\sigma}} = Z_{\widehat{\sigma}}(\sigma, \widehat{\sigma}, \widehat{\sigma}^*) = \left(\frac{\sigma}{\widehat{\sigma}^*} \right) \widehat{\sigma} \qquad (14.7)$$

is a GPQ for σ.

Generalized pivotal quantities for functions of μ and σ

As illustrated in the rest of this section, there are a number of important quantities of interest for which no PQ exists. In such cases one can construct a confidence interval that is based on a GPQ. As shown in Section F.2, a GPQ for a function of interest $g = g(\mu, \sigma)$ is obtained by substituting the GPQ (14.6) for μ and the GPQ (14.7) for σ into the function $g(\mu, \sigma)$. We provide examples in the rest of this section.

14.4.2 Confidence Intervals for Tail Probabilities for Location-Scale and Log-Location-Scale Distributions

There does not exist a PQ that can be used directly to define a confidence interval procedure for (log-)location-scale distribution tail probabilities. There is, however, a GPQ for this purpose. In particular, for a location-scale distribution (e.g., normal or SEV), a lower-tail probability is

$$p = \Pr(X \le x) = F(x) = \Phi\left(\frac{x - \mu}{\sigma} \right). \qquad (14.8)$$

For a log-location-scale distribution (e.g., lognormal or Weibull)

$$p = \Pr(T \le t) = F(t) = \Phi\left[\frac{\log(t) - \mu}{\sigma}\right]. \tag{14.9}$$

ML estimates of these probabilities can be obtained from the given data by evaluating the last two expressions using the parameter estimates $\widehat{\mu}$ and $\widehat{\sigma}$ in place of μ and σ. Substituting (14.6) for μ and (14.7) for σ into (14.8) or (14.9) and simplifying gives the GPQ for $p = F(t)$,

$$Z_{\widehat{F}} = \Phi\left[\left(\frac{\widehat{\sigma}^*}{\sigma}\right)\Phi^{-1}(\widehat{p}) + \frac{\widehat{\mu}^* - \mu}{\sigma}\right],$$

where $\widehat{p}$ is the ML estimate of p, based on the given data, and $\widehat{\mu}^*$ and $\widehat{\sigma}^*$ are the random variables defined by the sampling distribution of the ML estimators for μ and σ. In a manner similar to the usage in Sections 14.3.2–14.3.4 for obtaining a confidence interval from a PQ, we use the bootstrap ML estimates $\widehat{\mu}_j^*$ and $\widehat{\sigma}_j^*$, $j = 1, \ldots, B$, simulated from the assumed (log-)location-scale distribution with μ and σ replaced by $\widehat{\mu}$ and $\widehat{\sigma}$ to compute

$$z_{\widehat{F},j}^* = \Phi\left[\left(\frac{\widehat{\sigma}_j^*}{\widehat{\sigma}}\right)\Phi^{-1}(\widehat{p}) + \frac{\widehat{\mu}_j^* - \widehat{\mu}}{\widehat{\sigma}}\right]. \tag{14.10}$$

The $\alpha/2$ and $1 - \alpha/2$ quantiles of $z_{\widehat{F}}^*$ provide the endpoints of the $100(1 - \alpha)\%$ confidence interval for p.

In this case, unlike the case for GPQs in general, the confidence interval procedure is exact (see Section F.4.2 for technical details).

Example 14.7 Confidence Intervals for Pipeline Thickness Weibull Distribution Probabilities. The ML estimate of $p = F(0.15) = \Pr(\text{Thickness} \le 0.15)$ is obtained from the given data by substituting the ML estimates for μ and σ into (14.9), giving

$$\widehat{p} = \widehat{F}(0.15) = \Phi_{\text{SEV}}\left[\frac{\log(0.15) + 1.16406}{0.07003}\right] = 0.0000284.$$

To compute a confidence interval for $F(0.15)$, we proceed as in Examples 14.4–14.6. The bootstrap ML estimates $\widehat{\mu}_j^*$ and $\widehat{\sigma}_j^*$ from the SEV$(\widehat{\mu}, \widehat{\sigma})$ distribution from Example 14.3 and the ML estimate $\widehat{p}$ (previously calculated from the data) are used to compute the corresponding 200,000 values of $z_{\widehat{F},j}^*$ by substituting into (14.10). For a 95% confidence interval, the 0.025 and 0.975 quantiles of the resulting empirical distribution of $Z_{\widehat{F}(0.15)}$ are the 5,000th and 195,000th ordered $z_{\widehat{F}(0.15)}^*$ values. The resulting confidence interval for $F(0.15)$ is

$$\left[\underset{\sim}{F}(0.15), \quad \widetilde{F}(0.15)\right] = \left[z_{\widehat{F}(0.15)_{(0.025;17)}}, \quad z_{\widehat{F}(0.15)_{(0.975;17)}}\right] = [6.20 \times 10^{-7}, \quad 0.00182].$$

The preceding confidence interval, based on the exact GPQ-based simulation method, is compared in Table 14.3 with the intervals obtained (details not shown here) using the two approximate methods discussed in Chapter 12 (i.e., the likelihood-ratio and Wald $\widehat{z}$ methods). We recommend use of the GPQ method because the method is exact for this application. ∎

Although, as usual, we computed two-sided confidence intervals for $F(0.15)$, primary interest in this application would be in the upper endpoint of the chosen interval, which could be interpreted as a 97.5% upper confidence bound. It is interesting that the Wald-based upper bound is closer to the GPQ-based (exact method) bound than likelihood-based bound. This is an exception to what we usually see and is probably due to some combination of the small number

Method	Section	95% confidence interval
Exact parametric bootstrap method using GPQ	14.4.2	$[6.20 \times 10^{-7}, \ 0.00182]$
Likelihood-ratio method	12.5.4	$[3.22 \times 10^{-7}, \ 0.00111]$
Wald $\widehat{z}$ method	12.6.4	$[4.83 \times 10^{-7}, \ 0.00167]$

Table 14.3   95% confidence intervals for the pipeline thickness Weibull probability $F(0.15)$ using three different methods.

of observations, large amount of extrapolation (both having a negative effect on the adequacy of large-sample approximations), and happenstance.

The pointwise GPQ-based 95% confidence intervals for the entire cumulative distribution function for the pipeline thickness were calculated and are shown as dashed lines in Figure 14.1.

14.4.3   Confidence Intervals for the Mean of a Log-Location-Scale Distribution

There are no known exact confidence interval procedures for the mean (expected value) of log-location-scale distributions, such as the lognormal and Weibull. Procedures based on GPQs, however, have coverage probabilities that are close to the nominal confidence level (see the results of simulation studies reported in the references given in the Bibliographic Notes section at the end of this chapter). The mean of a lognormal distribution is

$$\mathrm{E}(T) = \exp\left(\mu + \sigma^2/2\right), \tag{14.11}$$

and the mean of a Weibull distribution is

$$\mathrm{E}(T) = \eta\Gamma\left(1 + \frac{1}{\beta}\right) = \exp(\mu)\Gamma(1 + \sigma). \tag{14.12}$$

Expressions for the means of other log-location-scale distributions are given in Section C.3.1.

Substituting (14.6) for μ and (14.7) for σ into (14.11) or (14.12) and simplifying gives the GPQs

$$Z_{\widehat{\mathrm{E}}(T)}(\mu, \sigma, \widehat{\mu}, \widehat{\sigma}, \widehat{\mu}^*, \widehat{\sigma}^*) = \exp\left[\widehat{\mu} + \left(\frac{\mu - \widehat{\mu}^*}{\widehat{\sigma}^*}\right)\widehat{\sigma} + \frac{1}{2}\left(\frac{\sigma}{\widehat{\sigma}^*}\right)^2\widehat{\sigma}^2\right] \tag{14.13}$$

for the lognormal mean and

$$Z_{\widehat{\mathrm{E}}(T)}(\mu, \sigma, \widehat{\mu}, \widehat{\sigma}, \widehat{\mu}^*, \widehat{\sigma}^*) = \exp\left[\widehat{\mu} + \left(\frac{\mu - \widehat{\mu}^*}{\widehat{\sigma}^*}\right)\widehat{\sigma}\right]\Gamma\left[1 + \left(\frac{\sigma}{\widehat{\sigma}^*}\right)\widehat{\sigma}\right] \tag{14.14}$$

for the Weibull mean. The empirical distributions of these GPQs, from Monte Carlo simulation, can be used to obtain approximate confidence intervals for $\mathrm{E}(T)$ for log-location-scale distributions. We note that these GPQs are not pivotal because they have distributions that depend on the observed parameter estimates $\widehat{\mu}$ and $\widehat{\sigma}$.

In particular, for bootstrap sample j, $j = 1, \ldots, B$, the bootstrap ML estimates $\widehat{\mu}_j^*$ and $\widehat{\sigma}_j^*$ are substituted into (14.13) or (14.14) to obtain

$$z_{\widehat{\mathrm{E}}(T),j}(\widehat{\mu}, \widehat{\sigma}, \widehat{\mu}_j^*, \widehat{\sigma}_j^*) = \exp\left[\widehat{\mu} + \left(\frac{\widehat{\mu} - \widehat{\mu}_j^*}{\widehat{\sigma}_j^*}\right)\widehat{\sigma} + \frac{1}{2}\left(\frac{\widehat{\sigma}}{\widehat{\sigma}_j^*}\right)^2\widehat{\sigma}^2\right] \tag{14.15}$$

or

$$z_{\widehat{\mathrm{E}}(T),j}(\widehat{\mu},\widehat{\sigma},\widehat{\mu}_j^*,\widehat{\sigma}_j^*) = \exp\left[\widehat{\mu} + \left(\frac{\widehat{\mu} - \widehat{\mu}_j^*}{\widehat{\sigma}_j^*}\right)\widehat{\sigma}\right]\Gamma\left[1 + \left(\frac{\widehat{\sigma}}{\widehat{\sigma}_j^*}\right)\widehat{\sigma}\right] \quad (14.16)$$

for the lognormal and Weibull distribution, respectively. Similar expressions can be derived for other log-location-scale distributions. Then an approximate $100(1-\alpha)\%$ confidence interval for the mean $\mathrm{E}(T)$ is

$$\left[\underline{\mathrm{E}(T)},\ \widetilde{\mathrm{E}(T)}\right] = \left[z_{\widehat{\mathrm{E}}(T),\alpha/2},\ z_{\widehat{\mathrm{E}}(T),1-\alpha/2}\right],$$

where $z_{\widehat{\mathrm{E}}(T),\gamma}$ is the γ quantile of the empirical distribution of the simulated values of $z_{\widehat{\mathrm{E}}(T),j}$, $j = 1,\ldots,B$.

Example 14.8 Confidence Interval for Tree Volume Mean Assuming a Weibull Distribution. The ML estimates of the SEV parameters fit to the natural log of the tree volume data are $\widehat{\mu} = -2.0964$ and $\widehat{\sigma} = 0.5475$. Substituting these and $\widehat{\mu}_j^*$ and $\widehat{\sigma}_j^*$, $j = 1,\ldots,B$, into (14.16) gives the simulated distribution of $z_{\widehat{\mathrm{E}}(T)}$. Then the endpoints of a 95% confidence interval for the Weibull distribution mean are given by the 0.025 and 0.975 quantiles of this empirical distribution, obtained from the 5,000th and 195,000th ordered $z_{\widehat{\mathrm{E}}(T)}$ values, resulting in the interval

$$\left[\underline{\mathrm{E}(T)},\ \widetilde{\mathrm{E}(T)}\right] = \left[z_{\widehat{\mathrm{E}}(T),0.025},\ z_{\widehat{\mathrm{E}}(T),0.975}\right] = [0.088,\ 0.138].$$

∎

14.4.4 Simplified Simulation and Confidence Interval Computation with PQs and GPQs

As mentioned earlier, when the data are complete (i.e., no censoring) or censored after a prespecified number of lower order statistics have been observed (known as Type 2 or failure censoring), the PQs like $Z_{\widehat{\mu}}$, $Z_{\widehat{\sigma}}$, and $Z_{\widehat{x}_p}$ have distributions that do not depend on the actual values of μ and σ. Also, GPQs such as $Z_{\widehat{F}}$ and $Z_{\widehat{\mathrm{E}}(T)}(\mu,\sigma,\widehat{\mu},\widehat{\sigma},\widehat{\mu}^*,\widehat{\sigma}^*)$ have distributions that, conditional on the observed data, do not depend on the actual values of μ and σ. In such cases, one can arbitrarily use any value of μ and $\sigma > 0$ in the Monte Carlo simulation.

Thus, it is possible to simulate using $\mu = 0$ and $\sigma = 1$ in place of $\mu = \widehat{\mu}$ and $\sigma = \widehat{\sigma}$. Then the computing formulas for the simulation simplify and it is possible to save the simulation results—which depend only on the assumed distribution and the sample size (and number of noncensored observations for Type 2 censoring)—so that these can be used subsequently for other purposes (without having to store $\widehat{\mu}$ and $\widehat{\sigma}$). For example, if $\mu = 0$ and $\sigma = 1$ are used in the simulation, (14.5) is replaced by

$$z_{\widehat{x}_p,j}^* = \left[-\frac{\widehat{\mu}_j^*}{\widehat{\sigma}_j^*} + \left(\frac{1}{\widehat{\sigma}_j^*} - 1\right)\Phi^{-1}(p)\right], \quad j = 1,\ldots,B,$$

and (14.10) can be replaced by

$$z_{\widehat{F},j}^* = \Phi\left[\widehat{\sigma}_j^*\Phi^{-1}(\widehat{p}) + \widehat{\mu}_j^*\right], \quad j = 1,\ldots,B.$$

Similar substitutions can be made for the other PQs and GPQs in this section.

14.5 SIMULATION-BASED TOLERANCE INTERVALS FOR LOCATION-SCALE OR LOG-LOCATION-SCALE DISTRIBUTIONS

Section 4.6 described and illustrated the use of tolerance intervals for a normal distribution. In this section we use parametric simulation-based methods to extend these methods to other location-scale and log-location-scale distributions.

14.5.1 Two-Sided Tolerance Intervals to Control the Center of a Distribution

From (B.14) in Appendix B, a two-sided control-the-center tolerance interval to contain at least a proportion β of a location-scale distribution with $100(1 - \alpha)\%$ confidence is given by

$$\left[\underset{\sim}{T_\beta}, \ \widetilde{T}_\beta\right] = [\widehat{\mu} + g_{L(1-\alpha;\beta,n)}\widehat{\sigma}, \ \widehat{\mu} + g_{U(1-\alpha;\beta,n)}\widehat{\sigma}].$$

For a log-location-scale distribution the corresponding tolerance interval is

$$\left[\underset{\sim}{T_\beta}, \ \widetilde{T}_\beta\right] = [\exp(\widehat{\mu} + g_{L(1-\alpha;\beta,n)}\widehat{\sigma}), \ \exp(\widehat{\mu} + g_{U(1-\alpha;\beta,n)}\widehat{\sigma})]. \tag{14.17}$$

For a symmetric distribution $g_{L(1-\alpha;\beta,n)} = -g_{U(1-\alpha;\beta,n)}$, $g_{L(1-\alpha;\beta,n)} = -g_{L(1-\alpha;1-\beta,n)}$, and $g_{U(1-\alpha;\beta,n)} = -g_{U(1-\alpha;1-\beta,n)}$.

The factors $g_{L(1-\alpha;\beta,n)}$ and $g_{U(1-\alpha;\beta,n)}$ are functions of quantiles of distributions of particular pivotal quantities, as shown for the normal distribution in Section E.5.1 (but the result is more generally true for location-scale distributions). For a symmetric distribution, these factors are obtained by choosing those values that make the coverage probability in (B.15) equal to the nominal confidence level. The coverage probability is evaluated using parametric bootstrap simulation, as shown in (B.16). For symmetric location-scale distributions, the values of $g_{L(1-\alpha;\beta,n)}$ and $g_{U(1-\alpha;\beta,n)}$ are determined, such that

$$\frac{1}{B}\sum_{j=1}^{B}\mathrm{I}\left[\Phi\left(z_{Uj}^*\right) - \Phi\left(z_{Lj}^*\right) > \beta\right] = 1 - \alpha, \tag{14.18}$$

$$z_{Lj}^* = \frac{\widehat{\mu}_j^* + g_{L(1-\alpha;\beta,n)}\widehat{\sigma}_j^* - \widehat{\mu}}{\widehat{\sigma}},$$

$$z_{Uj}^* = \frac{\widehat{\mu}_j^* + g_{U(1-\alpha;\beta,n)}\widehat{\sigma}_j^* - \widehat{\mu}}{\widehat{\sigma}},$$

subject to the symmetry constraints given above. Here $\mathrm{I}[A]$ is an indicator function which is equal to 1 when the statement A is true and equal to 0 otherwise. For a nonsymmetric distribution, the values of $g_{L(1-\alpha;\beta,n)}$ and $g_{U(1-\alpha;\beta,n)}$ are chosen subject to the constraint

$$\frac{1}{B}\sum_{j=1}^{B}\Phi\left(z_{Lj}^*\right) = 1 - \frac{1}{B}\sum_{j=1}^{B}\Phi\left(z_{Uj}^*\right). \tag{14.19}$$

This constraint assures that the error probabilities are equal for both the lower and the upper endpoints of the tolerance interval. Note that, as described in Section 14.4.4, (14.18) and (14.19) simplify somewhat if the simulation to obtain $\widehat{\mu}_j^*$ and $\widehat{\sigma}_j^*$, $j = 1, \ldots, B$, can be done using $\mu = 0$ and $\sigma = 0$ instead of $\mu = \widehat{\mu}$ and $\sigma = \widehat{\sigma}$.

Example 14.9 Two-Sided Control-the-Center 95% Tolerance Interval to Contain the Proportion 0.80 of the Carbon-Epoxy Fracture Strength Distribution. Example 14.2 introduced data giving fracture strength measured on 19 specimens of a carbon-epoxy composite material and suggested that a Weibull distribution provides an appropriate description of the fracture strength data for this process. We will use the data to construct a 95% tolerance interval to contain at least a proportion 0.80 of the distribution of fracture strengths.

The ML estimates for the Weibull distribution scale and shape parameters for the data are $\widehat{\eta} = 510.17855$ and $\widehat{\beta} = 18.86249$. The corresponding estimated SEV distribution location and scale parameters (that would be obtained by fitting an SEV distribution to the logs of the data) are $\widehat{\mu} = \log(\widehat{\eta}) = 6.23476$ and $\widehat{\sigma} = 1/\widehat{\beta} = 0.05302$.

Bootstrap ML estimates $\widehat{\mu}_j^*$ and $\widehat{\sigma}_j^*$, $j = 1, \ldots, B$, were computed from $B = 200{,}000$ samples of size $n = 19$ (the sample size of the fracture strength data) simulated from the standard SEV (i.e., $\mu = 0$ and $\sigma = 1$) distribution and used in (14.18) to determine $g_{L(0.95;0.80,19)} = -3.198916$ and $g_{U(0.95;0.80,19)} = 1.350704$, subject to the constraint in (14.19). Then the 95% control-the-center tolerance interval to contain at least a proportion 0.80 of the distribution of fracture strengths is obtained by substituting into (14.17), giving

$$[\exp(6.23476 - 3.198916 \times 0.05302), \ \exp(6.23476 + 1.350704 \times 0.05302)]$$
$$= [430.6, \ 548.1].$$

∎

14.5.2 Two-Sided Tolerance Intervals to Control Both Tails of a Distribution

From (B.14) in Appendix B, a two-sided tolerance interval to control both tails of a location-scale distribution to have no more than p_{tL} in the lower tail and no more than p_{tU} in the upper tail of the distribution is given by

$$\left[\underset{\sim}{T}_{p_{tL}}, \ \widetilde{T}_{p_{tU}} \right] = \left[\widehat{\mu} + g''_{L(1-\alpha;p_{tL},n)}\widehat{\sigma}, \ \widehat{\mu} + g''_{U(1-\alpha;p_{tU},n)}\widehat{\sigma} \right].$$

For a log-location-scale distribution the corresponding tolerance interval is

$$\left[\underset{\sim}{T}_{p_{tL}}, \ \widetilde{T}_{p_{tU}} \right] = \left[\exp\left(\widehat{\mu} + g''_{L(1-\alpha;p_{tL},n)}\widehat{\sigma} \right), \ \exp\left(\widehat{\mu} + g''_{U(1-\alpha;p_{tU},n)}\widehat{\sigma} \right) \right]. \tag{14.20}$$

As with the control-the-center tolerance interval described in Section 14.5.1, for a symmetric distribution $g''_{L(1-\alpha;p_{tL},n)} = -g''_{U(1-\alpha;p_{tU},n)}$, $g''_{L(1-\alpha;p_{tL},n)} = -g''_{L(1-\alpha;1-p_{tL},n)}$ and $g''_{U(1-\alpha;p_{tU},n)} = -g''_{U(1-\alpha;1-p_{tU},n)}$. The factors are obtained like those for the control-the-center tolerance interval described in Section 14.5.1, but using the coverage probability evaluation formula in (B.20). For symmetric location-scale distributions, the values of $g''_{L(1-\alpha;p_{tL},n)}$ and $g''_{U(1-\alpha;p_{tU},n)}$ are determined, such that

$$\frac{1}{B} \sum_{j=1}^{B} \mathrm{I}\left[z''^*_{Lj} \le \Phi^{-1}(p_{tL}) \text{ and } z''^*_{Uj} \ge \Phi^{-1}(p_{tU}) \right] = 1 - \alpha, \tag{14.21}$$

where

$$z''^*_{Lj} = \frac{\widehat{\mu}_j^* + g''_{L(1-\alpha;p_{tL},n)}\widehat{\sigma}_j^* - \widehat{\mu}}{\widehat{\sigma}}, \quad z''^*_{Uj} = \frac{\widehat{\mu}_j^* + g''_{U(1-\alpha;p_{tU},n)}\widehat{\sigma}_j^* - \widehat{\mu}}{\widehat{\sigma}},$$

subject to the above symmetry constraints. For a nonsymmetric distribution, the values of $g''_{L(1-\alpha;p_{tL},n)}$ and $g''_{U(1-\alpha;p_{tU},n)}$ are chosen in a nonsymmetric manner, again subject to a constraint similar to (14.19), using $g''_{L(1-\alpha;p_{tL},n)}$ and $g''_{U(1-\alpha;p_{tU},n)}$ instead of $g_{L(1-\alpha;p_{tL},n)}$ and $g_{U(1-\alpha;p_{tU},n)}$. Note that, as described in Section 14.4.4, (14.21) simplifies somewhat if the simulation to obtain $\widehat{\mu}_j^*$ and $\widehat{\sigma}_j^*$, $j = 1, \ldots, B$, is done using $\mu = 0$ and $\sigma = 0$ instead of $\mu = \widehat{\mu}$ and $\sigma = \widehat{\sigma}$.

Example 14.10   Two-Sided 95% Tolerance Interval to Control Both Tails of the Distribution of Fracture Strengths for Carbon-Epoxy Specimens so that Neither Tail has a Probability of More than 0.10. A control-both-tails tolerance interval is constructed in a manner similar to the control-the-center tolerance interval in Example 14.9. The same bootstrap ML estimates $\widehat{\mu}_j^*$ and $\widehat{\sigma}_j^*$, $j = 1, \ldots, B$, are used in (14.21) to determine $g''_{L(0.95;0.80,19)} = -3.669778$ and $g''_{U(0.95;0.80,19)} = 1.600181$, subject to the constraint in (14.19). Then a 95% tolerance interval to have no more than a proportion 0.10 in each tail of the distribution of fracture strengths is obtained by substituting into (14.20), giving

$$[\exp(6.23476 - 3.669778 \times 0.05302), \ \exp(6.23476 + 1.600181 \times 0.05302)]$$
$$= [419.97, \ 555.35].$$

This interval is wider than the control-the-center interval in Example 14.9, due to the more stringent constraint of controlling both tails of the distribution. ∎

14.5.3   One-Sided Tolerance Bounds

As described in Sections 2.4.2 and 4.6.3, a one-sided lower $100(1 - \alpha)\%$ tolerance bound to be exceeded by at least a proportion p of a distribution is equivalent to a one-sided lower $100(1 - \alpha)\%$ confidence bound for the $1 - p$ quantile of the distribution. Similarly, a one-sided upper $100(1 - \alpha)\%$ tolerance bound to exceed at least a proportion p of a distribution is equivalent to a one-sided upper $100(1 - \alpha)\%$ confidence bound for the p quantile of the distribution.

Therefore, a one-sided simulation-based tolerance bound is obtained using the method in Section 14.3.4. This would be done by taking the appropriate endpoint of a two-sided confidence interval for a distribution quantile, with a suitably adjusted confidence level. For example, to obtain a 95% lower tolerance bound to be exceeded by at least a proportion 0.90 of a population, one would use the lower endpoint of a 90% confidence interval for the 0.10 quantile.

14.6   SIMULATION-BASED PREDICTION INTERVALS AND ONE-SIDED PREDICTION BOUNDS FOR AT LEAST k OF m FUTURE OBSERVATIONS FROM LOCATION-SCALE OR LOG-LOCATION-SCALE DISTRIBUTIONS

Section 4.8 presented simultaneous prediction intervals to enclose at least k of m future observations from a normal distribution. This section extends these results to other location-scale and log-location-scale distributions, using simulation-based methods.

14.6.1   Simultaneous Two-Sided Prediction Intervals to Contain at Least k of m Future Observations

From (B.27) in Appendix B, a simultaneous two-sided $100(1 - \alpha)\%$ prediction interval to enclose at least k of m future observations from a particular location-scale distribution is

$$\left[\underset{\sim}{Y}_{k;m}, \ \widetilde{Y}_{k;m}\right] = [\widehat{\mu} + r_{L(1-\alpha;k,m,n)}\widehat{\sigma}, \ \widehat{\mu} + r_{U(1-\alpha;k,m,n)}\widehat{\sigma}].$$

The corresponding prediction interval for a log-location-scale distribution is

$$\left[\underset{\sim}{Y}_{k;m}, \ \widetilde{Y}_{k;m}\right] = [\exp(\widehat{\mu} + r_{L(1-\alpha;k,m,n)}\widehat{\sigma}), \ \exp(\widehat{\mu} + r_{U(1-\alpha;k,m,n)}\widehat{\sigma})]. \quad (14.22)$$

The factors $r_{L(1-\alpha;k,m,n)}$ and $r_{U(1-\alpha;k,m,n)}$ are quantiles of distributions of particular pivotal quantities, as shown in Section E.7 for the normal distribution, but the result is also true for other location-scale distributions. For a symmetric distribution, $r_{L(1-\alpha;k,m,n)} = -r_{U(1-\alpha;k,m,n)}$. The factors can be computed so that the coverage probability (B.28) in Appendix B is equal to the nominal confidence level.

For a nonsymmetric distribution, the values of $r_{L(1-\alpha;k,m,n)}$ and $r_{U(1-\alpha;k,m,n)}$ are chosen such that the coverage probability in (B.28) is equal to the nominal confidence level and the one-sided probabilities of noncoverage in each tail of the distribution are equal.

For symmetric location-scale distributions, the values of $r_{L(1-\alpha;k,m,n)}$ and $r_{U(1-\alpha;k,m,n)}$ are then determined, such that

$$\frac{1}{B}\sum_{j=1}^{B}\left[\sum_{i=k}^{m}\binom{m}{i}(p_j^*)^i(1-p_j^*)^{m-i}\right]=1-\alpha, \tag{14.23}$$

where

$$p_j^* = \Phi\left(\frac{\widehat{\mu}_j^* + r_{U(1-\alpha;k,m,n)}\widehat{\sigma}_j^* - \widehat{\mu}}{\widehat{\sigma}}\right) - \Phi\left(\frac{\widehat{\mu}_j^* + r_{L(1-\alpha;k,m,n)}\widehat{\sigma}_j^* - \widehat{\mu}}{\widehat{\sigma}}\right),$$

subject to the symmetry constraint given above. For a nonsymmetric distribution, the values of $r_{L(1-\alpha;k,m,n)}$ and $r_{U(1-\alpha;k,m,n)}$ are chosen (in a nonsymmetric manner) subject to the constraint

$$\frac{1}{B}\sum_{j=1}^{B}\left[\sum_{i=k}^{m}\binom{m}{i}(\omega_j^*)^i(1-\omega_j^*)^{m-i}\right]=\frac{1}{B}\sum_{j=1}^{B}\left[\sum_{i=k}^{m}\binom{m}{i}(\nu_j^*)^i(1-\nu_j^*)^{m-i}\right], \tag{14.24}$$

where

$$\omega_j^* = 1 - \Phi\left(\frac{\widehat{\mu}_j^* + r_{L(1-\alpha;k,m,n)}'\widehat{\sigma}_j^* - \widehat{\mu}}{\widehat{\sigma}}\right),$$

$$\nu_j^* = \Phi\left(\frac{\widehat{\mu}_j^* + r_{U(1-\alpha;k,m,n)}'\widehat{\sigma}_j^* - \widehat{\mu}}{\widehat{\sigma}}\right). \tag{14.25}$$

The constraint in (14.24) assures that the error probabilities are equal for both the lower and the upper endpoints of the simultaneous prediction interval. Note that, as described in Section 14.4.4, (14.25) can be simplified if the simulation to obtain $\widehat{\mu}_j^*$ and $\widehat{\sigma}_j^*$, $j = 1, \ldots, B$, is done using $\mu = 0$ and $\sigma = 0$ instead of $\mu = \widehat{\mu}$ and $\sigma = \widehat{\sigma}$.

Example 14.11 Simultaneous Two-Sided 95% Prediction Interval to Contain at Least k of m Future Fracture Strengths for Carbon-Epoxy Specimens. We use the carbon-epoxy fracture strength data from Example 14.2 to illustrate the computation of a simultaneous two-sided prediction interval to contain 4 out of 5 future observations from the fracture strength distribution. The procedure is similar to that used in Examples 14.9 and 14.10 for obtaining tolerance intervals using the ML estimates $\widehat{\mu} = 6.23476$ and $\widehat{\sigma} = 0.05302$ calculated from the data, and the bootstrap ML estimates $\widehat{\mu}_j^*$ and $\widehat{\sigma}_j^*$, $j = 1, \ldots, 2,000,000$. The bootstrap estimates are then used in (14.23) to determine $r_{L(0.95;4,5,19)} = -3.734$ and $r_{U(0.95;4,5,19)} = 1.429$, subject to the constraint (14.24). Thus, a simultaneous 95% prediction interval to contain 4 out of 5 future fracture strengths is obtained by substituting into (14.22), giving

$$[\exp(6.23476 - 3.734 \times 0.05302), \ \exp(6.23476 + 1.429 \times 0.05302)] = [418.6, \ 550.3].$$

Table 14.4 shows the preceding prediction interval, along with similar intervals to contain at least k out of m future observations from the fracture strength distribution for increasing values of k and m such that $k/m = 0.80$. The previously calculated 95% tolerance interval to contain a proportion 0.80 of the sampled distribution is also shown. This tabulation indicates that the prediction intervals become narrower as m and k increase, and ultimately converge to the tolerance interval to contain a proportion 0.80 of the sampled distribution. ∎

Interval type					Interval
Prediction	4	out of	5	future observations	[418.6, 550.3]
Prediction	8	out of	10	future observations	[420.7, 550.1]
Prediction	40	out of	50	future observations	[427.2, 548.8]
Prediction	80	out of	100	future observations	[428.8, 548.5]
Prediction	400	out of	500	future observations	[430.2, 548.2]
Tolerance		a proportion 0.80 of the distribution			[430.6, 548.1]

Table 14.4   Simultaneous two-sided 95% prediction intervals to contain at least k of m future fracture strengths with increasing k and m and a 95% tolerance interval to contain a proportion 0.80 of the fracture strength distribution, calculated from the carbon-epoxy data and based on a Weibull distribution fit.

14.6.2   Simultaneous One-Sided Prediction Bounds for k of m Future Observations

A simultaneous one-sided lower $100(1 - \alpha)\%$ prediction bound to be exceeded by at least k of m future observations from a location-scale distribution is

$$\underset{\sim}{Y}'_{k;m} = \widehat{\mu} + r'_{L(1-\alpha;k,m,n)}\widehat{\sigma}. \tag{14.26}$$

A simultaneous one-sided upper $100(1 - \alpha)\%$ prediction bound to exceed at least k of m future observations from a location-scale distribution is

$$\widetilde{Y}'_{k;m} = \widehat{\mu} + r'_{U(1-\alpha;k,m,n)}\widehat{\sigma}.$$

The values of $r'_{L(1-\alpha;k,m,n)}$ and $r'_{U(1-\alpha;k,m,n)}$ are determined such that (14.23) holds using

$$p_j^* = 1 - \Phi\left(\frac{\widehat{\mu}_j^* + r'_{L(1-\alpha;k,m,n)}\widehat{\sigma}_j^* - \widehat{\mu}}{\widehat{\sigma}}\right) \tag{14.27}$$

for the one-sided lower prediction bound and

$$p_j^* = \Phi\left(\frac{\widehat{\mu}_j^* + r'_{U(1-\alpha;k,m,n)}\widehat{\sigma}_j^* - \widehat{\mu}}{\widehat{\sigma}}\right) \tag{14.28}$$

for the one-sided upper prediction bound. For a log-location-scale distribution the corresponding prediction bounds are obtained by exponentiating $\underset{\sim}{Y}'_{k;m}$ and $\widetilde{Y}'_{k;m}$. Note that, as described in Section 14.4.4, (14.27) and (14.28) can be simplified if the simulation to obtain $\widehat{\mu}_j^*$ and $\widehat{\sigma}_j^*$, $j = 1, \ldots, B$, is done using $\mu = 0$ and $\sigma = 0$ instead of $\mu = \widehat{\mu}$ and $\sigma = \widehat{\sigma}$.

Example 14.12   Lower Prediction Bound to be Exceeded by at least k of m Future Fracture Strengths for Carbon-Epoxy Specimens. We again use the carbon-epoxy fracture strength data to illustrate the computation of a simultaneous one-sided lower 95% prediction bound to be exceeded by 4 out of 5 future observations from the fracture strength distribution. The procedure is similar to that used in Example 14.11 using the ML estimates $\widehat{\mu} = 6.23476$ and $\widehat{\sigma} = 0.05302$, calculated from the data, and the bootstrap ML estimates $\widehat{\mu}_j^*$ and $\widehat{\sigma}_j^*$, $j = 1, \ldots, 2,000,000$. The bootstrap estimates are then used in (14.23) with (14.27) to determine $r'_{L(0.95;4,5,19)} = -2.922$. Then the lower 95% prediction bound to be exceeded by 4 out

Bound type	To be exceeded by at least				Bound
Prediction	4	out of	5	future observations	437.0
Prediction	8	out of	10	future observations	438.9
Prediction	40	out of	50	future observations	444.9
Prediction	80	out of	100	future observations	446.4
Prediction	400	out of	500	future observations	447.8
Tolerance	a proportion 0.80 of the distribution				448.1

Table 14.5   Simultaneous lower 95% prediction bounds to be exceeded by at least k of m future fracture strengths with increasing k and m and a lower 95% tolerance bound to be exceeded by a proportion 0.80 of the fracture strength distribution, calculated from the carbon-epoxy data and based on a Weibull distribution fit.

of 5 future fracture strengths is obtained by substituting into (14.26) and then exponentiating, giving $\exp(6.23476 - 2.922 \times 0.05302) = 437.0$.

Table 14.5 compares the preceding lower prediction bound with similar bounds to be exceeded by at least k out of m future observations from the fracture strength distribution for increasing values of k and m such that $k/m = 0.80$. The lower 95% tolerance bound to be exceeded by the proportion 0.80 of the sampled population is also shown. Table 14.5 shows that the value of the prediction bound increases as k and m increase, ultimately converging to the lower 95% tolerance bound to be exceeded by a proportion 0.80 of the sampled distribution.

Note that, as described in Section 14.5.3, the lower 95% tolerance bound to be exceeded by a proportion 0.80 of the sampled distribution in the last line of Table 14.5 is equivalent to a 95% lower confidence bound for the 0.20 quantile of the distribution (or the lower endpoint of a two-sided 90% confidence interval for the 0.20 quantile of the distribution) and was obtained by using the methods described in Section 14.3.4.   ∎

14.7   OTHER SIMULATION AND BOOTSTRAP METHODS AND APPLICATION TO OTHER DISTRIBUTIONS AND MODELS

Hall (1988) commented that "There exists in the literature an almost bewildering array of bootstrap methods for constructing confidence intervals for a univariate parameter θ." Since 1988, the size of the array of bootstrap methods and applications seems to have grown explosively.

In this and the preceding chapter we have presented and illustrated the use of bootstrap and other simulation-based methods that are most commonly used in practice and those that we have found to be particularly useful in our own work. Some other important combinations, not explicitly presented previously in this chapter, also deserve mention here. Many more bootstrap/simulation-related methods are described in the textbooks and other references cited in the Bibliographic Notes section at the end of this chapter.

14.7.1   Resampling for Parametric Bootstrap Confidence Intervals

The parametric bootstrap methods in Sections 14.3–14.6 not only assumed a particular parametric distribution, but also took advantage of the theoretical properties of certain pivotal (or generalized pivotal) quantities. Where such pivotal quantities are not available, it is possible (and theoretically justifiable) to simply adapt the nonparametric methods to generate bootstrap samples in Section 13.4 to obtain parametric bootstrap intervals. For example, nonparametric resampling (Section 13.2.1) can be used to generate B bootstrap samples and for each of these bootstrap samples the assumed parametric model can be fit, giving B bootstrap estimates of a quantity of interest. Then any of the methods in Section 13.4 (simple percentile, BC percentile, BCa percentile, or percentile-t) could be adapted to construct the desired parametric bootstrap confidence interval, without regard to the existence of related pivotal quantities. In fitting a parametric distribution or model, the smoothness condition mentioned in Section 13.4.8 is met for most important inference needs because ML estimates, under the standard regularity conditions in parametric models, are smooth functions of the data. Thus, such methods generally work well for estimating distribution quantiles or probabilities. As with nonparametric bootstrap methods, the BCa percentile and percentile-t methods have superior theoretical properties and are to be preferred over the simple percentile and BC methods. Among parametric bootstrap methods, we recommend the bootstrap-t method because it requires only an estimate of the standard error from each bootstrap sample (usually readily available when using a parametric model) and does not require computation of an acceleration constant.

14.7.2   Random-Weight Bootstrap Sampling for Parametric Bootstrap Confidence Intervals

Although the random-weight bootstrap sampling method (Section 13.2.2) is nonparametric (because it does not require specification of a particular parametric distribution), the method is most commonly used for parametric inference problems. That is, at each bootstrap iteration, random weights are selected (without regard to any parametric assumption) and then a parametric distribution or model is fit using the weights, as described in Section 13.2.2, to obtain bootstrap estimates. As mentioned there, random-weight bootstrap sampling is particularly appealing when there is a nonnegligible probability that a bootstrap resample will not allow estimation of the parameters of the assumed parametric distribution or model or the quantity of interest. Another class of applications for which the random-weight bootstrap has strong appeal is when the original data or model have complications such as random censoring and truncation, where it would be difficult to develop parametric methods to simulate bootstrap samples.

As mentioned in Section 14.7.1, once the bootstrap samples and bootstrap estimates have been computed, any of the interval-construction methods in Section 13.4 (i.e., simple percentile, BC percentile, BCa percentile, or percentile-t) can be adapted to construct the desired bootstrap confidence intervals.

14.7.3   Bootstrap Methods with Other Distributions and Models

Up to this point, our presentation and examples of parametric bootstrap methods have been based on the assumption of a (log-)location-scale distribution. Parametric bootstrap methods can, however, be applied much more generally to other commonly used parametric distributions such as the gamma, generalized gamma, Birnbaum–Saunders, and inverse Gaussian or any other specified distribution. We use a further example to illustrate how bootstrap methods can be adapted to other distributions and kinds of data. Further examples and comparisons are given in Chapter 18. We also describe special considerations that may be needed when using bootstrap methods with censored data.

The following example illustrates the use of bootstrap methods for a more complicated problem.

Example 14.13   Parametric Bootstrap Confidence Intervals for Generalized Gamma Distribution Quantiles Fitted to the Ball Bearing Failure Data. As described in Section C.3.8,

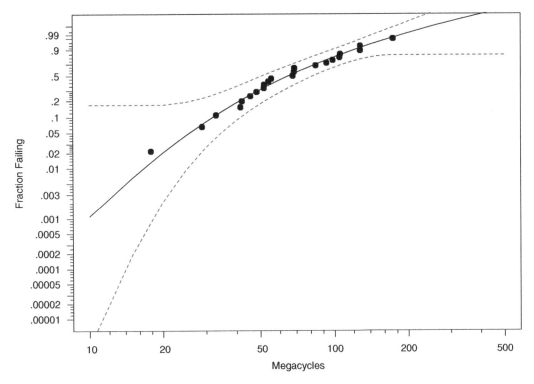

Figure 14.6 Weibull probability plot of the ball bearing failure data with the GNG ML estimate and a set of pointwise Wald 95% confidence intervals for the cdf.

the generalized gamma distribution has three parameters and includes the Weibull, lognormal, Fréchet, and gamma distributions as special cases. The following example is an application for which attempting to use the standard parametric bootstrap method caused difficulties but the random-weight bootstrap worked well.

Consider again the ball bearing fatigue failure data given in Table 4.1 and used in Example 4.15. We now extend the application of the generalized gamma (GNG) distribution used in Meeker and Escobar (1998, Example 11.2) by using bootstrap methods to compute confidence intervals for the 0.10 quantile of the bearing life distribution. The ML estimates of the GNG distribution parameters are $\hat{\mu} = 4.230$, $\hat{\sigma} = 0.5100$, and $\hat{\lambda} = 0.3077$. Figure 14.6 is a Weibull probability plot of the data, showing the ML estimate of the GNG cdf and a set of pointwise Wald 95% confidence intervals for the cdf. The ML estimate of the quantile $t_{0.10}$, obtained by substituting into (C.13), is $\hat{t}_{0.10} = 32.08$.

We initially tried to obtain bootstrap estimates with the traditional resampling method (Section 13.2.1), but found that the ML iterations failed to converge correctly in 193 of the 4,000 (4.8%) bootstrap resamples (in particular, the ML estimates of the shape parameter λ were, in these cases, on the boundary of the interval $[-12, 12]$ that is commonly used to constrain λ to practical values). That problem arose in only 8 of the 4,000 (0.2%) random-weight bootstrap samples (an occurrence rate that is small enough to ignore).

Table 14.6 compares six different methods for constructing a confidence interval for $t_{0.10}$. The likelihood-ratio and Wald methods were computed by direct application of the general methods described in Chapter 12. The four bootstrap intervals were all computed based on random-weight bootstrap ML estimates (Section 13.2.2) and applying the simple percentile and

Method	Section(s)	95% confidence interval
Random-weight bootstrap simple percentile	13.4.2	[23.4, 42.9]
Random-weight basic bootstrap	13.4.5	[21.2, 40.8]
Random-weight basic bootstrap (log transformation)	13.4.5	[24.0, 44.0]
Random-weight bootstrap-t	13.4.6	[20.9, 41.0]
Likelihood-ratio method	12.5.4	[18.4, 42.2]
Wald-approximation method	12.6.4	[20.1, 51.3]

Table 14.6 Parametric 95% confidence intervals for the ball bearing generalized gamma 0.10 quantile using six different methods.

bootstrap-t confidence interval methods (Sections 13.4.2 and 13.4.6, respectively), as well as the basic bootstrap method (Section 13.4.5) with and without a log transformation.

Table 14.6 shows that the upper bound of the Wald interval is appreciably higher than any of the other upper bounds, indicating skewness in the profile likelihood for $t_{0.10}$ (not shown here). The other intervals show only moderate differences. As in other examples, statistical theory suggests that the random-weight bootstrap-t method is more reliable than the other simpler alternatives. ■

14.7.4 Bootstrapping with Complicated Censoring

Methods for analyzing censored data were discussed in Section 12.2 and a more thorough presentation is given in Meeker and Escobar (1998). Bootstrap methods require special care when data are censored. In particular, bootstrap samples must be generated in a manner that accurately mimics the actual data-generating process. For simple Type 1 censoring (i.e., *time* censoring for which a laboratory life test is terminated at a fixed time for all units in a sample) or Type 2 censoring (i.e., *failure* censoring for which a laboratory life test is terminated when a specified number of units have failed), simulating bootstrap samples is straightforward (e.g., as described in Meeker and Escobar, 1998, Section 4.13). In more complicated situations, censoring can arise from a complicated process that may be difficult to model. This is especially true with field data where censoring arises due to a combination of, often random, factors that impact how units enter and leave service, for example, the withdrawal of units from service due to the occurrence of competing causes of failure in the analysis of a specific failure mode.

The classic nonparametric bootstrap resampling method (Section 13.2.1) provides a method of generating bootstrap samples that does not require explicit modeling of a censoring mechanism. Problems using this method can arise, however, because there is always a chance that a bootstrap sample will contain all censored observations, making it impossible to compute the needed bootstrap estimates. In such cases, the random-weight bootstrap sampling method (Section 13.2.2) avoids this problem because all of the original observations are represented in each bootstrap sample.

Example 14.14 Parametric Bootstrap Confidence Intervals for Evaluating Lifetime Quantiles of a Rocket Motor. The rocket motor life data in Table 14.7 were first presented in Olwell and Sorell (2001).

The US Navy had an inventory of approximately 20,000 missiles. Each included a rocket motor—one of five critical components. While in storage awaiting potential use, each of these rocket motors was subject to continuous thermal cycling. Only 1,940 of the systems had actually been put into flight over a period of time up to 18 years subsequent to their manufacture. At their

Years	Number of motors	Years	Number of motors	Years	Number of motors
> 1	105	> 8	211	> 14	14
> 2	164	> 9	124	> 15	5
> 3	153	> 10	90	> 16	3
> 4	236	> 11	72	< 8.5	1
> 5	250	> 12	53	< 14.2	1
> 6	197	> 13	30	< 16.5	1
> 7	230				

Table 14.7   Rocket motor life data (in years since manufacture).

time of flight, 1,937 of these motors performed satisfactorily; but there were three catastrophic launch failures. Responsible scientists and engineers believed that these failures were due to the thermal cycling to which the motors were exposed continually prior to deployment. In particular, it was believed that the thermal cycling resulted in failed bonds between the solid propellant and the missile casing.

The failures raised concern about the previously unanticipated possibility of a sharply increasing failure rate over time (i.e., rapid wearout) as the motors aged and were subjected to thermal cycling while in storage. If this were indeed the case, a possible—but highly expensive— remedial strategy might be to replace aged rocket motors with new ones. Thus, to assess the magnitude of the problem it was desired to quantify the rocket motor failure probability as a function of the amount of thermal cycling to which a motor was exposed and to obtain appropriate confidence bounds around such estimates based on the results for the 1,940 rocket motors—assuming these to be a random sample from the larger population (at least with regard to their failure time distribution).

Because no information was directly available on the thermal cycling history of the individual motors, the age of the motor (i.e., time since manufacture) at launch was used as a surrogate. This was not an ideal replacement because the thermal cycling rate, or rate of accumulation of other damage mechanisms, varied across the population of motors, depending on an individual missile's environmental storage history. The effect of such an imperfect time scale is to increase the variability in the observed lifetime response, as described in Meeker et al. (2009). The failure probability 20 years after manufacture was of particular interest.

The specific age at failure of each of the three failed motors was not known—all that was known was that failure, in each case, had occurred sometime prior to the time of launch—thus making the time since manufacture at launch left-censored observations of the actual failure times. Similarly, the information of (eventual) failure age for the 1,937 successful motors is right censored—all that is known is that the time to the yet-to-occur failure exceeds the calendar age at the time of launch. Thus, the available motor life data, which is shown in Table 14.7, contained only left- and right-censored observations—but *no* known exact failure times. Figure 14.7 is an event plot that further illustrates the structure of the data.

Because failure times are only loosely bounded and because of the very small number of known failures, the amount of information in the data is severely limited. Nevertheless, it is possible to estimate the Weibull distribution parameters from these data. The ML estimates of the location and scale parameters of the underlying SEV distribution are $\widehat{\mu} = 3.055$ and $\widehat{\sigma} = 0.123$. The corresponding Weibull parameter ML estimates are $\widehat{\eta} = 21.23$ and $\widehat{\beta} = 8.126$. The estimate of the Weibull shape parameter β is very large for a population that is known to have much variability. Figure 14.8 is a Weibull probability plot of the data containing a

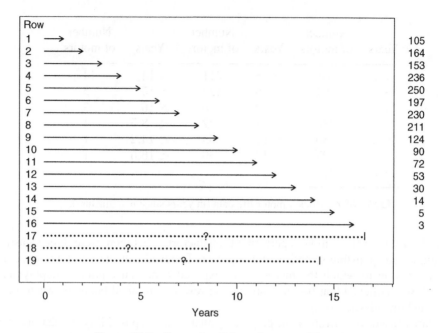

Figure 14.7   Event plot of the rocket motor life data.

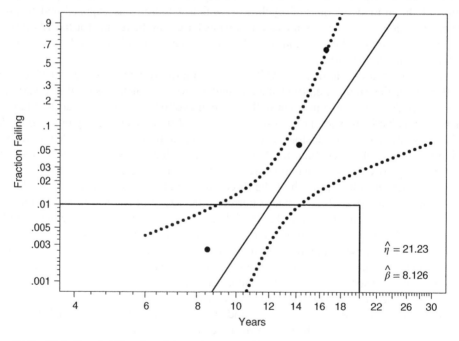

Figure 14.8   Weibull probability plot of the rocket motor life data showing a nonparametric estimate and the Weibull ML estimate of the cdf.

Method	Section(s)	95% confidence interval
Random-weight bootstrap simple percentile	14.4.2	[0.014, 1.0000000]
Likelihood ratio	12.5.4	[0.023, 0.9999957]
Wald approximation	12.6.4	[0.027, 0.9999988]

Table 14.8 95% confidence intervals for $F(20)$ for the Weibull distribution fit to the rocket motor data using different methods.

nonparametric estimate (the three plotted points) and the Weibull ML estimate of fraction failing as a function of years in service, along with pointwise Wald confidence intervals for the cdf. The inner rectangle indicates a target of no more than a fraction 0.01 failing in 20 years.

In this example, it would be difficult or impossible to model the censoring mechanism for parametric simulation of bootstrap samples. Also, traditional bootstrap resampling (Section 13.2.1) would not work because of the high probability of obtaining resamples that would be unable to estimate the Weibull parameters (e.g., the probability of obtaining all right-censored observations in a resample of size 1,940 is approximately 0.05). Thus we use the random-weight bootstrap method for which we were able to estimate $F(20)$ for all 4,000 bootstrap samples. Due to numerical difficulties, estimated standard errors were not always available. Thus we report bootstrap results only for the simple percentile method.

The ML estimate of $F(20)$, the fraction failing before 20 years after manufacture (or the probability that a missile of age 20 years will not successfully fire) is $\widehat{F}(20) = \Phi_{sev}[(\log(20) - 3.055)/0.123] = 0.46$, which appreciably exceeds the target value of 0.01.

Table 14.8 compares several different confidence intervals for $F(20)$. The Wald and likelihood procedures give similar results in this example. The simple percentile method results in an interval that is somewhat wider. These differences, however, are inconsequential when one looks at the width of the intervals. Each provides the same practical message: due to the limited amount of information in the data, useful bounds on $F(20)$ are not available from the data alone. In Section 18.7 we return to this example, supplementing the data with additional engineering information. ∎

BIBLIOGRAPHIC NOTES

Parametric Bootstrap

Hall (1988) presents a theoretical framework for comparing the properties of different bootstrap procedures. His results show that both the BCa percentile and the bootstrap-t methods are second-order correct as parametric bootstrap methods for the multivariate exponential family of distributions.

Intervals based on pivotal quantities

Thoman et al. (1970) show how to use PQs to construct confidence intervals for Weibull distribution failure probabilities and quantiles. Lawless (2003, Section 5.1.2 and Appendix E) provides PQs to construct exact confidence intervals for parameters, quantiles, and probabilities of location-scale (and log-location-scale) distributions for complete and (failure) censored samples. Monte Carlo simulation is generally required to obtain the distribution of the needed PQs.

Intervals based on generalized pivotal quantities

Tsui and Weerahandi (1989) introduced the concept of generalized p-values, which set the stage for generalized pivotal quantities (GPQs). Weerahandi (1993) shows how to use GPQs to construct approximate confidence intervals for applications for which no PQ exists, and Weerahandi (1995) describes a broad range of applications for making inferences (including confidence intervals) with GPQs. Hamada and Weerahandi (2000) apply GPQ methods to obtain confidence intervals for variance components and functions of variance components from gauge repeatability and reproducibility studies. Chiang (2001) independently developed the surrogate variable method to obtain confidence intervals on functions of the parameters of a random-effects model that is equivalent to the GPQ method. Weerahandi (2004) applies GPQs to problems involving repeated measures and other mixed effect models. Krishnamoorthy and Mathew (2009) describe GPQs for location-scale (and log-location-scale) distributions and how to compute confidence intervals for various applications. Again, Monte Carlo simulation is generally required to obtain the distributions of the needed GPQs. Li et al. (2009) provide an interesting application of GPQ confidence intervals.

Relationship between generalized pivotal quantities and generalized fiducial inference

Hannig et al. (2006) show the connection between GPQ based inference and extensions of classical fiducial inference, which is called generalized fiducial inference (GFI). They show that a wide range of GPQ confidence interval procedures could be obtained by using GFI methods. They also proved that confidence intervals based on GFI are asymptotically exact (i.e., have a coverage probability that approaches the nominal confidence level as the sample size increases) and gave a condition for GFI intervals to be exact in finite samples. In addition, they gave some general methods for constructing GPQs. Hannig (2009) provided a more general treatment of GFI ideas, including extensions to discrete distributions. Hannig (2013) shows that discretization of the data avoids problems of nonuniqueness of the fiducial distribution while maintaining asymptotic correctness of the GFI procedures. Hannig et al. (2016) provide a review of GFI methods.

Tolerance and prediction intervals

Yuan et al. (2017) provide a general simulation-based approach for constructing two-sided tolerance intervals based on data from a member of the (log-)location-scale family of distributions with complete or right-censored data. They treat both the control-the-center and the control-both-tails type of tolerance interval. The material in our Section 14.5 is based on this paper. Xie et al. (2017) provide a general simulation-based approach for constructing two-sided simultaneous prediction intervals to contain at least k out of m future observations and corresponding one-sided simultaneous prediction bounds, based on data from a member of the (log-)location-scale family of distributions with complete or right-censored data. The material in our Section 14.6 is based on this paper.

Chapter *15*

Introduction to Bayesian Statistical Intervals

OBJECTIVES AND OVERVIEW

This chapter presents the basic concepts behind the construction of Bayesian statistical intervals and the integration of prior information with data that Bayesian methods provide. The use of such methods has seen a rapid evolution over recent years. The development of the theory and application of Markov chain Monte Carlo (MCMC) methods and vast improvements in computational capabilities have made the use of such methods feasible.

There are, moreover, many applications for which practitioners have solid prior information on certain aspects of their applications based on knowledge of the physical-chemical mechanisms and/or relevant experience with a previously studied phenomenon. For example, engineers often have useful, but imprecise, knowledge about the effective activation energy in a temperature-accelerated life test or about the Weibull distribution shape parameter in the analysis of fatigue failure data or the amount of variability in a measurement process. In such applications, the use of Bayesian methods is compelling as it offers an appropriate compromise between assuming that such quantities are known and assuming that nothing is known.

We describe three specific methods for obtaining Bayesian intervals:

- A traditional approach based on conjugate distributions that is useful in a few particularly simple situations.

- A simple approach based on Monte Carlo simulation that provides intuition and insight into how Bayesian methods work.

- A general approach based on MCMC simulation that is recommended for most applications of Bayesian methods.

We then discuss the use of these methods for constructing Bayesian credible, tolerance, and prediction intervals. (Bayesian credible intervals are analogous to non-Bayesian confidence

Statistical Intervals: A Guide for Practitioners and Researchers, Second Edition.
William Q. Meeker, Gerald J. Hahn and Luis A. Escobar.
© 2017 John Wiley & Sons, Inc. Published 2017 by John Wiley & Sons, Inc.
Companion Website: www.wiley.com/go/meeker/intervals

intervals and a credible level is similar to a confidence level—see Section 1.15.) We apply these methods to the binomial, Poisson, and normal distributions in Chapter 16. Then we extend these methods in Chapter 17 to consider the construction of Bayesian intervals for the more complicated situation involving hierarchical models and provide other examples of the use of Bayesian methods in Chapter 18.

The topics discussed in this chapter are:

- An overview of the motivations for using Bayesian inference, how it differs from non-Bayesian likelihood inference, a statement of Bayes' theorem, and brief introductions to the specification of prior information and parameterization (Section 15.1).

- An example that illustrates a practical approach to making Bayesian inferences by introducing the specification of a prior distribution, the characterization of a posterior distribution using simulation, and the construction of Bayesian credible intervals from the results (Section 15.2).

- How to choose a prior distribution for a Bayesian analysis and the traditional use of conjugate distributions (Section 15.3).

- The basic ideas of using MCMC simulation to compute estimates and credible intervals in a Bayesian analysis (Section 15.4).

- The construction of Bayesian tolerance and prediction intervals (Section 15.5).

Some technical details about Bayesian inference methods are given in Appendix H.

15.1 BAYESIAN INFERENCE: OVERVIEW

15.1.1 Motivation

There are, depending on the application, three strong motivators for using Bayesian methods to construct statistical intervals:

- Bayesian methods provide a formal analytical framework for an analyst to incorporate prior information into a data analysis/modeling problem to supplement limited data, often providing important improvements in precision (and often resulting in cost savings).

- Bayesian methods can handle, with relative ease, complicated data-model combinations for which no software using classical non-Bayesian likelihood-based methods exists and for which implementing non-Bayesian methods would be difficult. For example, currently available software for doing Bayesian computations can analyze combinations of nonlinear relationships, random effects, and censored data that cannot be handled readily by currently available commercial software.

- When using Bayesian methods, it is generally easy to produce estimates and credible intervals for complicated functions of the model parameters, such as the probability of product failure or quantiles of a failure-time distribution, that might be extremely difficult using non-Bayesian methods.

15.1.2 Bayesian Inference versus Non-Bayesian Likelihood Inference

The top diagram in Figure 15.1 shows the components of a likelihood-based non-Bayesian inference procedure. Inputs are the data and a model for the data. The inference outputs are, for example, point estimates and confidence intervals for quantities of interest (e.g., a quantile

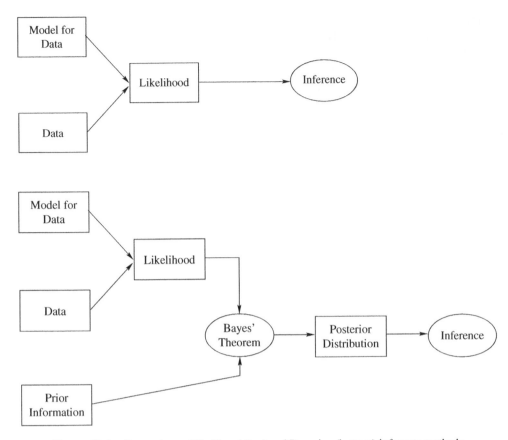

Figure 15.1   Comparison of likelihood (top) and Bayesian (bottom) inference methods.

or a failure probability associated with a failure-time distribution), or in some applications a tolerance or a prediction interval. The bottom diagram in Figure 15.1 is a similar diagram for the Bayesian inference procedure. In addition to the model and the data, one must also specify a joint prior distribution that describes one's knowledge about the unknown parameters of the model. Bayes' theorem is used to combine the prior information with the likelihood to produce a posterior distribution. Similar to non-Bayesian inference, outputs are point estimates and credible intervals, tolerance intervals, and prediction intervals.

15.1.3   Bayes' Theorem and Bayesian Data Analysis

Bayes' theorem is a probability rule that relates different kinds of conditional probabilities (or conditional probability density functions) to one another. This rule is also the basis for the Bayesian method of statistical inference which allows one to combine available data with prior information to obtain a posterior (or updated) distribution that can be used to make inferences about some vector $\boldsymbol{\theta}$ of unknown parameters or functions thereof. Bayes' theorem for continuous random variables can be written as

$$f(\boldsymbol{\theta}|\text{DATA}) = \frac{L(\text{DATA}|\boldsymbol{\theta})f(\boldsymbol{\theta})}{\int L(\text{DATA}|\boldsymbol{\theta})f(\boldsymbol{\theta})d\boldsymbol{\theta}} = \frac{R(\boldsymbol{\theta})f(\boldsymbol{\theta})}{\int R(\boldsymbol{\theta})f(\boldsymbol{\theta})d\boldsymbol{\theta}}, \tag{15.1}$$

where the joint prior distribution $f(\boldsymbol{\theta})$ quantifies the available prior information about the unknown parameters in $\boldsymbol{\theta}$. The output of (15.1) is $f(\boldsymbol{\theta}|\text{DATA})$, the joint posterior distribution for $\boldsymbol{\theta}$, reflecting knowledge of $\boldsymbol{\theta}$ after the information in the data and the prior distribution have been combined. $L(\text{DATA}|\boldsymbol{\theta})$ is the likelihood function and $R(\boldsymbol{\theta}) = L(\text{DATA}|\boldsymbol{\theta})/L(\text{DATA}|\widehat{\boldsymbol{\theta}})$ is the relative likelihood (introduced in Chapter 12), and the integral is computed over the region where $f(\boldsymbol{\theta}) > 0$. Also $L(\text{DATA}|\boldsymbol{\theta})$ is a function of the assumed model for the data and must be proportional to the probability of the data. It quantifies the information in the data. The vector $\boldsymbol{\theta}$ of unknown parameters is to be estimated. For discrete random variables, the integrals in (15.1) are replaced by summations.

In general, it is impossible to compute the integrals in (15.1) in closed form. Numerical methods can be computationally intensive or intractable when $\boldsymbol{\theta}$ has more than two or three elements, as is frequently the case in dealing with regression analyses and with the hierarchical models described in Chapter 17. In the past, this was an impediment to the use of Bayesian methods. Today, however, new statistical and numerical methods that take advantage of modern computing power are making it feasible to apply Bayesian methods to a much wider range of applications. Modern computing methods for Bayesian analysis make inferences on the basis of a large number of (relatively easy-to-compute) sample draws from the joint posterior distribution. This approach makes Bayesian inference computations relatively simple to conduct.

15.1.4   The Need for Prior Information

As we can see from (15.1), the use of Bayesian methods for statistical modeling and inference requires one to specify a joint prior distribution $f(\boldsymbol{\theta})$ to describe the prior knowledge that is available about the unknown parameters in $\boldsymbol{\theta}$. One reason why the use of Bayesian methods has been controversial in many applications is that it is possible that the joint prior distribution will have a strong influence on the resulting inferences, especially when the amount of data is limited, as is common in many applications. When the joint prior distribution $f(\boldsymbol{\theta})$ is diffuse (i.e., relatively flat over the range of $\boldsymbol{\theta}$ values for which the likelihood is nonnegligible) and if there is ample data, the data are likely to dominate the prior distribution. In such situations, one can expect the likelihood $L(\text{DATA}|\boldsymbol{\theta})$ to be approximately proportional to the joint posterior distribution. This will result in Bayesian inferences that are similar to what one would make using non-Bayesian methods like maximum likelihood (ML). The important topic of prior distribution selection is discussed more fully in Section 15.2.2.

15.1.5   Parameterization

Parametric statistical models have unknown parameters that are to be estimated from data (sometimes with the aid of prior information). For example, the Weibull distribution cumulative distribution function is often written as

$$\Pr(T \le t; \eta, \beta) = F(t; \eta, \beta) = 1 - \exp\left[-\left(\frac{t}{\eta}\right)^{\beta}\right], \quad t > 0, \tag{15.2}$$

where $\beta > 0$ is a unitless shape parameter and $\eta > 0$ is a scale parameter that has the same units as T. The scale parameter is (approximately) the 0.632 quantile of the distribution (e.g., in dealing with failure-time data, the time by which a proportion 0.632 of the population will fail). In Chapters 12 and 14, we found it useful to use the alternative parameterization $\mu = \log(\eta)$ and $\sigma = 1/\beta$, corresponding to the parameters of the smallest extreme value distribution (a member of the location-scale family) of the logarithms of the Weibull random variable. For more information about the Weibull and similar probability distributions, see Section C.3.1.

It might be, however, that neither (η, β) nor (μ, σ) is the best set of parameters to use in Bayesian applications. For example, in fitting a Weibull distribution to failure-time data, the use of an appropriate value of the quantile t_p for p other than 0.632, instead of η, together with β is generally a better choice for the following reasons:

- Usually, engineers are more likely to have prior information on a quantile of the failure-time distribution other than the 0.632 quantile, because one often observes only a small fraction failing in life tests or in field operation. It will, therefore, generally be easier to elicit prior information about a quantile in the lower tail of the distribution than to obtain prior information about η. In contrast, one would retain the shape parameter β because engineers often have some information about this parameter. In particular, the shape parameter is often indicative of a product's failure mechanism. Knowledge of the failure mechanism then often suggests an approximate plausible range for this parameter.

- When there is heavy censoring (i.e., only a small fraction failing), the likelihood surface for η and β will tend to have an elongated shape, reflecting the strong correlation between the ML estimators of η and β. This strong correlation can make the computation of ML (and also Bayesian) estimates more difficult by increasing the amount of computer time needed or the probability of algorithmic failure. As an alternative, we propose using what Ross (1970, 1990) calls *stable parameters* which generally correspond to quantities that one can readily identify in a plot of the data.

Although it is possible to use one parameterization for prior specification and a different parameterization for computing parameter estimates iteratively, we generally find that a single alternative parameterization usefully serves both purposes.

A useful reparameterization for the Weibull distribution replaces η with a particular distribution quantile that could be estimated nonparametrically directly from the available data. The p quantile of the Weibull distribution can be written as $t_p = \eta[-\log(1-p)]^{1/\beta}$. Replacing η with the equivalent expression $\eta = t_p/[-\log(1-p)]^{1/\beta}$ in (15.2) provides a reparameterized version of the Weibull distribution,

$$\Pr(T \le t; t_p, \beta) = F(t; t_p, \beta) = 1 - \exp\left[-\left(\frac{t}{t_p/[-\log(1-p)]^{1/\beta}} \right)^{\beta} \right]$$

$$= 1 - \exp\left[\log(1-p)\left(\frac{t}{t_p} \right)^{\beta} \right], \quad t > 0.$$

In addition, especially when there is heavy censoring (i.e., only a small fraction failing), estimation of (t_p, β) will be more stable than estimating (η, β) for some appropriately chosen value of p. Moreover, graphical estimates of the chosen t_p and β within the range of the data (e.g., estimated by fitting a simple linear regression line through the points on a probability plot) provide excellent starting values for either ML or Bayesian estimation. A useful rule of thumb is to choose parameters that are near the center of the data so that the parameters can be approximately identified from a plot of the data. For example, if the nonparametric estimate of the fraction failing at the largest failure time is 0.10, then choosing $t_{0.05}$ would be expected to work well.

15.2   BAYESIAN INFERENCE: AN ILLUSTRATIVE EXAMPLE

This section uses a relatively simple (but nontrivial, due to the censoring) example of the analysis of limited reliability field-failure data to illustrate the basic ideas and computational methods behind the use of Bayesian methods. In this and subsequent examples, we will compare:

- A non-Bayesian (ML) analysis
- A Bayesian analysis with a diffuse (approximately noninformative) prior distribution, and
- A Bayesian analysis with an informative prior distribution.

For the application in this section, the informative prior distribution will contain information about the Weibull shape parameter β. The comparison shows that the Bayesian analysis with diffuse prior information provides results that are similar to the non-Bayesian analysis, but that the Bayesian analysis with informative prior information provides more precise (i.e., narrower) statistical intervals than the preceding two analyses.

15.2.1   Example Using Weibull Distribution ML Analysis

To illustrate the basic ideas of Bayesian inference we will use a simple example of fitting a Weibull distribution to censored field failure data for an aircraft engine bearing cage. The following example introduces the data and starts with a non-Bayesian ML analysis.

Example 15.1   ML Estimation for the Bearing Cage Data. The data for this example were first given in Abernethy et al. (1983) and were also analyzed in Meeker and Escobar (1998, Chapters 8 and 14). Over time, 1,703 similar aircraft engines with a particular type of bearing cage had been introduced into service. The design life specification for the bearing cage was that the 0.10 quantile of bearing life (also known as B10 life) should be at least 8,000 hours of service. The longest running units had seen only 2,220 hours of service. At the time of the analysis of the data, there had been only six failures, and a preliminary Weibull distribution analysis of these limited data suggested that the reliability goal might not have been met. Management needed to know if a redesign of the bearing cage would be required and also wanted to predict how many spare parts would be needed over future years to keep the fleet of aircraft flight-ready.

Figure 15.2 is a Weibull probability plot of the bearing cage failure-time data. The top right-hand corner of the inner rectangle shows the bearing cage reliability goal of no more than a proportion 0.10 failing by 8,000 hours of service. The ML estimate of the Weibull distribution cdf at 8,000 hours lies appreciably above this proportion, suggesting that the reliability goal has most likely not been met. The dotted curves in Figure 15.2 are a set of pointwise approximate 95% confidence intervals for the bearing cage lifetime distribution. The lower endpoint of the confidence interval at 8,000 hours is about 0.03, suggesting the *possibility* that the proportion of the population failing by 8,000 hours could be that low. This provides an argument for postponing decision making until more information is available. At the same time, the upper endpoint of the confidence interval for the failure probability suggests that the actual proportion failing at 8,000 hours could be far more than 0.10. If this is, in fact, the case, waiting for more information could cause a bad situation to become much worse. The problem is that there is very little information in the available data. The use of Bayesian methods with informative prior information could provide more precision and a better basis for decision making. Fortunately, such prior information was available. ∎

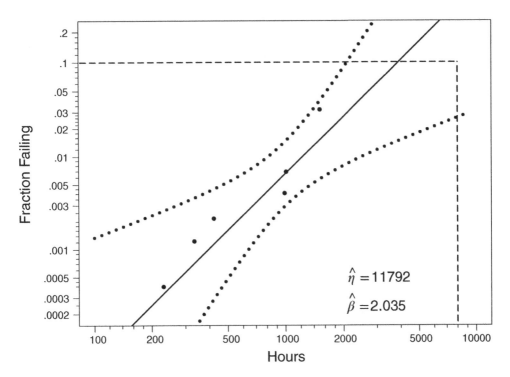

Figure 15.2   Weibull probability plot of the bearing cage data with ML estimates and a set of 95% pointwise confidence intervals.

15.2.2   Specification of Prior Information

For the sake of comparison, we will first do a Bayesian analysis with a *diffuse* joint prior distribution, followed by an analysis that uses a joint prior distribution with an *informative* prior distribution containing information about the Weibull shape parameter β. Uniform distributions, over a wide range of values, are often used to define a diffuse prior distribution for a parameter. A uniform prior distribution between a and b will be indicated by $\mathrm{UNIF}(a, b)$. A log-uniform distribution (see Section C.3.4) over a relatively wide range is often used to specify a diffuse prior distribution for a parameter that must be positive. The log-uniform distribution for a single parameter θ over a range from a to b is equivalent to a uniform distribution for $\log(\theta)$ between $\log(a)$ and $\log(b)$ and will be indicated by $\mathrm{LUNIF}(a, b)$.

Normal distributions are commonly used to specify an informative prior distribution. One method of eliciting prior information for an informative prior distribution for a particular parameter is to request a range of values that would, with 0.99 probability, contain the value of the parameter, with the understanding that the probability of being outside of the interval would be 0.005 on each side. We will denote such a prior distribution by $<\mathrm{NORM}>(a, b)$. Note the $<\ \ >$ indicates that the (a, b) values correspond to the above-defined "soft" endpoints and *not* the parameters of the distribution. Relating (a, b) to the mean and standard deviation of the normal distribution gives

$$<\mathrm{NORM}>(a, b) \equiv \mathrm{NORM}\left(\frac{a + b}{2}, \frac{b - a}{2 \times z_{0.995}}\right)$$

where $z_{0.995} = 2.5758$ is the 0.995 quantile of the standard normal distribution.

Analysis based on	Weibull distribution stable parameters	
	$t_{0.10}$	β
Diffuse prior	LUNIF$(1{,}000, 50{,}000)$	LUNIF$(0.30, 8.0)$
Informative prior	LUNIF$(1{,}000, 50{,}000)$	<LNORM>$(1.5, 3.0)$

Table 15.1   Prior distribution specification for the bearing cage data analyses.

Correspondingly, a lognormal distribution is often used to specify an informative prior distribution for a parameter that must be positive. Also, <LNORM>(a, b) will be used to indicate a lognormal distribution with probability 0.99 between a and b, again with the understanding that the probability of being outside of the interval would be 0.005 on each side. Relating (a, b) to the usual μ, σ parameters of the lognormal distribution gives

$$<\text{LNORM}>(a, b) \equiv \text{LNORM}\left(\frac{\log(a) + \log(b)}{2}, \frac{\log(b) - \log(a)}{2 \times z_{0.995}} \right).$$

Example 15.2   Choosing the Prior Distributions for the Bearing Cage Example. For the bearing cage example, as described in Section 15.1.5, we will specify both the diffuse and the informative prior distributions for the (more meaningful) 0.10 quantile of the failure-time distribution (i.e., $t_{0.10}$) in place of the scale parameter η. We will continue to use the Weibull shape parameter β as the second distribution parameter. We do this not only because, as previously suggested, it is easier to elicit prior information on these two parameters, but also because the information on them is more likely to be approximately independent (allowing the joint prior distribution to be specified more simply by two marginal distributions), and because the 0.10 quantile is the primary quantity of interest in this application. Table 15.1 summarizes the diffuse and informative prior distributions that we will use in this example.

Because there is little or no prior information available for $t_{0.10}$, we specify our prior (lack of) knowledge about $t_{0.10}$ by a log-uniform distribution over the wide range from 1,000 to 50,000 hours for both the diffuse and the informative prior distribution. Choosing an even wider range for this prior distribution would have little practical effect on the results.

For the diffuse-prior-distribution analysis, we use a log-uniform distribution between 0.30 and 8 to describe our lack of knowledge about the Weibull shape parameter β (i.e., a value of β outside of this range would not be expected). Again, choosing an even wider range for this prior distribution would have little practical effect on the final answers. For the informative-marginal prior distribution for β we use a lognormal distribution with 99% of its probability between 1.5 and 3.0 (denoted by <LNORM>$(1.5, 3.0)$). The justification for this informative prior distribution comes from previous field-data experience with fatigue failures in similar bearing cages, as well as an understanding of the underlying failure mechanism.   ■

15.2.3   Characterizing the Joint Posterior Distribution via Simulation

For a given likelihood and prior distribution, Bayes' theorem, as stated in (15.1), gives the joint posterior distribution of the parameters being estimated. The joint posterior distribution can be calculated explicitly using (15.1) only in special cases. In other situations, such computations are intractable. Instead, we characterize the joint posterior distribution using modern simulation methods—described further in this and subsequent sections—by obtaining a large number of "sample draws" from the joint posterior distribution. These sample draws are then used to construct Bayesian credible intervals for the quantities of interest, such as parameters, quantiles, and probabilities, as described in Section 15.2.6. Section 15.5 shows how to use the sample draws to compute Bayesian tolerance and prediction intervals.

In particular, the basic output of modern Bayesian analysis computational tools is generally a large number (B) of sample draws from the joint posterior distribution of the model parameters. These sample draws are usually organized as a matrix with columns corresponding to marginal posterior distributions for each of the unknown model parameters and the B rows corresponding to individual sample draws from the joint posterior distribution. Suppose that a model has q parameters. Then the matrix of sample draws would look like this:

$$\begin{pmatrix} \boldsymbol{\theta}_1^* \\ \boldsymbol{\theta}_2^* \\ \vdots \\ \boldsymbol{\theta}_B^* \end{pmatrix} = \begin{pmatrix} \theta_{11}^* & \theta_{12}^* & \cdots & \theta_{1q}^* \\ \theta_{21}^* & \theta_{22}^* & \cdots & \theta_{2q}^* \\ \vdots & \vdots & \vdots & \vdots \\ \theta_{B1}^* & \theta_{B2}^* & \cdots & \theta_{Bq}^* \end{pmatrix}.$$

Column j from this matrix is a set of B draws from the marginal posterior distribution of parameter θ_j and can be used to compute point estimates and a credible interval for θ_j, as described in Section 15.2.6 and subsequent examples. The preceding matrix would have only two columns for a (two-parameter) Weibull distribution, and would involve B rows (usually a large number). Additional columns for other quantities that are to be estimated (i.e., functions of the parameters $v(\boldsymbol{\theta})$ such as distribution quantiles or probabilities) can be added to the matrix by simply computing the function of interest for each row in the matrix. The following matrix shows two such additional columns:

$$\begin{pmatrix} \theta_{11}^* & \theta_{12}^* & \cdots & \theta_{1q}^* & v_1(\boldsymbol{\theta}_1^*) & v_2(\boldsymbol{\theta}_1^*) \\ \theta_{21}^* & \theta_{22}^* & \cdots & \theta_{2q}^* & v_1(\boldsymbol{\theta}_2^*) & v_2(\boldsymbol{\theta}_2^*) \\ \vdots & \vdots & \vdots & \vdots & \vdots & \vdots \\ \theta_{B1}^* & \theta_{B2}^* & \cdots & \theta_{Bq}^* & v_1(\boldsymbol{\theta}_B^*) & v_2(\boldsymbol{\theta}_B^*) \end{pmatrix}. \tag{15.3}$$

15.2.4 Comparison of Joint Posterior Distributions Based on Diffuse and Informative Prior Information on the Weibull Shape Parameter β

Example 15.3 Diffuse Prior Distribution and the Resulting Joint Posterior Distribution for the Bearing Cage Data. The points in the left-hand plot of Figure 15.3 are sample draws from the diffuse joint *prior* distribution (which, because of the assumed independence of the information sources, can be specified as the product of marginal distributions for both parameters). The contours in both plots in Figure 15.3 are relative likelihood contours, obtained by dividing the likelihood values by the value of the maximum of the likelihood. Because the likelihood is proportional to the probability of the data, we can, for example, say that the probability of the data at the ML estimate (where the relative likelihood is 1) is 100 times (10 times) larger than the probability of the data at any point on the 0.01 contour (the 0.10 contour). Also, as described in Section 12.12, the region enclosed by the 0.01 relative likelihood contour is an approximate 99% joint confidence region for the 0.10 quantile of the failure-time distribution and the Weibull shape parameter β. Similar statements can be made for any of the other relative likelihood contours. Figure 15.3 shows only points (each corresponding to a sample draw) with values of β below 5 to provide a better view of the interesting features of the interaction between the joint prior distribution and the relative likelihood contours.

The right-hand plot of Figure 15.3 shows sample draws from the diffuse-prior-analysis joint *posterior* distribution, along with the relative likelihood contours. As expected, the sample draws from the joint posterior distribution agree well with the likelihood contours because the joint posterior distribution is proportional to the likelihood function when the prior distribution is uniform (recall that the actual prior used here is uniform on the log scale). ∎

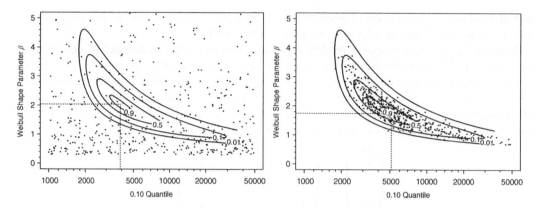

Figure 15.3   Sample draws from the diffuse joint prior distribution for $t_{0.10}$ and β with likelihood contours and ML parameter estimates (left) and the corresponding sample draws from the joint posterior distribution for $t_{0.10}$ and β with lines showing the Bayesian point estimates (right).

Example 15.4   Informative Prior Distribution and Resulting Joint Posterior Distribution for the Bearing Cage Data. The plots in Figure 15.4 are similar to those in Figure 15.3 but are based on the prior distribution that is informative for the Weibull shape parameter β. The points in the left-hand plot of Figure 15.4 are sample draws from the informative joint *prior* distribution, restricting the Weibull shape parameter β to be in the interval 1.5 to 3 with probability 0.99, according to a lognormal distribution. The right-hand plot of Figure 15.4 shows the corresponding sample draws from the joint *posterior* distribution. As expected, this plot shows that the joint posterior distribution is concentrated in the region where the joint prior distribution and the likelihood overlap.   ∎

There are a number of algorithms that can be used to generate the sample draws from the joint posterior distribution. In the next section, we use a particularly simple method that can be employed in situations for which the likelihood is easy to compute and there are only a few parameters. The more versatile MCMC methods described in Section 15.4 can be used both for simple models and for models that are much more complicated and that have a large number of parameters. The general methods that are used to compute Bayesian statistical intervals from

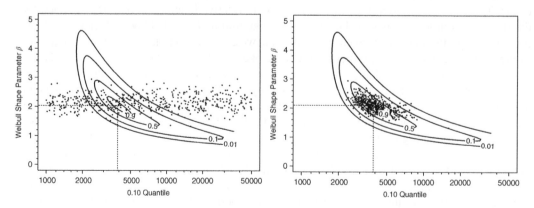

Figure 15.4   Sample draws from the informative (for the Weibull shape parameter β) joint prior distribution for $t_{0.10}$ and β with likelihood contours and ML parameter estimates (left) and the corresponding sample draws from the joint posterior distribution for $t_{0.10}$ and β with lines showing the Bayesian point estimates (right).

the sample draws do not, however, depend on the method that was used to compute the sample draws.

15.2.5 Generating Sample Draws with Simple Simulation

Sample draws from the posterior distribution are easy to generate in applications such as the preceding for which the likelihood is easy to compute and for which there is only a small number of parameters to be estimated. In particular, the sample draws from the joint posterior distribution are obtained by randomly filtering the sample draws from the joint prior distribution. A point from the joint prior distribution is accepted with a probability corresponding to the value of the relative likelihood at the point. For example, points on the 0.01 contour in the left-hand plot in Figure 15.3 (or 15.4) would be kept with probability 0.01 and points on the 0.90 contour would be kept with probability 0.90.

A total of 10,000–20,000 sample draws from the posterior distribution is sufficient for most practical purposes when the draws are independent and identically distributed (as they are in this simple simulation but not in the MCMC-type simulations described in Section 15.4 and used in subsequent examples in this chapter and Chapters 16–18). A much larger number of draws may be needed to be certain that a given number of significant digits has only a small probability of containing any errors. See the Bibliographic Notes section "How many draws."

15.2.6 Using the Sample Draws to Construct Bayesian Credible Intervals

Bayesian point estimates for parameters and other quantities of interest can be obtained by using a measure of central tendency for the marginal posterior distribution of the quantity of interest (represented by a given column in a matrix, such as (15.3). Theoretically, the mean of the marginal posterior distribution will provide (assuming a correct model) a Bayesian estimate that minimizes the squared-error loss, relative to the value of the actual quantity being estimated. The median of the marginal posterior distribution minimizes absolute-error loss (also known as linear-error loss). Also, the median is less affected than the mean by the long tail of a skewed marginal posterior distribution, and will generally agree better with the ML estimate when using a diffuse joint prior distribution. For these reasons, we will use the sample median of the marginal posterior distributions as a point estimate of the quantity of interest in all of our examples.

Bayesian credible intervals can be obtained by using the appropriate quantiles of the sample draws from the marginal posterior distribution for the quantity of interest. For example, a 95% credible interval is obtained by using the 0.025 and the 0.975 quantiles of the sample draws from the quantity's marginal posterior distribution. We illustrate these methods in the following examples by computing credible intervals for distribution cumulative probabilities and quantiles.

Example 15.5 Marginal Posterior Distributions and Credible Intervals for Bearing Cage Failure Probabilities. As an example of the matrix in (15.3), Table 15.2 shows the first ten sample draws (out of the $B = 100,000$ that were computed) from the posterior distribution for the parameters $\boldsymbol{\theta} = (t_{0.10}, \beta)$ (the 0.10 quantile and the Weibull shape parameter, respectively) for the bearing cage example, using the informative joint prior distributions. The table also shows the values that were computed subsequently for the Weibull scale parameter $v_1(\boldsymbol{\theta}^*) = \eta^* = t^*_{0.10}/[-\log(1-0.10)]^{1/\beta^*}$, and the cdf values, or fraction failing, at 5,000 hours of service (i.e., $v_2(\boldsymbol{\theta}^*) = F^*(5,000) = 1 - \exp[-(5,000/\eta^*)^{\beta^*}])$ and at 8,000 hours of service. The sample draws for η, $F(5,000)$, and $F(8,000)$ were computed row-wise for all of the B draws.

$t^*_{0.10}$	β^*	η^*	$F^*(5,000)$	$F^*(8,000)$
6,558.5	1.7567	23,612.4	0.063321	0.280081
7,475.8	1.8927	24,548.5	0.048018	0.184914
3,409.1	2.3598	8,846.9	0.229044	0.066405
3,597.5	2.3448	9,392.9	0.203867	0.241618
6,648.6	1.6411	26,198.3	0.063874	0.120376
4,589.2	2.3860	11,785.5	0.121268	0.087313
4,327.9	2.0856	12,731.4	0.132705	0.607407
5,171.0	1.7262	19,043.2	0.094635	0.094925
3,628.8	1.9332	11,623.0	0.177812	0.143343
4,880.5	1.7877	17,185.8	0.104179	0.145801

Table 15.2 Ten sample draws from the joint posterior distribution of the Weibull distribution parameters $t_{0.10}$ and β and the resulting calculated additional quantities of interest η, $F(5,000)$, and $F(8,000)$, based on the bearing cage data using the informative joint prior distribution.

For the examples in this section we will suppose that there is a matrix of sample draws, similar to those shown in Table 15.2, but with $B = 100,000$ rows, with columns named t0p10, beta, eta, F5000, and F8000. This matrix (actually an R object) has the name drawsBearingCageInformative. There is a similar matrix (R object) for draws from the joint posterior distribution that was computed using the *diffuse* joint prior distribution with name drawsBearingCageDiffuse.

Figure 15.5 shows histograms of the draws from the marginal posterior distribution of $F(5,000)$ (top) and $F(8,000)$ (bottom), the fraction failing at 5,000 and 8,000 hours of service, respectively, based on the diffuse prior distribution (left) and the informative prior distribution (right). The vertical lines in the histograms indicate the 0.025 and 0.975 quantiles of the empirical distribution of the marginal posterior distribution computed for each of the 100,000 draws from the joint posterior distribution. These quantiles define the 95% credible intervals for $F(5,000)$ and $F(8,000)$.

The following R commands, applied to the matrix of 100,000 sample draws, similar to those shown in Table 15.2, give

```
> quantile(drawsBearingCageDiffuse[,"F5000"], probs=c(0.025, 0.975))
[1]  0.01883178   0.6775634
> quantile(drawsBearingCageInformative[,"F5000"], probs=c(0.025,
  0.975))
[1]  0.05540326   0.4041599
> quantile(drawsBearingCageDiffuse[,"F8000"], probs=c(0.025, 0.975))
[1]  0.02994012   0.992824
> quantile(drawsBearingCageInformative[,"F8000"], probs=c(0.025,
  0.975))
[1]  0.1261585    0.8086649
```

providing the 95% credible intervals for $F(5,000)$ and $F(8,000)$ that are summarized in the bottom part of Table 15.3. We note that for the diffuse prior distribution, $F^*(8,000) \approx 1$ for an appreciable number of the draws, as shown in Figure 15.5, and thus the upper endpoint of the credible interval for $F(8,000)$ is close to 1.

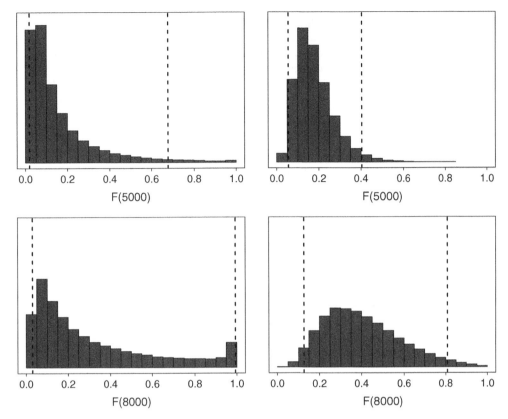

Figure 15.5 Marginal posterior distributions and 95% credible intervals for $F(5{,}000)$ (top) and F(8,000) (bottom) using the diffuse prior distribution (left) and the informative prior distribution (right) for the bearing cage data.

The credible bounds for this example are compared in Table 15.3 with the non-Bayesian likelihood-based confidence intervals that were obtained in Example 15.1. The comparison shows fairly good agreement (relative to the width of the intervals) between the credible intervals based on a diffuse prior distribution and the likelihood-based confidence intervals. As expected, the intervals obtained using the informative prior distribution are appreciably different from, and much narrower than, those obtained by the other two approaches. ■

As shown in previous applications in Chapters 4, 12, and 14, and Figure 15.2, it is generally useful to plot the parametric estimate of a cdf on a probability plot, along with a set of pointwise parametric confidence intervals. When doing a Bayesian analysis it is similarly useful to plot the Bayesian estimate and credible bounds for the cdf, as shown in the following example.

Example 15.6 Bayesian Probability Plot of the Bearing Cage Data Failure-Time cdf Based on the Joint Posterior Distribution. Figure 15.6 shows Weibull probability plots along with Bayesian estimates (median of the marginal posterior distribution) of the fraction failing as a function of time, based on the diffuse joint prior distribution (left) and the informative prior distribution (right). The estimates and credible intervals for $F(t)$ shown in these plots were computed using the same approach as in Example 15.5, for a large number of values of t.

Prior distribution	Interval/bound type	Interval for	Lower	Upper
None	Two-sided confidence	$t_{0.10}$	2,094	22,144
Diffuse	Two-sided credible	$t_{0.10}$	2,310	30,206
Informative	Two-sided credible	$t_{0.10}$	2,578	6,987
None	Two-sided confidence	β	0.971	3.580
Diffuse	Two-sided credible	β	0.891	3.203
Informative	Two-sided credible	β	1.682	2.582
None	Two-sided confidence	η	4,045	213,597
Diffuse	Two-sided credible	η	4,807	356,554
Informative	Two-sided credible	η	6,530	24,636
None	Two-sided confidence	$F(5,000)$	0.0223	0.8776
Diffuse	Two-sided credible	$F(5,000)$	0.0188	0.6776
Informative	Two-sided credible	$F(5,000)$	0.0554	0.4042
None	Two-sided confidence	$F(8,000)$	0.0365	0.99998
Diffuse	Two-sided credible	$F(8,000)$	0.0299	0.9928
Informative	Two-sided credible	$F(8,000)$	0.1262	0.8087

Table 15.3 Two-sided 95% confidence intervals and credible intervals based on diffuse and informative prior distributions for the Weibull distribution parameters and other quantities of interest for the bearing cage data.

The credible intervals in the left-hand plot (diffuse prior) are similar to those based on ML, shown in Figure 15.2. The width of the credible interval in the right-hand plot in Figure 15.6, based on the informative prior distribution for the Weibull shape parameter, is much narrower. The reason for this can be seen by looking at the sample draws from the informative joint prior and the resulting joint posterior distributions in Figure 15.4. In particular, focusing on the likelihood contours in the region where β is less than 1.5 and $t_{0.10}$ is larger than 8,000, we see that using the prior information that the Weibull shape parameter is larger than 1.5, the optimism that $t_{0.10}$ could be larger than 8,000 hours (or equivalently that the fraction failing by 8,000 hours could be less than 0.10) disappears. There is a similar, but less dramatic, change in the upper credible bound for $t_{0.10}$. ■

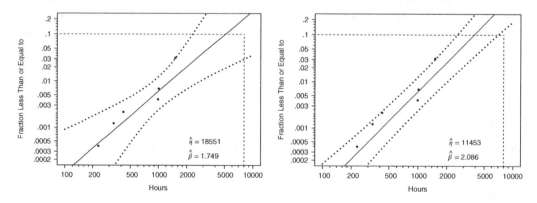

Figure 15.6 Weibull probability plots of the bearing cage failure data showing the Bayesian estimates for $F(t)$ and a set of pointwise 95% credible intervals for $F(t)$ based on the diffuse-prior-distribution analysis (left) and the informative-prior-distribution analysis (right).

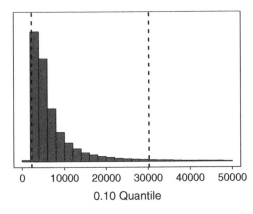
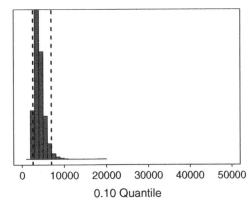

Figure 15.7   Marginal posterior distributions and 95% credible intervals for the 0.10 quantile of the bearing cage failure-time distribution using the diffuse-prior-distribution analysis (left) and the informative-prior-distribution analysis (right).

The methods described above for making inferences about $F(t)$ can also be used to compute estimates and credible bounds for quantiles or other functions of the model parameters, as illustrated in the following example.

Example 15.7   Marginal Posterior Distributions for the Bearing Cage Failure-Time Distribution 0.10 Quantile. Figure 15.7 shows histograms of the marginal posterior distribution for $t_{0.10}$ based on the diffuse-prior-distribution analysis (left) and the informative-prior-distribution analysis (right). The vertical lines on these plots show the lower and upper endpoints of the resulting 95% credible intervals for $t_{0.10}$.

The preceding credible intervals can be obtained easily from the sample draws by using the following R commands:

```
> quantile(drawsBearingCageDiffuse[,"t0p10"], probs=c(0.025, 0.975))
[1] 2309.784   30206.46
> quantile(drawsBearingCageInformative[,"t0p10"], probs=c(0.025,
  0.975))
[1] 2578.263    6986.95
```

These results are summarized and compared with the corresponding non-Bayesian likelihood-based confidence interval in Table 15.3. Again, our conclusion from the informative-prior-distribution analysis is that $t_{0.10}$ is less than 8,000 hours.   ■

15.3   MORE ABOUT SPECIFICATION OF A PRIOR DISTRIBUTION

Statistical models generally require a relatively small number of parameters (usually two if there are no explanatory variables). As explained in Section 15.1.5, reparameterization is often used to describe a model with quantities that are of particular interest to the analyst and we will, henceforth, also refer to these quantities of interest as parameters.

15.3.1   Diffuse versus Informative Prior Distributions

As mentioned in Section 15.1.1, one important reason for using Bayesian methods is that the analysis provides a formal mechanism for including prior information (i.e., knowledge beyond that provided by the data) into the analysis. Thus, if there is prior information for one or more of the model parameters and if the definition of parameters (i.e., the particular parameterization) has been chosen such that the information about each of these parameters is approximately mutually independent, then one can specify a joint prior distribution with separate marginal distributions for each parameter.

Informative marginal prior distributions can be used for those parameters for which there is appreciable prior information, while diffuse marginal prior distributions can be specified for the other parameters. A commonly used practice when there is little or no prior information about the parameters, or when there is need to present an objective analysis for which the results do not depend on prior information, is to specify diffuse marginal prior distributions for each of the parameters. A prior distribution that is flat (or uniform) over the entire parameter space is sometimes referred to as a noninformative prior distribution. A difficulty with this term is that a flat (noninformative) prior distribution for a parameter implies a nonflat (and thus informative) prior distribution for any nonlinear function of that parameter (e.g., a prior distribution that is noninformative (flat) for the standard deviation σ is informative (not flat) for the variance σ^2).

In applications for which there is little or no useful prior information on any of the parameters, one can specify a diffuse joint prior distribution (i.e., a distribution that is flat or approximately flat over the range of the parameters where the likelihood is nonnegligible). Commonly used diffuse prior distributions include uniform distributions with a wide (but finite) range or a normal distribution with a large variance. It is important to note that with limited data, the choice of a prior distribution (even a diffuse prior distribution) can have strong influence on the resulting inferences. When using a diffuse prior distribution, it is important to experiment with different specifications of the prior distribution to assess the sensitivity of the analysis results to the specification, especially when the information in the data is limited.

15.3.2 Whose Prior Distribution Should We Use?

A major reason why the use of Bayesian methods has been controversial is the need to specify a joint prior distribution for the unknown parameters. Analysts are faced with the question of which or whose prior information/distribution should be used in the analysis. One generally accepted principle for answering this question is that whoever is assuming the risks associated with decisions resulting from the Bayesian analysis should be allowed to choose, or at least have an important say in choosing, the prior distribution. If, however, different interested people, groups of people, or organizations have different risk functions, there is likely to be conflict. For example, in assessing whether or not a product is safe, customers who use the product and producers who benefit from its sale will have different risk functions. In such cases, it may be appropriate to use a compromise (and likely diffuse) joint prior distribution. In any case, we strongly recommend, especially when a compromise joint prior distribution cannot be readily agreed upon and/or one is dealing with limited data, that analyses be conducted using *different alternative* reasonable (or advocated) prior distributions and then the results compared to assess the sensitivity of the prior distribution assumption on the resulting inferences. In making this recommendation we fully recognize the added complexity that this presents. Also, it is likely to make it more difficult to explain findings to a nonstatistical audience—something that needs to be done with much care, recognizing that having to explain alternative credible intervals, in addition to the concept of credible intervals per se, will be challenging.

Finally, we need to recognize the ever-present danger that subjective prior information is contaminated with biases arising from the risks and rewards associated with decisions that are made on the basis of a Bayesian analysis. Pressure from top management and/or outside political considerations—especially in situations in which government oversight or funding is involved—can impact opinions on what prior distribution(s) to use. We must beware of such pressures, as well as wishful thinking masquerading as prior information.

15.3.3 Sources of Prior Information

In some applications solid prior information, based on a combination of physics or chemistry related to the phenomena being studied, combined with previous empirical experience, is available. This is particularly true in some engineering applications such as reliability, for which there are known, well-understood failure mechanisms. Thus, engineers working in specialized areas may well have previous experience with particular failure mechanisms and testing and product-use environments that will allow them, in some situations, to provide strong (but not precise) prior information about the failure-time model.

For a given failure mode (e.g., fracture due to fatigue crack growth), engineers will, for example, typically have some knowledge about the Weibull distribution shape parameter. In particular, if the primary failure mode for a component is a wearout-type failure mechanism, we immediately know that the Weibull distribution shape parameter is greater than 1. Previous field data with similar products (as in the bearing cage example) may, moreover, provide tighter specification of the parameter. Similar knowledge is often available for the shape parameter of a Weibull or a lognormal distribution when it is used to model the failure-time distribution of microelectronic devices that fail due to certain known causes. For example, if a component will fail only when it receives an external shock that arrives according to a homogeneous Poisson process, the failure-time distribution would be exponential (corresponding to a Weibull shape parameter equal to 1). Because such an assumption is an approximation, allowing for some uncertainty in the specification of the Weibull shape parameter (around 1) is appropriate.

Additionally, in product accelerated testing applications there is often available knowledge about the parameter describing the relationship between life and an accelerating variable. In the case of temperature acceleration of a particular chemical reaction, there is often strong knowledge about the effective activation energy in the Arrhenius relationship that is commonly used to describe how temperature affects the rate of the chemical reaction (or other similar mechanisms, such as diffusion). Indeed, some reliability handbooks on electronic reliability (e.g., Klinger et al., 1990, page 59) provide approximate values of the effective activation energy for different failure mechanisms (e.g., metalization, electromigration, or corrosion).

Sometimes, prior information is elicited from "expert opinion" panels. These typically consist of individuals or groups of individuals with knowledge about the process being studied that share and discuss their knowledge to, possibly, arrive at agreed-upon prior information. Methods used by such panels are described in various references given in the Bibliographic Notes section at the end of this chapter.

15.3.4 Implementing Bayesian Analyses Using Conjugate Distributions

For a few simple statistical models, the posterior distribution $f(\boldsymbol{\theta}|\text{DATA})$ is in the same family of distributions as the prior distribution. Such distributions are known as *conjugate distributions* and the particular form of the prior distribution is referred to as a *conjugate prior distribution*. With a conjugate prior distribution, Bayes' theorem can be used easily to update the parameters of the prior distribution to give the posterior distribution (of the same form). For example, when sampling from a binomial distribution, if a beta distribution (described in Section C.3.3) is used as the prior distribution for the binomial distribution parameter π, then the posterior distribution of π will also be a beta distribution with updated parameters (as shown in Section H.3.1).

Conjugate prior distributions generally afford advantages in the development of theory, ease of computation, and the possible closed-form expression for the posterior distribution. In spite of such advantages, the use of conjugate prior distributions can in many situations be limiting.

Before modern methods of computing for Bayesian inference, as described in this chapter, were developed, the use of Bayesian methods was, for the most part, constrained to simple models for which conjugate distributions were available. Although we no longer have such constraints and the main thrust of this chapter is much more general, the concept of conjugate distributions is still useful (mainly due to its simplicity) and will be illustrated in Sections 16.1.1 and 16.2.1. Section H.3 outlines technical details of conjugate distributions for estimation, and Section H.5 does the same for prediction.

15.3.5 Some Further Considerations in the Specification of Prior Distributions

As mentioned in Sections 15.1.5 and 15.3.1, if definitions of approximately independent model parameters are provided, it is possible to specify an appropriate joint prior distribution for these parameters by specifying individual marginal prior distributions for each unknown parameter. The parameters of these marginal prior distributions are known as hyperparameters.

Different textbooks and software packages use different parameterizations for the same distribution. Many textbooks, for example, characterize the normal (Gaussian) distribution in terms of its mean μ and variance σ^2. R uses the mean μ and standard deviation σ. OpenBUGS uses the mean μ and precision, which is defined to be $\tau = 1/\sigma^2$. One must take such differences into consideration in programming Bayesian estimation methods.

There are various ways that one can specify a marginal prior distribution to describe uncertainty in a model parameter. One simple method for a prior distribution with two parameters (marginal prior distributions in our examples will usually have two parameters) is to specify the form and the hyperparameters of the prior distribution using the distribution's usual parameterization. This approach usually is not user-friendly because some of the hyperparameters may not have an easy-to-understand interpretation. Thus, instead of requiring the user to specify the actual hyperparameters for a particular marginal prior distribution, one could require specification of the mean and standard deviation of the prior distribution. Such a specification may, however, not be meaningful when a prior distribution is highly skewed.

When a prior distribution has a finite range, an alternative is to specify the range of the distribution, together with a shape parameter or parameters. The beta distribution can be scaled to have any finite range and its shape parameters can be used to provide the desired prior distribution shape (the uniform distribution is a special case of the beta distribution). When the prior distribution does not have a finite range, another user-friendly alternative is to have the user specify a range of the distribution that contains some large proportion of the distribution's probability content. For example, the specified prior distribution range may be taken to be the 0.005 and 0.995 quantiles of the marginal prior distribution for that parameter.

The specification of diffuse prior distributions presents some additional challenges. When there is no longer a need to specify conjugate prior distributions, a uniform distribution over a wide range of potential parameter values (wide enough that the likelihood is near 0 at the extremes of the marginal prior distribution) often works well. For example, one could specify the marginal prior distribution for a distribution standard deviation by using a uniform distribution ranging from 10^{-5} to 10^4. If, however, the values of the profile likelihood for that parameter are essentially 0 outside the range from 10^{-1} to 10^3, then using a uniform distribution over this range instead would have little or no effect on the resulting joint posterior distribution and generally results in faster, more efficient computation of the joint posterior distribution. Another popular diffuse prior distribution alternative is to use a normal distribution with a large variance (poor precision).

For distributions with only one parameter, the Jeffreys prior distributions (Section H.4) generally provide Bayesian-based statistical intervals that have coverage probabilities close to the nominal credible level. Indeed, as mentioned in the Bibliographic Notes sections of Chapters 6 and 7, the Jeffreys confidence and prediction interval methods presented in those chapters arise by computing corresponding Bayesian intervals based on a Jeffreys prior distribution (see Sections H.4.1 and H.4.2 for technical details). Section H.4.3 describes a modification to the Jeffreys prior distribution that leads to Bayesian procedures for the normal distribution that result in statistical intervals that coincide with the exact methods presented in Chapter 4.

15.4 IMPLEMENTING BAYESIAN ANALYSES USING MARKOV CHAIN MONTE CARLO SIMULATION

15.4.1 Basic Ideas of MCMC Simulation

This section provides a general overview of the MCMC method. This is a powerful, versatile method of simulating sample draws from a particular joint posterior distribution (i.e., the joint posterior distribution resulting from a given model, data, and joint prior distribution). Because there is no other practicable method, MCMC methods are particularly important in inference problems that involve models with a large number of parameters. During the past 20 years there have been many developments and much has been written about MCMC. Rather than providing a detailed technical explanation of the method, we will treat it—and the computer software to execute it—principally as a useful "black box" to implement Bayesian inference. Further details and examples are provided in various books on Bayesian methods and MCMC, some of which are referenced in the Bibliographic Notes section at the end of this chapter.

A Markov chain is a well-known stochastic process model that can be used to characterize the probability of moving from one state to another. An important property of a Markov chain is that the probability of going from one state to another depends only on the current state and not on other history of the process. In the context of Bayesian inference, a Markov chain state corresponds to a point in the model's parameter space. MCMC simulates a sequence of jumps from one state to another (i.e., from one point in the parameter space to another). Numerous MCMC algorithms have been developed to simulate sample draws from a discrete-time continuous-space Markov chain such that after reaching a steady state, the sequence of sample draws constitutes a sample from the desired joint posterior distribution. The best-known methods are Gibbs sampling and the Metropolis–Hastings algorithm, although combinations of these two methods and other MCMC algorithms also exist.

Because the probability of being in a particular state at time i depends on the state at time $i - 1$, simulated sample draws from a Markov chain are not, in general, independent (as they were in the simple Monte Carlo simulation used in Section 15.2.5). Technically, this is not a problem, as estimators of marginal posterior distribution quantities computed from autocorrelated sample draws are still statistically consistent. The autocorrelation does, however, imply that a larger number of sample draws from the joint posterior distribution may be needed to adequately estimate the median and, especially, the more extreme quantiles (used to compute Bayesian credible interval endpoints) of the marginal posterior distributions for the parameters of interest. To address this, analysts will, in some cases, "thin" the sample draws by retaining every kth value in the sequence, where k should be larger if the autocorrelation is stronger. Thinned sample draws can be made to have little or no autocorrelation by making k sufficiently large. Moreover, one can use standard methods for independent samples to assess how long the chain needs to be to adequately estimate quantities of interest. Often the draws from a joint posterior distribution are saved after they are computed so that they can be used subsequently to compute estimates of functions of the parameters and statistical intervals. Thinning reduces the space required to store the results.

MCMC simulations generally require specification of starting values (a process that sometimes can be automated). Sample draws at the beginning of the MCMC sequence cannot be expected to represent draws from the desired posterior distribution. Thus it is common practice to drop some number (e.g., 1,000) of the initial sample draws, so that the remaining draws more accurately represent a sample from the limiting distribution Markov chain (i.e., after steady state has been reached). The discarded sample draws are referred to as "burn-in" draws.

Example 15.8 MCMC Output for the Bearing Cage Example. The plots in the top row and the bottom left in Figure 15.8, based on the bearing cage example from Section 15.2 with the informative prior distribution, show Metropolis–Hastings MCMC sample paths for three different relatively short chains (1,000 sample draws) for the joint posterior distribution of the Weibull shape parameter β and the 0.1 distribution quantile, using different starting values.

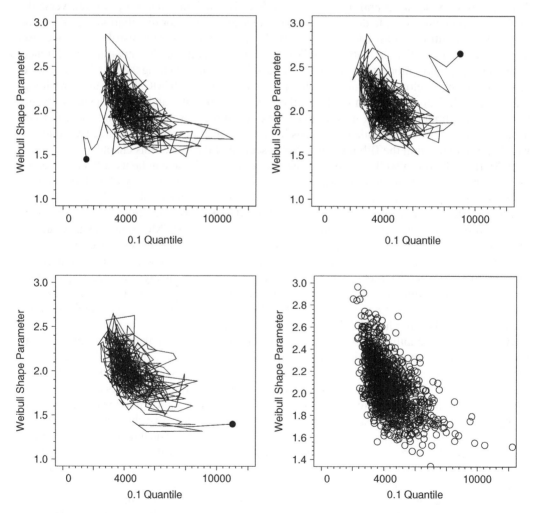

Figure 15.8 Illustrations of sample paths from a Markov chain with different starting values, generating samples from the joint posterior distribution of the Weibull distribution shape parameter and 0.1 quantile for the bearing cage failure-time data.

The plot in the bottom right-hand corner shows, on the same scales, a "final" set of sample draws obtained after initial "burn-in" samples have been discarded and the sample draws were thinned. For this plot, 20,000 samples were drawn, the first 2,000 were discarded, and then every 20th draw was retained. The resulting thinned samples are approximately independent and are easier to store and post-process. Additionally, because autocorrelation is generally small or nonexistent in the thinned sample draws, it is relatively easy to assess the amount of Monte Carlo error in estimates computed from the sample draws.   ■

15.4.2 Risks of Misuse and Diagnostics

Although modern MCMC methods are highly versatile and powerful, an inexperienced user can misapply them and obtain seriously incorrect results. Putting programming errors aside, problems in applying MCMC are especially likely to occur in dealing with diffuse prior distributions and limited data or when a poor parameterization (i.e., a parameterization for which the sample draws for different parameters are highly correlated with each other) is used. If an improper prior distribution (i.e., a prior distribution that does not integrate to a finite number) is used and the data are not sufficient to identify the unknown model parameters, the joint posterior distribution will be improper. In such cases, an MCMC algorithm will still give "answers," but they will generally be wrong. If a proper joint prior distribution is specified, then the joint posterior distribution will be proper. If, however, the proper prior distribution is diffuse and there is limited information in the data, then posterior inferences will usually be highly sensitive to the exact way in which the diffuse prior distribution was specified. Sensitivity analyses are recommended (i.e., try different diffuse joint prior distributions to assess their effect on posterior inferences).

If a proper joint posterior distribution exists for a given model, data, and joint prior distribution, then MCMC theory assures that eventually a properly chosen MCMC algorithm will converge and generate sample draws from the joint posterior distribution. In practice, however, there is no guarantee that a given MCMC simulation, run for a finite number of iterations, has converged. To gain some degree of assurance, it is necessary to use appropriate diagnostics to assess whether the sample draws from the chain represent draws from the limiting distribution and that a sufficient number of samples have been obtained to properly estimate quantities of interest. This assessment will be more difficult to make when the sequence of sample draws has high autocorrelation. Useful graphical diagnostics include trace (time series) and autocorrelation function (ACF) plots of the MCMC sample draws from the joint posterior distribution for each of the model parameters. It is common practice to generate three or four sample chains simultaneously, using different starting values, and then check (using plots and numerical diagnostics) that all of the chains have converged to the same distribution. There are also numerical summary diagnostic tools that complement the graphical approach. These are described in the books referenced in the Bibliographic Notes section at the end of this chapter.

A common, but difficult-to-answer, question is "How many MCMC sample draws do I need?" The determination of the appropriate number of MCMC sample draws depends on the strength of the autocorrelation in the sample draws and on the inherent variability in the MCMC output. The ACF for the sequence of draws will differ among different parameters and quantities of interest that are functions of the parameters. Displays of the ACF of the sample draws are helpful in assessing how long the simulated chain should be (after the burn-in sample draws have been removed). If, however, the MCMC sample draws are thinned sufficiently such that there is little remaining autocorrelation, then standard methods of sample size determination for independent samples can be used as a guideline. References addressing this topic are given in the Bibliographic Notes section at the end of this chapter.

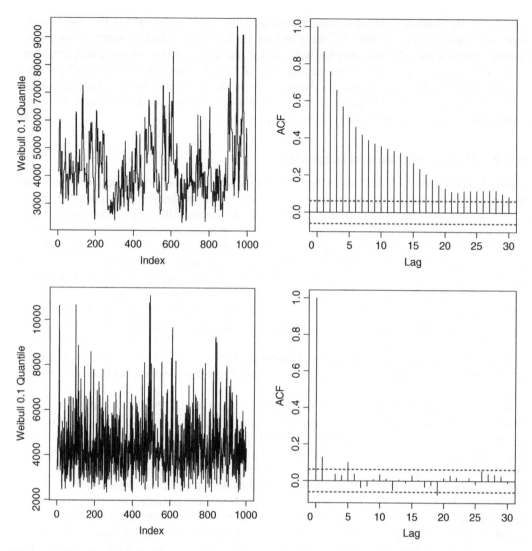

Figure 15.9   Trace plots (left) and ACF plots (right) comparing unthinned MCMC sample draws (top) and thinned MCMC sample draws (bottom) for the bearing cage Weibull distribution example.

Example 15.9   MCMC Diagnostics for the Bearing Cage Example. Figure 15.9 gives examples of some simple MCMC diagnostic plots from one chain corresponding to the bearing cage example and the MCMC output shown in Figure 15.8. The top left-hand plot of Figure 15.9 is a trace (time series) plot of the first 1,000 of the Weibull 0.10 quantile sample draws provided by the chain's output, after the initial 2,000 "burn-in" sample draws were discarded. Figure 15.9 suggests that the time series contains autocorrelation. Evidence of the strong autocorrelation can be seen more clearly in the plot of the ACF on the top right. The bottom part of the plot is similar, but shows the first 1,000 of the Weibull 0.10 quantile sample draws provided by the chain's output after the "burn-in" sample draws were discarded and after the output was "thinned" by keeping every 20th sample draw. The effect of the thinning is to provide sample draws that have little or no autocorrelation.   ∎

15.4.3   MCMC Summary

In summary, for given data, assumed model, and joint prior distribution, Bayesian inference typically involves the use of MCMC simulation implemented along the following lines.

- Using appropriate software, generate three long sequences of draws (perhaps 20,000) from the posterior distribution, using different starting values for each of the three sequences.

- The initial draws from each of the sequences of draws will have transient behavior, due to dependence on the starting values (as seen in the behavior shown in Figure 15.8). Therefore, one generally drops some number of draws (e.g., 1,000) from the beginning of each sequence.

- For each parameter look at a time series plot that superimposes the three different sequences to see if they are "mixing" well (i.e., whether the different sequences appear to be coming from the same marginal distribution). If the sequences are not mixing properly, it may be an indication of an inability to estimate the particular parameter or that a much larger number of draws will be required.

- If the different sequences seem to be mixing, then check to see if there is strong autocorrelation. If so, then it may be advisable to use "thinning" by retaining every 1 in 10 or 1 in 50 of the sample draws (depending on the strength of the autocorrelation). Thinned sequences will have less autocorrelation. Thinning will increase the number of draws that need to be generated but reduce the number of draws that need to be stored.

- Look at statistical summaries of the draws (generally combining the three sequences after the burn-in draws) to assess the precision with which Bayesian estimates and credible interval endpoints are being obtained. If the precision is not adequate, generate additional draws.

- Use the resulting empirically generated joint posterior distribution for the parameters of interest to obtain the desired Bayesian estimates and associated credible intervals, using the approach described in Section 15.2.6 and illustrated in the subsequent examples in this chapter and in Chapters 16–18.

15.4.4   Software for MCMC

Currently, there are only limited capabilities for computing Bayesian statistical intervals in commercial statistical software packages, but we expect that to change in the future. As described in Section 15.1.3, a posterior distribution (from which one can obtain Bayesian statistical intervals) is defined by Bayes' theorem for given data, assumed model, and prior distribution. Currently, many (if not most) users of Bayesian methods employ one of the available software packages that provides MCMC draws from a given posterior distribution. We used OpenBUGS for most of our examples. Other alternatives and references are given in the Bibliographic Notes section at the end of this chapter.

15.5   BAYESIAN TOLERANCE AND PREDICTION INTERVALS

This section provides methods for computing Bayesian two-sided tolerance intervals, two-sided simultaneous prediction intervals, and one-sided simultaneous prediction bounds (using the Bayesian terminology "credible level" instead of confidence level).

The methods used here parallel the non-Bayesian methods described in Sections 14.5 and 14.6. A key difference, however, is that in the non-Bayesian approach, inferences are based on the sampling distribution of the parameter estimates $\widehat{\theta}$, given the actual value of θ, using the B simulated values of $\widehat{\theta}^*$ to approximate the sampling distribution of $\widehat{\theta}$. Under the Bayesian method, inferences are instead based on the posterior distribution of θ, given the data. This is generally done by using the sample draws from the posterior distribution in a manner that is similar to, but different from, the use of the simulated $\widehat{\theta}^*$ values in the non-Bayesian approach.

15.5.1 One-Sided Bayesian Tolerance Bounds

Extending the discussion in Section 2.4.2 to Bayesian tolerance intervals, we note that a one-sided lower $100(1 - \alpha)\%$ Bayesian tolerance bound to be exceeded by at least a proportion p of a distribution is equivalent to a one-sided lower $100(1 - \alpha)\%$ credible bound for the $1 - p$ quantile of the distribution, and this equivalence can be used to construct Bayesian one-sided lower tolerance bounds. Similarly, a one-sided upper $100(1 - \alpha)\%$ Bayesian tolerance bound to exceed at least a proportion p of a distribution is equivalent to a one-sided upper $100(1 - \alpha)\%$ credible bound for the p quantile of the distribution, and this equivalence can be used to construct Bayesian one-sided upper tolerance bounds. For this reason, we discuss only two-sided Bayesian tolerance intervals in this chapter.

15.5.2 Two-Sided Bayesian Tolerance Intervals to Control the Center of a Distribution

A two-sided $100(1 - \alpha)\%$ control-the-center Bayesian tolerance interval to contain at least a proportion β of a distribution $F(x)$ is denoted by

$$\left[\underset{\sim}{T}_\beta, \ \widetilde{T}_\beta\right] = \left[\underset{\sim}{T}_\beta(\beta, 1 - \alpha), \ \widetilde{T}_\beta(\beta, 1 - \alpha)\right].$$

The interval endpoints depend on the data, the assumed distributional form $F(x)$, and the prior distribution, through the posterior distribution of the parameter vector $\boldsymbol{\theta}$. For the sake of simplicity, however, this dependency is not reflected in the notation. For a continuous distribution, the tolerance interval endpoints are determined such that

$$\Pr_{\boldsymbol{\theta}|\text{DATA}}\left[F(\widetilde{T}_\beta; \boldsymbol{\theta}) - F(\underset{\sim}{T}_\beta; \boldsymbol{\theta}) > \beta\right] = 1 - \alpha,$$

where $1 - \alpha$ is the specified credible level and $\boldsymbol{\theta}|\text{DATA}$ indicates that the probability is evaluated with respect to the joint posterior distribution of $\boldsymbol{\theta}$, given the data.

Operationally, the interval endpoints $\underset{\sim}{T}_\beta$ and $\widetilde{T}_\beta$ are chosen such that

$$\frac{1}{B}\sum_{j=1}^{B}\text{I}\left[F(\widetilde{T}_\beta; \boldsymbol{\theta}_j^*) - F(\underset{\sim}{T}_\beta; \boldsymbol{\theta}_j^*) > \beta\right] = 1 - \alpha \tag{15.4}$$

and

$$\frac{1}{B}\sum_{j=1}^{B}F(\widetilde{T}_\beta; \boldsymbol{\theta}_j^*) = \frac{1}{B}\sum_{j=1}^{B}\left[1 - F(\underset{\sim}{T}_\beta; \boldsymbol{\theta}_j^*)\right]. \tag{15.5}$$

The second constraint assures that the error probabilities are equal for both the lower and upper endpoints of the tolerance interval. Here $\text{I}[A]$ is an indicator function that is equal to 1 when the statement A is true and equal to 0 otherwise. For the important special case of a (log-)location-scale distribution,

$$F(\underset{\sim}{T_\beta}; \boldsymbol{\theta}_j^*) = \Phi\left(\frac{\underset{\sim}{T_\beta} - \mu_j^*}{\sigma_j^*}\right) \quad \text{and} \quad F(\widetilde{T}_\beta; \boldsymbol{\theta}_j^*) = \Phi\left(\frac{\widetilde{T}_\beta - \mu_j^*}{\sigma_j^*}\right). \tag{15.6}$$

For (log-)location-scale distributions, Yuan et al. (2017) describe an algorithm that can be used to solve (15.4) subject to (15.5) that requires only a simple root-finding procedure (such as uniroot in R). The method could, however, be adapted to non-(log-)location-scale distributions.

For discrete distributions (e.g., the binomial and Poisson distributions) the solution of (15.4) subject to (15.5) would have to be done with approximations, as will be illustrated in Example 16.5.

15.5.3 Two-Sided Bayesian Tolerance Intervals to Control Both Tails of a Distribution

A two-sided $100(1 - \alpha)\%$ control-both-tails Bayesian tolerance interval to contain at least a proportion β of a distribution $F(x)$ is given by

$$\left[\underset{\sim}{T_{p_t L}}, \ \widetilde{T}_{p_t U}\right] = \left[\underset{\sim}{T_{p_t L}}(\beta, 1 - \alpha), \ \widetilde{T}_{p_t U}(\beta, 1 - \alpha)\right].$$

These interval endpoints again depend on the data, the assumed distribution $F(x)$, and the prior distribution, through the posterior distribution of the parameter vector $\boldsymbol{\theta}$. For the sake of simplicity, however, this dependency is not reflected in the notation. For a continuous distribution, the tolerance interval endpoints are determined such that

$$\Pr_{\boldsymbol{\theta}|\text{DATA}}\left[F(\underset{\sim}{T_{p_t L}}; \boldsymbol{\theta}) < (1 - \beta)/2 \text{ and } 1 - F(\widetilde{T}_{p_t U}; \boldsymbol{\theta}) < (1 - \beta)/2\right] = 1 - \alpha,$$

where $1 - \alpha$ is the specified credible level and $\boldsymbol{\theta}|\text{DATA}$ indicates that the probability is evaluated with respect to the joint posterior distribution of $\boldsymbol{\theta}$.

Operationally, the interval endpoints $\underset{\sim}{T_{p_t L}}$ and $\widetilde{T}_{p_t U}$ are chosen such that

$$\frac{1}{B}\sum_{j=1}^{B} \text{I}\left[F(\underset{\sim}{T_{p_t L}}; \boldsymbol{\theta}_j^*) < (1 - \beta)/2 \text{ and } 1 - F(\widetilde{T}_{p_t U}; \boldsymbol{\theta}_j^*) < (1 - \beta)/2\right] = 1 - \alpha, \tag{15.7}$$

with a symmetry constraint similar to (15.5). For the important special case of a (log-)location-scale distribution, expressions similar to those in (15.6) are again used. The only changes needed are to replace $\underset{\sim}{T_\beta}$ and $\widetilde{T}_\beta$ with $\underset{\sim}{T_{p_t L}}$ and $\widetilde{T}_{p_t U}$, respectively. The algorithm described in Yuan et al. (2017) can also be used to find Bayesian control-both-tails tolerance intervals for continuous distributions.

As with the control-the-center Bayesian tolerance intervals in Section 15.5.2, for discrete distributions (e.g., the binomial and Poisson distributions), the solution of (15.7) will have to be obtained with approximations in a manner similar to that illustrated in Section 16.5.

Example 15.10   Bayesian Tolerance Intervals for the Bearing Cage Failure-Time Distribution. Following the approach described in Section 15.5.2, two-sided 95% Bayesian control-the-center tolerance intervals were computed using the draws from the joint posterior distribution, like those illustrated in Table 15.2, for both the diffuse and the informative prior distributions. Two-sided 95% Bayesian control-the-tails tolerance intervals were also computed following the approach described above. The results are given in Table 15.4, along with prediction intervals and bounds to be described in Section 15.5.4.

The upper endpoints of these particular tolerance intervals have little or no practical meaning (and are highly dependent on the assumed distributional model) due to the large amount of

Type of prior distribution	Interval/bound type	Interval for	Lower	Upper
Diffuse	Two-sided tolerance	Control center 0.90	2,298.1	920,576.1
Informative	Two-sided tolerance	Control center 0.90	2,552.3	40,883.5
Diffuse	Two-sided tolerance	Control tails 0.90	1,908.3	3,169,140.6
Informative	Two-sided tolerance	Control tails 0.90	1,986.5	105,957.8
Diffuse	Two-sided prediction	at least 9 of 10	1,872.4	705,701.8
Informative	Two-sided prediction	at least 9 of 10	1,830.1	38,809.7
Diffuse	One-sided prediction	at least 9 of 10	2,280.9	402,159.5
Informative	One-sided prediction	at least 9 of 10	2,212.6	31,890.4

Table 15.4   Two-sided 95% tolerance and prediction intervals and one-sided 95% prediction bounds for bearing cage life.

extrapolation needed to make inferences in the upper tail of the distribution (given that the nonparametric estimate of the fraction failing goes only up to 0.05). We note, however, that, as one would expect, the upper endpoints computed with the informative prior distribution are much closer to the center of the distribution than those calculated using the diffuse prior distribution. The lower endpoints of the tolerance intervals that were computed with the informative prior distribution are larger (i.e., closer to the center of the distribution) than those computed with the diffuse prior distribution. The differences, however, are not nearly as large as they were for the upper endpoints. This is because computing the lower endpoints does not involve extrapolation. ∎

15.5.4   Bayesian Simultaneous Prediction Intervals to Contain at Least k out of m Future Observations

In a manner that is similar to (but in some respects different from) the development in Section B.7, a two-sided $100(1 - \alpha)\%$ simultaneous prediction interval to contain at least k out of m future observations from a previously sampled distribution $F(x; \boldsymbol{\theta})$ is given by

$$\left[\underset{\sim}{Y}_{k;m}, \ \widetilde{Y}_{k;m} \right] = \left[\underset{\sim}{Y}_{k;m}(1 - \alpha; k, m), \ \widetilde{Y}_{k;m}(1 - \alpha; k, m) \right].$$

These interval endpoints depend also on the data, the assumed distribution $F(x; \boldsymbol{\theta})$, and the prior distribution, through the posterior distribution of the parameter vector $\boldsymbol{\theta}$. For the sake of simplicity, however, this dependency is not reflected in the notation. The prediction interval endpoints are determined such that

$$\mathrm{E}_{\boldsymbol{\theta}|\mathrm{DATA}} \left[\sum_{i=k}^{m} \binom{m}{i} p^i (1 - p)^{m-i} \right] = 1 - \alpha.$$

Here

$$p = \mathrm{Pr}\left[\underset{\sim}{Y}_{k;m}(1 - \alpha; k, m) < Y \leq \widetilde{Y}_{k;m}(1 - \alpha; k, m) \right]$$

$$= F\left[\widetilde{Y}_{k;m}(1 - \alpha; k, m); \boldsymbol{\theta} \right] - F\left[\underset{\sim}{Y}_{k;m}(1 - \alpha; k, m); \boldsymbol{\theta} \right],$$

where $F(y; \boldsymbol{\theta})$ is the cdf of Y, a single future observation from the previously sampled distribution, $1 - \alpha$ is the specified credible level, and $\boldsymbol{\theta}|\mathrm{DATA}$ indicates that the probability is evaluated with respect to the joint posterior distribution of $\boldsymbol{\theta}$.

The two-sided simultaneous prediction interval endpoints $\underset{\sim}{Y}_{k;m}$ and $\widetilde{Y}_{k;m}$ are chosen such that

$$\frac{1}{B}\sum_{j=1}^{B}\left[\sum_{i=k}^{m}\binom{m}{i}(p_j^*)^i(1-p_j^*)^{m-i}\right] = 1 - \alpha, \tag{15.8}$$

where

$$p_j^* = F\left[\widetilde{Y}_{k;m}(1-\alpha;k,m);\boldsymbol{\theta}_j^*\right] - F\left[\underset{\sim}{Y}_{k;m}(1-\alpha;k,m);\boldsymbol{\theta}_j^*\right],$$

and

$$\frac{1}{B}\sum_{j=1}^{B}\left[\sum_{i=k}^{m}\binom{m}{i}(\omega_j^*)^i(1-\omega_j^*)^{m-i}\right] = \frac{1}{B}\sum_{j=1}^{B}\left[\sum_{i=k}^{m}\binom{m}{i}(\nu_j^*)^i(1-\nu_j^*)^{m-i}\right], \tag{15.9}$$

where

$$\omega_j^* = 1 - F\left[\underset{\sim}{Y}_{k;m}(1-\alpha;k,m);\boldsymbol{\theta}_j^*\right],$$
$$\nu_j^* = F\left[\widetilde{Y}_{k;m}(1-\alpha;k,m);\boldsymbol{\theta}_j^*\right].$$

The constraint given by (15.9) assures that the error probabilities are equal for both the lower and upper endpoints of the two-sided prediction interval. For the important special case of a (log-)location-scale distribution,

$$F(\underset{\sim}{Y}_{k;m};\boldsymbol{\theta}_j^*) = \Phi\left(\frac{\underset{\sim}{Y}_{k;m}-\mu_j^*}{\sigma_j^*}\right) \quad\text{and}\quad F(\widetilde{Y}_{k;m};\boldsymbol{\theta}_j^*) = \Phi\left(\frac{\widetilde{Y}_{k;m}-\mu_j^*}{\sigma_j^*}\right).$$

For (log-)location-scale distributions, Xie et al. (2017) describe an algorithm to solve (15.8) subject to (15.9) that requires only a simple root-finding procedure (such as `uniroot` in R). The method could, however, be adapted to non-(log-)location-scale distributions.

One-sided simultaneous prediction bounds are defined in a similar manner, following the outline in Section B.7, but taking expectations with respect to the joint posterior distribution of $\boldsymbol{\theta}$ instead of the sampling distribution of the data. In particular, for one-sided prediction bounds one can use (15.8) with

$$p_j = \Pr\left[Y > \underset{\sim}{Y}_{k;m}(1-\alpha;k,m)\right] = 1 - F\left[\underset{\sim}{Y}_{k;m}(1-\alpha;k,m);\boldsymbol{\theta}_j^*\right]$$

for a one-sided lower prediction bound and

$$p_j = \Pr\left[Y \le \widetilde{Y}_{k;m}(1-\alpha;k,m)\right] = F\left[\widetilde{Y}_{k;m}(1-\alpha;k,m);\boldsymbol{\theta}_j^*\right]$$

for a one-sided upper prediction bound.

Example 15.11 Bayesian Prediction Intervals for a Bearing Cage Failure Time. Following the approach described above, two-sided 95% Bayesian prediction intervals to contain the failure times of at least $k = 9$ out of $m = 10$ future bearing cages from the same population were computed using the draws from the joint posterior distribution (like those illustrated in Table 15.2) for both the diffuse and the informative prior distributions. Corresponding one-sided prediction bounds were also computed. The results are given in Table 15.4.

As with the tolerance intervals discussed in Example 15.10, the upper endpoints of these prediction intervals have little or no practical meaning due to the large amount of extrapolation needed to make inferences in the upper tail of the distribution (given that the nonparametric estimate of fraction failing goes only up to 0.05).

As expected, the lower endpoints of the prediction intervals computed using the diffuse prior distribution are larger (i.e., closer to the center of the distribution) than those computed with the informative prior distribution.   ∎

15.5.5   An Alternative Method of Computing Bayesian Prediction Intervals

The Bayesian prediction interval methods described in Section 15.5.4 are based on using the draws from the joint posterior distribution to approximate the posterior predictive distribution for the random variable(s) to be predicted. That method is difficult to apply in situations with more complicated models for which the distribution of the random variable cannot be computed explicitly (e.g., hierarchical models to be considered in Chapter 17).

An alternative method is to use another layer of Monte Carlo simulation to generate draws from the posterior predictive distribution, simulating copies of the random variable (or variables) to be predicted for each of the draws from the joint posterior distribution. This approach is more general, but requires a much larger number of draws from the posterior predictive distribution, relative to the method described in Section 15.5.4.

Suppose that it is desired to find a prediction interval for a scalar random variable Y that has a cdf $F(y; \boldsymbol{\theta})$. Based on observed data, the specified distribution, and a prior distribution, draws $\boldsymbol{\theta}_j^*, j = 1, \ldots, B$, from the joint posterior distribution are computed (e.g., by using an MCMC algorithm as described in Section 15.4). Then B draws from the posterior predictive distribution of Y are obtained by simulating a draw from $F(y; \boldsymbol{\theta})$ for each value of $\boldsymbol{\theta}_j^*, j = 1, \ldots, B$. For continuous distributions, this can be done by using

$$Y_j^* = F^{-1}(U_j; \boldsymbol{\theta}_j^*), \quad j = 1, \ldots, B,$$

where $F^{-1}(p; \boldsymbol{\theta})$ is the quantile function of the random variable Y and $U_j, j = 1, \ldots, B$, are independent and identically distributed $\mathrm{UNIF}(0, 1)$ random variables.

For some distributions there may be a more direct method of generating the random draws from $F(y; \boldsymbol{\theta})$. Then two-sided prediction intervals and one-sided prediction bounds for Y can be obtained from the empirical distribution of the draws from the posterior predictive distribution of Y in a manner that is similar to the computation of credible intervals for parameters and functions of the parameters. For example, a two-sided $100(1 - \alpha)\%$ Bayesian prediction interval would be obtained as the $\alpha/2$ and $1 - \alpha/2$ quantiles of the empirical distribution of the draws from the posterior predictive distribution.

This simulation approach to computing draws from a posterior predictive distribution can be extended to more complicated prediction problems such as a prediction interval to contain at least k out of m future observations. In this case, the values for m future observations are computed for each draw from the joint posterior distribution of the parameters. For each such simulated sample, information about the location of the order statistics would have to be tabulated.

BIBLIOGRAPHIC NOTES

General literature

A large number of books have been written on the practical use of Bayesian methods in statistical analysis. These include Box and Tiao (1973), Casella and Robert (1999), Congdon (2007), Hoff (2009), Carlin and Louis (2009), Lunn et al. (2012), and Gelman et al. (2013).

Methods for eliciting prior distributions have been provided by Meyer and Booker (1991), O'Hagan (1998), O'Hagan and Oakley (2004), Garthwaite et al. (2005), and O'Hagan et al. (2006) and others.

The simple simulation method for generating sample draws from a posterior distribution used in Section 15.2.5 is a slight modification of the method described in Smith and Gelfand (1992).

Bayesian tolerance and prediction intervals

Aitchison (1964) defines Bayesian tolerance intervals and compares them with non-Bayesian tolerance intervals. Guttman (1970) describes both Bayesian and non-Bayesian tolerance and prediction intervals. Hamada et al. (2004) explain the differences between Bayesian tolerance and prediction intervals and illustrate both kinds of intervals with an example using a hierarchical linear model.

Software for MCMC

Software packages to generate MCMC draws from a given posterior distribution, based on given data, an assumed underlying statistical model for the data, and assumed prior distributions for the distribution parameters, include WinBUGS described in Lunn et al. (2000), JAGS described in Plummer (2003), and OpenBUGS described in Spiegelhalter et al. (2014). All three of these packages use similar model-specification languages and are available at no cost. Available packages for R (R Core Team, 2016) provide a convenient interface that allows one to use these MCMC packages directly from R. Lunn et al. (2012) provides an introduction to OpenBUGS. Stan (Stan Development Team, 2015a,b) is another open-source package that has powerful Bayesian modeling capabilities.

It is also possible to program MCMC algorithms directly. Numerous books provide guidance for doing this, including Albert (2007), Rizzo (2007), and Robert and Casella (2010).

How many draws?

An important question facing users of Bayesian methods is "How many MCMC draws are required?" The answer to this question requires an assessment of Monte Carlo error. Much of the literature on this subject concerns estimating the mean of a marginal posterior distribution. Examples include Flegal et al. (2008) and Koehler et al. (2009). Raftery and Lewis (1992) show how to choose the number of MCMC trials to estimate the probability in the tail of a marginal posterior distribution. Liu et al. (2016) describe a method to choose the number of MCMC draws needed to estimate quantiles of a marginal posterior distribution to a desired degree of accuracy. Their method was designed to make it possible to report a Bayesian credible interval with a specified number of correct significant digits.

Chapter *16*

Bayesian Statistical Intervals for the Binomial, Poisson, and Normal Distributions

OBJECTIVES AND OVERVIEW

This chapter describes the construction of Bayesian intervals for data generated from the distributions discussed in Chapters 3 and 4 (normal distribution), Chapter 6 (binomial distribution) and Chapter 7 (Poisson distribution). We extend these methods (for the same distributions) in Chapter 17 to consider the construction of Bayesian intervals for the more complicated situation involving hierarchical models.

The topics discussed in this chapter are:

- The construction of Bayesian intervals for the binomial distribution (Section 16.1).

- The construction of Bayesian intervals for the Poisson distribution (Section 16.2).

- The construction of Bayesian intervals for the normal distribution (Section 16.3).

In each section we show how to compute credible intervals for the distribution parameter(s), functions of the parameter(s), and Bayesian tolerance and prediction intervals.

As we saw in Chapter 15, Bayesian methods using appropriately chosen diffuse prior distributions result in credible intervals that are close to the confidence intervals obtained using non-Bayesian methods. For particular examples, in Chapters 6 and 7 we used non-Bayesian methods to construct confidence intervals for the parameters of the binomial and Poisson distributions, respectively (or functions thereof). One of these methods—and one which we recommended for practical use—was the Jeffreys approximate method. Moreover, in the Bibliographic Notes section at the end of Chapters 6 and 7, we observe that this method is derived from a Bayesian method based on a Jeffreys prior distribution. We will elaborate on this comment shortly.

Statistical Intervals: A Guide for Practitioners and Researchers, Second Edition.
William Q. Meeker, Gerald J. Hahn and Luis A. Escobar.
© 2017 John Wiley & Sons, Inc. Published 2017 by John Wiley & Sons, Inc.
Companion Website: www.wiley.com/go/meeker/intervals

In addition, as shown in Section H.4.3, for the normal distribution mean, using a modified Jeffreys prior results in Bayesian credible intervals that are *exactly* the same as the non-Bayesian confidence intervals presented in Chapter 4. Thus, because of the preceding equivalence (or near-equivalence) of results obtained by using non-Bayesian intervals and Bayesian intervals constructed using diffuse prior distributions, we will in this chapter focus mainly on comparing Bayesian credible intervals constructed using diffuse versus informative prior distributions. We also note that due to this equivalence (or near equivalence), we can refer to Bayesian intervals constructed from diffuse prior distributions as either "confidence intervals" or "credible intervals." We will use the term "credible interval" in this chapter.

16.1   BAYESIAN INTERVALS FOR THE BINOMIAL DISTRIBUTION

This section presents Bayesian methods for the binomial distribution that extend the (mostly) non-Bayesian methods given in Chapter 6. In some applications involving the binomial distribution, there is prior information on the parameter π. Using this information will generally result in a credible interval for π—and in some cases other statistical intervals of interest—that are narrower than the intervals that one would obtain without taking such prior information into consideration. We will illustrate the methods with the application introduced in Example 6.1. More specifically, we will compare the statistical intervals computed with a diffuse Jeffreys prior distribution (described in Sections 6.2.5 and H.4.1) with those obtained using an informative prior distribution. Bayesian statistical intervals based on a Jeffreys prior distribution generally have coverage probabilities that are close to the nominal credible level. In this section, the number of nonconforming (e.g., defective) units x out of n trials is the data and the actual proportion π (i.e., binomial distribution parameter) is the parameter of interest.

16.1.1   Binomial Distribution Conjugate Prior Distribution

The binomial distribution is one of the distributions that has a conjugate prior distribution. In particular, suppose that we have observed x nonconforming units out of n trials and that the prior distribution for π can be expressed as a beta distribution with parameters a and b (i.e., $\mathrm{BETA}(a, b)$). Then, as shown in Section H.3.1, $f(\pi|x)$, the posterior distribution for π, is a $\mathrm{BETA}(x + a, n + b - x)$ distribution (i.e., a beta distribution with parameters $x + a$ and $n + b - x$). We note that the prior distribution $\mathrm{BETA}(a, b)$ can be interpreted as having prior information about π that is equivalent to previous data consisting of $a - 1$ nonconforming units out of a total of $a + b - 2$ sampled units (as shown in Section H.3.1).

Example 16.1   Prior Distributions for the Defective Integrated Circuits Application. Continuing with the application introduced in Example 6.1, there were $x = 20$ defective (non-conforming) integrated circuits in a sample of size $n = 1,000$. To use Bayesian methods, we must specify a prior distribution for the binomial distribution parameter π. For the informative prior distribution, we use a conjugate $\mathrm{BETA}(25, 975)$ distribution (i.e., a beta distribution with parameters $a = 25$ and $b = 975$), which can be interpreted as information equivalent to having 24 defective integrated circuits in 998 past observations. For the diffuse prior distribution we use a Jeffreys prior distribution (Section H.4.1) which has a $\mathrm{BETA}(0.50, 0.50)$ density. In the examples we will illustrate the use of both the conjugate distribution approach and Markov chain Monte Carlo (MCMC) simulation to construct statistical intervals. For the MCMC examples, `drawsICJeffreys` and `drawsICInformative` are vectors (and R objects) containing $B = 80,000$ draws from the posterior distribution of π for the Jeffreys prior and the informative prior distributions, respectively. ∎

16.1.2   Credible Interval for the Binomial Distribution Parameter

Using the conjugate distribution approach, a $100(1 - \alpha)\%$ credible interval for π can be obtained from the $\alpha/2$ and $1 - \alpha/2$ quantiles of the beta posterior distribution. Alternatively, with the same inputs (i.e., data and prior distribution), an MCMC algorithm could be used to generate B draws from the posterior distribution $f(\pi|x)$. Then a two-sided $100(1 - \alpha)\%$ Bayesian credible interval for π is obtained as the $\alpha/2$ and $1 - \alpha/2$ quantiles of the empirical distribution of the B MCMC draws from the posterior distribution of π. The advantage of the MCMC approach over the conjugate prior distribution approach is that one can specify a prior distribution in a form other than a beta distribution. For example, the prior distribution for π might be specified to be a normal distribution with 99% of its probability between 0.4 and 0.6.

Example 16.2   Credible Interval for the Proportion of Defective Integrated Circuits. In Example 6.1, a confidence interval was calculated on the binomial distribution parameter π, based on $x = 20$ defective units from a random sample of $n = 1,000$ integrated circuits. The density functions for the Jeffreys prior distribution and the informative prior distribution are compared in the top row of Figure 16.1. The bottom row of Figure 16.1 shows histograms of the corresponding $B = 80,000$ sample draws from the posterior distributions for π, the proportion of defective integrated circuits. The vertical lines indicate the 95% credible interval.

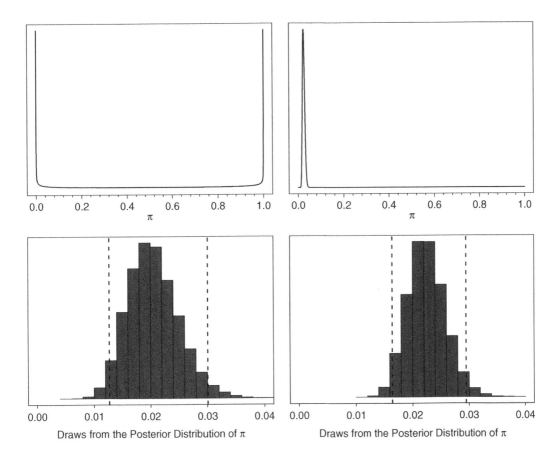

Figure 16.1   Comparison of prior distributions (top) and corresponding sample draws from the posterior distributions (bottom) of the proportion defective for the Jeffreys prior (left) and the informative prior (right) distributions for the integrated circuit data. The vertical lines indicate the 95% credible interval.

Type of prior	Interval/bound type	Interval for	Lower	Upper
Jeffreys	Two-sided credible	π	0.0127	0.0301
Informative	Two-sided credible	π	0.0165	0.0295
Jeffreys	One-sided credible	π	0.0137	0.0283
Informative	One-sided credible	π	0.0173	0.0282
Jeffreys	Two-sided credible	$\Pr(Y \leq 2)$, $m = 50$ ICs	0.8100	0.9743
Informative	Two-sided credible	$\Pr(Y \leq 2)$, $m = 50$ ICs	0.8175	0.9507
Jeffreys	Two-sided credible	$y_{0.90}$, $m = 50$ ICs	2	3
Informative	Two-sided credible	$y_{0.90}$, $m = 50$ ICs	2	3
Jeffreys	Two-sided tolerance	Control center 0.80, $m = 1{,}000$	9	34
Informative	Two-sided tolerance	Control center 0.80, $m = 1{,}000$	12	34
Jeffreys	Two-sided prediction	Y, $m = 50$ ICs	0	3
Informative	Two-sided prediction	Y, $m = 50$ ICs	0	4
Jeffreys	Two-sided prediction	Y, $m = 1{,}000$ ICs	10	34
Informative	Two-sided prediction	Y, $m = 1{,}000$ ICs	12	35

Table 16.1 Two-sided 95% credible intervals and one-sided 95% credible bounds for π, the binomial distribution proportion of defective integrated circuits, and other related intervals.

Using vectors of draws from the posterior distributions of π, the credible intervals for π can be obtained simply by using the following R commands:

```
> quantile(drawsICJeffreys, probs=c(0.025, 0.975))
[1] 0.0127 0.0301

> quantile(drawsICInformative, probs=c(0.025, 0.975))
[1] 0.0165 0.0295
```

These results and the results for all of the other examples in this section are summarized in Table 16.1.

Because we are using conjugate prior distributions for π, the 95% credible intervals for the Jeffreys prior and the informative prior distributions can also be obtained from the 0.025 and 0.975 quantiles of the $\mathrm{BETA}(20 + 0.5, 1000 + 0.5 - 20)$ and $\mathrm{BETA}(20 + 25, 1000 + 975 - 20)$ posterior distributions, respectively. Taking this approach, the credible intervals for π can be obtained by using the following simple R commands:

```
> qbeta(p=c(0.025, 0.975), shape1=20+0.5, shape2=1000+0.5-20)
[1] 0.0127 0.0301

> qbeta(p=c(0.025, 0.975), shape1=20+25, shape2=1000+975-20)
[1] 0.0165  0.0294
```

The slight difference in the upper bound using the informative prior distribution is due to Monte Carlo error for the $B = 80{,}000$ sample draws that were used for this example. This Monte Carlo error could be reduced by using a larger number of sample draws (we will see similar Monte Carlo error in some other examples in this chapter, but will not always point them out).

As expected (and previously noted in Chapter 6—see Table 6.2), the credible interval endpoints and bounds based on the Jeffreys prior distribution are close to those obtained using the non-Bayesian methods in Examples 6.4 and 6.5. In comparing the results using the Jeffreys prior distribution with those using the informative prior distribution, we find noticeable differences in the lower endpoints of the credible intervals (as well as the one-sided lower credible bounds), but little difference in the upper endpoints of the credible intervals (or the one-sided upper credible bounds). All in all, we conclude that even though the Jeffreys prior and the informative prior distributions are strikingly different (as seen in Figure 16.1), the prior distribution does not have a large effect on the credible intervals in this example. This is because there is a large amount of data and reasonably good agreement between the data and the prior distributions.   ∎

16.1.3 Credible Intervals for Functions of the Binomial Distribution Parameter

To find Bayesian credible intervals (or one-sided credible bounds) for a function of π, one can use the approach described in Section 15.2.6. That is, one can compute the function of interest for each sample draw from the posterior distribution of π to generate sample draws from the marginal posterior distribution of the quantity of interest. Then the $100(1 - \alpha)\%$ credible interval for the quantity of interest can be obtained from the $\alpha/2$ and $1 - \alpha/2$ quantiles of the empirical distribution of these draws. In the two examples below, we want to make inferences or predictions about a future sample from a binomial distribution having the same value of π but with a different sample size m. The random variable for this distribution will be denoted by Y.

If the function of interest is a nondecreasing (nonincreasing) function of π, then one can, as in Chapter 6, compute the lower and upper endpoints of the credible interval for the function of interest by just substituting the lower and upper (upper and lower) credible interval endpoints for π into the function of interest.

Example 16.3   Credible Interval for the Probability of Two or Fewer Defects in a Package of $m = 50$ Integrated Circuits. Figure 16.2 is a histogram of the $B = 80,000$ sample draws from the posterior distribution of the binomial probability $\Pr(Y \leq 2)$ for a package of $m = 50$ integrated circuits based on the Jeffreys prior distribution (left) and the informative prior distribution (right) for π. The vertical lines in Figure 16.2 indicate the endpoints of the 95%

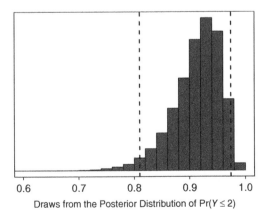
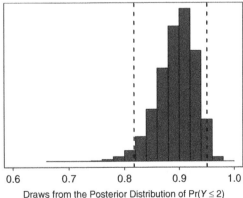

| 0.6 | 0.7 | 0.8 | 0.9 | 1.0 | | 0.6 | 0.7 | 0.8 | 0.9 | 1.0 |

Draws from the Posterior Distribution of Pr($Y \leq 2$)   Draws from the Posterior Distribution of Pr($Y \leq 2$)

Figure 16.2   Histograms of 80,000 draws from the posterior distribution of the binomial $\Pr(Y \leq 2)$ for a package of $m = 50$ integrated circuits based on a Jeffreys prior distribution (left) and an informative prior distribution (right). The vertical lines indicate the endpoints of the 95% Bayesian credible interval for $\Pr(Y \leq 2)$.

Bayesian credible interval for $\Pr(Y \le 2)$. These endpoints were obtained by taking the 0.025 and 0.975 quantiles of the empirical distribution of the $B = 80{,}000$ sample draws from the marginal posterior distributions of $\Pr(Y \le 2)$ shown in Figure 16.2. Using vectors of draws from the posterior distribution of π, one can compute the corresponding set of draws from the posterior distribution of $\Pr(Y \le 2)$ and the quantiles of that distribution by using the following R commands:

```
> quantile(pbinom(q=2, size=50, prob=drawsICJeffreys), p=c(0.025,
  0.975))
    2.5%   97.5%
 0.8100 0.9743
> quantile(pbinom(q=2, size=50, prob=drawsICInformative), p=c(0.025,
  0.975))
    2.5%   97.5%
 0.8175 0.9507
```

giving the credible intervals based on the Jeffreys prior and the informative prior distribution.

Alternatively, because there is only one unknown parameter in the model, we can, as in the non-Bayesian Example 6.9, substitute the credible interval endpoints from Example 16.2 into the function of interest. Thus, because $\Pr(Y \le 2)$ is a decreasing function of π (Casella and Berger, 2002, page 426), we substitute the upper (lower) endpoint of the credible interval for π to obtain the lower (upper) endpoint for the credible interval for $\Pr(Y \le 2)$. In particular, the interval endpoints can be obtained by using the following R commands for the Jeffreys prior and the informative prior distributions:

```
> pbinom(q=2, size=50, prob=qbeta(p=c(0.975, 0.025),
       shape1=20+0.5, shape2=1000+0.5-20))
[1] 0.8093 0.9745
> pbinom(q=2, size=50, prob=qbeta(p=c(0.975, 0.025),
       shape1=20+25, shape2=1000+975-20))
[1] 0.8179 0.9506
```

As expected, the credible interval computed with the informative prior distribution is narrower than the one computed from the Jeffreys prior distribution. The small difference between the intervals (e.g., $[0.8179, \; 0.9506]$ computed above using the exact conjugate posterior distribution of $\Pr(Y \le 2)$ and $[0.8175, \; 0.9507]$ computed from sample draws of the posterior distribution of $\Pr(Y \le 2)$) is due to Monte Carlo error, which could be reduced by using a larger value of B. ∎

Example 16.4   Credible Interval for the 0.90 Quantile of the Distribution of the Number of Defects in Packages of $m = 50$ Integrated Circuits. Figure 16.3 shows histograms of the $B = 80{,}000$ draws from the posterior distribution of $y_{0.90}$, the 0.90 quantile of the distribution of Y, the number of defects in packages of $m = 50$ integrated circuits, based on the Jeffreys prior and the assumed informative prior distribution, respectively. We note that the preponderance of draws in both cases resulted in either 2 or 3 defects. A small number of draws (hardly visible in Figure 16.3) yielded 1 or 4 defects.

A 95% credible interval for $y_{0.90}$, the 0.90 quantile of the distribution of Y, the number of defects in packages of $m = 50$ integrated circuits, can be obtained by using the 0.025 and 0.975 quantiles of the empirical distribution of the draws from the posterior distribution of $y_{0.90}$.

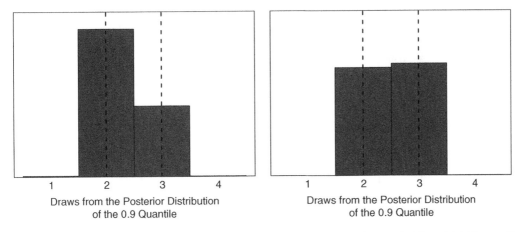

Figure 16.3 Histograms of 80,000 draws from the posterior distribution of $y_{0.90}$, the binomial 0.90 quantile of the distribution of the number of defects in packages of $m = 50$ integrated circuits based on the Jeffreys prior (left) and the informative prior distributions (right) for π.

The 95% credible intervals for $y_{0.90}$ can be computed for both prior distributions by using the following R commands:

```
> quantile(qbinom(p=0.90, size=50, prob=drawsICJeffreys),
  probs=c(0.025,0.975))
[1] 2   3
> quantile(qbinom(p=0.90, size=50, prob=drawsICInformative),
  probs=c(0.025,0.975))
    2   3
```

Alternatively, applying a conjugate prior (beta) distribution, one can use instead the quantiles from the analytically computed posterior distribution of $y_{0.90}$. For the informative prior distribution, the following R commands can be used:

```
> qbinom(p=0.90,size=50, prob=qbeta(p=c(0.025, 0.975),
      shape1=20+0.5, shape2=1000+0.5-20))
[1] 2 3
> qbinom(p=0.90,size=50, prob=qbeta(p=c(0.025, 0.975),
      shape1=20+25, shape2=1000+975-20))
[1] 2   3
```

In words, we are 95% confident that a proportion 0.90 of future packages of 50 integrated circuits will have either 2 or 3 defective units. For this example, there is no difference between the credible intervals using the Jeffreys prior and the informative prior distributions due to the discreteness of the distribution of $y_{0.90}$ and because there is little difference in the marginal posterior distributions for π (as seen in the bottom row of Figure 16.1). ∎

16.1.4   Tolerance Intervals for the Binomial Distribution

In this section we use a procedure that is similar to that used in Section 6.6.4 and explicitly illustrate the use of interval calibration suggested there and explained more generally in Section B.8.

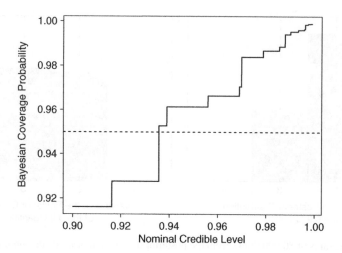

Figure 16.4   Bayesian coverage probability as a function of the input nominal credible level. The horizontal dashed line is the desired credible level.

In particular, as explained in Section D.7.4, an approximate $100(1 - \alpha)\%$ control-the-center tolerance interval to contain a proportion β can be obtained by combining a one-sided lower $100(1 - \alpha)\%$ credible bound on the $(1 - \beta)/2$ quantile of the distribution and a one-sided upper $100(1 - \alpha)\%$ credible bound on the $(1 + \beta)/2$ quantile of the distribution. Evaluation of the Bayesian coverage probability for such approximate tolerance intervals using (15.4) shows that the intervals tend to be somewhat conservative. Using the calibration approach, one can adjust the input nominal credible level $1 - \alpha$ to obtain an interval that has a Bayesian coverage that is closer to what is desired.

Example 16.5   Bayesian Tolerance Interval to Contain at Least a Proportion 0.80 of the Distribution of Defects in Batches of $m = 1{,}000$ Integrated Circuits with 95% Confidence.
Using the methods presented in Section 16.1.3, similar to Example 16.4, using the Jeffreys prior distribution draws, a one-sided lower 95% credible bound on the 0.10 quantile of the distribution is 9 and a one-sided upper 95% credible bound on the 0.90 quantile of the distribution is 35. The Bayesian coverage probability for this interval is computed from (15.4) to be 0.9613. Figure 16.4 shows the Bayesian coverage probability as a function of nominal credible interval input values between 0.90 and 0.99. We note from Figure 16.4 that the Bayesian coverage probability is a step function of the nominal credible level $1 - \alpha$. This is because of the discreteness of the interval endpoints (which results from the discreteness of the sample space). As $1 - \alpha$ increases, the lower endpoint will decrease and the upper endpoint will increase—but only at certain points. The larger jumps in the coverage probability function occur when the lower endpoint decreases. The coverage probability first crosses 0.95 when $1 - \alpha$ is approximately 0.936, as shown in Figure 16.4, and at that point the tolerance interval is [9, 34]. The following R commands illustrate the computations that were used to compute the initial interval endpoints (before calibration) and the associated Bayesian coverage probabilities for the intervals [9, 35] and [9, 34]:

```
> quantile(qbinom(p=0.10, size=1000, prob=drawsICJeffreys),
  probs=0.05)
5%
  9
```

```
> quantile(qbinom(p=0.90, size=1000, prob=drawsICJeffreys),
    probs=0.95)
95%
 35
> sum(pbinom(q=35, size=1000, prob=drawsICJeffreys)-
      pbinom(q=9, size=1000, prob=drawsICJeffreys) > 0.80)/
    length(drawsICJeffreys)
[1] 0.9613
> sum(pbinom(q=34, size=1000, prob=drawsICJeffreys)-
      pbinom(q=9, size=1000, prob=drawsICJeffreys) > 0.80)/
    length(drawsICJeffreys)
[1] 0.9528
```

The computations show that the calibration provides an alternative narrower interval $[9, \ 34]$ (instead of $[9, \ 35]$) that is still conservative. Similar computations using the informative prior distribution (with just a change in the object containing the MCMC draws) give a calibrated tolerance interval $[12, \ 34]$, providing a somewhat larger lower interval endpoint, when compared to the Jeffreys prior tolerance interval. ∎

16.1.5 Prediction Intervals for the Binomial Distribution

Either of the two general methods described in Sections 15.5.4 and 15.5.5 can be used to construct Bayesian prediction intervals for a future outcome of a binomial random variable involving m future independent and identically distributed Bernoulli trials. As with the binomial credible intervals described in Sections 16.1.2 and 16.1.3, using sample draws from the posterior predictive distribution (i.e., the distribution of the future random variable Y given the data) provides an easy-to-implement method that can be used with an arbitrary prior distribution.

If one uses a beta conjugate prior distribution for π, then a simple analytical expression for the posterior predictive distribution of Y can be used, as described in Section H.5.2. In this case, a Bayesian prediction interval is obtained by using the appropriate quantiles of the beta-binomial distribution (described in Section C.4.2). In particular, if the prior distribution for π is a $\mathrm{BETA}(a, b)$ distribution, then the prediction interval for the number nonconforming Y in m future trials is computed from

$$[\underset{\sim}{Y}, \ \widetilde{Y}] = [\text{qbetabinom}(\alpha/2; m, x + a, n - x + b),$$
$$\text{qbetabinom}(1 - \alpha/2; m, x + a, n - x + b)]. \qquad (16.1)$$

For a Jeffreys prior distribution one uses $a = b = 0.50$ (see Section H.6.1 and our earlier discussion), as illustrated in Section 6.7.4.

Example 16.6 Bayesian Prediction Interval for the Number of Defective Integrated Circuits. Using the alternative method described in Section 15.5.5 we return to the preceding example to compute prediction intervals for the number of defects in packages of $m = 50$ and $m = 1,000$ integrated circuits, assuming diffuse (Jeffreys) and informative prior distributions for π.

The top row of plots in Figure 16.5 gives scatter plots of the 1,000 draws from the posterior distribution for π versus the corresponding draws from the predictive distribution of Y, the number of defective components from a package of $m = 50$ integrated circuits based on the Jeffreys prior distribution on the left and the informative prior distribution on the right. A small amount of jitter was used around the integer prediction values so that one can see the density of the plotted points. The plots in the bottom row are similar to those in the top row, except that they

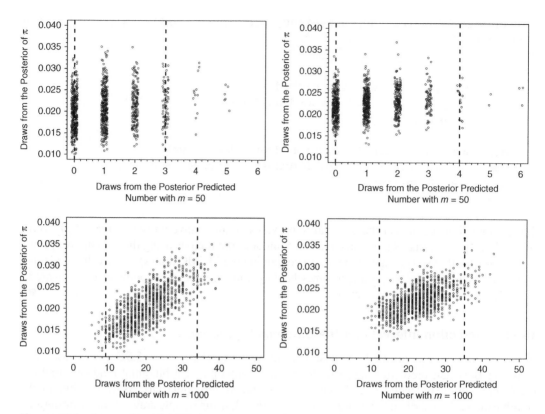

Figure 16.5 Scatter plots of 1,000 draws from posterior distribution of π versus the posterior predictive distributions of Y, the number of defects in packages of $m = 50$ (top) and $m = 1,000$ (bottom) integrated circuits, based on the Jeffreys prior (left) and the informative prior distributions (right). The vertical dashed lines indicate the Bayesian 95% prediction intervals for Y.

are for packages of $m = 1,000$ integrated circuits. The vertical lines indicate the 95% prediction interval endpoints (approximately 2.5% of the plotted points fall to the left (right) of the lower (upper) endpoint) for the number of defective units in the future sample. In the $m = 1,000$ plots in the bottom row, we can see the positive correlation (as expected) between the draws from the distribution of π and the corresponding prediction. There is also positive correlation for the $m = 50$ plot, but it is not as strong (and less evident in the plots) because the range of the draws from the predictive distribution of the future number of defective units is smaller.

The following R commands generate draws from the posterior predictive distribution of the number of defects in packages of $m = 50$ integrated circuits and then calculate the 0.025 and 0.975 quantiles of the empirical distribution of these draws, giving the corresponding 95% Bayesian prediction intervals using both the Jeffreys prior and the informative prior distributions:

```
> quantile(rbinom(n=length(drawsICJeffreys), size=50,
      prob=drawsICJeffreys), probs =c(0.025, 0.975))
 2.5% 97.5%
    0     3
> quantile(rbinom(n=length(drawsICInformative), size=50,
      prob=drawsICInformative), probs =c(0.025, 0.975))
 2.5% 97.5%
    0     4
```

The following R commands use the diffuse (Jeffreys) and informative beta conjugate prior distributions from the previous examples using (16.1) and the beta-binomial quantile function qbetabinom (in package StatInt), giving

```
> qbetabinom(p=c(0.025, 0.975), size=50, shape1=20+0.50, shape2=1000+
  0.5-20)
[1] 0 3
> qbetabinom(p=c(0.025, 0.975), size=50, shape1=20+25, shape2=1000+
  975-20)
[1] 0 4
```

The prediction interval results are summarized in the bottom part of Table 16.1. Note that, unlike the case for the credible interval for π, the Bayesian prediction interval using the informative prior distribution for π is, when compared to that obtained using the Jeffreys prior, only slightly narrower for the future sample size of $m = 1,000$. This is because, as we have seen for some other prediction interval examples, the width of the interval is due primarily to the variability in the future sample, as opposed to uncertainty in the model parameter π.

For the future sample size of $m = 50$ the prediction interval using the informative prior distribution for π is a little wider than that obtained using the Jeffreys prior. This is because the informative prior distribution suggested a slightly larger value of π than the data, causing the upper endpoint of the interval to increase from 3 to 4.   ∎

16.2 BAYESIAN INTERVALS FOR THE POISSON DISTRIBUTION

This section presents Bayesian methods for constructing statistical intervals for the Poisson distribution. These methods extend the non-Bayesian methods given in Chapter 7. In some applications involving the Poisson distribution, there is prior information on the parameter λ. Using this information will generally result in a credible interval for λ—and in other statistical intervals of interest—that are narrower than the intervals that one obtains without taking such information into consideration. We will illustrate the methods with the example introduced in Example 7.1.

For the binomial distribution examples in Section 16.1, we provided plots showing the shape of the marginal posterior distributions. Because the plots would be similar for the Poisson distribution examples, we do not present such plots here. They are, however, easy for the reader to construct using a computing/graphics environment like R.

Our discussion will proceed as in the preceding section and we will again compare the statistical intervals computed using a diffuse Jeffreys prior distribution (described in Section H.4.2) with those obtained using an informative prior distribution. As described in Chapter 7, the number of events x in n units of exposure is the data and the actual rate of occurrence of events per unit of exposure λ is the parameter of interest.

16.2.1 Poisson Distribution Conjugate Prior Distribution

The Poisson distribution is another distribution that has a conjugate prior distribution. Suppose that we have observed x events in n units of exposure and that the prior distribution for λ is a gamma distribution (described in Section C.3.5) with shape parameter a and rate parameter (the reciprocal of a scale parameter) b (i.e., $\text{GAMMA}(a, b)$). Then, as shown in Section H.3.2, the posterior distribution $f(\lambda|x)$ is a $\text{GAMMA}(x + a, n + b)$ distribution (i.e., a gamma distribution with shape parameter $x + a$ and rate parameter $n + b$). We note that the prior distribution GAMMA(a, b) can be interpreted as having prior information about λ that is equivalent to previous data consisting of $a - 1$ events in b units of exposure, as shown in Section H.3.2.

Example 16.7   Prior Distributions for the Unscheduled Computer Shutdown Application. In Example 7.1, there were $x = 24$ unscheduled shutdowns in $n = 5$ system-years of operation. To use Bayesian methods, we must specify a prior distribution for the Poisson parameter λ. For the informative prior distribution, we use a conjugate GAMMA$(32, 5)$ distribution (i.e., a gamma distribution with shape parameter $\alpha = 32$ and rate parameter $b = 5$). This can be interpreted as the information equivalent to having previously observed 31 unscheduled shutdowns in 5 system-years of exposure. For the diffuse prior distribution we use a Jeffreys prior distribution (Section H.4.2) that can be approximated by a conjugate GAMMA$(0.50, 0.00001)$ distribution. In the examples we will illustrate the use of both conjugate distributions and MCMC simulation to construct statistical intervals. For the MCMC examples, `drawsComputerShutDownJeffreys` and `drawsComputerShutDownInformative` are vectors (and R objects) containing $B = 80,000$ draws from the posterior distribution of λ for the Jeffreys prior and the informative prior distributions, respectively.   ∎

16.2.2   Credible Interval for the Poisson Event-Occurrence Rate

Using the conjugate distribution approach, a $100(1 - \alpha)\%$ credible interval for the Poisson event-occurrence rate λ can be obtained from the $\alpha/2$ and $1 - \alpha/2$ quantiles of the resulting gamma posterior distribution. Alternatively, with the same inputs (i.e., data and prior distribution), an MCMC algorithm could be used to generate B draws from the posterior distribution $f(\lambda|x)$. Then a two-sided $100(1 - \alpha)\%$ credible interval for λ is obtained as the $\alpha/2$ and $1 - \alpha/2$ quantiles of the empirical distribution of the B MCMC draws from the posterior distribution of λ. The advantage of the MCMC approach over the conjugate distribution approach is that one can specify a prior distribution in a form other than a gamma distribution. For example, the prior distribution for λ might be specified to be a normal distribution with 99% of its probability between two specified numbers.

Example 16.8   Credible Interval for the Rate of Occurrence of Unscheduled Shutdowns for a Group of Computing Systems. Here we illustrate the use of both the diffuse (Jeffreys) and the informative GAMMA$(32, 5)$ prior distributions in Example 16.7. We combine these prior distributions with the data (24 unscheduled shutdowns in 5 system-years of operation) to obtain the 95% credible intervals directly from the conjugate posterior distribution of λ by using the following R commands:

```
> qgamma(p=c(0.025, 0.975), shape=24+0.5, rate= 5)
[1] 3.155 7.022
> qgamma(p=c(0.025, 0.975), shape=24+32, rate=5+5)
[1] 4.230 7.159
```

Alternatively, the empirical quantiles of the sample of draws from the MCMC-computed posterior distribution of λ can be used to obtain the 95% credible intervals, using the following R commands:

```
> quantile(drawsComputerShutDownJeffreys,probs=c(0.025,0.975))
 2.5% 97.5%
3.151 7.021
> quantile(drawsComputerShutDownInformative, probs=c(0.025,0.975))
 2.5% 97.5%
4.233 7.181
```

Type of prior	Interval/bound type	Interval for	Lower	Upper
Jeffreys	Two-sided credible	λ	3.155	7.022
Informative	Two-sided credible	λ	4.230	7.159
Jeffreys	Two-sided credible	$\Pr(Y \leq 5)$, $m = 0.50$ years	0.8561	0.9943
Informative	Two-sided credible	$\Pr(Y \leq 5)$, $m = 0.50$ years	0.8469	0.9789
Jeffreys	Two-sided credible	$y_{0.90}$, $m = 1$ years	6	11
Informative	Two-sided credible	$y_{0.90}$, $m = 1$ years	7	11
Jeffreys	Two-sided tolerance	Control center 0.80, $m = 2$ years	3	18
Informative	Two-sided tolerance	Control center 0.80, $m = 2$ years	5	18
Jeffreys	One-sided prediction	Y, $m = 0.50$ years	0	5
Informative	One-sided prediction	Y, $m = 0.50$ years	0	6
Jeffreys	Two-sided prediction	Y, $m = 4$ years	9	32
Informative	Two-sided prediction	Y, $m = 4$ years	12	34

Table 16.2 Two-sided 95% credible intervals for λ, the Poisson distribution mean number of unscheduled shutdowns per year, and other related statistical intervals.

The small differences in results between the two methods are due to Monte Carlo error. These results and the results from other examples in this section are summarized in Table 16.2. ∎

16.2.3 Other Credible Intervals for the Poisson Distribution

To compute a Bayesian credible interval (or one-sided credible bound) for a function of λ one can use the approach described in Section 15.2.6. That is, one can use direct evaluation of the function of interest for each of the B sample draws from the posterior distribution of λ to generate sample draws from the posterior distribution of the function of interest. Then the $100(1 - \alpha)\%$ credible interval for the quantity of interest can be obtained from the $\alpha/2$ and $1 - \alpha/2$ quantiles of the empirical distribution of these draws.

If the function of interest is a nondecreasing (nonincreasing) function of λ, then one can, as in Chapter 7, compute the lower and upper endpoints of the credible interval of interest by just substituting the lower and upper (upper and lower) endpoints for λ into the function of interest.

Example 16.9 Credible Interval for the Probability of Five or Fewer Unscheduled Shutdowns in 6 Months of Operation ($m = 0.5$ **year**). Applying the MCMC approach, sample draws from the posterior distribution of $\Pr(Y \leq 5)$ can be computed from each of the $B = 80{,}000$ sample draws from the posterior distribution of λ. Then the desired 95% credible interval for $\Pr(Y \leq 5)$ in 6 months (i.e., half a year) of operation is obtained from the 0.025 and 0.975 quantiles of the empirical distribution of these draws. For the Jeffreys prior and the informative prior distributions, this is done with the following R commands:

```
> quantile(ppois(q=5, lambda=drawsComputerShutDownJeffreys*0.5),
        probs=c(0.025, 0.975))
  2.5%   97.5%
0.8562 0.9944
> quantile(ppois(q=5, lambda=drawsComputerShutDownInformative*0.5),
        probs=c(0.025, 0.975))
  2.5%   97.5%
0.8454 0.9789
```

Alternatively, using directly the credible intervals for λ, obtained as described in Section 16.2.2, the credible interval for $\Pr(Y \leq 5)$ in 6 months (i.e., half of a year) of operation is obtained, applying the conjugate distribution approach. This is done by using the following R commands:

```
> ppois(q=5, lambda=qgamma(p=c(0.975, 0.025), shape=24+0.5, rate=5)
  *0.50)
[1]  0.8561 0.9943
> ppois(q=5, lambda=qgamma(p=c(0.975, 0.025), shape=24+32, rate=5+5)
  *0.50)
[1]  0.8469 0.9789
```

The preceding results (using the conjugate distribution approach) are shown in Table 16.2. ■

Example 16.10   Credible Interval for the 0.90 Quantile of the Distribution of the Number of Unscheduled Computer Shutdowns in 1 Year of Operation ($m = 1$ year). In a manner similar to that used in the previous example, using the MCMC method, sample draws from the posterior distribution of the Poisson distribution quantile $y_{0.90}$—the 0.90 quantile of the distribution of the number of unscheduled shutdowns Y in 1 year of operation—can be computed for each of the $B = 80,000$ draws from the posterior distribution of λ. Then the desired 95% credible interval for $y_{0.90}$ is obtained from the 0.025 and 0.975 quantiles of the empirical distribution of these draws. For the Jeffreys prior and the informative prior distributions, this is done with the following R commands:

```
> quantile(qpois(p=0.90, lambda=drawsComputerShutDownJeffreys*1.0),
        probs=c(0.025, 0.975))
 2.5% 97.5%
    5    11
> quantile(qpois(p=0.90, lambda=drawsComputerShutDownInformative*1.0),
        probs=c(0.025, 0.975))
 2.5% 97.5%
    7    11
```

Alternatively, using the conjugate distribution approach, a credible interval for $y_{0.90}$, can be computed by evaluating the Poisson quantile function at the credible interval endpoints. For the Jeffreys prior and the informative prior distributions this is done with the following R commands:

```
> qpois(p=0.90, lambda=qgamma(p=c( 0.025, 0.975),
        shape=24+0.5, rate=5)*1.0)
[1]   6 11
> qpois(p=0.90, lambda=qgamma(p=c( 0.025, 0.975),
        shape=24+32, rate=5+5)*1.0)
[1]   7 11
```

The preceding results (using the conjugate distribution approach) are shown in Table 16.2. ■

16.2.4   Tolerance Intervals for the Poisson Distribution

This section uses a procedure that is similar to that used in Section 7.5.3 and Section 16.1.4. In particular, as explained in Section D.7.4, an approximate $100(1 - \alpha)\%$ control-the-center tolerance interval to contain a proportion β can be obtained by combining a one-sided lower $100(1 - \alpha)\%$ credible bound on the $(1 - \beta)/2$ quantile of the distribution and a one-sided upper $100(1 - \alpha)\%$ credible bound on the $(1 + \beta)/2$ quantile of the distribution. Evaluation of the Bayesian coverage probability for such approximate tolerance intervals using (15.4) shows that the intervals tend to be somewhat conservative.

Example 16.11 Bayesian Tolerance Interval to Contain at Least a Proportion 0.80 of the Distribution of the Number of Unscheduled Computer Shutdowns in 2 Years of Operation ($m = 2$ years). From the methods presented in Section 16.2.3, similar to Example 16.10, using the Jeffreys prior distribution draws, a one-sided lower 95% credible bound for the 0.10 quantile of the distribution is 4 and a one-sided upper 95% credible bound for the 0.90 quantile of the distribution is 18. The corresponding Bayesian coverage probability for the resulting two-sided tolerance interval, computed from (15.4), is 0.9463. By increasing the input coverage probability $1 - \alpha$ above 0.95, the lower endpoint decreases from 4 to 3 when $1 - \alpha \approx 0.956$, the Bayesian coverage probability increases sharply to 0.9867, and the tolerance interval becomes $[3, \ 18]$. The following R commands illustrate the computations that were used to obtain these results:

```
> quantile(qpois(p=0.10, lambda=drawsComputerShutDownJeffreys*2),
  probs=0.05)
5%
 4
> quantile(qpois(p=0.90, lambda=drawsComputerShutDownJeffreys*2),
  probs=0.95)
95%
 18
> sum(ppois(q=18, lambda=drawsComputerShutDownJeffreys*2)-
       ppois(q=4, lambda=drawsComputerShutDownJeffreys*2) > 0.80)/
       length(drawsComputerShutDownJeffreys)
[1] 0.9463
> sum(ppois(q=18, lambda=drawsComputerShutDownJeffreys*2)-
       ppois(q=3, lambda=drawsComputerShutDownJeffreys*2) > 0.80)/
       length(drawsComputerShutDownJeffreys)
[1] 0.9867
```

Similar computations (with just a change in the object containing the MCMC draws) give the calibrated tolerance interval for the informative prior distribution with the results $[5, \ 18]$, shown in Table 16.2. ∎

16.2.5 Prediction Intervals for the Poisson Distribution

Either of the two general methods described in Sections 15.5.4 and 15.5.5 can be used to construct Bayesian prediction intervals for a future outcome of a Poisson random variable involving m future units of exposure. As with the Poisson credible intervals described in Sections 16.2.2 and 16.2.3, using sample draws from the posterior predictive distribution provides an easy-to-implement method that can be used with an arbitrary prior distribution.

If one uses a gamma conjugate prior distribution for λ, then an analytical expression for the posterior predictive distribution can be used, as described at the end of Section H.5.3. In this case, a Bayesian prediction interval is obtained from the appropriate quantiles of the negative binomial distribution (described in Section C.4.3). In particular, if the prior distribution for λ

is GAMMA(a, b), then the prediction interval for the number events Y in m future units of exposure is computed as

$$[\underset{\sim}{Y}, \ \widetilde{Y}] = [\text{qnbinom}(\alpha/2; a + x, (b + n)/(b + n + m)),$$
$$\text{qnbinom}(1 - \alpha/2; a + x, (b + n)/(b + n + m))], \qquad (16.2)$$

where qnbinom$(p; k, \pi)$ is the p quantile of the negative binomial distribution with "stopping parameter" k and "proportion parameter" π. For a Jeffreys prior distribution one uses $a = 0.50$ and $b = 0$ (see Section H.6.2), as illustrated in Section 7.6.4. The following two examples illustrate the computation of Bayesian prediction intervals and bounds for the number of unscheduled computer shutdowns with a given amount of future exposure using both the Jeffreys prior and the informative prior distributions and both the simulation method and the conjugate prior prediction-distribution methods.

Example 16.12 Bayesian Prediction Interval for the Number of Unscheduled Computer Shutdowns in 4 Years of Operation ($m = 4$ years). Using the MCMC method described in Section 15.5.5, we will compute prediction intervals for Y, the number of unscheduled computer shutdowns in 4 years of operation. Thus, for each of the $B = 80,000$ draws from the posterior distribution of λ, multiplied by 4 to get the rate for 4 years of exposure, we generate a Poisson random variable. The resulting collection of B Poisson variates are draws from the posterior predictive distribution of Y. Then the Bayesian 95% prediction interval for Y is obtained from the 0.025 and 0.975 quantiles of the empirical distribution of these draws. For the Jeffreys prior and the informative prior distributions, this is done with the following R commands:

```
> quantile(rpois(n=length(drawsComputerShutDownJeffreys),
        lambda=drawsComputerShutDownJeffreys*4.0), probs=c
        (0.025, 0.975))
 2.5% 97.5%
    9    32
> quantile(rpois(n=length(drawsComputerShutDownJeffreys),
        lambda=drawsComputerShutDownInformative*4.0), probs=c
        (0.025, 0.975))
 2.5% 97.5%
   12    34
```

The conjugate predictive distribution approach in (16.2), using the prior distributions from the previous examples, can be implemented with the following R commands:

```
> qnbinom(p=c(0.025, 0.975), size=24+0.5, prob=5/(5+4))
[1]   9 32
> qnbinom(p=c(0.025, 0.975), size=24+32, prob=(5+5)/(5+5+4))
[1] 12 34
```

Again, the preceding results are shown in Table 16.2.   ■

Example 16.13 Bayesian Upper Prediction Bound for the Number of Unscheduled Computer Shutdowns in 6 Months of Operation ($m = 0.50$ year). Similarly to Example 16.12, we can use the MCMC method to construct a Bayesian *one-sided* upper prediction bound on the number of unscheduled shutdowns in the next 6 months of operation. The only difference is that we multiply the draws from the posterior distribution of λ by 0.50 instead of by 4, and use only the 0.95 quantile of the empirical distribution of the draws from the posterior predictive distribution of Y. For the Jeffreys prior and informative prior distributions used in previous examples in this section, this is done by using the following R commands:

```
> quantile(rpois(n=length(drawsComputerShutDownJeffreys),
        lambda=drawsComputerShutDownJeffreys*0.5), probs=0.95)
95%
  5
> quantile(rpois(n=length(drawsComputerShutDownInformative),
        lambda=drawsComputerShutDownInformative*0.5), probs=0.95)
95%
  6
```

The conjugate predictive distribution approach in (16.2), using the prior distributions from the previous examples, can be implemented with the following R commands:

```
> qnbinom(p=0.95, size=24+0.5, prob=5/(5+0.5)))
[1] 5
> qnbinom(p=0.95, size=24+32, prob=(5+5)/(5+5+0.5)))
[1] 6
```

■

16.3 BAYESIAN INTERVALS FOR THE NORMAL DISTRIBUTION

In some applications involving the normal distribution, there is prior information on the distribution parameters μ and σ. In this section, we describe the use of such information, using MCMC methods, to construct Bayesian statistical intervals.

16.3.1 Normal Distribution Conjugate Prior Distribution

The normal distribution is another distribution that has a conjugate prior distribution from which one can readily determine an appropriate (normal) conjugate posterior distribution. One can then compute statistical intervals directly from the resulting joint posterior distribution, from the parameters of that distribution, or (for complicated intervals) from iid draws from the distribution. Technical details are given in Section H.3.3. In this section, however, we will use the more general MCMC sample-draws method for constructing statistical intervals, which allows one to specify an arbitrary joint prior distribution.

16.3.2 MCMC Method

In what follows, the observed values $x_1, \ldots, x_n$ from a sample of size n is the data and the mean μ and standard deviation σ are the parameters. It will be assumed that, based on a given prior distribution and data, that B sample draws μ_j^* and $\sigma_j^*, j = 1, \ldots, B$, from the joint posterior distribution of the normal distribution parameters μ and σ have been generated, using an MCMC method.

Example 16.14 Prior Distributions and Draws from the Joint Posterior Distribution of the Normal Distribution Parameters for the Circuit Pack Output Voltage Application. The examples in this section are based on the circuit pack output voltage application introduced in Example 3.1. This application was used to illustrate the computation of statistical intervals based on an assumption of a normal distribution, as described in Chapters 3 and 4.

To obtain a diffuse prior distribution for the normal distribution mean μ, we follow common practice and use a flat (i.e., uniform) prior distribution over the entire real line (i.e., from $-\infty$ to ∞). Although this is an improper prior distribution, the joint posterior distribution for (μ, σ) will be proper (in general if there are at least two observations). We also follow common practice in defining the diffuse prior distribution for the standard deviation σ. In particular, the uncertainty in the reciprocal of the variance $1/\sigma^2$ (known as "precision") is described by a gamma distribution with a shape parameter $a = 0.001$ and a rate parameter $b = 0.001$. This distribution has a mean of 1 and a variance of 1,000 and is thus diffuse. This *implicit* diffuse prior for σ is popular because it is easy to specify in the commonly used Bayesian software packages and because it is approximately proportionate to the Jeffreys prior distribution for σ.

For the informative prior distribution, we continue to use a flat prior distribution for μ. To describe our prior knowledge about the standard deviation σ, however, we assume that there is strong belief that σ is between 1.0 and 1.6. and, therefore, use a lognormal prior distribution with probability 0.99 that σ is between 1.0 and 1.6. More precisely, the lognormal distribution is chosen such that the probability that σ is less than 1.0 is 0.005 and that it is greater than 1.6 is 0.005. This prior distribution is denoted by <LNORM>$(1.0, 1.6)$.

All of the Bayesian intervals computed in this section are based on matrices of sample draws from the joint posterior distribution of μ and σ, stored in R objects `drawsOutputVoltageDiffuse` and `drawsOutputVoltageInformative`, based on the assumed diffuse prior and informative prior distributions, respectively. Both of these objects contain $B = 60,000$ draws (i.e., 60,000 rows). The first two columns of Table 16.3 show the values of μ and σ for the first ten sample draws using the informative joint prior distributions taken from `drawsOutputVoltageInformative`. Table 16.3 also shows the first ten sample draws, that were computed subsequently, for $\Pr(Y > 48)$ and the 0.10 quantile of the distribution of voltage outputs, based on the informative joint prior distribution and to be discussed further subsequently.

μ^*	σ^*	$\Pr(Y > 48)^*$	$x_{0.10}^*$
49.36	1.1230	0.8871	47.92
49.61	0.9872	0.9485	48.34
49.54	1.1260	0.9143	48.10
50.18	0.8066	0.9966	49.15
49.37	1.1650	0.8802	47.88
49.91	1.0320	0.9679	48.59
50.01	1.0090	0.9768	48.72
49.86	1.1560	0.9462	48.38
50.01	1.1210	0.9635	48.57
50.04	1.1350	0.9639	48.59

Table 16.3   First ten sample draws from the joint posterior distribution of μ and σ and the marginal posterior distributions of $\Pr(Y > 48)$ and $x_{0.10}$ for the circuit pack voltage output data with the informative joint prior distribution for μ and σ.

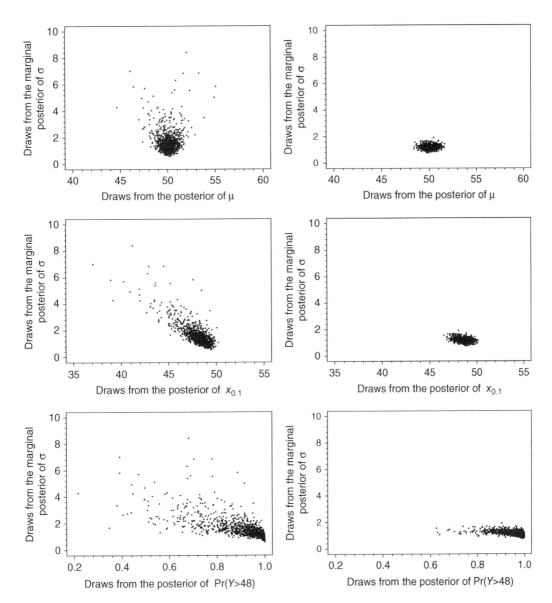

Figure 16.6 Scatter plots of draws from the joint posterior distribution of μ (top), $x_{0.10}$ (middle), and $\Pr(Y > 48)$ (bottom) versus σ for the circuit pack voltage output data with the diffuse (left) and informative (right) joint prior distributions.

Figure 16.6 shows scatter plots of the draws from the joint posterior distributions of μ and σ (top), of $x_{0.10}$ and σ (middle) and of $\Pr(Y > 48)$ and σ (bottom), using the Jeffreys (left) and informative (right) prior distributions for σ. These plots show the appreciable impact that the choice of the prior distribution for σ has on the inferences of interest. Table 16.4 summarizes the numerical results from this section, comparing the Bayesian methods results based on the diffuse and informative assumed prior distributions. The diffuse prior results are very close to (if not exactly the same as) the results that were obtained using the classical non-Bayesian methods in Chapter 4. This is as expected, given the theoretical results outlined in Sections H.4.3 and H.5.4. ■

Type of Prior Distribution	Interval/bound type	Interval for	Lower	Upper
Diffuse	Two-sided credible	μ	48.5	51.7
Informative	Two-sided credible	μ	49.0	51.1
Diffuse	Two-sided credible	σ	0.785	3.76
Informative	Two-sided credible	σ	0.890	1.51
Diffuse	Two-sided credible	$x_{0.10}$	44.6	49.6
Informative	Two-sided credible	$x_{0.10}$	47.4	49.6
Diffuse	One-sided credible	$x_{0.90}$	50.8	54.6
Informative	One-sided credible	$x_{0.90}$	50.7	52.6
Diffuse	Two-sided credible	$\Pr(X > 48)$	0.571	0.998
Informative	Two-sided credible	$\Pr(X > 48)$	0.797	0.998
Diffuse	Two-sided tolerance	Control center 0.90	44.4	55.7
Informative	Two-sided tolerance	Control center 0.90	47.4	52.8
Diffuse	Two-sided tolerance	Control tails 0.05	43.7	56.5
Informative	Two-sided tolerance	Control tails 0.05	46.9	53.2
Diffuse	Two-sided prediction	Y	46.1	54.1
Informative	Two-sided prediction	Y	47.5	52.6
Diffuse	Two-sided prediction	$\bar{Y}, m = 3$	47.4	52.7
Informative	Two-sided prediction	$\bar{Y}, m = 3$	48.4	51.8
Diffuse	Two-sided prediction	all 10	43.2	57.0
Informative	Two-sided prediction	all 10	46.4	53.9
Diffuse	One-sided prediction	all 10	44.3	55.9
Informative	One-sided prediction	all 10	46.7	53.5
Diffuse	Two-sided prediction	at least 9 of 10	44.9	55.4
Informative	Two-sided prediction	at least 9 of 10	47.3	53.0
Diffuse	Two-sided prediction	$S, m = 3$	0.211	4.32
Informative	Two-sided prediction	$S, m = 3$	0.181	2.32
Diffuse	One-sided prediction	$S, m = 3$	0.302	3.46
Informative	One-sided prediction	$S, m = 3$	0.258	2.08

Table 16.4 Bayesian two-sided 95% statistical intervals and some one-sided 95% statistical bounds for normal distribution quantities of interest for the circuit pack output voltage example.

16.3.3 Credible Intervals for the Normal Distribution Parameters

To find a Bayesian credible interval (or one-sided credible bound) for either of the normal distribution parameters, one can use the simple approach described in Section 15.2.6. That is, the $100(1 - \alpha)\%$ credible interval for the parameter of interest can be obtained from the $\alpha/2$ and $1 - \alpha/2$ quantiles of the sample draws from the posterior distribution for that parameter.

Example 16.15 Credible Interval for the Mean of the Distribution of Output Voltages.
A 95% credible interval for the mean of the distribution of output voltages in Example 3.1 is computed from the 0.025 and 0.975 quantiles of the empirical marginal posterior distribution

of μ. These quantiles are obtained for the diffuse prior and informative prior distributions by using the following R commands:

```
quantile(drawsOutputVoltageDiffuse[,"mu"], probs=c(0.025,0.975))
  2.5%    97.5%
48.45    51.72
> quantile(drawsOutputVoltageInformative[,"mu"], probs=c(0.025,0.975))
  2.5% 97.5%
49.05 51.14
```

As expected, the credible interval for μ, based on the diffuse prior distribution, is in good agreement with the non-Bayesian confidence interval found in Example 4.1. This type of agreement is also true for all of the other examples in this section when compared with the similar example in Chapter 3, so we will not repeat the statement for the other examples, and instead focus on the difference between the results from the diffuse and informative prior distributions.

The credible interval for μ, based on the informative prior distribution, is slightly narrower due to the assumed prior information about σ. This difference is as expected and can be explained and understood by comparing the two plots in the top row of Figure 16.6.   ∎

Example 16.16   Credible Interval for the Standard Deviation of the Distribution of Output Voltages. A 95% credible interval for the standard deviation of the distribution of output voltages is computed from the 0.025 and 0.975 quantiles of the empirical marginal posterior distribution of σ. These quantiles are obtained for the diffuse prior and informative prior distributions by using the following R commands:

```
> quantile(drawsOutputVoltageDiffuse[,"sigma"], probs=c(0.025, 0.975))
  2.5%    97.5%
0.7846 3.7630
> quantile(drawsOutputVoltageInformative[,"sigma"], probs=c(0.025,
  0.975))
  2.5% 97.5%
0.890 1.515
```

The credible interval for σ based on the informative prior distribution is considerably narrower than that for the diffuse prior distribution, as expected, due to the informative prior information that was used for σ. The reason for this difference can be explained and understood by comparing the pairs of plots in any of the rows of Figure 16.6.   ∎

16.3.4   Other Credible Intervals for the Normal Distribution

To find a Bayesian credible interval (or one-sided credible bound) for a function of normal distribution parameters, one can use the simple approach described in Section 15.2.6. That is, one uses direct evaluation of the function for each of the B sample draws from the joint posterior distribution of μ and σ to generate sample draws from the marginal posterior distribution of the function of interest. Then the $100(1 - \alpha)\%$ credible interval for the function can be obtained from the $\alpha/2$ and $1 - \alpha/2$ quantiles of the empirical distribution of these draws.

Example 16.17   Credible Interval for $x_{0.10}$, the 0.10 Quantile of the Distribution of Output Voltages. To compute a Bayesian credible interval (or one-sided credible bound) for $x_{0.10} = \mu - 1.645\sigma$, the 0.10 quantile of the distribution of output voltages, one computes draws from the marginal posterior distribution of $x_{0.10}$. This is done by first computing

$x_{0.10,j}^{*} = \mu_j^* - 1.645\sigma_j^*$, $j = 1, \ldots, B$. Then the 95% credible interval for $x_{0.10}$ is obtained from the 0.025 and the 0.975 quantiles of the empirical distribution of these draws. These are computed for the diffuse prior and informative prior distributions by using the following R commands:

```
quantile(drawsOutputVoltageDiffuse[,"mu"]+
        qnorm(0.10)*drawsOutputVoltageDiffuse[,"sigma"],
        probs=c(0.025, 0.975))
 2.5%    97.5%
44.62  49.59
> quantile(drawsOutputVoltageInformative[,"mu"]+
        qnorm(0.10)*drawsOutputVoltageInformative[,"sigma"],
        probs=c(0.025, 0.975))
 2.5% 97.5%
47.41 49.64
```

The credible interval for $x_{0.10}$ based on the informative prior distribution is considerably narrower than the diffuse prior distribution counterpart, as expected. More specifically, the lower endpoint of the informative prior credible interval is considerably larger than the corresponding endpoint of the diffuse prior interval (the upper endpoints are the same to three significant digits). The reason for this difference can be understood by comparing the two plots in the middle row of Figure 16.6. In particular, because the value of σ is constrained from above by the informative prior information about σ, the marginal posterior distribution of $x_{0.10}$ is constrained from above due to the negative correlation between σ and $x_{0.10}$ in the joint posterior seen in Figure 16.6.■

Example 16.18   One-Sided Upper Credible Bound for the 0.90 Quantile of the Distribution of Output Voltages. Similar to previous examples, a one-sided upper credible bound for $x_{0.90}$, the 0.90 quantile of the distribution of output voltages, can be obtained by computing sample draws from the marginal posterior distribution of $x_{0.90}$. This is done by first computing $x_{0.90,j}^* = \mu_j^* + 1.645\sigma_j^*$, $j = 1, \ldots, B$. Then the desired 95% upper credible bound for $x_{0.90}$ is the 0.95 quantile of the empirical distribution of these draws. This is calculated for the diffuse prior and informative prior distributions by using the following R commands:

```
quantile(drawsOutputVoltageDiffuse[,"mu"]+
        qnorm(0.90)*drawsOutputVoltageDiffuse[,"sigma"], probs=0.95)
  95%
54.55
> quantile(drawsOutputVoltageInformative[,"mu"]+
        qnorm(0.90)*drawsOutputVoltageInformative[,"sigma"], probs=0.95)
  95%
52.57
```

The upper credible bound for $x_{0.90}$ obtained by using the informative prior distribution is considerably smaller (i.e., closer to the center of the distribution) than the upper credible bound obtained by using the diffuse prior distribution. This is caused by an effect similar but opposite to the movement of the lower endpoint of the credible interval for $x_{0.10}$ in Example 16.17. The effect is due to a strong positive dependency between $x_{0.90}$ and σ and the restrictive informative prior distribution that was used for σ. That is, because σ is constrained to be small in the informative prior distribution, $x_{0.90}$ will be constrained to be small.   ■

Example 16.19   Credible Interval for $\Pr(X > 48)$ from the Distribution of Output Voltages. Similar to previous examples, a credible interval for $\Pr(X > 48)$ can be obtained from the distribution of output voltages obtained by computing draws from the marginal posterior distribution of $\Pr(X > 48)$. This can be done by computing $\Pr(X > 48)_j^* = \Phi_{\mathrm{norm}}[(\mu_j^* - 48)/\sigma_j^*]$, $j = 1, \ldots, B$. Then the 95% credible interval is given by the 0.025 and 0.975 quantiles of the empirical distribution of these draws. These are computed for the diffuse and informative prior distributions by the following R commands:

```
> quantile(pnorm(q=48, mean=drawsOutputVoltageDiffuse[,"mu"],
        sd=drawsOutputVoltageDiffuse[,"sigma"], lower.tail=FALSE),
        probs=c(0.025, 0.975))
  2.5%    97.5%
0.5709 0.9984
> quantile(pnorm(q=48, mean=drawsOutputVoltageInformative[,"mu"],
        sd=drawsOutputVoltageInformative[,"sigma"], lower.tail=FALSE),
        probs=c(0.025, 0.975))
  2.5%    97.5%
0.7975 0.9977
```

The lower endpoint of the credible interval for $\Pr(X > 48)$ based on the informative prior distribution is considerably larger than the lower endpoint based on the diffuse prior distribution. This is due to the dependency of $\Pr(X > 48)$ on σ and the informative prior information that was used for σ. The reason for this difference can be understood by comparing the two plots in the bottom row of Figure 16.6. ∎

16.3.5 Tolerance Intervals for the Normal Distribution

A Bayesian tolerance interval for a normal distribution can be obtained by using the general approach described in Sections 15.5.2 and 15.5.3. The basic input for both of these procedures is the sequence of draws from the joint posterior distribution of μ and σ used for the other statistical intervals in this section.

Example 16.20 Bayesian Control-the-Center Tolerance Interval to Contain at Least a Proportion 0.90 of the Distribution of Output Voltages. The results of implementing the procedure in Section 15.5.2 to construct control-the-center tolerance intervals for output voltage using both the diffuse prior and the informative prior distributions are shown in Table 16.4. The tolerance interval using an informative prior distribution is considerably narrower than the diffuse prior interval due to the informative prior information that was used for σ. ∎

Example 16.21 Bayesian Control-Both-Tails Tolerance Interval to Contain at Least a Proportion 0.90 of the Distribution of Output Voltages. The results of implementing the procedure in Section 15.5.3 to construct control-both-tails tolerance intervals for output voltage using both the diffuse prior and the informative prior distributions are shown in Table 16.4. The tolerance interval using the informative prior distribution is considerably narrower than the diffuse prior interval due to the informative prior information that was used for σ. ∎

16.3.6 Prediction Intervals for the Normal Distribution

Bayesian prediction intervals for a normal distribution can be obtained by using the general approaches described in Section 15.5.4 or 15.5.5. The basic input for both of these procedures is the sequence of MCMC draws from the joint posterior distribution of μ and σ used for the other statistical intervals in this section.

For situations in which it is relatively easy to simulate the random variable to be predicted, we will use the additional-layer-of-MC method described in Section 15.5.5 because it is easier to implement using simple R commands (even though this method generally requires a larger number of draws from the posterior predictive distribution to achieve the same amount of MC precision). For the simultaneous prediction interval/bounds applications, we will use the direct method described in Section 15.5.4.

Example 16.22 Bayesian Prediction Interval to Contain a Single Future Observation from the Distribution of Output Voltages. Draws from the posterior predictive distribution of Y can be obtained by generating one normal random variable for each of the B pairs of MCMC sample draws (μ^*, σ^*) from the joint posterior distribution of μ and σ. Then the Bayesian 95% prediction interval for Y is obtained from the 0.025 and 0.975 quantiles of the empirical distribution of these draws. For the diffuse prior and informative prior distributions, this is done using the following R commands:

```
> quantile(rnorm(n=nrow(drawsOutputVoltageDiffuse),
        mean=drawsOutputVoltageDiffuse[,"mu"],
        sd=drawsOutputVoltageDiffuse[,"sigma"]), probs=c(0.025, 0.975))
 2.5% 97.5%
46.07 54.11
> quantile(rnorm(n=nrow(drawsOutputVoltageInformative),
        mean=drawsOutputVoltageInformative[,"mu"],
        sd=drawsOutputVoltageInformative[,"sigma"]), probs=c(0.025,
          0.975))
 2.5% 97.5%
47.53 52.65
```

As expected, the Bayesian prediction interval using the informative prior distribution is narrower than the prediction interval using the diffuse prior distribution. ∎

Example 16.23 Bayesian Prediction Interval to Contain the Mean of $m = 3$ Future Observations from the Distribution of Output Voltages. Following the approach described in Section 15.5.5, sample draws from the posterior predictive distribution of the future sample mean $\bar{Y}$ can be obtained by generating a corresponding sample mean of $m = 3$ observations for each of the B pairs of MCMC sample draws (μ^*, σ^*) from the joint posterior distribution of μ and σ. Then the Bayesian 95% prediction interval for $\bar{Y}$ is obtained from the 0.025 and 0.975 quantiles of the empirical distribution of these draws. For the diffuse prior and informative prior distributions, this is done using the following R commands:

```
> quantile(apply(matrix(rnorm(n=3*nrow(drawsOutputVoltageDiffuse),
        mean=drawsOutputVoltageDiffuse[,"mu"],
        sd=drawsOutputVoltageDiffuse[,"sigma"]), ncol=3),
        MARGIN=1, FUN=mean), probs=c(0.025, 0.975))
 2.5% 97.5%
47.40 52.74
> quantile(apply(matrix(rnorm(n=3*nrow(drawsOutputVoltageInformative),
        mean=drawsOutputVoltageInformative[,"mu"],
        sd=drawsOutputVoltageInformative[,"sigma"]), ncol=3),
        MARGIN=1, FUN=mean), probs=c(0.025, 0.975))
 2.5% 97.5%
48.39 51.79
```

As expected, the Bayesian prediction interval based on the informative prior distribution is somewhat narrower than that based on the diffuse prior distribution. Moreover, as expected, the prediction interval for the mean of $m = 3$ future observations is narrower than the interval to contain a single future observation obtained in Example 16.22.   ∎

Example 16.24   Bayesian Prediction Interval to Contain All $m = 10$ Future Observations from the Distribution of Output Voltages. A 95% Bayesian prediction interval to contain all of $m = 10$ future observations from the distribution of output voltages is based on the B pairs of MCMC sample draws (μ^*, σ^*) from the joint posterior distribution of μ and σ and is computed as described in Section 15.5.4. The corresponding Bayesian upper (lower) prediction bound to exceed (be exceeded by) all $m = 10$ future observations from the distribution of output voltages is computed in a similar manner. The results are given in Table 16.4. As expected, the prediction interval based on the informative prior distribution is appreciably narrower than that based on the diffuse prior distribution. Also, as expected, the Bayesian prediction interval to contain all 10 future observations is considerably wider than the prediction interval to contain only one future observation in Example 16.22.   ∎

Example 16.25   Bayesian Prediction Interval to Contain at Least $k = 9$ of $m = 10$ Future Observations from the Distribution of Output Voltages. A 95% Bayesian prediction interval to contain at least $k = 9$ of $m = 10$ future observations from the distribution of output voltages is based on the B pairs of MCMC sample draws (μ^*, σ^*) from the joint posterior distribution of μ and σ and is computed as described in Section 15.5.4. The results are given in Table 16.4. As expected, the Bayesian prediction interval based on the informative prior distribution is appreciably narrower than that based on the diffuse prior distribution. Also, as expected, the Bayesian prediction interval to contain at least $k = 9$ of $m = 10$ future observations is a little narrower than the corresponding prediction interval to contain all $m = 10$ future observations in Example 16.24.   ∎

Example 16.26   Bayesian Prediction Interval to Contain the Sample Standard Deviation of $m = 3$ Future Observations from the Distribution of Output Voltages. Following the approach described in Section 15.5.5, a prediction interval for the future sample standard deviation S can be obtained from MCMC sample draws from the posterior predictive distribution of S. For a future sample of size m, S has the same distribution as

$$\sigma\sqrt{X^2_{(m-1)}/(m-1)},$$

where $X^2_{(m-1)}$ is a chi-square random variable with $m - 1$ degrees of freedom. Then sample draws from the posterior predictive distribution of S can be computed from $S^*_j = \sigma^*_j\sqrt{X^2_{(m-1)}/(m-1)}$, $j = 1, \ldots, B$, where independent values of $X^2_{(m-1)}/(m-1)$ need to be simulated for each value of j. Then the Bayesian 95% prediction interval for S is obtained from the 0.025 and 0.975 quantiles of the empirical distribution of these draws. For the diffuse prior and informative prior distributions, and a future sample of size $m = 3$, this is done using the following R commands:

```
> quantile(sqrt(rchisq(n=nrow(drawsOutputVoltageDiffuse),df=2)/2)*
      drawsOutputVoltageDiffuse[,"sigma"], probs=c(0.025,0.975))
  2.5%  97.5%
0.211 4.319
> quantile(sqrt(rchisq(n=nrow(drawsOutputVoltageInformative),df=2)/2)*
      drawsOutputVoltageInformative[,"sigma"], probs=c(0.025,0.975))
  2.5%  97.5%
0.181 2.322
```

When compared to the prediction interval for S using the diffuse prior distribution, the corresponding prediction interval using the informative prior distribution is narrower and shifted downwards. This is due to the dependency of the distribution of S on σ and the restrictive informative prior distribution that was used for σ.   ■

BIBLIOGRAPHIC NOTES

The technical results related to the conjugate prior distributions used in this chapter are given in Gelman et al. (2013) and summarized in Section H.3 of this book. Methods for computing MCMC draws from the posterior distribution for the distributions in this chapter are given in Lunn et al. (2012). Liu et al. (2016) describe a method to choose the number of MCMC draws needed to estimate quantiles of a marginal posterior distribution to a desired degree of accuracy. Their method was designed to make it possible to report a Bayesian credible interval with a known number of correct significant digits.

Chapter *17*

Statistical Intervals for Bayesian Hierarchical Models

OBJECTIVES AND OVERVIEW

This chapter extends the introductory discussion of Bayesian statistical models presented in Chapters 15 and 16, and shows how to compute statistical intervals for more complicated statistical models and/or data structures. It provides an introduction to the analysis of hierarchical (or multilevel) statistical models using Bayesian analysis—to be referred to as "Bayesian hierarchical models" for short. We describe basic concepts underlying such analyses and illustrate their use in several different applications. The following topics are discussed:

- The basic ideas for modeling data from multilevel (hierarchical) studies (Section 17.1).
- Hierarchical models for data that can be described by a normal distribution (Section 17.2).
- Hierarchical models for data that can be described by a binomial distribution (Section 17.3).
- Hierarchical models for data that can be described by a Poisson distribution (Section 17.4).
- The analysis of data when repeated measurement are taken on a sample of units over time, which can also be viewed as a hierarchical model (Section 17.5).

As indicated in Section 15.1.1 and illustrated by various examples in Chapters 15 and 16, one important motivation for using Bayesian methods is to incorporate prior information into a data analysis/modeling problem, especially to supplement limited data. In this chapter, however, the primary motivation is that Bayesian methods are particularly convenient for modeling data when there are multiple sources of variability, as in hierarchical models (see the references given in the Bibliographic Notes section at the end of this chapter). Indeed, all of the examples in this chapter use only diffuse prior distributions—although informative prior distributions could have been used if informative prior information had been available. In particular, the diffuse prior distributions for random effects used here and in Chapter 18 are those recommended by Gelman (2006b).

Statistical Intervals: A Guide for Practitioners and Researchers, Second Edition.
William Q. Meeker, Gerald J. Hahn and Luis A. Escobar.
© 2017 John Wiley & Sons, Inc. Published 2017 by John Wiley & Sons, Inc.
Companion Website: www.wiley.com/go/meeker/intervals

17.1   BAYESIAN HIERARCHICAL MODELS AND RANDOM EFFECTS

Figure 17.1 shows a traditional three-level hierarchical structure; the corresponding statistical model and analysis will be discussed in the next section. Hierarchical structures with varying shapes and size are encountered in practice, but are handled similarly. Also, we adopt a modern Bayesian-oriented interpretation of the concept of a hierarchical model in the examples in this chapter. Further examples are provided in the references given in the Bibliographic Notes section at the end of this chapter.

Figure 17.1 illustrates the structure of a hierarchical study of baseball players' batting averages. At the bottom of the hierarchy are batting averages for individual players "nested" within teams—which are at the second level. Each team belongs to a division, shown at the top level. The three levels (or tiers) in this example are, therefore, division, team and player. To provide a simple example, we consider data from the two US baseball divisions: American League Central (AL-C) and National League Central (NL-C).

Suppose that the primary goal of the study depicted in Figure 17.1 is to use the available data to estimate the mean batting average for each team, together with the associated statistical uncertainty of the estimate—to be quantified by a credible interval.

The simplest approach for estimating a team's mean batting average is to simply compute the mean of the individual batting averages of the team's players for each team. An alternative and more complex—but also potentially more informative—approach uses a hierarchical model and Bayesian methods, taking into consideration the common features of the data across the different teams in estimating the mean for each team. This approach implicitly recognizes some similarity among the teams and does what is sometimes referred to as "partial pooling" to capitalize on these similarities to provide better estimation precision.

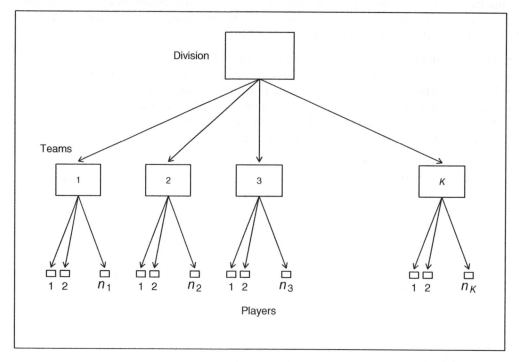

Figure 17.1   Illustration of a hierarchical study involving K baseball teams with n_k players' batting averages nested within team k, where $k = 1, 2, \ldots, K$.

We will illustrate and compare the preceding two approaches for constructing credible intervals for the baseball and other examples in the subsequent sections.

A third, and more extreme, approach would be to pool the batting averages of all the players, irrespective of team and, possibly, division to obtain a single mean batting average across all teams. This approach assumes that there is a single common distribution that describes all of the teams, without any team differentiation. We will not employ this approach for constructing a statistical interval in the subsequent sections, but will calculate the overall or pooled mean of the data.

17.2 NORMAL DISTRIBUTION HIERARCHICAL MODELS

This section illustrates the use of Bayesian methods for obtaining statistical intervals for characteristics of a hierarchical model when the resulting data are assumed to follow a normal distribution.

Example 17.1 Baseball batting averages for teams in the American League Central and the National League Central divisions. This example deals mainly with obtaining credible intervals on individual team batting averages for each of the ten AL-C and NL-C teams. A summary of the data is shown in Table 17.1.

| Team | Team sample mean $\bar{x}_k$ | Team standard error $s_{\bar{x}_k}$ | Number of batters | Division | 90% credible intervals for team mean θ_k | | | |
| | | | | | Individual | | Hierarchical | |
					Lower	Upper	Lower	Upper
Chicago White Sox (CHW)	0.281	0.0325	5	AL-C	0.20	0.32	0.24	0.30
Cleveland Indians (CLE)	0.271	0.0382	5	AL-C	0.21	0.33	0.25	0.30
Detroit Tigers (DET)	0.294	0.0304	5	AL-C	0.24	0.34	0.25	0.30
Kansas City Royals (KC)	0.270	0.0122	7	AL-C	0.22	0.31	0.25	0.30
Minnesota Twins (MIN)	0.266	0.0204	4	AL-C	0.22	0.33	0.25	0.30
Chicago Cubs (CHC)	0.268	0.0390	6	NL-C	0.21	0.34	0.25	0.29
Cincinnati Reds (CIN)	0.250	0.0445	7	NL-C	0.19	0.30	0.24	0.29
Milwaukee Braves (MIL)	0.262	0.0256	4	NL-C	0.23	0.31	0.25	0.29
Pittsburgh Pirates (PIT)	0.251	0.0229	6	NL-C	0.23	0.35	0.25	0.30
St. Louis Cardinals (STL)	0.289	0.0369	7	NL-C	0.24	0.31	0.25	0.29

Table 17.1 Batting averages in 2014 for teams in the American League Central (AL-C) and the National League Central (NL-C) divisions. The pooled point estimate of the players' mean batting average for both the AL-C and NL-C is 0.272.

The second and third columns in Table 17.1 show the sample mean $\bar{x}_k$ and the corresponding standard error $s_{\bar{x}_k}$ of the 2014 team batting averages for each team's batters who had at least 300 at-bats. The standard error for each team was computed using the well-known expression taking the sample standard deviation for the team divided by the square root of the number of eligible batters on the team (see Section 4.2).

As indicated in Section 17.1, one approach to characterizing the batting strength of the teams is to use the data from each team's players to estimate separately the team's mean batting average using the assumption

$$X_{ik} \sim \mathrm{NORM}(\theta_k, \sigma_k)$$

where X_{ik}, $i = 1, \ldots n_k$, is the batting average for team member i from team k, n_k is the number of batters in team k, while θ_k and σ_k are the unknown mean and standard deviations for the batting average for team k for $k = 1, 2, \ldots, 10$. (The normal distribution assumption is justified by the central limit theorem because each X_{ik} is the average of at least 300 at-bats for each player). Separate individual 90% credible intervals for the mean batting average θ_k for each team were computed using the methods in Section 16.3.3, with diffuse joint prior distributions for each θ_k and σ_k pair. The results are given in the right-hand columns in Table 17.1. This approach requires estimating a total of 20 parameters (θ_k and σ_k for $k = 1, \ldots, 10$).

Alternatively, to recognize the similarities among teams, we use the hierarchical model

$$\bar{x}_k \sim \mathrm{NORM}(\theta_k + \beta_1 \mathrm{AL}_k, s_{\bar{x}_k}), \quad \theta_k \sim \mathrm{NORM}(\mu_\theta, \sigma_\theta),$$

$k = 1, \ldots, 10$. Here θ_k is taken to be a "random effect" term that describes team-to-team variability within the two divisions, while μ_θ and σ_θ are known as hyperparameters, describing the distribution of the θ_k values. To take into consideration the difference between divisions, we define $\mathrm{AL}_k = 0.50$ for the American League Central and $\mathrm{AL}_k = -0.50$ for the National League Central, and use the parameter β_1 to quantify the estimated difference in batting averages between the two divisions. Thus, this model assumes that the batting averages for the two divisions may have different means, but are subject to the same team-to-team variability.

To conduct a Bayesian analysis using the preceding model, we then use the following independent diffuse prior distributions for the unknown model parameters:

$$\mu_\theta \sim \mathrm{UNIF}(-\infty, \infty), \quad \sigma_\theta \sim \mathrm{UNIF}(0, 20), \quad \beta_1 \sim \mathrm{UNIF}(-\infty, \infty).$$

As described in Section 15.1.4, use of this model and diffuse prior distributions generally provides results close to those that one would get in a non-Bayesian analysis.

An MCMC algorithm was used to generate 40,000 draws from the joint posterior distribution of μ_θ, σ_θ, β_1, and θ_k, $k = 1, \ldots, 10$, for this hierarchical model. The draws for the θ_k values were used to compute the 90% credible intervals for each team average, using the same approach as in the examples in Chapter 15. The results are shown in the last two columns of Table 17.1. The draws for β_1 were used to compute a 90% credible interval for β_1, giving $[-0.029, \quad 0.031]$, indicating that the observed difference between the two divisions is not statistically significant at the 10% level of significance (because the interval includes zero).

Figure 17.2 provides a graphical summary of the data analyses. The solid vertical lines show point estimates and 90% credible intervals for each team average, based on the hierarchical model; the dashed vertical lines are individual credible intervals based on the method from Section 4.2. The long horizontal dotted lines indicate the sample mean batting averages based on pooling all of the data within each of the two divisions. These two sample means, based on 26 batters in the AL-C and 30 batters in the NL-C, when rounded to three significant digits, are both equal to 0.272.

These plots show two (not unexpected) differences between the results from the data analyses using the individual Bayesian credible intervals and the Bayesian credible intervals based on

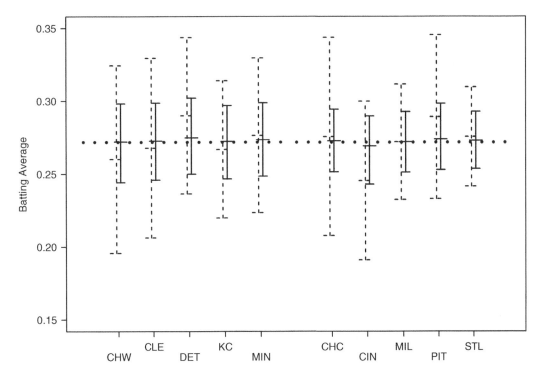

Figure 17.2 Individual 90% credible intervals (dashed vertical lines) and hierarchical model 90% credible intervals (solid vertical lines) for the mean batting averages for each of 10 teams. The horizontal dotted lines are the means for each division.

the hierarchical model. First, the credible intervals are considerably narrower than the intervals based on the individual analyses. Second, the hierarchical model point estimates (along with the intervals) are shifted toward the overall mean for the division. These shifts are known as "shrinkage toward the mean" and the effect is stronger for those teams that have wider intervals. This is because the teams with less information "borrow" more strength from the other teams. As described in references in the Bibliographic Notes section of this chapter (e.g., Gelman, 2006a), the shrunken estimators have better statistical properties than the individual-analysis estimators (assuming appropriateness of the assumed model) because the hierarchical model assumption allows the information in the data to be used more efficiently.   ■

17.3 BINOMIAL DISTRIBUTION HIERARCHICAL MODELS

This section illustrates the use of Bayesian methods for obtaining statistical intervals for characteristics of a hierarchical model when the resulting data are assumed to follow a binomial distribution.

Example 17.2 Assessing the Performance of Sales Agents. A study was conducted to assess the probability of making a successful sale for a particular product, based on "warm leads." A warm lead is a potential customer who is known to have an interest in purchasing the particular product. The company's sales force is divided into two different geographical regions, each having nine sales agents. The warm leads were randomly assigned to one of the sales agents in

Sales Agent	Region	Number of attempts n_k	Number of successes x_k	Sample proportion x_k/n_k	90% credible intervals Individual Lower	Upper	Hierarchical Lower	Upper
Agent A	Region 1	12	6	0.50	0.28	0.72	0.60	0.75
Agent B	Region 1	30	21	0.70	0.55	0.82	0.62	0.76
Agent C	Region 1	20	18	0.90	0.75	0.97	0.64	0.79
Agent D	Region 1	26	15	0.58	0.42	0.72	0.60	0.75
Agent E	Region 1	24	17	0.71	0.54	0.84	0.62	0.76
Agent F	Region 1	17	11	0.65	0.45	0.81	0.61	0.76
Agent G	Region 1	38	25	0.66	0.53	0.77	0.62	0.75
Agent H	Region 1	20	16	0.80	0.63	0.91	0.63	0.77
Agent I	Region 1	17	12	0.71	0.51	0.86	0.62	0.76
Agent J	Region 2	24	16	0.67	0.50	0.81	0.59	0.73
Agent K	Region 2	27	18	0.67	0.51	0.80	0.59	0.73
Agent L	Region 2	19	11	0.58	0.39	0.75	0.57	0.73
Agent M	Region 2	14	9	0.64	0.42	0.82	0.58	0.73
Agent N	Region 2	10	8	0.80	0.55	0.94	0.59	0.75
Agent O	Region 2	17	13	0.76	0.57	0.90	0.59	0.75
Agent P	Region 2	33	21	0.64	0.49	0.76	0.58	0.73
Agent Q	Region 2	28	20	0.71	0.56	0.84	0.59	0.74
Agent R	Region 2	14	7	0.50	0.29	0.71	0.56	0.72

Table 17.2   Sales successes based on warm leads from two different regions. The hierarchical model credible intervals are based on a logistic regression model. For region 1, the pooled point estimate of π is 0.69, and for region 2, the pooled point estimate of π is 0.66.

the appropriate region at the time of arrival. If the selected agent was not available (e.g., because of vacation or illness), the lead would be assigned to an agent who was available.

The third and fourth columns in Table 17.2 give the number of warm leads n_k in the assigned batch and the number of successful sales x_k for each of the 18 agents in the company. The fifth column gives the sample proportion x_k/n_k. We want to use these data to compute credible intervals to characterize the sales ability of each of the 18 different agents and for the two different sales regions.

One approach to carrying out the desired characterization of sales agents is to use the binomial distribution model

$$X_k \sim \text{BINOM}(n_k, \pi_k), \tag{17.1}$$

for $k = 1, \ldots, 18$, to *separately* estimate sales success probability π_k for each of the 18 agents. Using this approach, the individual 90% credible intervals in Table 17.2 were computed using the binomial distribution method described in Sections 6.2.5 and 16.1.2 based on a Bayesian approach with a Jeffreys diffuse prior distribution (see Section H.4.1). The results are shown in the sixth and seventh columns in Table 17.2.

A second extreme approach would be to pool the data within each region and compute an estimate of the probability of a successful sale within each region. We computed such estimates and present them as dotted horizontal lines in the plots given in Figure 17.3.

A compromise between the individual analyses and pooling is to use (17.1) in a hierarchical model where π_k is assumed to be a random quantity (called a "random effect") that is allowed to vary from sales agent to sales agent according to some probability distribution. Using the beta distribution

$$\pi_k \sim \text{BETA}(a, b), \quad k = 1, \ldots, 18, \tag{17.2}$$

is natural for this purpose because the beta distribution has outcomes that range between 0 and 1 with a wide range of shapes depending on the parameters a and b. The combination of (17.1) and (17.2) is known as the *beta-binomial* hierarchical model. The unknown model parameters a and b can be estimated from the available data. In particular, to implement Bayesian estimation for this model we specify diffuse prior distributions for the beta distribution parameters a and b. A common choice is to use gamma distributions with a very large variances, such as

$$a \sim \text{GAMMA}(0.001, 0.001), \quad b \sim \text{GAMMA}(0.001, 0.001).$$

We fit these models separately for both regions, based on the belief that the associated success probabilities could differ for the two regions, resulting in four parameters (two beta distribution parameters for each region). For each region, an MCMC algorithm was used to generate 10,000 draws from the joint posterior distribution of a, b, and π_k, $k = 1, \ldots, 9$, and these were used in the usual way (described and illustrated in the examples in Chapter 15) to compute point estimates (from the median of the draws) and 90% credible intervals (from the 0.05 and 0.95 quantiles of the empirical distribution of the draws) for each π_k. The results are shown in the last two columns of Table 17.2.

The top plot of Figure 17.3 compares the 90% credible intervals from the individual analyses (dashed vertical lines) with those from the hierarchical model in (17.2) (solid vertical lines) for both of the regions. The horizontal dotted lines indicate point estimates based on pooling all of the data within each region to estimate the overall proportion for that region. For the individual analyses, π_k is estimated separately for each sales agent/region combination. This is in contrast to the hierarchical model for which there is partial pooling within each region, linking the π_k through the beta distribution parameters a and b (separate values of a and b for each region). This partial pooling results in estimates that in all cases have better (and in some cases appreciably better) precision (i.e., narrower credible intervals). Also, the point estimates have shrunk toward the pooled estimate within each region.

An alternative hierarchical model again uses (17.1) but links the two different regions through a logistic regression model in which

$$\text{logit}(\pi_k) = \log\left(\frac{\pi_k}{1 - \pi_k}\right) = \beta_0 + \beta_1 \times \text{Region}_k + \alpha_k, \tag{17.3}$$

where $\text{Region}_k = 0.50$ for region 1 and $\text{Region}_k = -0.50$ for region 2. As a result, β_1 describes the region effect and the term

$$\alpha_k \sim \text{NORM}(0, \sigma)$$

describes the agent-to-agent variability. Again, we use diffuse independent prior distributions,

$$\beta_0 \sim \text{UNIF}(-\infty, \infty), \quad \beta_1 \sim \text{UNIF}(-\infty, \infty), \quad \sigma \sim \text{UNIF}(0, 20),$$

to describe the uncertainty in the unknown model parameters. An MCMC algorithm was used to generate 10,000 draws from the joint posterior distribution of β_0, β_1, σ, and α_k, $k = 1, \ldots, 18$. Then, using the inverse logit function, these draws were used to compute corresponding draws from the marginal posterior distributions of

$$\pi_k = \frac{\exp(\beta_0 + \beta_1 \times \text{Region}_k + \alpha_k)}{1 + \exp(\beta_0 + \beta_1 \times \text{Region}_k + \alpha_k)},$$

for $k = 1, \ldots, 18$. Point estimates and credible intervals for each of the π_k values are then obtained from the empirical quantiles of the respective set of draws.

The bottom plot in Figure 17.3 is similar to the top plot (the pooled estimate and the individual credible intervals are the same in the two plots) except that the bottom plot shows instead the hierarchical model credible intervals for the logistic regression model in (17.3). The credible

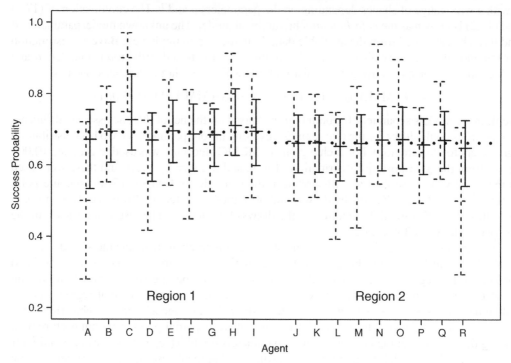

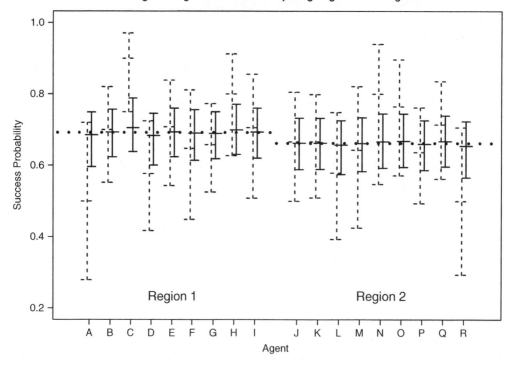

Figure 17.3   Individual 90% credible intervals (dashed vertical lines) and hierarchical model 90% credible intervals (solid vertical lines) for warm-lead sales success probability using different models. The horizontal dotted lines are the means for each region.

intervals from model (17.3) are somewhat narrower than those from model (17.2) because of the partial pooling across *all* 18 agents and because there are only three parameters in model (17.3) compared with four parameters in model (17.2). The draws for β_1 were used to compute a 90% credible interval for β_1, giving $[-0.24, \quad 0.52]$, indicating that the observed difference between region 1 and region 2 is not statistically significant (because the interval includes the value 0). With this result, one could justify combining the data from the two regions to fit a single model for both regions with just two parameters (the beta parameters in (17.2)) to describe the differences among the agents. ∎

17.4 POISSON DISTRIBUTION HIERARCHICAL MODELS

This section deals with using Bayesian methods for obtaining statistical intervals for characteristics of a hierarchical model when the resulting data are assumed to follow a Poisson distribution.

Example 17.3 Credit Card and ATM Fraud Rates in Rural Iowa. The second column in Table 17.3 gives x_k, the number of credit card and ATM fraud incidents reported in 2009 in a random sample of ten rural counties (defined as counties with fewer than 25,000 residents) in Iowa. The third column gives n_k, the county population size, stated in thousands. The fourth column gives x_k/n_k, the fraud-incident rate per 1,000 people in each of the sampled counties. We want to use these data to construct credible intervals for the fraud-incident rate, under the assumption of stationarity, for each of these ten counties. Such intervals quantify the uncertainty due to the random variability in the limited available sample data.

One approach for obtaining the desired statistical intervals is to use the Poisson distribution model

$$X_k \sim \text{POIS}(\lambda_k n_k)$$

and the data from each county to compute *individual* estimates of fraud-incident rates λ_k, $k = 1, \ldots, 10$. This analysis can be done using the methods in Chapters 7 or 16. In particular,

County	Reported fraud incidents x_k	Population ($\times 1,000$) n_k	Fraud-incident rate per 1,000 x_k/n_k	90% credible intervals for λ_k			
				Individual		Hierarchical	
				Lower	Upper	Lower	Upper
Adams	0	3.988	0.00	0.00049	0.48	0.012	0.33
Emmet	7	10.384	0.67	0.35	1.2	0.21	0.92
Hardin	1	17.193	0.06	0.010	0.23	0.022	0.23
Jones	1	20.331	0.05	0.0087	0.19	0.020	0.21
Lyon	2	8.740	0.23	0.068	0.65	0.060	0.44
O'Brien	2	13.927	0.14	0.041	0.40	0.046	0.32
Poweshiek	9	18.536	0.49	0.27	0.81	0.20	0.67
Taylor	0	6.218	0.00	0.00032	0.31	0.011	0.28
Washington	1	21.384	0.05	0.0082	0.18	0.019	0.20
Winneshiek	6	20.841	0.29	0.14	0.54	0.12	0.44

Table 17.3 Credit card and ATM fraud-incident counts reported in 2009 from a random sample of 10 rural counties in Iowa and credible intervals for the corresponding fraud-incident rates. The pooled point estimate of the fraud-incident rate λ is 0.21 per thousand population.

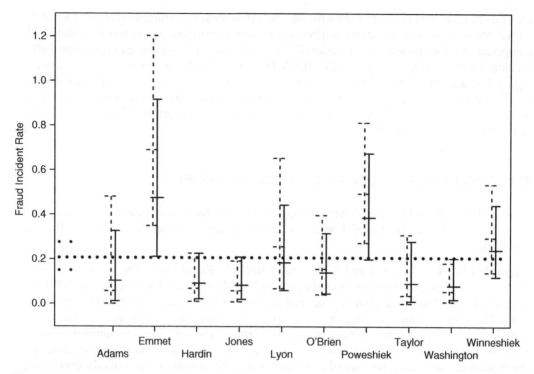

Figure 17.4   Individual 90% credible intervals (dashed vertical lines) and hierarchical model 90% credible intervals (solid vertical lines) for 2009 credit card and ATM fraud-incident rates in ten Iowa rural counties.

credible intervals were obtained, using a simple diffuse Jeffreys prior distribution, as described in Sections 7.2.5 and 16.2.2. The results are shown in the fifth and sixth columns of Table 17.3.

Another alternative extreme approach would be to pool the data from all 10 counties to estimate an overall fraud-incident rate for rural Iowa counties. The dotted horizontal line in Figure 17.4 indicates the pooled mean fraud-incident rate. Because such pooling can obscure important information in the data, we do not pursue this approach further and suggest, instead, the following analysis.

A third alternative is to use a compromise between the preceding two extremes that allows the fraud-incident rates in different rural Iowa counties to vary (as we know they do), using partial pooling through a Poisson distribution hierarchical model. In particular, let

$$X_k \sim \text{POIS}(\mu_k), \tag{17.4}$$

where

$$\mu_k = \lambda_k n_k = n_k \exp(\beta_0 + \theta_k), \tag{17.5}$$
$$\theta_k \sim \text{NORM}(0, \sigma_\theta),$$

for $k = 1, \ldots, 10$. In this model, μ_k is the 2009 fraud-incident rate for county k, $\lambda_k = \exp(\beta_0 + \theta_k)$ is the fraud-incident rate per $m = 1,000$ people in county k, and the θ_k values are random effects describing the county-to-county variability. We can interpret $\exp(\beta_0)$ to be a baseline (average) fraud-incident rate per $m = 1,000$ across all rural Iowa counties. Appropriate diffuse prior distributions for the model parameters are

$$\beta_0 \sim \text{UNIF}(-\infty, \infty), \quad \sigma_\theta \sim \text{UNIF}(0, 20).$$

An MCMC algorithm was used to generate 27,000 draws from the joint posterior distribution of β_0, σ_θ and θ_k, $k = 1, \ldots, 10$, and these were used to compute corresponding draws from the marginal posterior distributions of λ_k, the fraud-incident rates per 1000 people in county k for $k = 1, \ldots, 10$. Then these draws were used in the usual manner to compute credible intervals for each of the λ_k values shown in the last two columns of Table 17.3.

Figure 17.4 compares the 90% credible intervals from the individual analyses (dashed vertical lines) with those from the hierarchical model in (17.4) and (17.5) (solid vertical lines). The horizontal dotted line shows the average fraud incident rate for all ten sampled rural Iowa counties.

The widths of the credible intervals based on the individual analyses depend on both the number of fraud incidents and the population size. Because the expected value and the variance of a Poisson random variable are the same (see Section C.4.4), the counties with larger fraud-incident rates tend to have wider credible intervals. For counties with the same number of fraud incidents, those with larger populations have narrower credible bounds. For most of the counties, the hierarchical model point estimates and the associated credible intervals are shifted toward the estimated overall mean (as in other examples in this chapter), due to the partial pooling. Also, for more than half of the counties (especially those with high fraud-incident rates), the hierarchical model credible intervals are narrower than those obtained by the individual analyses; this is also due to the partial pooling that is implicit in the hierarchical model. ■

17.5 LONGITUDINAL REPEATED MEASURES MODELS

In this section we describe an example in which units are measured repeatedly over time. Such data are known as "longitudinal repeated measures data" and are typically represented by a hierarchical model with measurements over time nested within the observed units.

Example 17.4 Telecommunications Laser Degradation. Lasers used in telecommunications applications contain a feedback mechanism that will maintain nearly constant light output over the life of the laser, resulting in an increase in operating current as the laser degrades. When operating current becomes too high—a 10% increase in our application—the device is considered to have failed. Figure 17.5 is a plot of laser degradation data first presented in Meeker and Escobar (1998, Example 13.10). The percent increase in current (i.e., the degradation measurement) for each of 15 lasers was observed at 17 equally spaced times up to 4,000 hours. In this example we show, among other things, how to obtain a 95% credible interval on the probability of failure by 5,000 hours of operation and a 95% credible interval on the 0.10 quantile of the failure-time distribution.

As suggested by Figure 17.5, the model for laser degradation, relating laser exposure time t_j to percent increase in operating current $\mathcal{D}_{i,j}$, is linear with a zero intercept and a slope that varies from laser to laser; that is,

$$\mathcal{D}_{i,j} = b_i t_j$$

for the $i = 1, \ldots, 15$ lasers at $j = 1, \ldots, 17$ points in time, where b_i is the degradation slope (rate) for laser i. Then a model for $y_{i,j}$, the observed percent increase in measured current for laser i at time t_j hours, is

$$y_{i,j} = \mathcal{D}_{i,j} + \epsilon_{i,j} = b_i t_j + \epsilon_{i,j},$$

where $\epsilon_{i,j}$ denotes the random error term, assumed to be independent (for the different i and j) and having a normal distribution with mean 0 and a standard deviation σ_ϵ. The log of b_i is assumed to have a normal distribution (so b_i has a lognormal distribution) with a mean $\mu_{\log(b)}$

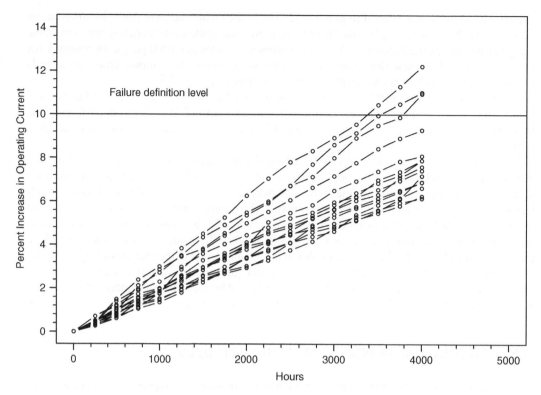

Figure 17.5   Percent increase in operating current over time for 17 lasers.

and a standard deviation $\sigma_{\log(b)}$. That is,

$$\log(b_i) \sim \text{NORM}(\mu_{\log(b)}, \sigma_{\log(b)}), \quad \epsilon_{i,j} \sim \text{NORM}(0, \sigma_\epsilon),$$

for $i = 1, \ldots, 15$ and $j = 1, \ldots, 17$. The assumption of independent residuals (from the fitted model) was justified in this application because the experiment was conducted under carefully controlled conditions and because there was no evidence of autocorrelation in the data.

References describing non-Bayesian methods for estimating the parameters and functions of the parameters for the preceding longitudinal repeated measures model are given in the Bibliographic Notes section at the end of this chapter. Here we use a Bayesian approach to construct credible intervals for functions of the parameters. The Bayesian approach is more straightforward than traditional likelihood-based methods when computing credible intervals for functions of the parameters. Also, with diffuse prior distributions, such procedures tend to have coverage probabilities that are close to the nominal credible level.

Diffuse prior distributions for the unknown parameters are taken to be

$$\mu_{\log(b)} \sim \text{UNIF}(-\infty, \infty), \quad \sigma_{\log(b)} \sim \text{UNIF}(0, 20), \quad \sigma_\epsilon^{-2} \sim \text{GAMMA}(0.001, 0.001).$$

The gamma prior distribution for the "precision" parameter σ_ϵ^{-2} is equivalent to having what is known as an "inverse-gamma" distribution for σ_ϵ^2 and is related to the conjugate inverse chi-square distribution for a normal distribution variance described in Section H.3.3. The GAMMA$(0.001, 0.001)$ distribution has a very large variance and is, for this reason, a commonly used diffuse prior distribution for an error variance.

In applications of longitudinal repeated measures models, statistical intervals are frequently desired for:

- Predictions about the future performance of the particular units in the data set or
- Estimates of functions of the parameters for the model describing the population or process from which the sample units were assumed to be randomly drawn.

In this example we will, as suggested earlier, focus on the latter application, considering estimates and associated intervals for the quantiles of the degradation distribution and on the probability of failure (i.e., the probability of having a current increase greater than $\mathcal{D}_{\mathrm{f}} = 10$) as a function of time.

An MCMC algorithm for this model was used to generate 27,000 draws from the joint posterior distribution of $\mu_{\log(b)}$, $\sigma_{\log(b)}$, σ_{ϵ}, and b_i, $i = 1, \ldots, 15$. The draws for the individual b_i values would be important for prediction of the future performance of individual units (the first issue stated above), but are not required for the present application. The cdf for the percent increase in operating current for a randomly selected laser at time t is

$$\Pr(\mathcal{D}(t) \leq d) = \Pr(bt \leq d) = \Pr[\log(bt) \leq \log(d)]$$
$$= \Phi_{\mathrm{norm}}\left[\frac{\log(d) - [\mu_{\log(b)} + \log(t)]}{\sigma_{\log(b)}}\right],$$

which can be seen to be a lognormal distribution with parameters $\mu_{\log(b)} + \log(t)$ and $\sigma_{\log(b)}$. Note that this distribution does not depend on σ_{ϵ}. The p quantile of the distribution of degradation at time t is

$$d_p(t) = t \times \exp\left[\mu_{\log(b)} + \Phi_{\mathrm{norm}}^{-1}(p)\sigma_{\log(b)}\right].$$

Figure 17.6 shows the 0.1, 0.5 and 0.9 quantiles for the fitted degradation distribution as a function of time and the lognormal density function for percent increase in operating current at several points in time. The probability of failure at a given time t can be visualized as the proportion of area of the density above the 10% failure definition at that value of t. More formally, the failure-time (i.e., the time to crossing $\mathcal{D}_{\mathrm{f}}$) cdf can be expressed as

$$F(t) = \Pr(T \leq t) = \Pr(\mathcal{D}(t) \geq \mathcal{D}_{\mathrm{f}}) = \Pr(b \geq \mathcal{D}_{\mathrm{f}}/t)$$
$$= 1 - \Phi_{\mathrm{norm}}\left[\frac{\log(\mathcal{D}_{\mathrm{f}}) - [\mu_{\log(b)} + \log(t)]}{\sigma_{\log(b)}}\right]$$
$$= \Phi_{\mathrm{norm}}\left[\frac{\log(t) - [\log(\mathcal{D}_{\mathrm{f}}) - \mu_{\log(b)}]}{\sigma_{\log(b)}}\right], \tag{17.6}$$

which again can be shown to be a lognormal distribution—now with parameters $\mu = \log(\mathcal{D}_{\mathrm{f}}) - \mu_{\log(b)}$ and $\sigma = \sigma_{\log(b)}$. The estimate and credible interval for the proportion of lasers failing by 5,000 hours can be obtained by substituting the pairs of draws for $\mu_{\log(b)}$ and $\sigma_{\log(b)}$ into (17.6) to obtain a set of draws from the marginal posterior distribution of $F(t)$ and finding the appropriate quantiles (0.5 for the point estimate and 0.05 and 0.95 for the 90% credible interval) of the empirical distribution of the resulting values. The point estimate of the proportion lasers failing by 5,000 hours is 0.503 and the credible interval is $[0.338, \ 0.667]$. Figure 17.7 shows the estimated fraction of lasers failing as a function of time, along with 90% pointwise credible intervals, plotted on lognormal probability axes. The horizontal lines indicate the point estimate and 90% credible interval at 5,000 hours.

The p quantile of the failure-time distribution is

$$t_p = \exp\left\{[\log(\mathcal{D}_{\mathrm{f}}) - \mu_{\log(b)}] + \Phi_{\mathrm{norm}}^{-1}(p)\sigma_{\log(b)}\right\}. \tag{17.7}$$

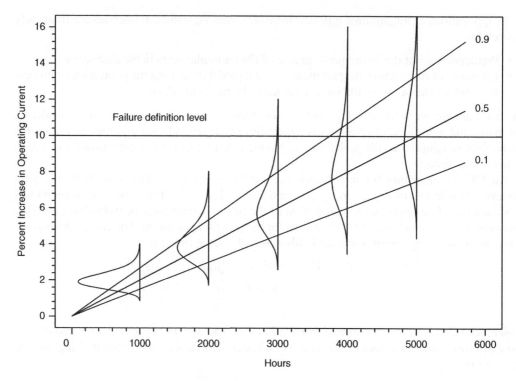

Figure 17.6   Plot of 0.1, 0.5 and 0.9 quantiles of fitted distribution for percent increase in laser operating current as a function of time and density functions at selected times.

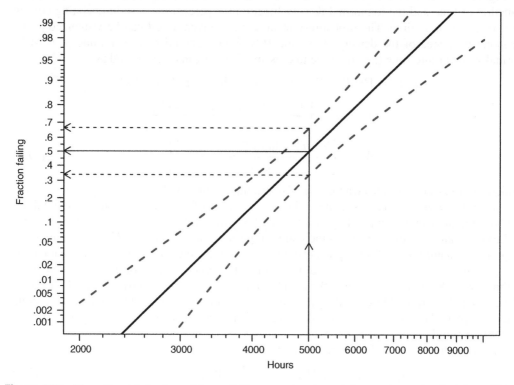

Figure 17.7   The estimated fraction of lasers failing as a function of time and associated 90% credible intervals, plotted on lognormal probability axes.

Estimates and Bayesian credible intervals for the p quantile can be found in a manner similar to that for the cdf by substituting the draws for $\mu_{\log(b)}$ and $\sigma_{\log(b)}$ into (17.7). Thus, the point estimate of the 0.10 quantile of the laser failure-time distribution is $\widehat{t}_{0.10} = 3{,}745$ hours and the corresponding 90% credible interval is $[\underset{\sim}{t}_{0.10}, \ \widetilde{t}_{0.10}] = [3{,}155, \ 4{,}187]$ hours. If a set of estimates of and credible intervals for the quantiles t_p (hours) had been plotted versus p (fraction failing), the resulting curves would be similar to but not exactly the same as those in Figure 17.7. ∎

BIBLIOGRAPHIC NOTES

General textbooks on Bayesian modeling

Most books on the application of Bayesian methods also discuss and illustrate (if not emphasize) Bayesian hierarchical models. These include Hoff (2009), Carlin and Louis (2009), Lunn et al. (2012), and Gelman et al. (2013).

Books and papers focusing on Bayesian hierarchical modeling

There are many books and papers that focus specifically on Bayesian hierarchical modeling. These include Raudenbush and Bryk (2002), Gelman and Hill (2006), Gelman (2006a), Congdon (2007), and Snijders (2011).

Specification of diffuse prior distributions in Bayesian hierarchical modeling

The specification of a diffuse joint prior distribution is an important part of the use of Bayesian methods in applications. As discussed in Section 15.3, when there is little information about a parameter, inferences about that parameter can be highly sensitive to the particular diffuse prior distribution that is used. Gelman (2006b) provides useful suggestions for specifying appropriate prior distributions for variance components in hierarchical models. Polson et al. (2012) discuss use of the half-Cauchy distribution as a prior distribution, and this distribution is also recommended by Gelman (2006b) for certain kinds of applications.

References for non-Bayesian multilevel/hierarchical estimation

This chapter has presented Bayesian inferential methods for hierarchical models. There are many references that present and use non-Bayesian methods for hierarchical models. These include Seber and Wild (1989, Chapter 7), Davidian and Giltinan (1995), Pinheiro and Bates (2000), Venables and Ripley (2002, Chapter 10), Singer and Willett (2003), Hox (2010), and Fitzmaurice et al. (2012). Browne and Draper (2006) compare Bayesian and non-Bayesian methods for hierarchical models; they show through simulation that Bayesian estimation with appropriately chosen diffuse joint prior distributions provides credible intervals that have coverage probabilities close to nominal, often performing better than commonly used non-Bayesian methods (such as Wald and likelihood-based confidence intervals).

## Chapter 18

Advanced Case Studies

OBJECTIVES AND OVERVIEW

This chapter contains seven advanced case studies that illustrate the broad applicability of the general methods presented in Chapters 15–17. The following applications are discussed:

- The case study in Section 18.1 shows how to construct, and contrasts, likelihood, bootstrap, and Wald-approximation confidence intervals on the proportion of defective integrated circuits from a manufacturing process, when the consequence of the defect is a product failure during the (early) life of the product operation and the available information is limited time-to-failure data.

- Gauge repeatability and reproducibility studies are used to assess the capability of a measurement process and to estimate components of variance attributable to different sources of variability. The case study in Section 18.2 shows the use of generalized pivotal quantity and Bayesian methods to compute confidence intervals for quantities of interest calculated from such studies.

- Naive computation of a tolerance interval when the data contain measurement errors will result in an interval that is too wide. The case study in Section 18.3 shows how to compute a tolerance interval on actual product performance that corrects for measurement error in the data, using parametric bootstrap and Bayesian methods.

- The case study in Section 18.4 illustrates the use of Bayesian and parametric bootstrap methods for computing a confidence interval on the probability of meeting a two-sided specification in product quality assessment applications.

- The case study in Section 18.5 shows how to compute a confidence interval on an estimate of the effect of a purchase-inducement strategy in a marketing study, using bootstrap and Bayesian methods, based on the results of a comparative study with random assignment of customers.

- The case study in Section 18.6 shows how to calculate confidence intervals on the probability of detecting material flaws as a function of flaw size using likelihood and Bayesian

Statistical Intervals: A Guide for Practitioners and Researchers, Second Edition.
William Q. Meeker, Gerald J. Hahn and Luis A. Escobar.
© 2017 John Wiley & Sons, Inc. Published 2017 by John Wiley & Sons, Inc.
Companion Website: www.wiley.com/go/meeker/intervals

methods, based on data from experiments involving the detection (or lack thereof) of flaws with known size. This application involves a binary regression.

- The case study in Section 18.7 returns to Example 14.14. This application dealt with obtaining a confidence interval on the probability of failure of a rocket motor as a function of age, based on scanty censored data and using Wald-approximation, likelihood, and bootstrap methods. All three of these methods resulted in intervals that were so wide that they were of little use. We now show how a more meaningful Bayesian credible interval can be obtained leveraging prior information based on engineering knowledge and/or past experience when such information is available.

Our main purpose in presenting these case studies is to demonstrate the broad applicability of the general, and mostly approximate, methods that we present in Chapters 12–17 and to help facilitate practitioners adopting them for yet other problems.

18.1 CONFIDENCE INTERVAL FOR THE PROPORTION OF DEFECTIVE INTEGRATED CIRCUITS

In some applications one needs to estimate the proportion of units in a population that possess a certain characteristic—but whether or not a particular unit possesses the characteristic becomes known only over time as a result of the occurrence (or non-occurrence) of some event, such as product failure. Thus, the information from which one estimates the population proportion possessing the characteristic is typically time-to-occurrence data. Moreover, the data may be incomplete in that some units have not experienced enough exposure time for the event occurrence to take place.

Example 18.1 Integrated Circuit Failure-Time Data. Meeker (1987) gives the results of an accelerated life test of $n = 4,156$ integrated circuits tested for 1,370 hours at 80°C and 80% relative humidity. A small proportion of the units had a manufacturing defect that would result in product failure early in life. The data are shown in Table 18.1. There were 25 failures in the first 100 hours, three more between 100 and 600 hours (with the last one occurring at 593 hours), and no more failures among the remaining 4,128 units by 1,370 hours, when the test was terminated. The data were also analyzed in Meeker and Escobar (1998, Section 11.5).

The primary purpose of the test was to estimate the proportion of defective units being manufactured by the production process and to determine how much initial in-house "burn-in" time each unit should receive so as to remove all, or at least most, of the defective units prior to shipment, assuming the proportion defective remains unchanged. The reliability engineers were also interested in determining whether it might be possible to obtain the needed information about the proportion defective in the future, using tests much shorter than 1,370 hours (say, 100 or 200 hours). ∎

0.10	0.10	0.15	0.60	0.80	0.80
1.20	2.50	3.00	4.00	4.00	6.00
10.00	10.00	12.50	20.00	20.00	43.00
43.00	48.00	48.00	54.00	74.00	84.00
94.00	168.00	263.00	593.00		

Table 18.1 Integrated circuit failure times in hours. When the test ended at 1,370 hours, there were 4,128 unfailed units.

We will illustrate the use of likelihood and Wald-approximation confidence interval methods from Chapter 12 (based on general theory outlined in Sections D.5.5 and D.5.6, respectively) and the bootstrap methods from Chapters 13 and 14 to estimate and construct confidence intervals for the proportion of defective units being generated by the manufacturing process.

18.1.1   The Limited Failure Population Model

The limited failure population (LFP) model, as used in this application, assumes that a proportion p of units from a population or process is defective and will fail according to a cumulative distribution $F(t; \mu, \sigma)$; the remaining proportion $1 - p$ will never fail—at least, due to the defect under consideration. This model has been found useful for describing integrated circuit infant mortality. Furthermore, if $F(t; \mu, \sigma)$ can be assumed to be a Weibull cdf, then the LFP failure-time model becomes

$$\Pr(T \leq t) = G(t; \mu, \sigma, p) = pF(t; \mu, \sigma) = p\Phi_{\mathrm{sev}}\left[\frac{\log(t) - \mu}{\sigma}\right], \qquad (18.1)$$

where $\Phi_{\mathrm{sev}}(z)$ is the standard smallest extreme value distribution cdf (see Section C.3.1). Note that as $t \to \infty$, $G(t; \mu, \sigma, p) \to p$. The lognormal LFP model is obtained by using Φ_{norm} instead of Φ_{sev} in (18.1).

18.1.2   Estimates and Confidence Intervals for the Proportion of Defective Units

Using the methodology of Chapter 12, the maximum likelihood estimates for the Weibull LFP model using the data available after 1,370 hours are $\widehat{\mu} = 3.35$, $\widehat{\sigma} = 2.02$, and $\widehat{p} = 0.0067$. The ML estimates using the data available after 100 hours are $\widehat{\mu} = 4.04$, $\widehat{\sigma} = 2.12$, and $\widehat{p} = 0.0083$. Figure 18.1 is a Weibull probability plot showing the nonparametric estimate of fraction failing as a function of time (the plotted points) and ML estimates of the Weibull LFP model cdf for the data available both after 100 hours and after 1,370 hours. The ML estimates of the Weibull model cdf were obtained by substituting the ML estimates of the parameters into (18.1) for values of t from 0.07 to 10,000 hours.

The two curves agree well up to 100 hours, but diverge after that. Focusing on the estimate for the proportion defective p, Figure 18.2 shows profile likelihood plots for both the 100-hour and the 1,370-hour data. As we saw in Chapter 12, the likelihood-based confidence intervals for a quantity of interest can be read from the profile likelihood plot. Figure 18.2 shows that one can obtain a reasonably precise estimate of p from the 1,370-hour data but that there is a huge amount of uncertainty when the estimate is based on the 100-hour data.

Table 18.2 summarizes the estimates and compares the confidence intervals for p using the likelihood and Wald-approximation methods as well as a bootstrap method based on random-weight bootstrap sampling and the simple percentile method to obtain the confidence intervals from the bootstrap estimates. There is good agreement between the likelihood and the bootstrap methods, but the Wald-approximation method deviates radically from the other two methods when using the 100-hour data.

There are two striking results from this example:

- Figure 18.2 shows that p can be estimated precisely from the 1,370-hour data. With the 100-hour data, however, the likelihood-based upper confidence bound for p is 1, implying that it is possible that all of the manufactured units are defective.

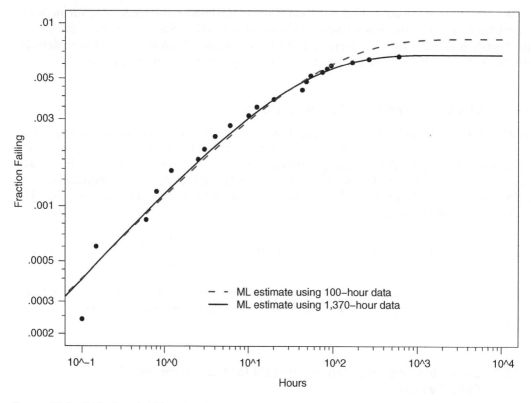

Figure 18.1   Weibull probability plot showing the LFP model estimates of the cdf for the 100-hour and 1,370-hour integrated circuit failure-time data.

- In sharp contrast to the upper endpoint of the likelihood and bootstrap confidence interval for p where $\widetilde{p} = 1$ for the 100-hour data, the corresponding upper endpoint of the Wald-approximation interval is only $\widetilde{p} = 0.0203$. Because the more trustworthy likelihood and bootstrap methods tell us that the upper confidence bound on p should be 1, the Wald-approximation interval gives a seriously inaccurate representation of the information in the data.

	1,370-hour data	100-hour data
ML estimate $\widehat{p}$	0.00674	0.00827
Standard error $\widehat{\text{se}}_{\widehat{p}}$	0.00127	0.00380
Approximate 95% confidence intervals for p		
Wald-approximation	[0.00466, 0.00975]	[0.0033, **0.0203**]
Likelihood-based	[0.00455, 0.00955]	[0.00463, **1.0000**]
Random-weight bootstrap	[0.00447, 0.00952]	[0.00464, **1.0000**]

Table 18.2   Comparison of Weibull LFP model estimates and confidence intervals for p, the proportion of defective integrated circuits for the 100-hour data and the 1,370-hour data.

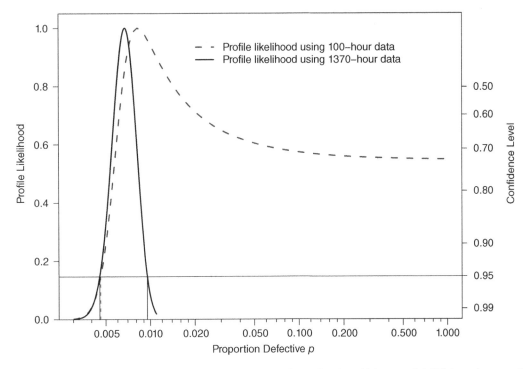

Figure 18.2 Profile likelihood for the proportion defective p for the 100-hour and 1,370-hour integrated circuit failure-time data.

The striking difference between the likelihood and bootstrap methods, when compared with the Wald-approximation methods, provides an extreme example in which the Wald approximation can be misleading and also supports the important practice of examining the likelihood in situations involving new models or data.

The contrast between the likelihoods for the 1,370-hour data and the 100-hour data also provides a partial answer the question of whether p can be estimated with a test of length 100 or 200 hours. The answer is no. Generally, one can obtain a reasonably precise estimate of p only if the nonparametric estimate of the fraction failing as a function of time (see the plotted points in Figure 18.1) levels off sufficiently by the end of the life test.

18.2 CONFIDENCE INTERVALS FOR COMPONENTS OF VARIANCE IN A MEASUREMENT PROCESS

Gauge repeatability and reproducibility (GR&R) studies are used to characterize the capability of measurement systems. In this case study we review the basic ideas of a GR&R study and present two modern methods for analysis of the resulting data—one based on a bootstrap generalized pivotal quantity approach and the other using Bayesian methods—to find confidence intervals for components of variance and functions of such components. Our application uses the common two-way random-effects model with interaction, but the methods can be readily adapted to other random-effects models.

Example 18.2 GR&R Study of Thermal Impedance Measurements on Power Modules. Houf and Berman (1988) describe a GR&R study to characterize a thermal impedance

Part	Operator 1	Operator 2	Operator 3
1	0.37, 0.38, 0.37	0.41, 0.41, 0.40	0.41, 0.42, 0.41
2	0.42, 0.41, 0.43	0.42, 0.42, 0.42	0.43, 0.42, 0.43
3	0.30, 0.31, 0.31	0.31, 0.31, 0.31	0.29, 0.30, 0.28
4	0.42, 0.43, 0.42	0.43, 0.43, 0.43	0.42, 0.42, 0.42
5	0.28, 0.30, 0.29	0.29, 0.30, 0.29	0.31, 0.29, 0.29
6	0.42, 0.42, 0.43	0.45, 0.45, 0.45	0.44, 0.46, 0.45
7	0.25, 0.26, 0.27	0.28, 0.28, 0.30	0.29, 0.27, 0.27
8	0.40, 0.40, 0.40	0.43, 0.42, 0.42	0.43, 0.43, 0.41
9	0.25, 0.25, 0.25	0.27, 0.29, 0.28	0.26, 0.26, 0.26
10	0.35, 0.34, 0.34	0.35, 0.35, 0.34	0.35, 0.34, 0.35

Table 18.3 Thermal impedance measurements on power modules in units of kelvin per watt.

measurement process applied to semiconductor power modules. Ten power modules were selected at random from a specific population of parts, and each was inspected three times by three different operators, in a randomized sequence. The three operators are also assumed to be a random sample from a larger population of operators. The resulting data are shown in Table 18.3. Analysis of variance (ANOVA) methods applied to the data allow one to estimate components of variance attributable to operator-to-operator variability, part-to-part variability, repeat variability, as well as other quantities of interest that are typically functions of these variance components. ∎

18.2.1 Two-Way Random-Effects Model

The two-way random-effects model with interaction for the observed thermal impedance measurement used by Hamada and Weerahandi (2000) is

$$Y_{ijk} = \mu + O_i + P_j + (OP)_{ij} + \epsilon_{ijk}, \tag{18.2}$$

where $i = 1, \ldots, o$; $j = 1, \ldots, p$; and $k = 1, \ldots, r$, $O_i \sim \text{NORM}(0, \sigma_O)$ is the effect for operator i, $P_j \sim \text{NORM}(0, \sigma_P)$ is the effect for part j, $(OP)_{ij} \sim \text{NORM}(0, \sigma_{OP})$ is the interaction effect between operator i and part j, and $\epsilon_{ijk} \sim \text{NORM}(0, \sigma_E)$ is the error term nested within operator i and part j. Also, O_i, P_j, OP_{ij}, and ϵ_{ijk} are assumed to be independent. The usual ANOVA table for this model is shown in Table 18.4. In this table,

$$\bar{y}_i = \sum_{j=1}^{p} \sum_{k=1}^{r} y_{ijk}/(pr), \quad \bar{y}_j = \sum_{i=1}^{o} \sum_{k=1}^{r} y_{ijk}/(or),$$

$$\bar{y}_{ij} = \sum_{k=1}^{r} y_{ijk}/r, \quad \bar{y} = \sum_{i=1}^{o} \sum_{j=1}^{p} \sum_{k=1}^{r} y_{ijk}/(opr)$$

are the sample means for each operator, each part, each operator-part combination, and the overall mean, respectively.

The E(MS) expected mean squares values given in Table 18.4 are denoted by $(\theta_P, \theta_O, \theta_{OP}, \theta_E)$ and can be thought of as an alternative set of parameters for the random-effects model in (18.2); these alternative parameters can be estimated directly (and independently, for balanced data) by the sample mean squares computed from the ANOVA table (e.g., Table 18.6).

Source	Degrees of freedom (df)	Sum of squares (SS)	Expected mean square (MS)
P	$p-1$	$or \sum_{j=1}^{p}(\bar{y}_j - \bar{y})^2$	$\theta_P = \sigma_E^2 + r\sigma_{OP}^2 + or\sigma_P^2$
O	$o-1$	$pr \sum_{i=1}^{o}(\bar{y}_i - \bar{y})^2$	$\theta_O = \sigma_E^2 + r\sigma_{OP}^2 + pr\sigma_O^2$
OP	$(o-1)(p-1)$	$r \sum_{i=1}^{o}\sum_{j=1}^{p}(y_{ij} - \bar{y}_i - \bar{y}_j + \bar{y})^2$	$\theta_{OP} = \sigma_E^2 + r\sigma_{OP}^2$
E	$op(r-1)$	$\sum_{i=1}^{o}\sum_{j=1}^{p}\sum_{k=1}^{r}(y_{ijk} - \bar{y}_{ij})^2$	$\theta_E = \sigma_E^2$
T	$opr-1$	$\sum_{i=1}^{o}\sum_{j=1}^{p}\sum_{k=1}^{r}(y_{ijk} - \bar{y})^2$	

Table 18.4 ANOVA table for two-factor random-effects model with interaction. For each row, the sample mean square (MS) is the mean sum of squares (i.e., SS/df).

Table 18.5 shows some quantities of interest (functions of the components of variance model parameters) in a GR&R study that will be estimated in this case study.

Example 18.3 ANOVA Table for the Thermal Impedance Measurement GR&R Study. In this application we have $p = 10$ parts, $o = 3$ operators, and $r = 3$ repeat observations by each operator on each part. The sample mean of all of the thermal impedance measurements is $\hat{\mu} = \bar{y} = 0.3580$. Table 18.6 gives the ANOVA table for the analysis of the data from this GR&R study. As will be shown in Section 18.2.2, entries from this table can be used to compute estimates and confidence intervals for the quantities of interest given in Table 18.5. ∎

18.2.2 Bootstrap (GPQ) Method

This section provides a more advanced application of the GPQ method described in Sections 14.4 and Appendix F. The development follows the approach given in Hamada and Weerahandi (2000). Table 18.7 shows GPQs for the expected mean square parameters for the ANOVA in Table 18.4. Note that in Table 18.7, s_P^2, s_O^2, s_{OP}^2, and s_E^2 are sample mean squares for the variance components for P, O, OP and E, respectively. Thus the numerators of the GPQ ratios are the *sample sums of squares* in Table 18.4. The $X_{(\nu)}^2$ values in the denominators are independent chi-square random variables with ν degrees of freedom, where ν equals $p-1$, $o-1$, $(o-1)(p-1)$, and $op(r-1)$, respectively, for the four rows in Table 18.7.

Variability type	GR&R name	Quantity
Part-to-part	Process	$\gamma_P = \sigma_P^2$
Measurement system	Gauge	$\gamma_M = \sigma_O^2 + \sigma_{OP}^2 + \sigma_E^2$
Reproducibility	Reproducibility	$\gamma_R = \sigma_O^2 + \sigma_{OP}^2$
Error	Repeatability	$\gamma_E = \sigma_E^2$
Total	Total	$\gamma_T = \sigma_O^2 + \sigma_P^2 + \sigma_{OP}^2 + \sigma_E^2$
Proportion of variability due to part-to-part variation		$\rho_P = \gamma_P/\gamma_T$
Proportion of variability due to the measurement system variation		$\rho_M = \gamma_M/\gamma_T$

Table 18.5 GR&R quantities of interest.

Source	Degrees of freedom (df)	Sum of squares (SS)	Mean square (MS)
P	9	0.394	0.0437
O	2	0.00393	0.00196
OP	18	0.00485	0.000270
E	60	0.00307	5.11×10^{-5}
T	89	0.405	0.00456

Table 18.6 ANOVA table for the thermal impedance measurement GR&R study. For each row, the sample mean square is the mean sum of squares (i.e., SS/df).

The GPQs for the other quantities of interest in Table 18.5 can be shown, using the substitution method in Section F.2, to be simple functions of the GPQs in Table 18.7 and are given by

$$Z_{\widehat{\gamma}_P} = \left(Z_{\widehat{\theta}_P} - Z_{\widehat{\theta}_{OP}} \right) / (or)$$
$$Z_{\widehat{\gamma}_M} = \left[Z_{\widehat{\theta}_O} + (p-1)Z_{\widehat{\theta}_{OP}} + p(r-1)Z_{\widehat{\theta}_E} \right] / (pr)$$
$$Z_{\widehat{\gamma}_R} = \left[Z_{\widehat{\theta}_O} + (p-1)Z_{\widehat{\theta}_{OP}} - pZ_{\widehat{\theta}_E} \right] / (pr) \tag{18.3}$$
$$Z_{\widehat{\gamma}_T} = \left[pZ_{\widehat{\theta}_P} + oZ_{\widehat{\theta}_O} + (op-p-o)Z_{\widehat{\theta}_{OP}} + op(r-1)Z_{\widehat{\theta}_E} \right] / (opr)$$
$$Z_{\widehat{\rho}_P} = Z_{\widehat{\gamma}_P} / Z_{\widehat{\gamma}_T}, \qquad Z_{\widehat{\rho}_M} = Z_{\widehat{\gamma}_M} / Z_{\widehat{\gamma}_T}.$$

Confidence intervals for the quantities of interest in Table 18.5 can then be obtained by using the expressions for the GPQs for the E(MS) values in Table 18.7, substituted into the appropriate expressions in (18.3) and applying the GPQ method described in Appendix F and illustrated for other applications in Section 14.4. The results for the thermal impedance measurement GR&R study are presented in the following example.

Example 18.4 GPQ Confidence Intervals for the Thermal Impedance Measurement GR&R Study. GPQ confidence intervals for quantities of interest, such as those given in Table 18.5, can be computed by generating a large number of draws (e.g., 1 million) from the distribution of the corresponding GPQ. The draws can be obtained by substituting the sum of squares from Table 18.6 into the GPQ expression for the quantity of interest in Table 18.7 and/or (18.3) and independently simulating the values of the different chi-square random variables. Then an approximate $100(1 - \alpha)\%$ confidence interval for the quantity of interest is obtained

E(MS)	GPQ for E(MS)
$\theta_P = \sigma_E^2 + r\sigma_{OP}^2 + or\sigma_P^2$	$Z_{\widehat{\theta}_P} = (p-1)s_P^2 / X_{(p-1)}^2$
$\theta_O = \sigma_E^2 + r\sigma_{OP}^2 + pr\sigma_O^2$	$Z_{\widehat{\theta}_O} = (o-1)s_O^2 / X_{(o-1)}^2$
$\theta_{OP} = \sigma_E^2 + r\sigma_{OP}^2$	$Z_{\widehat{\theta}_{OP}} = (o-1)(p-1)s_{OP}^2 / X_{((o-1)(p-1))}^2$
$\theta_E = \sigma_E^2$	$Z_{\widehat{\theta}_E} = op(r-1)s_E^2 / X_{(op(r-1))}^2$

Table 18.7 GPQs for the E(MS) values in Table 18.4.

from the $\alpha/2$ and $1 - \alpha/2$ quantiles of the resulting empirical distribution of the draws. The median of the empirical distribution provides a point estimate for the quantity of interest.

For example, to obtain a point estimate and approximate 95% confidence interval for γ_P one can use the R command

```
quantile((0.394/rchisq(1.e7, df=9) -
        0.00485/rchisq(1.e7, df=18))/(3*3),
        probs=c(0.50, 0.025, 0.975))
       50%         2.5%        97.5%
0.005212903 0.002268804 0.016186309
```

where `1.e7` is to be read as 10^7. Similarly, to obtain a point estimate and approximate 95% confidence interval for γ_M one can use the R command

```
quantile((0.00393/rchisq(1.e7, df=2) +
        (10-1)*0.00485/rchisq(1.e7, df=18) +
        10*(3-1)*0.00307/rchisq(1.e7, df=60))/(10*3),
        probs=c(0.50, 0.025, 0.975))
       50%          2.5%         97.5%
0.0002272204 0.0001186154 0.0027125311
```

Table 18.8 summarizes the results for all of the quantities of interest in Table 18.5, along with similar results using the Bayesian method, to be described in Section 18.2.3. ∎

This example has dealt with point estimates and confidence intervals for variances and ratios of variances. For some applications, we need to present instead estimates and intervals for

Method	Quantity of interest	Estimate	Interval endpoints Lower	Upper	Ratio upper/lower endpoints
GPQ	γ_P	0.0052	0.0023	0.016	7.0
Bayesian	γ_P	0.0059	0.0024	0.020	8.2
GPQ	γ_M	0.00023	0.00012	0.0027	23
Bayesian	γ_M	0.00035	0.00013	0.016	121
GPQ	γ_R	0.00017	6.7×10^{-5}	0.0027	40
Bayesian	γ_R	0.00028	6.6×10^{-5}	0.016	237
GPQ	θ_E	5.2×10^{-5}	3.7×10^{-5}	7.6×10^{-5}	2.1
Bayesian	θ_E	6.3×10^{-5}	4.5×10^{-5}	9.2×10^{-5}	2.0
GPQ	γ_T	0.0056	0.0025	0.018	7.0
Bayesian	γ_T	0.0069	0.0029	0.030	10
GPQ	ρ_P	0.96	0.64	0.99	1.5
Bayesian	ρ_P	0.94	0.26	0.99	3.8
GPQ	ρ_M	0.044	0.012	0.36	31
Bayesian	ρ_M	0.059	0.012	0.74	64

Table 18.8 95% confidence intervals and 95% credible intervals for quantities of interest for the thermal inductance GR&R study.

standard deviations (which are in general easier to interpret because they are in the same units as the original response variable). These are easily obtained by simply taking the square root of the values corresponding to the variance. For example, a GPQ 95% confidence interval for the measurement error standard deviation σ_E^2 is

$$\left[\sqrt{\underset{\sim}{\gamma_M}}, \ \sqrt{\tilde{\gamma}_M}\right] = \left[\sqrt{0.00012}, \ \sqrt{0.0027}\right] = [0.011, \ 0.052].$$

18.2.3 Bayesian Method

The construction of confidence intervals for the GR&R quantities of interest (Table 18.5) presented in this section follows the Bayesian approach given in Example 2 of Weaver et al. (2012). This approach uses the model given in Section 18.2.1, but also requires one to specify prior distributions for the unknown model parameters. Then an MCMC algorithm is used to generate a large number of draws from the resulting joint posterior distribution of the model parameters; these are, in turn, used to compute draws from the marginal posterior distributions of the quantities of interest needed to compute Bayesian point estimates and credible intervals for the quantities of interest.

Example 18.5 Bayesian Credible Intervals for the Thermal Impedance Measurement GR&R Study. Because no physically or empirically based prior information was available for this application, and to have results that are comparable with the non-Bayesian GPQ method in Section 18.2.2, we follow the advice in the literature (e.g., Gelman, 2006b, Weaver et al., 2012) and use the diffuse prior distributions $\mu \sim \text{UNIF}(-\infty, \infty)$, $\sigma_P \sim \text{UNIF}(0, 10,000)$, $\sigma_{OP} \sim \text{UNIF}(0, 10,000)$, $\sigma_E \sim \text{UNIF}(0, 10,000)$, and $\sigma_O \sim \text{HCAUCHY}(0.20)$. The half-Cauchy prior distribution is recommended for a random-effect parameter when there are only a small number of units (and thus a small number of degrees of freedom) available to estimate the random effect (there were only three operators).

An MCMC algorithm was used to generate 1 million draws from the joint posterior distribution of μ, σ_P, σ_{OP}, σ_E, and σ_O. These draws were then used to generate the corresponding 1 million draws for the marginal posterior distributions of the GR&R quantities of interest in Table 18.5, using the formulas on the right-hand column of that table. Then, as described in Section 15.2.6, these sample draws were used to compute Bayesian point estimates and credible intervals for the quantities of interest. The results are shown in Table 18.8. ∎

18.2.4 Comparison of Results Using GPQ and Bayesian Methods and Recommendations

We note that in Table 18.8, the upper endpoints for the intervals for γ_R and $\gamma_M = \gamma_R + \sigma_E^2$ agree because σ_E^2 is negligible in magnitude relative to γ_R. The GPQ and Bayesian intervals for $\gamma_P = \sigma_P$ are similar because there is a reasonable amount of data to estimate σ_P (i.e., there were 10 parts) and, therefore, the analysis did not rely heavily on the assumed prior distribution for σ_P. On the other hand, there were only three operators, and thus σ_O^2 and functions of σ_O^2 (see Table 18.5) such as γ_M and γ_R are not estimated precisely (note the last column in Table 18.8, in which larger numbers indicate less precision). Because there is relatively little information in the data about γ_M and γ_R, the Bayesian inferences are highly dependent on the particular forms of the specified diffuse prior distributions, probably contributing to the differences between the GPQ and Bayesian intervals.

In applications with random effects and limited (but balanced) data for which the specification of different diffuse prior distributions can have a strong effect on the resulting inferences (because the number of degrees of freedom to estimate variance components of interest is small), we recommend the GPQ method over the Bayesian method. This is because simulation studies (see the references in the Bibliographic Notes section at the end of this chapter) have shown that GPQ methods tend to have coverage probabilities close to the nominal confidence level. An exception would be situations for which good prior information exists and there is no objection to its use.

If the data are unbalanced, the GPQ method does not apply directly. In this case, alternatives approximate parametric bootstrap/simulation procedures can be developed. If a Bayesian approach (which does not require balanced data) is used, it is important to compare the results obtained using different diffuse prior distributions so as to assess the sensitivity of the results.

18.3 TOLERANCE INTERVAL TO CHARACTERIZE THE DISTRIBUTION OF PROCESS OUTPUT IN THE PRESENCE OF MEASUREMENT ERROR

Chapters 4 and 16 show how to compute tolerance intervals for a normal distribution and a general (log-)location-scale distribution, respectively. When there is measurement error, however, using the methods in Chapters 4 and 16 will result in tolerance intervals that overstate the actual product or process variability. On the other hand, the naive interval $\widehat{\mu} \mp z_{(1+\beta)/2}\widehat{\sigma}_P$ (where $\widehat{\sigma}_P$ is the estimate of the part standard deviation, introduced in Section 18.2.1) is likely be too narrow because it ignores the statistical uncertainty in the estimates $\widehat{\mu}$ and $\widehat{\sigma}_P$. The next two subsections provide improved approximate tolerance intervals for the actual part or process distribution without the measurement error when the available data are subject to measurement error. We will continue to use the notation and data from the thermal impedance example, described in Section 18.2. Under the model (18.2), the distribution of the particular part characteristic is $\mathrm{NORM}(\mu, \sigma_P)$ and a control-the-center tolerance interval is desired for this distribution.

18.3.1 Bootstrap (GPQ) Method

The method used here is a special case of the more general method given in Liao et al. (2005), but has been simplified and uses our notation. Tolerance intervals for the responses of other random-effects models (e.g., when there are only two sources of variability in a response) can be constructed similarly.

First, note that $\sigma_{\bar{y}}^2 = \gamma_T/(opr)$ is the variance of $\widehat{\mu} = \bar{y}$ (see Table 18.5). Then $\sigma_{\mathrm{Tol}} = (\sigma_{\bar{y}}^2 + \sigma_P^2)^{1/2}$ can be interpreted as a kind of tolerance interval standard deviation, accounting for both variability in the $\mathrm{NORM}(\mu, \sigma_P)$ distribution and the uncertainty in the estimates of the model parameters. A GPQ method can then be used to account for the parameter uncertainty to construct a control-the-center tolerance interval, in a manner similar to that for the confidence intervals given in Section 18.2.2. Specifically, an approximate $100(1 - \alpha)\%$ control-the-center tolerance interval for the $\mathrm{NORM}(\mu, \sigma_P)$ distribution is

$$\left[\underset{\sim}{T}_\beta(\boldsymbol{y}, 1 - \alpha), \ \widetilde{T}_\beta(\boldsymbol{y}, 1 - \alpha) \right] = \widehat{\mu} \mp z_{(1+\beta)/2}\, \widetilde{\sigma}_{\mathrm{Tol}}, \tag{18.4}$$

where $\widetilde{\sigma}_{\mathrm{Tol}}$ is an approximate $100(1 - \alpha)\%$ one-sided upper confidence bound for σ_{Tol}.

In a manner similar to that used to find the GPQs in (18.3), the GPQ for $\widehat{\sigma}_{\mathrm{Tol}}^2$ is

$$Z_{\widehat{\sigma}_{\mathrm{Tol}}^2} = \left[\frac{p}{(opr)^2} + \frac{1}{or} \right] Z_{\widehat{\theta}_P} + \frac{o}{(opr)^2} Z_{\widehat{\theta}_O} + \left[\frac{op - p - o}{(opr)^2} - \frac{1}{or} \right] Z_{\widehat{\theta}_{OP}} + \frac{op(r-1)}{(opr)^2} Z_{\widehat{\theta}_E}. \tag{18.5}$$

Then $\widetilde{\sigma}_{\mathrm{Tol}}$ is obtained as the square root of the $1 - \alpha$ quantile of the empirical distribution of B (a large number, such as 1 million) draws from the distribution of $Z_{\widehat{\sigma}^2_{\mathrm{Tol}}}$, computed using the B independent draws from the distributions of the GPQs in Table 18.7.

Example 18.6   GPQ Tolerance Interval for the Distribution of Thermal Impedance Values.
As in Example 18.4, GPQ intervals can be computed by generating draws from the distribution of the GPQ for the quantity of interest ($\widehat{\sigma}^2_{\mathrm{Tol}}$ for the tolerance interval in (18.4)). For this example, taking the values of the sums of squares and degrees of freedom from Table 18.6, recalling from Example 18.2 that $p = 10$, $o = 3$, $r = 3$, substituting into (18.5), and using the R command

```
sqrt(quantile((10/90^2+1/9)*0.394/rchisq(1.e7, df=9) +
        (3/90^2)*0.00393/rchisq(1.e7, df=2) +
        (18/90^2-1/9)*0.00485/rchisq(1.e7, df=18) +
        (60/90^2)*0.00307/rchisq(1.e7, df=60), probs=0.95))
     95%
0.1152904
```

provides $\widetilde{\sigma}_{\mathrm{Tol}} = 0.1153$. Substituting $\widetilde{\sigma}_{\mathrm{Tol}}$ into (18.4) gives an approximate 95% tolerance interval to contain 90% of the thermal impedance values as

$$\left[\underset{\sim}{T}_{\beta}, \ \widetilde{T}_{\beta}\right] = 0.3580 \mp z_{(0.95)} \, 0.1153 = [0.168, \ 0.548].$$

Using R as a calculator to compute the tolerance interval, we obtain

```
0.3580 + c(-1,1)*qnorm(0.95)*0.1153
```

$\blacksquare$

18.3.2   Bayesian Method

We now apply the general Bayesian approach given in Section 15.5.2 to construct a control-the-center tolerance interval for a normal distribution when the observations are subject to measurement error. Because of the symmetry of the normal distribution, the desired tolerance interval has the form

$$\left[\underset{\sim}{T}_{\beta}(\boldsymbol{y}, 1 - \alpha), \ \widetilde{T}_{\beta}(\boldsymbol{y}, 1 - \alpha)\right] = \widehat{\mu} \mp k_{\mathrm{Tol}}.$$

Then, adapting (15.4), k_{Tol} is chosen such that

$$\frac{1}{B} \sum_{j=1}^{B} \mathrm{I}\left[\Phi_{\mathrm{norm}}\left(\frac{\widetilde{T}_{\beta} - \mu^*_j}{\sigma^*_{P_j}}\right) - \Phi_{\mathrm{norm}}\left(\frac{\underset{\sim}{T}_{\beta} - \mu^*_j}{\sigma^*_{P_j}}\right) > \beta \right] = 1 - \alpha, \qquad (18.6)$$

where the indicator function $\mathrm{I}[A]$ is equal to 1 when the statement A is true and equal to 0 otherwise, and μ^*_j and $\sigma^*_{P_j}$ are calculated from draw j of B draws from the marginal posterior distributions of μ and σ_P, respectively, for $j = 1, \ldots, B$.

Example 18.7   Bayesian Tolerance Interval for the Distribution of Thermal Impedance Values. In this example we use the same 1 million draws from the joint posterior distribution of μ and σ_P as in Example 18.5. The median of the posterior distribution of μ gives $\widehat{\mu} = 0.3581$, and the value of k_{Tol} satisfying (18.6) for $\beta = 0.90$ and $1 - \alpha = 0.95$ is 0.2456, giving

$$\left[\underset{\sim}{T}_{\beta}, \ \widetilde{T}_{\beta}\right] = 0.3581 \mp 0.2456 = [0.113, \ 0.604].$$

This interval is similar to but a little wider than the interval computed with the GPQ method in Example 18.4. $\blacksquare$

18.4   CONFIDENCE INTERVAL FOR THE PROPORTION OF PRODUCT CONFORMING TO A TWO-SIDED SPECIFICATION

Determining the probability of meeting specifications, or equivalently, the proportion of product conforming to specifications—a subject introduced in Section 2.2.3—is desired in many practical applications. When there is only a *one-sided upper specification limit U*, the proportion of product meeting (i.e., falling below) such a limit for a location-scale distribution is

$$p_U = \Pr(X \le x_U) = F(x_U) = \Phi\left(\frac{x_U - \mu}{\sigma}\right), \qquad (18.7)$$

where x_U on the right-hand side is replaced by $\log(x_U)$ for a log-location-scale distribution. When one has to rely on data to estimate μ and σ—as is usually the case—the ML estimate for p_U is obtained by substituting the ML estimates for μ and σ into (18.7). A similar approach is used to determine the proportion of product p_L meeting (i.e., exceeding) a *one-sided lower specification limit L*. In Section 4.5 we provided an exact method for constructing confidence intervals for the probabilities p_L or p_U of meeting such *one-sided* specification limits for a normal distribution. Section 14.4.2 provides a more general approach for any (log-)location-scale distribution.

This case study shows how to obtain a confidence interval for the probability of meeting a *two-sided* specification limit (e.g., $x_L \le X \le x_U$) for a normal distribution (but the method extends easily to any (log-)location-scale distribution). In particular, we show how to compute a confidence interval for

$$p_I = \Pr(x_L \le X \le x_U) = p_U - p_L = \Phi\left(\frac{x_U - \mu}{\sigma}\right) - \Phi\left(\frac{x_L - \mu}{\sigma}\right), \qquad (18.8)$$

where again x_i on the right-hand side is replaced by $\log(x_i)$, $i = L, U$, for a log-location-scale distribution and the ML estimate of p_I is obtained by substituting the ML estimates of μ and σ into (18.8).

Unlike the case for one-sided specification limits, we know of no exact methods for finding a confidence interval for p_I. Thus, we will present approximate methods based on the GPQ (Chapter 14) and Bayesian methods (Chapter 15). The methods presented in this case study apply directly to any location-scale or log-location-scale distribution and can be extended to distributions outside these families. They can also be applied for obtaining approximate confidence intervals for one-sided specification limits for distribution families other than (log-)location-scale distributions.

18.4.1   Bootstrap Simulation (GPQ) Method

A GPQ for p_I is obtained by substituting GPQs for μ and σ (given in (14.6) and (14.7), respectively) into (18.8), giving

$$Z_{\hat{p}_I} = \Phi\left[\left(\frac{\hat{\sigma}^*}{\sigma}\right)\Phi^{-1}(\hat{p}_U) + \frac{\hat{\mu}^* - \mu}{\sigma}\right] - \Phi\left[\left(\frac{\hat{\sigma}^*}{\sigma}\right)\Phi^{-1}(\hat{p}_L) + \frac{\hat{\mu}^* - \mu}{\sigma}\right]. \qquad (18.9)$$

The preceding expression can be simplified when, as described in Section 14.4.4, the data are complete (i.e., no censoring) or censoring occurs after a prespecified number of lower order statistics have been observed. In these cases it is possible to simulate using $\mu = 0$ and $\sigma = 1$, resulting in the simplified form of (18.9),

$$Z_{\hat{p}_I} = \Phi\left[\hat{\sigma}^*\Phi^{-1}(\hat{p}_U) + \hat{\mu}^*\right] - \Phi\left[\hat{\sigma}^*\Phi^{-1}(\hat{p}_L) + \hat{\mu}^*\right].$$

Example 18.8   GPQ Confidence Interval for the Proportion of Conforming Circuit Pack Output Voltages. Assume that in Example 3.1 systems using the circuit pack will operate satisfactorily as long as their output is between 48 and 52 volts. It is desired to construct a

95% confidence interval to contain the proportion of such "conforming" units (i.e., to obtain a confidence interval for $p_I = \Pr(48 \leq X \leq 52)$).

The ML estimate of p_I is computed by substituting the ML estimates $\widehat{\mu} = 50.10$ and $\widehat{\sigma} = 1.1713$ into (18.8), giving

$$\widehat{p}_I = \widehat{p}_U - \widehat{p}_L = \Phi_{\text{norm}}\left(\frac{52 - 50.10}{1.1713}\right) - \Phi_{\text{norm}}\left(\frac{48 - 50.10}{1.1713}\right)$$

$$= 0.9476086 - 0.03649902 = 0.9111.$$

Given a matrix VoltageMLdraws of draws ($B = 200{,}000$ in this example) from the joint sampling distribution of $\widehat{\mu}$ and $\widehat{\sigma}$, one can use the following R command to generate an approximate 95% confidence interval for p_I:

```
quantile(
    pnorm(VoltageMLdraws[,"sigma"]*qnorm(0.9476086)+VoltageMLdraws
[,"mu"])-
    pnorm(VoltageMLdraws[,"sigma"]*qnorm(0.0364990)+VoltageMLdraws
[,"mu"]),
    probs=c(0.025, 0.975))
```

This gives the 95% confidence interval $[p_I, \ \widetilde{p}_I] = [0.366, \ 0.984]$. The left-hand plot of Figure 18.3 shows the empirical density function of the generated values of $Z_{\widehat{p}_I}$ from which this confidence interval was obtained. ∎

18.4.2 Bayesian Method

To obtain the marginal posterior distribution for p_I, one first substitutes the marginal posterior distributions for μ and σ into (18.8). Then, in manner similar to that in the examples in Chapter 14, a credible interval can be obtained from the appropriate quantiles of the marginal posterior distribution for p_I.

Example 18.9 Bayesian Credible Interval for the Proportion of Conforming Circuit Pack Output Voltages. We obtain credible intervals for the proportion conforming in Example 3.1

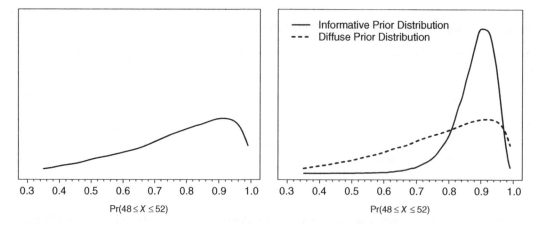

Figure 18.3 The empirical density of the GPQ $Z_{\widehat{p}_I}$ (left) and the Bayesian marginal posterior distributions of $p_I = \Pr(48 \leq X \leq 52)$ based on diffuse and informative prior distributions (right) for the circuit pack output voltage application.

using Bayesian methods with the diffuse and informative prior distributions for μ and σ that were used in Example 16.15. In particular, draws from the posterior distributions for μ and σ were used to generate draws from the marginal posterior distributions for p_I. The right-hand plot in Figure 18.3 shows the resulting marginal posterior distributions under the assumptions of diffuse and informative prior distributions, respectively. It is interesting (and reassuring) to note that the marginal posterior distribution of p_I using the diffuse prior information has almost exactly the same shape as the GPQ distribution shown on the left-hand plot.

The 95% credible interval for p_I based on diffuse prior information can be computed from the corresponding marginal distribution for p_I with the R command

```
quantile(pnorm(q=52, mean=drawsOutputVoltageDiffuse[,"mu"],
               sd=drawsOutputVoltageDiffuse[,"sigma"])-
         pnorm(q=48, mean=drawsOutputVoltageDiffuse[,"mu"],
               sd=drawsOutputVoltageDiffuse[,"sigma"]),
         probs=c(0.50, 0.025, 0.975))
     50%       2.5%       97.5%
0.7984249 0.3558718 0.9840402
```

giving $[p_I, \; \widetilde{p}_I] = [0.356, \; 0.984]$ which is in close agreement with the GPQ interval. The corresponding interval based on the informative prior distribution is $[p_I, \; \widetilde{p}_I] = [0.716, \; 0.969]$ which is, as expected, appreciably narrower, reflecting the impact of the assumed prior information. ∎

18.5 CONFIDENCE INTERVAL FOR THE TREATMENT EFFECT IN A MARKETING CAMPAIGN

18.5.1 Background

Marketing managers often conduct experiments to evaluate the effect that a treatment (e.g., a direct mail offer of a discount) will have on the probability that a customer will purchase a product.

Example 18.10 Direct Mail Marketing Experiment. Table 18.9 shows the results of a marketing experiment in which $n_T = 300,000$ potential customers were randomly selected to receive, by direct mail, a special inducement to purchase a particular product. The other $n_C = 1,300,000$ potential customers were not offered the inducement and are considered the control group.

Of those potential customers in the treatment group, $x_T = 810$ (proportion 0.0027) made a purchase. Of the 1,300,000 potential customers in the control group, $x_C = 1,950$ (proportion 0.0015) made a purchase. The difference between the treatment and the control proportions is $0.0027 - 0.0015 = 0.0012$. The results are usually reported in terms of the percentage increase arising from the marketing campaign:

$$\widehat{R}_{\text{Total}} = \frac{\dfrac{x_T}{n_T} - \dfrac{x_C}{n_C}}{\dfrac{x_C}{n_C}} = \frac{\dfrac{810}{300,000} - \dfrac{1,950}{1,300,000}}{\dfrac{1,950}{1,300,000}} \qquad (18.10)$$

$$= \frac{0.0027 - 0.0015}{0.0015} = \frac{0.0012}{0.0015} = 0.80,$$

	Potential customers	Number purchased	Proportion purchased
Treatment	300,000	810	0.0027
Control	1,300,000	1,950	0.0015
Increment			0.0012

Table 18.9   Marketing experiment results.

suggesting an 80% "lift" (i.e., 80% increase in the proportion of customers). Although the sample sizes were large, the response proportions were small. Thus, a statement about the statistical uncertainty in $\widehat{R}_{\text{Total}}$, in the form of a confidence interval, was desired for presentation to management.   ■

18.5.2   Bootstrap (Simulation) Method

The model for the data from the marketing experiment is

$$X_{\text{T}} \sim \text{BINOM}(n_{\text{T}}, p_{T}), \quad X_{\text{C}} \sim \text{BINOM}(n_{\text{C}}, p_{C}), \tag{18.11}$$

where X_{T} and X_{C} are assumed to be mutually independent. The parameters are estimated by $\widehat{p}_{\text{T}} = x_{\text{T}}/n_{\text{T}}$ and $\widehat{p}_{\text{C}} = x_{\text{C}}/n_{\text{C}}$.

To compute the bootstrap intervals for given n_{T}, x_{T}, n_{C}, and x_{C}, we simulated $B = 50{,}000$ realizations $(x_{\text{T}}^{*}, x_{\text{C}}^{*})$ according to (18.11) and computed the corresponding bootstrap estimates $(\widehat{p}_{\text{T}}^{*}, \widehat{p}_{\text{C}}^{*}$ and $\widehat{R}_{\text{Total}}^{*})$. Then a bootstrap approximate 95% confidence interval for R_{Total} is obtained from the 0.025 and 0.975 quantiles of the empirical distribution of $\widehat{R}_{\text{Total}}^{*}$, giving $[\underset{\sim}{R}_{\text{Total}}, \ \widetilde{R}_{\text{Total}}] = [0.66, \ 0.95]$ or $[66\%, \ 95\%]$. It can be claimed, with 95% confidence, that the lift in sales provided by the marketing campaign is between 66% and 95%.

18.5.3   Bayesian Method

The Bayesian method also uses the model in (18.11), but has to be supplemented with prior distributions for the parameters. In this example, we will use Jeffreys prior distributions to represent diffuse prior information; that is,

$$p_{T} \sim \text{BETA}(0.50, 0.50), \quad p_{C} \sim \text{BETA}(0.50, 0.50).$$

Because the number of trials and the number of positive responses are large for each of the binomial distributions in this example, the results are not expected to be highly sensitive to the prior distribution specification.

An MCMC algorithm was used to generate $B = 297{,}000$ draws from the joint posterior distribution of (p_{T}, p_{C}). These draws were then substituted into (18.10) to compute the marginal posterior distribution of R_{Total}. Then a Bayesian 95% credible interval for R_{Total} is obtained from the 0.025 and 0.975 quantiles of the marginal posterior distribution of R_{Total}, giving $[\underset{\sim}{R}_{\text{Total}}, \ \widetilde{R}_{\text{Total}}] = [0.66, \ 0.95]$, the same result as the simulation method given in Section 18.5.2.

18.6   CONFIDENCE INTERVAL FOR THE PROBABILITY OF DETECTION WITH LIMITED HIT/MISS DATA

Nondestructive evaluation (NDE), also known as nondestructive inspection, is commonly used to detect flaws (e.g., cracks or voids) in physical components such as tubes in nuclear power plant heat exchangers, fan blades in aircraft engines, and pipeline welds. Some inspection methods provide quantitative information about flaw size. Others give only binary "hit/miss" results. This is often the case when inspectors (or radiologists in the case of medical inspection) assess images to decide whether or not a flaw (or other characteristic of interest) exists. In this case study, we show two approaches for constructing confidence intervals on the probability of detecting material flaws of different sizes, based on the results of a study to assess the probability of detection of flaws of known sizes.

Example 18.11   Hit/Miss Inspection Data. A study was conducted to estimate the probability of detection (POD) as a function of flaw size in specimens of a particular material. In total, 48 specimens were prepared and a flaw was seeded into each specimen. Four different flaw sizes (5, 10, 15 or 20 mils) were used with 12 specimens randomly assigned for fabrication for each of the four flaw sizes. Each specimen was then subjected to an X-ray inspection. The resulting images were inspected in random order by a trained technician to determine whether or not a flaw was detected. The results are shown in Table 18.10.   ■

18.6.1   Logistic Regression Model

A commonly used model for binary (hit/miss) inspection data is the logistic regression model in which the probability of detection of a flaw of size x is

$$\text{POD}(x) = \Phi_{\text{logis}}(\beta_0 + \beta_1 x) = \frac{\exp(\beta_0 + \beta_1 x)}{1 + \exp(\beta_0 + \beta_1 x)},$$

where β_1 is generally greater than 0, in which case $p \to 0$ as $x \to -\infty$ and $p \to 1$ as $x \to \infty$. In NDE applications, x is often defined to be the log of the flaw size. An alternative parameterization uses

$$\text{POD}(x) = \Phi_{\text{logis}}\left(\frac{x - \mu}{\sigma}\right), \tag{18.12}$$

where $\sigma = 1/\beta_1 > 0$ is known as the POD slope because it controls the steepness of the increasing POD curve, and $\mu = -\beta_0/\beta_1$ is the POD median (i.e., the (log) flaw size at which the POD is 0.50). Constraining σ to be positive assures that $\text{POD}(x)$ is a monotone increasing function of x.

When x is log flaw size, $a_{50} = \exp(\mu)$ is the flaw size at which the POD is 0.50 and $a_{90} = \exp[\mu + \Phi_{\text{logis}}^{-1}(0.90)\sigma]$ is the flaw size at which the POD is 0.90, where

Flaw size (mils)	Misses	Hits
5	12	0
10	6	6
15	0	12
20	0	12

Table 18.10   Hit/miss inspection results for different flaw sizes.

$\Phi_{\text{logis}}^{-1}(p) = \log[p/(1-p)]$ is the p quantile of the standard logistic distribution. A one-sided upper 95% confidence bound on a_{90}, commonly denoted by $a_{90/95}$ in the NDE literature, is a widely used metric for NDE inspection capability, indicating the largest flaw that might be missed in an inspection (in NDE applications it is often stated that the primary concern is the largest flaw that might be missed, not the smallest flaw that can be detected).

18.6.2 Likelihood Method

Analysts attempted to fit model (18.12) to the data in Table 18.10 using two different statistical software packages. The two packages gave different answers and one of them gave warnings of instability. Neither of the software packages provided sensible confidence intervals for the parameters. The problem is that the maximum of the likelihood is not unique because there is only one flaw size that has a mix of hits and misses and the software packages did not detect this data problem.

Figure 18.4 illustrates the problem by showing profile likelihood plots (described in Section 12.5.2) for $a_{50} = \exp(\mu)$, the flaw size having POD equal to 0.50 (left), and the POD slope parameter σ (right). These plots show that the data provide good information about a_{50} but not about σ. The profile likelihood plot for σ, however, shows that values of σ between 0 and 0.05 are nearly equally likely and thus it is possible (although not expected in practice) for the data to have arisen from a model with value of σ that is close to 0 (which would imply that the POD function is close to a step function). From these plots, we obtain the likelihood-based 90% confidence interval for a_{50} to be $[\underset{\sim}{a}_{50}, \ \widetilde{a}_{50}] = [9.0, \ 10.8]$ mils and the likelihood-based confidence interval for σ to be $[\underset{\sim}{\sigma}, \ \widetilde{\sigma}] = [0+, \ 0.17]$, where $0+$ indicates a positive value that is close to 0 (e.g., 10^{-5}).

The left-hand plot in Figure 18.5 shows the POD estimates obtained by fixing the POD slope parameter σ to be 10^{-5} (as an approximation to the smallest possible value for σ) and 0.1706 (the one-sided upper 95% confidence bound for σ). These two POD curve estimates help provide some insight into the effect that σ has on the shape of the estimated POD function and help in understanding the likelihood-based confidence intervals for POD to follow.

Although there is not a unique ML estimate for σ, we have been able to use the likelihood to obtain an upper confidence bound on σ. Correspondingly, we can use the likelihood to compute profile likelihoods and obtain one-sided confidence bounds for POD for given values of flaw size a.

The right-hand plot in Figure 18.5 shows a set of pointwise likelihood-based approximate 90% confidence intervals for the POD as a function of a. Note that for some small flaw sizes

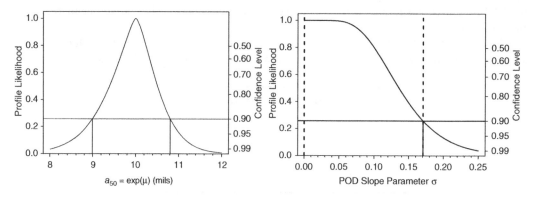

Figure 18.4 Profile likelihood plots for a_{50} (left) and the POD slope parameter σ (right) for the hit/miss inspection data.

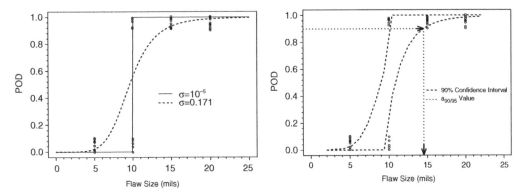

Figure 18.5   POD estimates with POD slope parameter σ set to 10^{-5} and 0.1706 (left) and likelihood-based pointwise 90% confidence intervals for POD (right) for the hit/miss inspection data.

the lower endpoint of the confidence interval is approaching 0 and for some large flaw sizes the upper endpoint is approaching 1. These limiting behaviors correspond to the step function in the left-hand plot in Figure 18.5 with σ approaching 0. Also, the upper endpoint of this two-sided 90% confidence interval for POD can be taken as a one-sided upper 95% confidence bound for POD. Thus the dotted-line rectangle in the plot indicates that the one-sided upper 95% confidence bound on a_{90} is $a_{90/95} = \widetilde{a}_{90} = 14.5$.

18.6.3   Bayesian Method

Because the data provide inadequate information about the POD slope parameter σ, it was decided to improve the estimate of the POD function by using (fortunately) available prior information about this parameter, based on previous experience using similar X-ray images to detect similar flaws in a similar material. In particular, past experience indicated that σ would most likely be between 0.05 and 0.15, suggesting the use of an accordingly informative prior distribution for σ. It was also decided to use a diffuse prior distribution for μ. In particular, the prior distributions were chosen to be $\mu \sim \mathrm{UNIF}(-\infty, \infty)$ and $\sigma \sim {<}\mathrm{LNORM}{>}(0.05, 0.15)$. As described in Section 15.2.2, ${<}\mathrm{LNORM}{>}(0.05, 0.15)$ implies a lognormal distribution for σ that has 0.05 as its 0.005 quantile and 0.15 as its 0.995 quantile (implying that 99% of the probability is between 0.05 and 0.15).

Similar to previous Bayesian-analysis examples, an MCMC algorithm was used to generate 57,000 draws from the joint posterior distribution of μ and σ. The left-hand plot in Figure 18.6 compares the prior and marginal posterior distributions for σ, showing that they are nearly identical. This indicates that the prior distribution dominates the small amount of information in the data in determining the marginal posterior distribution for σ. The right-hand plot in Figure 18.6 shows a sample of 1,000 draws from the joint posterior distribution of $a_{50} = \exp(\mu)$ and σ. This plot shows that the marginal posterior distribution of $a_{50} = \exp(\mu)$ ranges from approximately 8.5 to 12 (recall that the prior distribution for $a_{50} = \exp(\mu)$ ranges from 0 to ∞), indicating that the data have had a large effect on the marginal posterior distribution of $a_{50} = \exp(\mu)$. The 90% credible interval for $a_{50} = \exp(\mu)$ is $[9.12, \quad 10.82]$, which is similar to the likelihood interval that was given in Section 18.6.2. The similarity is because the prior distribution for μ (and thus for $a_{50} = \exp(\mu)$) is diffuse.

The left-hand plot in Figure 18.7 shows the Bayesian estimates and pointwise two-sided 90% credible intervals for POD over a range of flaw sizes. For each flaw size, these were computed, as in previous examples, by substituting the 57,000 draws from the joint posterior

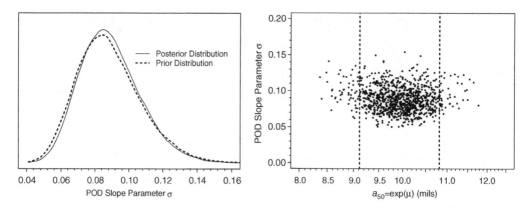

Figure 18.6 Comparison of the prior and the marginal posterior distributions for the POD slope parameter σ (left) and 1,000 draws from the joint posterior distribution for σ and the POD median parameter $a_{50} = \exp(\mu)$ (right) for the hit/miss inspection data. The vertical dashed lines indicate the 90% credible interval for $a_{50} = \exp(\mu)$.

distribution of μ and σ into (18.12) and using the median of the posterior distribution draws as the point estimate and the 0.05 and 0.95 quantiles as the endpoints of the two-sided 90% credible interval. In comparison with the right-hand plot in Figure 18.5, the Bayesian analysis using an informative prior distribution for σ has substantially improved the POD estimation precision, compared to the likelihood method. The $a_{90/95}$ value for the Bayesian analysis, indicated by the dotted-line rectangle in the plot, is $a_{90/95} = \tilde{a}_{90} = 13.5$ (compared with 14.5 for the non-Bayesian analysis). The right-hand plot in Figure 18.7 shows the marginal posterior distribution of a_{90}, with the vertical lines indicating the 90% credible interval for a_{90} (and again, the upper endpoint of this interval $a_{90/95} = \tilde{a}_{90}$ is the one-sided upper 95% confidence bound on a_{90}).

18.7 USING PRIOR INFORMATION TO ESTIMATE THE SERVICE-LIFE DISTRIBUTION OF A ROCKET MOTOR

18.7.1 Rocket Motor Example Revisited

Example 14.14 used likelihood and bootstrap methods to estimate the lifetime distribution of a rocket motor of a missile system and especially $F(20)$, the fraction of rocket motors that

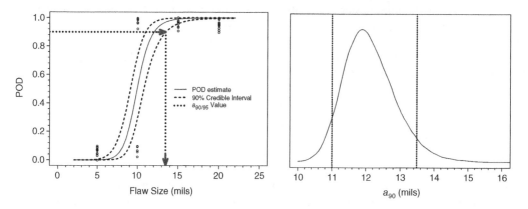

Figure 18.7 Bayesian estimates and pointwise 90% credible intervals for POD (left) and marginal posterior distribution for a_{90} showing 90% credible interval (right), for the hit/miss inspection data.

would be defective by the time they have been in the stockpile for 20 years. Because the data consisted of only censored observations (three left-censored observations corresponding to the failures, and 1,937 right-censored observations corresponding to successful launches) the resulting estimates of distribution characteristics in general and $F(20)$ in particular were so wide as to be essentially useless. Fortunately, there was relevant engineering information pertaining to the Weibull shape parameter that could be used to develop a prior distribution for a Bayesian analysis.

18.7.2   Rocket Motor Prior Information

As described in Olwell and Sorell (2001), based on conversations with engineers, values of the Weibull shape parameter β between 1 and 5 were felt to be plausible, but values outside this range seemed to be out of line with past experience and/or engineering theory. Thus, we assume the (relatively) informative prior distribution $\beta \sim <$LNORM$>(1, 5)$ for β. Because there was no other relevant prior information available, we will—using the type of reparameterization suggested in Section 15.1.5—assume a diffuse prior distribution for $t_{0.10}$, the Weibull 0.10 quantile. In particular, we will use $t_{0.10} \sim$ LUNIF$(5, 400)$ years. For the sake of comparison, we will also consider the diffuse prior distribution $\beta \sim$ UNIF$(0.20, 30)$ for β, with the same diffuse prior distribution as before for $t_{0.10}$.

18.7.3   Rocket Motor Bayesian Estimation Results

Similar to previous examples of Bayesian estimation, an MCMC algorithm was used to generate 297,000 draws from the joint posterior distribution of $t_{0.10}$ and β. The left-hand plot in Figure 18.8 compares the diffuse prior distribution for β with the marginal posterior distribution for β (computed using the diffuse prior distribution for β). We can see from the plots that the data provide new information about β, but the 95% credible interval on this parameter is $[\underset{\sim}{\beta}, \ \widetilde{\beta}] = [2.3, \ 14.7]$, indicating only limited knowledge about its true value. The right-hand plot in Figure 18.8 provides a similar comparison between the prior and the marginal posterior distribution for β using the assumed informative prior distribution for this parameter. In this case, there is good agreement between the two distributions; this is because the informative prior dominates the limited information about β in the data. The 95% credible interval on β

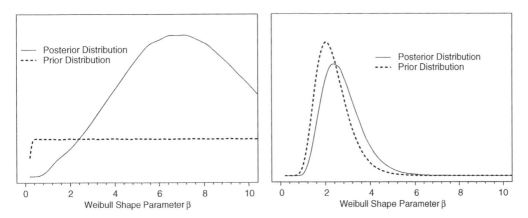

Figure 18.8   Comparison of the posterior and prior distributions for the rocket motor Weibull shape parameter β with diffuse prior information (left) and informative prior information (right) for β.

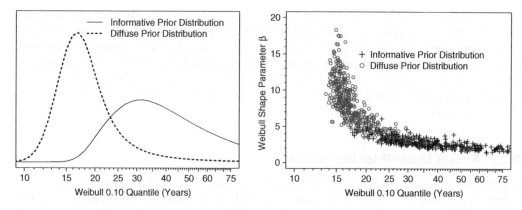

Figure 18.9   Marginal posterior distributions for the 0.10 quantile of the rocket motor lifetime distribution (left) and draws from the joint posterior distributions (right) using diffuse and informative prior distributions for β.

with the informative prior distribution is $[\underset{\sim}{\beta},\ \widetilde{\beta}] = [1.42,\ 4.77]$, considerably narrower than the interval with the diffuse prior distribution.

The left-hand plot of Figure 18.9 comparing the marginal posterior distributions of $t_{0.10}$, the rocket motor lifetime distribution 0.10 quantile, using the diffuse and the informative prior distributions for β, shows strikingly different results. Interestingly, the marginal posterior distribution for β using the informative prior distribution is wider than that using the diffuse prior distribution, resulting in 95% credible intervals for $t_{0.10}$ of $[\underset{\sim}{t}_{0.10},\ \widetilde{t}_{0.10}] = [19.3,\ 163]$ years and $[\underset{\sim}{t}_{0.10},\ \widetilde{t}_{0.10}] = [14.3,\ 48.6]$ years, respectively. The reason for this is apparent from examining the draws from the joint posterior distributions for $t_{0.10}$ and β in the two situations, compared in the right-hand plot of Figure 18.9. The informative prior distribution constrains β to have almost all of its values below 5 years. With the associated strong concentration of values with $t_{0.10}$ less than 20 years largely eliminated (by using the informative in place of the diffuse prior distribution for β), the total probability is redistributed to the larger values of $t_{0.10}$.

18.7.4   Credible Interval for the Proportion of Healthy Rocket Motors after 20 or 30 Years in the Stockpile

The left-hand plot of Figure 18.10 compares marginal posterior distributions for the Weibull $F(20)$ using the diffuse and informative prior distributions. Table 18.11 is an extension of Table 14.8, adding Bayesian credible intervals to the comparison of several non-Bayesian confidence intervals for $F(20)$. As in most previous similar comparisons, the credible interval

Method	Section	95% confidence interval
Wald approximation	12.6.4	[0.027, 0.9999988]
Likelihood ratio	12.5.4	[0.023, 0.999996]
Random-weight bootstrap simple percentile	14.4.2	[0.014, 1.0000000]
Bayesian with diffuse prior distribution	15.2.6	[0.012, 0.9998]
Bayesian with informative prior distribution	15.2.6	[0.0035, 0.116]

Table 18.11   95% confidence and credible intervals for the rocket motor Weibull $F(20)$ using different methods.

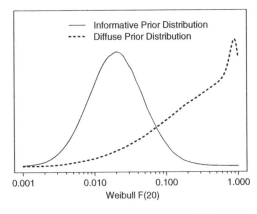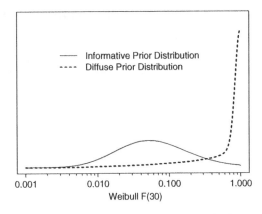

Figure 18.10 Comparison of the marginal posterior distributions of the rocket motor lifetime $F(20)$ (left) and $F(30)$ (right) using diffuse and informative prior distributions.

using the diffuse prior distribution is close to the non-Bayesian intervals. The Bayesian analysis using the informative prior distribution resulted in a credible interval for $F(20)$ that has a much narrower and more meaningful interval length, showing an important advantage of using the Bayesian method when reliable informative prior information is available.

The right-hand side of Figure 18.10 provides a similar comparison for $F(30)$. These plots again show the dramatic effect that the informative prior has on providing more precise inferences. The 95% credible interval for $F(30)$, based on the informative prior distribution, is $[\underset{\sim}{F}(30), \; \widetilde{F}(30)] = [0.0072, \; 0.53]$, which is much wider than the interval for $F(20)$, and indicates, as expected, that reliability could be seriously deteriorated after 30 years in the stockpile.

BIBLIOGRAPHIC NOTES

The LFP model

In addition to the microelectronic example in Section 18.1, first presented in Meeker (1987), the LFP model (also known as the "cure model" and the "defective-subpopulation model") has been used in addressing a variety of problems, including estimating the proportion of treated patients that are cured (e.g., Boag, 1949) and the recidivism (i.e., relapse) rate of those released from prison (e.g., Maltz and McCleary, 1977). Also, Maller and Zhou (1996) provide a detailed description of "cure models" for a number of biomedical applications. Trindade (1991) gave another microelectronic example.

Random-effects models

Hamada and Weerahandi (2000) conducted a small simulation study that showed (for the cases considered) that the GPQ method has coverage probabilities for variance components and functions of variance components that are generally close to their nominal confidence levels. Chiang (2001) independently developed a similar method that he called the "surrogate variable method" for obtaining confidence intervals for variance components. Liao et al. (2005) show how to use GPQ methods to compute two-sided tolerance intervals and one-sided tolerance bounds for general balanced mixed models and unbalanced one-way random-effects models.

Confidence interval for the proportion in the center of a distribution

Corresponding to the GPQ procedure used in Section 18.4, Hannig et al. (2006) show how to use the GPQ method to compute a confidence interval for the proportion in a specified two-sided interval of a distribution (one of many applications in the paper) and show that the procedure has asymptotically correct coverage probability. The coverage properties of the procedure were evaluated for finite sample sizes in Patterson et al. (2004) and shown to be satisfactory for most practical applications.

Epilogue

The concept of this book was born about 40 years ago. After agreeing to go for it with Wiley, our publishers, it took two of us 12 years to put together the first edition, published in 1991. After a 14-year hiatus, it took another 12 years for the three of us to get this second edition into print. Why so long? Admittedly, our day jobs had to be our first priority. But another reason is that there is so much to say and limited space to say it precisely and understandably. Statistical intervals, in fact, illustrate the proverbial "bottomless pit" with seemingly endless opportunities for adding to the exposition. We have tried to provide both practitioners and researchers the specific tools they need to quantify uncertainty in their data via statistical intervals for commonly occurring situations and to provide added guidance in constructing such intervals in more complex and less frequently occurring situations. Thus, we will conclude by reemphasizing some of our major ideas, and adding a few anecdotes about:

- The importance of calculating the "right" statistical interval.
- The role of statistical intervals versus other forms of inference.
- The limitations of statistical inference.

STATISTICAL INTERVALS: VIVE LA DIFFÉRENCE!

We have presented a wide variety of statistical intervals, and tried to explain the situations for which each is appropriate, how each is calculated, and the underlying assumptions. Some of these—such as confidence intervals for the population or process mean or standard deviation (assuming a normal distribution) or for a population or process proportion (assuming a binomial distribution)—are well known to users of statistical methods. Others, such as a confidence interval for the proportion below or above a threshold, or a prediction interval to contain one

Statistical Intervals: A Guide for Practitioners and Researchers, Second Edition.
William Q. Meeker, Gerald J. Hahn and Luis A. Escobar.
© 2017 John Wiley & Sons, Inc. Published 2017 by John Wiley & Sons, Inc.
Companion Website: www.wiley.com/go/meeker/intervals

or more future observations, are surprisingly unfamiliar, even to many professional statisticians. We say "surprisingly" because, as evidenced by the case studies in Chapters 11 and 18, these intervals are frequently needed in applications. We have suggested that the reasons for this unfamiliarity include tradition, the relatively advanced nature of the underlying mathematics (generally *not* needed to *use* the intervals), and the complexity of some of the calculations (becoming increasingly less relevant with the rapid development of needed computer algorithms and computational capabilities).

In this regard, we recount our experience in interviewing, for positions in industry, promising recent PhDs in statistics. As part of the screening process, we often ask the following two questions, typical of those we are asked by our clients:

- An appliance, built in large quantities, is required to have a noise level of less than 50 decibels. A sample of eight units has resulted in the following readings:

$$46.4, 46.7, 46.9, 47.0, 47.0, 47.2, 47.6, 48.1.$$

What can one conclude with a "high degree of assuredness" about the percentage of units manufactured during the year that fail to meet the 50-decibel threshold? What important assumptions, that are implicit in our inferences, do we need emphasize?

- Consider again the preceding noise measurements. However, now assume that a single added appliance is to be selected. What can one say, with a high degree of confidence, about the maximum noise that one may reasonably expect from this ninth appliance, and what added assumptions does this require?

We find that the great majority of interviewees either are unable to tell us how they would go about constructing the desired intervals or, worse still, answer incorrectly (e.g., by proposing a confidence interval for the mean in response to the second question). In contrast, diligent readers of this book will have no trouble passing our test!

THE ROLE OF STATISTICAL INTERVALS

We feel strongly that, before reporting or using any form of statistical inference, an analyst should carefully examine and plot the data. Numerous methods and software for exploratory data analysis are available for this purpose and should be applied before one proceeds to more sophisticated evaluations, should these seem necessary.

However, when we do try to draw formal conclusions about a population or process from an appropriately selected random sample, statistical intervals play a central role in quantifying uncertainty, and provide an important supplement to point estimates. In our experience, such intervals are much more useful than significance or hypothesis tests. As previously indicated, we believe that this is because few statistical hypotheses hold exactly. Moreover, one can reject almost any statistical hypotheses by taking a sufficiently large sample—and avoid disproving a hypothesis by having a small enough sample, or even no sample at all.

THE LIMITATIONS OF STATISTICAL INFERENCE

Starting with Chapter 1, and throughout this book, we have stressed the basic assumptions underlying statistical inferences about a sampled population or process, especially in analytic studies. We conclude with a recent example.

In a mail survey, the 730 member families of a congregation were asked the following question: "On an overall basis, do you feel the minister is doing a good job (answer yes or no)?" Among the 105 respondents, 58 answered "yes," and 47 said "no." Because the results were from "sample data," one of us was asked to make a statement that incorporated the "statistical uncertainty" about the proportion of families in the congregation that favored the minister. In this example, there was a well-defined population (the 730 families) which also comprised the sample frame. In fact, if all families had responded, one would have had data from the entire population and there would be no statistical uncertainty, at least with regard to the stated viewpoints at the time of the survey. Moreover, *if* the respondents could be considered as randomly selected from the congregation, one could apply the methods presented in Chapter 6 for drawing inferences about the population proportion (perhaps including an appropriate finite population adjustment for the fact that the sample constituted an appreciable part of the population).

In reality, however, the sample was far from random; the respondents, in fact, were self-selected. Thus, those who felt most strongly about the issue, and, perhaps, those most active in the congregation (and in organizing the survey), were the ones who were most likely to respond, or urge their similarly viewed friends to do so. Without further study, little can be said about how representative this nonrandom sample really is of the population as a whole. Therefore, we felt that it would be misleading to calculate a statistical interval to contain the proportion of all congregants favoring the minister. Instead, we proposed that the results be presented as they stand, with appropriate comments concerning their possible inadequacy and encouragement for recanvassing nonrespondents in a follow-up study.

Where does this leave us? Despite our enthusiasm for statistical intervals, we feel that there are numerous situations where the practitioner is better served by not calculating such intervals and by emphasizing instead the limitations of the available information, perhaps suggesting how improved data can be obtained. Moreover, when statistical intervals are calculated in such situations, one needs to stress that they provide only a lower bound on the total uncertainty.

Learning is an iterative process. Sometimes, the available data are sufficient to draw meaningful conclusions; statistical intervals may then play an important role in quantifying uncertainty. Often, however, current information provides only a stepping stone to further study. In such cases, statistical intervals are frequently useful in describing what is known (or, indeed, unknown). This understanding can help practitioners decide on the next step in their investigations.

Appendix *A*

Notation and Acronyms

This appendix outlines most of the notation that is used in this book. Some symbols, when they are only used within a particular section, are not listed here but are defined where used. Generally, we have tried to use notation that is most commonly used in the statistical literature. In some cases we needed to use the same symbol for more than one purpose. These are explained in the following list, and in such usage we have been careful to make sure that the meaning of the notation is clear from the context of the usage.

Notation for probability distribution cdfs and quantiles is given in the following list when such notation is widely used in the book chapters (e.g., pnorm and qnorm refer, respectively, to the normal distribution cdf and quantile function). Technical details about these and other distributions are given in Appendix C and summarized in Table C.1.

$*$	Used to indicate a bootstrap sample value (e.g., $\widehat{\theta}_j^*$ is a bootstrap estimate of θ computed from the jth bootstrap sample). Also, in Bayesian inferences, it is used to indicate a sample draw from a posterior distribution.
$\square$	Indicates planning value for a parameter; used in choosing an appropriate sample size (e.g., $\sigma^\square$ is used in computing a sample size when σ is unknown).
$\dagger$	Indicates a specified limit for a parameter (e.g., one might need to demonstrate that $y_p \geq y_p^\dagger$) or a value that has been adjusted (e.g., $\widehat{\pi}^\dagger$ in Chapter 6).
$\widehat{}$	Denotes an estimator or estimate (e.g., $\widehat{\lambda}$ is an estimate of λ).
$\widetilde{}$	Indicates endpoints of a statistical interval or one-sided statistical bound (e.g., $[\underset{\sim}{\lambda},\ \widetilde{\lambda}]$ is a confidence interval for λ; $\underset{\sim}{\lambda}$ and $\widetilde{\lambda}$ are one-sided lower and upper confidence bounds for λ, respectively).

Statistical Intervals: A Guide for Practitioners and Researchers, Second Edition.
William Q. Meeker, Gerald J. Hahn and Luis A. Escobar.
© 2017 John Wiley & Sons, Inc. Published 2017 by John Wiley & Sons, Inc.
Companion Website: www.wiley.com/go/meeker/intervals

–	$F(z^-)$ is equal to $F(z)$ for continuous random variables and equal to the limit of $F(w)$ when w approaches z from below for discrete random variables.
$\sim$	Distributed as (e.g., $W \sim \chi^2(r)$ indicates that W has a chi-square distribution with r degrees of freedom).
$\overset{\cdot}{\sim}$	Approximately distributed as (e.g., $W \overset{\cdot}{\sim} \chi^2(r)$).
$\approx$	Approximately equal to (e.g., $n \approx 11$).
$100(1-\alpha)\%$	Confidence level (percent).
β	Probability content of a tolerance interval. Also the shape parameter for a Weibull distribution and the rate parameter for a gamma distribution.
β_i	Regression model parameter in Chapters 4, 17, and 18.
$\mathcal{B}(a,\,b)$	Beta function; $\mathcal{B}(a,\,b) = \int_0^1 w^{a-1}(1-w)^{b-1}\,dw = \Gamma(a)\Gamma(b)/\Gamma(a+b), a > 0, b > 0.$
γ	The power parameter of a power transformation in Chapter 4 (e.g., x^γ is power transformation of x).
$\Gamma(z)$	Gamma function; $\Gamma(z) = \int_0^\infty t^{z-1} \exp(-t)\,dt,\ z > 0.$ $\Gamma(z) = (z-1)!$, where z is a nonnegative integer with the convention $0! = 1$.
δ	$\delta = 1 - p_{\mathrm{dem}}$ in Chapter 9. Also the noncentrality parameter of a noncentral t-distribution.
θ	Generic scalar parameter. Also the mean of an exponential distribution.
$\boldsymbol{\theta}$	Generic parameter vector.
λ	Exponential and Poisson distribution rate parameter.
μ	Location parameter of a location-scale distribution (e.g., the mean of a normal or logistic distribution). Also, $\eta = \exp(\mu)$ is the scale parameter of the corresponding log-location-scale distribution.
ν	$\nu - 1$ is the number of extreme observations to be removed from the upper (or lower) end of the sample of size n to obtain the order statistic that provides the desired one-sided upper (lower) confidence bound for x_p. Also, $\nu - 2$ is the total number extreme observations to be removed from the upper and lower ends of the sample of size n to obtain the order statistics that provide the desired two-sided tolerance interval.
η	Scale parameter of a log-location-scale distribution (e.g., for a Weibull distribution).
π	The probability of a single randomly selected unit being nonconforming or the probability of some other particular event.
ρ	Correlation coefficient between two random variables.
σ	Scale parameter for a location-scale distribution; also the standard deviation of a normal distribution.
τ	Precision $\tau = 1/\sigma^2$, where σ^2 is the variance.
$\phi(z)$	The pdf of a standardized location-scale distribution (i.e., $\phi(z) = d\Phi(z)/dz$). See Table C.2.

$\Phi(z)$	The cdf for a standardized location-scale distribution (i.e., $\Pr(Z \leq z) = \Phi(z)$, where $Z = (X - \mu)/\sigma$ and X has a location-scale distribution with location parameter μ and scale parameter σ). See Table C.2.
$\Phi^{-1}(p)$	p quantile of a standardized location-scale distribution. See Table C.2.
$\chi^2(r)$	Chi-square distribution with r degrees of freedom.
$\chi^2_{(p;r)}$	p quantile of a chi-square distribution with r degrees of freedom; equivalent to `qchisq(p;r)`.
ω	Weight for linear interpolation in Chapter 5. Also an observation weight in Chapter 13.
a	A beta distribution shape parameter, the gamma distribution shape parameter, a general interval endpoint, and also an acceleration constant in Chapter 13.
ACF	Autocorrelation function.
ANOVA	Analysis of variance.
b	A beta distribution shape parameter, a gamma distribution rate parameter, and also a general interval endpoint.
B	Number of bootstrap samples. Also the number of sample draws from a posterior distribution.
BC	Bias-corrected bootstrap procedure in Chapter 13.
BCa	Bias-corrected and accelerated bootstrap procedure in Chapter 13.
$\text{BETA}(a, b)$	Beta distribution with shape parameters a and b.
$\text{BINOM}(n, \pi)$	Binomial distribution for the number of nonconforming (or conforming) units in a random sample of size n and probability π of observing a nonconforming (or conforming) unit in a single trial.
$c_{(\cdot)}, c'_{(\cdot)}, c_{L(\cdot)}, c_{U(\cdot)}, c'_{L(\cdot)}, c'_{U(\cdot)}$	Factors for computing statistical intervals for a normal distribution in Chapter 3.
c	Critical number of nonconforming items in an attribute demonstration test. If the observed number nonconforming is less than or equal to c, the demonstration is successful.
cdf	Cumulative distribution function for a random variable (i.e., $\Pr(X \leq x)$).
$\text{CP}(\boldsymbol{\theta})$	Coverage probability as a function of the parameter(s) $\boldsymbol{\theta}$.
$\text{CPKM}(n, \ell, u, m, k)$	The distribution-free probability that the interval defined by the order statistics $[x_{(\ell)}, x_{(u)}]$ from a random sample of size n will contain at least k observations from a subsequent independent random sample of size m from the same distribution. For one-sided lower (upper) bounds, set $u = n + 1$ ($\ell = 0$).
$\text{CPTI}(n, \ell, u, \beta)$	The distribution-free probability that the interval defined by the order statistics $[x_{(\ell)}, x_{(u)}]$ from a random sample of size n will cover at least a proportion β of the sampled distribution. For one-sided lower (upper) bounds, set $u = n + 1$ ($\ell = 0$).
$\text{CPXP}(n, \ell, u, p)$	The distribution-free probability that the interval defined by the order statistics $[x_{(\ell)}, x_{(u)}]$ from a random sample of size n will contain the p quantile of the distribution. For one-sided lower (upper) bounds, set $u = n + 1$ ($\ell = 0$).

$\text{CPYJ}(n, \ell, u, m, j)$	The distribution-free probability that the interval defined by the order statistics $[x_{(\ell)}, x_{(u)}]$ from a random sample of size n will contain $Y_{(j)}$, the jth largest observation from a subsequent independent random sample of size m from the same distribution. For one-sided lower (upper) bounds, set $u = n + 1$ ($\ell = 0$).
d	Desired confidence interval half-width.
	Degrees of freedom.
D	Number of nonconforming units in a finite population of size N.
DATA	A data set (including a response and other information such as explanatory variables, a censoring indicator, or a frequency count).
DATA^*	A simulated or bootstrap data set.
DATA_i^*	The ith simulated or bootstrap data set.
$\text{EXP}(\theta)$	Exponential distribution with mean parameter θ.
$f(\boldsymbol{\theta})$	Prior probability density function of a random variable.
$f(\boldsymbol{\theta}|\text{DATA})$	Posterior distribution of $\boldsymbol{\theta}$ given a prior distribution $f(\boldsymbol{\theta})$ and DATA.
$f(\text{DATA}|\boldsymbol{\theta})$	Likelihood for the DATA given the parameters $\boldsymbol{\theta}$. See also $L(\text{DATA}|\boldsymbol{\theta})$.
$F(x) = \Pr(X \le x)$	The cdf for a random variable X.
$F_{(p; r_1, r_2)}$	p quantile of Snedecor's F-distribution with r_1 numerator and r_2 denominator degrees of freedom; equivalent to $\texttt{qf}(p; r_1, r_2)$.
$g_{(1-\alpha; p, n)}$	Factors from Tables J.5a–J.5b used to compute $100(1 - \alpha)\%$ two-sided tolerance intervals to control the center of a normal distribution.
$g''_{(1-\alpha; p, n)}$	Factors from Tables J.6a–J.6b used to compute $100(1 - \alpha)\%$ two-sided tolerance intervals to control both tails of a normal distribution.
$g'_{(\gamma; p, n)}$	Factors from Tables J.7a–J.7d used to compute two-sided confidence intervals or one-sided confidence bounds for quantiles from a normal distribution; also used to compute one-sided tolerance bounds for a normal distribution.
$\text{GAMMA}(a, b)$	Gamma distribution with shape parameter a and rate parameter b.
GFI	Generalized fiducial inference.
$\text{GNG}(\mu, \sigma, \lambda)$	Generalized gamma distribution with scale parameter $\exp(\mu)$ and shape parameters σ and λ.
GPQ	Generalized pivotal quantity.
$h(\widehat{\boldsymbol{\theta}}; \boldsymbol{\theta})$	Sampling distribution pdf for the estimator $\widehat{\boldsymbol{\theta}}$.
$\text{HYPER}(n, D, N)$	Hypergeometric distribution, for which n is sample size and N is the population size which initially contains D nonconforming units.
iid	Independent and identically distributed.
$\text{I}[A]$	Indicator function for statement A. The function $\text{I}[A]$ is equal to 1 when the statement A is true and 0 otherwise.
ℓ, u	Indices for particular ordered observations.

$L(\text{DATA}|\boldsymbol{\theta})$	The likelihood of the DATA as a function of $\boldsymbol{\theta}$. See also $f(\text{DATA}|\boldsymbol{\theta})$.
$\text{LNORM}(\mu, \sigma)$	Lognormal distribution with scale parameter $\exp(\mu)$ and shape parameter σ.
$<\text{LNORM}>(a, b)$	A lognormal (prior) distribution that has 99% of its probability between a and b, with $0 < a < b$.
$\log(a)$	Natural (base e) logarithm of a for $a > 0$.
$\text{LOGIS}(\mu, \sigma)$	Logistic distribution with location parameter μ and scale parameter σ.
$\text{LLOGIS}(\mu, \sigma)$	Log-logistic distribution with scale parameter $\exp(\mu)$ and shape parameter σ.
$\text{LUNIF}(a, b)$	Log-uniform distribution with location parameters a and b.
m	Sample size of a future sample.
MC	Monte Carlo.
MCMC	Markov chain Monte Carlo.
ML	Maximum likelihood.
n	Sample size of a previous sample.
normCenterTI	A StatInt R function to compute a control-the-center tolerance interval for a normal distribution.
normTailCI	A StatInt R function to compute confidence intervals for a normal distribution tail probability.
N	Size of a finite population.
$\text{NORM}(\mu, \sigma)$	Normal distribution with mean μ and standard deviation σ.
$<\text{NORM}>(a, b)$	A normal (prior) distribution that has 99% of its probability between a and b.
pdf	Probability density function for a continuous random variable.
pmf	Probability mass function for a discrete random variable.
p_{dem}	Probability of successful demonstration.
p_{GT}	Probability that a random variable X is greater than a specified number x (i.e., $p_{GT} = \Pr(X > x)$).
p_{LE}	Probability that a random variable X is less or equal than a specified number x (i.e., $p_{LE} = \Pr(X \leq x)$).
$\text{POIS}(n\lambda)$	Poisson distribution with exposure amount n and rate parameter λ.
PQ	Pivotal quantity.
p_{tL}	Lower tail probability for a control-both-tails tolerance interval.
p_{tU}	Upper tail probability for a control-both-tails tolerance interval.
$\text{pbeta}(x; a, b)$	The cdf for a beta distribution with shape parameters a and b.
$\text{pbinom}(x; n, \pi)$	The cdf for a binomial distribution with probability parameter π and sample size n.
$\text{pchisq}(p; r)$	The cdf for a chi-square distribution with r degrees of freedom.
$\text{pf}(x; r_1, r_2)$	The cdf for Snedecor's F-distribution with r_1 numerator and r_2 denominator degrees of freedom.
$\text{pgamma}(x; a, b)$	The cdf for a gamma distribution with shape parameter a and rate parameter b.
$\text{phyper}(x; n, D, N)$	The cdf for a hypergeometric distribution. The sample size is n; N is the finite population size which initially contains D nonconforming units.

$\texttt{plnorm}(x; \mu, \sigma)$	The cdf for a lognormal distribution with scale parameter $\exp(\mu)$ and shape parameter σ.
$\texttt{pnhyper}(x; k, D, N)$	The cdf for a negative hypergeometric distribution, where x is the number of conforming units observed in sequential sampling without replacement until exactly k nonconforming units have been drawn from a population of size N which initially contains D nonconforming units.
$\texttt{pnorm}(p; \mu, \sigma)$	The cdf for a normal distribution with mean μ and standard deviation σ.
$\texttt{ppois}(x; n\lambda)$	The cdf for a Poisson distribution with given exposure n and rate parameter λ.
$\texttt{pt}(x; r)$	The cdf for Student's t-distribution with r degrees of freedom.
$\texttt{pt}(x; r, \delta)$	The cdf for a noncentral t-distribution with r degrees of freedom and noncentrality parameter δ.
q	Length of a parameter vector.
$\texttt{qbeta}(p; a, b)$	p quantile for a beta distribution with shape parameters a and b.
$\texttt{qbinom}(p; n, \pi)$	p quantile for a binomial distribution with probability parameter π and sample size n.
$\texttt{qchisq}(p; r)$	p quantile for a chi-square distribution with r degrees of freedom; equivalent to $\chi^2_{(p;r)}$.
$\texttt{qf}(p; r_1, r_2)$	p quantile for Snedecor's F-distribution with r_1 numerator and r_2 denominator degrees of freedom; equivalent to $F_{(p; r_1, r_2)}$.
$\texttt{qgamma}(p; a, b)$	p quantile for a gamma distribution with shape parameter a and rate parameter b.
$\texttt{qhyper}(p; n, D, N)$	p quantile for a hypergeometric distribution. The sample size is n; N is the finite population size which initially contains D nonconforming units.
$\texttt{qnhyper}(p; k, D, N)$	p quantile for the negative hypergeometric distribution, involving sequential sampling without replacement until exactly k nonconforming units have been drawn from a population of size N which initially contains D nonconforming units.
$\texttt{qnorm}(p; \mu, \sigma)$	p quantile for a normal distribution with mean μ and standard deviation σ; equivalent to $z_{(p)}$.
$\texttt{qpois}(x; n\lambda)$	p quantile for a Poisson distribution with given exposure n and rate parameter λ.
$\texttt{qt}(p; r)$	p quantile for Student's t-distribution with r degrees of freedom; equivalent to $t_{(p;r)}$.
$\texttt{qt}(p; r, \delta)$	p quantile for a noncentral t-distribution with r degrees of freedom and noncentrality parameter δ; equivalent to $t_{(p; r, \delta)}$.
$r_{(1-\alpha; m, m, n)}$	Factors from Table J.8 used to compute $100(1 - \alpha)\%$ two-sided simultaneous prediction intervals to contain m out of m future observations from a normal distribution.
$r'_{(1-\alpha; m, m, n)}$	Factors from Table J.9 used to compute $100(1 - \alpha)\%$ one-sided simultaneous prediction bounds to contain m out of m future observations from a normal distribution.
$R(\theta)$	Profile relative likelihood for parameter θ.
s or s_n	Sample standard deviation based on a previous sample of size n; an estimate of the normal distribution σ.

S_m	Sample standard deviation of a future sample of size m (estimate for σ).
$[\underset{\sim}{S}_m,\ \widetilde{S}_m]$	Prediction interval or bounds to contain the sample standard deviation computed from m future observations.
se	Standard error.
$\mathrm{SEV}(\mu, \sigma)$	Smallest extreme value distribution with location parameter μ and scale parameter σ.
$t_{(p;r)}$	p quantile for Student's t-distribution with r degrees of freedom; equivalent to $\mathrm{qt}(p; r)$.
$t_{(p;r,\delta)}$	p quantile for a noncentral t-distribution with r degrees of freedom and noncentrality parameter δ; equivalent to $\mathrm{qt}(p; r, \delta)$.
$[\underset{\sim}{T}_\beta,\ \widetilde{T}_\beta]$	Tolerance interval to contain at least a proportion β of the sampled distribution.
$\underset{\sim}{T}'_\beta$	One-sided lower tolerance bound to be exceeded by at least a proportion β of the sampled distribution; equivalent to $\underset{\sim}{x}_{(1-\beta)}$, a one-sided lower confidence bound on the $1-\beta$ quantile of the previously sampled distribution.
$\widetilde{T}'_\beta$	One-sided upper tolerance bound to exceed at least a proportion β of the previously sampled distribution; equivalent to $\widetilde{x}_\beta$, a one-sided upper confidence bound on the β quantile of the previously sampled distribution.
$[\underset{\sim}{T}_{p_{tL}},\ \widetilde{T}_{p_{tU}}]$	Tolerance interval to control both tails of a distribution. See definitions for p_{tL} and p_{tL}.
U	A $[0, 1]$ uniform random variable (in most places, but also used as a general random vector in Appendix D).
$\mathrm{UNIF}(a, b)$	Uniform distribution over the range $[a, b]$.
$v(\boldsymbol{\theta})$	A general scalar function of a parameter vector $\boldsymbol{\theta}$.
$\underset{\sim}{v}(\boldsymbol{x}, 1-\alpha)$ and $\widetilde{v}(\boldsymbol{x}, 1-\alpha)$	Endpoints of a $100(1-\alpha)\%$ confidence interval for $v(\boldsymbol{\theta})$; that is, $[\underset{\sim}{v}(\boldsymbol{x}, 1-\alpha),\ \widetilde{v}(\boldsymbol{x}, 1-\alpha)]$ is a $100(1-\alpha)\%$ confidence interval for $v(\boldsymbol{\theta})$.
$\underset{\sim}{v}'(\boldsymbol{x}, 1-\alpha)$	One-sided lower $100(1-\alpha)\%$ confidence bound for $v(\boldsymbol{\theta})$.
$\widetilde{v}'(\boldsymbol{x}, 1-\alpha)$	One-sided upper $100(1-\alpha)\%$ confidence bound for $v(\boldsymbol{\theta})$.
W	Prediction interval relative width.
x	Single observed quantity (e.g., the number of binomial nonconforming units in a sample of size n or the number of Poisson occurrences in n units of exposure).
$x_1, \ldots, x_n$	Previous observations from a sample of size n.
$x_{(1)}, \ldots, x_{(n)}$	Order statistics; previous sample observations, ordered from smallest $x_{(1)}$ to largest $x_{(n)}$.
$\bar{x}$	Sample mean of a previous sample; estimate of the mean of a sampled distribution.
x_p	p quantile of the distribution of X.
$\widehat{x}_p$	Estimate for the p quantile of the distribution of X.
$[\underset{\sim}{x}_p,\ \widetilde{x}_p]$	Confidence interval or one-sided confidence bounds for x_p.
$\widehat{X}_p$	Estimator for the p quantile of the distribution of X.
$[\underset{\sim}{\widehat{X}}_p,\ \widetilde{\widehat{X}}_p]$	Prediction interval or bounds to contain $\widehat{X}_p$.

$X^2_{(r)}$	A chi-square random variable with r degrees of freedom.
y_p	p quantile of the distribution of Y.
$\widehat{y}_p$	Estimate for the p quantile of the distribution of Y.
$[\underset{\sim}{y}_p, \ \widetilde{y}_p]$	Confidence interval or one-sided confidence bounds for y_p.
$y^{(\gamma)}$	Transformed data value.
$[\underset{\sim}{y}^{(\gamma)}, \ \widetilde{y}^{(\gamma)}]$	Statistical interval on a transformed data scale.
Y	Single future random variable (e.g., number of binomial nonconforming units in a future sample of size m).
$Y_1, \ldots, Y_m$	Future sample observations.
$Y_{(1)}, \ldots, Y_{(m)}$	Future sample observations, ordered from smallest to largest (order statistics).
$[\underset{\sim}{Y}, \ \widetilde{Y}]$	Prediction interval or bounds to contain a single future observation Y.
$\bar{Y}_m$	Sample mean of a future sample of m observations.
$[\underset{\sim}{\bar{Y}}_m, \ \widetilde{\bar{Y}}_m]$	Prediction interval or bounds to contain the mean of a future sample of m observations.
$\widehat{Y}_p$	Estimator for the p quantile of the distribution of Y.
$[\underset{\sim}{\widehat{Y}}_p, \ \widetilde{\widehat{Y}}_p]$	Prediction interval or bounds to contain $\widehat{Y}_p$.
$[\underset{\sim}{Y}_{(j)}, \ \widetilde{Y}_{(j)}]$	Prediction interval or bounds to contain the jth largest observation (the jth order statistic) from a future sample of m observations.
$[\underset{\sim}{Y}_{k;m}, \ \widetilde{Y}_{k;m}]$	Simultaneous prediction intervals or one-sided bounds to contain at least k out of m future observations.
$z_{(p)}$	p quantile of the standard normal distribution; equivalent to $\texttt{qnorm}(p)$.
$z_{\widehat{\theta}_{(\gamma)}}$	γ quantile of the distribution of $Z_{\widehat{\theta}}$.
$Z_{\widehat{\theta}}$	A pivotal quantity or a generalized pivotal quantity for θ (a parameter or other quantity of interest).

Appendix *B*

Generic Definition of Statistical Intervals and Formulas for Computing Coverage Probabilities

B.1 INTRODUCTION

This appendix provides formal definitions of two-sided confidence intervals, tolerance intervals, prediction intervals, and the corresponding one-sided bounds. A procedure to compute a statistical interval (or bound) defines a computational method by which interval endpoint(s) are obtained from observed data. This appendix also provides methods to compute the "coverage probability" (CP) associated with a procedure for calculating a statistical interval, the CP being is the probability that the interval obtained using the procedure actually contains what it is claimed to contain, as a function of the procedure's definition. General expressions are followed by some specific examples that can be programmed directly for purposes of computation. Both analytical and simulation-based methods are described.

Knowledge of the CP of an interval is useful for several purposes:

- In some cases, exact (or approximate) statistical interval procedures can be obtained by finding interval endpoints that result in a CP that is equal to (or approximates) the desired nominal confidence level.

- To evaluate the adequacy of an approximate interval procedure.

- To calibrate a procedure (i.e., to improve the approximation). This topic is discussed in Section B.8 and applied in Chapters 6 and 7.

Statistical Intervals: A Guide for Practitioners and Researchers, Second Edition.
William Q. Meeker, Gerald J. Hahn and Luis A. Escobar.
© 2017 John Wiley & Sons, Inc. Published 2017 by John Wiley & Sons, Inc.
Companion Website: www.wiley.com/go/meeker/intervals

This appendix describes the definition and CPs for:

- Confidence intervals and one-sided confidence bounds (Section B.2).

- Two-sided control-the-center tolerance intervals (Section B.3).

- Two-sided tolerance intervals to control both tails of a distribution (Section B.4).

- One-sided tolerance bounds (Section B.5).

- Two-sided prediction intervals and one-sided prediction bounds (Section B.6).

- Two-sided simultaneous prediction intervals and one-sided simultaneous prediction bounds (Section B.7).

The appendix concludes with a brief introduction to the concept of statistical interval calibration (Section B.8).

B.2   TWO-SIDED CONFIDENCE INTERVALS AND ONE-SIDED CONFIDENCE BOUNDS FOR DISTRIBUTION PARAMETERS OR A FUNCTION OF PARAMETERS

B.2.1   Two-Sided Confidence Interval Definition

In statistical studies we are often interested in estimating and constructing two-sided confidence intervals for quantities such as (a) distribution parameters, (b) distribution quantiles, or (c) probabilities of events or other functions of the distribution parameters.

First, we introduce some generic notation (in contrast to the specific notation used in the main parts of this book) for a two-sided confidence interval. Let

$$\left[\underset{\sim}{v}(\boldsymbol{x}, 1 - \alpha), \ \widetilde{v}(\boldsymbol{x}, 1 - \alpha) \right]$$

denote a confidence interval for $v(\boldsymbol{\theta})$, a scalar function of a parameter vector $\boldsymbol{\theta}$. Here $\boldsymbol{x}$ denotes the observed data used to construct the confidence interval, and $100(1 - \alpha)\%$ is the nominal confidence level, expressed as a percentage. To simplify the presentation, in the rest of this section we will suppress the dependency of the interval endpoints on $1 - \alpha$. Then, the particular functions $\underset{\sim}{v}(\boldsymbol{x})$ and $\widetilde{v}(\boldsymbol{x})$ define what we will call a *confidence interval procedure*.

An *exact* confidence interval procedure is one for which the CP is exactly equal to the nominal confidence level $100(1 - \alpha)\%$. That is,

$$\mathrm{CP}(\boldsymbol{\theta}) = \Pr\left[\underset{\sim}{v}(\boldsymbol{X}) \leq v(\boldsymbol{\theta}) \leq \widetilde{v}(\boldsymbol{X}) \right] = 1 - \alpha, \tag{B.1}$$

where $\boldsymbol{X}$ denotes the not-yet-observed data that will be used to construct the confidence interval. Note that in this expression $v(\boldsymbol{\theta})$ is fixed but $\underset{\sim}{v}(\boldsymbol{X})$ and $\widetilde{v}(\boldsymbol{X})$ are random because they depend on the random data $\boldsymbol{X}$.

In some situations (e.g., when the data are discrete) an exact confidence interval procedure may not be available. Then one can use an approximate procedure for which the CP in (B.1) is approximately equal to the nominal confidence level $1 - \alpha$. In some applications it may be desirable to have a confidence interval procedure that is conservative in the sense that the CP is greater than or equal to the nominal confidence level.

Chapters 3–7 show how to compute a confidence interval $\left[\underset{\sim}{v}(\boldsymbol{x}), \ \widetilde{v}(\boldsymbol{x}) \right]$ for given data $\boldsymbol{x}$ and different assumed probability distributions. Chapters 12–15 and Appendices D–F describe general approaches for defining confidence interval procedures to determine the interval endpoints $\underset{\sim}{v}(\boldsymbol{x})$ and $\widetilde{v}(\boldsymbol{x})$.

B.2.2 One-Sided Confidence Bound Definition

In other situations, one needs to compute one-sided confidence bounds. As described in Section 2.7, it is possible to combine two one-sided confidence bounds to create a two-sided confidence interval. Let $\underset{\sim}{v}'(\boldsymbol{x})$ and $\widetilde{v}'(\boldsymbol{x})$ denote, respectively, one-sided lower and upper confidence bounds for $v(\boldsymbol{\theta})$. The probability statement defining an exact one-sided lower confidence bound procedure is

$$\Pr\left[\underset{\sim}{v}'(\boldsymbol{X}) \le v(\boldsymbol{\theta})\right] = 1 - \alpha_L. \tag{B.2}$$

The corresponding one-sided upper confidence bound procedure is

$$\Pr[\widetilde{v}'(\boldsymbol{X}) \ge v(\boldsymbol{\theta})] = 1 - \alpha_U. \tag{B.3}$$

There are similar definitions for approximate and conservative confidence bound procedures.

Assuming exact procedures and $0 < \alpha < 0.50$, $\underset{\sim}{v}'(\boldsymbol{x})$ and $\widetilde{v}'(\boldsymbol{x})$ can be combined to obtain the two-sided CP,

$$\Pr\left[\underset{\sim}{v}'(\boldsymbol{X}) \le v(\boldsymbol{\theta}) \le \widetilde{v}'(\boldsymbol{X})\right] = 1 - \alpha_L - \alpha_U.$$

This shows that, in general, one can obtain a two-sided confidence interval by combining separate one-sided lower and upper confidence bounds. As described in Section 2.7, it is desirable to have $\alpha_L = \alpha_U$. If the one-sided confidence bounds are approximate (conservative), then combining them results in an approximate (conservative) two-sided confidence interval.

B.2.3 A General Expression for Computing the Coverage Probability of a Confidence Interval Procedure

The CP for a two-sided confidence interval procedure can be expressed as

$$\begin{aligned} \mathrm{CP}(\boldsymbol{\theta}) &= 1 - \Pr\left[\underset{\sim}{v}(\boldsymbol{X}) > v(\boldsymbol{\theta})\right] - \Pr[\widetilde{v}(\boldsymbol{X}) < v(\boldsymbol{\theta})] \\ &= 1 - \mathrm{E}_{\boldsymbol{X}}\left\{\mathrm{I}\left[\underset{\sim}{v}(\boldsymbol{X}) > v(\boldsymbol{\theta})\right]\right\} - \mathrm{E}_{\boldsymbol{X}}\{\mathrm{I}[\widetilde{v}(\boldsymbol{X}) < v(\boldsymbol{\theta})]\}, \end{aligned} \tag{B.4}$$

where the indicator function $\mathrm{I}[A]$ is equal to 1 when the statement A is true and equal to 0 otherwise. We have expressed the CP in the manner given in (B.4) because, motivated by the discussion in Section 2.7, we find it advisable to examine separately the probability of the two possible events that will lead to the interval not covering the quantity of interest $v(\boldsymbol{\theta})$ (i.e., the interval can be entirely to the left or to the right of $v(\boldsymbol{\theta})$). Note that in some cases (especially when no exact confidence interval procedure is available), the CP will depend on the parameter(s) $\boldsymbol{\theta}$. The expectations with respect to $\boldsymbol{X}$ in (B.4) sometimes have simple closed-form expressions. In other cases these expectations can be expressed as functions of well-known probability distributions or the distribution of a pivotal quantity (as, for example, described in Appendix E). When simple and/or easy-to-compute expressions are not available, Monte Carlo simulation can be used to perform the needed evaluations.

B.2.4 Computing the Coverage Probability for a Confidence Interval Procedure When Sampling from a Discrete Probability Distribution

When a confidence interval is to be computed for a parameter (or some function of a parameter) of a univariate discrete distribution, such as the binomial or the Poisson distribution, instead of the data $\boldsymbol{x}$ we have a scalar x and easy-to-compute expressions are available for the CP of the procedure. In particular,

$$\mathrm{CP}(\boldsymbol{\theta}) = 1 - \sum \mathrm{I}\left[\underset{\sim}{v}(x) > v(\boldsymbol{\theta})\right]\Pr(X = x) - \sum \mathrm{I}[\widetilde{v}(x) < v(\boldsymbol{\theta})]\Pr(X = x). \tag{B.5}$$

Say, for example, that we wish to compute the CP for a confidence interval procedure to contain the parameter π of the binomial distribution based on a sample of n trials, as described in Section 6.2. Letting $\underset{\sim}{\pi} = \underset{\sim}{\pi}(\boldsymbol{x})$ and $\widetilde{\pi} = \widetilde{\pi}(\boldsymbol{x})$, (B.5) reduces to

$$
\mathrm{CP}(\pi) = 1 - \sum_{x=0}^{n} \mathrm{I}[\underset{\sim}{\pi} > \pi] \binom{n}{x} \pi^x (1-\pi)^{n-x} - \sum_{x=0}^{n} \mathrm{I}[\widetilde{\pi} < \pi] \binom{n}{x} \pi^x (1-\pi)^{n-x}
$$

$$
= 1 - \sum_{x=0}^{n} \mathrm{I}[\underset{\sim}{\pi} > \pi] \texttt{dbinom}(x; \pi, n) - \sum_{x=0}^{n} \mathrm{I}[\widetilde{\pi} < \pi] \texttt{dbinom}(x; \pi, n), \qquad \text{(B.6)}
$$

where the interval endpoints $\underset{\sim}{\pi}$ and $\widetilde{\pi}$ depend on n and x, as described in Section 6.2.

The CP for a confidence interval to contain the probability that a future binomial random variable Y from m trials will be less than or equal to j, given the results of n trials from the same distribution, would be the same as (B.6) because $\mathrm{I}\left[\underset{\sim}{p}_{LE} > p_{LE}\right] = \mathrm{I}[\underset{\sim}{\pi} > \pi]$ and $\mathrm{I}[\widetilde{p}_{LE} < p_{LE}] = \mathrm{I}[\widetilde{\pi} < \pi]$.

Finally, for a confidence interval to contain the mean event-occurrence rate parameter λ of the Poisson distribution based on an exposure of n units, (B.5) reduces to

$$
\mathrm{CP}(\lambda) = 1 - \sum_{x=0}^{\infty} \mathrm{I}[\underset{\sim}{\lambda} > \lambda] \frac{\exp(-n\lambda)(n\lambda)^x}{x!} - \sum_{x=0}^{\infty} \mathrm{I}\left[\widetilde{\lambda} < \lambda\right] \frac{\exp(-n\lambda)(n\lambda)^x}{x!}
$$

$$
= 1 - \sum_{x=0}^{\infty} \mathrm{I}[\underset{\sim}{\lambda} > \lambda] \texttt{dpois}(x; n\lambda) - \sum_{x=0}^{\infty} \mathrm{I}\left[\widetilde{\lambda} < \lambda\right] \texttt{dpois}(x; n\lambda), \qquad \text{(B.7)}
$$

where the interval endpoints $\underset{\sim}{\lambda}$ and $\widetilde{\lambda}$ depend on n and x, as described in Section 7.2.

B.2.5 Computing the Coverage Probability for a Confidence Interval Procedure When Sampling from a Continuous Probability Distribution

Two-sided confidence intervals

When a confidence interval is to be computed for one of the parameters (or a scalar function of the parameters) of a univariate continuous distribution, such as the normal distribution or the Weibull distribution, the CP can be expressed as

$$
\mathrm{CP}(\boldsymbol{\theta}) = 1 - \int \mathrm{I}[\underset{\sim}{v}(\boldsymbol{x}) > v(\boldsymbol{\theta})] f(\boldsymbol{x}; \boldsymbol{\theta}) d\boldsymbol{x} - \int \mathrm{I}[\widetilde{v}(\boldsymbol{x}) < v(\boldsymbol{\theta})] f(\boldsymbol{x}; \boldsymbol{\theta}) d\boldsymbol{x},
$$

where $f(\boldsymbol{x}; \boldsymbol{\theta})$ is the joint density function of the data $\boldsymbol{X}$ and the integration is over the region of $\boldsymbol{x}$ values for which $f(\boldsymbol{x}; \boldsymbol{\theta}) > 0$. When the confidence interval endpoints can be expressed as a function of the model parameter estimates $\widehat{\boldsymbol{\theta}}$, the CP can be computed from

$$
\mathrm{CP}(\boldsymbol{\theta}) = 1 - \int \mathrm{I}\left[\underset{\sim}{v}(\widehat{\boldsymbol{\theta}}) > v(\boldsymbol{\theta})\right] h(\widehat{\boldsymbol{\theta}}; \boldsymbol{\theta}) d\widehat{\boldsymbol{\theta}} - \int \mathrm{I}\left[\widetilde{v}(\widehat{\boldsymbol{\theta}}) < v(\boldsymbol{\theta})\right] h(\widehat{\boldsymbol{\theta}}; \boldsymbol{\theta}) d\widehat{\boldsymbol{\theta}}, \qquad \text{(B.8)}
$$

where $h(\widehat{\boldsymbol{\theta}}; \boldsymbol{\theta})$ is the joint density function (sampling distribution) of the parameter estimator $\widehat{\boldsymbol{\theta}}$ and the integration is over the region of $\widehat{\boldsymbol{\theta}}$ values for which $h(\widehat{\boldsymbol{\theta}}; \boldsymbol{\theta}) > 0$. If the confidence interval procedure is based on a Wald statistic (described in Sections 12.6 and D.5.6), then the integrals in (B.8) reduce to integration by parts of an ellipsoid and simplifications may be possible. In particular, when the underlying distribution of $\boldsymbol{X}$ is a member of the location-scale or log-location-scale families of distributions and the data are complete (i.e., not censored or truncated), it is generally possible to replace the distribution of $\widehat{\theta}$ with the distribution of a pivotal quantity. Some illustrations are given in Appendix E.

When no simplifications are available, Monte Carlo simulation can be used to evaluate the CP of a given procedure using

$$\mathrm{CP}(\boldsymbol{\theta}) \approx 1 - \frac{1}{B} \sum_{j=1}^{B} \mathrm{I}\big[\underset{\sim}{v}(\boldsymbol{x}_j^*) > v(\boldsymbol{\theta})\big] - \frac{1}{B} \sum_{j=1}^{B} \mathrm{I}\big[\widetilde{v}(\boldsymbol{x}_j^*) < v(\boldsymbol{\theta})\big],$$

where $\boldsymbol{x}_j^*, j = 1, \ldots, B$, are simulated samples from the density $f(\boldsymbol{x}; \boldsymbol{\theta})$. The approximation is due to Monte Carlo error which can be made arbitrarily small by using a sufficiently large value of B. When the confidence interval endpoints can be expressed as a function of the model parameter estimates, $\widehat{\boldsymbol{\theta}}$, the CP can be computed from

$$\mathrm{CP}(\boldsymbol{\theta}) \approx 1 - \frac{1}{B} \sum_{j=1}^{B} \mathrm{I}\big[\underset{\sim}{v}(\widehat{\boldsymbol{\theta}}_j^*) > v(\boldsymbol{\theta})\big] - \frac{1}{B} \sum_{j=1}^{B} \mathrm{I}\big[\widetilde{v}(\widehat{\boldsymbol{\theta}}_j^*) < v(\boldsymbol{\theta})\big],$$

where $\widehat{\boldsymbol{\theta}}_j^*$ is the estimate of $\boldsymbol{\theta}$ computed from the sample $\boldsymbol{x}_j^*$, simulated from the density $f(\boldsymbol{x}; \boldsymbol{\theta})$, $j = 1, \ldots, B$.

An important special case arises when X comes from a location-scale (or log-location-scale) distribution. For example, by applying an extension of the notation used in Section 4.6.1, a two-sided confidence interval for a location-scale distribution quantile $x_p = \mu + \Phi^{-1}(p)\sigma$ (where $\Phi^{-1}(p)$ is the p quantile of the corresponding standard location-scale distribution) is

$$\underset{\sim}{v}(\boldsymbol{x}) = \underset{\sim}{v}(\widehat{\mu}, \widehat{\sigma}) = \widehat{\mu} + g'_{L(1-\alpha/2;p,n)}\widehat{\sigma}, \tag{B.9}$$
$$\widetilde{v}(\boldsymbol{x}) = \widetilde{v}(\widehat{\mu}, \widehat{\sigma}) = \widehat{\mu} + g'_{U(1-\alpha/2;p,n)}\widehat{\sigma},$$

where for a symmetric distribution $g'_{L(1-\alpha/2;p,n)} = -g'_{U(1-\alpha/2;p,n)}$, $g'_{L(1-\alpha/2;p,n)} = -g'_{L(1-\alpha/2;1-p,n)}$ and $g'_{U(1-\alpha/2;p,n)} = -g'_{U(1-\alpha/2;1-p,n)}$, for any $0 < 1 - \alpha/2 < 1$.

One-sided confidence bounds

One-sided confidence bounds for x_p can be expressed in a similar manner. In particular, a one-sided lower confidence bound is

$$\underset{\sim}{v}'(\boldsymbol{x}) = \underset{\sim}{v}'(\widehat{\mu}, \widehat{\sigma}) = \widehat{\mu} + g'_{L(1-\alpha;p,n)}\widehat{\sigma}$$

and a one-sided upper confidence bound is

$$\widetilde{v}'(\boldsymbol{x}) = \widetilde{v}'(\widehat{\mu}, \widehat{\sigma}) = \widehat{\mu} + g'_{U(1-\alpha;p,n)}\widehat{\sigma}.$$

Computation of coverage probability

It follows from (B.4) that the CP for the two-sided confidence interval procedure defined by (B.9) is

$$\mathrm{CP}(\mu, \sigma) = 1 - \mathrm{E}_{\widehat{\mu}, \widehat{\sigma}}\big\{\mathrm{I}\big[\widehat{\mu} + g'_{L(1-\alpha/2;p,n)}\widehat{\sigma} > x_p\big]\big\} - \mathrm{E}_{\widehat{\mu}, \widehat{\sigma}}\big\{\mathrm{I}\big[\widehat{\mu} + g'_{U(1-\alpha/2;p,n)}\widehat{\sigma} < x_p\big]\big\},$$

where the expectation is with respect to the distribution of the estimators $(\widehat{\mu}, \widehat{\sigma})$. With complete data from a normal distribution, it is possible to evaluate the expectations by numerical integration (see the references in the Bibliographic Notes section at the end of Chapter 4). Otherwise,

Monte Carlo simulation can be used to evaluate the CP. In particular, for sufficiently large B,

$$\mathrm{CP}(\mu, \sigma) \approx 1 - \frac{1}{B} \sum_{j=1}^{B} \mathrm{I}\left[\widehat{\mu}_j^* + g'_{L(1-\alpha/2;p,n)}\widehat{\sigma}_j^* > \widehat{x}_p\right] - \frac{1}{B} \sum_{j=1}^{B} \mathrm{I}\left[\widehat{\mu}_j^* + g'_{U(1-\alpha/2;p,n)}\widehat{\sigma}_j^* < \widehat{x}_p\right],$$

where $\widehat{x}_p = \widehat{\mu} + \Phi^{-1}(p)\widehat{\sigma}$ and $\widehat{\mu}_j^*$ and $\widehat{\sigma}_j^*$ are the ML estimates based on sample j simulated from a normal distribution with $\mu = \widehat{\mu}$ and $\sigma = \widehat{\sigma}$, $j = 1, \ldots, B$.

B.3　TWO-SIDED CONTROL-THE-CENTER TOLERANCE INTERVALS TO CONTAIN AT LEAST A SPECIFIED PROPORTION OF A DISTRIBUTION

B.3.1　Control-the-Center Tolerance Interval Definition

In many applications, one is interested in estimating, with a high level of confidence, the extent of a distribution (as opposed to obtaining a confidence interval to contain a scalar descriptor, such as a distribution mean or quantile). A tolerance interval to contain a proportion of the distribution (i.e., to control the center of a distribution) is used for this purpose.

Let x denote the observed data and let $\left[\underset{\sim}{T}_\beta(x, 1-\alpha), \; \widetilde{T}_\beta(x, 1-\alpha)\right]$ be a tolerance interval that will contain, with $100(1-\alpha)\%$ confidence, at least a proportion β of the randomly sampled distribution. To simplify the presentation, in the rest of this section we will suppress the dependency of the interval endpoints on $1 - \alpha$.

In what follows, $F(z^-)$ is equal to $F(z)$ for continuous random variables and equal to the limit of $F(w)$ when w approaches z from below for discrete random variables. Note that for any realization of $X = x$, the population content of the tolerance interval is $\Delta F(x, \theta) = F\left[\widetilde{T}_\beta(x); \theta\right] - F\left[\underset{\sim}{T}_\beta(x); \theta\right]$, where $F(x; \theta)$ is the distribution of the random variable X from which the not-yet-observed data X will be taken. One is interested in the probability with which the random quantity $\Delta F(X, \theta)$ exceeds β. The functions $\underset{\sim}{T}_\beta(x)$ and $\widetilde{T}_\beta(x)$ define a *control-the-center tolerance interval procedure*, often referred to more briefly as just a *tolerance interval procedure*.

An *exact* tolerance interval procedure is one for which the CP is exactly equal to the nominal confidence level $100(1-\alpha)\%$. That is,

$$\begin{aligned}
\mathrm{CP}(\theta) &= \Pr\left\{\Pr\left[\underset{\sim}{T}_\beta(x) \leq X \leq \widetilde{T}_\beta(x) | X = x\right] > \beta\right\} \\
&= \Pr\left\{\Pr\left[F[\widetilde{T}_\beta(x); \theta] - F\left[\underset{\sim}{T}_\beta(x); \theta\right] | X = x\right] > \beta\right\} \\
&= \Pr\left\{F\left[\widetilde{T}_\beta(X); \theta\right] - F\left[\underset{\sim}{T}_\beta(X); \theta\right] > \beta\right\} = 1 - \alpha, \quad (\text{B.10})
\end{aligned}$$

where X is a single observation from the randomly sampled distribution and independent of the data X. The inside probability in the first two lines in (B.10) is computed with respect to the single observation X. Note that $\underset{\sim}{T}_\beta(X)$ and $\widetilde{T}_\beta(X)$ are random because they depend on the random data X. The other probabilities in (B.10) are computed with respect to the distribution of the data X. Generally there are not unique values of $\underset{\sim}{T}_\beta(X)$ and $\widetilde{T}_\beta(X)$ that satisfy (B.10) and an equal-tail probability constraint like

$$\Pr\left\{F\left[\underset{\sim}{T}_\beta(X); \theta\right]\right\} = 1 - \Pr\left\{F\left[\widetilde{T}_\beta(X); \theta\right]\right\}$$

is a natural constraint to determine the interval endpoints uniquely.

When an exact control-the-center tolerance interval procedure is not available (e.g., when the data are discrete), one can use an approximate procedure for which the CP in (B.10) is approximately equal to the nominal confidence level $1 - \alpha$. In some applications it may be desirable to have a tolerance interval procedure that is conservative in the sense that the CP is greater than or equal to the nominal confidence level.

Chapters 3–7 and 14 show how to compute tolerance intervals like $\left[\underset{\sim}{T}_\beta(\boldsymbol{x}), \ \widetilde{T}_\beta(\boldsymbol{x})\right]$ for given data $\boldsymbol{x}$ and different assumed probability distributions in the (log-)location-scale family of distributions. Section E.5.1 describes numerical methods for computing the CP and how to determine the interval endpoints $\underset{\sim}{T}_\beta(\boldsymbol{x})$ and $\widetilde{T}_\beta(\boldsymbol{x})$ for the normal distribution.

B.3.2 A General Expression for Computing the Coverage Probability of a Control-the-Center Tolerance Interval Procedure

The CP for a two-sided tolerance interval procedure can be expressed as

$$\mathrm{CP}(\boldsymbol{\theta}) = \Pr\left\{ F\left[\widetilde{T}_\beta(\boldsymbol{X}); \boldsymbol{\theta}\right] - F\left[\underset{\sim}{T}_\beta(\boldsymbol{X}); \boldsymbol{\theta}\right] > \beta \right\} \tag{B.11}$$
$$= \mathrm{E}_{\boldsymbol{X}}\left(\mathrm{I}\left\{ F\left[\widetilde{T}_\beta(\boldsymbol{X}); \boldsymbol{\theta}\right] - F\left[\underset{\sim}{T}_\beta(\boldsymbol{X}); \boldsymbol{\theta}\right] > \beta \right\} \right),$$

where $F(x)$ is the cdf of the random variable X and the indicator function $\mathrm{I}[A]$ is equal to 1 when the statement A is true and is equal to 0 otherwise. Note that in some cases (especially when no exact tolerance interval procedure is available), the CP will depend on the parameter(s) $\boldsymbol{\theta}$ through the probability distribution of the data $\boldsymbol{X}$.

B.3.3 Computing the Coverage Probability for a Control-the-Center Tolerance Interval Procedure When Sampling from a Discrete Distribution

For a univariate discrete distribution, such as a binomial or a Poisson distribution, instead of the data $\boldsymbol{x}$ we have a scalar x and easy-to-compute expressions are available for the CP of the tolerance interval procedure. In particular,

$$\mathrm{CP}(\boldsymbol{\theta}) = \sum \mathrm{I}\left\{ F\left[\widetilde{T}_\beta(x)\right] - F\left[\underset{\sim}{T}_\beta(x)\right] > \beta \right\} \Pr(X = x), \tag{B.12}$$

where the summation is over all values of x such that $\Pr(X = x) > 0$. If, for example, we wish to compute the CP for a tolerance interval procedure to contain a proportion β of an m-trial binomial distribution based on a past sample of n trials, as described in Section B.4, (B.12) reduces to

$$\mathrm{CP}(\pi) = \sum_{x=0}^{n} \mathrm{I}\left[\sum_{k=\underset{\sim}{T}_\beta}^{\widetilde{T}_\beta} \binom{m}{k} \pi^k (1-\pi)^{m-k} > \beta \right] \binom{n}{x} \pi^x (1-\pi)^{n-x}$$
$$= \sum_{x=0}^{n} \left\{ \mathrm{I}\left[\mathrm{pbinom}(\widetilde{T}_\beta; \pi, m) - \mathrm{pbinom}(\underset{\sim}{T}_\beta - 1; \pi, m) \right] > \beta \right\} \mathrm{dbinom}(x; \pi, n),$$

where the interval endpoints $\underset{\sim}{T}_\beta$ and $\widetilde{T}_\beta$ depend on n, x, m, β, and $1 - \alpha$ as described in Section 6.4.

For a tolerance interval to contain a proportion β of a Poisson distribution with m exposure units and mean occurrence rate parameter λ, based on past data x from an exposure amount of

n units, (B.12) reduces to

$$\mathrm{CP}(\lambda) = \sum_{x=0}^{\infty} \mathrm{I}\left[\sum_{k=T_{\beta}}^{\widetilde{T}_{\beta}} \frac{\exp(-m\lambda)(m\lambda)^k}{k!} > \beta\right] \frac{\exp(-n\lambda)(n\lambda)^x}{x!}$$

$$= \sum_{x=0}^{\infty} \mathrm{I}\left\{\left[\mathrm{ppois}(\widetilde{T}_{\beta}; m\lambda) - \mathrm{ppois}(\underset{\sim}{T}_{\beta} - 1; m\lambda)\right] > \beta\right\} \mathrm{dpois}(x; n\lambda),$$

where the interval endpoints $\underset{\sim}{T}_{\beta}$ and $\widetilde{T}_{\beta}$ depend on n, x, m, β, and $1 - \alpha$ as described in Section 7.5.

B.3.4 Computing the Coverage Probability for a Control-the-Center Tolerance Interval Procedure When Sampling from a Continuous Probability Distribution

When a tolerance interval is to be computed for a univariate continuous distribution, such as a normal or Weibull distribution, the CP can be expressed as

$$\mathrm{CP}(\boldsymbol{\theta}) = \int \mathrm{I}\left\{F\left[\widetilde{T}_{\beta}(\boldsymbol{x}); \boldsymbol{\theta}\right] - F[\underset{\sim}{T}_{\beta}(\boldsymbol{x}); \boldsymbol{\theta}] > \beta\right\} f(\boldsymbol{x}; \boldsymbol{\theta}) d\boldsymbol{x},$$

where $f(\boldsymbol{x}; \boldsymbol{\theta})$ is the joint density function of the data $\boldsymbol{X}$ and the integration is over the region of $\boldsymbol{x}$ values for which $f(\boldsymbol{x}; \boldsymbol{\theta}) > 0$. When the tolerance interval endpoints can be expressed as a function of model parameter estimates $\widehat{\boldsymbol{\theta}}$, the CP can be computed as

$$\mathrm{CP}(\boldsymbol{\theta}) = \int \mathrm{I}\left\{F\left[\widetilde{T}_{\beta}(\widehat{\boldsymbol{\theta}}); \boldsymbol{\theta}\right] - F\left[\underset{\sim}{T}_{\beta}(\widehat{\boldsymbol{\theta}}); \boldsymbol{\theta}\right] > \beta\right\} h(\widehat{\boldsymbol{\theta}}; \boldsymbol{\theta}) d\widehat{\boldsymbol{\theta}},$$

where $h(\widehat{\boldsymbol{\theta}}; \boldsymbol{\theta})$ is the joint density function (sampling distribution) of the parameter estimator $\widehat{\boldsymbol{\theta}}$ and the integration is over the region of $\widehat{\boldsymbol{\theta}}$ values for which $h(\widehat{\boldsymbol{\theta}}; \boldsymbol{\theta}) > 0$.

For some distributions (e.g., the normal distribution), simplification may be possible. For example, when the underlying distribution of $\boldsymbol{X}$ is a member of the location-scale or log-location-scale families of distributions and the data are complete (i.e., not censored or truncated), it is generally possible to replace the distribution of $\widehat{\boldsymbol{\theta}}$ with the distribution of a pivotal quantity. Some examples are given in Section E.6.

When no simplifications are available, Monte Carlo simulation can be used to evaluate the CP of a given procedure using

$$\mathrm{CP}(\boldsymbol{\theta}) \approx \frac{1}{B} \sum_{j=1}^{B} \mathrm{I}\left\{F\left[\widetilde{T}_{\beta}(\boldsymbol{x}_j^*); \boldsymbol{\theta}\right] - F\left[\underset{\sim}{T}_{\beta}(\boldsymbol{x}_j^*); \boldsymbol{\theta}\right] > \beta\right\}, \tag{B.13}$$

where $\boldsymbol{x}_j^*$, $j = 1, \ldots, B$, are simulated samples from the density $f(\boldsymbol{x}; \boldsymbol{\theta})$. The approximation in (B.13) is due to Monte Carlo error which can be made arbitrarily small by using a sufficiently large value of B.

When the tolerance interval endpoints can be expressed as a function of the model parameter estimates, the CP can be computed from

$$\mathrm{CP}(\boldsymbol{\theta}) \approx \frac{1}{B} \sum_{j=1}^{B} \mathrm{I}\left\{F\left[\widetilde{T}_{\beta}(\widehat{\boldsymbol{\theta}}_j^*); \boldsymbol{\theta}\right] - F\left[\underset{\sim}{T}_{\beta}(\widehat{\boldsymbol{\theta}}_j^*); \boldsymbol{\theta}\right] > \beta\right\},$$

where $\widehat{\boldsymbol{\theta}}_j^*$ is the estimate of $\boldsymbol{\theta}$ computed from the sample $\boldsymbol{x}_j^*$, simulated from the density $f(\boldsymbol{x}; \boldsymbol{\theta})$, $j = 1, \ldots, B$.

An important special case arises when X comes from a continuous location-scale (or log-location-scale) distribution. In this case, using an extension of the notation used in Section 4.6.1, for a two-sided tolerance interval to control the center of the distribution,

$$\underline{T}_\beta(\boldsymbol{x}) = \underline{T}_\beta(\widehat{\mu}, \widehat{\sigma}) = \widehat{\mu} + g_{L(1-\alpha;\beta,n)}\widehat{\sigma}, \tag{B.14}$$

$$\widetilde{T}_\beta(\boldsymbol{x}) = \widetilde{T}_\beta(\widehat{\mu}, \widehat{\sigma}) = \widehat{\mu} + g_{U(1-\alpha;\beta,n)}\widehat{\sigma},$$

where $g_{L(1-\alpha;\beta,n)} = -g_{U(1-\alpha;\beta,n)}$, $g_{L(1-\alpha;\beta,n)} = -g_{L(1-\alpha;1-\beta,n)}$ and $g_{U(1-\alpha;\beta,n)} = -g_{U(1-\alpha;1-\beta,n)}$ for a symmetric distribution.

Following from (B.11), the CP for a two-sided tolerance interval to control the center of the distribution can be expressed as

$$\text{CP}(\mu, \sigma) = \text{E}_{\widehat{\mu},\widehat{\sigma}} \left\{ \text{I}\left[\Phi\left(\frac{\widetilde{T}_\beta(\widehat{\mu}, \widehat{\sigma}) - \mu}{\sigma} \right) - \Phi\left(\frac{\underline{T}_\beta(\widehat{\mu}, \widehat{\sigma}) - \mu}{\sigma} \right) > \beta \right] \right\} \tag{B.15}$$

$$= \text{E}_{\widehat{\mu},\widehat{\sigma}} \left\{ \text{I}\left[\Phi\left(\frac{\widehat{\mu} + g_{U(1-\alpha;\beta,n)}\widehat{\sigma} - \mu}{\sigma} \right) - \Phi\left(\frac{\widehat{\mu} + g_{L(1-\alpha;\beta,n)}\widehat{\sigma} - \mu}{\sigma} \right) > \beta \right] \right\},$$

where $\Phi(z)$ is the cdf of a particular standard location-scale distribution and the expectation is with respect to the joint distribution of the estimators $(\widehat{\mu}, \widehat{\sigma})$.

With complete data from a normal distribution, it is possible to evaluate the expectation using numerical integration (see the references in the Bibliographic Notes section at the end of Chapter 4 and Section E.5.1). Otherwise, Monte Carlo simulation can be used to evaluate the CP. In particular, for sufficiently large B,

$$\text{CP}(\mu, \sigma) \approx \frac{1}{B} \sum_{j=1}^{B} \text{I}\left[\Phi\left(\frac{\widehat{\mu}_j^* + g_{U(1-\alpha;\beta,n)}\widehat{\sigma}_j^* - \widehat{\mu}}{\widehat{\sigma}} \right) - \Phi\left(\frac{\widehat{\mu}_j^* + g_{L(1-\alpha;\beta,n)}\widehat{\sigma}_j^* - \widehat{\mu}}{\widehat{\sigma}} \right) > \beta \right],$$

$$\tag{B.16}$$

where $\widehat{\mu}_j^*$ and $\widehat{\sigma}_j^*$ are the ML estimates based on sample j simulated from a normal distribution with $\mu = \widehat{\mu}$ and $\sigma = \widehat{\sigma}$, $j = 1, \dots, B$. This approach is used in Section 14.5.1.

B.4 TWO-SIDED TOLERANCE INTERVALS TO CONTROL BOTH TAILS OF A DISTRIBUTION

B.4.1 Control-Both-Tails Tolerance Interval Definition

In some applications, in contrast to the control-the-center tolerance interval described in the preceding section, one might wish to construct a more stringent tolerance interval that simultaneously controls the amount of probability *in both tails* of the distribution. The "control-both-tails" tolerance interval specifies the maximum probability in each tail. Often the *same* probability is specified in each distribution tail, and this procedure is sometimes referred to as an "equal-tail tolerance interval." Our presentation allows for unequal tail probabilities.

In particular, let $\left[\underline{T}_{p_{tL}}(\boldsymbol{x}, p_{tL}, 1 - \alpha), \ \widetilde{T}_{p_{tU}}(\boldsymbol{x}, p_{tU}, 1 - \alpha) \right]$ denote a control-both-tails tolerance interval procedure that will, with $100(1 - \alpha)\%$ confidence, have no more than a proportion p_{tL} less than $\underline{T}_{p_{tL}}$ and no more than p_{tU} greater than $\widetilde{T}_{p_{tU}}$. To simplify the presentation, in the rest of this section we will suppress the dependency of the interval endpoints on $1 - \alpha$ and p_{tL} or p_{tU}.

Note that in the upper tail, the requirement is $1 - F\left[\widetilde{T}_{p_{tU}}(x); \theta\right] \leq p_{tU}$, which is equivalent to $F\left[\widetilde{T}_{p_{tU}}(x); \theta\right] \geq 1 - p_{tU}$. Here x denotes the observed data used to construct the control-both-tails tolerance interval. Typically p_{tL} and p_{tU} will be between 0 and 0.50, but all that is required is that these probabilities be positive and have a sum less than 1.

An *exact* control-both-tails tolerance interval procedure is one for which the CP is exactly equal to the nominal confidence level $100(1 - \alpha)\%$. That is,

$$\mathrm{CP}(\boldsymbol{\theta}) = \Pr\left\{ F\left[\underset{\sim}{T}_{p_{tL}}^{-}(\boldsymbol{X}); \boldsymbol{\theta}\right] \leq p_{tL} \text{ and } F\left[\widetilde{T}_{p_{tU}}(\boldsymbol{X}); \boldsymbol{\theta}\right] \geq 1 - p_{tU} \right\} = 1 - \alpha, \quad \text{(B.17)}$$

where $\boldsymbol{X}$ is the not-yet-observed data that will be used to construct the control-both-tails tolerance interval, and the probability is computed with respect to the distribution of $\boldsymbol{X}$. Note that $\underset{\sim}{T}_{p_{tL}}(\boldsymbol{X})$ and $\widetilde{T}_{p_{tU}}(\boldsymbol{X})$ are random because they depend on the random data $\boldsymbol{X}$.

In some situations (e.g., when the data are discrete), it is impossible to obtain an exact control-both-tails tolerance interval procedure. Then one can use an approximate procedure for which the CP in (B.17) is approximately equal to the nominal confidence level $1 - \alpha$. In some applications it may be desirable to have a control-both-tails tolerance interval procedure that is conservative in the sense that the CP is greater than or equal to the nominal confidence level.

Section 4.6.2 shows how to compute a control-both-tails tolerance interval $\left[\underset{\sim}{T}_{p_{tL}}(x), \ \widetilde{T}_{p_{tU}}(x)\right]$ for a normal distribution. Section 14.5.2 does the same for general (log-)location-scale distributions.

B.4.2   A General Expression for Computing the Coverage Probability of a Control-Both-Tails Tolerance Interval Procedure

This section provides an expression for the CP of a control-both-tails tolerance interval procedure. First note that the random event

$$F\left[\underset{\sim}{T}_{p_{tL}}^{-}(\boldsymbol{X}); \boldsymbol{\theta}\right] \leq p_{tL} \text{ and } F\left[\widetilde{T}_{p_{tU}}(\boldsymbol{X}); \boldsymbol{\theta}\right] \geq 1 - p_{tU},$$

indicating the probability of correctness in both tails of a control-both-tails tolerance interval, is equivalent to the event

$$\underset{\sim}{T}_{p_{tL}}(\boldsymbol{X}) \leq x_{p_{tL}} \text{ and } \widetilde{T}_{p_{tU}}(\boldsymbol{X}) \geq x_{(1-p_{tU})},$$

which implies that the endpoints of the tolerance interval $\left[\underset{\sim}{T}_{p_{tL}}, \ \widetilde{T}_{p_{tU}}\right]$ enclose the unknown interval $[x_{p_{tL}}, \ x_{(1-p_{tU})}]$ where $x_{p_{tL}}$ and $x_{(1-p_{tU})}$ are the corresponding quantiles of the sampled distribution. That is, the interval is correct if $\underset{\sim}{T}_{p_{tL}} \leq x_{p_{tL}} < x_{(1-p_{tU})} \leq \widetilde{T}_{p_{tU}}$. Then, the CP for a two-sided control-both-tails tolerance interval procedure can be expressed as

$$\begin{aligned}
\mathrm{CP}(\boldsymbol{\theta}) &= \Pr\left\{ F\left[\underset{\sim}{T}_{p_{tL}}^{-}(\boldsymbol{X}); \boldsymbol{\theta}\right] \leq p_{tL} \text{ and } F\left[\widetilde{T}_{p_{tU}}(\boldsymbol{X}); \boldsymbol{\theta}\right] \geq 1 - p_{tU} \right\} \\
&= \Pr\left[\underset{\sim}{T}_{p_{tL}}(\boldsymbol{X}) \leq x_{p_{tL}} \text{ and } \widetilde{T}_{p_{tU}}(\boldsymbol{X}) \geq x_{(1-p_{tU})}\right] \qquad \text{(B.18)} \\
&= \mathrm{E}_{\boldsymbol{X}}\left\{ \mathrm{I}\left[\underset{\sim}{T}_{p_{tL}}(\boldsymbol{X}) \leq x_{p_{tL}} \text{ and } \widetilde{T}_{p_{tU}}(\boldsymbol{X}) \geq x_{(1-p_{tU})}\right]\right\},
\end{aligned}$$

where $\mathrm{E}_{\boldsymbol{X}}$ is the expectation with respect to the distribution of $\boldsymbol{X}$ and the indicator function $\mathrm{I}[A]$ is equal to 1 when the statement A is true and is equal to 0 otherwise. Note that in some cases (especially when no exact control-both-tails tolerance interval procedure is available), the CP will depend on the parameter(s) $\boldsymbol{\theta}$ through the probability distribution of the data $\boldsymbol{X}$.

B.4.3 Computing the Coverage Probability for a Control-Both-Tails Tolerance Interval Procedure When Sampling from a Discrete Probability Distribution

When a control-both-tails tolerance interval is to be computed for a univariate discrete distribution, such as a binomial or a Poisson distribution, instead of the data $\boldsymbol{x}$ we have a scalar x and easy-to-compute expressions are available for its CP. In particular,

$$\mathrm{CP}(\boldsymbol{\theta}) = \sum \mathrm{I}\left[\underset{\sim}{T}_{p_{tL}}(x) \leq x_{p_{tL}} \text{ and } \widetilde{T}_{p_{tU}}(x) \geq x_{(1-p_{tU})}\right] \Pr(X = x), \tag{B.19}$$

where the summation is over all values of x for which $\Pr(X = x) > 0$.

Say, for example, that we wish to compute the CP for a control-both-tails tolerance interval procedure to have no more than a proportion p_{tL} in the lower tail and no more than a proportion p_{tU} in the upper tail of an m-trial binomial distribution based on a past sample of n trials. Then (B.19) reduces to

$$\mathrm{CP}(p) = \sum_{x=0}^{n} \mathrm{I}\left[\underset{\sim}{T}_{p_{tL}}(x) \leq x_{p_{tL}} \text{ and } \widetilde{T}_{p_{tU}}(x) \geq x_{(1-p_{tU})}\right] \binom{n}{x} p^x (1-p)^{n-x},$$

where the interval endpoints $\underset{\sim}{T}_{p_{tL}}(x)$ and $\widetilde{T}_{p_{tU}}(x)$ depend on n, x, m, p_{tL} or p_{tU}, and $1-\alpha$ in a manner similar to that described in Section B.3.3.

As a second example, say we desire a control-both-tails tolerance interval to have no more than a proportion p_{tL} in the lower tail and no more than a proportion p_{tU} in the upper tail of a Poisson distribution with m exposure units and mean occurrence rate λ, based on past data of exposure amount of n units. For this case, (B.19) reduces to

$$\mathrm{CP}(\lambda) = \sum_{x=0}^{\infty} \mathrm{I}\left[\underset{\sim}{T}_{p_{tL}}(x) \leq x_{p_{tL}} \text{ and } \widetilde{T}_{p_{tU}}(x) \geq x_{(1-p_{tU})}\right] \frac{\exp(-n\lambda)(n\lambda)^x}{x!},$$

where the interval endpoints $\underset{\sim}{T}_{p_{tL}}(x)$ and $\widetilde{T}_{p_{tU}}(x)$ depend on n, x, m, p_{tL} or p_{tU}, and $1-\alpha$ in a manner similar to that described in Section B.3.3.

B.4.4 Computing the Coverage Probability for a Control-Both-Tails Tolerance Interval Procedure When Sampling from a Continuous Probability Distribution

When a control-both-tails tolerance interval is to be computed for a univariate continuous distribution, such as a normal or Weibull distribution, the CP can be expressed as

$$\mathrm{CP}(\boldsymbol{\theta}) = \int \mathrm{I}\left[\underset{\sim}{T}_{p_{tL}}(\boldsymbol{x}) \leq x_{p_{tL}} \text{ and } \widetilde{T}_{p_{tU}}(\boldsymbol{x}) \geq x_{(1-p_{tU})}\right] f(\boldsymbol{x}; \boldsymbol{\theta}) d\boldsymbol{x},$$

where $f(\boldsymbol{x}; \boldsymbol{\theta})$ is the joint density function of the data $\boldsymbol{X}$ and the integration is over the region of $\boldsymbol{x}$ values for which $f(\boldsymbol{x}; \boldsymbol{\theta}) > 0$. When the control-both-tails tolerance interval endpoints can be expressed as a function of the model parameter estimates $\widehat{\boldsymbol{\theta}}$, the CP can be computed from

$$\mathrm{CP}(\boldsymbol{\theta}) = \int \mathrm{I}\left[\underset{\sim}{T}_{p_{tL}}(\widehat{\boldsymbol{\theta}}) \leq x_{p_{tL}} \text{ and } \widetilde{T}_{p_{tU}}(\widehat{\boldsymbol{\theta}}) \geq x_{(1-p_{tU})}\right] h(\widehat{\boldsymbol{\theta}}; \boldsymbol{\theta}) d\widehat{\boldsymbol{\theta}},$$

where $h(\widehat{\boldsymbol{\theta}}; \boldsymbol{\theta})$ is the joint density function (sampling distribution) of the parameter estimator $\widehat{\boldsymbol{\theta}}$ and the integration is over the region of $\widehat{\boldsymbol{\theta}}$ values for which $h(\widehat{\boldsymbol{\theta}}; \boldsymbol{\theta}) > 0$.

For some distributions (e.g., the normal distribution), simplifications may be possible. When the underlying distribution of $\boldsymbol{X}$ is a member of the location-scale or log-location-scale families of distributions and data are complete (i.e., not censored or truncated), it is generally possible to replace the distribution of $\widehat{\boldsymbol{\theta}}$ with the distribution of a pivotal quantity.

When no simplifications are available, Monte Carlo simulation can be used to evaluate the CP of a given procedure using

$$\mathrm{CP}(\boldsymbol{\theta}) \approx \frac{1}{B} \sum_{j=1}^{B} \mathrm{I}\left[\underset{\sim}{T}_{p_{tL}}\left(\boldsymbol{x}_j^*\right) \le x_{p_{tL}} \text{ and } \widetilde{T}_{p_{tU}}\left(\boldsymbol{x}_j^*\right) \ge x_{(1-p_{tU})}\right],$$

where $\boldsymbol{x}_j^*$, $j = 1, \ldots, B$, are simulated samples from the density $f(\boldsymbol{x}; \boldsymbol{\theta})$. When the control-both-tails tolerance interval endpoints can be expressed as a function of the model parameter estimates $\widehat{\boldsymbol{\theta}}$, the CP can be computed from

$$\mathrm{CP}(\boldsymbol{\theta}) \approx \frac{1}{B} \sum_{j=1}^{B} \mathrm{I}\left[\underset{\sim}{T}_{p_{tL}}\left(\widehat{\boldsymbol{\theta}}_j^*\right) \le x_{p_{tL}} \text{ and } \widetilde{T}_{p_{tU}}\left(\widehat{\boldsymbol{\theta}}_j^*\right) \ge x_{(1-p_{tU})}\right],$$

where $\widehat{\boldsymbol{\theta}}_j^*$ is the estimate of $\boldsymbol{\theta}$ computed from the sample $\boldsymbol{x}_j^*$, simulated from the density $f(\boldsymbol{x}; \boldsymbol{\theta})$, $j = 1, \ldots, B$.

An important special case arises when X comes from a location-scale (or log-location-scale) distribution. In this case, using an extension of the notation in Section 4.6.2, for a two-sided tolerance interval to control the probability in both tails of the distribution,

$$\underset{\sim}{T}_{p_{tL}}(\boldsymbol{x}) = \underset{\sim}{T}_{p_{tL}}(\widehat{\mu}, \widehat{\sigma}) = \widehat{\mu} + g''_{L(1-\alpha;p_{tL},n)}\widehat{\sigma},$$

$$\widetilde{T}_{p_{tU}}(\boldsymbol{x}) = \widetilde{T}_{p_{tU}}(\widehat{\mu}, \widehat{\sigma}) = \widehat{\mu} + g''_{U(1-\alpha;p_{tU},n)}\widehat{\sigma},$$

where for a symmetric distribution $g''_{L(1-\alpha;p_{tL},n)} = -g''_{U(1-\alpha;p_{tU},n)}$, $g''_{L(1-\alpha;p_{tL},n)} = -g''_{L(1-\alpha;1-p_{tL},n)}$, and $g''_{U(1-\alpha;p_{tU},n)} = -g''_{U(1-\alpha;1-p_{tU},n)}$ for any $0 < 1 - \alpha < 1$.

Following from (B.18), the CP for a two-sided tolerance interval to control the probability in both tails of the distribution can be expressed as

$$\mathrm{CP}(\mu, \sigma) = \mathrm{E}_{\boldsymbol{X}}\left\{\mathrm{I}\left[\underset{\sim}{T}_{p_{tL}}(\boldsymbol{X}) \le x_{p_{tL}} \text{ and } \widetilde{T}_{p_{tU}}(\boldsymbol{X}) \ge x_{(1-p_{tU})}\right]\right\}$$

$$= \mathrm{E}_{\widehat{\mu}, \widehat{\sigma}}\left\{\mathrm{I}\left[\widehat{\mu} + g''_{L(1-\alpha;p_{tL},n)}\widehat{\sigma} \le x_{p_{tL}} \text{ and } \widehat{\mu} + g''_{U(1-\alpha;p_{tU},n)}\widehat{\sigma} \ge x_{(1-p_{tU})}\right]\right\},$$

where $x_p = \mu + \Phi^{-1}(p)\sigma$ is the p quantile of the location-scale distribution and the expectation is with respect to the joint distribution of the estimators $(\widehat{\mu}, \widehat{\sigma})$.

With complete data from a normal distribution, it is possible to evaluate the expectation using numerical integration (see the references in the previously cited Bibliographic Notes section at the end of Chapter 4 and Section E.5.2). Otherwise, Monte Carlo simulation can be used to evaluate the CP. In particular, for sufficiently large B,

$$\mathrm{CP}(\mu, \sigma) \approx \frac{1}{B} \sum_{j=1}^{B} \mathrm{I}\left[z_{Lj}''^* \le \Phi^{-1}(p_{tL}) \text{ and } z_{Uj}''^* \ge \Phi^{-1}(p_{tU})\right] = 1 - \alpha, \qquad \text{(B.20)}$$

$$z_{Lj}''^* = \frac{\widehat{\mu}_j^* + g''_{L(1-\alpha;p_{tL},n)}\widehat{\sigma}_j^* - \widehat{\mu}}{\widehat{\sigma}},$$

$$z_{Uj}''^* = \frac{\widehat{\mu}_j^* + g''_{U(1-\alpha;p_{tU},n)}\widehat{\sigma}_j^* - \widehat{\mu}}{\widehat{\sigma}},$$

where $\widehat{\mu}_j^*$ and $\widehat{\sigma}_j^*$ are the ML estimates based on bootstrap sample j simulated from a normal distribution with $\mu = \widehat{\mu}$ and $\sigma = \widehat{\sigma}$, $i = 1, \ldots, B$. This approach is used in Section 14.5.2.

B.5 ONE-SIDED TOLERANCE BOUNDS

A $100(1 - \alpha)\%$ one-sided lower tolerance bound to be exceeded by a proportion β of the population, denoted by $\underset{\sim}{T'_\beta}(\boldsymbol{x})$, is equivalent to a $100(1 - \alpha)\%$ lower confidence bound on $x_{(1-\beta)}$, the $1 - \beta$ quantile of the sampled distribution. To see this, we modify the coverage statement in (B.11) to be a one-sided lower tolerance bound CP. That is,

$$\text{CP}(\boldsymbol{\theta}) = \Pr\left\{1 - F\left[\underset{\sim}{T'^{-}_\beta}(\boldsymbol{X})\right] > \beta\right\} = \Pr\left[\underset{\sim}{T'_\beta}(\boldsymbol{X}) \leq x_{(1-\beta)}\right]$$
$$= \Pr\left[\underset{\sim}{v'}(\boldsymbol{X}) \leq x_{(1-\beta)}\right],$$

which is in agreement with the one-sided confidence interval statement in (B.2) when $v(\boldsymbol{\theta}) = x_{(1-\beta)}$.

Similarly, a $100(1 - \alpha)\%$ one-sided upper tolerance bound to exceed a proportion β of the sampled distribution, denoted by $\widetilde{T}'_\beta(\boldsymbol{x})$, is equivalent to a $100(1 - \alpha)\%$ upper confidence bound on x_β, the β quantile of the sampled distribution. To see this, we now modify the coverage statement in (B.11) to be a one-sided upper tolerance bound CP. That is,

$$\text{CP}(\boldsymbol{\theta}) = \Pr\left\{F\left[\widetilde{T}'_\beta(\boldsymbol{X})\right] > \beta\right\} = \Pr\left[\widetilde{T}'_\beta(\boldsymbol{X}) \geq x_\beta\right]$$
$$= \Pr[\widetilde{v}'(\boldsymbol{X}) \geq x_\beta],$$

which is in agreement with the one-sided confidence interval statement in (B.3) when $v(\boldsymbol{\theta}) = x_\beta$.

B.6 TWO-SIDED PREDICTION INTERVALS AND ONE-SIDED PREDICTION BOUNDS FOR FUTURE OBSERVATIONS

B.6.1 Prediction Interval Definition

One often needs to assess the uncertainty associated with the prediction of one or more future observations or a function of future observations. This is done by constructing a *prediction interval*. First, we introduce some generic notation for a prediction interval (in contrast to the specific notation used in the main parts of this book). Define $V = v(\boldsymbol{Y})$ as a scalar function of future observations $\boldsymbol{Y}$. Let $\left[\underset{\sim}{V}(\boldsymbol{x}, 1 - \alpha), \ \widetilde{V}(\boldsymbol{x}, 1 - \alpha)\right]$ denote a prediction interval for V. Here $\boldsymbol{x}$ denotes the observed data used to construct the prediction interval, and $100(1 - \alpha)\%$ is the nominal confidence level. To simplify the presentation, in the rest of this section we will suppress the dependency of the interval endpoints on $1 - \alpha$. Then the functions $\underset{\sim}{V}(\boldsymbol{x})$ and $\widetilde{V}(\boldsymbol{x})$ define a *prediction interval procedure*.

The probability statement defining an exact prediction interval procedure (i.e., a procedure for which the CP of the procedure is exactly equal to its nominal confidence level $100(1 - \alpha)\%$) is

$$\Pr\left[\underset{\sim}{V}(\boldsymbol{X}) \leq v(\boldsymbol{Y}) \leq \widetilde{V}(\boldsymbol{X})\right] = 1 - \alpha, \tag{B.21}$$

where $\boldsymbol{X}$ is the not-yet-observed data that will be used to construct the prediction interval and $\boldsymbol{Y}$ is the future data from the sampled population. Note that in this expression $v(\boldsymbol{Y})$, $\underset{\sim}{V}(\boldsymbol{X})$, and $\widetilde{V}(\boldsymbol{X})$ are random because they depend on the random data $\boldsymbol{Y}$ and $\boldsymbol{X}$.

In some situations (e.g., when the data are discrete), it is impossible to obtain an exact prediction interval procedure. Then one can use an approximate procedure for which the CP in (B.21) is approximately equal to the nominal confidence level $1 - \alpha$. In some applications it may be desirable to have a prediction interval procedure that is conservative in the sense that the CP is greater than or equal to the nominal confidence level.

Chapters 3–7 show how to compute a prediction interval $\left[\underline{V}(\boldsymbol{x}),\ \widetilde{V}(\boldsymbol{x})\right]$ for given data $\boldsymbol{x}$ and different assumed probability distributions. Chapters 12–15 describe general approaches for defining prediction interval procedures to determine the prediction interval endpoints $\underline{V}(\boldsymbol{x})$ and $\widetilde{V}(\boldsymbol{x})$. Section E.6 gives theory, based on pivotal quantities, for prediction intervals related to the normal distribution. The methods given there can be adapted to other (log-)location-scale distributions.

B.6.2   One-Sided Prediction Bound Definition

As described in Section 2.7, it is possible to combine two one-sided prediction bounds for a future scalar random variable to create a two-sided prediction interval. Namely, let $\underline{V}'(\boldsymbol{x})$ and $\widetilde{V}'(\boldsymbol{x})$ denote, respectively, lower and upper prediction bounds for $v(\boldsymbol{Y})$. The probability statement defining an "exact" lower prediction bound procedure is then

$$\Pr\left[\underline{V}'(\boldsymbol{X}) \leq v(\boldsymbol{Y})\right] = 1 - \alpha_L$$

and the corresponding upper prediction bound procedure is

$$\Pr\left[\widetilde{V}'(\boldsymbol{X}) \geq v(\boldsymbol{Y})\right] = 1 - \alpha_U.$$

There are similar definitions for approximate and conservative prediction bound procedures.

Assuming exact procedures and $0 < \alpha < 0.50$, we use $\underline{V}'(\boldsymbol{x})$ and $\widetilde{V}'(\boldsymbol{x})$ to obtain the two-sided CP,

$$\Pr\left[\underline{V}'(\boldsymbol{X}) \leq v(\boldsymbol{Y}) \leq \widetilde{V}'(\boldsymbol{X})\right] = 1 - \alpha_L - \alpha_U.$$

Thus, one can combine separate lower and upper prediction bounds for a scalar random variable to obtain a two-sided prediction interval. As described in Section 2.7, it is desirable to have $\alpha_L = \alpha_U$. Note, however, that this result holds only for prediction intervals to contain a *single* random quantity, but not for the simultaneous prediction intervals described in Section B.7.

B.6.3   A General Expression for Computing the Coverage Probability of a Prediction Interval Procedure

There are two kinds of CP for prediction intervals and prediction bounds: the conditional CP and the unconditional CP. First, for given data $\boldsymbol{x}$, we have a CP for the corresponding interval, $\left[\underline{V}(\boldsymbol{x}),\ \widetilde{V}(\boldsymbol{x})\right]$, conditional on $\boldsymbol{X} = \boldsymbol{x}$, that can be expressed as

$$\mathrm{CP}(\boldsymbol{\theta}|\boldsymbol{X} = \boldsymbol{x}) = 1 - \Pr\left[v(\boldsymbol{Y}) < \underline{V}(\boldsymbol{x})\right] - \Pr\left[v(\boldsymbol{Y}) > \widetilde{V}(\boldsymbol{x})\right],$$

where $\boldsymbol{\theta}$ is a vector of parameters of the distribution of $\boldsymbol{X}$ and $\boldsymbol{Y}$. This conditional CP is *unknown* because it depends on the unknown parameter(s) $\boldsymbol{\theta}$. Also, this conditional probability depends on the observed data $\boldsymbol{x}$ and in this sense is random (before the data are obtained). For these reasons, the conditional CP is *not* useful for describing the properties of a prediction interval procedure.

The unconditional CP of the prediction interval procedure can be expressed as

$$\mathrm{CP}(\boldsymbol{\theta}) = 1 - \Pr\left[v(\boldsymbol{Y}) < \underline{V}(\boldsymbol{X})\right] - \Pr\left[v(\boldsymbol{Y}) > \widetilde{V}(\boldsymbol{X})\right]$$

$$= 1 - \mathrm{E}_{\boldsymbol{X}}\left\{\Pr\left[v(\boldsymbol{Y}) < \underline{V}(\boldsymbol{x})|\boldsymbol{X} = \boldsymbol{x}\right]\right\} - \mathrm{E}_{\boldsymbol{X}}\left\{\Pr\left[v(\boldsymbol{Y}) > \widetilde{V}(\boldsymbol{x})|\boldsymbol{X} = \boldsymbol{x}\right]\right\},$$

(B.22)

where $\mathrm{E}_{\boldsymbol{X}}$ is the expectation with respect to the distribution of the data $\boldsymbol{X}$. Note that both $\boldsymbol{X}$ and $\boldsymbol{Y}$ are random. As with the other statistical interval procedures, we have expressed the CP in this form because, as motivated by the discussion in Section 2.7, we generally find it advisable to examine separately the probability of the two possible events that will lead to the interval not covering the quantity of interest $v(\boldsymbol{Y})$ (i.e., when the interval ends up being entirely to the left or to the right of $v(\boldsymbol{Y})$). Note that in some cases (especially when no exact prediction interval procedure is available), the CP will depend on the parameter(s) $\boldsymbol{\theta}$.

Sometimes the expectations in (B.22) have simple closed-form expressions. In other cases they can be expressed as functions of well-known probability distributions or the distribution of a pivotal quantity (e.g., as described in Section E.6). When simple easy-to-compute expressions are not available, Monte Carlo simulation can be used for the needed evaluations.

B.6.4   Computing the Coverage Probability for a Prediction Interval Procedure When Sampling from a Discrete Probability Distribution

When a prediction interval is to be computed for a function of future observations from a univariate discrete distribution, such as the binomial or the Poisson distribution, the observed data x and the future observations $\boldsymbol{Y}$ are scalars (denoted by x and Y, respectively). In this case, easy-to-compute expressions are generally available for the CP of a given prediction interval procedure. Let $\underset{\sim}{Y} = \underset{\sim}{V}(x)$ and $\widetilde{Y} = \widetilde{V}(x)$. Then

$$\mathrm{CP}(\boldsymbol{\theta}) = 1 - \sum_x \left[\Pr(X = x) \sum_{y < \underset{\sim}{Y}} \Pr(Y = y) \right] - \sum_x \left[\Pr(X = x) \sum_{y > \widetilde{Y}} \Pr(Y = y) \right],$$

(B.23)

where the summation is over all values of x such that $\Pr(X = x) > 0$.

For example, for a prediction interval to contain a future outcome Y from m (future) trials from a binomial distribution based on a previous sample of n trials, (B.23) reduces to

$$\mathrm{CP}(\pi) = \sum_{x=0}^{n} \left[\texttt{pbinom}(\widetilde{Y}; m, \pi) - \texttt{pbinom}(\underset{\sim}{Y} - 1; m, \pi) \right] \texttt{dbinom}(x; \pi, n), \quad (\text{B}.24)$$

where the interval endpoints $\underset{\sim}{Y}$ and $\widetilde{Y}$ depend on n, the observed value of x, m, and $1 - \alpha$ as described in Section 6.7.

Also, for a prediction interval to contain the future outcome from a Poisson distribution with m units of exposure, based on a previous sample with n units of exposure, (B.23) reduces to

$$\mathrm{CP}(\lambda) = \sum_{x=0}^{\infty} \left[\texttt{ppois}(\widetilde{Y}; m\lambda) - \texttt{ppois}(\underset{\sim}{Y} - 1; m\lambda) \right] \texttt{dpois}(x; n\lambda), \quad (\text{B}.25)$$

where the interval endpoints $\underset{\sim}{Y}$ and $\widetilde{Y}$ depend on n, the observed value of x, m, and $1 - \alpha$ as described in Section 7.6.

B.6.5   Computing the Coverage Probability for a Prediction Interval Procedure When Sampling from a Continuous Probability Distribution

When a prediction interval is to be computed for a function of future observations from a univariate continuous distribution, such as the normal distribution or the Weibull distribution, the CP can be expressed as

$$\mathrm{CP}(\boldsymbol{\theta}) = 1 - \int \Pr \left[v(\boldsymbol{Y}) < \underset{\sim}{V}(\boldsymbol{x}) \right] f(\boldsymbol{x}; \boldsymbol{\theta}) d\boldsymbol{x} - \int \Pr \left[v(\boldsymbol{Y}) > \widetilde{V}(\boldsymbol{x}) \right] f(\boldsymbol{x}; \boldsymbol{\theta}) d\boldsymbol{x},$$

where $f(\boldsymbol{x}; \boldsymbol{\theta})$ is the joint density function of the data $\boldsymbol{X}$ and the integration is over the region of $\boldsymbol{x}$ values for which $f(\boldsymbol{x}; \boldsymbol{\theta}) > 0$. When the prediction interval endpoints can be expressed as a function of the model parameter estimates $\widehat{\boldsymbol{\theta}}$, the CP can be computed from

$$\mathrm{CP}(\boldsymbol{\theta}) = 1 - \int \mathrm{Pr}\big[v(\boldsymbol{Y}) < \underset{\sim}{V}(\boldsymbol{x})\big] h(\widehat{\boldsymbol{\theta}}; \boldsymbol{\theta}) d\widehat{\boldsymbol{\theta}} - \int \mathrm{Pr}\big[v(\boldsymbol{Y}) > \widetilde{V}(\boldsymbol{x})\big] h(\widehat{\boldsymbol{\theta}}; \boldsymbol{\theta}) d\widehat{\boldsymbol{\theta}},$$

$$\text{(B.26)}$$

where $h(\widehat{\boldsymbol{\theta}}; \boldsymbol{\theta})$ is the joint density function (sampling distribution) of the parameter estimator $\widehat{\boldsymbol{\theta}}$ and the integration is over the region of $\widehat{\boldsymbol{\theta}}$ values for which $h(\widehat{\boldsymbol{\theta}}; \boldsymbol{\theta}) > 0$.

If the prediction interval procedure is based on a Wald-like procedure (described in Sections 12.6 and D.5.6), the integrals in (B.26) reduce to integration over parts of an ellipsoid and simplification may be possible. For example, when the underlying distribution of $\boldsymbol{X}$ is a member of the location-scale or log-location-scale families of distributions and the data are complete (i.e., not censored or truncated), it is generally possible to replace the distribution of $\widehat{\boldsymbol{\theta}}$ with the distribution of a pivotal quantity. Some examples are given in Section E.6.

When no simplifications are available, Monte Carlo simulation can be used to evaluate the CP of a given procedure using

$$\mathrm{CP}(\boldsymbol{\theta}) \approx 1 - \frac{1}{B} \sum_{j=1}^{B} \mathrm{Pr}\big[v(\boldsymbol{Y}) < \underset{\sim}{V}(\boldsymbol{x}_j^*)\big] - \frac{1}{B} \sum_{j=1}^{B} \mathrm{Pr}\big[v(\boldsymbol{Y}) > \widetilde{V}(\boldsymbol{x}_j^*)\big],$$

where $\boldsymbol{x}_j^*$, $j = 1, \ldots, B$, are simulated samples from the density $f(\boldsymbol{x}; \boldsymbol{\theta})$. When the prediction interval endpoints can be expressed as a function of the model parameter estimates, $\widehat{\boldsymbol{\theta}}$, CP can be computed from

$$\mathrm{CP}(\boldsymbol{\theta}) \approx 1 - \frac{1}{B} \sum_{j=1}^{B} \mathrm{Pr}\big[v(\boldsymbol{Y}) < \underset{\sim}{V}(\widehat{\boldsymbol{\theta}}_j^*)\big] - \frac{1}{B} \sum_{j=1}^{B} \mathrm{Pr}\big[v(\boldsymbol{Y}) > \widetilde{V}(\widehat{\boldsymbol{\theta}}_j^*)\big],$$

where $\widehat{\boldsymbol{\theta}}_j^*$ is the bootstrap estimate of $\boldsymbol{\theta}$ computed from the bootstrap sample $\boldsymbol{x}_j^*$, simulated from the density $f(\boldsymbol{x}; \boldsymbol{\theta})$, $j = 1, \ldots, B$.

B.7   TWO-SIDED SIMULTANEOUS PREDICTION INTERVALS AND ONE-SIDED SIMULTANEOUS PREDICTION BOUNDS

Let $\boldsymbol{Y}$ denote a vector of m independent future observations from a previously sampled population. A common statistical problem is to obtain a *simultaneous* prediction interval

$$\big[\underset{\sim}{Y}_{k;m}(\boldsymbol{x}, k, m, 1-\alpha), \ \widetilde{Y}_{k;m}(\boldsymbol{x}, k, m, 1-\alpha)\big]$$

to contain at least k of the m components of $\boldsymbol{Y}$ with $100(1-\alpha)\%$ confidence. Other situations call for obtaining one-sided simultaneous prediction bounds. To simplify the presentation, in the rest of this section we will suppress the dependency of the interval endpoints on k, m, and $1-\alpha$.

The CP for a two-sided simultaneous prediction interval is

$$\mathrm{CP}(\boldsymbol{\theta}) = \mathrm{Pr}\big[\text{at least } k \text{ of } m \text{ values lie between } \underset{\sim}{Y}_{k;m}(\boldsymbol{X}) \text{ and } \widetilde{Y}_{k;m}(\boldsymbol{X})\big]$$

$$= \mathrm{E}_{\boldsymbol{X}}\left[\sum_{i=k}^{m} \binom{m}{i} p^i (1-p)^{m-i}\right],$$

where the expectation is with respect to the distribution of $\boldsymbol{X}$. Conditional on a fixed $\boldsymbol{X} = \boldsymbol{x}$,

$$p = p(\boldsymbol{x}) = \Pr\left[\underset{\sim}{Y}_{k;m}(\boldsymbol{x}) \leq Y \leq \widetilde{Y}_{k;m}(\boldsymbol{x})|\boldsymbol{X} = \boldsymbol{x}\right]$$
$$= F\left[\widetilde{Y}_{k;m}(\boldsymbol{x})\right] - F\left[\underset{\sim}{Y}_{k;m}^{-}(\boldsymbol{x})\right],$$

where $F(y)$ is the cdf of Y, a single future observation from the previously sampled distribution. One can make similar CP statements for one-sided simultaneous prediction bounds. In particular, for a lower simultaneous prediction bound,

$$\mathrm{CP}(\boldsymbol{\theta}) = \Pr\left[\text{at least } k \text{ of } m \text{ values are greater than or equal to } \underset{\sim}{Y}_{k;m}'(\boldsymbol{X})\right]$$
$$= \mathrm{E}_{\boldsymbol{X}}\left[\sum_{i=k}^{m}\binom{m}{i}p^i(1-p)^{m-i}\right],$$

where, conditional on a fixed $\boldsymbol{X} = \boldsymbol{x}$,

$$p = p(\boldsymbol{x}) = \Pr[Y \geq \underset{\sim}{Y}_{k;m}'(\boldsymbol{x})|\boldsymbol{X} = \boldsymbol{x}] = 1 - F\left[\underset{\sim}{Y}_{k;m}'^{-}(\boldsymbol{x})\right].$$

Similarly, for a one-sided upper simultaneous prediction bound,

$$\mathrm{CP}(\boldsymbol{\theta}) = \Pr\left[\text{at least } k \text{ of } m \text{ values are less than or equal to } \widetilde{Y}_{k;m}'(\boldsymbol{X})\right]$$
$$= \mathrm{E}_{\boldsymbol{X}}\left[\sum_{i=k}^{m}\binom{m}{i}p^i(1-p)^{m-i}\right],$$

where, conditional on a fixed $\boldsymbol{X} = \boldsymbol{x}$,

$$p = p(\boldsymbol{x}) = \Pr\left[Y \leq \widetilde{Y}_{k;m}'(\boldsymbol{x})|\boldsymbol{X} = \boldsymbol{x}\right] = F\left[\widetilde{Y}_{k;m}'(\boldsymbol{x})\right].$$

An important special case arises when X and Y come from a location-scale (or log-location-scale) distribution. In this case, using an extension of the notation used in Section 4.8, for a two-sided simultaneous prediction interval,

$$\underset{\sim}{Y}_{k;m}(\boldsymbol{x}) = \underset{\sim}{Y}_{k;m}(\widehat{\mu}, \widehat{\sigma}) = \widehat{\mu} + r_{L(1-\alpha;k,m,n)}\widehat{\sigma}, \tag{B.27}$$
$$\widetilde{Y}_{k;m}(\boldsymbol{x}) = \widetilde{Y}_{k;m}(\widehat{\mu}, \widehat{\sigma}) = \widehat{\mu} + r_{U(1-\alpha;k,m,n)}\widehat{\sigma},$$

where $r_{L(1-\alpha;k,m,n)} = -r_{U(1-\alpha;k,m,n)}$ for a symmetric distribution. For a nonsymmetric distribution, we suggest that the values of $r_{L(1-\alpha;k,m,n)}$ and $r_{U(1-\alpha;k,m,n)}$ be chosen to have equal one-sided probabilities of noncoverage in each tail of the distribution.

One-sided simultaneous prediction bounds can be expressed in a similar manner. In particular, the one-sided lower simultaneous prediction bound is

$$\underset{\sim}{Y}_{k;m}'(\boldsymbol{x}) = \underset{\sim}{Y}_{k;m}'(\widehat{\mu}, \widehat{\sigma}) = \widehat{\mu} + r_{L(1-\alpha;k,m,n)}'\widehat{\sigma}$$

and the one-sided upper simultaneous prediction bound is

$$\widetilde{Y}_{k;m}'(\boldsymbol{x}) = \widetilde{Y}_{k;m}'(\widehat{\mu}, \widehat{\sigma}) = \widehat{\mu} + r_{U(1-\alpha;k,m,n)}'\widehat{\sigma}.$$

The CP for two-sided prediction intervals and one-sided prediction bounds can be expressed as

$$\mathrm{CP}(\mu, \sigma) = \mathrm{E}_{\widehat{\mu}, \widehat{\sigma}}\left[\sum_{i=k}^{m}\binom{m}{i}p^i(1-p)^{m-i}\right],$$

where p is given by the following expressions and the expectation is with respect to the joint distribution of the estimators $(\widehat{\mu}, \widehat{\sigma})$: for a *two-sided prediction interval*,

$$p = p(\widehat{\mu}, \widehat{\sigma}) = \Pr\left[\underset{\sim}{Y}_{k;m}(\widehat{\mu}, \widehat{\sigma}) \leq Y \leq \widetilde{Y}_{k;m}(\widehat{\mu}, \widehat{\sigma})\right]$$
$$= \Phi(\widehat{\mu} + r_{U(1-\alpha;k,m,n)}\widehat{\sigma}) - \Phi(\widehat{\mu} + r_{L(1-\alpha;k,m,n)}\widehat{\sigma});$$

for a *one-sided lower prediction bound*,

$$p = p(\widehat{\mu}, \widehat{\sigma}) = \Pr\left[Y \geq \underset{\sim}{Y}'_{k;m}(\widehat{\mu}, \widehat{\sigma})\right] = 1 - \Phi(\widehat{\mu} + r'_{L(1-\alpha;k,m,n)}\widehat{\sigma});$$

and for a *one-sided upper prediction bound*,

$$p = p(\widehat{\mu}, \widehat{\sigma}) = \Pr\left[Y \leq \widetilde{Y}'_{k;m}(\widehat{\mu}, \widehat{\sigma})\right] = \Phi(\widehat{\mu} + r'_{U(1-\alpha;k,m,n)}\widehat{\sigma}),$$

where $\Phi(z)$ is the cdf of a particular standard location-scale distribution and $\widehat{\mu}$ and $\widehat{\sigma}$ are the ML estimates of μ and σ, respectively.

With complete data from a normal distribution, it is possible to evaluate the expectation using numerical integration (see the references cited in the Bibliographic Notes section at the end of Chapter 4 and Section E.6.4). Otherwise, Monte Carlo simulation can be used to evaluate the CP. In particular, for sufficiently large B,

$$CP(\mu, \sigma) \approx \frac{1}{B} \sum_{j=1}^{B} \left[\sum_{i=k}^{m} \binom{m}{i} (p_j^*)^i (1 - p_j^*)^{m-i} \right], \tag{B.28}$$

where p_j is given by the following expressions in which $\widehat{\mu}_j^*$ and $\widehat{\sigma}_j^*$ are the ML estimates based on sample j simulated from a normal distribution with $\mu = 0$ and $\sigma = 1$, $j = 1, \ldots, B$: for a *two-sided prediction interval*,

$$p_j^* = p(\widehat{\mu}_j^*, \widehat{\sigma}_j^*) = \Pr\left[\underset{\sim}{Y}_{k;m}(\widehat{\mu}_j^*, \widehat{\sigma}_j^*) \leq Y \leq \widetilde{Y}_{k;m}(\widehat{\mu}_j^*, \widehat{\sigma}_j^*)\right]$$
$$= \Phi(\widehat{\mu}_j^* + r_{U(1-\alpha;k,m,n)}\widehat{\sigma}_j^*) - \Phi(\widehat{\mu}_j^* + r_{L(1-\alpha;k,m,n)}\widehat{\sigma}_j^*);$$

for a *one-sided lower prediction bound*,

$$p_j^* = p(\widehat{\mu}_j^*, \widehat{\sigma}_j^*) = \Pr\left[Y \geq \underset{\sim}{Y}'_{k;m}(\widehat{\mu}_j^*, \widehat{\sigma}_j^*)\right] = 1 - \Phi(\widehat{\mu}_j^* + r'_{L(1-\alpha;k,m,n)}\widehat{\sigma}_j^*);$$

and for a *one-sided upper prediction bound*,

$$p_j^* = p(\widehat{\mu}_j^*, \widehat{\sigma}_j^*) = \Pr\left[Y \leq \widetilde{Y}'_{k;m}(\widehat{\mu}_j^*, \widehat{\sigma}_j^*)\right] = \Phi(\widehat{\mu}_j^* + r'_{U(1-\alpha;k,m,n)}\widehat{\sigma}_j^*),$$

where $\Phi(z)$ is the cdf of a particular standard location-scale distribution. This approach is used in Section 14.6.

B.8 CALIBRATION OF STATISTICAL INTERVALS

Up to this point in this appendix we have notationally suppressed the dependency of the coverage probability $CP(\boldsymbol{\theta})$ on the nominal confidence level $1 - \alpha$. In this section, however, we will write $CP(\boldsymbol{\theta}, 1 - \alpha)$. Then for an exact interval, $CP(\boldsymbol{\theta}, 1 - \alpha) = 1 - \alpha$.

As described in several places in this appendix, in applications for which an exact statistical interval procedure is not available, we often resort to the use of an approximate interval procedure, in which case $CP(\boldsymbol{\theta}, 1 - \alpha) \approx 1 - \alpha$. In situations for which $CP(\boldsymbol{\theta}, 1 - \alpha)$ is consistently greater than or consistently less than the nominal confidence level $1 - \alpha$ for all possible values of the parameter(s) $\boldsymbol{\theta}$, one may be able to use the method of calibration to find a procedure that has $\mathrm{CP}(\boldsymbol{\theta}, 1 - \alpha)$ closer to $1 - \alpha$ than the simple approximation alone. We describe calibration for confidence bounds and intervals only. However, this method can be used for any kind of statistical interval,

For a two-sided confidence interval, because it is desirable to have equal probability of being outside the interval on each side of a two-sided confidence interval (as described in Sections 2.7 and B.2.2), we would do separate calibrations for one-sided lower and upper $100(1 - \alpha)\%$ confidence bounds and then combine them to form an approximate $100(1 - 2\alpha)\%$ confidence interval. Thus we describe the calibration of one-sided confidence bound procedures.

To obtain a one-sided lower confidence bound with nominal confidence level $1 - \alpha$, we use $\underset{\sim}{v}'(\boldsymbol{X}, 1 - \alpha_{cl})$ and choose $1 - \alpha_{cl}$ such that the CP is $\mathrm{CP}(\widehat{\boldsymbol{\theta}}, 1 - \alpha_{cl}) = 1 - \alpha$. Similarly, to obtain a one-sided upper confidence bound with nominal confidence level $1 - \alpha$, we use $\widetilde{v}'(\boldsymbol{X}, 1 - \alpha_{cu})$ and choose $1 - \alpha_{cu}$ such that the CP is $\mathrm{CP}(\widehat{\boldsymbol{\theta}}, 1 - \alpha_{cu}) = 1 - \alpha$.

In simple special cases it may be possible to do the calibration analytically or to compute CPs directly. Frequently, however, simulation is required. For a given set of data and a specified model, a single simulation generally can be used to compute $\mathrm{CP}(\widehat{\boldsymbol{\theta}}, 1 - \alpha_{cl})$ as a function of $1 - \alpha_{cl}$. This provides a so-called calibration curve that can be used to determine the value of $1 - \alpha_{cl}$ that will give the desired $1 - \alpha$. If $\mathrm{CP}(\widehat{\boldsymbol{\theta}}, 1 - \alpha_{cl})$ does not depend on the value of $\widehat{\boldsymbol{\theta}}$, the calibrated procedure is exact. Otherwise, the procedure is approximate. A similar approach is used to calibrate a one-sided upper confidence bound.

The ideas behind calibration and theory to support its use have been described in numerous works, including Loh (1987, 1991) and Zheng and Loh (1995). Beran (1990) treats calibration for prediction intervals. Calibration is closely related to the bootstrap/simulation methods in Chapters 13 and 14.

Useful Probability Distributions

INTRODUCTION

This appendix provides notation, expressions for the density or the probability mass function (pmf), the cumulative distribution function, the quantiles, the mean, and the variance for distributions that are important in setting different kinds of statistical intervals. Most of the presentation is descriptive, but when needed we provide some technical details.

The topics discussed in this appendix are:

- The relationship between the symbols used in our formulas in this book and the arguments used by the R software package (R Core Team, 2016) for the pdf/pmf, cdf, and quantile functions (Section C.1).

- Information about important functions related to the probability distributions used in this book, namely the cdf, pdf, and quantile functions (Section C.2).

- Continuous distributions used in the book (Section C.3).

- Discrete distributions used in the book (Section C.4).

C.1 PROBABILITY DISTRIBUTIONS AND R COMPUTATIONS

In some places in this book, we show how to do simple computations where the software package R is used as a sophisticated calculator that can not only perform the usual scientific hand calculator functions (e.g., add, subtract, multiply, divide, log, antilog, and square root), but also compute probabilities and quantiles of particular distributions used in constructing confidence intervals (replacing tables that were utilized before the advent of personal computers). To aid readers who are new to R, Table C.1 shows the relationship between the symbols used in our formulas and the arguments used by R.

For each probability distribution in Table C.1, R provides functions to compute the pdf (or pmf), cdf, the quantile function, and to generate random variates. The names of these functions

Statistical Intervals: A Guide for Practitioners and Researchers, Second Edition.
William Q. Meeker, Gerald J. Hahn and Luis A. Escobar.
© 2017 John Wiley & Sons, Inc. Published 2017 by John Wiley & Sons, Inc.
Companion Website: www.wiley.com/go/meeker/intervals

Distribution	R root name	Distribution parameters	R function parameter names	Section number
Normal	norm	μ, σ	mean, std	C.3.2
Lognormal	lnorm	μ, σ	meanlog, sdlog	C.3.2
Smallest extreme value	sev[†]	μ, σ	location, scale	C.3.2
Weibull	weibull	β, η	shape, scale	C.3.2
Largest extreme value	lev[†]	μ, σ	location, scale	C.3.2
Fréchet	frechet[†]	β, η	shape, scale	C.3.2
Logistic	logis	μ, σ	location, scale	C.3.2
Log-logistic	llogis[†]	μ, σ	locationlog, scalelog	C.3.2
Cauchy	cauchy	μ, σ	location, scale	C.3.2
Log-Cauchy	lcauchy[†]	μ, σ	locationlog, scalelog	C.3.2
Beta	beta	a, b	shape1, shape2	C.3.3
Log-uniform	lunif[†]	a, b	location1, location2	C.3.4
Gamma	gamma	a, b	shape, rate	C.3.5
Chi-square	chisq	r	df	C.3.6
Exponential	exp	λ	rate	C.3.7
Noncentral t	t	r, δ	df, ncp	C.3.9
Student's t	t	r	df	C.3.10
Snedecor's F	f	r_1, r_2	df1, df2	C.3.11
Binomial	binom	n, π	size, prob	C.4.1
Beta-binomial	betabinom[†]	n, a, b	size, shape1, shape2	C.4.2
Negative binomial	nbinom	n, π	size, prob	C.4.3
Poisson	pois	λ	lambda	C.4.4
Hypergeometric	hyper2[†]	n, D, N	size, D, N	C.4.5
Negative hypergeometric	nhyper[†]	k, D, N	k, D, N	C.4.6

Table C.1 Relationship between mathematical notation and R parameter names for probability distributions. [†] Indicates functions that require R package StatInt.

are obtained by prefixing the root name with the letters d, p, q, and r, respectively. The first argument for each of these five functions indicates the point where the pdf (or pmf) is to be evaluated (x), the point where the cdf is to be evaluated (q), the particular quantile (p, $0 \le p \le 1$), and the number of random observations to generate ($n \ge 1$). For example, using the binomial distribution functions, R gives the following results:

```
> dbinom(x=5, size=20, prob=0.10)
[1] 0.03192136
> pbinom(q=5, size=20, prob=0.10)
[1] 0.9887469
> qbinom(p=0.50, size=20, prob=0.10)
[1] 2
> rbinom(n=4, size=20, prob=0.10)
[1] 0 3 1 2
```

for `dbinom`$(x = 5; n = 20, \pi = 0.10)$, `pbinom`$(q = 5; n = 20, \pi = 0.10)$, `qbinom`$(p = 0.50; n = 20, \pi = 0.10)$ and four random observations from the $\mathrm{BINOM}(n = 20, \pi = 0.10)$ distribution. Functions in package `StatInt` are for distributions that either do not exist in basic R (smallest and largest extreme value, log-logistic, log-Cauchy, log-uniform, beta-binomial, and negative hypergeometric) or have been redefined to use our more standard parameterization (hypergeometric).

C.2   IMPORTANT CHARACTERISTICS OF RANDOM VARIABLES

In this book we describe distributions for two distinct types of random variables: continuous and discrete. The support of a random variable is the collection of values for which the random variable has positive probability.

Continuous random variables used here have their support on a finite interval (e.g., the $\mathrm{UNIF}(0, 1)$ distribution has support over the interval $(0, 1))$ or on an infinite interval (e.g., the $\mathrm{NORM}(\mu, \sigma)$ distribution has support over the entire real line).

Discrete random variables have support on countable sets. For example, the binomial distribution with $n = 1$ assigns positive probability to the points 0 and 1, while the $\mathrm{POIS}(\lambda)$ distribution assigns positive probability to each nonnegative integer.

Next we define some important functions and metrics associated with a random variable.

C.2.1   Density and Probability Mass Functions

A continuous and nonnegative function $f(x; \boldsymbol{\theta})$ is a density (pdf) for a *continuous* random variable X if, for all real values between a and b (with $a < b$),

$$\Pr(a < X \le b) = \int_a^b f(x; \boldsymbol{\theta})\, dx,$$

where the vector $\boldsymbol{\theta}$ contains the distribution parameters. A nonnegative function $f(x; \boldsymbol{\theta})$ is a pmf for a *discrete* random variable X if $f(x_i; \boldsymbol{\theta}) = \Pr(X = x_i)$ for each point x_i in the support of the distribution, where $\boldsymbol{\theta}$ is the vector of distribution parameters.

C.2.2   Cumulative Distribution Function

For a *continuous* random variable X, the cumulative distribution function (cdf) is defined by

$$F(x; \boldsymbol{\theta}) = \Pr(X \le x) = \int_{-\infty}^x f(w; \boldsymbol{\theta})\, dw, \quad -\infty < x < \infty,$$

where the real-valued function $f(w; \boldsymbol{\theta})$ is nonnegative and continuous for all w. This definition implies that for all x the density $f(x; \boldsymbol{\theta})$ is the derivative of $F(x; \boldsymbol{\theta})$ with respect to x. The cdf $F(x; \boldsymbol{\theta})$ is commonly called an absolutely continuous distribution, but we simply call it a continuous distribution.

For a *discrete* random variable X, the cdf is defined by

$$F(x; \boldsymbol{\theta}) = \Pr(X \le x) = \sum_{x_i \le x} f(x_i),$$

where the sum is over all the x_i values in the support of the distribution that are less than or equal to x. Then $F(x; \boldsymbol{\theta})$ is a step function that increases at each point in the support of the distribution and that remains constant between adjacent points in the support. That is, if

$x_i < x_{i+1}$ are two adjacent points in the support of the distribution, $F(x; \boldsymbol{\theta}) = F(x_i; \boldsymbol{\theta})$ for all $x \in [x_i, x_{i+1})$ and $F(x_i; \boldsymbol{\theta}) < F(x_{i+1}; \boldsymbol{\theta})$.

C.2.3   Quantile Function

The quantile x_p is the inverse of the cumulative distribution function. Formally, for $0 < p < 1$, the p quantile of the cdf $F(x; \boldsymbol{\theta})$ is defined as

$$x_p = F^{-1}(p; \boldsymbol{\theta}) = \inf\{x | p \le F(x; \boldsymbol{\theta})\}. \tag{C.1}$$

For all practical purposes in this book, it suffices to interpret (C.1) as $x_p = \min\{x | p \le F(x; \boldsymbol{\theta})\}$. When $F(x; \boldsymbol{\theta})$ is continuous and strictly increasing as a function of x, there is a unique value of x_p such that $F(x_p; \boldsymbol{\theta}) = p$, or equivalently $x_p = F^{-1}(p; \boldsymbol{\theta})$. For example, for the exponential distribution with rate parameter $\lambda = 1$, the distribution function is $F(x) = \texttt{pexp}(x) = 1 - \exp(-x)$ for $x > 0$ (see Section C.3.7 for details on this distribution), and the p quantile function of the cdf is $x_p = \texttt{qexp}(p) = -\log(1 - p)$ for $0 < p < 1$. Then the 0.5 quantile (which is the median of the distribution) is $x_{0.5} = -\log(0.5) \approx 0.6931$. A direct computation shows that $\texttt{pexp}(0.6931) = 0.4999 \approx 0.5$, which verifies the computed value for $x_{0.5}$.

When $F(t; \boldsymbol{\theta})$ is continuous but not strictly monotone increasing, or when $F(x; \boldsymbol{\theta})$ is discrete, there may not be a unique value of x such that $F(x; \boldsymbol{\theta}) = p$. In such cases, x_p is the smallest of the x values with the property that $F(x; \boldsymbol{\theta}) \ge p$. For example, suppose one is interested in the $p = 0.25$ quantile, $x_{0.25}$, of the $\text{BINOM}(n = 5, \pi = 0.5)$ distribution (see Section C.4.1 for properties of this distribution). In this case, $\boldsymbol{\theta} = (n, \pi) = (5, 0.5)$ and $F(x; \boldsymbol{\theta}) = \texttt{pbinom}(x; 5, 0.5)$. It can be verified that $0.25 < \texttt{pbinom}(x; 5, 0.5)$ for $x \ge 2$ and $0.25 > \texttt{pbinom}(x; 5, 0.5)$ for $x < 2$, which shows that $x_{0.25} = \texttt{qbinom}(0.25; 5, 0.5) = 2$.

For a continuous cdf $F(x; \boldsymbol{\theta})$, the quantile function (C.1) is useful in the generation of random samples from $F(x; \boldsymbol{\theta})$. That is, to generate a random sample from $F(x; \boldsymbol{\theta})$, first generate a random sample from a $\text{UNIF}(0, 1)$ distribution, say u_i, $i = 1, \ldots, n$. Then $x_i = F^{-1}(u_i; \boldsymbol{\theta})$, $i = 1, \ldots, n$, is a random sample from $F(x; \boldsymbol{\theta})$ (see Example D.2 for an explanation of this result).

Behavior of quantiles under monotone transformations

The quantile function of a distribution has the important property of predictable behavior under monotone increasing (or decreasing) transformations as follows. Suppose that $X \sim F(x)$ and x_α (where $0 < \alpha < 1$) is the α quantile of $F(x)$. Define $Y = h(X)$, where $h(x)$ is a monotone function, and let $G(y)$ denote the cdf of Y. The following results hold:

1. For a continuous (or discrete) random variable X and a monotone increasing function $h(x)$, $y_\alpha = h(x_\alpha)$ (i.e., the quantiles of $G(y)$ are the corresponding quantiles of $F(x)$ transformed through $h(x)$). This result is explained by the fact that $G[h(x_\alpha)] = F(x_\alpha) \ge \alpha$ and $G[h(x))] = F(x) < \alpha$ for every $x < x_\alpha$.

2. For a continuous random variable X and a monotone decreasing function $h(x)$, the relationship between the quantiles is $y_\alpha = h[x_{(1-\alpha)}]$. This result is explained by the fact that $G[h(x_\alpha)] = 1 - F(x_\alpha) = 1 - \alpha$. As explained next, this result is not true when X is a discrete random variable.

3. Consider a discrete random variable X and a monotone decreasing function $h(x)$, and let i be the observation index for the $1 - \alpha$ quantile (i.e., $x_{(1-\alpha)} = x_i$). Then

$$y_\alpha = \begin{cases} h(x_{i+1}) & \text{if } F(x_i) = 1 - \alpha \\ h(x_i) = h[x_{(1-\alpha)}] & \text{otherwise,} \end{cases} \tag{C.2}$$

where x_{i+1} is the smallest point in the support of the distribution of X with the property that $F(x_{i+1}) > 1 - \alpha$. There is always a point with this property when $F(x_i) = 1 - \alpha$ because $\lim_{x \to \infty} F(x) = 1$.

Suppose that $F(x_i) = 1 - \alpha$. Thus, because $h(x)$ is decreasing, using (D.9) gives

$$G[h(x_{i+1})] = 1 - F(x_i) = \alpha.$$

This shows that $h(x_{i+1}) = y_\alpha$. Note that $y_\alpha < h(x_i)$ because $h(x)$ is a decreasing function of x.

Now suppose that $F(x_i) > 1 - \alpha$. Then $F(x) < 1 - \alpha$ for $x < x_i$. Consequently,

$$G[h(x_i)] = 1 - \lim_{x \uparrow x_i} F(x) \geq \alpha,$$
$$G[h(x_{i+1})] = 1 - \lim_{x \uparrow x_{i+1}} F(x) = 1 - F(x_i) < \alpha. \tag{C.3}$$

The two inequalities in (C.3) show that $h(x_i)$ is the α quantile of $Y = h(X)$. The relationship in (C.2) is used in (C.18) and (C.29) to determine the quantiles for the distribution of the number of conforming units as a function of the quantiles of the number of nonconforming units in binomial and hypergeometric distributions. These results were also needed to establish the computational methods presented in Chapter 5.

Example C.1 Quantiles of a Monotone Decreasing Function. Consider the monotone decreasing function $h(x) = 4 - x$, for $0 \leq x \leq 4$. Suppose that $X \sim \text{BINOM}(4, 1/2)$ and define $Y = h(X) = 4 - X$. We show the computation of two quantiles $\alpha_1 = 1/16$ and $\alpha_2 = 1/2$ for the distribution of Y.

Using the relationship (C.2) with $\alpha = \alpha_1 = 1/16$, we have $1 - \alpha_1 = 15/16$ and the quantile function qbinom gives $x_{(1-\alpha_1)} = \text{qbinom}(1 - \alpha_1; 4, 1/2) = 3$. This implies that $x_i = 3$ and $x_{i+1} = 4$. From the first row in (C.2), $y_{(1-\alpha_1)} = y_{(1/16)} = h(x_{i+1}) = 4 - 4 = 0$.

Now for the quantile $\alpha = \alpha_2 = 1/2$, $x_{(1-\alpha_2)} = \text{qbinom}(1 - \alpha_2, 4, 1/2) = 2$. This implies $x_i = 2$ and $\text{pbinom}(x_i, 4, 1/2) = 0.6875 \neq 1/2$. Thus from the second row in (C.2), $y_\alpha = h(x_i) = 4 - 2 = 2$. In this example, $Y \sim \text{BINOM}(4, 1/2)$ and the quantile function of this distribution gives $\text{qbinom}(\alpha_1, 4, 1/2) = 0$ and $\text{qbinom}(\alpha_2, 4, 1/2) = 2$, which verifies the computations done using (C.2). ■

C.3 CONTINUOUS DISTRIBUTIONS

C.3.1 Location-Scale and Log-Location-Scale-Distributions

The location-scale and log-location-scale families of distributions contain the most important continuous probability distributions used in practical applications, including the normal, lognormal, and Weibull distributions.

Location-scale distributions

A random variable X has a location-scale distribution with location parameter μ and scale parameter σ if its pdf and cdf are given by

$$f(x; \mu, \sigma) = \frac{1}{\sigma} \phi\left(\frac{x - \mu}{\sigma}\right) \quad \text{and} \quad F(x; \mu, \sigma) = \Phi\left(\frac{x - \mu}{\sigma}\right). \tag{C.4}$$

Here $-\infty < \mu < \infty$, $\sigma > 0$, $-\infty < x < \infty$, $\phi(z)$ is a continuous pdf that does not depend on unknown parameters and $\Phi(z)$ is the cdf corresponding to $\phi(z)$. Note that the density $f(x; \mu, \sigma)$ is completely determined by $\phi(z)$ and the parameters μ and σ. Similarly, $F(x; \mu, \sigma)$

	pdf	cdf	Quantile
Distribution	$\phi(z)$	$\Phi(z)$	$\Phi^{-1}(p)$
NORM	$\phi_{\text{norm}}(z) = \dfrac{\exp(-z^2/2)}{\sqrt{2\pi}}$	$\Phi_{\text{norm}}(z) = \int_{-\infty}^{z} \phi_{\text{norm}}(w)\,dw$	$\Phi_{\text{norm}}^{-1}(p)$
SEV	$\phi_{\text{sev}}(z) = \exp[z - \exp(z)]$	$\Phi_{\text{sev}}(z) = 1 - \exp[-\exp(z)]$	$\log[-\log(1-p)]$
LEV	$\phi_{\text{lev}}(z) = \exp[-z - \exp(-z)]$	$\Phi_{\text{lev}}(z) = \exp[-\exp(-z)]$	$-\log[-\log(p)]$
LOGIS	$\phi_{\text{logis}}(z) = \dfrac{\exp(z)}{[1 + \exp(z)]^2}$	$\Phi_{\text{logis}}(z) = \dfrac{\exp(z)}{1 + \exp(z)}$	$\log[p/(1-p)]$
CAUCHY	$\phi_{\text{cauchy}}(z) = \dfrac{1}{\pi(1 + z^2)}$	$\Phi_{\text{cauchy}}(z) = \dfrac{1}{2} + \dfrac{1}{\pi}\arctan(z)$	$\tan\left[\pi\left(p - \dfrac{1}{2}\right)\right]$

Table C.2   Standardized pdfs, cdfs, and quantile functions for commonly used location-scale and log-location-scale distributions.

is completely determined by $\Phi(z)$, μ, and σ. From the cdf in (C.4), the quantile function for the distribution $F(x; \mu, \sigma)$ is $x_p = \mu + \sigma\Phi^{-1}(p)$, where $\Phi^{-1}(p)$ is the p quantile of $\Phi(z)$.

The location-scale cdf representation in (C.4) leads to certain general properties of the location-scale family. In particular, let $Z = (X - \mu)/\sigma$. Then

$$\text{E}(X) = \mu + \sigma\text{E}(Z) \quad \text{and} \quad \text{Var}(X) = \sigma^2\text{Var}(Z).$$

Note that $\text{E}(X)$ is finite only when $\text{E}(Z)$ is finite. Similarly, $\text{Var}(X)$ is finite only when $\text{Var}(Z)$ is finite. Table C.2 lists five standardized pdfs $\phi(z)$ that characterize five location-scale distributions used in this book. In Section C.3.2, we summarize general properties of these five distributions.

Log-location-scale distributions

For each random variable X that has a location-scale distribution, there is a corresponding log-location-scale random variable defined by $T = \exp(X)$. Using transformation of random variables, as explained in Section D.1, the pdf and cdf of T are

$$f(t) = \frac{1}{\sigma t}\phi\left[\frac{\log(t) - \mu}{\sigma}\right] \quad \text{and} \quad F(t) = \Phi\left[\frac{\log(t) - \mu}{\sigma}\right], \tag{C.5}$$

where $t > 0$. The quantile function for $F(t)$ is $t_p = \exp[\mu + \sigma\Phi^{-1}(p)]$.

The five corresponding log-location-scale pdfs related to the location-scale distributions derived from Table C.2 are the lognormal distribution (using $\phi_{\text{norm}}(z)$), Weibull distribution (using $\phi_{\text{sev}}(z)$), Fréchet distribution (using $\phi_{\text{lev}}(z)$), log-logistic distribution (using $\phi_{\text{logis}}(z)$), and log-Cauchy distribution (using $\phi_{\text{cauchy}}(z)$).

C.3.2   Examples of Location-Scale and Log-Location-Scale Distributions

In this section we describe the five location-scale and the corresponding log-location-scale distributions derived from the standardized distributions in Table C.2.

Normal distribution

A random variable that has a normal distribution, denoted by $X \sim \text{NORM}(\mu, \sigma)$, is a member of the location-scale family and has a pdf $\text{dnorm}(x; \mu, \sigma)$ and cdf $\text{pnorm}(x; \mu, \sigma)$. The expressions for the pdf and cdf are obtained by replacing the functions $\phi(z)$ and $\Phi(z)$ in (C.4) with $\phi_{\text{norm}}(z)$ and $\Phi_{\text{norm}}(z)$ from Table C.2, respectively. That is,

$$\text{dnorm}(x; \mu, \sigma) = \frac{1}{\sigma}\phi_{\text{norm}}\left(\frac{x - \mu}{\sigma}\right) \quad \text{and} \quad \text{pnorm}(x; \mu, \sigma) = \Phi_{\text{norm}}\left(\frac{x - \mu}{\sigma}\right).$$

The normal distribution quantile function is $\text{qnorm}(p; \mu, \sigma) = \mu + \sigma \Phi_{\text{norm}}^{-1}(p)$, where $\Phi_{\text{norm}}^{-1}(p)$ is given in Table C.2. The mean and variance of the $\text{NORM}(\mu, \sigma)$ distribution are $\text{E}(X) = \mu$ and $\text{Var}(X) = \sigma^2$.

The normal distribution cdf has the following important symmetry relationship. For any value x,

$$\text{pnorm}(\mu + x; \mu, \sigma) + \text{pnorm}(\mu - x; \mu, \sigma) = 1. \tag{C.6}$$

Letting $\mu = 0$ and $\sigma = 1$ in (C.6) gives

$$\Phi_{\text{norm}}(z) = 1 - \Phi_{\text{norm}}(-z). \tag{C.7}$$

The latter result is useful when tabulating the cdf of the $\text{NORM}(0, 1)$ distribution because it suffices to tabulate $\Phi_{\text{norm}}(z)$ for $z \leq 0$. The values of $\Phi_{\text{norm}}(z)$ for $z > 0$ are obtained from (C.7). The relationship in (C.7) also implies that the quantiles of the $\text{NORM}(0, 1)$ distribution are symmetric in the sense that $z_{(p)} = -z_{(1-p)}$. Thus in tabulating the quantiles of the $\text{NORM}(0, 1)$ distribution, it suffices to tabulate $z_{(p)}$ for $0 < p \leq 0.5$.

Lognormal distribution

A random variable that has a lognormal distribution, denoted by $T \sim \text{LNORM}(\mu, \sigma)$, has the property that $\log(T) \sim \text{NORM}(\mu, \sigma)$ and is thus a member of the log-location-scale family. Let $\text{dlnorm}(t; \mu, \sigma)$ and $\text{plnorm}(t; \mu, \sigma)$ denote the pdf and cdf of the distribution, respectively. The expressions for the pdf and cdf are obtained by replacing the functions $\phi(z)$ and $\Phi(z)$ in (C.5) with $\phi_{\text{norm}}(z)$ and $\Phi_{\text{norm}}(z)$ from Table C.2, respectively. That is,

$$\text{dlnorm}(t; \mu, \sigma) = \frac{1}{\sigma t}\phi_{\text{norm}}\left[\frac{\log(t) - \mu}{\sigma}\right] \quad \text{and} \quad \text{plnorm}(t; \mu, \sigma) = \Phi_{\text{norm}}\left[\frac{\log(t) - \mu}{\sigma}\right],$$

where $t > 0$. In this case, μ is the mean of $\log(T)$, $\exp(\mu)$ is the median (and a scale parameter) of the distribution of T, σ is a shape parameter, and σ^2 is the variance of $\log(T)$.

The lognormal distribution quantile function is $\text{qlnorm}(p; \mu, \sigma) = \exp[\mu + \sigma \Phi_{\text{norm}}^{-1}(p)]$. The mean and variance of the $\text{LNORM}(\mu, \sigma)$ distribution are

$$\text{E}(T) = \exp\left(\mu + \frac{\sigma^2}{2}\right) \quad \text{and} \quad \text{Var}(T) = \exp(2\mu + \sigma^2)\left[\exp(\sigma^2) - 1\right].$$

Smallest extreme value distribution

A random variable that has a smallest extreme value distribution, denoted by $X \sim \text{SEV}(\mu, \sigma)$, is a member of the location-scale family and has a pdf and cdf given by $\text{dsev}(x; \mu, \sigma)$ and $\text{psev}(x; \mu, \sigma)$, respectively. The expressions for the pdf and cdf are obtained by replacing the functions $\phi(z)$ and $\Phi(z)$ in (C.4) with $\phi_{\text{sev}}(z)$ and $\Phi_{\text{sev}}(z)$ from Table C.2, respectively. The quantile function of the $\text{SEV}(\mu, \sigma)$ distribution is $\text{qsev}(p; \mu, \sigma) = \mu + \sigma \Phi_{\text{sev}}^{-1}(p)$, where Φ_{sev}^{-1}

is given in Table C.2. The mean and variance of the $SEV(\mu, \sigma)$ distribution are

$$E(X) = \mu - \sigma\gamma \quad \text{and} \quad Var(X) = \frac{\sigma^2\pi^2}{6},$$

where $\gamma \approx 0.5772$ is Euler's constant.

Weibull distribution

A random variable that has a Weibull distribution, denoted by $T \sim WEIBULL(\eta, \beta)$, has the property that $\log(T) \sim SEV(\mu, \sigma)$, where $\mu = \log(\eta)$ and $\sigma = 1/\beta$. Thus, the Weibull distribution is a member of the log-location-scale family and replacing the functions $\phi(z)$ and $\Phi(z)$ with $\phi_{sev}(z)$ and $\Phi_{sev}(z)$, respectively, from Table C.2 in (C.5) gives the Weibull pdf and cdf as

$$\texttt{dweibull}(t; \eta, \beta) = \frac{1}{\sigma t}\phi_{sev}\left[\frac{\log(t) - \mu}{\sigma}\right] = \left(\frac{\beta}{t}\right)\left(\frac{t}{\eta}\right)^{\beta}\exp\left[-\left(\frac{t}{\eta}\right)^{\beta}\right],$$

$$\texttt{pweibull}(t; \eta, \beta) = \Phi_{sev}\left[\frac{\log(t) - \mu}{\sigma}\right] = 1 - \exp\left[-\left(\frac{t}{\eta}\right)^{\beta}\right],$$

(C.8)

where $\eta > 0$ is a scale parameter, $\beta > 0$ is a shape parameter, and $t > 0$.

Note that (C.8) gives two different parameterizations for the Weibull distribution. The (η, β) parameterization is common in applications and the (μ, σ) parameterization is useful for graphical purposes, regression models, and theoretical work (see Meeker and Escobar, 1998, Chapters 4 and 6, for details).

The quantile function of the Weibull distribution is

$$\texttt{qweibull}(p; \eta, \beta) = \exp\left[\mu + \sigma\Phi_{sev}^{-1}(p)\right] = \eta[-\log(1 - p)]^{1/\beta}.$$

The mean and variance of the $WEIBULL(\eta, \beta)$ distribution are

$$E(T) = \eta\Gamma\left(1 + \frac{1}{\beta}\right) \quad \text{and} \quad Var(T) = \eta^2\left[\Gamma\left(1 + \frac{2}{\beta}\right) - \Gamma^2\left(1 + \frac{1}{\beta}\right)\right],$$

where $\Gamma(z)$ is the gamma function defined in Appendix A.

Largest extreme value distribution

A random variable that has a largest extreme value distribution, denoted by $X \sim LEV(\mu, \sigma)$, is a member of the location-scale family and has a pdf and cdf given by $\texttt{dlev}(x; \mu, \sigma)$ and $\texttt{plev}(x; \mu, \sigma)$, respectively. The expressions for the pdf and cdf are obtained by replacing the functions $\phi(z)$ and $\Phi(z)$ in (C.4) with $\phi_{lev}(z)$ and $\Phi_{lev}(z)$ from Table C.2, respectively. The quantile function for this distribution is $\texttt{qlev}(p; \mu, \sigma) = \mu + \sigma\Phi_{lev}^{-1}(p)$, where the function $\Phi_{lev}^{-1}(p)$ is given in Table C.2. The mean and variance of the $LEV(\mu, \sigma)$ distribution are

$$E(X) = \mu + \sigma\gamma \quad \text{and} \quad Var(X) = \frac{\sigma^2\pi^2}{6},$$

where $\gamma \approx 0.5772$ is Euler's constant.

Fréchet distribution

A random variable that has a Fréchet distribution, denoted by $T \sim \text{FREC}(\eta, \beta)$, has the property that $\log(T) \sim \text{LEV}(\mu, \sigma)$, where $\mu = \log(\eta)$ and $\sigma = 1/\beta$. Thus, T is a member of the log-location-scale family with pdf $\texttt{dfrechet}(t; \eta, \beta)$ and cdf $\texttt{pfrechet}(t; \eta, \beta)$. Using (C.5) and replacing the functions $\phi(z)$ and $\Phi(z)$ with $\phi_{\text{lev}}(z)$ and $\Phi_{\text{lev}}(z)$, respectively, from Table C.2 gives

$$
\begin{aligned}
\texttt{dfrechet}(t; \eta, \beta) &= \frac{1}{\sigma t} \phi_{\text{lev}} \left[\frac{\log(t) - \mu}{\sigma} \right] = \left(\frac{\beta}{t} \right) \left(\frac{\eta}{t} \right)^{\beta} \exp\left[-\left(\frac{\eta}{t} \right)^{\beta} \right], \\
\texttt{pfrechet}(t; \eta, \beta) &= \Phi_{\text{lev}} \left[\frac{\log(t) - \mu}{\sigma} \right] = \exp\left[-\left(\frac{\eta}{t} \right)^{\beta} \right],
\end{aligned}
\tag{C.9}
$$

where $\eta > 0$ is a scale parameter, $\beta > 0$ is a shape parameter, and $t > 0$.

Note that (C.9) gives two equivalent parameterizations for the Fréchet distribution. The (μ, σ) parameterization is useful for graphical purposes, regression models, and theoretical work. The (η, β) parameterization is more common in other applications.

The p quantile of the Fréchet distribution is

$$
\texttt{qfrechet}(p; \eta, \beta) = \exp\left[\mu + \sigma \Phi_{\text{lev}}^{-1}(p) \right] = \frac{\eta}{[-\log(p)]^{1/\beta}}.
$$

The mean and variance of the $\text{FREC}(\eta, \beta)$ distribution are

$$
\text{E}(T) = \eta \Gamma \left(1 - \frac{1}{\beta} \right) \quad \text{and} \quad \text{Var}(T) = \eta^2 \left[\Gamma \left(1 - \frac{2}{\beta} \right) - \Gamma^2 \left(1 - \frac{1}{\beta} \right) \right],
$$

where $\text{E}(T)$ is finite if $\beta > 1$ and $\text{Var}(T)$ is finite if $\beta > 2$. The gamma function $\Gamma(z)$ is defined in Appendix A.

Logistic distribution

A random variable that has a logistic distribution, denoted by $X \sim \text{LOGIS}(\mu, \sigma)$, is a member of the location-scale family and has a pdf and cdf given by $\texttt{dlogis}(x; \mu, \sigma)$ and $\texttt{plogis}(x; \mu, \sigma)$, respectively. The expressions for the pdf and cdf are obtained by replacing the functions $\phi(z)$ and $\Phi(z)$ in (C.4) with $\phi_{\text{logis}}(z)$ and $\Phi_{\text{logis}}(z)$ from Table C.2, respectively. The logistic distribution quantile function is $\texttt{qlogis}(x; \mu, \sigma) = \mu + \sigma \Phi_{\text{logis}}^{-1}(p)$, where $\Phi_{\text{logis}}^{-1}(p)$ is given in Table C.2. The mean and variance of the $\text{LOGIS}(\mu, \sigma)$ distribution are $\text{E}(X) = \mu$ and $\text{Var}(X) = \sigma^2 \pi^2 / 3$.

Log-logistic distribution

A random variable that has a log-logistic distribution, denoted by $T \sim \text{LLOGIS}(\mu, \sigma)$, has property that $\log(T) \sim \text{LOGIS}(\mu, \sigma)$ and is thus a member of the log-location-scale family. Let $\texttt{dllogis}(t; \mu, \sigma)$ and $\texttt{pllogis}(t; \mu, \sigma)$ denote, respectively, the pdf and cdf of the distribution. The expressions for the pdf and cdf are obtained by replacing the functions $\phi(z)$ and $\Phi(z)$ in (C.5) with $\phi_{\text{logis}}(z)$ and $\Phi_{\text{logis}}(z)$ from Table C.2, respectively. The log-logistic distribution quantile function is

$$
\texttt{qllogis}(p; \mu, \sigma) = \exp\left[\mu + \sigma \Phi_{\text{logis}}^{-1}(p) \right] = \exp(\mu) \left(\frac{p}{1-p} \right)^{\sigma},
$$

where $\Phi_{\text{logis}}^{-1}(z)$ is given in Table C.2. For this distribution, $\exp(\mu)$ is the median of the distribution (and a scale parameter) and σ is a shape parameter. The mean and variance of the LLOGIS(μ, σ) distribution are

$$E(T) = \exp(\mu)\Gamma(1 + \sigma)\Gamma(1 - \sigma),$$
$$\text{Var}(T) = \exp(2\mu)\left[\Gamma(1 + 2\sigma)\Gamma(1 - 2\sigma) - \Gamma^2(1 + \sigma)\Gamma^2(1 - \sigma)\right],$$

where $\Gamma(z)$ is the gamma function defined in Appendix A. Note that $E(T)$ is finite only if $0 < \sigma < 1$ and $\text{Var}(T)$ is finite only if $0 < \sigma < 1/2$.

Cauchy distribution

A random variable that has a Cauchy distribution, denoted by $X \sim \text{CAUCHY}(\mu, \sigma)$, is a member of the location-scale family and has a pdf given by $\texttt{dcauchy}(x; \mu, \sigma)$ and cdf $\texttt{pcauchy}(x; \mu, \sigma)$. The expressions for the pdf and cdf are obtained by replacing the functions $\phi(z)$ and $\Phi(z)$ in (C.4) with $\phi_{\text{cauchy}}(z)$ and $\Phi_{\text{cauchy}}(z)$ from Table C.2, respectively. The Cauchy distribution quantile function is $\texttt{qcauchy}(p; \mu, \sigma) = \mu + \sigma\Phi_{\text{cauchy}}^{-1}(p)$, where $\Phi_{\text{cauchy}}^{-1}(p)$ is given in Table C.2. That is,

$$x_p = \texttt{qcauchy}(p; \mu, \sigma) = \mu + \sigma\tan\left[\pi\left(p - \frac{1}{2}\right)\right],$$

where $0 < p < 1$ and $\tan(z)$ is the tangent function for the angle z in radians. In particular, $\mu - \sigma$, μ, and $\mu + \sigma$ are the first, second, and third quartiles of the distribution, respectively. That is, $\texttt{pcauchy}(\mu - \sigma; \mu, \sigma) = 0.25$, $\texttt{pcauchy}(\mu; \mu, \sigma) = 0.5$, and $\texttt{pcauchy}(\sigma; \mu + \sigma) = 0.75$.

The Cauchy distribution does not have a finite mean or variance. When the location parameter μ is equal to 0 and the scale parameter σ is equal to 1, the Cauchy distribution is a Student's t-distribution with 1 degrees of freedom (see Section C.3.10 for more details).

Log-Cauchy distribution

A random variable that has a log-Cauchy distribution, denoted by $T \sim \text{LCAUCHY}(\mu, \sigma)$, has the property that $\log(T) \sim \text{CAUCHY}(\mu, \sigma)$ and is thus a member of the log-location-scale family. Let $\texttt{dlcauchy}(t; \mu, \sigma)$ and $\texttt{plcauchy}(t; \mu, \sigma)$ denote the pdf and cdf of the distribution, respectively. The expressions for the pdf and cdf of the distribution are obtained by replacing the functions $\phi(z)$ and $\Phi(z)$ in (C.5) with $\phi_{\text{cauchy}}(z)$ and $\Phi_{\text{cauchy}}(z)$ from Table C.2, respectively. For this distribution, $\exp(\mu)$ is the median of the distribution and σ is a shape parameter. The log-Cauchy distribution quantile function is $\texttt{qlcauchy}(p; \mu, \sigma) = \exp\left[\mu + \sigma\Phi_{\text{cauchy}}^{-1}(p)\right]$, where $\Phi_{\text{cauchy}}^{-1}(z)$ is given in Table C.2. The log-Cauchy distribution does not have a finite mean or variance.

C.3.3 Beta Distribution

The pdf and cdf of the beta random variable, denoted by $X \sim \text{BETA}(a, b)$, are

$$\texttt{dbeta}(x; a, b) = \frac{1}{\mathcal{B}(a, b)} x^{a-1}(1 - x)^{b-1} \quad \text{and} \quad \texttt{pbeta}(x; a, b) = \int_0^x \texttt{dbeta}(w; a, b)\, dw,$$

where $a > 0$ and $b > 0$ are shape parameters, $0 < x < 1$, and $\mathcal{B}(a, b)$ is the beta function defined in Appendix A. The beta distribution quantile function $\texttt{qbeta}(p; a, b)$ does not have a closed-form expression. The mean and variance of the $\mathrm{BETA}(a, b)$ distribution are

$$\mathrm{E}(X) = \frac{a}{a+b} \quad \text{and} \quad \mathrm{Var}(X) = \frac{ab}{(a+b)^2(a+b+1)}.$$

The $\mathrm{UNIF}(0, 1)$ is the special case of the beta distribution when $a = b = 1$.

Monotone decreasing behavior of a single parameter beta distribution.

Consider the special case of a single-parameter beta cdf given by $\mathtt{pbeta}(x, a, n - a + 1)$, where $a > 0$ is a shape parameter and n is a given positive quantity. This cdf is a monotone decreasing function of a in the range $0 < a < n + 1$. Let δ be such that $0 < \delta < n + 1 - a$. Direct computations give

$$\frac{\mathtt{dbeta}(x; a+\delta, n+1-a-\delta)}{\mathtt{dbeta}(x; a, n+1-a)} = \frac{\Gamma(a)\Gamma(n-a+1)}{\Gamma(a+\delta)\Gamma(n-a-\delta+1)} \frac{x^{a+\delta-1}(1-x)^{n-a-\delta}}{x^{a-1}(1-x)^{n-a}}$$

$$= \frac{\Gamma(a)\Gamma(n-a+1)}{\Gamma(a+\delta)\Gamma(n-a-\delta+1)} \left(\frac{x}{1-x}\right)^{\delta}. \qquad \text{(C.10)}$$

Because (C.10) is a monotone increasing function of x for $0 < x < 1$, for fixed x and n the cdf $\mathtt{pbeta}(x; a, n - a + 1)$ is a monotone decreasing function of a (see details in Lehmann, 1986, pages 85). Equivalently, for fixed p and n, $\mathtt{qbeta}(p; a, n - a + 1)$ is increasing in a. This result is important, for example, in determining the smallest sample size to have a positive probability of successful demonstration test for a binomial parameter, as done in Section I.2.2.

Beta probabilities and quantiles as function of Snedecor's F-distribution probabilities and quantiles

Relationships to compute beta quantiles and probabilities as a function of Snedecor's F-distribution probabilities and quantiles are:

$$\mathtt{pbeta}(x; a, b) = \mathtt{pf}\left[\frac{bx}{a(1-x)}; 2a, 2b\right] = 1 - \mathtt{pf}\left[\frac{a(1-x)}{bx}; 2b, 2a\right],$$

$$\mathtt{qbeta}(p; a, b) = \frac{a}{a + b/\mathtt{qf}(p; 2a, 2b)} = \frac{a}{a + b\,\mathtt{qf}(1 - p; 2b, 2a)}, \qquad \text{(C.11)}$$

where $0 \le x < 1$ and $\mathtt{qf}(\gamma; r_1, r_2) = F_{(\gamma; r_1, r_2)}$ is the γ quantile for Snedecor's F-distribution with (r_1, r_2) degrees of freedom (see Example D.3 for details on these relationships). For example, using R as a calculator gives

```
> qbeta(p=0.3, shape1=2, shape2=3)
[1]  0.2723839
> 2/(2+3*qf(p=1-0.3, df1=2*3, df2=2*2))
[1]  0.2723839
> 2/(2+3/qf(p=0.3, df1=2*2, df2=2*3))
[1]  0.2723839
> pbeta(q=0.27238, shape1=2, shape2=3)
[1]  0.2999932
> pf((q=3/2)*0.27238/(1-0.27238), df1=2*2, df2=2*3)
[1]  0.2999932
```

The relationship between the beta quantiles and Snedecor's F-distribution quantiles in (C.11) is used in Sections 6.2.2 and 6.2.5 to express, respectively, the conservative and the Jeffreys confidence interval methods for the binomial parameter π in terms of F-distribution quantiles.

C.3.4 Log-Uniform Distribution

A random variable that has a log-uniform distribution, denoted by $T \sim \text{LUNIF}(a, b)$, has the property that $\log(T) \sim \text{UNIF}[\log(a), \log(b)]$. Thus T has pdf and cdf given by

$$\texttt{dlunif}(t; a, b) = \frac{1}{t\,[\log(b) - \log(a)]} \quad \text{and} \quad \texttt{plunif}(t; a, b) = \frac{\log(t) - \log(a)}{\log(b) - \log(a)},$$

where $0 < a < b$ and $a \leq t \leq b$. The quantile function of the distribution is $t_p = \texttt{qlunif}(p; a, b) = a(b/a)^p$, where $0 < p < 1$. The mean and variance of the $\text{LUNIF}(a, b)$ distribution are

$$\text{E}(T) = \frac{b - a}{\log(b) - \log(a)},$$

$$\text{Var}(T) = \frac{(b^2 - a^2)[\log(b) - \log(a)] - 2\,(b - a)^2}{2\,[\log(b) - \log(a)]^2}.$$

C.3.5 Gamma Distribution

The pdf and cdf for the gamma random variable, denoted by $T \sim \text{GAMMA}(a, b)$, are

$$\texttt{dgamma}(t; a, b) = \frac{b^a\, t^{a-1}}{\Gamma(a)}\exp(-bt) \quad \text{and} \quad \texttt{pgamma}(t; a, b) = \int_0^t \texttt{dgamma}(w; a, b)\, dw,$$

where $a > 0$ is a shape parameter, $b > 0$ is a rate parameter, $t > 0$, and $\Gamma(a)$ is the gamma function defined in Appendix A. The gamma distribution quantile function $\texttt{qgamma}(p; a, b)$ does not have a closed-form expression. The mean and variance of the $\text{GAMMA}(a, b)$ distribution are $\text{E}(T) = a/b$ and $\text{Var}(T) = a/b^2$.

Sometimes the gamma distribution is parameterized using a scale parameter $\eta = 1/b$. In this case, to obtain the pdf, cdf, quantile function, mean, and variance of the distribution, replace b with $1/\eta$ in the corresponding expressions above.

C.3.6 Chi-Square Distribution

A chi-square distribution with r degrees of freedom, denoted by $\chi^2(r)$, is a special case of the $\text{GAMMA}(a, b)$ distribution where $a = r/2$ and $b = 1/2$. Observe that r can take noninteger values. Using the expressions for the $\text{GAMMA}(a, b)$ distribution with $a = r/2$ and $b = 1/2$ gives the chi-square pdf and cdf

$$\texttt{dchisq}(t; r) = \frac{t^{r/2-1}}{2^{r/2}\Gamma(r/2)}\exp\left(-\frac{t}{2}\right) \quad \text{and} \quad \texttt{pchisq}(t; r) = \int_0^t \texttt{dchisq}(w; r)\, dw,$$

where $\Gamma(z)$ is the gamma function defined in Appendix A and $t > 0$. The chi-square quantile function $\texttt{qchisq}(p; r) = \chi^2_{(p;r)}$ does not have a closed-form expression. A chi-square random variable with r degrees of freedom is denoted by $X^2_{(r)}$. The mean and variance of the $\chi^2(r)$ distribution are $\text{E}\left(X^2_{(r)}\right) = r$ and $\text{Var}\left(X^2_{(r)}\right) = 2r$.

C.3.7 Exponential Distribution

An exponential random variable, denoted by $T \sim \text{EXP}(\lambda)$, is a $\text{GAMMA}(a, \lambda)$ with $a = 1$. Then using the $\text{GAMMA}(1, \lambda)$ pdf and cdf expressions gives the exponential pdf and cdf as

$$\text{dexp}(t; \lambda) = \lambda \exp(-\lambda t) \quad \text{and} \quad \text{pexp}(t; \lambda) = 1 - \exp(-\lambda t),$$

where $\lambda > 0$ is a rate parameter and $t > 0$. The exponential distribution quantile function is $\text{qexp}(p; \lambda) = -(1/\lambda) \log(1 - p)$. The mean and variance of the $\text{EXP}(\lambda)$ distribution are $\text{E}(T) = 1/\lambda$ and $\text{Var}(T) = 1/\lambda^2$.

A common *alternative* parameterization of the exponential distribution is in terms of the scale parameter $\theta = 1/\lambda$ which is also the mean of the distribution. In this case, the pdf, cdf, and quantile functions are obtained from the expressions given above with $\lambda = 1/\theta$. In particular, $F(t; \theta) = 1 - \exp(-t/\theta)$. This parameterization shows that the exponential distribution is also a Weibull distribution with shape parameter $\beta = 1$ and scale parameter θ. This parameterization in terms of the mean parameter is used in Section 12.3 to illustrate likelihood-based confidence intervals for a single-parameter distribution.

C.3.8 Generalized Gamma Distribution

The generalized gamma (GNG) distribution (which has also been known as the extended generalized gamma distribution) has three parameters and contains the Weibull, lognormal, Fréchet, and gamma distributions as special cases. The GNG cdf is

$$\Pr(T \le t) = F(t; \mu, \sigma, \lambda) = \begin{cases} \Phi_{\text{lg}}[\lambda\omega + \log(\lambda^{-2}); \lambda^{-2}] & \text{if } \lambda > 0 \\ \Phi_{\text{norm}}(\omega) & \text{if } \lambda = 0 \\ 1 - \Phi_{\text{lg}}[\lambda\omega + \log(\lambda^{-2}); \lambda^{-2}] & \text{if } \lambda < 0, \end{cases} \quad (\text{C.12})$$

where $t > 0$, $\omega = [\log(t) - \mu]/\sigma$, $-\infty < \mu < \infty$, $-\infty < \lambda < \infty$, $\sigma > 0$, and

$$\Phi_{\text{lg}}(z; a) = \Gamma_{\text{I}}[\exp(z); a]$$

with $a > 0$ and

$$\Gamma_{\text{I}}(v; a) = \frac{\int_0^v x^{a-1} \exp(-x)\, dx}{\Gamma(a)}, \quad v > 0,$$

where $\Phi_{\text{lg}}(z; a)$ is known as the incomplete gamma function. Inverting the cdf in (C.12) gives the p quantile of the GNG distribution,

$$t_p = \exp[\mu + \sigma \times \omega(p; \lambda)], \quad (\text{C.13})$$

where $\omega(p; \lambda)$ is the p quantile of $[\log(T) - \mu]/\sigma$ given by

$$\omega(p; \lambda) = \begin{cases} \left(\dfrac{1}{\lambda}\right) \log\left[\lambda^2 \Gamma_{\text{I}}^{-1}(p; \lambda^{-2})\right] & \text{if } \lambda > 0 \\ \Phi_{\text{norm}}^{-1}(p) & \text{if } \lambda = 0 \\ \left(\dfrac{1}{\lambda}\right) \log\left[\lambda^2 \Gamma_{\text{I}}^{-1}(1 - p; \lambda^{-2})\right] & \text{if } \lambda < 0. \end{cases}$$

From (C.12), we conclude that $\exp(\mu)$ is a scale parameter and σ and λ are shape parameters. For any given fixed value of λ, the GNG distribution is a log-location-scale distribution. The GNG distribution has the following important special cases:

- If $\lambda = 1$, T has a Weibull distribution with $\eta = \exp(\mu)$ and $\beta = 1/\sigma$.

- If $\lambda = 0$, T has a lognormal distribution with parameters μ and σ.

- If $\lambda = -1$, T has a Fréchet distribution of *maxima* (which is equivalent to the distribution of the reciprocal of a Weibull random variable).

For more information about the GNG distribution, see Meeker and Escobar (1998, Section 5.4).

C.3.9   Noncentral t-Distribution

A random variable T with a noncentral t-distribution is denoted by $T \sim t(r, \delta)$ and is defined by the ratio

$$T = \frac{Z + \delta}{\sqrt{X_{(r)}^2/r}},$$

where $Z \sim \text{NORM}(0, 1)$ and $X_{(r)}^2 \sim \chi^2(r)$ are independent random variables, r is a positive real value, and δ is a real value. The pdf of this distribution (see Example D.4 for a detailed derivation) is

$$\text{dt}(t; r, \delta) = \frac{r^{r/2} \exp(-r\delta^2/[2(r + t^2)])}{\sqrt{\pi}\,\Gamma(r/2)\,(r + t^2)^{(r+1)/2}} \int_0^\infty z^{(r-1)/2} \exp\left[-\left(\sqrt{z} - \frac{t\delta}{\sqrt{2(r + t^2)}}\right)^2\right] dz,$$

where $-\infty < t < \infty$. There are no simple formulas for the noncentral t-distribution pdf, cdf, or quantile function. The mean and variance of the noncentral t-distribution are

$$\text{E}(T) = \frac{\delta\,\Gamma[(r - 1)/2]}{\Gamma(r/2)} \sqrt{\frac{r}{2}} \quad \text{and} \quad \text{Var}(T) = \frac{(1 + \delta^2)\,r}{r - 2} - [\text{E}(T)]^2.$$

Note that $\text{E}(T)$ is finite only if $r > 1$ and $\text{Var}(T)$ is finite only if $r > 2$. Some important properties of the noncentral t-distribution are as follows.

A symmetry relationship of the noncentral t-distribution cdf

For given $r > 0$ and real-valued quantity δ, $\text{pt}(-t; r, -\delta) + \text{pt}(t; r, \delta) = 1$ for any value t. This is a consequence of the fact that if $T \sim t(r, \delta)$ then $W = -T \sim t(r, -\delta)$. Thus, using the transformation of variables $W = g(T) = -T$ and (D.3) gives $\Pr(W \le w) = 1 - \Pr(T \le -w)$. That is, $\text{pt}(w; r; -\delta) = 1 - \text{pt}(-w; r, \delta)$. Equivalently, with $w = -t$,

$$\text{pt}(t; r, \delta) + \text{pt}(-t; r, -\delta) = 1, \tag{C.14}$$

which is the proposed result. This result is useful in establishing an important property of the `normTailCI` function defined in (E.9). A direct consequence of (C.14) is that the quantiles of the distributions $\text{pt}(t; r, \delta)$ and $\text{pt}(w; r, -\delta)$ are related through $t_{(\gamma;r,\delta)} = -t_{(1-\gamma;r,-\delta)}$, for every $0 < \gamma < 1$ and real value δ. In particular, if $z_{(p)}$ is the p quantile of a $\text{NORM}(0, 1)$ distribution and $n > 1$ is a given sample size, we have

$$t_{(\gamma;n-1,\sqrt{n}\,z_{(p)})} = -t_{(1-\gamma;n-1,\sqrt{n}\,z_{(1-p)})}. \tag{C.15}$$

This result is useful when tabulating the factors in Tables J.7a–J.7d because it implies $g'_{(\gamma;p,n)} = -g'_{(1-\gamma;1-p,n)}$ and thus it suffices to tabulate the factors for $0 < p < 0.5$. The factors for $0.5 \le p < 1$ are obtained using the relationship in (C.15).

The noncentral t-distribution cdf is a monotone decreasing function of δ

$\mathrm{pt}(t; r, \delta)$ is monotone decreasing function of δ. This monotone decreasing property follows from the result given in Lehmann (1986, page 295). This result is useful in obtaining pivotal-based confidence intervals for tail probabilities of a normal distribution in Sections E.3.4 and 4.5.

C.3.10   Student's t-Distribution

Student's t-distribution is a special case of the noncentral t-distribution when the noncentrality parameter is equal to zero (i.e., $\delta = 0$). Using the convention of omitting the noncentrality parameter δ when this parameter is equal to zero, Student's t-distribution is denoted by the simpler notation $T \sim t(r)$. The pdf, mean, and variance of Student's t-distribution are obtained by substituting $\delta = 0$ into the corresponding expressions for the noncentral t-distribution. This gives

$$\mathrm{dt}(t; r) = \frac{\Gamma[(r+1)/2]}{\Gamma(r/2)} \frac{1}{\sqrt{r\pi}} \frac{1}{(1 + t^2/r)^{(r+1)/2}}, \quad -\infty < t < \infty,$$

$\mathrm{E}(T) = 0$ for $r > 1$, and $\mathrm{Var}(T) = r/(r-2)$ for $r > 2$. The pdf, cdf, and quantile functions of Student's t-distribution are denoted by $\mathrm{dt}(t; r)$, $\mathrm{pt}(t; r)$, and $\mathrm{qt}(p; r) = t_{(p;r)}$, respectively. Because of the symmetry of the distribution, the quantiles of the distribution follow the relationship $\mathrm{qt}(p; r) = -\mathrm{qt}(1 - p; r)$. This is useful because most tables of Student's t-distribution quantiles have entries only for values of $p \ge 0.5$.

When $r = 1$, Student's t-distribution is known as the standardized Cauchy distribution (see Section C.3.2 for details about the Cauchy distribution).

C.3.11   Snedecor's F-Distribution

A random variable X that has Snedecor's F-distribution with (r_1, r_2) degrees of freedom, denoted by $X \sim F(r_1, r_2)$, is defined by the ratio $X = (X^2_{(r_1)}/r_1)/(X^2_{(r_2)}/r_2)$, where $X^2_{(r_1)} \sim \chi^2(r_1)$ and $X^2_{(r_2)} \sim \chi^2(r_2)$ are independent random variables. Using transformation of random variables, as explained in Section D.1, gives the following pdf and cdf for X:

$$\mathrm{df}(x; r_1, r_2) = \frac{1}{\mathcal{B}(r_1/2, r_2/2)} \left(\frac{r_1}{r_2}\right)^{r_1/2} \frac{x^{(r_1-2)/2}}{[1 + (r_1/r_2)x]^{(r_1+r_2)/2}},$$

$$\mathrm{pf}(x; r_1, r_2) = \int_0^x \mathrm{df}(w; r_1, r_2)\, dw,$$

where $r_1 > 0$, $r_2 > 0$, $\mathcal{B}(a, b)$ is the beta function defined in Appendix A, and $x \ge 0$.

The Snedecor's F-distribution quantile function $\mathrm{qf}(p; r_1, r_2) = F_{(p; r_1, r_2)}$ does not have a closed-form expression. The mean and variance of the $F(r_1, r_2)$ distribution are

$$\mathrm{E}(X) = \frac{r_2}{r_2 - 2} \quad \text{and} \quad \mathrm{Var}(X) = 2\left(\frac{r_2}{r_2 - 2}\right)^2 \frac{(r_1 + r_2 - 2)}{r_1(r_2 - 4)}.$$

Note that $\mathrm{E}(X)$ is finite if $r_2 > 2$ and $\mathrm{Var}(X)$ is finite if $r_2 > 4$.

Snedecor's F-distribution probabilities and quantiles as a function of beta probabilities and quantiles

Relationships to compute Snedecor's F-distribution probabilities and quantiles as a function of beta probabilities and quantiles are:

$$\text{pf}(x; r_1, r_2) = \text{pbeta}[r_1\, x/(r_2 + r_1 x); r_1/2, r_2/2],$$

$$\text{qf}(p; r_1, r_2) = \frac{r_2\, \text{qbeta}(p;\, r_1/2, r_2/2)}{r_1[1 - \text{qbeta}(p;\, r_1/2, r_2/2)]},$$

where $0 \le x < \infty$ and $\text{qbeta}(\gamma; a, b)$ is the γ quantile for a $\text{BETA}(a, b)$ (see Example D.3 for details about these relationships).

A symmetry property of Snedecor's F-distribution cdf

The relationship

$$\text{qf}(p;\, r_1, r_2) = 1/\text{qf}(1 - p;\, r_2, r_1)$$

is useful because most available tables for F-distribution quantiles contain entries only for $p \ge 0.50$.

C.3.12   Scale Half-Cauchy Distribution

A scale half-Cauchy distribution is related to a $\text{CAUCHY}(0, \sigma)$ as follows. If $X \sim \text{CAUCHY}(0, \sigma)$ then $H = |X|$ has a scale half-Cauchy distribution, which is denoted by $H \sim \text{HCAUCHY}(\sigma)$. This distribution is also known as a folded-CAUCHY distribution. The pdf and cdf of scale half-Cauchy are

$$\text{dhcauchy}(h; \sigma) = 2\, \text{dcauchy}(h; 0, \sigma) \quad \text{and} \quad \text{phcauchy}(h; \sigma) = 2\, \text{pcauchy}(h; 0, \sigma) - 1,$$

where σ is a scale parameter and $h \ge 0$. The quantiles for the $\text{HCAUCHY}(\sigma)$ distribution are

$$h_p = \text{qhcauchy}(p; \sigma) = \text{qcauchy}[(1 + p)/2; 0, \sigma] = \sigma \tan\left(\frac{\pi p}{2}\right).$$

Note that σ is the median of the $\text{HCAUCHY}(\sigma)$ distribution (i.e., $h_{0.5} = \sigma$). The scale half-Cauchy distribution does not have a finite mean or variance. The $\text{HCAUCHY}(\sigma)$ distribution is used in Chapter 18 to specify a diffuse prior for a random effect parameter.

C.4   DISCRETE DISTRIBUTIONS

C.4.1   Binomial Distribution

A binomial random variable, denoted by $X \sim \text{BINOM}(n, \pi)$, arises as the number of nonconforming units in n independent binary trials, when at each trial the probability of observing a nonconforming unit is π. The pmf and cdf for the binomial distribution are

$$\text{dbinom}(x; n, \pi) = \frac{n!}{x!\,(n - x)!}\, \pi^x (1 - \pi)^{n-x} \quad \text{and} \quad \text{pbinom}(x; n, \pi) = \sum_{i=0}^{x} \text{dbinom}(i; n, \pi),$$

$$(\text{C.16})$$

where n is a positive integer, $0 \le \pi \le 1$, and x is nonnegative integer such that $0 \le x \le n$.

The binomial quantile function $\texttt{qbinom}(p, n, \pi)$ does not have a closed-form expression. The mean and variance of the $\mathrm{BINOM}(n, \pi)$ distribution are $\mathrm{E}(X) = n\pi$ and $\mathrm{Var}(X) = n\pi(1 - \pi)$.

The distribution of the number of conforming units in a random sample

If $X \sim \mathrm{BINOM}(n, \pi)$ then $Y = (n - X) \sim \mathrm{BINOM}(n, 1 - \pi)$. This result follows directly from switching the role of the conforming and nonconforming units. Then Y is the number of conforming units in a sample of size n when at each trial the probability of observing a conforming unit is $1 - \pi$. From the definition of the binomial distribution, $Y \sim \mathrm{BINOM}(n, 1 - \pi)$. Because $Y = n - X$, we can obtain the cdf of Y from its own distribution (i.e., $\mathrm{Pr}(Y \leq y) = \texttt{pbinom}(y; n, 1 - \pi)$) or as a function of the distribution of X (i.e., $\mathrm{Pr}(Y \leq y) = 1 - \texttt{pbinom}(n - y - 1; n, \pi)$). These two equivalent expressions for $\mathrm{Pr}(Y \leq y)$ imply the identity

$$\texttt{pbinom}(y; n, 1 - \pi) = 1 - \texttt{pbinom}(n - y - 1; n, \pi), \tag{C.17}$$

where y is a nonnegative integer and $y \leq n$. This relationship is used in (5.4) to obtain a simple expression for the distribution-free coverage probability of a lower confidence bound for a distribution quantile.

Also, using (C.2), the quantiles of Y and X are related as follows:

$$y_\alpha = \begin{cases} n - x_{(1-\alpha)} - 1 & \text{if } \texttt{pbinom}(x_{(1-\alpha)}; n, \pi) = 1 - \alpha \\ n - x_{(1-\alpha)} & \text{otherwise.} \end{cases} \tag{C.18}$$

The binomial distribution cdf as a function of the beta distribution cdf

Two relationships to compute binomial probabilities using beta probabilities are

$$\begin{aligned} \mathrm{Pr}(X \leq x; n, \pi) &= \texttt{pbinom}(x; n, \pi) = 1 - \texttt{pbeta}(\pi; x + 1, n - x), \\ \mathrm{Pr}(X \geq x; n, \pi) &= 1 - \texttt{pbinom}(x - 1; n, \pi) = \texttt{pbeta}(\pi; x, n - x + 1), \end{aligned} \tag{C.19}$$

where $x = 0, 1, \ldots, n$. These beta probability expressions are particularly useful to compute binomial probabilities when n is large. Note that, by continuity, $\texttt{pbeta}(\pi; n + 1, 0) = 0$ and $\texttt{pbeta}(\pi; 0, n + 1) = 1$. For the proof of (C.19) see Arnold et al. (2008, page 13). The relationships in (C.19) are used in Section G.2 to express the coverage probabilities of distribution-free confidence intervals and bounds for a quantile in terms of binomial probabilities.

The binomial distribution cdf as a function of Snedecor's F-distribution cdf

The relationship

$$\texttt{pbinom}(x; n, \pi) = \texttt{pf}\left[\frac{1 - \pi}{k\pi}; 2(n - x), 2(x + 1)\right],$$

shows that binomial distribution cdf values can be expressed a function of Snedecor's F-distribution cdf values, where $k = (n - x)/(x + 1)$ and $x = 0, 1, \ldots, n - 1$. This relationship follows from (C.19) and (C.11).

The binomial distribution cdf is a monotone decreasing function of π

Taking derivatives with respect to π on both sides of the cdf relationship in (C.19) gives

$$\frac{d}{d\pi}\text{pbinom}(x; n, \pi) = \frac{d}{d\pi}[1 - \text{pbeta}(\pi; x + 1, n - x)]$$

$$= -\frac{d}{d\pi}\text{pbeta}(\pi; x + 1, n - x)$$

$$= -\text{dbeta}(\pi; x + 1, n - x) < 0. \tag{C.20}$$

This shows that the binomial cdf $\text{pbinom}(x; n, \pi)$ is a decreasing function of the probability parameter π. This implies that for fixed p and n, the quantile function $\text{qbinom}(p; n, \pi)$ is nondecreasing in π. The result in (C.20) is used in Section 6.4 to determine a confidence interval for the probability that the number of nonconforming units in a sample is less than or equal to (or greater than) a specified number.

The binomial distribution cdf is a nonincreasing function of n

Directly from (C.16),

$$\frac{\text{dbinom}(x; n + 1, \pi)}{\text{dbinom}(x; n, \pi)} = \begin{cases} \dfrac{n + 1}{n + 1 - x} \times (1 - \pi) & \text{if } 0 \le x \le n, \\ \infty & \text{if } x = n + 1. \end{cases}$$

Because this ratio is monotone increasing in x, the binomial pmf $\text{dbinom}(x; n, \pi)$ is a monotone increasing function of x. This implies that the binomial cdf $\text{pbinom}(x; n, \pi)$ is nonincreasing as a function of n (see details in Lehmann, 1986, page 85). This result is important in obtaining the prediction interval procedure given in Section 7.6.1.

C.4.2   Beta-Binomial Distribution

The beta-binomial distribution is a $\text{BETA}(a, b)$ distribution mixture of binomial distributions. Formally, suppose that $X|\pi \sim \text{BINOM}(n, \pi)$ and $\pi \sim \text{BETA}(a, b)$. Then the marginal distribution of X is a beta-binomial distribution with parameters n, a, and b. The beta-binomial distribution pmf and cdf are

$$\text{dbetabinom}(x; n, a, b) = \frac{1}{(n + 1)\,\mathcal{B}(x + 1, n - x + 1)} \times \frac{\mathcal{B}(x + a, n - x + b)}{\mathcal{B}(a, b)},$$

$$\text{pbetabinom}(x; n, a, b) = \sum_{i=0}^{x} \text{dbetabinom}(i; n, a, b),$$

where n is a positive integer, $a > 0$, $b > 0$, $\mathcal{B}(u, v)$ is the beta function defined in Appendix A, and x is a nonnegative integer such that $0 \le x \le n$. The quantile function of the beta-binomial distribution, $\text{qbetabinom}(p; n, a, b)$, does not have a closed-form expression. The mean and variance of the beta-binomial distribution are

$$\text{E}(X) = \frac{na}{a + b} \quad \text{and} \quad \text{Var}(X) = \frac{nab(a + b + n)}{(a + b)^2(a + b + 1)}.$$

C.4.3 Negative Binomial Distribution

The negative binomial random variable, denoted by $X \sim \text{NBINOM}(k, \pi)$, has a pmf and cdf given by

$$\text{dnbinom}(x; k, \pi) = \frac{\Gamma(x+k)}{\Gamma(k)\,\Gamma(x+1)} \pi^k (1-\pi)^x \quad \text{and}$$

$$\text{pnbinom}(x; k, \pi) = \sum_{i=0}^{x} \text{dnbinom}(i; k, \pi),$$

where $0 < \pi \le 1$, k is a positive quantity (not necessarily an integer), and x is a nonnegative integer.

The quantile function of the negative binomial distribution $\text{qnbinom}(p)$ does not have a closed-form expression. The mean and variance of the $\text{NBINOM}(k, \pi)$ distribution are $\text{E}(X) = k(1-\pi)/\pi$ and $\text{Var}(X) = k(1-\pi)/\pi^2$.

When k is a positive integer, the negative binomial random variable has the following interpretation. In a sequence of independent binary trials with constant probability π of observing a nonconforming unit, X is the number of *conforming* units observed at the time that k *nonconforming* units are obtained. Note that $X + k$ is the total number of binary trials.

The negative binomial distribution cdf as function of the beta distribution cdf

A relationship to compute negative binomial cdf values using the beta distribution cdf is

$$\text{pnbinom}(x; k, \pi) = \text{pbeta}(\pi, k, x+1), \tag{C.21}$$

where $x = 0, 1, \dots$. This relationship is particularly useful to compute negative binomial probabilities when either p or k is large.

The negative binomial distribution cdf is a monotone increasing function of π

Taking derivatives with respect to π on both sides of the relationship in (C.21) gives

$$\frac{d}{d\pi}\text{pnbinom}(x; k, \pi) = \text{dbeta}(\pi, k, x+1) > 0, \tag{C.22}$$

where $x = 0, 1, \dots$. This shows that, for fixed $k > 0$ and $0 < \pi < 1$, the negative binomial cdf $\text{pnbinom}(x; r, \pi)$ is an increasing function of the probability parameter π. This result is useful to obtain a confidence interval for the negative binomial distribution probability parameter in Example D.16.

C.4.4 Poisson Distribution

A Poisson distribution random variable, denoted by $X \sim \text{POIS}(\lambda)$, has a pmf and cdf given by

$$\text{dpois}(x; \lambda) = \frac{\lambda^x \exp(-\lambda)}{x!} \quad \text{and} \quad \text{ppois}(x; \lambda) = \sum_{i=0}^{x} \text{dpois}(i; \lambda),$$

where $\lambda > 0$ and x is a nonnegative integer.

The quantile function for the Poisson distribution $\text{qpois}(p; \lambda)$ does not have a closed-form expression. The mean and variance of the $\text{POIS}(\lambda)$ distribution are $\text{E}(X) = \lambda$ and $\text{Var}(X) = \lambda$.

Modeling rate of occurrence data with a Poisson distribution

The Poisson distribution is often used to model the number of events X occurring during n units of exposure when the rate of events per unit of exposure is λ. The assumptions are that $n > 0$ (not necessarily an integer) is given and that the rate is constant during the exposure time. In this setting,

$$\texttt{dpois}(x; n\lambda) = \frac{(n\lambda)^x \exp(-n\lambda)}{x!} \quad \text{and} \quad \texttt{ppois}(x; n\lambda) = \sum_{i=0}^{x} \texttt{dpois}(i; n\lambda),$$

where $\lambda > 0$ and x is a nonnegative integer. The quantile function of the Poisson distribution using this parameterization is $\texttt{qpois}(p; n\lambda)$. The mean and variance are $E(X) = n\lambda$ and $\text{Var}(X) = n\lambda$.

This model is convenient because one can combine independent observations with different amounts of exposure to estimate the rate λ. Another application of the model arises when X is the cumulative number of occurrences in n independent processes that have the same constant rate λ.

Poisson distribution probabilities as a function of chi-square distribution probabilities

Relationships to compute Poisson distribution probabilities using chi-square distribution probabilities are

$$\begin{aligned} \Pr(X \le x; n\lambda) = \texttt{ppois}(x; n\lambda) = 1 - \texttt{pchisq}[2n\lambda; 2(x+1)], \\ \Pr(X \ge x; n\lambda) = 1 - \texttt{ppois}(x-1; n\lambda) = \texttt{pchisq}(2n\lambda; 2x), \end{aligned} \tag{C.23}$$

where $X \sim \text{POIS}(n\lambda)$, $\lambda > 0$, $n > 0$, and x is a nonnegative integer. The chi-square distribution expressions are particularly useful to compute Poisson distribution probabilities when n is large.

The Poisson distribution cdf is a monotone decreasing function of λ

Taking derivatives with respect to λ on both sides of the cdf relationship in (C.23) gives

$$\frac{d}{d\lambda}\texttt{ppois}(x; n\lambda) = -2n\,\texttt{dchisq}[2n\lambda; 2(x+1)] < 0. \tag{C.24}$$

This shows that, for given x, the Poisson distribution cdf $\texttt{ppois}(x; n\lambda)$ is a decreasing function of the parameter λ. This result is used in Section D.6.2 to define the confidence interval for λ using the cdf pivotal method.

The Poisson distribution quantile function is a nondecreasing function of λ

This follows directly from (C.24). This property of the Poisson distribution is used in Section 7.3 to obtain confidence intervals for the probability that the number of events in a specified amount of exposure is less than or equal to (or greater than) a specified number.

C.4.5   Hypergeometric Distribution

Consider sampling without replacement from a population of size N that initially had D nonconforming units and $N - D$ conforming units. Define X as the random variable that counts the number of *nonconforming* units observed in a sample of size n. Then X is said to have a hypergeometric distribution, denoted by $X \sim \text{HYPER}(n, D, N)$. The pdf and cdf of X are

$$\text{dhyper}(x; n, D, N) = \frac{\binom{D}{x}\binom{N-D}{n-x}}{\binom{N}{n}} \quad \text{and}$$

$$\text{phyper}(x; n, D, N) = \sum_{i=0}^{x} \text{dhyper}(i; n, D, N), \tag{C.25}$$

where $0 < D < N$, $0 < n \leq N$, and x is a nonnegative integer such that $\max\{0, n + D - N\} \leq x \leq \min\{n, D\}$.

The quantile function of this distribution, $\text{qhyper}(p; n, D, N)$, does not have a closed-form expression. The mean and variance of the $\text{HYPER}(n, D, N)$ distribution are

$$\text{E}(X) = \frac{nD}{N} \quad \text{and} \quad \text{Var}(X) = \frac{nD}{N}\left(1 - \frac{D}{N}\right)\left(\frac{N-n}{N-1}\right).$$

Equivalence of the HYPER(n, D, N) and HYPER(D, n, N) distributions

To show the equivalence, it suffices to show that $\text{dhyper}(x; n, D, N) = \text{dhyper}(x; D, n, N)$ for $\max\{0, n + D - N\} \leq x \leq \min\{n, D\}$. Using (C.25) and expanding the terms in the numerator and denominator gives

$$\text{dhyper}(x; n, D, N) = \frac{D!}{(D-x)!\,x!} \times \frac{(N-D)!}{(N-D-n+x)!\,(n-x)!} \times \frac{(N-n)!\,n!}{N!}$$

$$= \frac{n!}{(n-x)!\,x!} \times \frac{(N-n)!}{(N-D-n+x)!\,(D-x)!} \times \frac{(N-D)!\,D!}{N!}$$

$$= \frac{\binom{n}{x}\binom{N-n}{D-x}}{\binom{N}{D}} = \text{dhyper}(x; D, n, N), \tag{C.26}$$

which shows the equivalence of the two pdf expressions for the hypergeometric distribution. This result is used in Section 6.7 to obtain prediction intervals for the number of nonconforming units in a future sample from a binomial distribution.

The hypergeometric distribution cdf is a nonincreasing function of D or n

Directly from (C.25),

$$\frac{\text{dhyper}(x; n, D+1, N)}{\text{dhyper}(x; n, D, N)} = \begin{cases} \dfrac{D+1}{N-D} \times \dfrac{N-D-n+x}{D+1-x} & \text{if } n+D+1-N \leq x \leq D, \\ 0 \text{ or } \infty & \text{if } x = n+D \text{ or } x = D+1. \end{cases}$$

This shows that, for fixed values of n, D, and N, the hypergeometric pmf $\text{dhyper}(x; n, D, N)$ is a decreasing function of x. This implies that the hypergeometric cdf $\text{phyper}(x; n, D, N)$ is a nonincreasing function of D (this is related to what is known as the monotone likelihood-ratio property discussed, for example, in Lehmann, 1986, pages 80 and 85). Because of the relationship $\text{phyper}(x; n, D, N) = \text{phyper}(x; D, n, N)$, it follows that $\text{phyper}(x; n, D, N)$ is also nonincreasing as a function of n. This result is used in Section 6.7 to obtain prediction intervals for the number of nonconforming units in a future sample from a distribution.

C.4.6   Negative Hypergeometric Distribution

Consider sampling without replacement from a population of size N which initially has D nonconforming units and $N - D$ conforming units. The sampling is sequential, without replacement, and the sampling ends as soon as k nonconforming units have been observed. Define $X \sim \text{NHYPER}(k, D, N)$ as the random variable that counts the number of *conforming* units observed in the sample by the time that exactly k *nonconforming* units of interest are obtained. Note that the total number of observed units in the sample is $X + k$ and that X takes positive probabilities in the set $\{0, 1, 2, \ldots, N - D\}$.

The pmf and cdf of the negative hypergeometric distribution are

$$\texttt{dnhyper}(x; k, D, N) = \frac{\binom{x+k-1}{x} \binom{N-x-k}{N-D-x}}{\binom{N}{D}},$$

$$\texttt{pnhyper}(x; k, D, N) = \sum_{i=0}^{x} \texttt{dnhyper}(i; k, D, N), \tag{C.27}$$

where $0 < D < N$, $1 \le k \le D$, and $x = 0, \ldots, N - D$. The quantile function of the negative hypergeometric distribution $\texttt{qnhyper}(p; k, D, N)$ does not have a closed-form expression. The mean and variance of the $\text{NHYPER}(k, D, N)$ distribution are

$$\text{E}(X) = \frac{k(1-\pi)}{\pi + 1/N} \quad \text{and} \quad \text{Var}(X) = \frac{k(1-\pi)}{(\pi + 1/N)^2} \frac{(1 + 1/N)(D + 1 - k)}{(D + 2)},$$

where $\pi = D/N$.

A symmetry relationship for the negative hypergeometric distribution

Here we show that if $X \sim \text{NHYPER}(k, D, N)$ then $Y = (N - D - X) \sim \text{NHYPER}(D - k + 1, D, N)$. Suppose that $g(y)$ is the pmf of Y. Using transformation of variables gives

$$g(y) = \Pr(Y = y) = \Pr(X = N - D - y)$$

$$= \frac{\binom{N-D-y+k-1}{N-D-y} \binom{D+y-k}{y}}{\binom{N}{D}} = \frac{\binom{y+(D-k+1)-1}{y} \binom{N-(D-k+1)-y}{N-D-y}}{\binom{N}{D}}$$

$$= \frac{\binom{y+k^*-1}{y} \binom{N-k^*-y}{N-D-y}}{\binom{N}{D}} = \texttt{dnhyper}(y; k^*, D, N),$$

where $k^* = D - k + 1$.

In particular, with $D = m$, $N = m + n$, and $k = m + j - 1$, we get $Y = n - X$, $k^* = j$, $X \sim \text{NHYPER}(m - j + 1, m, m + n)$, and $Y \sim \text{NHYPER}(j, m, m + n)$. Because $Y = n - X$, the cdf of Y can be obtained from its own distribution $\Pr(Y \le y) = \texttt{pnhyper}(y; j, m, m + n)$ or as a function of the distribution of X (i.e., $\Pr(Y \le y) = 1 - \texttt{pnhyper}(n - y - 1; m - j + 1, m, m + n)$). This gives the identity

$$\texttt{pnhyper}(y; j, m, m + n) = 1 - \texttt{pnhyper}(n - y - 1; m - j + 1, m, m + n), \tag{C.28}$$

where $0 \le y \le n$. Also, using (C.2), the quantiles of Y and X are related as follows:

$$y_\alpha = \begin{cases} n - x_{(1-\alpha)} - 1 & \text{if } \texttt{pnhyper}(x_{(1-\alpha)}; m - j + 1, m, m + n) = 1 - \alpha \\ n - x_{(1-\alpha)} & \text{otherwise.} \end{cases} \tag{C.29}$$

The negative hypergeometric pmf as a beta distribution mixture of binomial distribution probabilities

The $\mathtt{dnhyper}(i; k, D, N)$ pmf can be expressed as a mixture of binomial distribution probabilities as follows:

$$\mathtt{dnhyper}(i; k, D, N) = \int_0^1 \mathtt{dbinom}(i; N - D, v)\mathtt{dbeta}(v; k, D - k + 1)\, dv. \quad \text{(C.30)}$$

We use this result to show the equivalence between the negative hypergeometric distribution cdf and a beta distribution mixture of beta distribution probabilities as given in (C.31). Now we prove (C.30):

$$\int_0^1 \mathtt{dbinom}(i; N - D, v)\,\mathtt{dbeta}(v; k, D - k + 1)\, dv$$

$$= \frac{\binom{N-D}{i} \int_0^1 v^{i+k-1}(1 - v)^{N-i-k}\, dv}{\mathcal{B}(k, D - k + 1)} = \frac{\binom{N-D}{i} \mathcal{B}(i + k, N - i - k + 1)}{\mathcal{B}(k, D - k + 1)}$$

$$= \frac{\binom{i+k-1}{i}\binom{N-i-k}{N-D-i}}{\binom{N}{D}} = \mathtt{dnhyper}(i; k, D, N).$$

The negative hypergeometric distribution cdf as a beta distribution mixture of beta distribution probabilities

The negative hypergeometric distribution cdf can be expressed as a beta distribution mixture of beta distribution probabilities as follows:

$$\mathtt{pnhyper}(x - 1; k, D, N) = 1 - \int_0^1 \mathtt{pbeta}(v; x, N - D - x + 1)\mathtt{dbeta}(v; k, D - k + 1)\, dv,$$

$$\text{(C.31)}$$

for $k = 1, \ldots, D$ and $x = 1, \ldots, N - D$. To verify this result, we use (C.30) in the proof below:

$$\mathtt{pnhyper}(x - 1; k, D, N) = \sum_{i=0}^{x-1} \mathtt{dnhyper}(i; k, D, N)$$

$$= \int_0^1 \left[\sum_{i=0}^{x-1} \mathtt{dbinom}(i; N - D, v)\right]\mathtt{dbeta}(v; k, D - k + 1)\, dv$$

$$= \int_0^1 \mathtt{pbinom}(x - 1; N - D, v)\,\mathtt{dbeta}(v; k, D - k + 1)\, dv$$

$$= \int_0^1 [1 - \mathtt{pbeta}(v; x, N - D - x + 1)]\mathtt{dbeta}(v; k, D - k + 1)\, dv$$

$$= 1 - \int_0^1 \mathtt{pbeta}(v; x, N - D - x + 1)\mathtt{dbeta}(v; k, D - k + 1)\, dv,$$

where the fourth line follows from the third by (C.19). This result is used in Sections G.4 and G.5 to obtain coverage probabilities for distribution-free prediction intervals given in Sections 5.4 and 5.5.

Appendix *D*

General Results from Statistical Theory and Some Methods Used to Construct Statistical Intervals

INTRODUCTION

This appendix provides some useful tools and results from statistical theory. These tools facilitate the justification and extension of much of the methodology in the book.

The topics discussed in this appendix are:

- Basic theory for transformation of random variables (Section D.1).

- The delta method to obtain expressions for approximate variances of random quantities as a function of the variances and covariances of the function arguments (Section D.2).

- Expected and observed information matrices (Section D.3).

- Some general regularity conditions assumed in most of the book and needed for certain technical results (Section D.3.2).

- A definition of convergence in distribution of random variables, with examples of its use in this book (Section D.4).

- An outline of general maximum likelihood theory relevant to applications in this book (Section D.5).

- The cdf pivotal method for constructing confidence intervals, their coverage probabilities, and examples for continuous and discrete distributions (Section D.6).

Statistical Intervals: A Guide for Practitioners and Researchers, Second Edition.
William Q. Meeker, Gerald J. Hahn and Luis A. Escobar.
© 2017 John Wiley & Sons, Inc. Published 2017 by John Wiley & Sons, Inc.
Companion Website: www.wiley.com/go/meeker/intervals

- Bonferroni approximate statistical intervals with application to simultaneous confidence intervals as well as the construction of tolerance and simultaneous prediction intervals (Section D.7).

D.1   THE CDFS AND PDFS OF FUNCTIONS OF RANDOM VARIABLES

This section reviews the procedure for obtaining the cdf and pdf (or pmf) of a one-to-one function of a random variable(s) that has known distribution. Most of the development is for scalar monotone increasing (or decreasing) functions of a given random variable. All the examples in this section are illustrations of transformations used in this book.

D.1.1   Transformation of Continuous Random Variables

This section shows how to obtain expressions for the pdf and cdf of functions of random variables. Let V be a k-dimensional continuous random vector with pdf $h(v)$. We consider a k-dimensional transformation $U = g(V)$ with the following properties:

1. The function $u = g(v) = [g_1(v), \ldots, g_k(v)]$ is a one-to-one transformation.

2. The inverse function $v = g^{-1}(u) = [g_1^{-1}(u), \ldots, g_k^{-1}(u)]$ has continuous first partial derivatives with respect to u.

3. The Jacobian $J(u)$ of $g^{-1}(u)$ is nonzero, where

$$
J(u) = \det \begin{bmatrix} \dfrac{\partial g_1^{-1}(u)}{\partial u_1} & \cdots & \dfrac{\partial g_k^{-1}(u)}{\partial u_1} \\ \vdots & \vdots & \vdots \\ \dfrac{\partial g_1^{-1}(u)}{\partial u_k} & \cdots & \dfrac{\partial g_k^{-1}(u)}{\partial u_k} \end{bmatrix}. \tag{D.1}
$$

Then the pdf and cdf of U are

$$
f(u) = h\left[g^{-1}(u)\right]|J(u)|,
$$

$$
F(u) = \int_{z \leq u} h\left[g^{-1}(z)\right]|J(z)|\,dz,
$$

where $u = (u_1, \ldots, u_k)$ and the integral is evaluated for all values $z = (z_1, \ldots, z_k)$ such that $z_i \leq u_i$, $i = 1, \ldots, k$.

For the scalar case (i.e., $k = 1$) the formulas simplify to

$$
f(u) = h[g^{-1}(u)]\left|\frac{dg^{-1}(u)}{du}\right|, \tag{D.2}
$$

for the pdf of the transformed random variable and

$$
F(u) = \int_{-\infty}^{u} h[g^{-1}(z)]\left|\frac{dg^{-1}(z)}{dz}\right|dz
$$

$$
= \begin{cases} H[g^{-1}(u)] & \text{if } g \text{ is increasing} \\ 1 - H[g^{-1}(u)] & \text{if } g \text{ is decreasing,} \end{cases} \tag{D.3}
$$

for the cdf of the transformed random variable. For illustration, consider the following examples.

Example D.1 Log-Location-Scale Transformation. Suppose that V has a location-scale distribution with location parameter $-\infty < \mu < \infty$, scale parameter $\sigma > 0$, and pdf and cdf given by

$$f(v; \mu, \sigma) = \phi\left(\frac{v - \mu}{\sigma}\right) \quad \text{and} \quad F(v; \mu, \sigma) = \Phi\left(\frac{v - \mu}{\sigma}\right),$$

where $\infty < v < \infty$, $\phi(z)$ is a completely specified cdf, and $\Phi(z)$ is the cdf corresponding to $\phi(z)$.

Consider the transformation $U = g(V) = \exp(V)$. Then $g^{-1}(u) = \log(u)$ and $J(u) = 1/u$. Consequently, from (D.2) and (D.3),

$$f(u; \mu, \sigma) = h[g^{-1}(u)] = \frac{1}{\sigma t} \phi\left[\frac{\log(u) - \mu}{\sigma}\right], \tag{D.4}$$

$$F(u; \mu, \sigma) = H[g^{-1}(u)] = \Phi\left[\frac{\log(u) - \mu}{\sigma}\right], \tag{D.5}$$

for $0 < u < \infty$.

The family of distributions $F(u; \mu, \sigma)$ is known as the log-location-scale family with parameters (μ, σ). Note that μ and σ are the location and scale parameters for the distribution of $\log(U) = V$.

The results in (D.4) and (D.5) justify all of the log-location-scale distributions discussed in Section C.3.1.   ∎

Example D.2 The cdf Transform for Continuous Distributions. Suppose that V has a continuous and strictly monotone increasing cdf $H(v)$. Consider the transformation $U = g(V) = H(V)$. That is, V is transformed using its own cdf. In this case $g^{-1}(u) = H^{-1}(u)$ and $J(u) = (h[H^{-1}(u)])^{-1}$. Thus,

$$f(u) = \frac{h[H^{-1}(u)]}{h[H^{-1}(u)]} = 1 \quad \text{and} \quad F(u) = H[H^{-1}(u)] = u,$$

where $0 < u < 1$. This implies that $U \sim \text{UNIF}(0, 1)$.

This result is useful in the generation of random numbers for a continuous random variable as follows. The relationship $U = H(V)$ gives $V = H^{-1}(U)$, and if $u_1, \ldots, u_n$ is a random sample from the UNIF$(0, 1)$ then $[v_1, \ldots, v_n] = [H^{-1}(u_1), \ldots, H^{-1}(u_n)]$ is a random sample from V. The cdf transform is used in Section D.6 to derive confidence intervals based on pivotal quantities.   ∎

Example D.3 A Snedecor's F Random Variable as a Function of a Beta Random Variable and Vice Versa. Suppose that $V \sim \text{BETA}(a, b)$ and let $r = a/b$. We show that the random variable defined by

$$U = g(V) = \frac{V}{r(1 - V)} \tag{D.6}$$

has Snedecor's $F(2a, 2b)$ distribution. Note that the degrees of freedom need not be integers.

The transformation $g(v)$ in (D.6) is a monotone increasing function in v with inverse function $g^{-1}(u) = ru/(1+ru)$ and Jacobian $J = r/(1+ru)^2$. Thus the density of U in (D.6) is

$$
\begin{aligned}
f(u; a, b) &= \frac{r}{(1+ru)^2} \; \mathtt{dbeta}\left(\frac{ru}{1+ru}; a, b\right) \\
&= \frac{\Gamma(a+b)}{\Gamma(a)\,\Gamma(b)} \frac{r}{(1+ru)^2} \left(\frac{ru}{1+ru}\right)^{a-1} \left(1 - \frac{ru}{1+ru}\right)^{b-1} \\
&= \frac{\Gamma(a+b)}{\Gamma(a)\,\Gamma(b)} \frac{r^a u^{a-1}}{(1+ru)^{a+b}} = \frac{\Gamma(a+b)}{\Gamma(a)\,\Gamma(b)} \left(\frac{a}{b}\right)^a \frac{u^{a-1}}{[1+(a/b)u]^{a+b}} \\
&= (u; 2a, 2b).
\end{aligned}
$$

Because U is a monotone increasing function of V, the quantiles $\mathtt{qf}(p; 2a, 2b)$ of U are related to the quantiles $\mathtt{qbeta}(p; a, b)$ of V through (D.6), which gives the relationship

$$
\mathtt{qf}(p; 2a, 2b) = \frac{\mathtt{qbeta}(p; a, b)}{r[1 - \mathtt{qbeta}(p; a, b)]}.
$$

Now, solving for $\mathtt{qbeta}(p; a, b)$ in this relationship gives

$$
\mathtt{qbeta}(p; a, b) = \frac{1}{a + b/\mathtt{qf}(p; 2a, 2b)}.
$$

Similarly, because the transformation (D.6) is one-to-one, one can get a $\mathrm{BETA}(a, b)$ random variable as a transformation of a Snedecor's $F(r_1, r_2)$ random variable. Specifically, suppose that $V \sim F(r_1, r_2)$. Define $\xi = r_1/r_2$ and consider the transformation

$$
U = g(V) = \frac{\xi V}{1 + \xi V}.
$$

Then $U \sim \mathrm{BETA}(r_1/2, r_2/2)$. The proof of this result is similar to the proof of (D.6). ∎

Example D.4   A Noncentral t Random Variable as a Function of a $\mathrm{NORM}(\delta, 1)$ Random Variable and an Independent Chi-Square Random Variable with r Degrees of Freedom. Suppose that X and W are independent, $X \sim \mathrm{NORM}(\delta, 1)$, and $W \sim \chi^2(r)$. A noncentral t-distribution random variable is defined by the ratio

$$
T = \frac{X}{\sqrt{W/r}},
$$

where the scalar δ is a noncentrality parameter of the distribution of T and $r > 0$ is the degrees of freedom (which need not be an integer).

Because of the independence of X and W, the joint distribution of $\boldsymbol{V} = (X, W)$ is a product of a normal density and a chi-square density. To obtain the distribution of T, consider the one-to-one transformation $\boldsymbol{u} = g(x, w) = (t, w)$, where $t = x/\sqrt{w/r}$. The inverse of the transformation is $g^{-1}(t, w) = (t\sqrt{w/r}, w)$ and the Jacobian of the transformation is $J = \sqrt{w/r}$. Thus, the joint distribution of $\boldsymbol{U} = (T, W)$ is

$$
\begin{aligned}
f(t, w) &= \sqrt{w/r} \times \mathtt{dnorm}(t\sqrt{w/r}; \delta, 1) \times \mathtt{dchisq}(w; r) \\
&= \sqrt{\frac{w}{r}} \frac{1}{\sqrt{2\pi}} \exp\left[-\frac{1}{2}\left(t\sqrt{\frac{w}{r}} - \delta\right)^2\right] \frac{w^{r/2-1}}{2^{r/2}\Gamma(r/2)} \exp\left(-\frac{w}{2}\right) \\
&= \frac{w^{(r-1)/2}}{2^{(r+1)/2} \sqrt{r\pi}\,\Gamma(r/2)} \exp\left[-\frac{1}{2}\left(t\sqrt{\frac{w}{r}} - \delta\right)^2 - \frac{w}{2}\right].
\end{aligned}
$$

The marginal pdf for T, denoted by $\mathrm{dt}(t; r, \delta)$, is obtained by integrating $f(t, w)$ with respect to w. This gives

$$\mathrm{dt}(t; r, \delta) = \int_0^\infty \frac{w^{(r-1)/2}}{2^{(r+1)/2}\sqrt{r\pi}\,\Gamma(r/2)}\exp\left[-\frac{1}{2}\left(t\sqrt{\frac{w}{r}} - \delta\right)^2 - \frac{w}{2}\right]dw.$$

Changing the integration variable to $z = w(r + t^2)/(2r)$,

$$\mathrm{dt}(t; r, \delta) = \frac{r^{r/2}\exp(-r\delta^2/[2(r+t^2)])}{\sqrt{\pi}\,\Gamma(r/2)\,(r+t^2)^{(r+1)/2}}\int_0^\infty z^{(r-1)/2}\exp\left[-\left(\sqrt{z} - \frac{t\delta}{\sqrt{2(r+t^2)}}\right)^2\right]dz.$$

(D.7)

When the noncentrality parameter is equal to zero, the distribution is known as Student's t-distribution. In this case, the integral in (D.7) is equal to $\Gamma[(r + 1)/2]$, and the pdf is

$$\mathrm{dt}(t; r) = \frac{\Gamma[(r+1)/2]}{\Gamma(r/2)}\frac{1}{\sqrt{r\pi}}\frac{1}{(1+t^2/r)^{(r+1)/2}},$$

where $-\infty < t < \infty$.   ∎

D.1.2   Transformation of Discrete Random Variables

This section shows how to obtain expressions for the pmf and cdf of functions of discrete random variables. Let $\boldsymbol{V}$ be a k-dimensional discrete random vector with pmf $h(\boldsymbol{v})$. We consider a one-to-one transformation k-dimensional transformation $\boldsymbol{U} = \boldsymbol{g}(\boldsymbol{V})$. Then the pmf and cdf of $\boldsymbol{U}$ are

$$f(\boldsymbol{u}) = h\left[\boldsymbol{g}^{-1}(\boldsymbol{u})\right] \quad \text{and} \quad F(\boldsymbol{u}) = \sum_{\boldsymbol{z} \le \boldsymbol{u}} h\left[\boldsymbol{g}^{-1}(\boldsymbol{z})\right],$$

where $\boldsymbol{u} = (u_1, \ldots, u_k)$ and the summation is evaluated for all values $\boldsymbol{z} = (z_1, \ldots, z_k)$ such that $h\left[\boldsymbol{g}^{-1}(\boldsymbol{z})\right] > 0$ and $z_i \le u_i$, $i = 1, \ldots, k$.

For the scalar case (i.e., $k = 1$), the pdf and cdf of the transformed random variable are

$$f(u) = h[g^{-1}(u)]$$ (D.8)

and

$$F(u) = \sum_{v \le g^{-1}(u)} h(v) = \begin{cases} H[g^{-1}(u)] & \text{if } g \text{ is increasing} \\ 1 - H^-[g^{-1}(u)] & \text{if } g \text{ is decreasing,} \end{cases}$$ (D.9)

where $H^-[g^{-1}(u)] = \lim_{v \uparrow g^{-1}(u)} H(v)$ is the limiting value of $H(v)$ when v approaches $g^{-1}(u)$ from below. For example, if V takes just integer values, for integer $g^{-1}(u)$ the limit from below for $H[g^{-1}(u)]$ is $H^-[g^{-1}(u)] = H[g^{-1}(u) - 1]$.

Example D.5   The Distribution of V_1 Conditional on the Sum $V_1 + V_2$ when V_1 and V_2 are Independent and Poisson Distributed. Suppose that $\boldsymbol{V} = (V_1, V_2)$, where $V_1 \sim \mathrm{POIS}(n\lambda)$ and $V_2 \sim \mathrm{POIS}(m\lambda)$. The joint distribution of (V_1, V_2) is

$$f(v_1, v_2) = \frac{(n\lambda)^{v_1}(m\lambda)^{v_2}}{v_1!\,v_2!}\exp[-\lambda(n + m)],$$

where v_1 and v_2 are nonnegative integers.

Consider the transformation of variables $U = (U_1, U_2) = g(V)$, where $g(v_1, v_2) = (u_1, u_2)$ with

$$u_1 = v_1 \quad \text{and} \quad u_2 = v_1 + v_2.$$

The inverse $g^{-1}(u_1, u_2)$ of the transformation is

$$v_1 = u_1 \quad \text{and} \quad v_2 = u_2 - u_1.$$

The purpose is to obtain the conditional distribution of $V_1 | (V_1 + V_2)$, or equivalently the conditional distribution of $U_1 | U_2$. Note that the support of the distribution of U has the restrictions that u_1 and u_2 are nonnegative integers with $u_1 \leq u_2$. From (D.8) and using the joint distribution $f(v_1, v_2)$ of (V_1, V_2) given above, the joint distribution of (U_1, U_2) is

$$f(u_1, u_2) = \frac{(n\lambda)^{u_1} (m\lambda)^{u_2 - u_1}}{u_1! (u_2 - u_1)!} \exp[-\lambda(n + m)], \quad 0 \leq u_1 \leq u_2.$$

The marginal distribution for U_2 is

$$f(u_2) = \sum_{u_1=0}^{u_2} f(u_1, u_2) = \lambda^{u_2} \exp[-\lambda(n + m)] \sum_{u_1=0}^{u_2} \frac{n^{u_1} m^{u_2 - u_1}}{u_1! (u_2 - u_1)!}$$

$$= \frac{[\lambda(n + m)]^{u_2}}{u_2!} \exp[-\lambda(n + m)].$$

That is, $U_2 \sim \text{POIS}[\lambda(n + m)]$. Thus, the conditional distribution of $U_1 | U_2$ is

$$f(u_1 | u_2) = \frac{f(u_1, u_2)}{f(u_2)} = \frac{u_2!}{u_1! (u_2 - u_1)!} \left(\frac{n}{n + m}\right)^{u_1} \left(1 - \frac{n}{n + m}\right)^{u_2 - u_1},$$

where $u_1 = 0, \ldots, u_2$. That is, $U_1 | U_2 = V_1 | (V_1 + V_2) \sim \text{BINOM}(u_2, \pi)$, where $\pi = n/(n + m)$. This result is used in Section 7.6.1 to obtain a Poisson prediction interval. ∎

D.2 STATISTICAL ERROR PROPAGATION—THE DELTA METHOD

This section shows how to compute approximate expected values, variances, and covariances of functions of parameter estimators. Let $g(\theta)$ be a real-valued function of the parameters $\theta = (\theta_1, \ldots, \theta_r)'$ and let $\widehat{\theta} = (\widehat{\theta}_1, \ldots, \widehat{\theta}_r)'$ and $g(\widehat{\theta})$ be estimates of θ and $g(\theta)$, respectively. The objective is to obtain expressions or approximate expressions for $\text{E}\left[g(\widehat{\theta})\right]$ and $\text{Var}\left[g(\widehat{\theta})\right]$ as a function of $\text{E}(\widehat{\theta}_i)$, $\text{Var}(\widehat{\theta}_i)$, and $\text{Cov}(\widehat{\theta}_i, \widehat{\theta}_j)$.

The simplest case is when $g(\widehat{\theta})$ is a linear function of the $\widehat{\theta}_i$, say, $g(\widehat{\theta}) = a_0 + \sum_{i=1}^r a_i \widehat{\theta}_i$, where the a_i are constants. To facilitate the development, express $g(\widehat{\theta})$ as

$$g(\widehat{\theta}) = a_0 + \sum_{i=1}^r a_i \widehat{\theta}_i = b_0 + \sum_{i=1}^r b_i \left[\widehat{\theta}_i - \text{E}(\widehat{\theta}_i)\right], \tag{D.10}$$

where $b_0 = a_0 + \sum_{i=1}^{r} a_i \mathrm{E}(\widehat{\theta}_i)$ and $b_i = a_i$, $i = 1, \ldots, r$. In this case, simple computations with expectations and variances give

$$\mathrm{E}\left[g(\widehat{\boldsymbol{\theta}})\right] = b_0,$$

$$\mathrm{Var}\left[g(\widehat{\boldsymbol{\theta}})\right] = \sum_{i=1}^{r} b_i^2 \, \mathrm{Var}(\widehat{\theta}_i) + \sum_{i=1}^{r} \sum_{\substack{j=1 \\ j \neq i}}^{r} b_i \, b_j \, \mathrm{Cov}(\widehat{\theta}_i, \widehat{\theta}_j).$$

When $g(\widehat{\boldsymbol{\theta}})$ is a smooth nonlinear function of the $\widehat{\theta}_i$ values and $g(\widehat{\boldsymbol{\theta}})$ can be approximated by a linear function of the $\widehat{\theta}_i$ values in the region with nonnegligible likelihood, it is still possible to apply the methodology above. The general procedure is known as the *delta method* or *statistical error propagation*, and here we describe a simplified version of the methodology. For a more detailed account, see Hahn and Shapiro (1967, page 228) or Stuart and Ord (1994, page 350).

When $g(\boldsymbol{\theta})$ has continuous second partial derivatives with respect to $\boldsymbol{\theta}$, a first-order (i.e., keeping linear terms only) Taylor series expansion of $g(\widehat{\boldsymbol{\theta}})$ about $\boldsymbol{\mu} = \left[\mathrm{E}(\widehat{\theta}_1), \ldots, \mathrm{E}(\widehat{\theta}_r)\right]$ is given by

$$g(\widehat{\boldsymbol{\theta}}) \approx g(\boldsymbol{\mu}) + \sum_{i=1}^{r} \frac{\partial g(\boldsymbol{\theta})}{\partial \theta_i} \left[\widehat{\theta}_i - \mathrm{E}(\widehat{\theta}_i)\right], \tag{D.11}$$

where the partial derivatives in (D.11) are evaluated at $\boldsymbol{\mu}$.

Observe that (D.11) looks like (D.10) with

$$b_0 = g(\boldsymbol{\mu}) \quad \text{and} \quad b_i = \frac{\partial g(\boldsymbol{\theta})}{\partial \theta_i}, \quad i = 1, \ldots, r.$$

Consequently,

$$\mathrm{E}\left[g(\widehat{\boldsymbol{\theta}})\right] \approx g(\boldsymbol{\mu}),$$

$$\mathrm{Var}\left[g(\widehat{\boldsymbol{\theta}})\right] \approx \sum_{i=1}^{r} \left[\frac{\partial g(\boldsymbol{\theta})}{\partial \theta_i}\right]^2 \mathrm{Var}(\widehat{\theta}_i) + \sum_{i=1}^{r} \sum_{\substack{j=1 \\ j \neq i}}^{r} \left[\frac{\partial g(\boldsymbol{\theta})}{\partial \theta_i}\right] \left[\frac{\partial g(\boldsymbol{\theta})}{\partial \theta_j}\right] \mathrm{Cov}(\widehat{\theta}_i, \widehat{\theta}_j). \tag{D.12}$$

When the $\widehat{\theta}_i$ values are uncorrelated or when the covariances $\mathrm{Cov}(\widehat{\theta}_i, \widehat{\theta}_j)$, $i \neq j$, are small when compared with the variances $\mathrm{Var}(\widehat{\theta}_i)$, the last term on the right of (D.12) is usually omitted from the approximation.

The same ideas apply to vector-valued functions. For example, if $g_1(\boldsymbol{\theta})$ and $g_2(\boldsymbol{\theta})$ are two real-valued functions then

$$\mathrm{Cov}\left[g_1(\widehat{\boldsymbol{\theta}}), g_2(\widehat{\boldsymbol{\theta}})\right] \approx \sum_{i=1}^{r} \left[\frac{\partial g_1(\boldsymbol{\theta})}{\partial \theta_i}\right] \left[\frac{\partial g_2(\boldsymbol{\theta})}{\partial \theta_i}\right] \mathrm{Var}(\widehat{\theta}_i)$$

$$+ \sum_{i=1}^{r} \sum_{\substack{j=1 \\ j \neq i}}^{r} \left[\frac{\partial g_1(\boldsymbol{\theta})}{\partial \theta_i}\right] \left[\frac{\partial g_2(\boldsymbol{\theta})}{\partial \theta_j}\right] \mathrm{Cov}(\widehat{\theta}_i, \widehat{\theta}_j).$$

In general, for a vector-valued function $g(\boldsymbol{\theta})$ of the parameters such that all the second partial derivatives with respect to the elements of $\boldsymbol{\theta}$ are continuous,

$$\mathrm{Var}\left[g(\widehat{\boldsymbol{\theta}})\right] \approx \left[\frac{\partial g(\boldsymbol{\theta})}{\partial \boldsymbol{\theta}}\right]' \mathrm{Var}(\widehat{\boldsymbol{\theta}})\left[\frac{\partial g(\boldsymbol{\theta})}{\partial \boldsymbol{\theta}}\right],$$

where $\partial g(\boldsymbol{\theta})/\partial \boldsymbol{\theta} = [\partial g_1(\boldsymbol{\theta})/\partial \boldsymbol{\theta}, \ldots, \partial g_r(\boldsymbol{\theta})/\partial \boldsymbol{\theta}]$ is the matrix of gradient vectors of first partial derivatives of $g(\boldsymbol{\theta})$ with respect to $\boldsymbol{\theta}$, and

$$\mathrm{Var}(\widehat{\boldsymbol{\theta}}) = \begin{bmatrix} \mathrm{Var}(\widehat{\theta}_1) & \mathrm{Cov}(\widehat{\theta}_1, \widehat{\theta}_2) & \cdots & \mathrm{Cov}(\widehat{\theta}_1, \widehat{\theta}_r) \\ & \mathrm{Var}(\widehat{\theta}_2) & \cdots & \mathrm{Cov}(\widehat{\theta}_2, \widehat{\theta}_r) \\ & & \ddots & \vdots \\ \text{symmetric} & & & \mathrm{Var}(\widehat{\theta}_r) \end{bmatrix}$$

are both evaluated at $\widehat{\boldsymbol{\theta}}$.

The delta method can provide good approximations for $\mathrm{E}[g(\widehat{\boldsymbol{\theta}})]$ and $\mathrm{Var}[g(\widehat{\boldsymbol{\theta}})]$. However, one needs to exercise caution in applying this method because the adequacy of the approximation depends on the validity of the Taylor series approximation and the size of the remainder in the approximation. Simulation can be used to check the adequacy of the approximation.

D.3 LIKELIHOOD AND FISHER INFORMATION MATRICES

This section presents expected and observed information matrices in the context of likelihood estimation. Expected information matrices used in the book are given to facilitate the presentation when they are used.

D.3.1 Information Matrices

Let $\mathcal{L}(\boldsymbol{\theta}) = \sum_{i=1}^{n} \mathcal{L}_i(\boldsymbol{\theta})$ denote the total log-likelihood for a given model and data that will consist of n independent but not necessarily identically distributed observations. Here it is understood that $\mathcal{L}_i(\boldsymbol{\theta})$ is the contribution of the ith observation to the total log-likelihood. Let $\widehat{\boldsymbol{\theta}}$ be the ML estimator of $\boldsymbol{\theta}$ with a sample of size n. This $\widehat{\boldsymbol{\theta}}$, when it exists, is the value of $\boldsymbol{\theta}$ that maximizes $\mathcal{L}(\boldsymbol{\theta})$. Let $\mathcal{I}(\boldsymbol{\theta})$ denote the large-sample (or limiting) average amount of information per observation. That is,

$$\mathcal{I}(\boldsymbol{\theta}) = \lim_{n \to \infty} \left\{ \frac{1}{n} \sum_{i=1}^{n} \mathrm{E}\left[-\frac{\partial^2 \mathcal{L}_i(\boldsymbol{\theta})}{\partial \boldsymbol{\theta}\, \partial \boldsymbol{\theta}'} \right] \right\}, \tag{D.13}$$

where the expectation is with respect to the as-yet-unobserved data. Then in general, for large samples, the matrix $I_{\boldsymbol{\theta}} = n\mathcal{I}(\boldsymbol{\theta})$ approximately quantifies the amount of information that we "expect" to obtain from the future data. Intuitively, larger second derivatives of $\mathcal{L}(\boldsymbol{\theta})$ indicate more curvature in the likelihood, implying that the likelihood is more concentrated about its maximum. For a large class of model situations satisfying regularity conditions that ensure consistency of the ML estimator of $\boldsymbol{\theta}$ (see Boos and Stefanski, 2013, Chapter 6, for details), $I_{\boldsymbol{\theta}}$ simplifies to the well-known Fisher information matrix for $\boldsymbol{\theta}$,

$$I_{\boldsymbol{\theta}} = \mathrm{E}\left[\frac{\partial \mathcal{L}(\boldsymbol{\theta})}{\partial \boldsymbol{\theta}}\frac{\partial \mathcal{L}(\boldsymbol{\theta})}{\partial \boldsymbol{\theta}'}\right] = \mathrm{E}\left[-\frac{\partial^2 \mathcal{L}(\boldsymbol{\theta})}{\partial \boldsymbol{\theta}\, \partial \boldsymbol{\theta}'}\right] = \sum_{i=1}^{n} \mathrm{E}\left[-\frac{\partial^2 \mathcal{L}_i(\boldsymbol{\theta})}{\partial \boldsymbol{\theta}\, \partial \boldsymbol{\theta}'}\right]. \tag{D.14}$$

I_{θ} is often known as the Fisher information or "expected information" matrix for $\boldsymbol{\theta}$. When data are available, one can compute the "local" (or "observed information") matrix for $\boldsymbol{\theta}$ as

$$\widehat{I}_{\theta} = -\frac{\partial^2 \mathcal{L}(\boldsymbol{\theta})}{\partial \boldsymbol{\theta}\, \partial \boldsymbol{\theta}'} = \sum_{i=1}^{n} \left[-\frac{\partial^2 \mathcal{L}_i(\boldsymbol{\theta})}{\partial \boldsymbol{\theta}\, \partial \boldsymbol{\theta}'} \right], \tag{D.15}$$

where the derivatives are evaluated at $\boldsymbol{\theta} = \widehat{\boldsymbol{\theta}}$.

D.3.2 Fisher Information for a One-to-One Transformation of θ

Consider the one-to-one transformation $\boldsymbol{\nu} = \boldsymbol{h}(\boldsymbol{\theta}) = (\nu_1, \ldots, \nu_k)'$ of $\boldsymbol{\theta}$. Then $\boldsymbol{\theta} = \boldsymbol{g}(\boldsymbol{\nu}) = [g_1(\boldsymbol{\nu}), \ldots, g_k(\boldsymbol{\nu})]'$, where $\boldsymbol{g}(\boldsymbol{\nu}) = \boldsymbol{h}^{-1}(\boldsymbol{\nu})$. When the interest is in the Fisher information matrix for $\boldsymbol{\nu}$, we use the chain rule for the derivative of a scalar function with respect a vector to show that

$$\frac{\partial \mathcal{L}(\boldsymbol{\nu})}{\partial \boldsymbol{\nu}} = \left[\frac{\partial \boldsymbol{g}(\boldsymbol{\nu})}{\partial \boldsymbol{\nu}} \right] \left[\frac{\partial \mathcal{L}[\boldsymbol{g}(\boldsymbol{\nu})]}{\partial \boldsymbol{g}(\boldsymbol{\nu})} \right],$$

where

$$\left[\frac{\partial \boldsymbol{g}(\boldsymbol{\nu})}{\partial \boldsymbol{\nu}} \right] = \begin{bmatrix} \dfrac{\partial g_1(\boldsymbol{\nu})}{\partial \nu_1} & \cdots & \dfrac{\partial g_k(\boldsymbol{\nu})}{\partial \nu_1} \\ \vdots & \vdots & \vdots \\ \dfrac{\partial g_1(\boldsymbol{\nu})}{\partial \nu_k} & \cdots & \dfrac{\partial g_k(\boldsymbol{\nu})}{\partial \nu_k} \end{bmatrix} \quad \text{and} \quad \frac{\partial \mathcal{L}[\boldsymbol{g}(\boldsymbol{\nu})]}{\partial \boldsymbol{g}(\boldsymbol{\nu})} = \left. \frac{\partial \mathcal{L}(\boldsymbol{\theta})}{\partial \boldsymbol{\theta}} \right|_{\boldsymbol{\theta}=\boldsymbol{g}(\boldsymbol{\nu})}.$$

Then from $\partial \mathcal{L}(\boldsymbol{\nu})/\partial \boldsymbol{\nu}$ and using (D.14), we obtain

$$I_{\nu} = \left[\frac{\partial \boldsymbol{g}(\boldsymbol{\nu})}{\partial \boldsymbol{\nu}} \right] I_{\theta} \left[\frac{\partial \boldsymbol{g}(\boldsymbol{\nu})}{\partial \boldsymbol{\nu}} \right]', \tag{D.16}$$

where I_{θ} is evaluated at $\boldsymbol{\theta} = \boldsymbol{g}(\boldsymbol{\nu})$.

In Section D.5.1, we explain that, under the standard regularity conditions, $n\Sigma_{\widehat{\theta}} = n(I_{\theta})^{-1}$ is the covariance matrix for the asymptotic (large-sample) distribution of $\sqrt{n}\,(\widehat{\boldsymbol{\theta}} - \boldsymbol{\theta})$ and an estimate of I_{θ}^{-1} can be used to estimate the sampling variability in $\widehat{\boldsymbol{\theta}}$.

Each technical asymptotic result, such as the asymptotic distribution of an estimator, or a specific asymptotic property of an estimator, requires its own set of regularity conditions on the model. For example, under a certain set of conditions it is possible to show that ML estimators are asymptotically normal. With additional conditions, it can be shown that ML estimators are also asymptotically efficient. The model, in this case, includes the underlying probability model for the process (e.g., a failure-time process) and for the observation process, such as sampling or inspections (when there is not continuous inspection) and characteristics of any censoring processes. Lehmann (1983, Chapter 6), for example, gives precise regularity conditions in the context of "continuous inspection." Rao (1973, Section 5e) does the same assuming an underlying discrete multinomial observation scheme. Although censoring is not explicitly treated in either of these references, the same asymptotic results hold under the standard kinds of noninformative censoring mechanisms as long as the average amount of information per sample (elements of $\mathcal{I}(\boldsymbol{\theta})$) does not decrease substantially as the sample size increases. For a rigorous treatment of the asymptotic properties of ML estimators based on Type 2 censored data, see Bhattacharyya (1985). See also the Bibliographic Notes section at the end of Chapter 12 for sources of more detailed technical information.

Example D.6 Fisher Information for the Binomial Distribution Parameter π**.** For $X \sim \mathrm{BINOM}(n, \pi)$, the likelihood is $L(\pi) = \mathtt{dbinom}(x; \pi)$. Using the expression for $\mathtt{dbinom}(x; \pi)$ in Section C.4.1 and taking its logarithm gives the log-likelihood

$$\mathcal{L}(\pi) = \log(C) + X \log(\pi) + (n - X) \log(1 - \pi),$$

where the constant $C = n!/[X!(n - X)!]$ does not depend on π. Direct computations, and using the result that $\mathrm{E}(X) = n\pi$, give

$$I_\pi = \mathrm{E}\left[-\frac{\partial^2 \mathcal{L}(\pi)}{\partial \pi^2}\right] = \mathrm{E}\left(\frac{X}{\pi^2} + \frac{n - X}{(1 - \pi)^2}\right) = \frac{n}{\pi(1 - \pi)}. \tag{D.17}$$

This information matrix is used in determining the Jeffreys prior distribution for the $\mathrm{BINOM}(n, \pi)$ distribution in Section H.4.1. ∎

Example D.7   Fisher Information for the Binomial Distribution Log-Odds Parameter. For the binomial model $X \sim \mathrm{BINOM}(n, \pi)$, sometimes one is interested in the log-odds parameter $\nu = \log[\pi/(1 - \pi)]$. Then, as a function of ν, $\pi = g(\nu) = 1/[1 + \exp(-\nu)]$ and

$$\frac{\partial g(\nu)}{\partial \nu} = \frac{\exp(-\nu)}{[1 + \exp(-\nu)]^2}.$$

Using this derivative and substituting (D.17) into (D.16) gives the Fisher information for ν,

$$I_\nu = \frac{n \exp(-\nu)}{[1 + \exp(-\nu)]^2}.$$ ∎

Example D.8 Fisher Information for the Poisson Distribution Rate Parameter λ**.** For $X \sim \mathrm{POIS}(n\lambda)$, where n is amount of exposure and λ is an unknown rate of occurrence parameter, the likelihood is $L(\lambda) = \mathtt{dpois}(x; n\lambda)$. Using the expression for $\mathtt{dpois}(x; n\lambda)$ given in Section C.4.4 and taking its logarithm gives the log-likelihood

$$\mathcal{L}(\lambda) = \log(C) + X \log(\lambda) - n \log(\lambda),$$

where the constant $C = n^X/X!$ does not depend on λ. Direct computations, and using the result that $\mathrm{E}(X) = n\lambda$, give

$$I_\lambda = \mathrm{E}\left[-\frac{\partial^2 \mathcal{L}(\lambda)}{\partial \lambda^2}\right] = \mathrm{E}\left(\frac{X}{\lambda^2}\right) = \frac{n}{\lambda}. \tag{D.18}$$

This information matrix is used in determining the Jeffreys prior distribution for the Poisson distribution model in Section H.4.2. ∎

Example D.9   Fisher Information Matrix for Normal Distribution Parameters. In this case the log-likelihood is

$$\mathcal{L}(\mu, \sigma) = -\frac{n}{2} \log(2\pi) - n \log(\sigma) - \frac{1}{2\sigma^2} \sum_{i=1}^{n} (X_i - \mu)^2. \tag{D.19}$$

1. **Fisher information for** μ **when** σ **is given.** Differentiating (D.19) twice with respect to μ and computing the expectation gives

$$\frac{\partial^2 \mathcal{L}(\mu, \sigma)}{\partial \mu^2} = -\frac{n}{\sigma^2}$$

and thus

$$I_\mu = \mathrm{E}\left(-\frac{\partial^2 \mathcal{L}}{\partial \mu^2}\right) = \frac{n}{\sigma^2}. \tag{D.20}$$

2. **Fisher information for** σ **and** σ^2 **when** μ **is given.** Differentiating (D.19) twice with respect to σ and computing the expectation gives

$$\frac{\partial^2 \mathcal{L}(\mu, \sigma)}{\partial \sigma^2} = \frac{n}{\sigma^2} - \frac{3}{\sigma^4} \sum_{i=1}^{n} (X_i - \mu)^2$$

and

$$I_\sigma = \mathrm{E}\left(-\frac{\partial^2 \mathcal{L}(\mu, \sigma)}{\partial \sigma^2} \right) = \frac{2n}{\sigma^2}. \tag{D.21}$$

A similar computation gives

$$I_{\sigma^2} = \frac{n}{2\sigma^4}. \tag{D.22}$$

3. **Fisher information for** (μ, σ) **and** (μ, σ^2)**.** Differentiating (D.19) twice with respect to (μ, σ) and computing the expectations gives

$$I_{(\mu, \sigma)} = \mathrm{E}\left[-\frac{\partial^2 \mathcal{L}(\boldsymbol{\theta})}{\partial \boldsymbol{\theta} \partial \boldsymbol{\theta}'} \right] = \frac{n}{\sigma^2} \begin{bmatrix} 1 & 0 \\ 0 & 2 \end{bmatrix}, \tag{D.23}$$

where $\boldsymbol{\theta} = (\mu, \sigma)$. A similar computation gives

$$I_{(\mu, \sigma^2)} = \frac{n}{\sigma^2} \begin{bmatrix} 1 & 0 \\ 0 & \frac{1}{2\sigma^2} \end{bmatrix}.$$

∎

D.4 CONVERGENCE IN DISTRIBUTION

In this section we use a subscript n to identify explicitly an estimator or quantity with properties that depend on the sample size n. Considering the sequence for increasing n facilitates the description of these properties when n gets large (i.e., when $n \to \infty$).

Convergence in distribution is an important concept for describing the behavior of estimators in large samples. For example, one is often interested in the statistical properties of the ML estimators $\widehat{\theta}_n$ of the scalar θ when the sample size n increases. In this case a common approach is to consider the distribution of the studentized ratios

$$Z_n = Z_n(\theta) = \frac{\widehat{\theta}_n - \theta}{\widehat{\mathrm{se}}_{\widehat{\theta}_n}}, \quad n = 2, \ldots,$$

where $\widehat{\mathrm{se}}_{\widehat{\theta}_n}$ is a consistent estimator of $\mathrm{se}_{\widehat{\theta}_n}$. In general, the exact distribution of Z_n is complicated, depending on the model, the actual parameter values, and the sample size. But under the regularity conditions of Section D.3.2, if $Z_n(\theta)$ is evaluated at the actual value of $\boldsymbol{\theta}$, then for all z,

$$\lim_{n \to \infty} F_{Z_n}(z) = \Phi_{\mathrm{norm}}(z).$$

Thus, for finite n, one can use the approximation

$$\Pr[z_{(\alpha/2)} < Z_n \leq z_{(1-\alpha/2)}] = F_{Z_n}[z_{(1-\alpha/2)}] - F_{Z_n}[z_{(\alpha/2)}]$$
$$\approx \Phi_{\mathrm{norm}}[z_{(1-\alpha/2)}] - \Phi_{\mathrm{norm}}[z_{(\alpha/2)}] = 1 - \alpha.$$

The adequacy of this approximation has to be studied (e.g., by simulation) for each individual application, but in general it works well for a large class of problems and moderate to large sample sizes.

More generally, we say that the sequence of scalars Z_n converges in distribution to the continuous random variable V if

$$\lim_{n \to \infty} F_{Z_n}(z) = F_V(z), \quad \text{for all } z,$$

where $F_V(z)$ is the cdf of V. Thus one can use the limiting distribution F_V to approximate the probabilities for finite n as follows:

$$\Pr(a < Z_n \le b) = F_{Z_n}(b) - F_{Z_n}(a) \approx F_V(b) - F_V(a),$$

where a and b are specified constants. This approximation improves as n increases. These ideas of convergence in distribution generalize to vector random variables, (see Billingsley, 2012, page 402, for details).

For other examples, let $\widehat{\boldsymbol{\theta}} = (\widehat{\boldsymbol{\theta}}_1, \widehat{\boldsymbol{\theta}}_2)$ be the ML estimator of a vector $\boldsymbol{\theta} = (\boldsymbol{\theta}_1, \boldsymbol{\theta}_2)$ with a sample of size n, where $\boldsymbol{\theta}_1$ is a parameter(s) of interest and $\boldsymbol{\theta}_2$ contains nuisance parameters. Although the distribution of $\widehat{\boldsymbol{\theta}}$ and its components depend on n, for simplicity in the notation, we do not make that dependence explicit in the notation in what follows.

Likelihood-ratio statistic

The profile likelihood of $\boldsymbol{\theta}_1$ is

$$R(\boldsymbol{\theta}_1) = \max_{\boldsymbol{\theta}_2} \left[\frac{L(\boldsymbol{\theta}_1, \boldsymbol{\theta}_2)}{L(\widehat{\boldsymbol{\theta}})} \right].$$

The corresponding parameter subset log-likelihood-ratio statistic is $\mathrm{LLR}(\boldsymbol{\theta}_1) = -2\log[R(\boldsymbol{\theta}_1)]$. This statistic, when evaluated at the actual value $\boldsymbol{\theta}_1$, converges in distribution to a chi-square distribution with r_1 degrees of freedom, where r_1 is the number of parameters in $\boldsymbol{\theta}_1$.

Wald statistic

For the parameter subset $\boldsymbol{\theta}_1$ the Wald statistic is

$$W(\boldsymbol{\theta}_1) = (\widehat{\boldsymbol{\theta}}_1 - \boldsymbol{\theta}_1)' \left(\widehat{\Sigma}_{\widehat{\boldsymbol{\theta}}_1} \right)^{-1} (\widehat{\boldsymbol{\theta}}_{1n} - \boldsymbol{\theta}_1).$$

When evaluated at the actual value of $\boldsymbol{\theta}_1$, $W(\boldsymbol{\theta}_1)$ converges in distribution to a chi-square random variable with r_1 degrees of freedom.

Score statistic

The score function is defined by the derivative of the log-likelihood

$$S(\boldsymbol{\theta}) = \frac{\partial \mathcal{L}(\boldsymbol{\theta})}{\partial \boldsymbol{\theta}}.$$

For given $\boldsymbol{\theta}_1$, Rao's score statistic for $\boldsymbol{\theta}_1$ is

$$V(\boldsymbol{\theta}_1) = \left[S(\breve{\boldsymbol{\theta}}) \right]' \left(\breve{\Sigma}_{\widehat{\boldsymbol{\theta}}} \right) \left[S(\breve{\boldsymbol{\theta}}) \right], \tag{D.24}$$

where $\breve{\boldsymbol{\theta}} = (\boldsymbol{\theta}_1, \breve{\boldsymbol{\theta}}_2)$, and $\breve{\boldsymbol{\theta}}_2$ maximizes $\mathcal{L}(\boldsymbol{\theta}_1, \boldsymbol{\theta}_2)$ with respect to $\boldsymbol{\theta}_2$ for fixed $\boldsymbol{\theta}_1$. The derivatives are evaluated at $\breve{\boldsymbol{\theta}}$, and $\breve{\Sigma}_{\hat{\boldsymbol{\theta}}}$ is the covariance matrix $\Sigma_{\hat{\boldsymbol{\theta}}}$ evaluated at $\boldsymbol{\theta} = \breve{\boldsymbol{\theta}}$.

In (D.24), $\boldsymbol{\theta}_1$ could be the full parameter vector, a single element of $\boldsymbol{\theta}$, or some other subset of $\boldsymbol{\theta}$. When $\boldsymbol{\theta}_1$ is a subset of the entire parameter vector and $\breve{\boldsymbol{\theta}}_2$ maximizes the likelihood with respect to $\boldsymbol{\theta}_2$, with $\boldsymbol{\theta}_1$ held constant, we have

$$S(\breve{\boldsymbol{\theta}}) = \begin{bmatrix} S_1(\breve{\boldsymbol{\theta}}) \\ S_2(\breve{\boldsymbol{\theta}}) \end{bmatrix} = \begin{bmatrix} S_1(\breve{\boldsymbol{\theta}}) \\ 0 \end{bmatrix},$$

where $S_i(\breve{\boldsymbol{\theta}}) = \partial \mathcal{L} / \partial \boldsymbol{\theta}_i$ and all the derivatives are evaluated at $\breve{\boldsymbol{\theta}}$. Consequently, the score statistic simplifies to

$$V(\boldsymbol{\theta}_1) = \left[S_1(\breve{\boldsymbol{\theta}}) \right]' \left[\breve{\Sigma}_{11} \right] \left[S_1(\breve{\boldsymbol{\theta}}) \right], \tag{D.25}$$

where the matrices $\breve{\Sigma}_{ij}$ are defined by the partitioned matrix

$$\breve{\Sigma}_{\hat{\boldsymbol{\theta}}} = \begin{bmatrix} \breve{\Sigma}_{11} & \breve{\Sigma}_{12} \\ \breve{\Sigma}_{21} & \breve{\Sigma}_{22} \end{bmatrix}.$$

Note that $\breve{\Sigma}_{ii}$, is a square matrix with row dimension equal to the length of $\boldsymbol{\theta}_i$, for $i = 1$ and $i = 2$.

When $n \rightarrow \infty$, the score statistic $V(\boldsymbol{\theta}_1)$, evaluated at the actual $\boldsymbol{\theta}_1$, converges in distribution to a chi-square distribution with r_1 degrees of freedom, where r_1 is the length of $\boldsymbol{\theta}_1$.

Example D.10 Score Statistic for the Poisson Distribution Rate Parameter. Suppose that X is the number of events during n units of exposure from a Poisson process with rate λ. Then $X \sim \text{POIS}(n\lambda)$. The data are x, the number of events observed during the exposure period, and the likelihood is

$$L(\lambda) = \Pr(X = x) = \frac{(n\lambda)^x}{x!} \exp(-n\lambda).$$

The log-likelihood is $\mathcal{L}(\lambda) = x \log(n\lambda) - n\lambda - \log(x!)$. Then the score function is

$$S(\lambda) = \frac{\partial \mathcal{L}(\lambda)}{\partial \lambda} = \frac{x}{\lambda} - n.$$

Solving $S(\lambda) = 0$ gives the ML estimator for λ, which is $\widehat{\lambda} = x/n$. The Fisher information is obtained from (D.18), which gives $I_\lambda = n/\lambda$. In this case $\boldsymbol{\theta}_1 = \boldsymbol{\theta} = \lambda$ and the score statistic is obtained directly from (D.24), giving

$$V(\lambda) = \left(\frac{x}{\lambda} - n \right) \frac{\lambda}{n} \left(\frac{x}{\lambda} - n \right) = \frac{(\lambda - \widehat{\lambda})^2}{\lambda/n}. \tag{D.26}$$

∎

D.5 OUTLINE OF GENERAL MAXIMUM LIKELIHOOD THEORY

This section is a summary of basic results for ML estimators. The presentation is descriptive with little development but relevant references for technical details. The focus is on results needed to justify large-sample approximations for the distribution of statistics used in obtaining some statistical intervals in this book.

D.5.1   Asymptotic Distribution of ML Estimators

In this section we assume that $\widehat{\boldsymbol{\theta}}$ is the ML estimator of $\boldsymbol{\theta}$ based on n observations and that the regularity conditions given in Section D.3.2 hold. Then it can be shown that $\sqrt{n}(\widehat{\boldsymbol{\theta}} - \boldsymbol{\theta})$ converges in distribution to a multivariate normal with mean zero and covariance matrix $\mathcal{I}^{-1}(\boldsymbol{\theta})$ defined in (D.13). In a convenient casual wording, we say that $\widehat{\boldsymbol{\theta}}$ is approximately normal with mean $\boldsymbol{\theta}$ and covariance matrix $\Sigma_{\widehat{\theta}} = I_{\theta}^{-1}$, where $I_{\theta} = n\mathcal{I}(\boldsymbol{\theta})$. In large samples, statistical theory shows that, under the standard regularity conditions, the elements of $\Sigma_{\widehat{\theta}}$ are of the order of n^{-1}. This can be seen by noting that $n\Sigma_{\widehat{\theta}}$ does not depend on n, following from the definition of $\mathcal{I}(\boldsymbol{\theta})$ in (D.13).

Similarly, using the observed information, it can be shown that $\widehat{\boldsymbol{\theta}}$ is approximately multivariate normal with mean $\boldsymbol{\theta}$ and covariance matrix $\widehat{\Sigma}_{\widehat{\theta}}$. This follows from convergence in distribution of $\sqrt{n}(\widehat{\boldsymbol{\theta}} - \boldsymbol{\theta})$ and the fact that $n\widehat{\Sigma}_{\widehat{\theta}}$ is a consistent estimator of $n\Sigma_{\widehat{\theta}}$.

D.5.2   Asymptotic Distribution of Functions of ML Estimators via the Delta Method

In general, one is interested in inferences on functions of $\boldsymbol{\theta}$. For example, consider a vector function $\boldsymbol{g}(\boldsymbol{\theta})$ of the parameters such that all the second derivatives with respect to the elements of $\boldsymbol{\theta}$ are continuous. The ML estimator of $\boldsymbol{g}(\boldsymbol{\theta})$ is $\widehat{\boldsymbol{g}} = \boldsymbol{g}(\widehat{\boldsymbol{\theta}})$. In large samples, $\boldsymbol{g}(\widehat{\boldsymbol{\theta}})$ is approximately normally distributed with mean $\boldsymbol{g}(\boldsymbol{\theta})$ and covariance matrix

$$\Sigma_{\widehat{g}} = \left[\frac{\partial \boldsymbol{g}(\boldsymbol{\theta})}{\partial \boldsymbol{\theta}}\right]' \Sigma_{\widehat{\theta}} \left[\frac{\partial \boldsymbol{g}(\boldsymbol{\theta})}{\partial \boldsymbol{\theta}}\right]. \tag{D.27}$$

This delta-method approximation is based on the assumption that $\boldsymbol{g}(\widehat{\boldsymbol{\theta}})$ is approximately linear in $\boldsymbol{\theta}$ in the region near to $\widehat{\boldsymbol{\theta}}$. The approximation is better in large samples because then the variation in $\widehat{\boldsymbol{\theta}}$ is smaller and thus the region over which $\boldsymbol{\theta}$ varies is correspondingly smaller. If this region is small enough, the linear approximation will be adequate. See Section D.2 for more details.

For scalar g and θ the formula simplifies to

$$\text{Avar}[g(\widehat{\theta})] = \left[\frac{\partial g(\theta)}{\partial \theta}\right]^2 \text{Avar}(\widehat{\theta}),$$

where Avar is the large-sample approximate variance function. For example, if θ is positive and $g(\theta)$ is the log function, the asymptotic variance of $\log(\widehat{\theta})$ is $\text{Avar}[\log(\widehat{\theta})] = \text{Avar}(\widehat{\theta})/\theta^2$.

D.5.3   Estimating the Variance-Covariance Matrix of ML Estimates

Under mild regularity conditions (see Section D.3.2), $\widehat{\Sigma}_{\widehat{\theta}} = (\widehat{I}_{\theta})^{-1}$ is a consistent estimator of $\Sigma_{\widehat{\theta}}$, where $\widehat{I}_{\theta}$ is defined in (D.15). This "local" estimate of $\Sigma_{\widehat{\theta}}$ is obtained by estimating the "expected curvature" in (D.14) by the "observed curvature" in (D.15). It is possible to estimate $\widehat{\Sigma}_{\widehat{\theta}}$ directly by evaluating (D.14) at $\theta = \widehat{\theta}$, but this approach is rarely used because it is more complicated and has no clear advantage.

The "local" estimate of the covariance matrix of $\widehat{g} = g(\widehat{\theta})$ can be obtained by substituting $\widehat{\Sigma}_{\widehat{\theta}}$ for $\Sigma_{\widehat{\theta}}$ in (D.27), giving

$$\widehat{\Sigma}_{\widehat{g}} = \left[\frac{\partial g(\theta)}{\partial \theta}\right]' \widehat{\Sigma}_{\widehat{\theta}} \left[\frac{\partial g(\theta)}{\partial \theta}\right], \tag{D.28}$$

where the derivatives are again evaluated at $\theta = \widehat{\theta}$. For scalar g and θ the formula simplifies to

$$\widehat{\mathrm{Var}}[g(\widehat{\theta})] = \left[\frac{\partial g(\theta)}{\partial \theta}\right]^2 \widehat{\Sigma}_{\widehat{\theta}} = \left[\frac{\partial g(\theta)}{\partial \theta}\right]^2 \widehat{\mathrm{Var}}(\widehat{\theta}).$$

For example, if θ is positive and $g(\theta)$ is the log function, the local estimate of the variance of $\log(\widehat{\theta})$ is $\widehat{\mathrm{Var}}[\log(\widehat{\theta})] = \widehat{\mathrm{Var}}(\widehat{\theta})/\widehat{\theta}^2$ and $\widehat{\mathrm{se}}[\log(\widehat{\theta})] = \widehat{\mathrm{se}}(\widehat{\theta})/\widehat{\theta}$.

D.5.4 Likelihood Ratios and Profile Likelihoods

Suppose we want to estimate θ_1 from the partition $\theta = (\theta_1, \theta_2)$. Let r_1 denote the length of θ_1. The profile likelihood for θ_1 is

$$R(\theta_1) = \max_{\theta_2} \left[\frac{L(\theta_1, \theta_2)}{L(\widehat{\theta})}\right]. \tag{D.29}$$

When the length of θ_2 is 0 (as in the exponential distribution in Example 12.12), expression (D.29) is a relative likelihood for $\theta = \theta_1$. Otherwise we have a "maximized relative likelihood" for θ_1. In either case, $R(\theta_1)$ is commonly known as a "profile likelihood" because it provides a view of the profile of $L(\theta)$ as viewed along a line that is perpendicular to the axes of θ_1.

- When θ_1 is of length 1, $R(\theta_1)$ is a curve projected onto a plane.

- When θ_1 is of length 2 or more, $R(\theta_1)$ is a surface projected onto a three-dimensional hyperplane.

In either case the projection is in a direction perpendicular to the coordinate axes for θ_1. When θ_1 is of length 1 or 2, it is useful to display $R(\theta_1)$ graphically.

Asymptotically, $\mathrm{LLR}(\theta_1) = -2\log[R(\theta_1)]$, when evaluated at the actual θ_1, has a chi-square distribution with r_1 degrees of freedom. To do a likelihood-ratio significance test, we would reject the null hypothesis that $\theta = \theta_0$, at the α level of significance, if

$$\mathrm{LLR}(\theta_1) = -2\log[R(\theta_0)] > \chi^2_{(1-\alpha;r_1)}.$$

D.5.5 Approximate Likelihood-Ratio-Based Confidence Regions or Confidence Intervals for the Model Parameters

An approximate $100(1 - \alpha)\%$ likelihood-ratio-based confidence region for θ_1 is the set of all values of θ_1 such that $\mathrm{LLR}(\theta_1) = -2\log[R(\theta_1)] < \chi^2_{(1-\alpha;r_1)}$ or, equivalently, $R(\theta_1) > \exp\left[-\chi^2_{(1-\alpha;r_1)}/2\right]$. Here θ_1 could be the full parameter vector, a single element of θ, or some other subset of θ. If one is interested in a scalar function $g(\theta)$, these same ideas can be applied after a reparameterization such that $g(\theta)$ is one of the parameters. Simulation studies for different applications and models (e.g., Meeker, 1987; Ostrouchov and Meeker, 1988; Vander Wiel and Meeker, 1990; Jeng and Meeker, 2000) have shown that in terms of closeness to the nominal confidence level, the likelihood-based intervals have important advantages over the standard normal-approximation intervals discussed in Section D.5.6, especially when there are only a small number of failures in the data.

D.5.6 Approximate Wald-Based Confidence Regions or Confidence Intervals for the Model Parameters

The large-sample normal approximation for the distribution of ML estimators can be used to obtain approximate confidence intervals (regions) for scalar (vector) functions of $\boldsymbol{\theta}$. In particular, an approximate $100(1 - \alpha)\%$ confidence region for $\boldsymbol{\theta}$ is the set of all values of $\boldsymbol{\theta}$ inside the ellipsoid

$$(\widehat{\boldsymbol{\theta}} - \boldsymbol{\theta})' \left(\widehat{\Sigma}_{\widehat{\boldsymbol{\theta}}} \right)^{-1} (\widehat{\boldsymbol{\theta}} - \boldsymbol{\theta}) \leq \chi^2_{(1-\alpha;r)},$$

where r is the length of $\boldsymbol{\theta}$. This is sometimes known as the "Wald method" or the "normal-approximation" method. This confidence region (or interval) is based on the distributional result that, asymptotically, when evaluated at the actual value of $\boldsymbol{\theta}$, the Wald statistic

$$W(\boldsymbol{\theta}) = (\widehat{\boldsymbol{\theta}} - \boldsymbol{\theta})' \left(\widehat{\Sigma}_{\widehat{\boldsymbol{\theta}}} \right)^{-1} (\widehat{\boldsymbol{\theta}} - \boldsymbol{\theta})$$

has a chi-square distribution with r degrees of freedom.

More generally, let $\boldsymbol{g}(\boldsymbol{\theta})$ be a vector function of $\boldsymbol{\theta}$. An approximate $100(1 - \alpha)\%$ normal-approximation confidence region for an r_1-dimensional subset $\boldsymbol{g}_1 = \boldsymbol{g}_1(\boldsymbol{\theta})$, from the partition $\boldsymbol{g}(\boldsymbol{\theta}) = [\boldsymbol{g}_1(\boldsymbol{\theta}), \boldsymbol{g}_2(\boldsymbol{\theta})]$, is the set of all the $\boldsymbol{g}_1$ values inside the ellipsoid

$$(\widehat{\boldsymbol{g}}_1 - \boldsymbol{g}_1)' \left(\widehat{\Sigma}_{\widehat{\boldsymbol{g}}_1} \right)^{-1} (\widehat{\boldsymbol{g}}_1 - \boldsymbol{g}_1) \leq \chi^2_{(1-\alpha;r_1)},$$

where $\widehat{\boldsymbol{g}}_1 = \boldsymbol{g}_1(\widehat{\boldsymbol{\theta}})$ is the ML estimator of $\boldsymbol{g}_1(\boldsymbol{\theta})$ and $\widehat{\Sigma}_{\widehat{\boldsymbol{g}}_1}$ is the local estimate of the covariance matrix of $\widehat{\boldsymbol{g}}_1$. The estimate $\widehat{\Sigma}_{\widehat{\boldsymbol{g}}_1}$ can be obtained from the local estimate of $\Sigma_{\widehat{\boldsymbol{g}}}$ in (D.28). This confidence region (or interval) is based on the distributional result that the Wald subset statistic

$$W(\boldsymbol{g}_1) = (\widehat{\boldsymbol{g}}_1 - \boldsymbol{g}_1)' \left(\widehat{\Sigma}_{\widehat{\boldsymbol{g}}_1} \right)^{-1} (\widehat{\boldsymbol{g}}_1 - \boldsymbol{g}_1),$$

when evaluated at the actual $\boldsymbol{g}_1$, has, asymptotically, a chi-square distribution with r_1 degrees of freedom. As shown in Meeker and Escobar (1995), this normal-approximation confidence region (or interval) can be viewed as a quadratic approximation for the log profile likelihood of $\boldsymbol{g}_1(\boldsymbol{\theta})$ at $\widehat{\boldsymbol{g}}_1$.

When $r_1 = 1$, $g_1 = g_1(\boldsymbol{\theta})$ is a scalar function of $\boldsymbol{\theta}$, and an approximate $100(1 - \alpha)\%$ Wald interval is obtained from the familiar formula

$$[\underset{\sim}{g_1}, \quad \widetilde{g}_1] = \widehat{g}_1 \pm z_{(1-\alpha/2)} \widehat{\mathrm{se}}_{\widehat{g}_1},$$

where $\widehat{\mathrm{se}}_{\widehat{g}_1} = \sqrt{\widehat{\mathrm{Var}}[g_1(\widehat{\boldsymbol{\theta}})]}$ is the local estimate for the standard error of $\widehat{g}_1$ and $z_{(1-\alpha/2)}$ is the $1 - \alpha/2$ quantile of the standard normal distribution.

Wald-based (normal-approximation) confidence intervals tend to have coverage probabilities that are less than the nominal confidence levels. Likelihood-ratio-based confidence intervals tend to have coverage probabilities that are much closer to the nominal. See also Meeker and Escobar (1995).

D.5.7   Approximate Score-Based Confidence Regions or Confidence Intervals for Model Parameters

Consider the parameter vector $\boldsymbol{\theta} = (\boldsymbol{\theta}_1, \boldsymbol{\theta}_2)$. An approximate $100(1 - \alpha)\%$ score-based confidence region for the subset $\boldsymbol{\theta}_1$ is the set of all values of $\boldsymbol{\theta}_1$ inside the ellipsoid defined by

$$V(\boldsymbol{\theta}_1) = \left[S(\breve{\boldsymbol{\theta}})\right]'\left(\breve{\Sigma}_{\widehat{\boldsymbol{\theta}}}\right)\left[S(\breve{\boldsymbol{\theta}})\right] \leq \chi^2_{(1-\alpha;r_1)}, \tag{D.30}$$

where $\breve{\boldsymbol{\theta}} = (\boldsymbol{\theta}_1, \breve{\boldsymbol{\theta}}_{2n})$, and $\breve{\boldsymbol{\theta}}_{2n}$ maximizes $\mathcal{L}(\boldsymbol{\theta}_1, \boldsymbol{\theta}_2)$ with respect to $\boldsymbol{\theta}_2$ (see (D.25) for an equivalent expression for $V(\boldsymbol{\theta}_1)$). This confidence region (or interval) is based on the distributional result that, asymptotically, when evaluated at the actual $\boldsymbol{\theta}_1$, the score statistic in (D.25) is distributed as $\chi^2(r_1)$.

If one is interested in a scalar function $g(\boldsymbol{\theta})$, these same ideas can be applied after a reparameterization such that $g(\boldsymbol{\theta})$ is one of the model parameters (see Boos and Stefanski, 2013, Chapter 3, for a general discussion of the score statistic).

Example D.11   Score Based Confidence Interval for the Poisson Distribution Rate Parameter. Using (D.30) with the score statistic for the λ parameter in (D.26), an approximate $100(1 - \alpha)\%$ confidence region for λ is

$$\frac{(\lambda - \widehat{\lambda})^2}{\lambda/n} \leq \chi^2_{(1-\alpha;1)}.$$

Using the fact that $\chi^2_{(1-\alpha;1)} = z^2_{(1-\alpha/2)}$, the confidence region contains the values of λ that satisfy the relationship $(\lambda - \widehat{\lambda})^2 - z^2_{(1-\alpha/2)}(\lambda/n) \leq 0$. Or equivalently, grouping the quadratic and linear terms for λ in the relationship, the confidence region contains values of λ that satisfy the relationship

$$\lambda^2 - \left(2\widehat{\lambda} + z^2_{(1-\alpha/2)}/n\right)\lambda + \widehat{\lambda}^2 \leq 0. \tag{D.31}$$

Because the left-hand side in the inequality (D.31) is a quadratic function that takes positive values at $\lambda = 0$ and for large values of λ, the confidence region is the interval given by the two roots of the quadratic equation

$$\lambda^2 - \left(2\widehat{\lambda} + z^2_{(1-\alpha/2)}/n\right)\lambda + \widehat{\lambda}^2 = 0.$$

A simple computation gives the two roots which determine the score-based confidence interval for λ,

$$\left[\underset{\sim}{\lambda}, \ \widetilde{\lambda}\right] = \widehat{\widehat{\lambda}} \mp z_{(1-\alpha/2)}\frac{1}{\sqrt{n}}\left(\widehat{\lambda} + \frac{z^2_{(1-\alpha/2)}}{4n}\right)^{1/2},$$

where $\widehat{\widehat{\lambda}} = (x + z^2_{(1-\alpha/2)}/2)/n$. This is the basis of the score interval in Section 7.2.4   ∎

D.6   THE CDF PIVOTAL METHOD FOR OBTAINING CONFIDENCE INTERVALS

This section describes the cdf *pivotal* method to obtain confidence intervals for scalar parameters. The method is very useful and provides exact intervals for parameters from some continuous distributions and conservative intervals for parameters from some discrete distribution.

As mentioned in Section 2.7, it is possible to combine a one-sided lower $100(1 - \alpha_L)\%$ confidence bound and a one-sided upper $100(1 - \alpha_L)\%$ confidence bound to obtain a two-sided $100(1 - 2\alpha_L)\%$ confidence interval (see Section B.2.2 for a technical demonstration of this result). To simplify the presentation, the development in this section will focus on two-sided confidence intervals, but there is a parallel set of results for both lower and one-sided upper confidence bounds.

D.6.1   Continuous Distributions

Consider a completely specified scalar function of the data $W = g(\boldsymbol{X})$; for example, $W = \sum_{i=1}^{n} X_i^2$. Suppose that W has a cdf $F(w; \theta) = \Pr(W \leq w; \theta)$ which is continuous in w, where θ is a scalar parameter. In addition, suppose that for all w, the function $F(w; \theta)$ is continuous and strictly monotone *decreasing* in θ.

Because of the cdf transform (see Example D.2 in Section D.1), the random quantity $F(W; \theta)$ has a $\mathrm{UNIF}(0, 1)$ distribution. This implies that $F(W; \theta)$ is pivotal. Thus

$$\Pr\left(\frac{\alpha}{2} \leq F(W; \theta) \leq 1 - \frac{\alpha}{2}\right) = 1 - \alpha.$$

Using the pivotal procedure in Section E.1, an exact $100(1 - \alpha)\%$ confidence interval $[\underset{\sim}{\theta}, \ \widetilde{\theta}]$ for θ is obtained from the solutions to

$$F(w; \underset{\sim}{\theta}) = 1 - \frac{\alpha}{2} \quad \text{and} \quad F(w; \widetilde{\theta}) = \frac{\alpha}{2},$$

where $w = g(\boldsymbol{x})$ is the sample value of W. Equivalently, obtain $\underset{\sim}{\theta}$ and $\widetilde{\theta}$ as the solutions to

$$\Pr(W \geq w; \underset{\sim}{\theta}) = \frac{\alpha}{2} \quad \text{and} \quad \Pr(W \leq w; \widetilde{\theta}) = \frac{\alpha}{2}. \tag{D.32}$$

When $F(w; \theta)$ is continuous and strictly monotone *increasing* on θ, the computation of the confidence interval for θ is similar. In particular, the interval $[\underset{\sim}{\theta}, \ \widetilde{\theta}]$ endpoints are obtained from to the solutions to

$$\Pr(W \geq w; \widetilde{\theta}) = \frac{\alpha}{2} \quad \text{and} \quad \Pr(W \leq w; \underset{\sim}{\theta}) = \frac{\alpha}{2}. \tag{D.33}$$

Note that it is implicitly assumed that there are solutions $\underset{\sim}{\theta}$ and $\widetilde{\theta}$ for (D.32), which need not be the case in certain special cases.

Example D.12   Pivoting a CDF to Obtain a Confidence Interval for a Normal Distribution Variance Parameter. Consider a random sample $X_1, \ldots, X_n$ from a $\mathrm{NORM}(0, \sigma)$ distribution. A sufficient statistic for σ^2 is $W = \sum_{i=1}^{n} X_i^2 / n$. Let $x_1, \ldots, x_n$ be the sample values. The ML estimate for σ^2 is $\widehat{\sigma}^2 = \sum_{i=1}^{n} x_i^2 / n$, which is the observed value of W.

Using the fact that $(X_i / \sigma)^2 \sim \chi^2(1)$ and that the X_i are independent, then $W \sim (\sigma^2 / n)\chi^2(n)$. Consequently, the sampling distribution of $\widehat{\sigma}^2$ is

$$F(\widehat{\sigma}^2; \sigma^2) = \Pr(W \leq \widehat{\sigma}^2) = \Pr\left(X_{(n)}^2 \leq \frac{n\widehat{\sigma}^2}{\sigma^2}\right) = \texttt{pchisq}\left(\frac{n\widehat{\sigma}^2}{\sigma^2}; n\right). \tag{D.34}$$

Because

$$\frac{d}{d\sigma^2} F(\widehat{\sigma}^2; \sigma^2) = -\frac{n\widehat{\sigma}^2}{\sigma^4} \texttt{dchisq}\left(\frac{n\widehat{\sigma}^2}{\sigma^2}, n\right) < 0,$$

for any fixed value of $\widehat{\sigma}^2$, $F(\widehat{\sigma}^2; \sigma^2)$ is a continuous monotone decreasing function of σ^2. Using (D.32) and (D.34), $\Pr(W \geq \widehat{\sigma}^2; \underset{\sim}{\sigma}^2) = \alpha/2$, which implies that $\Pr(W \leq \widehat{\sigma}^2; \underset{\sim}{\sigma}^2) =$

$\texttt{pchisq}(n\widehat{\sigma}^2/\underset{\sim}{\sigma}^2; n) = 1 - \alpha/2$. Equivalently, $\underset{\sim}{\sigma}^2 = n\widehat{\sigma}^2/\chi^2_{(1-\alpha/2;n)}$. Similarly, $\Pr(W \geq \widehat{\sigma}^2; \widetilde{\sigma}^2) = \alpha/2$ gives $\widetilde{\sigma}^2 = n\widehat{\sigma}^2/\chi^2_{(\alpha/2;n)}$. Thus an exact $100(1 - \alpha)\%$ confidence interval for σ^2 is

$$[\underset{\sim}{\sigma}^2, \ \widetilde{\sigma}^2] = \left[\frac{n\widehat{\sigma}^2}{\chi^2_{(1-\alpha/2;n)}}, \ \frac{n\widehat{\sigma}^2}{\chi^2_{(\alpha/2;n)}} \right].$$

■

Example D.13 Pivoting a CDF to Obtain a Confidence Interval for an Exponential Distribution Rate Parameter. Consider a random sample $T_1, \ldots, T_n$ from an $\text{EXP}(\theta)$ distribution; the objective is to obtain a confidence interval for the rate parameter $\lambda = 1/\theta$. A sufficient statistic for λ is $W = \sum_{i=1}^{n} T_i/n$. Let $t_1, \ldots, t_n$ be the sample values and let $\bar{t}$ be the sample mean of the data. Note that $\bar{t}$ is a sample value for W. Then the ML estimate for λ is $\widehat{\lambda} = 1/\bar{t}$. It can be verified that $2\lambda X_i \sim \text{EXP}(2) = \chi^2(2)$. Thus $2n\lambda W \sim \chi^2(2n)$ or, equivalently, $W \sim \chi^2(2n)/(2n\lambda)$. Then

$$F(\bar{t}; \lambda) = \Pr(W \leq \bar{t}) = \Pr(X^2_{(2n)} \leq 2n\lambda\bar{t}) = \texttt{pchisq}(2n\lambda\bar{t}; 2n). \tag{D.35}$$

Because

$$\frac{d}{d\lambda}F(\bar{t}; \lambda) = 2n\bar{t}\,\texttt{dchisq}(2n\lambda\bar{t}; 2n) > 0,$$

for any value of $\bar{t}$, $F(\bar{t}; \lambda)$ is a continuous monotone increasing function in λ. Using (D.33) and (D.35) with $\theta = \underset{\sim}{\lambda}$ gives $\Pr(W \leq \bar{t}; \underset{\sim}{\lambda}) = \alpha/2$, which implies that $\Pr(X^2_{(2n)} \leq 2n\underset{\sim}{\lambda}\bar{t}) = \alpha/2$. Thus $2n\underset{\sim}{\lambda}\bar{t} = \chi^2_{(\alpha/2;2n)}$ and $\underset{\sim}{\lambda} = (1/\bar{t})\,[\chi^2_{(\alpha/2;2n)}/(2n)]$. Similarly, $\Pr(W \leq \bar{t}; \widetilde{\lambda}) = \alpha/2$ gives $\widetilde{\lambda} = (1/\bar{t})\,[\chi^2_{(1-\alpha/2;2n)}/(2n)]$. Thus an exact $100(1 - \alpha)\%$ confidence interval for λ is

$$[\underset{\sim}{\lambda}, \ \widetilde{\lambda}] = \left[\frac{\widehat{\lambda}\chi^2_{(\alpha/2;2n)}}{2n}, \ \frac{\widehat{\lambda}\chi^2_{(1-\alpha/2;2n)}}{2n} \right].$$

■

D.6.2 Discrete Distributions

To focus on the discrete distributions of interest in this book, and avoid pathological situations, we consider discrete distributions with support on a finite set (e.g., binomial distribution, beta-binomial distribution) or a subset of the nonnegative integers (i.e., the Poisson distribution).

When $W = g(\boldsymbol{X})$ is a discrete random variable with cdf $F(w; \theta)$, the distribution of $F(W; \theta)$ is not pivotal because it depends on θ. Nevertheless, we can still use the cdf inversion method given in (D.32) and (D.33). As shown in Section D.6.3, this approach provides conservative confidence interval procedures for θ.

Example D.14 Pivoting a CDF to Obtain a Confidence Interval for a Binomial Distribution Probability Parameter. Consider a single observation X from a $\text{BINOM}(n, \pi)$ distribution. Then $W = X \sim \text{BINOM}(n, \pi)$. Using $w = x$ for the sample value, the derivative in (C.20) shows that, for each x, $F(x; \pi)$ is a continuous and decreasing function of π.

Using (D.32) shows that $\Pr(X \geq x; n, \underset{\sim}{\pi}) = \alpha/2$. From the relationship between the beta and binomial distributions in (C.19), $\Pr(X \geq x; n, \underset{\sim}{\pi}) = \texttt{pbeta}(\underset{\sim}{\pi}, x, n - x + 1) = \alpha/2$. Thus $\underset{\sim}{\pi} = \texttt{qbeta}(\alpha/2, x, n - x + 1)$. Similarly, $\Pr(X \leq x; n, \widetilde{\pi}) = \alpha/2$ gives $\widetilde{\pi} = \texttt{qbeta}(1 - \alpha/2, x + 1, n - x)$. Thus a conservative $100(1 - \alpha)\%$ confidence interval for π is

$$[\underset{\sim}{\pi}, \ \widetilde{\pi}] = [\texttt{qbeta}(\alpha/2; x, n - x + 1), \ \texttt{qbeta}(1 - \alpha/2; x + 1, n - x)],$$

which agrees with (6.1).

■

Example D.15 Pivoting a CDF to Obtain a Confidence Interval for the Poisson Rate Parameter. Now consider a single observation X from a $\mathrm{POIS}(n\lambda)$ distribution where $n > 0$ is the amount of exposure and λ is the rate of occurrences per unit of exposure. Then $W = X \sim \mathrm{POIS}(n\lambda)$. Using $w = x$ for the sample value, the derivative in (C.24) shows that, for each x, $F(x; \lambda)$ is a continuous and decreasing function of λ.

Using (D.32) shows that $\Pr(X \geq x; n\underset{\sim}{\lambda}) = \alpha/2$. Using, the relationship between the chi-square and the Poisson distribution cdf in (C.23), $\Pr(X \geq x; n\underset{\sim}{\lambda}) = \Pr(X_{(2x)}^2 \leq 2n\underset{\sim}{\lambda}) = \alpha/2$. Thus $\underset{\sim}{\lambda} = (0.5/n)\chi_{(\alpha/2;2x)}^2$. Similarly, $\Pr(X \leq x; n\widetilde{\lambda}) = \alpha/2$ gives $\widetilde{\lambda} = (0.5/n)\chi_{(1-\alpha/2;2x+2)}^2$. Thus a conservative $100(1-\alpha)\%$ confidence interval for the rate parameter λ is

$$[\underset{\sim}{\lambda}, \ \widetilde{\lambda}] = \left[\frac{0.5\,\chi_{(\alpha/2;2x)}^2}{n}, \ \frac{0.5\,\chi_{(1-\alpha/2;2x+2)}^2}{n} \right],$$

which agrees with (7.1).   ∎

Example D.16 Pivoting a CDF to Obtain a Confidence Interval for a Negative Binomial Distribution Probability Parameter. Now consider a single observation X from an $\mathrm{NBINOM}(k, \pi)$ distribution where X is the number of *conforming* units observed before the kth nonconforming unit is observed, and π is the probability of observing a nonconforming unit at each trial. Then $W = X \sim \mathrm{NBINOM}(k, \pi)$. Using $w = x$ for the sample value, the derivative in (C.22) shows that, for each x, $F(x; k, \pi)$ is a continuous and increasing function of π.

Using (D.33), $\Pr(X \leq x; \underset{\sim}{\pi}) = \alpha/2$. Then using the relationship between the beta and the negative binomial cdfs in (C.21) gives $\Pr(X \leq x; \underset{\sim}{\pi}) = \mathtt{pbeta}(\underset{\sim}{\pi}, k, x+1) = \alpha/2$, which implies that $\underset{\sim}{\pi} = \mathtt{qbeta}(\alpha/2, k, x+1)$. Similarly, $\Pr(X \geq x; k, \widetilde{\pi}) = \alpha/2$ implies $\widetilde{\pi} = \mathtt{qbeta}(1 - \alpha/2, k, x)$. Thus a conservative $100(1-\alpha)\%$ confidence interval for π is

$$\left[\underset{\sim}{\pi}, \ \widetilde{\pi}\right] = [\mathtt{qbeta}(\alpha/2, k, x+1), \ \mathtt{qbeta}(1 - \alpha/2, k, x)].$$ ∎

Example D.17 Pivoting a CDF to Obtain a Confidence Interval for a Hypergeometric Distribution Parameter. Consider a sample of size n without replacement from a population of size N containing D nonconforming units and $N - D$ conforming units, where N and n are known and D is unknown. In this case the parameter of interest is $\theta = D$. Define X as the random variable that counts the number of *nonconforming* units observed in the sample (without replacement) of size n. As shown later, for fixed x, $H(x; n, D, N) = \Pr(X \leq x; D)$ is a nonincreasing step function of D, for $0 \leq D \leq N$.

Because of the discontinuities in $H(x; n, D, N)$ with respect to D, $\widetilde{D}$ is the largest value such that $\Pr(X \leq x; \widetilde{D}) > \alpha/2$. Similarly, $\underset{\sim}{D}$ is the smallest value such that $\Pr(X \geq x; \underset{\sim}{D}) > \alpha/2$. Then, as shown in Section D.6.3, $[\underset{\sim}{D}, \ \widetilde{D}]$ is a conservative $100(1-\alpha)\%$ confidence interval for D.   ∎

D.6.3 Coverage Probability of the Intervals Derived from the cdf Pivotal Method

Here we consider the coverage probability for intervals obtained from pivoting a cdf of a function $W = g(\boldsymbol{X})$ of the data. The general assumptions are as follows:

(a) $W = g(\boldsymbol{X})$ is a continuous (or discrete) random variable with cdf $F(w; \theta) = \Pr(W \leq w)$ and survival function $S(w; \theta) = \Pr(W \geq w)$. In this section, for convenience, the survival function $S(w; \theta)$ is defined as the probability that W takes values equal to or larger than w.

(b) For each w, $F(w;\theta)$ is a continuous and monotone decreasing (or increasing) function of θ.

(c) For a given observed value of w, the interval endpoints $\widetilde{\theta}$ and $\underset{\sim}{\theta}$ are defined to be values of θ such that $F(w;\widetilde{\theta}) = \alpha/2$ and $S(w;\underset{\sim}{\theta}) = \alpha/2$.

Continuous distribution

When $W = g(\boldsymbol{X})$ is a continuous random variable with cdf $F(w;\theta)$, because of the cdf transform $F(W;\theta) \sim \mathrm{UNIF}(0,1)$ and also $S(W;\theta) \sim \mathrm{UNIF}(0,1)$. From (D.32) the coverage probability of the one-sided lower confidence bound is

$$
\begin{aligned}
\Pr\left(\underset{\sim}{\theta} \le \theta\right) &= 1 - \Pr(\theta < \underset{\sim}{\theta}) = 1 - \Pr[S(W;\theta) \le S(W;\underset{\sim}{\theta})] \\
&= 1 - \Pr[S(W;\theta) \le \alpha/2] = 1 - \alpha/2.
\end{aligned}
\tag{D.36}
$$

This shows that the one-sided lower confidence bound is exact (i.e., the coverage probability is equal to the nominal confidence level).

Similarly, for the one-sided upper confidence bound,

$$
\begin{aligned}
\Pr\left(\theta \le \widetilde{\theta}\right) &= 1 - \Pr(\widetilde{\theta} < \theta) = 1 - \Pr[F(W;\theta) \le F(W;\widetilde{\theta})] \\
&= 1 - \Pr[F(W;\theta) \le \alpha/2] = 1 - \alpha/2,
\end{aligned}
\tag{D.37}
$$

which shows that the one-sided upper confidence bound is exact.

Now consider the coverage probability for the two-sided confidence interval for θ. First notice that $\Pr(\theta < \underset{\sim}{\theta}) + \Pr\left(\underset{\sim}{\theta} \le \theta \le \widetilde{\theta}\right) + \Pr(\widetilde{\theta} < \theta) = 1$. Using this result, (D.36), and (D.37) gives

$$
\begin{aligned}
\Pr\left(\underset{\sim}{\theta} \le \theta \le \widetilde{\theta}\right) &= 1 - \Pr(\theta < \underset{\sim}{\theta}) - \Pr(\widetilde{\theta} < \theta) \\
&= 1 - \alpha/2 - \alpha/2 = 1 - \alpha.
\end{aligned}
$$

Thus, the two-sided confidence intervals are also exact. This implies that the confidence intervals in Examples D.12 and D.13 are exact.

Discrete distribution

Now we consider the coverage probability for the intervals when $W = g(\boldsymbol{X})$ is a discrete random variable. To avoid pathological situations, we consider discrete distributions whose set of discontinuities is either finite (e.g., binomial and beta-binomial distributions) or a subset of the nonnegative integers (e.g., the Poisson distribution). This includes all the discrete distributions used in this book.

We now show that the cdf pivotal method, under assumptions (b) and (c) above, as applied in Section D.6.2, gives conservative one-sided confidence bounds and conservative two-sided confidence intervals. To prove the result, recall that the cdf of a $\mathrm{UNIF}(0,1)$ distribution is given by the identity function $\Pr(U \le v) = v$, where $U \sim \mathrm{UNIF}(0,1)$. We first show that for each given θ, $F(W;\theta)$ is stochastically ordered with respect to the $\mathrm{UNIF}(0,1)$ cdf in the sense that $\Pr[F(W;\theta) \le v] \le v$ for all $0 < v < 1$ and $\Pr[F(W;\theta) \le v] < v$ for at least some values of v. Let $J = \{w_0 < \ldots < w_k < \ldots\}$ be the set of discontinuity points of the cdf $F(w;\theta)$. For each w_i in J, define $v_i = F(w_i;\theta) = \Pr(W \le w_i;\theta)$. Note that the v_i values are monotone increasing (i.e., $v_0 < v_1 < \ldots$). Using the fact that $F(w;\theta)$ is nondecreasing in w,

$$
v_i = \Pr(W \le w_i;\theta) = \Pr[F(W;\theta) \le v_i].
\tag{D.38}
$$

This shows that $\Pr[F(W; \theta) \leq v_i] = v_i$. Thus the cdf of $F(W; \theta)$ agrees with the cdf of a UNIF$(0, 1)$ distribution at the v_i points. Now, $\Pr[F(W; \theta) \leq v]$ is right-continuous and constant in the interval $[v_i, \ v_{i+1})$. Then

$$\Pr[F(W; \theta) \leq v] = v_i < v, \quad \text{when } v_i < v < v_{i+1}. \tag{D.39}$$

This shows that at each v value in interval $(v_i, \ v_{i+1})$ the cdf of $F(W; \theta)$ evaluated at v is smaller than the cdf of the UNIF$(0, 1)$ distribution. From (D.38) and (D.39), we conclude that

$$\Pr[F(W; \theta) \leq v] \leq v, \quad \text{for all } 0 \leq v \leq 1, \tag{D.40}$$

where the equality between the UNIF$(0, 1)$ cdf and the cdf $\Pr[F(W; \theta) \leq v]$ occur at and only at the discontinuity points v_i of this distribution.

First we show that the one-sided *upper* $100(1 - \alpha/2)\%$ confidence bound is conservative. Hereafter it is assumed that $F(w; \theta)$ is decreasing in θ (the proof when the cdf is increasing in θ is similar). The upper confidence bound $\widetilde{\theta}$ is determined by $F(w; \widetilde{\theta}) = \alpha/2$. Using (D.40) gives

$$\Pr\left(\theta \leq \widetilde{\theta}\right) = 1 - \Pr(\widetilde{\theta} < \theta) \geq 1 - \Pr(\widetilde{\theta} \leq \theta)$$
$$= 1 - \Pr[F(W; \theta) \leq F(W; \widetilde{\theta})] = 1 - \Pr[F(W; \theta) \leq \alpha/2] \geq 1 - \alpha/2. \tag{D.41}$$

This shows that the one-sided upper confidence bound $\widetilde{\theta}$ is conservative.

The survival function $S(W; \theta)$ is also stochastically ordered with respect to the UNIF$(0, 1)$. To prove this result, note that as a function of s, the function $S(w; \theta)$ is nonincreasing, continuous from the left, with discontinuities at the points in the set J. For each w_i in J, define $s_i = S(w_i; \theta) = \Pr(W \geq w_i; \theta)$. Note that the s_i values are monotone decreasing (i.e., $s_0 > s_1 > \ldots$). Then because $S(w; \theta)$ is nonincreasing in w,

$$s_i = \Pr(W \geq w_i; \theta) = \Pr[S(W; \theta) \leq S(w_i; \theta)] = \Pr[S(W; \theta) \leq s_i]. \tag{D.42}$$

This shows that $\Pr[S(W; \theta) \leq s_i] = s_i$, which is the same as the value of the UNIF$(0, 1)$ cdf at s_i. Now, as a function of s, $\Pr[S(W; \theta) \leq s]$ is right-continuous and constant in the interval $[s_{i+1}, \ s_i)$. Then

$$\Pr[S(W; \theta) \leq s] = s_{i+1} < s, \quad \text{when } s_{i+1} < s < s_i. \tag{D.43}$$

From (D.42) and (D.43),

$$\Pr[S(W; \theta) \leq s] \leq s, \quad \text{for all } 0 \leq s \leq 1, \tag{D.44}$$

where the equality between the UNIF$(0, 1)$ cdf and the cdf $\Pr[S(W; \theta) \leq s]$ occur at and only at the discontinuity points s_i of this distribution.

To show that the one-sided *lower* $100(1 - \alpha/2)\%$ confidence bound is conservative, using (D.44) gives

$$\Pr\left(\underset{\sim}{\theta} \leq \theta\right) = 1 - \Pr(\theta < \underset{\sim}{\theta}) < \theta) \geq 1 - \Pr(\theta \leq \underset{\sim}{\theta})$$
$$= 1 - \Pr[S(W; \theta) \leq S(W; \underset{\sim}{\theta})] = 1 - \Pr[S(W; \theta) \leq \alpha/2] \geq 1 - \alpha/2. \tag{D.45}$$

This shows that the one-sided lower bound $\underset{\sim}{\theta}$ is also conservative.

Now consider the two-sided $100(1 - \alpha)\%$ confidence interval $[\underset{\sim}{\theta}, \ \widetilde{\theta}]$. The interval endpoints are determined from $F(w; \widetilde{\theta}) = S(w; \underset{\sim}{\theta}) = \alpha/2$. Because the three events $\{\theta < \underset{\sim}{\theta}\}$, $\{\underset{\sim}{\theta} \leq \theta \leq \widetilde{\theta}\}$, and $\{\widetilde{\theta} < \theta\}$ are disjoint and their union is equal to $(-\infty, \ \infty)$,

$$\Pr\left(\theta < \underset{\sim}{\theta}\right) + \Pr\left(\underset{\sim}{\theta} \leq \theta \leq \widetilde{\theta}\right) + \Pr\left(\widetilde{\theta} < \theta\right) = 1.$$

Solving this equation for $\Pr\left(\underset{\sim}{\theta} \le \theta \le \widetilde{\theta}\right)$ and using (D.41)–(D.45) gives

$$\Pr\left(\underset{\sim}{\theta} \le \theta \le \widetilde{\theta}\right) = 1 - \Pr\left(\theta < \underset{\sim}{\theta}\right) - \Pr\left(\widetilde{\theta} < \theta\right) \ge 1 - \Pr\left(\theta \le \underset{\sim}{\theta}\right) - \Pr\left(\widetilde{\theta} \le \theta\right)$$
$$\ge 1 - \alpha/2 - \alpha/2 = 1 - \alpha. \tag{D.46}$$

This shows that the confidence interval $[\underset{\sim}{\theta},\ \widetilde{\theta}]$ is conservative.

In summary, the confidence intervals based on the CDF pivotal methods for discrete distributions that satisfy assumptions (a)–(c) are conservative. In particular, the intervals in Examples D.14–D.16 meet the assumptions and they are conservative.

In the case of Example D.17 where $W = X$, the conditions in items (b)–(c) above are not satisfied. The difficulty is caused by the discontinuity of $F(w; D)$ as a function of D because D is a discrete parameter. The interval in this example, however, is also conservative, as we show next. The cdf $F(w; D)$ is nonincreasing in D. This result is a consequence of the fact that for fixed x, N and n, the hypergeometric distribution cdf $\texttt{phyper}(x; n, D, N)$ is a nonincreasing function of D (see Section C.4.5). Recall that in Example D.17 the bounds were specified as $\widetilde{D}$, the largest value such that $F(w; \widetilde{D}) > \alpha/2$, and $\underset{\sim}{D}$, the smallest value such that $S(w; \underset{\sim}{D}) > \alpha/2$.

With $\underset{\sim}{D}$ and $\widetilde{D}$ as given in Example D.17, using $F(w; \widetilde{D}) > \alpha/2$, $S(w; \underset{\sim}{D}) > \alpha/2$, and proceeding as in deriving (D.46) gives

$$\Pr\left(\underset{\sim}{D} \le D \le \widetilde{D}\right) \ge 1 - \Pr[S(W; D) \le S(W; \underset{\sim}{D})] - \Pr[F(W; D) \le F(W; \widetilde{D})]$$
$$\ge 1 - \Pr[S(W; D) \le \alpha/2] - \Pr[F(W; D) \le \alpha/2] \tag{D.47}$$
$$\ge 1 - \frac{\alpha}{2} - \frac{\alpha}{2} = 1 - \alpha.$$

This shows that the interval $\left[\underset{\sim}{D},\ \widetilde{D}\right]$ is conservative because its coverage probability is larger than or equal to the nominal $1 - \alpha$ confidence level. The one-sided lower $\underset{\sim}{D}$ and upper $\widetilde{D}$ confidence bounds are also conservative. The proof of this is a simplified version of the result in (D.47).

D.7 BONFERRONI APPROXIMATE STATISTICAL INTERVALS

D.7.1 The Bonferroni Inequality

Given a collection of m sets $A_1, \ldots, A_m$, the Bonferroni inequality provides a bound for the probability of the intersection of m sets, say $\bigcap_{i=1}^{m} A_i = A_1 \cap A_2 \cap \cdots \cap A_m$, as a function of the probabilities of the A_i or the probability of the A_i^c. To be precise,

$$\Pr\left(\bigcap_{i=1}^{m} A_i\right) \ge \sum_{i}^{m} \Pr(A_i) - (m - 1) = 1 - \sum_{i}^{m} \Pr(A_i^c).$$

To prove the inequality for $m = 2$, note that

$$1 \ge \Pr(A_1 \cup A_2) = \Pr(A_1) + \Pr(A_2) - \Pr(A_1 \cap A_2). \tag{D.48}$$

From the inequality between the far left and far right terms in (D.48), after rearranging some terms, we obtain

$$\Pr(A_1 \cap A_2) \ge \Pr(A_1) + \Pr(A_2) - 1 = 1 - [\Pr(A_1^c) + \Pr(A_2^c)],$$

where A_i^c is the complement of the set A_i for $i = 1, 2$. This proves the Bonferroni inequality for $m = 2$. The general Bonferroni inequality is obtained using induction in m.

D.7.2   Bonferroni Conservative Approach for Simultaneous Statistical Intervals

Suppose A_i is a statement that a confidence interval contains a parameter θ_i of interest, where $i = 1, \ldots, k$. Also suppose that, individually, the coverage probabilities of the confidence intervals are $\Pr(A_1) = 1 - \alpha_1, \ldots, \Pr(A_k) = 1 - \alpha_k$. Then using Bonferroni, the combined simultaneous confidence region $\bigcap_{i=1}^{k} A_i$ has a conservative coverage probability of

$$\Pr\left(\bigcap_{i=1}^{k} A_i\right) \geq 1 - (\alpha_1 + \alpha_2 + \cdots + \alpha_k).$$

For example, if $[a_i, \; b_i]$ is an exact 95% confidence level interval for θ_i for each $i = 1, 2$, then the square with corners $\{(a_1, a_2), (b_1, a_2), (b_1, b_2), (a_1, b_2)\}$ is a conservative simultaneous 90% confidence level region for the pair (θ_1, θ_2). Section 2.9 discuss the combination of individual confidence intervals to obtain simultaneous confidence intervals.

D.7.3   Bonferroni Conservative Approach for Tolerance Intervals

One-sided tolerance bounds can be combined to obtain *approximate* two-sided tolerance intervals. For example, suppose that $\underset{\sim}{T'_{p_L}}$ is a one-sided lower tolerance bound for T that one can claim with $100(1 - \alpha_L)\%$ confidence is exceeded by at least $100p_L \%$ of the population. Also suppose that $\widetilde{T}'_{p_U}$ is a one-sided upper tolerance bound for T that one can claim with $100(1 - \alpha_U)\%$ confidence exceeds at least $100p_U \%$ of the population. Then $[\underset{\sim}{T'_{p_L}}, \; \widetilde{T}'_{p_U}]$ is an approximate two-sided tolerance interval that one can claim with $100(1 - \alpha_L - \alpha_U)\%$ confidence encloses at least $100(p_L + p_U - 1)\%$ of the sampled population. The coverage probability for this two-sided tolerance interval is greater than $100(1 - \alpha_L - \alpha_U)\%$, and thus the interval is conservative (i.e., wider than it needs to be). This is an application of the Bonferroni approximation, as follows.

Define the sets

$$A_1 = \left\{ \Pr\left(\underset{\sim}{T'_{p_L}} < T \mid \boldsymbol{X}\right) > p_L \right\} = \left\{ 1 - F(\underset{\sim}{T'_{p_L}}) > p_L \right\},$$
$$A_2 = \left\{ \Pr\left(T \leq \widetilde{T}'_{p_U} \mid \boldsymbol{X}\right) > p_U \right\} = \left\{ F(\widetilde{T}'_{p_U}) > p_U \right\}.$$

Note that A_1 and A_2 depend on $\boldsymbol{X}$ through $\underset{\sim}{T'_{p_L}}$ and $\widetilde{T}'_{p_U}$, respectively. Also $\Pr_{\boldsymbol{X}}(A_1) \geq 1 - \alpha_L$ and $\Pr_{\boldsymbol{X}}(A_2) \geq 1 - \alpha_U$. Define the set B as

$$B = \left\{ \Pr\left(\underset{\sim}{T'_{p_L}} < T \leq \widetilde{T}'_{p_U} \mid \boldsymbol{X}\right) > p_L + p_U - 1 \right\} = \left\{ F(\widetilde{T}'_{p_U}) - F(\underset{\sim}{T'_{p_L}}) > p_L + p_U - 1 \right\}.$$

Note that B depends on $\boldsymbol{X}$ through $\underset{\sim}{T'_{p_L}}$ and $\widetilde{T}'_{p_U}$. Because the inequalities $1 - F(\underset{\sim}{T'_{p_L}}) > p_L$ and $F(\widetilde{T}'_{p_U}) > p_U$ imply the inequality $F(\widetilde{T}'_{p_U}) - F(\underset{\sim}{T'_{p_L}}) > p_L + p_U - 1$, then $A_1 \cap A_2$ is a subset of B which implies $\Pr(B) \geq \Pr_{\boldsymbol{X}}(A_1 \cap A_2)$. Thus

$$\begin{aligned} \Pr\left(\underset{\sim}{T'_{p_L}} < T \leq \widetilde{T}'_{p_U}\right) &= \Pr_{\boldsymbol{X}}(B) \geq \Pr_{\boldsymbol{X}}(A_1 \cap A_2) \\ &\geq \Pr_{\boldsymbol{X}}(A_1) + \Pr_{\boldsymbol{X}}(A_2) - 1 \qquad \text{(D.49)} \\ &\geq 1 - (\alpha_L + \alpha_U), \end{aligned}$$

where the inequality in the second line of (D.49) is obtained by applying the Bonferroni inequality to $\Pr_{\boldsymbol{X}}(A_1 \cap A_2)$.

When an exact two-sided tolerance interval is available (e.g., for the normal and exponential distributions), such an exact interval is more precise (i.e., narrower) than the two-sided tolerance interval from the preceding approximation.

D.7.4   Alternative Approximate Tolerance Interval Method

Simulation studies (like those reported in Sections 6.6.4 and 7.5.4) have shown that, for discrete distributions, the one-sided tolerance bound procedures tend to be conservative, and thus applying the method in Section D.7.3 results in additional conservatism (resulting in overly wide tolerance intervals). Thus, for discrete distributions we will suggest a modification of the approach in Section D.7.3. To obtain an approximate two-sided $100(1 - \alpha)\%$ tolerance interval to contain at least a proportion β of the distribution, one combines a one-sided lower $100(1 - \alpha)\%$ confidence bound on the $(1 - \beta)/2$ quantile for the lower endpoint and a one-sided upper $100(1 - \alpha)\%$ confidence bound on the $(1 + \beta)/2$ quantile for the upper endpoint. That is $[\underset{\sim}{T}_\beta, \ \widetilde{T}_\beta] = [\underset{\sim}{y}_{(1-\beta)/2}, \ \widetilde{y}_{(1+\beta)/2}]$. Note that this procedure is not guaranteed to be conservative (but evaluations in Sections 6.6.4 and 7.5.4 show that they are).

D.7.5   Two-Sided Prediction Intervals Based on Two One-Sided Prediction Bounds and a Bonferroni Approximation

Section 4.8 gives the simultaneous two-sided $100(1 - \alpha)\%$ prediction interval to contain the values of all m future randomly selected observations from a previously sampled normal distribution

$$[\underset{\sim}{Y}_{m:m}, \ \widetilde{Y}_{m:m}] = \bar{x} \mp r_{(1-\alpha;m,m,n)} s,$$

where $r_{(1-\alpha;m,m,n)}$ is computed by the methods shown in Section E.6.4. For $m = 1$, $r_{(1-\alpha;1,1,n)} = (1 + 1/n)^{1/2} t_{(1-\alpha/2;n-1)}$ (see Section 4.7 for details). We now show that a conservative approximation for $r_{(1-\alpha;m,m,n)}$ is

$$r^a_{(1-\alpha;m,m,n)} = \left(1 + \frac{1}{n}\right)^{1/2} t_{(1-\alpha/(2m);n-1)}. \tag{D.50}$$

Suppose that Y_i, $i = 1, 2, \ldots, m$, are the future observations. Define $A_i = \left\{\underset{\sim}{Y}^a_{m:m}, \leq Y_i \leq \widetilde{Y}^a_{m:m}\right\}$. Note that for a *single observation* (i.e., $m = 1$), the prediction interval $\left[\underset{\sim}{Y}^a_{m:m}, \ \widetilde{Y}^a_{m:m}\right]$ is exact with coverage probability equal to $\Pr(A_i) = 1 - \alpha/m$, for each i. Then using the Bonferroni inequality gives

$$\Pr\left(\bigcap_{i=1}^m A_i\right) \geq 1 - \Pr\left(\sum_{i=1}^m A_i^c\right) = 1 - \left(\frac{\alpha}{m} + \cdots + \frac{\alpha}{m}\right) = 1 - \alpha.$$

Because $\Pr(\bigcap_{i=1}^m A_i)$ is the probability that the prediction interval contains all of the m future observations, the approximation $r^a_{(1-\alpha;m,m,n)} \approx r_{(1-\alpha;m,m,n)}$ gives a conservative prediction interval.

The conservative approximation in (D.50) was suggested by Chew (1968). See the Bibliographic Notes section at the end of Chapter 4 for additional information.

Appendix *E*

Pivotal Methods for Constructing Parametric Statistical Intervals

INTRODUCTION

This appendix outlines some of the most useful methods for defining particular statistical interval procedures. In particular, we present pivotal quantities that lead to well-known interval procedures for the normal distribution. We also present pivotal quantities for the more general location-scale distributions and indicate how methods for location-scale distributions can usually be applied directly to log-location-scale distributions and, in an approximate manner, to obtain statistical intervals and bounds for other parametric distributions or when the data are censored.

The topics discussed in this appendix are:

- Pivotal quantities and their use to construct confidence intervals (Section E.1).

- Examples of normal distribution pivotal quantities (Section E.2).

- How to obtain confidence intervals for the mean, the standard deviation, quantiles, and tail probabilities of a normal distribution using pivotal quantities (Section E.3).

- Examples of pivotal methods to construct confidence intervals to compare means and sample variances for data coming from two normal distributions (Section E.4).

- The use of pivotal quantities to obtain tolerance intervals for normally distributed data (Section E.5).

- Examples of pivotal methods to construct prediction intervals to predict the mean or the sample standard deviation for data coming from a normal distribution. Also, examples of pivotal methods to construct prediction bounds and intervals to bound or contain at least

Statistical Intervals: A Guide for Practitioners and Researchers, Second Edition.
William Q. Meeker, Gerald J. Hahn and Luis A. Escobar.
© 2017 John Wiley & Sons, Inc. Published 2017 by John Wiley & Sons, Inc.
Companion Website: www.wiley.com/go/meeker/intervals

473

k out of m future observations for data coming from a normal or a log-location-scale distribution (Section E.6).

- How to obtain confidence intervals for the location, scale, and quantiles for a location-scale and log-location-scale distribution using pivotal quantities (Section E.7).

E.1 GENERAL DEFINITION AND EXAMPLES OF PIVOTAL QUANTITIES

Consider a vector of random variables $\boldsymbol{X} = (X_1, \ldots, X_k)$ with joint distribution $f(\boldsymbol{x}; \boldsymbol{\theta})$. A function of the observations $\boldsymbol{X}$ and the parameters $\boldsymbol{\theta}$, say $g(\boldsymbol{X}, \boldsymbol{\theta})$, is a pivotal quantity if the distribution of $g(\boldsymbol{X}, \boldsymbol{\theta})$ does not depend on $\boldsymbol{\theta}$. As illustrated later, the definition of pivotal quantities does not require that the $X_1, \ldots, X_n$ are independent and identically distributed. The vector of parameters $\boldsymbol{\theta}$ may have one or more components. There is the potential to use vector pivotal quantities, but in this book we use only scalar pivotal quantities.

Pivotal quantities are useful to construct confidence intervals. The general process is as follows.

- Suppose that $g(\boldsymbol{X}, \boldsymbol{\theta})$ is continuous and that g_{α_1} and $g_{(1-\alpha_2)}$ are the α_1 and $1 - \alpha_2$ quantiles of its distribution, where $\alpha_1 + \alpha_2 = \alpha < 1$ and usually $\alpha_1 = \alpha_2 = \alpha/2$. Then

$$\Pr[g_{\alpha_1} \leq g(\boldsymbol{X}, \boldsymbol{\theta}) \leq g_{(1-\alpha_2)}] = 1 - \alpha. \qquad (\text{E.1})$$

- Now suppose that $\boldsymbol{\theta} = \theta$ is a scalar and $g(\boldsymbol{X}, \theta)$ is a scalar monotone decreasing function of θ. Let $\underset{\sim}{\theta}$ and $\widetilde{\theta}$ be the solutions for θ to the equations $g(\boldsymbol{X}, \theta) = g_{(1-\alpha_2)}$, and $g(\boldsymbol{X}, \theta) = g_{\alpha_1}$, respectively. Then

$$\Pr(\underset{\sim}{\theta} \leq \theta \leq \widetilde{\theta}) = 1 - \alpha$$

defines an exact $100(1 - \alpha)\%$ confidence interval for θ. Note that $\underset{\sim}{\theta}$ and $\widetilde{\theta}$ are functions of $\boldsymbol{X}$ (e.g., $\bar{X}$ and S^2). The procedure generating the interval $[\underset{\sim}{\theta}, \ \widetilde{\theta}]$ satisfies the definition of an exact confidence interval because the coverage probability is equal to the nominal confidence level of $100(1 - \alpha)\%$.

- There are similar results when $g(\boldsymbol{X}, \theta)$ is monotone increasing in θ.

- For a nonmonotonic function $g(\boldsymbol{X}, \theta)$ or for a vector $\boldsymbol{\theta}$, the probability statement in (E.1) still applies. The final confidence interval for $\boldsymbol{\theta}$ may, however, have a more complicated form, but a similar procedure can be applied to obtain the interval.

E.2 PIVOTAL QUANTITIES FOR THE NORMAL DISTRIBUTION

This section gives pivotal quantities for data from a normal distribution. The derivation of the distributions $g_1, \ldots, g_6$ below is material covered in courses and textbooks dealing with transformation of random variables. See, for example, Casella and Berger (2002, Chapter 5) for derivation of the distributions of some of these pivotal quantities.

E.2.1 Pivotal Quantities from a Single Normal Distribution Sample

Suppose that the data $\boldsymbol{X} = (X_1, \ldots, X_n)$ are n independent observations from a $\mathrm{NORM}(\mu, \sigma)$ distribution and $\bar{Y}$ is the mean of m independent future observations from the same distribution. In the notation of Section E.1, $\boldsymbol{\theta} = (\mu, \sigma)$ and the components of $\boldsymbol{X}$ are iid. Let $\bar{X}$ and S be the sample mean and sample standard deviation of the observations (see Section 3.1.1). In this case, the following random quantities are pivotal quantities, with the distribution indicated in each case:

$$g_1(\boldsymbol{X}, \boldsymbol{\theta}) = \frac{\sqrt{n}(\bar{X} - \mu)}{S} \sim t(n-1), \tag{E.2}$$

$$g_2(\boldsymbol{X}, \boldsymbol{\theta}) = \frac{(n-1)S^2}{\sigma^2} \sim \chi^2(n-1), \tag{E.3}$$

$$g_3(\boldsymbol{X}, \boldsymbol{\theta}) = \frac{\sqrt{n}(\bar{X} - \mu)/\sigma - \sqrt{n}\, z_p}{S/\sigma} \sim t(n-1, \delta), \tag{E.4}$$

$$g_4(\boldsymbol{X}, \boldsymbol{\theta}) = \frac{\bar{Y} - \bar{X}}{(1/n + 1/m)^{1/2}\, S} \sim t(n-1), \tag{E.5}$$

where z_p is the p quantile of the $\mathrm{NORM}(0, 1)$ distribution, $t(n-1)$ is a Student's t-distribution with $n-1$ degrees of freedom, and $t(n-1, \delta)$ is a noncentral t-distribution with $n-1$ degrees of freedom and noncentrality parameter $\delta = -\sqrt{n}\, z_p$ (see Section C.3.9).

E.2.2   Pivotal Quantities Involving Data from Two Normal Distribution Samples

Suppose $\boldsymbol{X}$ is a vector of n independent observations from a $\mathrm{NORM}(\mu_X, \sigma_X)$, $\boldsymbol{Y}$ is a vector of m independent observations from a $\mathrm{NORM}(\mu_Y, \sigma_Y)$, and that $\boldsymbol{X}$ and $\boldsymbol{Y}$ are independent. In this case the vector of unknown parameters is $\boldsymbol{\theta} = (\mu_X, \mu_Y, \sigma_X, \sigma_Y)$. Let $(\bar{X}, S_X)$ be the sample mean and sample standard deviation of $\boldsymbol{X}$ and let $(\bar{Y}, S_Y)$ be the corresponding sample values for $\boldsymbol{Y}$. Then

$$g_5(\boldsymbol{X}, \boldsymbol{Y}, \boldsymbol{\theta}) = \frac{S_Y^2/\sigma_Y^2}{S_X^2/\sigma_X^2} \sim F(m-1, n-1) \tag{E.6}$$

is a pivotal quantity. With the constraint that $\sigma_X = \sigma_Y$,

$$g_6(\boldsymbol{X}, \boldsymbol{Y}, \boldsymbol{\theta}) = \frac{\bar{X} - \bar{Y} - (\mu_X - \mu_Y)}{(1/n + 1/m)^{1/2}\, S_p} \sim t(n+m-2) \tag{E.7}$$

is also a pivotal quantity, where $S_p = \{[(n-1)S_X^2 + (m-1)S_Y^2]/(n+m-2)\}^{1/2}$ is the pooled sample standard deviation computed from $(\boldsymbol{X}, \boldsymbol{Y})$.

E.3   CONFIDENCE INTERVALS FOR A NORMAL DISTRIBUTION BASED ON PIVOTAL QUANTITIES

In this section we consider the situation when the data are a random sample $\boldsymbol{X}$ of size n from a $\mathrm{NORM}(\mu, \sigma)$ distribution where the mean μ and the standard deviation σ are unknown.

E.3.1   Confidence Interval for the Mean of a Normal Distribution

From the pivotal quantity $g_1(\boldsymbol{X}, \boldsymbol{\theta})$ in (E.2),

$$\Pr\left[t_{(\alpha/2; n-1)} \leq \frac{\sqrt{n}(\bar{X} - \mu)}{S} \leq t_{(1-\alpha/2; n-1)} \right] = 1 - \alpha.$$

Symmetry of the t-distribution, $t_{(\alpha/2;n-1)} = -t_{(1-\alpha/2;n-1)}$, gives

$$\Pr\left[-t_{(1-\alpha/2;n-1)} \le \frac{\sqrt{n}(\bar{X} - \mu)}{S} \le t_{(1-\alpha/2;n-1)}\right] = 1 - \alpha.$$

Solving for μ in the center gives

$$\Pr\left[\bar{X} - t_{(1-\alpha/2;n-1)}) \frac{S}{\sqrt{n}} \le \mu \le \bar{X} + t_{(1-\alpha/2;n-1)} \frac{S}{\sqrt{n}}\right] = 1 - \alpha.$$

Thus an exact $100(1-\alpha)\%$ confidence interval for μ is

$$[\underset{\sim}{\mu}, \ \widetilde{\mu}] = \bar{x} \mp t_{(1-\alpha/2;n-1)} \frac{s}{\sqrt{n}},$$

as given in Section 4.2.

E.3.2   Confidence Interval for the Standard Deviation of a Normal Distribution

From the pivotal quantity $g_2(\boldsymbol{X}, \boldsymbol{\theta})$ in (E.3),

$$\Pr\left[\chi^2_{(\alpha/2;n-1)} \le \frac{(n-1)S^2}{\sigma^2} \le \chi^2_{(1-\alpha/2;n-1)}\right] = 1 - \alpha.$$

Solving for σ in the center gives

$$\Pr\left[S\left(\frac{n-1}{\chi^2_{(1-\alpha/2;n-1)}}\right)^{1/2} \le \sigma \le S\left(\frac{n-1}{\chi^2_{(\alpha/2;n-1)}}\right)^{1/2}\right].$$

Thus, an exact $100(1-\alpha)\%$ confidence interval for σ is

$$[\underset{\sim}{\sigma}, \ \widetilde{\sigma}] = \left[s\left(\frac{n-1}{\chi^2_{(1-\alpha/2;n-1)}}\right)^{1/2}, \ s\left(\frac{n-1}{\chi^2_{(\alpha/2;n-1)}}\right)^{1/2}\right],$$

as given in Section 4.3.

E.3.3   Confidence Interval for the Quantile of a Normal Distribution

From the pivotal quantity $g_3(\boldsymbol{X}, \boldsymbol{\theta})$ in (E.4),

$$\Pr\left[t_{(\alpha/2;n-1,\delta)} \le \frac{\sqrt{n}(\bar{X} - \mu)/\sigma - \sqrt{n}\,z_p}{S/\sigma} \le t_{(1-\alpha/2;n-1,\delta)}\right] = 1 - \alpha,$$

where $\delta = -\sqrt{n}z_{(p)}$ is the t-distribution noncentrality parameter. Solving for the p quantile $x_p = \mu + \sigma z_p$ in the center gives

$$\Pr\left[\bar{X} - t_{(1-\alpha/2;n-1,\delta)} \frac{S}{\sqrt{n}} \le x_p \le \bar{X} - t_{(\alpha/2;n-1,\delta)} \frac{S}{\sqrt{n}}\right] = 1 - \alpha.$$

Thus an exact $100(1 - \alpha)\%$ confidence interval for x_p is

$$[\underset{\sim}{x}_p, \ \widetilde{x}_p] = \left[\bar{x} - t_{(1-\alpha/2;n-1,\delta)} \frac{s}{\sqrt{n}}, \ \bar{x} - t_{(\alpha/2;n-1,\delta)} \frac{s}{\sqrt{n}} \right],$$

as given in Section 4.4.

E.3.4   Confidence Intervals for Tail Probabilities of a Normal Distribution

First, we consider a lower tail probability. The probability that an observation from a normal distribution, with given mean μ and standard deviation σ, is less than a specified value x, is given by $p_{LT} = \Pr(X \leq x) = \Phi_{\text{norm}}[(x - \mu)/\sigma]$. The interest is in a confidence interval for p_{LT}.

Let $K = (x - \bar{X})/S$, where $\bar{X}$ and S are the sample mean and sample standard deviation. Define

$$Z = \frac{\sqrt{n}(\mu - \bar{X})}{\sigma}, \quad \delta = \frac{\sqrt{n}(x - \mu)}{\sigma}, \quad W = \frac{(n-1)S^2}{\sigma^2},$$

where n is the sample size. Note that Z and W are independent, $Z \sim \text{NORM}(0,1)$, $W \sim \chi^2(n-1)$, Then

$$\frac{\delta + Z}{S/\sigma} = \frac{\delta + Z}{\sqrt{W/(n-1)}} \sim t(n-1, \delta).$$

Define $G(k) = \Pr(K \leq k)$. Direct computation gives

$$G(k) = \Pr(K \leq k) = \Pr\left(\frac{x - \bar{X}}{S} \leq k \right)$$

$$= \Pr\left[\frac{\sqrt{n}(x - \mu)/\sigma + \sqrt{n}(\mu - \bar{X})/\sigma}{S/\sigma} \leq k\sqrt{n} \right]$$

$$= \Pr\left(\frac{\delta + Z}{S/\sigma} \leq k\sqrt{n} \right) = \text{pt}(k\sqrt{n}; n-1, \delta),$$

where $\text{pt}(k\sqrt{n}; n-1, \delta)$ is the noncentral t-distribution cdf with $n-1$ degrees of freedom and noncentrality parameter δ. Thus, $G(k)$ is completely specified by n and δ. To make explicit the dependence of $G(k)$ on the parameters (n, δ), we write $G(k) = G(k; n-1, \delta)$. That is,

$$\Pr(K \leq k) = G(k; n-1, \delta) = \text{pt}(k\sqrt{n}; n-1, \delta). \tag{E.8}$$

For specified k and n, $\text{pt}(k\sqrt{n}; n-1, \delta)$ is decreasing in δ, as discussed in Section C.3.9. Thus, for specified k and n, $G(k; n-1, \delta)$ is decreasing in δ. Using the cdf pivot method discussed in Section D.6, we proceed to obtain a confidence interval for δ as follows.

An exact $100(1 - \alpha)\%$ confidence interval $[\underset{\sim}{\delta}, \ \widetilde{\delta}]$ for δ is given by the solutions to the equations $G(k; n-1, \underset{\sim}{\delta}) = 1 - \alpha/2$ and $G(k; n-1, \widetilde{\delta}) = \alpha/2$. Equivalently, in terms of the noncentral t-distribution cdf, the confidence interval is provided by the solutions to the equations

$$\text{pt}\left(k\sqrt{n}; n-1, \underset{\sim}{\delta} \right) = 1 - \alpha/2 \quad \text{and} \quad \text{pt}\left(k\sqrt{n}; n-1, \widetilde{\delta} \right) = \alpha/2.$$

Because $p_{LT} = \Phi_{\text{norm}}(\delta/\sqrt{n})$ is a monotone increasing function of δ, a $100(1 - \alpha)\%$ confidence interval for p_{LT} is

$$[\underset{\sim}{p}_{LT}, \ \widetilde{p}_{LT}] = \left[\Phi_{\text{norm}}(\underset{\sim}{\delta}/\sqrt{n}), \ \Phi_{\text{norm}}(\widetilde{\delta}/\sqrt{n})\right]$$

$$= [\text{normTailCI}(\alpha/2; k, n), \ \text{normTailCI}(1 - \alpha/2; k, n)],$$

where $k = (\bar{x} - y)/s$ is the observed value for K. For $0 < \gamma < 1$, the function $\text{normTailCI}(\gamma; k, n)$ provides $\Phi_{\text{norm}}(\delta/\sqrt{n})$, where δ is the solution to $\text{pt}(k\sqrt{n}; n - 1, \delta) = 1 - \gamma$. The normTailCI function has the important property that

$$\text{normTailCI}(\gamma; k, n) = 1 - \text{normTailCI}(1 - \gamma; -k, n). \tag{E.9}$$

This is explained as follows. Suppose that

$$\text{normTailCI}(\gamma; k, n) = \Phi_{\text{norm}}(\delta/\sqrt{n}) \quad \text{and} \quad \text{normTailCI}(1 - \gamma; -k, n)$$

$$= \Phi_{\text{norm}}(\delta^*/\sqrt{n}). \tag{E.10}$$

These assumptions imply

$$\text{pt}(k\sqrt{n}; n - 1, \delta) = 1 - \gamma \quad \text{and} \quad \text{pt}(-k\sqrt{n}; n - 1, \delta^*) = \gamma.$$

These two equalities and the noncentral t-distribution property that (see the details in Section C.3.9)

$$\text{pt}(-k\sqrt{n}; n - 1, \delta^*) = 1 - \text{pt}(k\sqrt{n}; n - 1, -\delta^*)$$

imply that $\text{pt}(k\sqrt{n}; n - 1, \delta) = \text{pt}(k\sqrt{n}; n - 1, -\delta^*)$. Because of the monotonicity of $\text{pt}(k; n - 1, \delta)$ with respect to the noncentrality parameter, it must be the case that $\delta = -\delta^*$. Thus, evaluating (E.10) with the restriction that $\delta = -\delta^*$ gives the result in (E.9).

We now consider a confidence interval and bounds for the probability of an observation being greater than a specified value y. That is, $p_{GT} = \Pr(Y > y) = 1 - \Phi_{\text{norm}}[(y - \mu)/\sigma] = 1 - p_{LT}$. Because p_{GT} is a monotone decreasing function of p_{LT}, it follows that an exact $100(1 - \alpha)\%$ confidence interval for p_{GT} is given by

$$[\underset{\sim}{p}_{GT}, \ \widetilde{p}_{GT}] = [1 - \widetilde{p}_{LT}, \ 1 - \underset{\sim}{p}_{LT}]$$

$$= [1 - \text{normTailCI}(1 - \alpha/2; k, n), \ 1 - \text{normTailCI}(\alpha/2; k, n)]$$

$$= [\text{normTailCI}(\alpha/2; -k, n), \ \text{normTailCI}(1 - \alpha/2; -k, n)],$$

as given in Section 4.5.

E.4 CONFIDENCE INTERVALS FOR TWO NORMAL DISTRIBUTIONS BASED ON PIVOTAL QUANTITIES

In this section we consider situations for which $\boldsymbol{X}$ is a vector of n independent observations from a $\text{NORM}(\mu_X, \sigma_X)$ distribution, $\boldsymbol{Y}$ is a vector of m independent observations from a $\text{NORM}(\mu_Y, \sigma_Y)$ distribution, and $\boldsymbol{X}$ and $\boldsymbol{Y}$ are independent. The vector of unknown parameters is $\boldsymbol{\theta} = (\mu_X, \mu_Y, \sigma_X, \sigma_Y)$. Let $(\bar{X}, S_X)$ be the sample mean and sample standard deviation of $\boldsymbol{X}$ and let $(\bar{Y}, S_Y)$ be the corresponding sample values for $\boldsymbol{Y}$.

E.4.1   Confidence Interval to Compare Two Sample Variances

From the pivotal quantity $g_5(\boldsymbol{X}, \boldsymbol{Y}, \boldsymbol{\theta})$ in (E.6),

$$\Pr\left(\frac{S_X^2}{S_Y^2} F_{(\alpha/2; m-1, n-1)} \leq \frac{\sigma_X^2}{\sigma_Y^2} \leq \frac{S_X^2}{S_Y^2} F_{(1-\alpha/2; m-1, n-1)}\right) = 1 - \alpha.$$

Then using the relationship $F_{(\alpha; r_1, r_2)} = 1/F_{(1-\alpha; r_2, r_1)}$ in Section C.3.11, an exact $100(1 - \alpha)\%$ confidence interval for $\nu = \sigma_X^2/\sigma_Y^2$ is

$$[\underset{\sim}{\nu}, \ \widetilde{\nu}] = \left[\frac{s_X^2}{s_Y^2} \frac{1}{F_{(1-\alpha/2; n-1, m-1)}}, \ \frac{s_X^2}{s_Y^2} F_{(1-\alpha/2; m-1, n-1)}\right].$$

E.4.2   Confidence Interval for the Difference between Two Normal Distribution Means

Here we assume that $\sigma_X = \sigma_Y = \sigma$. From the pivotal quantity $g_6(\boldsymbol{X}, \boldsymbol{Y}, \boldsymbol{\theta})$ in (E.7),

$$\Pr\left[-t_{(1-\alpha/2; n+m-2)} \leq \frac{\bar{X} - \bar{Y} - (\mu_X - \mu_Y)}{(1/n + 1/m)^{1/2} S_p} \leq t_{(1-\alpha/2; n+m-2)}\right] = 1 - \alpha,$$

where $\bar{X}$ and $\bar{Y}$ are sample means and S_p^2 is the pooled sample variance of σ. Solving for $\delta_\mu = \mu_X - \mu_Y$ and using $g'''_{(1-\alpha; n, m)} = t_{(1-\alpha/2; n+m-2)}(1/m + 1/n)^{1/2}$, we obtain

$$\Pr\left[\bar{X} - \bar{Y} - g'''_{(1-\alpha; n, m)} S_p \leq \delta_\mu \leq \bar{X} - \bar{Y} + g'''_{(1-\alpha; n, m)} S_p\right] = 1 - \alpha.$$

Thus, an exact $100(1 - \alpha)\%$ confidence interval for the difference δ_μ between the two normal distribution means is

$$[\underset{\sim}{\delta_\mu}, \ \widetilde{\delta_\mu}] = (\bar{x} - \bar{y}) \mp g'''_{(1-\alpha; n, m)} s_p,$$

where $(\bar{x}, s_X^2)$ and $(\bar{y}, s_Y^2)$ are the observed sample mean and sample variance from $\boldsymbol{X}$ and $\boldsymbol{Y}$, respectively. The pooled variance estimate is $s_p^2 = [(n-1)s_X^2 + (m-1)s_Y^2]/(n+m-2)$.

E.5   TOLERANCE INTERVALS FOR A NORMAL DISTRIBUTION BASED ON PIVOTAL QUANTITIES

E.5.1   Tolerance Intervals to Control the Center

Section 4.6.1 gives an exact $100(1 - \alpha)\%$ tolerance interval to contain at least a proportion β of a normal distribution,

$$[\underset{\sim}{T_\beta}, \ \widetilde{T}_\beta] = \bar{x} \mp g_{(1-\alpha; \beta, n)} s.$$

The purpose here is to provide details for the computations of the factors $g_{(\gamma; \beta, n)}$ given in Tables J.5a and J.5b. To facilitate the development below, we used the compact notation $g = g_{(1-\alpha; \beta, n)}$. We also will use the notation

$$A(\bar{x}, w) = \Phi_{\text{norm}}(\bar{x} + w) - \Phi_{\text{norm}}(\bar{x} - w).$$

Note that for given w, $A(\bar{x}, w) = A(-\bar{x}, w)$, and for given $\bar{x}$,

$$\frac{\partial A(\bar{x}, w)}{\partial w} = \phi_{\mathrm{norm}}(\bar{x} + w) + \phi_{\mathrm{norm}}(\bar{x} - w) > 0,$$

where $\phi_{\mathrm{norm}}(z)$ is the $\mathrm{NORM}(0, 1)$ pdf. Then $A(\bar{x}, w)$ is monotone increasing in w. For given $\bar{x}$, we denote by $r(\bar{x}, \beta)$ the unique solution to the equation $A(\bar{x}, w) = \beta$. That is,

$$A[\bar{x}, r(\bar{x}, \beta)] = \beta. \tag{E.11}$$

The $\mathrm{CP}(\mu, \sigma)$ for the tolerance interval is

$$\mathrm{CP}(\mu, \sigma) = \Pr\left[\Pr_X\left(\underset{\sim}{T}_\beta < X \le \widetilde{T}_\beta\right) > \beta\right]$$

$$= \Pr\left[\Phi_{\mathrm{norm}}\left(\frac{\widetilde{T}_\beta - \mu}{\sigma}\right) - \Phi_{\mathrm{norm}}\left(\frac{\underset{\sim}{T}_\beta - \mu}{\sigma}\right) > \beta\right],$$

where the outside probability is computed with respect to the joint distribution of $(\bar{X}, S)$, that is, the distribution of the sufficient statistics for the data.

Note that $(\widetilde{T}_\beta - \mu)/\sigma = (\bar{X} - \mu + gS)/\sigma$ is a pivotal quantity. Similarly, $[(\underset{\sim}{T}_\beta - \mu)/\sigma]$ is also a pivotal quantity. Thus, we can arbitrarily use $\mu = 0$ and $\sigma = 1$ in computing the coverage probability, and also we write $\mathrm{CP}(0, 1)$ without loss of generality. In this case $\bar{X} \sim \mathrm{NORM}(0, 1/\sqrt{n})$ with pdf $f(\bar{x}) = \mathrm{dnorm}(\bar{x}; 0, 1/\sqrt{n})$, and S has density $h(s; n - 1)$ such that $(n - 1)S^2 \sim \chi^2(n - 1)$, that is

$$h(s; k) = \frac{k^{k/2}\, s^{k-2}\, \exp(-k\, s^2/2)}{2^{k/2-1}\, \Gamma(k/2)}, \tag{E.12}$$

where $k = n - 1$ and $s > 0$.

Using the fact that $\bar{X}$ and S are independent and the result that $A(\bar{x}, gs) > \beta$ if and only if $s > r(\bar{x}, \beta)/g$, and the indicator function $\mathrm{I}[A(\bar{x}, gs) > \beta]$, gives

$$\mathrm{CP}(0, 1) = \Pr\left[\Phi_{\mathrm{norm}}\left(\bar{X} + gS\right) - \Phi_{\mathrm{norm}}\left(\bar{X} - gS\right) > \beta\right] = \Pr\left[A(\bar{X}, gS) > \beta\right]$$

$$= \int_{-\infty}^{\infty} \int_{0}^{\infty} \mathrm{I}[A(\bar{x}, gs) > \beta]\, h(s) f(\bar{x})\, ds\, d\bar{x}$$

$$= 2 \int_{0}^{\infty} \int_{r(\bar{x})/g}^{\infty} h(s) f(\bar{x})\, ds\, d\bar{x}$$

$$= 2 \int_{0}^{\infty} \Pr[S > r(\bar{x}, \beta)/g]\, f(\bar{x})\, d\bar{x}$$

$$= 2 \int_{0}^{\infty} (1 - \Pr[S \le r(\bar{x}, \beta)/g])\, f(\bar{x})\, d\bar{x}$$

$$= 2\left(1 - \int_{0}^{\infty} \Pr[S \le r(\bar{x}, \beta)/g]\, f(\bar{x})\, d\bar{x}\right).$$

Finding g such that $\mathrm{CP}(0, 1) = 1 - \alpha$ requires solving

$$2\left(1 - \int_{0}^{\infty} \Pr[S \le r(\bar{x}, \beta)/g]\, f(\bar{x})\, d\bar{x}\right) = 1 - \alpha$$

for g. This is equivalent to solving $\int_{0}^{\infty} \Pr[S \le r(\bar{x}, \beta)/g]\, f(\bar{x})\, d\bar{x} = (1 + \alpha)/2$ for g. Expressed more simply,

$$\int_{0}^{\infty} \mathrm{pchisq}[(n - 1)r^2(\bar{x}, \beta)/g^2; n - 1]\, \mathrm{dnorm}(\bar{x}; 0, 1/\sqrt{n})\, d\bar{x} = (1 + \alpha)/2$$

can be solved for g where, for given $\bar{x}$ and β, the solution $r(\bar{x}, \beta)$ is given by (E.11). The presentation here is similar to that in Wald and Wolfowitz (1946) and Odeh and Owen (1980).

E.5.2   Tolerance Intervals to Control Both Tails

The distribution theory for tolerance intervals to control both tails is somewhat more complicated than for control-the-center tolerance intervals. Such intervals were first given in Owen (1964), with corrections noted in Owen (1965, 1966). Odeh and Owen (1980) presents tables, theory, and computation methods. Owen and Frawley (1971) also present tables and an example. Krishnamoorthy and Mathew (2009, Section 2.3.2) present the theory for such intervals (which they call "equal-tailed tolerance intervals" and describe a computing method that uses two-dimensional integration.

The simulation method presented in Section 14.5.2 and by Yuan et al. (2017) is relatively simple to implement and can be applied for any (log-)location-scale distribution. Without using a variance-reduction technique, however, such simulations require much computer time to compute factors that have three or four significant digits of accuracy, except in the complete-data normal distribution case.

E.6   NORMAL DISTRIBUTION PREDICTION INTERVALS BASED ON PIVOTAL QUANTITIES

In this section we consider situations for which $\boldsymbol{X}$ is a vector of n independent observations and $\boldsymbol{Y}$ is a vector of m independent observations both from the same $\text{NORM}(\mu, \sigma)$ distribution and $\boldsymbol{X}$ and $\boldsymbol{Y}$ are independent. The vector of unknown parameters is $\boldsymbol{\theta} = (\mu, \sigma)$. Let $(\bar{X}, S_X)$ be the sample mean and sample standard deviation of $\boldsymbol{X}$ and let $(\bar{Y}, S_Y)$ be the corresponding sample values for $\boldsymbol{Y}$.

E.6.1   Prediction Interval to Contain the Mean of m Future Observations from a Previously Sampled Normal Distribution

From the pivotal quantity $g_4(\boldsymbol{X}, \boldsymbol{\theta})$ in (E.5),

$$\Pr\left[-t_{(1-\alpha/2;n-1)} \leq \frac{(\bar{Y} - \bar{X})}{(1/n + 1/m)^{1/2}\, S_X} \leq t_{(1-\alpha/2;n-1)}\right] = 1 - \alpha.$$

Solving for $\bar{Y}$, one obtains the prediction interval for $\bar{Y}$,

$$\Pr\left[\bar{X} - r''_{(1-\alpha;m,n)} S_X \leq \bar{Y} \leq \bar{X} + r''_{(1-\alpha;m,n)} S_X\right] = 1 - \alpha,$$

where $r''_{(1-\alpha;m,n)} = t_{(1-\alpha/2;n-1)}(1/m + 1/n)^{1/2}$. Thus an exact $100(1 - \alpha)\%$ prediction interval for $\bar{Y}$ is

$$[\underset{\sim}{\bar{Y}}_m,\ \widetilde{\bar{Y}}_m] = \bar{x}_n \mp r''_{(1-\alpha;m,n)} s_n,$$

where $\bar{x}$ and s_n^2 are the observed sample mean and sample variance for the data $\boldsymbol{X}$. This prediction interval is given in Section 4.7.

E.6.2 Prediction Interval for the Sample Standard Deviation from a Sample of m Future Observations from a Previously Sampled Normal Distribution

From the pivotal quantity $g_5(\boldsymbol{X}, \boldsymbol{Y}, \boldsymbol{\theta})$ in (E.6), and noting that $\sigma_Y = \sigma_X$ because the future observations are from the same distribution, gives

$$\Pr\left(F_{(\alpha/2;m-1,n-1)} \leq \frac{S_Y^2}{S_X^2} \leq F_{(1-\alpha/2;m-1,n-1)}\right) = 1 - \alpha.$$

Solving for S_Y gives

$$\Pr\left[S_X\left(F_{(\alpha/2;m-1,n-1)}\right)^{1/2} \leq S_Y \leq S_X\left(F_{(1-\alpha/2;m-1,n-1)}\right)^{1/2}\right] = 1 - \alpha.$$

Then, using the relationship $F_{(\alpha;r_1,r_2)} = 1/F_{(1-\alpha;r_2,r_1)}$ in Section C.3.11, an exact $100(1-\alpha)\%$ prediction interval for S_Y is

$$[\underline{s}_m, \ \widetilde{s}_m] = \left[s\left(\frac{1}{F_{(1-\alpha/2;n-1,m-1)}}\right)^{1/2}, \ s\left(F_{(1-\alpha/2;m-1,n-1)}\right)^{1/2}\right], \tag{E.13}$$

where s is the sample standard deviation from the data $\boldsymbol{X}$. This prediction interval is given in Section 4.9.

E.6.3 One-Sided Prediction Bound to Exceed (be Exceeded by) at Least k Out of m Future Observations from a Previously Sampled Normal Distribution

This section provides background information for the one-sided prediction bounds in Section 4.8. First we consider the one-sided upper prediction bound to exceed at least k of m future observations from the same normal distribution. There is not a simple pivotal quantity that would directly provide this prediction bound. Pivotal quantities, however, facilitate the computation of the bounds using either simulation or numerical integration, as we indicate next.

A one-sided upper prediction bound exceeds at least k of the m future observations if and only if the kth largest order statistic from the m future observations, say $Y_{(k)}$, is exceeded by the bound. Consider an upper prediction bound of the type $\bar{x} + r'_{(1-\alpha;k,m,n)}s$. Using $r' = r'_{(1-\alpha;k,m,n)}$, the coverage probability that the upper prediction bound $\bar{X} + r'S$ exceeds at least k of the m future observations is

$$\mathrm{CP}_U(r') = \Pr(Y_{(k)} \leq \bar{X} + r'S) = \Pr\left[\frac{Y_{(k)} - \mu}{\sigma} \leq \frac{\bar{X} - \mu}{\sigma} + r'\frac{S}{\sigma}\right]$$

$$= \Pr(Z_{(k)} \leq Z + r'W), \tag{E.14}$$

where $Z_{(k)}$, Z and W are independent. The random variable $Z_{(k)}$ is the kth order statistic from a $\mathrm{NORM}(0,1)$ distribution, Z has a $\mathrm{NORM}(0, 1/\sqrt{n})$ distribution with pdf denoted by $f(z;n)$, and W has a pdf $h(w; n-1)$ such that $(n-1)W^2 \sim \chi^2(n-1)$; see (E.12).

Consequently, for given r', the probability on the right-hand side of (E.14) does not depend on unknown parameters. Using the distribution function of $Z_{(k)}$ as given in Arnold et al. (2008,

Chapter 1), an equivalent expression for the coverage probability in (E.14) is

$$
\begin{aligned}
\mathrm{CP}_U(r') &= \mathrm{E}_{Z,W}\left[\Pr(Z_{(k)} \le Z + r'W | Z, W)\right] \\
&= \int_{-\infty}^{\infty} \int_0^{\infty} \Pr(Z_{(k)} \le z + r'w)\, h(w; n-1) f(z; n)\, dw\, dz \\
&= \int_{-\infty}^{\infty} \int_0^{\infty} \left[\sum_{i=k}^m \binom{m}{i} p_U^i (1 - p_U)^{m-i}\right] h(w; n-1) f(z; n)\, dw\, dz \quad (\mathrm{E.15}) \\
&= \int_{-\infty}^{\infty} \int_0^{\infty} \mathtt{pbeta}(p_U; k, m-k+1)\, h(w; n-1) f(z; n)\, dw\, dz, \quad (\mathrm{E.16})
\end{aligned}
$$

where $p_U = \Phi_{\mathrm{norm}}(z + r'w)$ and the last equality was obtained using the relationship $\mathtt{pbeta}(p_U; k, m-k+1) = 1 - \mathtt{pbinom}(k-1; m, p_U)$ in (C.19).

To find the value of r' that provides an exact $100(1-\alpha)\%$ prediction bound, one solves for r' in the relationship $\mathrm{CP}_U(r') = 1 - \alpha$. For this purpose, one can use (E.15) or (E.16). Using the latter, r' is the solution to

$$
\int_{-\infty}^{\infty} \int_0^{\infty} \mathtt{pbeta}(p_U; k, m-k+1)\, h(w; n-1) f(z; n)\, dw\, dz = 1 - \alpha.
$$

Due to the symmetry of the normal distribution, the one-sided lower prediction bound to be exceeded by at least k of m future observations is $\bar{x} - r's$. In this case the coverage probability of the lower prediction bound is

$$
\mathrm{CP}_L(-r') = \int_{-\infty}^{\infty} \int_0^{\infty} \mathtt{pbeta}(1 - p_L; k, m-k+1)\, h(w; n-1) f(z; n)\, dw\, dz,
$$

where $p_L = \Phi_{\mathrm{norm}}(z - r'w)$. The procedure here to compute the factors r' is similar, and numerically equivalent, to that in Fertig and Mann (1977).

E.6.4 Simultaneous Prediction Interval to Contain at Least k Out of m Future Observations from a Previously Sampled Normal Distribution

This section provides background information for the simultaneous prediction intervals in Section 4.8. The derivation of the simultaneous prediction interval $[\bar{x} - r_{(1-\alpha;k,m,n)}s, \ \bar{x} + r_{(1-\alpha;k,m,n)}s]$ to contain at least k out of m future observations is similar to the derivation of the one-side prediction bound in Section E.6.3

Let B_k be the set corresponding to "at least k of the m future observations are contained in the prediction interval." Thus, using the notation $r = r_{(1-\alpha;k,m,n)}$, the coverage probability for the prediction interval is

$$
\begin{aligned}
\mathrm{CP}(r, -r) &= \Pr(B_k) \\
&= \mathrm{E}_{Z,W}\left[\Pr(B_k | Z, W)\right] \\
&= \int_{-\infty}^{\infty} \int_0^{\infty} \left[\sum_{i=k}^m \binom{m}{i} (p_U - p_L)^i (1 - p_U + p_L)^{m-i}\right] h(w; n-1) f(z; n)\, dw\, dz \\
&= \int_{-\infty}^{\infty} \int_0^{\infty} \mathtt{pbeta}(p_U - p_L; k, m-k+1]\, h(w; n-1) f(z; n)\, dw\, dz,
\end{aligned}
$$

where $p_U = \Phi_{\mathrm{norm}}(z + rw)$ and $p_L = \Phi_{\mathrm{norm}}(z - rw)$. To find the value of r that provides an exact $100(1-\alpha)\%$ simultaneous prediction interval that contains at least k of m future observations, one solves for r in the relationship $\mathrm{CP}(r, -r) = 1 - \alpha$.

The procedure here to compute the factors r for $k = m$ is similar, and numerically equivalent, to that in Hahn (1969, 1970a). Odeh (1990) gives a procedure to compute the factors r when $k \leq m$. His procedure requires only integration in one dimension but it does not seem to generalize to the related prediction problem when the data are from a general location-scale family (see Section B.7).

E.7   PIVOTAL QUANTITIES FOR LOG-LOCATION-SCALE DISTRIBUTIONS

In this section we describe some pivotal quantities for log-location-scale families. The cdf for the location-scale family (see Section C.3 for a description of the location-scale family) is

$$F(x) = \Pr(X \leq x) = \Phi\left(\frac{x - \mu}{\sigma}\right),$$

where μ is the location parameter and σ is the scale parameter of the distribution of X. The p quantile of the distribution is $x_p = \mu + \sigma z_p$, where $z_p = \Phi^{-1}(p)$ is the p quantile of the standardized distribution $\Phi(z)$. For the log-location-scale family the cdf is $F(t) = \Phi[(\log(t) - \mu)/\sigma]$ and the p quantile is $t_p = \exp(\mu + \sigma z_p)$.

E.7.1   Pivotal Quantities Involving Data from a Location-Scale Distribution

Suppose that the data, $\boldsymbol{X}$, are from a location-scale distribution (see Section C.3.1) and are complete (i.e., no censoring) or censored after a prespecified number of lower order statistics (Type 2 censoring). Then the likelihood of the data is $L(\mu, \sigma)$, which is a special case of the likelihood in (12.7). The parameter values of μ and σ that maximize the likelihood are denoted by $\widehat{\mu}$ and $\widehat{\sigma}$, respectively.

In this case, the following random quantities are pivotal quantities (see Lawless, 2003, Appendix A for details):

$$g_7(\boldsymbol{X}, \boldsymbol{\theta}) = Z_{\widehat{\mu}} = \frac{\mu - \widehat{\mu}}{\widehat{\sigma}}, \tag{E.17}$$

$$g_8(\boldsymbol{X}, \boldsymbol{\theta}) = Z_{\widehat{\sigma}} = \frac{\sigma}{\widehat{\sigma}},$$

$$g_9(\boldsymbol{X}, \boldsymbol{\theta}) = Z_{\widehat{x}_p} = \frac{x_p - \widehat{x}_p}{\widehat{\sigma}} = \frac{\mu - \widehat{\mu}}{\widehat{\sigma}} + \left(\frac{\sigma}{\widehat{\sigma}} - 1\right)\Phi^{-1}(p). \tag{E.18}$$

If the data $\boldsymbol{T}$ are from a log-location-scale distribution, then with $\boldsymbol{X} = \log(\boldsymbol{T})$ the relationships in $g_7(\boldsymbol{X}, \boldsymbol{\theta})$, $g_8(\boldsymbol{X}, \boldsymbol{\theta})$, and $g_9(\boldsymbol{X}, \boldsymbol{\theta})$ still hold, where now x_p is the logarithm of the p quantile t_p of the log-location-scale distribution. That is, $x_p = \log(t_p)$.

E.7.2   Confidence Interval for a Location Parameter

To obtain an exact $100(1 - \alpha)\%$ confidence interval for μ, let $z_{\widehat{\mu}(\alpha/2;n)}$ and $z_{\widehat{\mu}(1-\alpha/2;n)}$ be the $\alpha/2$ and $1 - \alpha/2$ quantiles of $g_7(\boldsymbol{X}, \boldsymbol{\theta}) = Z_{\widehat{\mu}}$, respectively. Thus

$$\Pr\left[z_{\widehat{\mu}(\alpha/2;n)} \leq \frac{\mu - \widehat{\mu}}{\widehat{\sigma}} \leq z_{\widehat{\mu}(1-\alpha/2;n)}\right] = 1 - \alpha. \tag{E.19}$$

Rearranging terms to isolate μ at the center of the inequalities in (E.19) gives

$$\Pr\left[\widehat{\mu} + z_{\widehat{\mu}(\alpha/2;n)}\,\widehat{\sigma} \leq \mu \leq \widehat{\mu} + z_{\widehat{\mu}(1-\alpha/2;n)}\,\widehat{\sigma}\right] = 1 - \alpha.$$

Thus, an exact $100(1 - \alpha)\%$ confidence interval for μ is

$$\left[\underset{\sim}{\mu}, \; \widetilde{\mu}\right] = \left[\widehat{\mu} + z_{\widehat{\mu}_{(\alpha/2;n)}} \; \widehat{\sigma}, \; \widehat{\mu} + z_{\widehat{\mu}_{(1-\alpha/2;n)}} \; \widehat{\sigma}\right].$$

If the data are from a log-location-scale distribution, one might be interested in the scale parameter $\eta = \exp(\mu)$ (e.g., if the data are from a Weibull distribution). In this case an exact $100(1 - \alpha)\%$ confidence interval for η is

$$\left[\underset{\sim}{\eta}, \; \widetilde{\eta}\right] = \left[\exp(\underset{\sim}{\mu}), \; \exp(\widetilde{\mu})\right].$$

E.7.3 Confidence Interval for a Scale Parameter

To obtain an exact $100(1 - \alpha)\%$ confidence interval for σ, let $z_{\widehat{\sigma}_{(\alpha/2;n)}}$ and $z_{\widehat{\sigma}_{(1-\alpha/2;n)}}$ be the $\alpha/2$ and $1 - \alpha/2$ quantiles of $g_8(\boldsymbol{X}, \boldsymbol{\theta}) = Z_{\widehat{\sigma}}$, respectively. Then

$$\Pr\left[z_{\widehat{\sigma}_{(\alpha/2;n)}} \leq \frac{\sigma}{\widehat{\sigma}} \leq z_{\widehat{\sigma}_{(1-\alpha/2;n)}}\right] = 1 - \alpha. \tag{E.20}$$

Rearranging terms to isolate σ at the center of the inequalities in (E.20) gives

$$\Pr\left[z_{\widehat{\sigma}_{(\alpha/2;n)}} \; \widehat{\sigma} \leq \sigma \leq z_{\widehat{\sigma}_{(1-\alpha/2;n)}} \; \widehat{\sigma}\right] = 1 - \alpha.$$

Thus, an exact $100(1 - \alpha)\%$ confidence interval for σ is

$$\left[\underset{\sim}{\sigma}, \; \widetilde{\sigma}\right] = \left[z_{\widehat{\sigma}_{(\alpha/2;n)}} \; \widehat{\sigma}, \; z_{\widehat{\sigma}_{(1-\alpha/2;n)}} \; \widehat{\sigma}\right].$$

If the data are from a log-location-scale distribution, one might be interested in the shape parameter $\beta = 1/\sigma$ (e.g., if the data are from a Weibull distribution). In this case an exact $100(1 - \alpha)\%$ confidence interval for β is

$$\left[\underset{\sim}{\beta}, \; \widetilde{\beta}\right] = \left[\frac{1}{\widetilde{\sigma}}, \; \frac{1}{\underset{\sim}{\sigma}}\right].$$

E.7.4 Confidence Interval for a Distribution Quantile

To obtain an exact $100(1 - \alpha)\%$ confidence interval for $x_p = \mu + \Phi^{-1}(p)\sigma$, let $z_{\widehat{x}_p(1-\alpha/2;n)}$ and $z_{\widehat{x}_p(1-\alpha/2;n)}$ be the $\alpha/2$ and $1 - \alpha/2$ quantiles of $g_9(\boldsymbol{X}, \boldsymbol{\theta}) = Z_{\widehat{x}_p}$, respectively. Thus,

$$\Pr\left[z_{\widehat{x}_p(\alpha/2;n)} \leq \frac{x_p - \widehat{x}_p}{\widehat{\sigma}} \leq z_{\widehat{x}_p(1-\alpha/2;n)}\right] = 1 - \alpha, \tag{E.21}$$

where $\widehat{x}_p = \widehat{\mu} + \Phi^{-1}(p)\widehat{\sigma}$. Rearranging terms to isolate x_p in the center of the inequalities in (E.21) gives

$$\Pr\left[\widehat{x}_p + z_{\widehat{x}_p(\alpha/2;n)} \; \widehat{\sigma} \leq x_p \leq \widehat{x}_p + z_{\widehat{x}_p(1-\alpha/2;n)} \; \widehat{\sigma}\right] = 1 - \alpha.$$

Thus, an exact $100(1 - \alpha)\%$ confidence interval for x_p is

$$\left[\underset{\sim}{x}_p, \; \widetilde{x}_p\right] = \left[\widehat{x}_p + z_{\widehat{x}_p(\alpha/2;n)} \; \widehat{\sigma}, \; \widehat{x}_p + z_{\widehat{x}_p(1-\alpha/2;n)} \; \widehat{\sigma}\right].$$

If the data are from a log-location-scale distribution, one might be interested in the p quantile $t_p = \exp[\mu + \Phi^p \sigma]$ of the distribution (e.g., the data are from a lognormal distribution). In this case an exact $100(1 - \alpha)\%$ confidence interval for t_p is

$$\left[\underline{t}_p, \ \widetilde{t}_p\right] = \left[\exp(\underline{x}_p), \ \exp(\widetilde{x}_p)\right].$$

E.7.5 One-Sided Simultaneous Prediction Bound to Exceed (be Exceeded by) at Least k Out of m Future Observations from a Previously Sampled (Log-)Location-Scale Distribution

The results Section E.6.3 extend readily to (log-)location-scale distributions in the case where the data are complete (i.e., no censoring) or censored after a prespecified number of lower order statistics have been observed (Type 2 censoring). First, we consider the case for which the data are from a location-scale distribution with cdf $\Phi[(x - \mu)/\sigma]$, where the cdf $\Phi(z)$ does not depend on unknown parameters. Let $(\widehat{\mu}, \widehat{\sigma})$ denote the maximum likelihood estimators of the parameters. Then a one-sided upper prediction bound, to exceed at least k future observations of a previous sampled distribution, has the form $\widehat{\mu} + r'_U \widehat{\sigma}$. The coverage probability for this one-sided upper prediction bound is

$$\mathrm{CP}_U(r'_U) = \int_{-\infty}^{\infty} \int_{0}^{\infty} \mathrm{pbeta}(p_U; k, n - k + 1)\, g(z, w)\, dw\, dz, \qquad (\text{E.22})$$

where $p_U = \Phi(z + r'_U w)$ and $g(z, w)$ is the joint density of $[Z, W] = [(\widehat{\mu} - \mu)/\sigma, \widehat{\sigma}/\sigma]$ in random samples of size n. Note that $g(z, w)$ does not depend on unknown parameters. To obtain an exact $100(1 - \alpha)\%$ one-sided upper prediction bound, set $\mathrm{CP}_U(r'_U) = 1 - \alpha$ and solve for r'_U.

For a nonsymmetric distribution $\Phi(z)$, the factor r'_L for the one-sided lower prediction bound $\widehat{\mu} + r'_L \widehat{\sigma}$ is not equal to $-r'_U$. The coverage probability of this one-sided lower prediction bound is

$$\mathrm{CP}_L(r'_L) = \int_{-\infty}^{\infty} \int_{0}^{\infty} \mathrm{pbeta}(1 - p_L; k, n - k + 1)\, g(z, w)\, dw\, dz, \qquad (\text{E.23})$$

where $p_L = \Phi(z + r'_L w)$. To obtain an exact $100(1 - \alpha)\%$ one-sided lower prediction bound, set $\mathrm{CP}_L(r'_L) = 1 - \alpha$ and solve for r'_L.

When the data are from a log-location-scale distribution, the exact $100(1 - \alpha)\%$ one-sided upper prediction bound is $\exp(\widehat{\mu} + r'_U \widehat{\sigma})$, where r'_U is obtained from (E.22). Similarly, the exact $100(1 - \alpha)\%$ one-sided lower prediction bound is $\exp(\widehat{\mu} + r'_L \widehat{\sigma})$, where r'_L is obtained from (E.23).

For situations where the joint density $g(z, w)$ cannot be expressed in simple terms, the integration in (E.22) and (E.23) can be accomplished by using Monte Carlo simulation, as described in Section B.7 and illustrated in Section 14.6.

E.7.6 Two-Sided Simultaneous Prediction Interval to Contain at Least k Out of m Future Observations from a Previously Sampled (Log-)Location-Scale Distribution

The results in Section E.6.4 extend to general location-scale distributions, as follows. A prediction interval to contain at least k of m future observations from a previously sampled location-scale distribution is given by $[\widehat{\mu} + r_L \widehat{\sigma}, \ \widehat{\mu} + r_U \widehat{\sigma}]$. The coverage probability for this prediction interval is

$$\mathrm{CP}(r_L, r_U) = \int_{-\infty}^{\infty} \int_{0}^{\infty} \mathrm{pbeta}(p_U - p_L; k, m - k + 1]\, g(z, w)\, dw\, dz,$$

where $p_U = \Phi(z + r_U w)$, $p_L = \Phi(z + r_L w)$, and $g(z, w)$ is the joint density of $[Z, W] = [(\widehat{\mu} - \mu)/\sigma, \widehat{\sigma}/\sigma]$ in random samples of size n. Note that $g(z, w)$ does not depend on unknown parameters.

To obtain an exact $100(1 - \alpha)\%$ prediction interval, solve

$$CP(r_L, r_U) = 1 - \alpha \tag{E.24}$$

for r_L and r_U. Because there are multiple solutions (r_L, r_U) to (E.24), to ensure a unique solution, a plausible constraint for finding the roots r_L and r_U is to require that the prediction interval be balanced in the sense that

$$CP_U(r_U) = CP_L(r_L),$$

where $CP_U(r_U)$ and $CP_L(r_L)$ are the one-sided coverage probabilities given in (E.22) and (E.23), respectively. The presentation here for the location-scale family is similar to that in Xie et al. (2017).

For situations where the joint density $g(z, w)$ cannot be expressed in simple terms, the integration in (E.24) can be accomplished by using Monte Carlo simulation, as described in Section B.7 and illustrated in Section 14.6.

E.7.7 Pivotal Quantities Involving Data from Two Similar (Log-)Location-Scale Distributions

Pivotal quantities similar to the normal distribution pivotal quantities g_5 and g_6 in Section E.2.2 can be used to obtain exact statistical interval procedures similar to those given in Sections E.4.1, E.4.2, and E.6.2. Details are not provided here, but are easy to work out following the approaches given in those sections.

Appendix F

Generalized Pivotal Quantities

INTRODUCTION

This appendix provides a general definition of a generalized pivotal quantity (GPQ), a method to obtain GPQs as a function of other GPQs, and a set of conditions that when satisfied ensure exact confidence intervals based on a GPQ.

The topics discussed in the appendix are:

- Definition of a GPQ (Secton F.1).

- A substitution method to obtain GPQs (Section F.2).

- Examples of GPQs for functions of parameters from location-scale distributions (Section F.3).

- Conditions for exact intervals derived from GPQs (Section F.4).

F.1 DEFINITION OF A GENERALIZED PIVOTAL QUANTITY

Let S be a vector function of the observed data x (i.e., a vector of parameter estimators). Suppose that the cdf of S is $F(\cdot; \nu)$, where ν is a vector of unknown parameters. Denote an observation from S by s and let S^* be an independent copy of S. That is, S and S^* are independent and have the same distribution. A scalar function $Z_s = Z(S^*; s, \nu)$ is a GPQ for a scalar parameter $\theta = \theta(\nu)$ if it satisfies the following two conditions:

1. For given s, the distribution of Z_s does not depend on unknown parameters.

2. Evaluating Z_s at $S^* = s$ gives $Z_s = Z(s; s, \nu) = \theta$.

Note that Hannig et al. (2006, Definition 2) refer to this as a "fiducial generalized pivotal quantity."

Statistical Intervals: A Guide for Practitioners and Researchers, Second Edition.
William Q. Meeker, Gerald J. Hahn and Luis A. Escobar.
© 2017 John Wiley & Sons, Inc. Published 2017 by John Wiley & Sons, Inc.
Companion Website: www.wiley.com/go/meeker/intervals

489

For the purposes of implementation, we can consider s as the estimate of the parameters ν and that the objective is to obtain a confidence interval for θ based on s and the information that the distribution of S^* is $F(\cdot; \nu)$. Note that S^* is not observable; we just know that S^* has the same distribution as S (which we denote by $S^* \sim S$). In this setting, we will use GPQs to obtain confidence intervals for some functions of the unknown parameters.

Although the GPQ methods are more general (see the application in Section 18.2), consider, for example, sampling from a location-scale family $\Phi[(x - \mu)/\sigma]$. Let $x = (x_1, \ldots, x_n)$ be the data. Obtain the maximum likelihood (ML) estimates $(\widehat{\mu}, \widehat{\sigma})$ of (μ, σ). Define $\nu = (\mu, \sigma)$, $s = (\widehat{\mu}, \widehat{\sigma})$, and $S^* = (\widehat{\mu}^*, \widehat{\sigma}^*)$, where $(\widehat{\mu}^*, \widehat{\sigma}^*)$ have the same distribution as $(\widehat{\mu}, \widehat{\sigma})$. We derive GPQs for functions of (μ, σ) and use those GPQs to obtain GPQ-based confidence intervals using s and the fact that the joint distribution of the GPQs does not depend on unknown parameters.

F.2   A SUBSTITUTION METHOD TO OBTAIN GENERALIZED PIVOTAL QUANTITIES

An important property of GPQs is the following:

Result F.1 *Suppose that θ_1 and θ_2 are parameters of interest. Suppose that $Z_{\widehat{\theta}_1} = Z_{\widehat{\theta}_1}(S^*; s, \nu)$ and $Z_{\widehat{\theta}_2} = Z_{\widehat{\theta}_2}(S^*; s, \nu)$ are GPQs for θ_1 and θ_2, respectively, and that, given s, the joint distribution of $(Z_{\widehat{\theta}_1}, Z_{\widehat{\theta}_2})$ does not depend on unknown parameters. Consider a function of θ_1, θ_2, say $h(\theta_1, \theta_2)$. Then $h(Z_{\widehat{\theta}_1}, Z_{\widehat{\theta}_2})$ is a GPQ for $h(\theta_1, \theta_2)$.*

This result is verified as follows. Given s, the distribution of $h(Z_{\widehat{\theta}_1}, Z_{\widehat{\theta}_2})$ does not depend on unknown parameters because it is determined by the function $h(\cdot, \cdot)$ and the joint distribution of $(Z_{\widehat{\theta}_1}, Z_{\widehat{\theta}_2})$, which does not depend on unknown parameters. When $S^* = s$, we have that $Z_{\widehat{\theta}_1} = \theta_1$ and $Z_{\widehat{\theta}_2} = \theta_2$ and then $h(Z_{\widehat{\theta}_1}, Z_{\widehat{\theta}_2}) = h(\theta_1, \theta_2)$. Thus $h(Z_{\widehat{\theta}_1}, Z_{\widehat{\theta}_2})$ has the two properties required to be a GPQ.

Note that the conditions for Result F.1 are satisfied, for example, when the GPQs $Z_{\widehat{\theta}_1}$ and $Z_{\widehat{\theta}_2}$ are independent because in this case the joint distribution is just the product of the marginals which do not depend on unknown parameters. In (F.3) and (F.6) the generalized pivotal quantities $Z_{\widehat{\theta}_1}$ and $Z_{\widehat{\theta}_2}$ are not independent but Result F.1 still applies because the joint distribution of $Z_{\widehat{\theta}_1}$ and $Z_{\widehat{\theta}_2}$ does not depend on any unknown parameters.

F.3   EXAMPLES OF GENERALIZED PIVOTAL QUANTITIES FOR FUNCTIONS OF LOCATION-SCALE DISTRIBUTION PARAMETERS

F.3.1   GPQ Function for μ

Here the interest is in $\theta = \mu$. Then $\nu = (\mu, \sigma)$, where σ is a nuisance parameter. Define

$$Z_{\widehat{\mu}} = Z_{\widehat{\mu}}(S^*; s, \nu) = Z_{\widehat{\mu}}[(\widehat{\mu}^*, \widehat{\sigma}^*); (\widehat{\mu}, \widehat{\sigma}), \nu] = \widehat{\mu} + \left(\frac{\mu - \widehat{\mu}^*}{\widehat{\sigma}^*}\right)\widehat{\sigma}. \qquad (F.1)$$

Thus:

1. Given the data x and the corresponding values $s = (\widehat{\mu}, \widehat{\sigma})$, $Z_{\widehat{\mu}}$ has a distribution that does not depend on unknown parameters. This is the case because $(\widehat{\mu}^* - \mu)/\widehat{\sigma}^*$ is a PQ and, given the data, $\widehat{\mu}$ and $\widehat{\sigma}$ are known.

Note that this is a conditional argument in the sense that the distribution of $Z_{\widehat{\mu}}$ depends on the values of $\widehat{\mu}$ and $\widehat{\sigma}$, but for purposes of constructing the confidence interval for μ those quantities are known and computed from the data.

2. Substituting $(\widehat{\mu}^*, \widehat{\sigma}^*)$ for $(\widehat{\mu}, \widehat{\sigma})$ in (F.1) gives

$$Z_{\widehat{\mu}} = Z_{\widehat{\mu}}(S^*; s, \nu) = Z_{\widehat{\mu}}[(\widehat{\mu}, \widehat{\sigma}); (\widehat{\mu}, \widehat{\sigma}), \nu] = \widehat{\mu} + \left(\frac{\mu - \widehat{\mu}}{\widehat{\sigma}} \right) \widehat{\sigma} = \mu.$$

Because of the results in items 1 and 2 above $Z_{\widehat{\mu}}$ is a GPQ for μ.

F.3.2   GPQ Function for σ

Here the interest is in $\theta = \sigma$. Then $\nu = (\mu, \sigma)$, where μ is a nuisance parameter. Define

$$Z_{\widehat{\sigma}} = Z_{\widehat{\sigma}}(S^*; s, \nu) = Z_{\widehat{\sigma}}[(\widehat{\mu}^*, \widehat{\sigma}^*); (\widehat{\mu}, \widehat{\sigma}), \nu] = \left(\frac{\sigma}{\widehat{\sigma}^*} \right) \widehat{\sigma}. \tag{F.2}$$

Thus:

1. Given the data x, one obtains the ML parameter estimates, say $s = (\widehat{\mu}, \widehat{\sigma})$. The conditional distribution of $Z_{\widehat{\sigma}}$, given s, does not depend on unknown parameters because $\widehat{\sigma}$ is known and $\sigma/\widehat{\sigma}^*$ is a PQ.

2. Substituting $(\widehat{\mu}^*, \widehat{\sigma}^*)$ for $(\widehat{\mu}, \widehat{\sigma})$ in (F.2) gives

$$Z_{\widehat{\sigma}} = Z_{\widehat{\sigma}}[(\widehat{\mu}, \widehat{\sigma}); (\widehat{\mu}, \widehat{\sigma}), \nu] = \left(\frac{\sigma}{\widehat{\sigma}} \right) \widehat{\sigma} = \sigma.$$

Because of the results in items 1 and 2 above $Z_{\widehat{\sigma}}$ is a GPQ for σ.

F.3.3   GPQ Function for a Tail Probability $F(x; \mu, \sigma)$

Here we consider a GPQ for the probability $p = F(x)$, where

$$F(x) = F(x; \mu, \sigma) = \Phi\left(\frac{x - \mu}{\sigma} \right),$$

and x is fixed and given. The ML estimate of p is

$$\widehat{p} = \Phi\left(\frac{x - \widehat{\mu}}{\widehat{\sigma}} \right).$$

To obtain a GPQ for p, say $Z_{\widehat{p}}$, note that the GPQs for μ and σ in (F.1) and (F.2) have a joint distribution that does not depend on unknown parameters. Then, using Result F.1,

$$Z_{\widehat{p}} = Z_{\widehat{p}}[(\widehat{\mu}^*, \widehat{\sigma}^*); (\widehat{\mu}, \widehat{\sigma}), \nu] = \Phi\left(\frac{x - Z_{\widehat{\mu}}}{Z_{\widehat{\sigma}}} \right)$$

$$= \Phi\left[\left(\frac{\widehat{\sigma}^*}{\sigma} \right) \left(\frac{x - \widehat{\mu}}{\widehat{\sigma}} \right) + \frac{\widehat{\mu}^* - \mu}{\sigma} \right]$$

$$= \Phi\left[\left(\frac{\widehat{\sigma}^*}{\sigma} \right) \Phi^{-1}(\widehat{p}) + \frac{\widehat{\mu}^* - \mu}{\sigma} \right]. \tag{F.3}$$

The confidence interval for $F(x; \mu, \sigma)$ is obtained from the $\alpha/2$ and $1 - \alpha/2$ quantiles of (F.3), considering $\widehat{p}$ fixed.

For example, for observed data $\boldsymbol{x}$ from a $\mathrm{NORM}(\mu, \sigma)$ distribution,

$$\widehat{\mu} = \bar{x}, \quad \widehat{\sigma}^2 = \frac{(n-1)s^2}{n}, \quad \Phi_{\mathrm{norm}}^{-1}(\widehat{p}) = \frac{x - \widehat{\mu}}{\widehat{\sigma}}, \tag{F.4}$$

where $\bar{x}$ and s^2 are the sample mean and variance, respectively. Also

$$\frac{\widehat{\mu}^* - \mu}{\sigma} \sim \mathrm{NORM}\left(0, \frac{1}{\sqrt{n}}\right) \quad \text{and} \quad \frac{\widehat{\sigma}^*}{\sigma} \sim \sqrt{\frac{X_{(n-1)}^2}{n}}, \tag{F.5}$$

where $(\widehat{\mu}^* - \mu)/\sigma$ and $\widehat{\sigma}^*/\sigma$ are independent. Using (F.3), with $\Phi(z) = \Phi_{\mathrm{norm}(z)}$ and the information in (F.4) and (F.5), one obtains the distribution of $Z_{\widehat{p}}$. The α and $1 - \alpha/2$ quantiles of this distribution provide the GPQ-based confidence interval for $F(x; \mu, \sigma)$.

The GPQ-based confidence interval for $F(x; \mu, \sigma)$ obtained from (F.3) is exact in the sense that its coverage probability over the sampling distribution of $\boldsymbol{X}$ is equal to $1 - \alpha$; see Sections F.4.1 and F.4.2 for details.

F.3.4 GPQ Function for Probability Content of an Interval

An inference of interest is the probability content of an interval $[x_L, \ x_U]$. This can be expressed as $p_I = F(x_U) - F(x_L) = p_U - p_L$, where $F(x)$ is a location-scale distribution and $p_i = \Phi[(x_i - \mu)/\sigma]$, $i = L, U$. A GPQ for p_I is obtained from Result F.1, with $h(Z_{\widehat{p}_U}, Z_{\widehat{p}_L}) = Z_{\widehat{p}_U} - Z_{\widehat{p}_L}$, where the $Z_{\widehat{p}_i}$ are obtained from (F.3) with $\widehat{p} = \widehat{p}_i$ and $i = L, U$. Thus,

$$Z_{\widehat{p}_I} = \Phi\left[\left(\frac{\widehat{\sigma}^*}{\sigma}\right)\Phi^{-1}(\widehat{p}_U) + \frac{\widehat{\mu}^* - \mu}{\sigma}\right] - \Phi\left[\left(\frac{\widehat{\sigma}^*}{\sigma}\right)\Phi^{-1}(\widehat{p}_L) + \frac{\widehat{\mu}^* - \mu}{\sigma}\right]. \tag{F.6}$$

F.4 CONDITIONS FOR EXACT CONFIDENCE INTERVALS DERIVED FROM GENERALIZED PIVOTAL QUANTITIES

In this section we first establish and prove a necessary and sufficient condition to ensure that a GPQ interval procedure has an exact coverage probability from the frequentist point of view. We also establish and prove an alternative sufficient condition for a GPQ interval procedure to have exact coverage. This sufficient condition may be easier to check in some special cases.

F.4.1 A Necessary and Sufficient Condition for an Exact GPQ Confidence Interval Procedure

Result F.2 *Consider a GPQ for a scalar parameter θ, say $Z_{\widehat{\theta}} = Z_{\widehat{\theta}}(\boldsymbol{S}^*; \boldsymbol{s}, \boldsymbol{\nu})$, where $\boldsymbol{s}$ is obtained from the data $\boldsymbol{x}$, $\boldsymbol{\nu}$ is a vector of unknown parameters, and $\boldsymbol{S}^*$ is random variable independent of the sampling process generating $\boldsymbol{s}$ and with a distribution equal to the sampling distribution of $\boldsymbol{s}$. Consider the $100(1 - \alpha)\%$ confidence interval for θ given by $[Z_{\widehat{\theta};\alpha/2}, \ Z_{\widehat{\theta};1-\alpha/2}]$ where $Z_{\widehat{\theta};\gamma}$ is the γ quantile of Z_θ. This GPQ-based confidence interval procedure is exact if*

$$\Pr_{\boldsymbol{S}}\left(Z_{\widehat{\theta};\alpha/2} \leq \theta \leq Z_{\widehat{\theta};1-\alpha/2}\right) = 1 - \alpha$$

for all θ.

A necessary and sufficient condition for an exact GPQ-based confidence interval procedure is that the random variable $U(S; \theta)$ defined by

$$U(S; \theta) = \Pr_{S^*|S}[Z_{\widehat{\theta}}(S^*; S, \nu) \leq \theta | S] \tag{F.7}$$

has a $\mathrm{UNIF}(0, 1)$ *distribution.*

This result is verified as follows. Because the confidence interval is obtained from putting together two one-sided confidence bounds, it suffices to prove the result for the one-sided upper bound. Suppose that $0 < \gamma < 1$ and that the one-sided bound procedure $Z_{\widehat{\theta};\gamma}$ has exact coverage γ. Thus,

$$\Pr(\theta \leq Z_{\widehat{\theta};\gamma}) = \Pr_{S}\left[\Pr_{S^*|S}\left(\theta \leq Z_{\widehat{\theta};\gamma} \mid S\right)\right] = \gamma.$$

Using the definition of the $U(S; \theta)$ function in (F.7), one obtains the following two properties: $U(s; Z_{\widehat{\theta};\gamma}) = \gamma$ and $U(s; \theta)$ is nondecreasing as a function of θ. Then, using these two properties, we have

$$\Pr(\theta \leq Z_{\widehat{\theta};\gamma}) = \Pr_{S}\left[\Pr_{S^*|S}(\theta \leq Z_{\widehat{\theta};\gamma} \mid S)\right] = \Pr_{S}\left\{\Pr_{S^*|S}\left[U(S, \theta) \leq U(S, Z_{\widehat{\theta};\gamma}) \mid S\right]\right\}$$

$$= \Pr_{S}\left\{\Pr_{S^*|S}[U(S, \theta) \leq \gamma \mid S]\right\} = \Pr_{S}[U(S, \theta) \leq \gamma] = \gamma. \tag{F.8}$$

The last equality in (F.8) holds because the procedure is exact. Also (F.8) shows that $U(S; \theta) \sim \mathrm{UNIF}(0, 1)$.
Now suppose that $U(S, \theta) \sim \mathrm{UNIF}(0, 1)$. Then

$$\Pr(\theta \leq Z_{\widehat{\theta};\gamma}) = \Pr_{S}\left[\Pr_{S^*|S}(\theta \leq Z_{\widehat{\theta};\gamma} \mid S)\right] = \Pr_{S}\left\{\Pr_{S^*|S}\left[U(S, \theta) \leq U(S, Z_{\widehat{\theta};\gamma}) \mid S\right]\right\}$$

$$= \Pr_{S}\left\{\Pr_{S^*|S}[U(S, \theta) \leq \gamma \mid S]\right\} = \Pr_{S}[U(S, \theta) \leq \gamma] = \gamma. \tag{F.9}$$

The last equality in (F.9) holds because $U(S; \theta) \sim \mathrm{UNIF}(0, 1)$. Result F.2 appears in Hannig et al. (2006, Remark 7) without proof.

Example F.1 Checking that a Tail Probability Confidence Interval is Exact. We now illustrate the use of Result F.2. Consider the GPQ in (F.3) used to derive a confidence interval for the tail probability $p = F(x)$, where x is given and the parameters μ and σ are unknown. In this case

$$s = (\widehat{\mu}, \widehat{\sigma}), \qquad\qquad S^* = (\widehat{\mu}^*, \widehat{\sigma}^*),$$

$$p = \Phi\left(\frac{x - \mu}{\sigma}\right), \qquad\qquad \widehat{p} = \Phi\left(\frac{x - \widehat{\mu}}{\widehat{\sigma}}\right),$$

$$Z_{\widehat{p}} = \Phi\left[\frac{\widehat{\sigma}^*}{\sigma}\Phi^{-1}(\widehat{p}) + \frac{\widehat{\mu}^* - \mu}{\sigma}\right].$$

Thus,

$$
\begin{aligned}
U(\boldsymbol{S}, p) &= \Pr_{\boldsymbol{S}^*|\boldsymbol{S}}[Z_{\widehat{p}}(\boldsymbol{S}^*; \boldsymbol{S}, \boldsymbol{\nu}) \leq p|\boldsymbol{S}] \\
&= \Pr_{\boldsymbol{S}^*|\boldsymbol{S}}\{Z_{\widehat{p}}[(\widehat{\mu}^*, \widehat{\sigma}^*); (\widehat{\mu}, \widehat{\sigma}), \boldsymbol{\nu}] \leq p|\boldsymbol{S}\} \\
&= \Pr_{\boldsymbol{S}^*|\boldsymbol{S}}\left\{\Phi\left[\frac{\widehat{\sigma}^*}{\sigma}\Phi^{-1}(\widehat{p}) + \frac{\widehat{\mu}^* - \mu}{\sigma}\right] \leq p \middle| \boldsymbol{S}\right\} \\
&= \Pr_{\boldsymbol{S}^*|\boldsymbol{S}}\left[\frac{\widehat{\sigma}^*}{\sigma}\Phi^{-1}(\widehat{p}) + \frac{\widehat{\mu}^* - \mu}{\sigma} \leq \Phi^{-1}(p) \middle| \boldsymbol{S}\right] \\
&= \Pr_{\boldsymbol{S}^*|\boldsymbol{S}}\left(\frac{x - \widehat{\mu}}{\widehat{\sigma}} \leq \frac{x - \widehat{\mu}^*}{\widehat{\sigma}^*} \middle| \boldsymbol{S}\right).
\end{aligned}
$$

Let $G(z)$ be the cdf of $(x - \widehat{\mu}^*)/\widehat{\sigma}^*$. Then

$$
\begin{aligned}
\Pr_{\boldsymbol{S}}[U(\boldsymbol{S}, p) \leq w] &= \Pr_{\boldsymbol{S}}\left[\Pr_{\boldsymbol{S}^*|\boldsymbol{S}}\left(\frac{x - \widehat{\mu}}{\widehat{\sigma}} \leq \frac{x - \widehat{\mu}^*}{\widehat{\sigma}^*} \middle| \boldsymbol{S}\right) \leq w\right] \\
&= \Pr_{\boldsymbol{S}}\left[1 - \Pr_{\boldsymbol{S}^*|\boldsymbol{S}}\left(\frac{x - \widehat{\mu}^*}{\widehat{\sigma}^*} \leq \frac{x - \widehat{\mu}}{\widehat{\sigma}} \middle| \boldsymbol{S}\right) \leq w\right] \\
&= \Pr_{\boldsymbol{S}}\left[1 - G\left(\frac{x - \widehat{\mu}}{\widehat{\sigma}}\right) \leq w\right] \\
&= 1 - \Pr_{\boldsymbol{S}}\left[G\left(\frac{x - \widehat{\mu}}{\widehat{\sigma}}\right) \leq 1 - w\right] = w.
\end{aligned}
$$

The last equality above follows from the probability integral transform for continuous variables which shows that $G\{[x - \widehat{\mu}]/\widehat{\sigma}\}$ has a $\mathrm{UNIF}(0, 1)$ distribution. ∎

F.4.2 A Sufficient Condition for an Exact GPQ Confidence Interval Procedure

Here we establish a sufficient condition which ensures that a GPQ-based confidence procedure is exact. In some cases, it might be easier to verify this condition than the necessary and sufficient condition in Result F.2.

Result F.3 *The $U(\boldsymbol{S}; \theta)$ function in (F.7) has a $\mathrm{UNIF}(0, 1)$ distribution if the inequality $Z_{\widehat{\theta}}(\boldsymbol{S}^*; s, \nu) \leq \theta$ is equivalent to the inequality $h(\boldsymbol{S}, \boldsymbol{\nu}) \leq h(\boldsymbol{S}^*, \boldsymbol{\nu})$ for some $h(u, v)$ function.*

To show the result, suppose that the two inequalities are equivalent. Let $G(w)$ be the cdf of $h(\boldsymbol{S}, \boldsymbol{\nu})$. Note that $G(w)$ is also the distribution of $h(\boldsymbol{S}^*, \boldsymbol{\nu})$ because $\boldsymbol{S} \sim \boldsymbol{S}^*$. Also $G[h(\boldsymbol{S}, \boldsymbol{\nu})]$

and $G[h(\boldsymbol{S}^*, \boldsymbol{\nu})]$ have $\mathrm{UNIF}(0, 1)$ distribution because of the probability integral transform for continuous variables. Then

$$
\begin{aligned}
\Pr_{\boldsymbol{S}}[U(\boldsymbol{S}, \theta) \leq w] &= \Pr_{\boldsymbol{S}}\left[\Pr_{\boldsymbol{S}^*|\boldsymbol{S}}\left[Z_{\widehat{\theta}}(\boldsymbol{S}^*; \boldsymbol{S}, \boldsymbol{\nu}) \leq \theta|\boldsymbol{S}\right] \leq w\right] \\
&= \Pr_{\boldsymbol{S}^*}\left[\Pr_{\boldsymbol{S}|\boldsymbol{S}^*}\left[Z_{\widehat{\theta}}(\boldsymbol{S}^*; \boldsymbol{S}, \boldsymbol{\nu}) \leq \theta|\boldsymbol{S}^*\right] \leq w\right] \\
&= \Pr_{\boldsymbol{S}^*}\left[\Pr_{\boldsymbol{S}|\boldsymbol{S}^*}\left[h(X, \boldsymbol{\nu}) \leq h(X^*, \boldsymbol{\nu})|\boldsymbol{S}^*\right] \leq w\right] \\
&= \Pr_{\boldsymbol{S}^*}\left(\Pr_{\boldsymbol{S}|\boldsymbol{S}^*}\left[G[h(X, \boldsymbol{\nu})] \leq G[h(X^*, \boldsymbol{\nu})] \mid \boldsymbol{S}^*\right] \leq w\right) \\
&= \Pr_{\boldsymbol{S}^*}\{G[h(X^*, \boldsymbol{\nu})] \leq w\} = w.
\end{aligned}
$$

This shows that $U(\boldsymbol{S}, \theta) \sim \mathrm{UNIF}(0, 1)$, which implies that the GPQ-based confidence-interval procedure in Section F.3.3 is exact. Result F.3 appears in Hannig et al. (2006, Remark 7) without proof.

Example F.2   A Simpler Check that a Tail Probability Confidence Interval is Exact. We now illustrate the use of Result F.3. We again consider the GPQ in (F.3) used to derive a confidence interval for $p = F(x)$, where x is fixed.

It can be show that the following inequalities are equivalent:

$$
Z_{\widehat{p}}(\boldsymbol{S}^*; \boldsymbol{S}, \boldsymbol{\nu}) \leq p,
$$

$$
Z_{\widehat{p}}[(\widehat{\mu}^*, \widehat{\sigma}^*); (\widehat{\mu}, \widehat{\sigma}), \boldsymbol{\nu}] \leq p,
$$

$$
\Phi\left[\frac{\widehat{\sigma}^*}{\sigma}\Phi^{-1}(\widehat{p}) + \frac{\widehat{\mu}^* - \mu}{\sigma}\right] \leq p,
$$

$$
\Phi^{-1}(\widehat{p}) + \frac{\widehat{\mu}^* - \mu}{\sigma} \leq \Phi^{-1}(p),
$$

$$
\frac{x - \widehat{\mu}}{\widehat{\sigma}} \leq \frac{x - \widehat{\mu}^*}{\widehat{\sigma}^*}.
$$

Therefore, the inequality $Z_{\widehat{p}}(\boldsymbol{S}^*; \boldsymbol{S}, \boldsymbol{\nu}) \leq p$ is equivalent to the inequality $h[(\widehat{\mu}, \widehat{\sigma}), (\mu, \sigma)] \leq h[(\widehat{\mu}^*, \widehat{\sigma}^*), (\mu, \sigma)]$, for all (μ, σ), where $h[(a, b), (\mu, \sigma)] = (x - a)/b$. This shows that the sufficient condition in Result F.3 is met and thus the GPQ-based confidence interval procedure in Section F.3.3 is exact. ∎

Appendix *G*

Distribution-Free Intervals Based on Order Statistics

INTRODUCTION

This appendix provides technical explanations for the distribution-free statistical intervals based on order statistics, which are described in Chapter 5. The purpose is to provide information such that an interested reader could understand the theoretical basis for the methodology.

The topics discussed in this appendix are:

- Some basic results used to develop and evaluate the coverage probabilities of distribution-free statistical intervals (Section G.1).

- Justification of the distribution-free confidence intervals (and corresponding one-sided confidence bounds) for a distribution quantile (Section G.2).

- Justification of the distribution-free tolerance intervals to contain a given proportion of a distribution (Section G.3).

- Justification of the distribution-free statistical intervals to contain a specified ordered observation in a future sample (Section G.4).

- Justification of the distribution-free statistical intervals to contain at least a given number of observations from a future sample and the relationship of this prediction problem to the prediction of a future ordered observation (Section G.5).

G.1 BASIC STATISTICAL RESULTS USED IN THIS APPENDIX

In this appendix we use the following results:

(a) If X is a continuous random variable with cdf $F(x)$ and one defines $W = F(X)$ then $W \sim \text{UNIF}(0, 1)$ (see Example D.2 for a proof of this result).

Statistical Intervals: A Guide for Practitioners and Researchers, Second Edition.
William Q. Meeker, Gerald J. Hahn and Luis A. Escobar.
© 2017 John Wiley & Sons, Inc. Published 2017 by John Wiley & Sons, Inc.
Companion Website: www.wiley.com/go/meeker/intervals

(b) If $W_{(i)}$ is the ith order statistic in a sample of size n from a $\text{UNIF}(0, 1)$ distribution then $W_{(i)} \sim \text{BETA}(i, n - i + 1)$.

(c) Suppose that ℓ and u are positive integers with $\ell < u$ and define $d = u - \ell$. If $W_{(\ell)}$ and $W_{(u)}$ are the ℓth and uth order statistics from a $\text{UNIF}(0, 1)$ distribution then $W_{(u)} - W_{(\ell)} \sim \text{BETA}(d, n - d + 1)$.

(d) The binomial cdf is related to a beta cdf as follows: $\text{pbinom}(x - 1; n, \pi) = 1 - \text{pbeta}(\pi; x, n - x + 1)$, where x and n are positive integers, and $x \leq n$. See the discussion following (C.19) for details about this relationship and a reference for its proof.

(e) Negative hypergeometric cumulative probabilities, denoted by $\text{pnhyper}(x - 1; j, m, m + n)$, can be expressed as a $\text{dbeta}(v; j, m - j + 1)$ mixture of $\text{pbeta}(v; x, n - x + 1)$ probabilities as follows:

$$\text{pnhyper}(x - 1; j, m, m + n) = 1 - \int_0^1 \text{pbeta}(v; x, n - x + 1)$$
$$\text{dbeta}(v; j, m - j + 1)\, dv, \qquad \text{(G.1)}$$

where j, m, n, x are positive integers with $j \leq m$, $x \leq n$ (see Section C.4.6 for details about this relationship).

For proofs of items (b) and (c) above, see David and Nagaraja (2003, page 14).

G.2 DISTRIBUTION-FREE CONFIDENCE INTERVALS AND BOUNDS FOR A DISTRIBUTION QUANTILE

G.2.1 Distribution-Free Confidence Intervals for Quantiles

For a continuous and strictly increasing cdf $F(x)$, the p quantile (where $0 < p < 1$) of the distribution is $x_p = F^{-1}(p)$. Suppose that $x_{(1)} < \cdots < x_{(n)}$ are the order statistics in a random sample of size n from a continuous cdf $F(x)$. A distribution-free confidence interval for the p quantile x_p is

$$[\underset{\sim}{x_p},\ \tilde{x}_p] = [x_{(\ell)},\ x_{(u)}].$$

In general, the interval endpoints $x_{(\ell)}$ and $x_{(u)}$ are chosen such that the procedure has a coverage probability equal or approximately equal to the nominal confidence level $100(1 - \alpha)\%$.

Because $F(x)$ is continuous, the coverage probability of the procedure to obtain the confidence interval $[\underset{\sim}{x_p},\ \tilde{x}_p]$ for a quantile is

$$\begin{aligned}
\text{CPXP}(n, \ell, u, p) &= \Pr(X_{(\ell)} \leq x_p \leq X_{(u)}) = \Pr(X_{(\ell)} \leq x_p) - \Pr(X_{(u)} \leq x_p) \\
&= \Pr[F(X_{(\ell)}) \leq p] - \Pr[F(X_{(u)}) \leq p] = \Pr(W_{(\ell)} \leq p) \\
&\quad - \Pr(W_{(u)} \leq p) \\
&= \text{pbeta}(p; \ell, n - \ell + 1) - \text{pbeta}(p; u, n - u + 1) \qquad \text{(G.2)} \\
&= \text{pbinom}(u - 1; n, p) - \text{pbinom}(\ell - 1; n, p). \qquad \text{(G.3)}
\end{aligned}$$

Note that (G.2) and (G.3) are equivalent forms for computing the coverage probability of the confidence interval obtained using the relationship between the beta cdf and the binomial cdf as explained in (C.19).

G.2.2 Distribution-Free One-Sided Confidence Bounds for Quantiles

A one-sided upper confidence bound for the quantile x_p is $\widetilde{x}_p = x_{(u)}$. The coverage probability of this upper one-sided confidence bound procedure is

$$
\begin{aligned}
\mathrm{CPXP}(n,0,u,p) &= \Pr(X_{(u)} \geq x_p) = 1 - \Pr(X_{(u)} \leq x_p) \\
&= 1 - \Pr[F(X_{(u)}) \leq p] = 1 - \Pr(W_{(u)} \leq p) \\
&= 1 - \mathtt{pbeta}(p; u, n - u + 1) = \mathtt{pbinom}(u - 1; n, p).
\end{aligned}
$$

Operationally $\mathrm{CPXP}(n, 0, u, p)$ is obtained from $\mathrm{CPXP}(n, \ell, u, p)$ with $\ell = 0$.

To determine u to provide a conservative $100(1 - \alpha)\%$ upper bound for x_p, we need to find the smallest integer u satisfying $\mathrm{CPXP}(n, 0, u, p) = \mathtt{pbinom}(u - 1; n, p) \geq 1 - \alpha$. This inequality implies $u - 1 = \mathtt{qbinom}(1 - \alpha; n, p)$. Thus

$$
u = \mathtt{qbinom}(1 - \alpha; n, p) + 1.
$$

A one-sided lower confidence bound for the quantile x_p is $\underset{\sim}{x}_p = x_{(\ell)}$. The coverage probability for the lower-bound procedure is

$$
\begin{aligned}
\mathrm{CPXP}(n, \ell, n + 1, p) &= \Pr(X_{(\ell)} \leq x_p) = \Pr[F(X_{(\ell)}) \leq p] \\
&= \Pr(W_{(\ell)} \leq p) = \mathtt{pbeta}(p; \ell, n - \ell + 1) \\
&= 1 - \mathtt{pbinom}(\ell - 1; n, p).
\end{aligned}
$$

Operationally $\mathrm{CPXP}(n, \ell, n + 1, p)$ is obtained from $\mathrm{CPXP}(n, \ell, u, p)$ with $u = n + 1$. Furthermore, using the identity (C.17), we obtain the following equivalent expression for $\mathrm{CPXP}(n, \ell, n + 1, p)$:

$$
\mathrm{CPXP}(n, \ell, n + 1, p) = \mathtt{pbinom}(n - \ell; n, 1 - p).
$$

This expression is convenient for obtaining a conservative lower bound ℓ for x_p, as follows. A conservative $100(1 - \alpha)\%$ lower bound ℓ for x_p is the largest integer ℓ satisfying $\mathrm{CPXP}(n, \ell, n + 1, p) = \mathtt{pbinom}(n - \ell; n, 1 - p) \geq 1 - \alpha$. This inequality implies that $n - \ell = \mathtt{qbinom}(1 - \alpha; n, 1 - p)$. Thus,

$$
\ell = n - \mathtt{qbinom}(1 - \alpha; n, 1 - p).
$$

Note that $\mathrm{CPXP}(n, \ell, u, p) = \mathrm{CPXP}(n, \ell, n + 1, p) + \mathrm{CPXP}(n, 0, u, p) - 1$. This shows that a $100(1 - \alpha_L - \alpha_U)\%$ confidence interval for a particular quantile can be obtained by putting together a one-sided lower $100(1 - \alpha_L)\%$ confidence bound and a one-sided upper $100(1 - \alpha_U)\%$ confidence bound for the same quantile.

G.3 DISTRIBUTION-FREE TOLERANCE INTERVALS TO CONTAIN A GIVEN PROPORTION OF A DISTRIBUTION

G.3.1 Distribution-Free Tolerance Intervals

A distribution-free tolerance interval to contain at least a proportion β of the sampled population, based on a random sample of size n from the same population, is

$$
[\underset{\sim}{T}_\beta, \ \widetilde{T}_\beta] = [x_{(\ell)}, \ x_{(u)}], \tag{G.4}
$$

where the order statistics $x_{(\ell)}$ and $x_{(u)}$ are chosen such that the interval has a coverage probability equal or approximately equal to the nominal confidence level $100(1 - \alpha)\%$.

For samples from a continuous random variable, the coverage probability for interval (G.4) is

$$
\begin{aligned}
\text{CPTI}(n, \ell, u, \beta) &= \Pr[F(X_{(u)}) - F(X_{(\ell)}) \geq \beta] \\
&= \Pr[(W_{(u)} - W_{(\ell)}) \geq \beta] = 1 - \Pr[(W_{(u)} - W_{(\ell)}) \leq \beta] \\
&= 1 - \texttt{pbeta}(\beta; d, n - d + 1) = \texttt{pbinom}(d - 1; n, \beta), \qquad \text{(G.5)}
\end{aligned}
$$

where $d = u - \ell$, and $W_{(u)} = F(X_{(u)})$, $W_{(\ell)} = F(X_{(\ell)})$ are the uth and ℓth order statistics in a sample of size n from a $\text{UNIF}(0, 1)$ distribution, respectively.

To have tolerance intervals that have equal error probabilities in both tails, we recommend the symmetric interval $[x_{(n-u+1)}, \ x_{(u)}]$, with $u > (n + 1)/2$. Such a symmetric interval has the property that $\Pr[F(X) > F(X_{(u)})] = \Pr[F(X) < F(X_{(n-u+1)})] = (n - u + 1)/(n + 1)$. This property implies that the expected proportion of the population less than $\underset{\sim}{T}_\beta$ is equal to the expected proportion of the population greater than $\widetilde{T}_\beta$.

Because $\text{CPTI}(n, \ell, u, \beta)$ in (G.5) depends on ℓ and u only through $d = u - \ell$, there are, in general, multiple choices for ℓ and u that achieve the same $\text{CPTI}(n, \ell, u, \beta)$. The narrowest tolerance interval will remove the largest number of ordered observations from the ends of the ordered sample, subject to the constraint $\text{CPTI}(n, \ell, u, \beta) < 1 - \alpha$, leading to a conservative tolerance interval $[x_{(\ell)}, \ x_{(u)}]$ that can be obtained as follows. Let $\nu - 2$ be the largest number of observations to be removed from the extremes of the ordered sample to obtain a conservative tolerance interval that contains at least a proportion β of the distribution. Then, $n = (\nu - 2) + (u - \ell + 1)$ or equivalently, $n - \nu = u - \ell - 1$ and it is required that $\text{CPTI}(n, \ell, u, \beta) = \texttt{pbinom}(n - \nu; n, \beta) \geq 1 - \alpha$. This implies that $n - \nu = \texttt{qbinom}(1 - \alpha; n, \beta)$. Thus,

$$
\nu = n - \texttt{qbinom}(1 - \alpha; n, \beta). \qquad \text{(G.6)}
$$

G.4 DISTRIBUTION-FREE PREDICTION INTERVAL TO CONTAIN A SPECIFIED ORDERED OBSERVATION FROM A FUTURE SAMPLE

G.4.1 Distribution-Free Prediction Interval for $Y_{(j)}$ in a Future Sample

A distribution-free prediction interval to contain $Y_{(j)}$, the jth ordered observation in a future sample of size m from the previously sampled distribution, is

$$
[\underset{\sim}{Y}_{(j)}, \ \widetilde{Y}_{(j)}] = [x_{(\ell)}, \ x_{(u)}], \qquad \text{(G.7)}
$$

where $1 \leq \ell < u \leq n$, and $x_{(\ell)}$ and $x_{(u)}$ are the ℓth and uth order statistics from the original sample of size n. These order statistics are chosen such that the interval has a coverage probability equal or approximately equal to the nominal confidence level $100(1 - \alpha)\%$.

The coverage probability for the prediction interval procedure $[\underset{\sim}{Y}_{(j)}, \ \widetilde{Y}_{(j)}]$ given in (G.7) is

$$
\begin{aligned}
\text{CPYJ}(n, \ell, u, m, j) &= \int_0^1 [\texttt{pbeta}(v; \ell, n - \ell + 1) \\
&\quad - \texttt{pbeta}(v; u, n - u + 1)]\texttt{dbeta}(v; j, m - j + 1)\, dv \\
&= \texttt{pnhyper}(u - 1; j, m, m + n) - \texttt{pnhyper}(\ell - 1; j, m, m + n),
\end{aligned}
$$

where $\mathrm{pnhyper}(x; j, m, m + n)$ is the cdf for a negative hypergeometric random variable (see Section C.4.6 for details) and

$$\mathrm{pnhyper}(x; k, D, N) = \sum_{i=0}^{x} \mathrm{dnhyper}(i; k, D, N) = \sum_{i=0}^{x} \frac{\binom{i+k-1}{i}\binom{N-i-k}{N-D-i}}{\binom{N}{D}},$$

where $0 < D < N, 1 \le k \le D, 0 \le x \le N - D$, k is a positive integer, and x is a nonnegative integer.

G.4.2 Distribution-Free One-Sided Prediction Bounds for $Y_{(j)}$ in a Future Sample

The coverage probability for the one-sided upper prediction bound $\widetilde{Y}_{(j)} = x_{(u)}$ to exceed $Y_{(j)}$, the jth order statistic in a future sample of size m, is

$$\mathrm{CPYJ}(n, 0, u, m, j) = 1 - \int_0^1 \mathrm{pbeta}(s; u, n - u + 1)\mathrm{dbeta}(s; j, m - j + 1)\, ds$$

$$= \mathrm{pnhyper}(u - 1; j, m, m + n). \tag{G.8}$$

To determine u to provide a conservative $100(1 - \alpha)\%$ upper prediction bound for $Y_{(j)}$, we need to find the smallest integer u satisfying $\mathrm{CPYJ}(n, 0, u, m, j) = \mathrm{pnhyper}(u - 1; j, m, m + n) \ge 1 - \alpha$. This implies $u - 1 = \mathrm{qnhyper}(1 - \alpha; j, m, m + n)$. Thus,

$$u = \mathrm{qnhyper}(1 - \alpha; j, m, m + n) + 1. \tag{G.9}$$

The coverage probability for the one-sided lower prediction bound $\underset{\sim}{Y}_{(j)} = x_{(\ell)}$ to be exceeded by $Y_{(j)}$, the jth order statistic in a future sample of size m, is

$$\mathrm{CPYJ}(n, \ell, n + 1, m, j) = \int_0^1 \mathrm{pbeta}(r; \ell, n - \ell + 1)\mathrm{dbeta}(r; j, m - j + 1)\, dr$$

$$= 1 - \mathrm{pnhyper}(\ell - 1; j, m, m + n).$$

Alternatively, using the identity (C.28) with $y = \ell - 1$, we can express this coverage probability as

$$\mathrm{CPYJ}(n, \ell, n + 1, m, j) = \mathrm{pnhyper}(n - \ell; m - j + 1, m, m + n). \tag{G.10}$$

To determine ℓ to provide a conservative $100(1 - \alpha)\%$ lower prediction bound for $Y_{(j)}$, ℓ must be the largest integer ℓ satisfying $\mathrm{CPYJ}(n, \ell, n + 1, m, j) = \mathrm{pnhyper}(n - \ell; m - j + 1, m, m + n) \ge 1 - \alpha$. This inequality implies $n - \ell = \mathrm{qnhyper}(1 - \alpha; m - j + 1, m, m + n)$. Thus,

$$\ell = n - \mathrm{qnhyper}(1 - \alpha; m - j + 1, m, m + n). \tag{G.11}$$

Note that $\mathrm{CPYJ}(n, \ell, u, m, j) = \mathrm{CPYJ}(n, \ell, n + 1, m, j) + \mathrm{CPYJ}(n, 0, u, m, j) - 1$. This shows that a $100(1 - \alpha)\%$ prediction interval to contain a future ordered observation can be obtained by combining a one-sided lower $100(1 - \alpha_L)\%$ prediction bound and a one-sided upper $100(1 - \alpha_U)\%$ prediction bound for the future ordered observation. Then the coverage probability for the two-sided prediction interval is $1 - \alpha_L - \alpha_U$.

G.4.3   Prediction Interval to Contain a Specified Ordered Observation $Y_{(j)}$ in a Future Sample (Technical Details)

The coverage probability of the prediction interval procedure given in (G.7) is

$$\text{CPYJ}(n, \ell, u, m, j) = \Pr(\underset{\sim}{Y}_{(j)} \leq Y_{(j)} \leq \widetilde{Y}_{(j)}) = \Pr(X_{(\ell)} \leq Y_{(j)} \leq X_{(u)}). \qquad \text{(G.12)}$$

Using the cdf transformation, it follows that $W_{(\ell)} = F(X_{(\ell)})$ and $W_{(u)} = F(X_{(u)})$ are the lth and uth order statistics in samples of size n from a $\text{UNIF}(0, 1)$ distribution, and $V_{(j)} = F(Y_{(j)})$ is the jth order statistic in samples of size m from a $\text{UNIF}(0, 1)$ distribution. Note that $V_{(j)}$ is independent of the pair $(W_{(\ell)}, W_{(u)})$ and that $V_{(j)} \sim \text{BETA}(j, m - j + 1)$.
   From (G.12),

$$\begin{aligned} \text{CPYJ}(n, \ell, u, m, j) &= \Pr[F(x_{(\ell)}) \leq F(Y_{(j)}) \leq F(X_{(u)})] = \Pr(W_{(\ell)} \leq V_{(j)} \leq W_{(u)}) \\ &= \Pr(W_{(\ell)} \leq V_{(j)}) - \Pr(W_{(u)} \leq V_{(j)}). \end{aligned}$$

Conditioning on $V_{(j)} = v$, we obtain

$$\begin{aligned} &\text{CPYJ}(n, \ell, u, m, j) \\ &= \int_0^1 [\Pr(W_{(\ell)} \leq v) - \Pr(W_{(u)} \leq v)] \texttt{dbeta}(v; j, m - j + 1)\, dv \\ &= \int_0^1 [\texttt{pbeta}(v; \ell, n - \ell + 1) - \texttt{pbeta}(v; u, n - u + 1)] \texttt{dbeta}(v; j, m - j + 1)\, dv \quad \text{(G.13)} \\ &= \texttt{pnhyper}(u - 1; j, m, m + n) - \texttt{pnhyper}(\ell - 1; j, m, m + n). \qquad \text{(G.14)} \end{aligned}$$

In the transition from (G.13) to (G.14), we use twice the relationship (G.1).
   Similarly, for the one-sided lower prediction bound, the coverage probability of the procedure is

$$\begin{aligned} \text{CPYJ}(n, \ell, n + 1, m, j) &= \Pr[F(x_{(\ell)}) \leq F(Y_{(j)})] = \Pr(W_{(\ell)} \leq V_{(j)}) \\ &= \int_0^1 \Pr(W_{(\ell)} \leq v) \texttt{dbeta}(v; j, m - j + 1)\, dv \\ &= \int_0^1 \texttt{pbeta}(v; \ell, n - \ell + 1) \texttt{dbeta}(v; j, m - j + 1)\, dv \\ &= 1 - \texttt{pnhyper}(\ell - 1; j, m, m + n). \end{aligned}$$

For the upper prediction bound procedure, the coverage probability is

$$\begin{aligned} \text{CPYJ}(n, 0, u, m, j) &= \Pr[F(Y_{(j)}) \leq F(X_{(u)})] = \Pr(V_{(j)} \leq W_{(u)}) \\ &= 1 - \Pr(W_{(u)} \leq V_{(j)}) \\ &= 1 - \int_0^1 \Pr(W_{(u)} \leq v) \texttt{dbeta}(v; j, m - j + 1)\, dv \\ &= 1 - \int_0^1 \texttt{pbeta}(v; u, n - u + 1) \texttt{dbeta}(v; j, m - j + 1)\, dv \\ &= \texttt{pnhyper}(u - 1; j, m, m + n). \end{aligned}$$

G.5 DISTRIBUTION-FREE PREDICTION INTERVALS AND BOUNDS TO CONTAIN AT LEAST k OF m FUTURE OBSERVATIONS FROM A FUTURE SAMPLE

G.5.1 Prediction Intervals to Contain at Least k of m Future Observations

A distribution-free $100(1 - \alpha)\%$ prediction interval to contain at least k of m future observations $Y_1, \ldots, Y_m$ from the previously sampled population, based on a random sample x of size n, is

$$\left[\underset{\sim}{Y}_{k:m}, \ \widetilde{Y}_{k:m} \right] = [x_{(\ell)}, \ x_{(u)}], \tag{G.15}$$

where $1 \leq \ell < u \leq n$, $x_{(\ell)}$, $x_{(u)}$ are the ℓth and uth order statistics from the original sample and they are chosen such that the interval has a coverage probability equal or approximately equal to the nominal confidence level $100(1 - \alpha)\%$.

Letting $d = u - \ell$, the coverage probability of the prediction interval procedure giving $\left[\underset{\sim}{Y}_{k:m}, \ \widetilde{Y}_{k:m} \right]$ is

$$\begin{aligned} \mathrm{CPKM}(n, \ell, u, k, m) &= \int_0^1 \mathtt{pbeta}(v; k, m - k + 1)\mathtt{dbeta}(v; d, n - d + 1) \, dv \\ &= 1 - \int_0^1 \mathtt{pbeta}(v; d, n - d + 1)\mathtt{dbeta}(v; k, m - k + 1) \, dv \\ &= \mathtt{pnhyper}(d - 1; k, m, m + n), \end{aligned} \tag{G.16}$$

where $\mathtt{pnhyper}(d - 1; k, m, m + n)$ is the negative hypergeometric cdf, explained in Section C.4.6, evaluated at $x = d - 1$, with parameters $k = k$, $D = m$, and $N = n + m$, with $d = u - \ell$ as defined earlier.

Let $\nu - 2$ be the total number of observations to be removed from the extremes of the ordered sample to obtain a conservative prediction interval that contains at least k of m future observations. Then $n - \nu = u - \ell - 1$ and it is required that $\mathrm{CPKM}(n, \ell, u, k, m) = \mathtt{pnhyper}(n - \nu; m, m + n) \geq 1 - \alpha$. This implies that $n - \nu = \mathtt{qnhyper}(1 - \alpha; k, m, m + n)$, which gives

$$\nu = n - \mathtt{qnhyper}(1 - \alpha; k, m, m + n). \tag{G.17}$$

G.5.2 One-Sided Prediction Bounds to Exceed (or Be Exceeded by) at Least k of m Future Observations

For a one-sided upper prediction bound, $\widetilde{Y}_{k:m} = x_{(u)}$, to exceed at least k of m future observations from a previously sampled population with continuous cdf $F(x)$, the coverage probability is given by (G.16) with $\ell = 0$. That is,

$$\begin{aligned} \mathrm{CPKM}(n, 0, u, k, m) &= \int_0^1 \mathtt{pbeta}(v; k, m - k + 1)\mathtt{dbeta}(v; u, n - u + 1) \, dv \\ &= 1 - \int_0^1 \mathtt{pbeta}(v; u, n - u + 1)\mathtt{dbeta}(v; k, m - k + 1) \, dv \\ &= \mathtt{pnhyper}(u - 1; k, m, m + n). \end{aligned} \tag{G.18}$$

Operationally, $\mathrm{CPKM}(n, 0, u, k, m) = \mathrm{CPKM}(n, u, \ell, k, m)$ with $\ell = 0$. Also note that $\mathrm{CPKM}(n, 0, u, k, m) = \mathrm{CPYJ}(n, 0, u, m, j)$ when $j = k$ in (G.8). To obtain a one-sided upper conservative prediction bound that exceeds at least k of m future observations, it is required that $\mathrm{CPKM}(n, 0, u, k, m) = \mathrm{pnhyper}(u - 1; k, m, m + n) \geq 1 - \alpha$. Thus

$$u = \mathrm{qnhyper}(1 - \alpha; k, m, m + n) + 1. \tag{G.19}$$

For a one-sided lower prediction bound, $\underset{\sim}{Y}_{k:m} = x_{(\ell)}$, to be exceeded by at least k of m future observations from a previously sampled population, the coverage probability is given by (G.16) with $u = n + 1$. That is,

$$\mathrm{CPKM}(n, \ell, n + 1, k, m) = \int_0^1 \mathrm{pbeta}(v; k, m - k + 1)\mathrm{dbeta}(v; n - \ell + 1, \ell)\, dv$$

$$= 1 - \int_0^1 \mathrm{pbeta}(v; n - \ell + 1, \ell)\mathrm{dbeta}(v; k, m - k + 1)\, dv$$

$$= \mathrm{pnhyper}(n - \ell; k, m, m + n). \tag{G.20}$$

Operationally, $\mathrm{CPKM}(n, \ell, n + 1, k, m) = \mathrm{CPKM}(n, \ell, u, k, m)$ with $u = n + 1$.

To obtain a one-sided lower conservative prediction bound that is exceeded by at least k of m future observations, it is required that $\mathrm{CPKM}(n, \ell, n + 1, k, m) = \mathrm{pnhyper}(n - \ell; k, m, m + n) \geq 1 - \alpha$. Thus,

$$\ell = n - \mathrm{qnhyper}(1 - \alpha; k, m, m + n). \tag{G.21}$$

G.5.3   Special Cases, Limiting Behavior, and Relationship to Predicting a Future Ordered Observation

Special cases

- A situation of interest is finding the coverage probability of the prediction interval to contain all m of m future observations and defined by the smallest and largest observations in the sample. That is, the prediction interval is given by $[\underset{\sim}{Y}_{m:m}, \widetilde{Y}_{m:m}] = (x_{(1)}, \ x_{(n)})$. In this case $d = u - \ell = n - 1$ and $k = m$. Because $\mathrm{pnhyper}(n; m, m, m + n) = 1$, using the negative hypergeometric density in (C.27) with $k = m$, $D = m$, $N = m + n$, and the convention $0! = 1$,

$$\mathrm{CPKM}(n, 1, n, m, m) = \mathrm{pnhyper}(n - 2; m, m, m + n)$$
$$= 1 - \mathrm{dnhyper}(n - 1; m, m, m + n)$$
$$\quad - \mathrm{dnhyper}(n; m, m, m + n)$$
$$= \frac{n(n - 1)}{(m + n)(m + n - 1)}.$$

- Another prediction problem of interest is the coverage probability of the one-sided *lower* prediction bound using the smallest observation $x_{(1)}$ to be exceeded by all m future observations. In this case $\ell = 1$ and (G.20) gives $\mathrm{CPKM}(n, 1, n + 1, m, m) = \mathrm{pnhyper}(n - 1; m, m, m + n) = n/(m + n)$.

  Using a symmetry argument shows that the one-sided *upper* prediction bound $u_{(n)}$ to exceed all m future observations has $\mathrm{CPKM}(n, 0, n, m, m) = \mathrm{CPKM}(n, 1, n + 1, m, m) = n/(m + n)$.

A limiting behavior of prediction intervals

Prediction intervals to contain at least k of m future observations with increasing k and m behave like a tolerance interval to contain a proportion k/m of the $F(x)$ distribution. Formally, consider the prediction interval $[\underset{\sim}{Y}_{k_r:m_r}, \; \widetilde{Y}_{k_r:m_r}] = [x_{(\ell)}, \; x_{(u)}]$, given in (G.15), to contain at least k_r out of m_r future observations in a sequence, $r = 1, 2, \ldots$, of prediction cases in which as $r \to \infty$, $m_r \to \infty$ and $(k_r/m_r) \to \beta$ (where $0 < \beta < 1$). In this situation, in the limit, the prediction interval is equivalent to the tolerance interval $[\underset{\sim}{T}_\beta, \; \widetilde{T}_\beta] = [x_{(\ell)}, \; x_{(u)}]$ to contain at least a proportion β of the distribution $F(x)$.

The practical implication of this result is that k is large and $0 < k/m < 1$, there is no important difference between the prediction interval $[\underset{\sim}{Y}_{k:m}, \; \widetilde{Y}_{k:m}] = [x_{(\ell)}, \; x_{(u)}]$ to contain at least k of m future observations and the tolerance interval $[\underset{\sim}{T}_\beta, \; \widetilde{T}_\beta] = [x_{(\ell)}, \; x_{(u)}]$ to contain a fraction $0 < k/m < 1$ of the distribution $F(x)$.

Relationship to predicting a future ordered observation

A one-sided upper prediction bound $\widetilde{Y}_{(j)} = x_{(u)}$ to exceed the jth ordered observation $Y_{(j)}$ in a future sample of size m is equivalent to a one-sided upper prediction bound $\widetilde{Y}_{k:m} = x_{(u)}$ to exceed at least $k = j$ of m future observations from the same distribution. For this result, note that the coverage probability in (G.8) and the u in (G.9) are the same as the coverage probability in (G.18) and the u in (G.19) when $k = j$.

Similarly, $\underset{\sim}{Y}_{(j)} = x_{(\ell)}$ is equivalent to a one-sided lower prediction bound $\underset{\sim}{Y}_{k:m}$ to be exceeded by at least $k = m - j + 1$ of m future observations from the same distribution. For this result, note that the coverage probability in (G.10) and ℓ in (G.11) are the same as the coverage probability in (G.20) and ℓ in (G.21) when $k = m - j + 1$.

G.5.4 Distribution-Free Prediction Intervals and Bounds to Contain at Least k of m Future Observations: Coverage Probabilities (Technical Details)

Suppose that the data and the future observations are independent simple random samples from the same continuous cdf $F(x)$.

The prediction interval case

The probability that a single future observation Y is within the random prediction interval is

$$\Pr(\underset{\sim}{Y}_{k:m} \leq Y \leq \widetilde{Y}_{k:m}) = \Pr(X_{(\ell)} \leq Y \leq X_{(u)})$$
$$= \Pr[F(X_{(\ell)}) \leq F(Y) \leq F(X_{(u)})]$$
$$= \Pr(W_{(\ell)} \leq W \leq W_{(u)}),$$

where $W_{(\ell)} = F(X_{(\ell)})$, $W_{(u)} = F(X_{(u)})$ are the ℓth and uth order statistics in a sample of size n from a $\text{UNIF}(0, 1)$ distribution; and $W = F(Y)$ has a $\text{UNIF}(0, 1)$ distribution independent of $(W_{(\ell)}, W_{(u)})$.

Define K as the number of future observations contained in the prediction interval. Then $\Pr(K = i; \ell, u)$ is the probability that the prediction interval $[W_{(\ell)}, W_{(u)}]$ contains exactly i of m independent observations from a $\text{UNIF}(0, 1)$ distribution. Let $g(w_\ell, w_u)$ be the joint pdf of $(W_{(\ell)}, W_{(u)})$. Using conditional probabilities,

$$\Pr(K = i; \ell, u) = \int_0^1 \int_0^{w_u} \Pr(K = i; \ell, u | W_{(\ell)} = w_\ell, W_{(u)} = w_u) g(w_\ell, w_u) d\,w_\ell d\,w_u.$$

Because $\Pr(K = i; \ell, u | W_{(\ell)} = w_\ell, W_{(u)} = w_u)$ is the probability of observing exactly i of m independent uniform random variables in the interval $[w_\ell, \; w_u]$, it follows from binomial sampling that

$$\Pr(K = i; \ell, u | W_{(\ell)} = w_\ell, W_{(u)} = w_u) = \binom{m}{i} (w_u - w_\ell)^i [1 - (w_u - w_\ell)]^{m-i}.$$

Thus,

$$
\begin{aligned}
\Pr(K = i; \ell, u) &= \mathrm{E}[\Pr(K = i; \ell, u | W_{(\ell)} = w_\ell, W_{(u)} = w_u)] \\
&= \binom{m}{i} \mathrm{E}\left\{ [W_{(u)} - W_{(\ell)}]^i [1 - (W_{(u)} - W_{(\ell)})]^{m-i} \right\} \\
&= \binom{m}{i} \int_0^1 v^i (1 - v)^{m-i} \mathtt{dbeta}(v; d, n - d + 1)\, dv, \qquad (\text{G.22})
\end{aligned}
$$

where $d = u - \ell$, $v = w_u - w_\ell$, and the pdf of $W_{(u)} - W_{(\ell)}$ is $\mathtt{dbeta}(v; d, n - d + 1)$. Consequently,

$$
\begin{aligned}
\mathrm{CPKM}(n, \ell, u, k, m) &= \int_0^1 \sum_{i=k}^m \binom{m}{i} v^i (1 - v)^{m-i} \mathtt{dbeta}(v; d, n - d + 1)\, dv \\
&= \int_0^1 \mathtt{pbeta}(v; k, m - k + 1) \mathtt{dbeta}(v; d, n - d + 1)\, dv \\
&= 1 - \int_0^1 \mathtt{pbeta}(v; d, n - d + 1) \mathtt{dbeta}(v; k, m - k + 1)\, dv \\
&= \mathtt{pnhyper}(d - 1; k, m, m + n). \qquad (\text{G.23})
\end{aligned}
$$

One-sided lower and upper prediction bounds

For the one-sided prediction bounds, the formulas in (G.16) still apply.

- For the one-sided lower prediction bound, $\widetilde{Y}_{k:m} = x_{(\ell)}$, it suffices to use $u = n + 1$ in the formulas as shown next. Using conditional expectations with respect to $W_{(\ell)} \sim g(w_\ell) = \mathtt{dbeta}(w_\ell; \ell, n - \ell + 1)$,

$$
\begin{aligned}
\Pr(K = i; l) &= \mathrm{E}[\Pr(K = i; l | W_{(\ell)} = w_\ell)] = \binom{m}{i} \mathrm{E}\left[(1 - W_{(\ell)})^i W_{(\ell)}^{m-i} \right] \\
&= \binom{m}{i} \int_0^1 v^i (1 - v)^{m-i} \mathtt{dbeta}(v; n - \ell + 1, \ell)\, dv. \qquad (\text{G.24})
\end{aligned}
$$

Note that (G.24) is a special case of (G.22) when $u = n + 1$. Similar to the derivation of (G.23), here (G.24) implies

$$
\begin{aligned}
\mathrm{CPKM}(n, \ell, n + 1, k, m) &= \int_0^1 \mathtt{pbeta}(v; k, m - k + 1) \mathtt{dbeta}(v; n - \ell + 1, \ell)\, dv \\
&= 1 - \int_0^1 \mathtt{pbeta}(v; n - \ell + 1, \ell) \mathtt{dbeta}(v; k, m - k + 1)\, dv \\
&= \mathtt{pnhyper}(n - \ell; k, m, m + n).
\end{aligned}
$$

Thus, operationally, the coverage probabilities for the lower prediction bound are obtained from (G.23), setting $u = n + 1$.

- Similarly, for the upper-prediction bound $\widetilde{Y}_{k:m} = x_{(u)}$, set $\ell = 0$ in (G.23) to get $\Pr(K = i; u) = \mathtt{dnhyper}(i; u, n, m + n)$ and

$$
\begin{aligned}
\mathrm{CPKM}(n, 0, u, k, m) &= \int_0^1 \mathtt{pbeta}(v; k, m - k + 1)\mathtt{dbeta}(v; u, n - u + 1)\, dv \\
&= 1 - \int_0^1 \mathtt{pbeta}(v; u, n - u + 1)\mathtt{dbeta}(v; k, m - k + 1)\, dv \\
&= \mathtt{pnhyper}(u - 1; k, m, m + n).
\end{aligned}
$$

Basic Results from Bayesian Inference Models

INTRODUCTION

This appendix describes and provides background information and results used mainly in the construction of Bayesian intervals in Chapters 15–17. The appendix contains:

- Some useful technical results that are used to obtain the posterior distributions and posterior predictive distributions in this appendix (Section H.1).

- A formal presentation of Bayes' theorem (Section H.2).

- The definition of conjugate prior distributions with examples for the binomial, Poisson, and normal distributions (Section H.3).

- The definition of Jeffreys prior distributions with examples for the binomial, Poisson, and normal distributions. Also a modified Jeffreys prior distribution for the normal distribution (Section H.4).

- The definition of posterior predictive distributions with examples, using conjugate prior distributions, for the binomial, Poisson, and normal distributions (Section H.5).

- Posterior predictive distributions for the binomial and the Poisson distributions using Jeffreys prior distributions. Also a posterior predictive distribution for the normal distributions using a modified Jeffreys prior distribution (Section H.6).

Statistical Intervals: A Guide for Practitioners and Researchers, Second Edition.
William Q. Meeker, Gerald J. Hahn and Luis A. Escobar.
© 2017 John Wiley & Sons, Inc. Published 2017 by John Wiley & Sons, Inc.
Companion Website: www.wiley.com/go/meeker/intervals

H.1   BASIC RESULTS USED IN THIS APPENDIX

In this appendix we use the following results:

(a) We frequently need to compute the marginal distribution corresponding to a mixture of two normal distributions. The following result is useful in making those computations.

Consider the continuous normal-normal mixture $(Y|\mu) \sim \text{NORM}(\mu, \sigma)$ and $\mu \sim \text{NORM}(\tau, \sigma/\sqrt{\nu})$, where $f(y|\mu) = \text{dnorm}(y; \mu, \sigma)$, $f(\mu) = \text{dnorm}(\mu; \tau, \sigma/\nu)$, and τ, σ, and ν are constants. The interest is in the marginal distribution of Y. The joint pdf of (Y, μ) is $f(y|\mu)f(\mu)$ and the marginal pdf of Y is

$$
\begin{aligned}
f(y; \tau, \sigma, \nu) &= \int_{-\infty}^{\infty} f(y|\mu)\, f(\mu) d\mu \\
&= \int_{-\infty}^{\infty} \text{dnorm}(y; \mu, \sigma)\, \text{dnorm}(\mu; \tau, \sigma/\nu)\, d\mu \\
&= \text{dnorm}\big[y; \tau, \sigma(1 + 1/\nu)^{1/2}\big]
\end{aligned}
\tag{H.1}
$$

(see Villa and Escobar, 2006, for details). Note that the marginal distribution of Y is a normal distribution with mean equal to the expected value of μ and variance equal to the sum of the variances $\text{Var}(Y|\mu)$ and $\text{Var}(\mu)$.

(b) The following two identities are integrals that occur several times in this appendix and are given here to simplify the development. These identities can be verified directly after making the change of variables $w = 1/\sigma^2$ in evaluating the integrals. Suppose that D is positive and does not depend on σ. Then

$$
\int_0^{\infty} \frac{1}{\sigma^r} \exp\left(-\frac{D}{\sigma^2}\right) d\sigma = \frac{\Gamma[(r-1)/2]}{2\, D^{(r-1)/2}},
\tag{H.2}
$$

where $r > 1$. Also

$$
\int_0^{\infty} \frac{1}{\sigma^r} \exp\left(-\frac{D}{\sigma^2}\right) d\sigma^2 = \frac{\Gamma[(r-2)/2]}{D^{(r-2)/2}},
\tag{H.3}
$$

where $r > 2$.

H.2   BAYES' THEOREM

Bayes' theorem (also known as Bayes' rule) is the basis for the Bayesian method of statistical inference. The rule allows one to combine available data with prior information (expressed as a probability distribution) to obtain a posterior (or updated) distribution as follows.

Suppose that $f(x|\theta)$ is the joint distribution of the data x as a function of the parameters θ and that $f(\theta)$ is a completely specified joint prior pdf of θ. Then using Bayes' theorem, the joint posterior distribution of θ as a function of the data is

$$
f(\theta|x) = \frac{f(x|\theta)f(\theta)}{\int f(x|\theta)f(\theta)d\theta}.
\tag{H.4}
$$

The integral in the denominator of (H.4) is a "normalizing constant" which is computed over the region where $f(\theta) > 0$. Bayes' theorem also applies when $f(\theta)$ is a discrete distribution. In that case the integral in the denominator of (H.4) is replaced with $\sum f(x|\theta)f(\theta)$, where the sum is over all values of θ for which $f(\theta) > 0$. In computing $f(\theta|x)$, the most challenging task is usually in the computation of the normalizing constant.

H.3 CONJUGATE PRIOR DISTRIBUTIONS

In principle, the parametric form of $f(\theta)$ should be chosen based on past experience, knowledge of the process being studied, or a plausible subjective argument. Sometimes, however, $f(\theta)$ is chosen for convenience of the computations, for ease of interpreting the prior information as additional data, or because the choice provides the basis for a simple but useful approximation of more realistic prior distributions. For these purposes, conjugate prior distributions are useful.

Informally, $f(\theta)$ is a conjugate prior distribution for $f(x|\theta)$ if the posterior distribution $f(\theta|x)$ has the same parametric form as $f(\theta)$. That is, if $f(x|\theta)$ is a class of joint distributions for x as a function of θ and $f(\theta)$ is a class of joint prior distributions for θ. Then $f(\theta)$ is conjugate for $f(x|\theta)$ if $f(\theta|x)$ is an element of the class $f(x|\theta)$ for all choices of $f(x|\theta)$ and $f(\theta)$, respectively. Next we describe some well-known conjugate prior distributions.

H.3.1 Conjugate Prior Distribution for the Binomial Distribution

Suppose that $X \sim \mathrm{BINOM}(n, \pi)$ is the number of conforming units in n independent Bernoulli trials. Then $\theta = \pi$ and the distribution of number of conforming units is

$$\Pr(X = x) = f(x|\pi) = \binom{n}{x} \pi^x (1 - \pi)^{n-x}, \tag{H.5}$$

where $0 < \pi < 1$, n is a given positive integer, and $0 \leq x \leq n$.

We now show that the beta prior distribution,

$$f(\pi) = \frac{\Gamma(a+b)}{\Gamma(a)\,\Gamma(b)} \pi^{a-1} (1 - \pi)^{b-1},$$

is a conjugate prior distribution for $f(x|\pi)$ in (H.5). Here $a > 0$ and $b > 0$ are given hyper-parameters. The joint distribution for (π, x) is $f(\pi, x) = f(x|\pi)f(\pi)$. Then the posterior distribution is

$$f(\pi|x) = \frac{f(\pi, x)}{f(x)} = C\,\pi^{x+a-1}(1 - \pi)^{n-x+b-1},$$

where C is a constant that does not depend on π. We find C using the constraint that $\int_0^1 f(\pi|x)\,d\pi = 1$ giving the posterior distribution,

$$f(\pi|x) = \frac{\Gamma(n+a+b)}{\Gamma(x+a)\,\Gamma(n-x+b)} \pi^{x+a-1}(1 - \pi)^{n-x+b-1}$$
$$= \mathtt{dbeta}(\pi; x + a, n - x + b). \tag{H.6}$$

This shows that the family of beta prior distributions $\mathrm{BETA}(a, b)$ is conjugate for the binomial $\mathrm{BINOM}(n, \pi)$ distribution. Note that $f(\pi|x)$ in (H.6) is proportional to a binomial probability in which there are $x + a - 1$ conforming units and $n - x + b - 1$ nonconforming units. Thus the posterior distribution can be interpreted as a binomial distribution likelihood after observing $x + a - 1$ nonconforming units in $n + a + b - 2$ trials.

The beta prior distribution $f(\pi)$ is also conjugate for the negative binomial distributions in Section C.4.3.

H.3.2 Conjugate Prior Distribution for the Poisson Distribution

Suppose that $X \sim \text{POIS}(n\lambda)$, where X is the number of events during a specified amount of exposure n, with $n > 0$ (not necessarily an integer) and $\lambda > 0$. Then $\boldsymbol{\theta} = \lambda$ and the distribution of the number of events is

$$\Pr(X = x) = f(x|\lambda) = \frac{(n\lambda)^x}{x!} \exp(-n\lambda), \tag{H.7}$$

where x is a nonnegative integer.

We now show that the prior distribution $\lambda \sim \text{GAM}(a, b)$ is a conjugate prior distribution for the Poisson distribution in (H.7). The gamma prior distribution pdf is

$$f(\lambda) = \frac{b^a \, \lambda^{a-1}}{\Gamma(a)} \exp(-b\lambda),$$

where $a > 0$, $b > 0$, and $\Gamma(z)$ is the gamma function defined in Appendix A.

The joint distribution for (λ, x) is $f(\lambda, x) = f(x|\lambda)f(\lambda)$. Then the posterior distribution for λ is

$$f(\lambda|x) = \frac{f(\lambda, x)}{f(x)} = C \, \lambda^{x+a-1} \exp[-(n+b)\lambda],$$

where C is a constant that does not depend on λ. We find C using the constraint that $\int_0^\infty f(\lambda|x) \, d\lambda = 1$. This gives the posterior distribution

$$f(\lambda|x) = \frac{(n+b)^{x+a} \, \lambda^{x+a-1}}{\Gamma(x+a)} \exp[-(n+b)\lambda] = \texttt{dgamma}(\lambda; x + a, n + b). \tag{H.8}$$

This posterior distribution can be interpreted as a Poisson distribution likelihood with an occurrence rate λ after observing $x + a - 1$ events during $n + b$ units of exposure.

H.3.3 Conjugate Prior Distribution for the Normal Distribution

For a random sample $\boldsymbol{x} = (x_1, \ldots, x_n)$ from a $\text{NORM}(\mu, \sigma)$ distribution, the likelihood is

$$f(\boldsymbol{x}|\mu, \sigma) = \left(\frac{1}{\sqrt{2\pi\sigma^2}} \right)^n \exp\left(-\frac{1}{\sigma^2} \left[\sum_{i=1}^n (x_i - \mu)^2 \right] \right).$$

Using the identity $\sum_{i=1}^n (x_i - \mu)^2 = n(\bar{x} - \mu)^2 + (n-1)s^2$, we can write the normal distribution likelihood as a function of the sufficient statistics (sample mean $\bar{x}$ and sample variance s^2):

$$f(\boldsymbol{x}|\mu, \sigma) = \left(\frac{1}{\sqrt{2\pi\sigma^2}} \right)^n \exp\left(-\frac{1}{\sigma^2} \left[\frac{n(\bar{x} - \mu)^2 + (n-1)s^2}{2} \right] \right). \tag{H.9}$$

A conjugate prior distribution for (μ, σ^2) is

$$\mu|\sigma^2 \sim \text{NORM}(\mu_0, \sigma/\sqrt{n_0}) \quad \text{and} \quad \sigma^2 \sim \text{Inv-}\chi^2(r_0, \sigma_0^2), \tag{H.10}$$

where μ_0, n_0, r_0, and σ_0 are given constants (see Gelman et al., 2013, Chapter 3, for details). The pdf of $\mu|\sigma^2$ is $f(\mu|\sigma^2) = \texttt{dnorm}(\mu; \mu_0, \sigma/\sqrt{n_0})$, and $\text{Inv-}\chi^2(r_0, \sigma_0^2)$ is the distribution of the scaled ratio $\sigma_0^2 r_0 / X_{(r_0)}^2$, where $X_{(r_0)}^2 \sim \chi^2(r_0)$. The pdf of the $\text{Inv-}\chi^2(r_0, \sigma_0^2)$ distribution is

$$f(\sigma^2) = \frac{1}{\Gamma(r_0/2)} \frac{1}{\sigma^2} \left(\frac{r_0 \sigma_0^2}{2\sigma^2} \right)^{r_0/2} \exp\left(-\frac{r_0 \sigma_0^2}{2\sigma^2} \right). \tag{H.11}$$

In (H.10), the prior distribution specification n_0 can be interpreted as the effective number of observations providing information about μ, and r_0 as the effective number of degrees of freedom providing information about σ^2.

The joint prior distribution for (μ, σ^2) is $f(\mu, \sigma^2) = f(\mu|\sigma^2) f(\sigma^2)$. The joint posterior distribution is given by $f(\mu, \sigma^2|\boldsymbol{x}) = f(\mu|\sigma^2, \boldsymbol{x}) f(\sigma^2|\boldsymbol{x})$, where

$$f(\mu|\sigma^2, \boldsymbol{x}) = \mathrm{dnorm}(\mu; \mu_n, \sigma/\sqrt{n_n}),$$

$$f(\sigma^2|\boldsymbol{x}) = \frac{1}{\Gamma(r_n/2)} \frac{1}{\sigma^2} \left(\frac{r_n \sigma_n^2}{2\sigma^2} \right)^{r_n/2} \exp\left(-\frac{r_n \sigma_n^2}{2\sigma^2} \right) \tag{H.12}$$

(see Gelman et al., 2013, page 68, for details), and the updated parameters are

$$\mu_n = \frac{n_0}{n_0 + n}\mu_0 + \frac{n}{n_0 + n}\bar{x}, \quad n_n = n_0 + n, \quad r_n = r_0 + n,$$

$$\sigma_n^2 = \frac{1}{r_n} \left[n_0 \sigma_0^2 + (n-1)s^2 + \frac{n_0 n}{n_0 + n}(\bar{x} - \mu_0)^2 \right]. \tag{H.13}$$

In summary,

$$\mu|\sigma^2, \boldsymbol{x} \sim \mathrm{NORM}(\mu_n, \sigma/\sqrt{n_n}), \quad \sigma^2|\boldsymbol{x} \sim \mathrm{Inv}\text{-}\chi^2\left(r_n, \sigma_n^2\right),$$

$$\left. \frac{\mu - \mu_n}{\sigma_n/\sqrt{n_n}} \right| \boldsymbol{x} \sim t(r_n). \tag{H.14}$$

Note that the marginal posterior distribution for μ is a location-scale distribution with location μ_n, scale $\sigma_n/\sqrt{n_n}$, and with a Student's t-distribution with r_n degrees of freedom as the standardized distribution.

Consequently, using (H.14), to draw a random pair (μ^*, σ^*) from the joint posterior distribution $(\mu, \sigma|\boldsymbol{x})$, we proceed as follows: First, draw w from a $\chi^2(r_n)$ distribution and compute $\sigma^* = (r_n \sigma_n^2/w)^{1/2}$. Then draw μ^* from a $\mathrm{NORM}(\mu_n, \sigma^*/\sqrt{n_n})$ distribution.

H.4 JEFFREYS PRIOR DISTRIBUTIONS

Suppose that $f(\boldsymbol{x}|\boldsymbol{\theta})$ is the likelihood of the data. Consider the Fisher information matrix, $I_{\boldsymbol{\theta}}$, defined in (D.14). The Jeffreys diffuse prior distribution (referred to henceforth as the Jeffreys prior distribution) is

$$f(\boldsymbol{\theta}) \propto [\det(I_{\boldsymbol{\theta}})]^{1/2},$$

where $\det(I_{\boldsymbol{\theta}})$ is the determinant of $I_{\boldsymbol{\theta}}$. Jeffreys prior distributions for single-parameter models are important because they generally lead to statistical interval procedures that have frequentist coverage probabilities close (and in some cases exactly equal) to the nominal credible level.

The Jeffreys prior distribution is invariant to one-to-one transformations of $\boldsymbol{\theta}$ in the following sense. Suppose that $\boldsymbol{\nu} = \boldsymbol{h}(\boldsymbol{\theta})$ is a one-to-one transformation of $\boldsymbol{\theta}$ and that $f(\boldsymbol{\theta})$ is the prior distribution for $\boldsymbol{\theta}$. We can use this distribution to obtain a prior distribution for $\boldsymbol{\nu}$ in two alternative but equivalent ways: (a) Let $\boldsymbol{g}(\boldsymbol{\nu})$ be the inverse function of $\boldsymbol{h}(\boldsymbol{\theta})$. Then using transformation of variables, the implied prior distribution for $\boldsymbol{\nu}$ is proportional to

$$|J(\boldsymbol{\nu})|f[\boldsymbol{g}(\boldsymbol{\nu})] = |J(\boldsymbol{\nu})|[\det(I_{\boldsymbol{\nu}})]^{1/2}, \tag{H.15}$$

where $J(\boldsymbol{\nu})$ is the Jacobian for $\boldsymbol{\nu}$ as defined in (D.1). (b) Compute the Fisher information for $\boldsymbol{\nu}$, say $I_{\boldsymbol{\nu}}$, using (D.16). Then, using a well-known result about the determinant of a product of square matrices (e.g., Harville, 1997, page 187), the Jeffreys prior distribution for $\boldsymbol{\nu}$ is proportional to

$$\{\det[J(\boldsymbol{\nu})\,I_{\boldsymbol{\theta}}\,J'(\boldsymbol{\nu})]\}^{1/2} = |J(\boldsymbol{\nu})|[\det(I_{\boldsymbol{\nu}})]^{1/2}. \tag{H.16}$$

Because of the equality of (H.15) and (H.16), the Jeffreys prior distribution is invariant to one-to-transformations of the parameters.

H.4.1   Jeffreys Prior Distribution for the Binomial Distribution

For the binomial distribution $\mathrm{BINOM}(n, \pi)$, the Fisher information is $n/[\pi(1-\pi)]$, as given in (D.17). Then the Jeffreys prior distribution for π is $f(\pi) \propto \pi^{-1/2}(1-\pi)^{-1/2}$. Or equivalently,

$$f(\pi) = C\,\pi^{-1/2}(1-\pi)^{-1/2} = \mathtt{dbeta}(1/2, 1/2),$$

where $C = 1/\mathcal{B}(1/2, 1/2)$ and $\mathcal{B}(a,\,b)$ is the beta function defined in Appendix A. As shown in (H.6), $\mathtt{dbeta}(1/2, 1/2)$ is a conjugate prior distribution for the $\mathtt{dbinom}(x; n, \pi)$ distribution. Thus using (H.6), the posterior distribution for π is $f(\pi|x) = \mathtt{dbeta}(\pi; x + 1/2, n - x + 1/2)$. This provides the basis for the Jeffreys confidence interval method in Section 6.2.5.

H.4.2   Jeffreys Prior Distribution for the Poisson Distribution

For Poisson data $X \sim \mathrm{POIS}(n\lambda)$, the Fisher information is n/λ, as given in (D.18). Then the Jeffreys prior distribution for λ is $f(\lambda) \propto 1/\sqrt{\lambda}$. In this case, the Jeffreys prior distribution is an improper prior distribution because $\int_0^\infty \lambda^{-1/2}\,d\lambda = \infty$. Using Bayes' theorem, however, the posterior distribution is proper. That is,

$$f(\lambda|\boldsymbol{x}) \propto f(\boldsymbol{x}|\lambda)f(\lambda) = C\,\lambda^{x-1/2}\exp(-n\lambda).$$

The constant C is obtained from the constraint $\int_0^\infty f(\lambda|\boldsymbol{x})\,d\lambda = 1$, giving $C = n^{x+1/2}/\Gamma(x+1/2)$. Thus, the posterior distribution for λ is $f(\lambda|\boldsymbol{x}) = \mathtt{dgamma}(\lambda; x + 1/2, n)$. This provides the basis for the Jeffreys confidence interval method in Section 7.2.5.

H.4.3   Jeffreys Prior Distribution for the Normal Distribution

The following are examples of Jeffreys prior distributions for the normal distribution.

Example H.1   Jeffreys Prior Distribution for μ with a $\mathrm{NORM}(\mu, \sigma)$ Distribution and Given σ. From the Fisher information $I_\mu = n/\sigma^2$ in (D.20) and the fact that n and σ are given, the Jeffreys prior distribution is

$$f(\mu) \propto 1$$

which is an improper prior distribution. Using Bayes' theorem, however, the posterior distribution is proper, as shown next. Because σ is given, the likelihood is

$$f(\boldsymbol{x}|\mu) \propto \exp\left[-\frac{n}{2\sigma^2}(\bar{x} - \mu)^2\right].$$

Then the joint distribution of $(\mu, \boldsymbol{x})$ is $f(\mu, \boldsymbol{x}) \propto f(\mu)f(\boldsymbol{x}|\mu)$ and $f(\mu|\boldsymbol{x}) = Cf(\mu, \boldsymbol{x})$, where the constant C is chosen such that $f(\mu|\boldsymbol{x})$ is a pdf for μ. This gives $C = \sqrt{n}/(\sigma\sqrt{2\pi})$. That is, the posterior distribution for μ is

$$f(\mu|\boldsymbol{x}) = \mathtt{dnorm}\big(\mu; \bar{x}, \sigma/\sqrt{n}\big). \qquad \blacksquare$$

Example H.2 Jeffreys Prior Distribution for σ^2 with a $\mathrm{NORM}(\mu, \sigma)$ Distribution and Given μ. From the Fisher information $I_{\sigma^2} = n/(2\sigma^4)$ given in (D.22) and the fact that n is given, the Jeffreys prior distribution is

$$f(\sigma^2) \propto \frac{1}{\sigma^2}$$

which is an improper prior distribution. Using Bayes' theorem, however, the posterior distribution is proper, as shown next. Because μ is given, the likelihood is

$$f(\boldsymbol{x}|\sigma^2) \propto \frac{1}{\sigma^n} \exp\left[-\frac{1}{2\sigma^2} \sum_{i=1}^{n} (x_i - \mu)^2\right] \propto \frac{1}{\sigma^n} \exp\left(-\frac{n\widehat{\sigma}^2}{2\sigma^2}\right), \qquad (\text{H.17})$$

where $\widehat{\sigma}^2 = \sum_{i=1}^{n}(x_i - \mu)^2/n$ is the ML estimate of σ^2. Then the posterior distribution for σ^2 is $f(\sigma^2|\boldsymbol{x}) = f(\sigma^2)f(\boldsymbol{x}|\sigma^2)$. That is,

$$f(\sigma^2|\boldsymbol{x}) = \frac{C}{\sigma^{n+2}} \exp\left(-\frac{n\widehat{\sigma}^2}{2\sigma^2}\right).$$

Using the constraint that $f(\sigma^2|\boldsymbol{x})$ must integrate to 1 and (H.3) with $r = n + 2$ and using $D = n\widehat{\sigma}^2/2$ to compute C leads to

$$f(\sigma^2|\boldsymbol{x}) = \frac{1}{\Gamma(n/2)} \frac{1}{\sigma^2} \left(\frac{n\widehat{\sigma}^2}{2\sigma^2}\right)^{n/2} \exp\left(-\frac{n\widehat{\sigma}^2}{2\sigma^2}\right). \qquad (\text{H.18})$$

This shows that $\sigma^2|\boldsymbol{x} \sim \mathrm{Inv}\text{-}\chi^2(n, \widehat{\sigma}^2)$ (see (H.11)). ∎

Example H.3 Jeffreys Prior Distribution for σ with a $\mathrm{NORM}(\mu, \sigma)$ Distribution and Given μ. From the information $I_\sigma = 2n/\sigma^2$ given in (D.21), the Jeffreys prior distribution for σ is $f(\sigma) \propto 1/\sigma$ which is an improper prior distribution. Using Bayes' theorem, however, the posterior distribution is proper, as shown next. The likelihood is, as in (H.17),

$$f(\boldsymbol{x}|\sigma) \propto \frac{1}{(\sigma^2)^{n/2}} \exp\left(-\frac{n\widehat{\sigma}^2}{2\sigma^2}\right),$$

where $\widehat{\sigma} = \sum_{i=1}^{n}(x_i - \mu)^2/n$. Because of the Jeffreys invariance property discussed in Section H.4, we obtain the posterior distribution $f(\sigma|\boldsymbol{x})$ using $f(\sigma^2|\boldsymbol{x})$ in (H.18) and the transformation $\sigma = \sqrt{\sigma^2}$. This gives

$$f(\sigma|\boldsymbol{x}) = \frac{2}{\Gamma(n/2)} \frac{1}{\sigma} \left(\frac{n\widehat{\sigma}^2}{2\sigma^2}\right)^{n/2} \exp\left(-\frac{n\widehat{\sigma}^2}{2\sigma^2}\right). \qquad ∎$$

Example H.4 Jeffreys Prior Distribution for (μ, σ) with a $\mathrm{NORM}(\mu, \sigma)$ Distribution. From the Fisher matrix $I_{(\mu, \sigma)}$ given in (D.23), the Jeffreys prior distribution is

$$f(\mu, \sigma) \propto \frac{1}{\sigma^2}$$

which is an improper prior distribution. The posterior distribution is, as shown next, a proper distribution if $n \geq 2$. The likelihood is

$$f(\boldsymbol{x}|\mu, \sigma) \propto \frac{1}{\sigma^n} \exp\left(-\frac{A}{\sigma^2}\right),$$

where $A = [n(\bar{x} - \mu)^2 + (n-1)s^2]/2$. Consequently, the joint distribution of $(\boldsymbol{x}, \mu, \sigma)$ is

$$f(\boldsymbol{x}, \mu, \sigma) \propto \frac{1}{\sigma^{n+2}} \exp\left(-\frac{A}{\sigma^2}\right).$$

From this joint distribution, the marginal posterior distributions are

$$f(\mu|\boldsymbol{x}, \sigma) = C_1 \frac{1}{\sigma} \exp\left[-\frac{n}{2\sigma^2}(\mu - \bar{x})^2\right], \quad f(\sigma|\boldsymbol{x}) = C_2 \frac{1}{\sigma^{n+1}} \exp\left[-\frac{(n-1)s^2}{2\sigma^2}\right],$$

where the constants C_1 and C_2 are chosen such that the corresponding posterior distribution integrates to 1. Directly, one gets $C_1 = \sqrt{n}/\sqrt{2\pi}$. From (H.2) with $r = n + 1$ and using $D = (n-1)s^2/2$ gives $C_2 = 2[(n-1)s^2/2]^{n/2}$. Thus,

$$f(\mu|\boldsymbol{x}, \sigma) = \frac{\sqrt{n}}{\sigma\sqrt{2\pi}} \exp\left[-\frac{n}{2\sigma^2}(\mu - \bar{x})^2\right],$$

$$f(\sigma|\boldsymbol{x}) = \frac{2}{\Gamma(n/2)} \left[\frac{(n-1)s^2}{2}\right]^{n/2} \frac{1}{\sigma^{n+1}} \exp\left[-\frac{(n-1)s^2}{2\sigma^2}\right]. \tag{H.19}$$

Also,

$$f(\mu|\boldsymbol{x}) \propto \int_0^\infty \frac{1}{\sigma^{n+2}} \exp\left(-\frac{A}{\sigma^2}\right) d\sigma.$$

This last integral is evaluated using (H.2), which leads to

$$f(\mu|\boldsymbol{x}) = \frac{\sqrt{n}}{s\sqrt{(n-1)/n}} \frac{\Gamma[(n+1)/2]}{\Gamma(n/2)\sqrt{\pi n}} \left(1 + \frac{1}{n}\left[\frac{\sqrt{n}(\mu - \bar{x})}{s\sqrt{(n-1)/n}}\right]^2\right)^{-(n+1)/2}.$$

That is, given the data, the marginal posterior distribution $f(\mu|\boldsymbol{x})$ implies that

$$\frac{\sqrt{n}(\mu - \bar{x})}{s\sqrt{(n-1)/n}} \sim t(n).$$

That is, $f(\mu|\boldsymbol{x})$ is a location-scale distribution with location $\bar{x}$, scale $(s/\sqrt{n})\sqrt{(n-1)/n}$, and a Student's t-distribution with n degrees of freedom as the standardized distribution.

The fact that the marginal distribution $f(\mu|\boldsymbol{x})$ has n degrees of freedom, rather than $n-1$ as suggested by other statistical considerations, was one of the reasons for Jeffreys to recommend against the use of this prior distribution (i.e., $f(\mu, \sigma) \propto 1/\sigma^2$) for the normal distribution (see Jeffreys, 1946, for details). See Kass and Wasserman (1996) for a related discussion of the issues with the Jeffreys prior distribution for the normal distribution. We do not use this Jeffreys prior distribution for the normal distribution in this book. ∎

Example H.5   A Modified Jeffreys Prior Distribution for (μ, σ^2) with a $\mathrm{NORM}(\mu, \sigma)$ Distribution. Here we consider the prior distribution $f(\mu, \sigma) = (1/\sigma)$ or equivalently $f(\mu, \sigma^2) = 1/\sigma^2$, which was suggested by Jeffreys to address the deficiencies encountered with the original Jeffreys prior distribution, discussed earlier in Example H.4. Note that $f(\mu, \sigma^2) \propto 1/\sigma^2$ is the product of the Jeffreys prior distributions $f(\mu) \propto 1$ and $f(\sigma^2) \propto 1/\sigma^2$. In this case, as shown in (H.9), the likelihood is

$$f(\boldsymbol{x}|\mu, \sigma^2) = \left(\frac{1}{\sqrt{2\pi\sigma^2}}\right)^n \exp\left(-\frac{A}{\sigma^2}\right),$$

where $A = [n(\bar{x} - \mu)^2 + (n-1)s^2]/2$. Consequently, the joint distribution of $(\boldsymbol{x}, \mu, \sigma^2)$ is

$$f(\boldsymbol{x}, \mu, \sigma^2) \propto \frac{1}{\sigma^{n+2}} \exp\left(-\frac{A}{\sigma^2}\right).$$

Similar to the development leading to (H.19), we obtain the posterior distributions

$$f(\mu|\boldsymbol{x}, \sigma^2) = \frac{\sqrt{n}}{\sigma\sqrt{2\pi}} \exp\left[-\frac{n}{2\sigma^2}(\mu - \bar{x})^2\right],$$

$$f(\sigma^2|\boldsymbol{x}) = \frac{1}{\Gamma[(n-1)/2]} \left[\frac{(n-1)s^2}{2\sigma^2}\right]^{(n-1)/2} \frac{1}{\sigma^2} \exp\left[-\frac{(n-1)s^2}{2\sigma^2}\right].$$

(H.20)

Thus

$$f(\mu|\boldsymbol{x}) = C \int_0^\infty \frac{1}{\sigma^{n+2}} \exp\left(-\frac{A}{\sigma^2}\right) d\sigma^2,$$

(H.21)

where

$$C = \frac{\sqrt{n}}{\sqrt{2\pi}} \frac{1}{\Gamma[(n-1)/2]} \left[\frac{(n-1)s^2}{2}\right]^{(n-1)/2}.$$

The integral in (H.21) is evaluated using (H.3), which leads to

$$f(\mu|\boldsymbol{x}) = \frac{1}{s/\sqrt{n}} \frac{\Gamma(n/2)}{\Gamma[(n-1)/2] \sqrt{\pi(n-1)}} \left[1 + \frac{1}{n-1}\left(\frac{\mu - \bar{x}}{s/\sqrt{n}}\right)^2\right]^{(-n/2)}.$$

In summary, using (H.11),

$$\mu|\sigma^2, \boldsymbol{x} \sim \text{NORM}\left(\bar{x}, \sigma/\sqrt{n}\right), \quad \sigma^2|\boldsymbol{x} \sim \text{Inv-}\chi^2\left(n-1, s^2\right).$$

(H.22)

Credible interval for μ

Combining the results in (H.22) gives

$$\frac{\mu - \bar{x}}{s/\sqrt{n}} \,\bigg|\, \boldsymbol{x} \sim t(n-1),$$

which implies

$$\Pr\left(t_{(\alpha/2, n-1)} \leq \frac{\mu - \bar{x}}{s/\sqrt{n}} \leq t_{(1-\alpha/2, n-1)}\right) = 1 - \alpha.$$

(H.23)

Then solving (H.23) for μ in the middle leads to

$$[\underset{\sim}{\mu}, \ \widetilde{\mu}] = \bar{x} \mp t_{(1-\alpha/2; n-1)} \frac{s}{\sqrt{n}}.$$

This $100(1 - \alpha)\%$ credible interval is the same as the non-Bayesian confidence interval for μ in Section 4.2.

Credible interval for σ

To obtain a credible interval for σ, we use the $1 - \alpha/2$ quantile, $(n-1)s^2/\chi^2_{(1-\alpha/2;n-1)}$, and the $\alpha/2$ quantile, $(n-1)s^2/\chi^2_{(\alpha/2;n-1)}$, of the posterior distribution $(\sigma^2|x)$ in (H.22). Solving for σ gives

$$[\underset{\sim}{\sigma},\ \widetilde{\sigma}] = \left[s\left(\frac{n-1}{\chi^2_{(1-\alpha/2;n-1)}} \right)^{1/2}, \quad s\left(\frac{n-1}{\chi^2_{(\alpha/2;n-1)}} \right)^{1/2} \right].$$

This $100(1-\alpha)\%$ credible interval is the same as the non-Bayesian confidence interval for σ in Section 4.3.

Sample draws from the joint posterior distribution of μ and σ

Note that $f(\mu,\sigma^2|x) = f(\mu|x,\sigma^2)\, f(\sigma^2|x)$, where the conditional distributions for $\mu|\sigma^2, x$ and $\sigma^2|x$ are given in (H.22). Thus to draw a random pair (μ^*, σ^*) from the posterior distribution $\mu, \sigma|x$, we proceed as follows: First, draw w at random from a $\chi^2(n-1)$ distribution and compute $\sigma^* = [(n-1)s^2/w]^{1/2}$. Then draw μ^* at random from a $\mathrm{NORM}(\bar{x}, \sigma^*/\sqrt{n})$ distribution. Such draws will be useful computing more complicated Bayesian intervals such as tolerance intervals (see Sections 15.5.2 and 15.5.3) and k-out-of-m prediction intervals (see Section 15.5.4). When compared with the MCMC method, these draws are easier to obtain and are iid. ∎

H.5   POSTERIOR PREDICTIVE DISTRIBUTIONS

H.5.1   General Results

In Bayesian inference, we have data $x = (x_1, \ldots, x_n)$ from $X \sim f(x|\theta)$, a prior distribution $f(\theta)$ for θ, and future observations $y = (y_1, \ldots, y_m)$ from $Y \sim f(y|\theta)$. In many applications, it is plausible to assume that, for given θ, the data X and the future observation Y are independent. Of interest is the posterior predictive distribution of the future observation Y given the data x. For this purpose (assuming that θ has a continuous distribution and ignoring irrelevant constants) the joint distribution of (Y, X, θ) is

$$f(y, x, \theta) = f(y, x|\theta)\, f(\theta) = f(y|\theta)\, f(x|\theta)\, f(\theta) = f(y|\theta)\, f(\theta|x) f(x). \quad \text{(H.24)}$$

Integrating both sides of (H.24) with respect to θ gives $f(y, x) = f(x) \int f(y|\theta)\, f(\theta|x)\, d\theta$. Consequently, the posterior predictive distribution $f(y|x)$ for $Y|X = x$ is

$$f(y|x) = \frac{f(y, x)}{f(x)} = \int f(y|\theta)\, f(\theta|x)\, d\theta. \quad \text{(H.25)}$$

When it is difficult to sample directly from $f(y|x)$, but one can sample from both $f(y|\theta)$ and $f(\theta|x)$, one can sample indirectly from $f(y|x)$ using a Monte Carlo procedure.

Note that $f(y|x) = \mathrm{E}[f(y|\theta)]$, where the expectation is with respect to the posterior distribution $f(\theta|x)$. Then, if the interest is in $\Pr(Y = y|x)$ for a specific value y, one can proceed as follows. Let $\theta_1^*, \ldots, \theta_B^*$ be approximately independent draws from $f(\theta|x)$. Then the mean $\sum_j^B f(y|\theta_j^*)/B$ approximates $f(y|x)$ (see Hoff, 2009, Chapter 4, for details). This is the method described in Section 15.5.5 and used to obtain prediction intervals in Sections 16.1.5, 16.2.5, and 16.3.6.

The following three subsections provide simple expressions for the posterior predictive distributions using conjugate prior distributions for the binomial, Poisson, and normal distributions. Section H.6 gives posterior predictive distributions using Jeffreys conjugate prior distributions.

H.5.2 Posterior Predictive Distribution for the Binomial Distribution Based on a Conjugate Prior Distribution

The relationship in (H.6) shows that for $X \sim \text{BINOM}(n; \pi)$ and $\pi \sim \text{BETA}(a, b)$, the conjugate posterior distribution for π is $f(\pi|x) = \texttt{dbeta}(\pi; x + a, n - x + b)$. Thus, given x, the posterior predictive distribution for a future observation from $Y \sim \text{BINOM}(m, \pi)$ is a beta-binomial distribution with pmf given by

$$
\begin{aligned}
f(y|x) &= \int_0^1 \texttt{dbinom}(y; m, \pi)\, \texttt{dbeta}(\pi; x + a, n - x + b)\, d\pi \\
&= \frac{m!}{y!(m - y)!} \int_0^1 \frac{\pi^{y+x+a-1}(1 - \pi)^{m-y+n-x+b-1}}{\mathcal{B}(x + a, n - x + b)}\, d\pi \\
&= \frac{1}{(m + 1)\, \mathcal{B}(y + 1, m - y + 1)} \frac{\mathcal{B}(y + x + a, m - y + n - x + b)}{\mathcal{B}(x + a, n - x + b)} \\
&= \texttt{dbetabinom}(y; m, x + a, n - x + b),
\end{aligned} \tag{H.26}
$$

where y is a nonnegative integer and $y \leq m$.

H.5.3 Posterior Predictive Distribution for the Poisson Distribution Based on a Conjugate Prior Distribution

The relationship in (H.8) shows that for $X \sim \text{POIS}(n\lambda)$ and $\lambda \sim \text{GAM}(a, b)$, the conjugate posterior distribution for λ is $f(\lambda|x) = \texttt{dgamma}(\lambda; x + a, n + b)$. Thus, given x, the posterior predictive distribution for the number of events $Y \sim \text{POIS}(m\lambda)$ during a specified future amount of exposure $m > 0$ for events occurring at a rate of λ per exposure unit is a negative binomial distribution with pmf given by

$$
\begin{aligned}
f(y|x) &= \int_0^1 \texttt{dpois}(y; m\lambda)\, \texttt{dgamma}(\lambda; x + a, n + b)\, d\lambda \\
&= \frac{(n + b)^{x+a} m^y}{\Gamma(x + a)\, y!} \int_0^1 \lambda^{y+x+a-1} \exp[-(n + b + m)\lambda]\, d\lambda.
\end{aligned}
$$

Using a change of variable $w = (n + b + m)\lambda$ gives

$$
\begin{aligned}
f(y|x) &= \frac{(n + b)^{x+a} m^y}{\Gamma(x + a)\, y!} \frac{\Gamma(y + x + a)}{(n + b + m)^{y+x+a}} \\
&= \frac{\Gamma(y + x + a)}{\Gamma(x + a)\, \Gamma(y + 1)} \left(\frac{n + b}{n + b + m}\right)^{x+a} \left(\frac{m}{n + b + m}\right)^y \\
&= \texttt{dnbinom}[y; x + a, (n + b)/(n + b + m)],
\end{aligned}
$$

where y is a nonnegative integer.

H.5.4 Posterior Predictive Distribution for the Normal Distribution Based on a Conjugate Prior Distribution

Here the joint prior distribution for the normal distribution parameters is the modified Jeffreys prior distribution $f(\mu, \sigma^2) = 1/\sigma^2$, used in Example H.5. To obtain the posterior predictive posterior for the normal distribution, use the identity $f(\mu, \sigma^2|\boldsymbol{x}) = f(\sigma^2|\boldsymbol{x})f(\mu|\sigma^2, \boldsymbol{x})$. Using (H.25) gives

$$f(y|\boldsymbol{x}) = \int_0^\infty f(\sigma^2|\boldsymbol{x}) \left[\int_{-\infty}^\infty f(y|\mu, \sigma^2)f(\mu|\sigma^2, \boldsymbol{x}) \, d\mu \right] d\sigma^2. \qquad \text{(H.27)}$$

Using (H.12), the inner integral on the right-hand side of (H.27) is the pdf of normal-normal mixture with $f(y|\mu, \sigma^2) = \mathtt{dnorm}(y; \mu, \sigma)$ and $f(\mu|\sigma^2, \boldsymbol{x}) = \mathtt{dnorm}(\mu; \mu_n, \sigma/\sqrt{n_n})$. Thus applying (H.1) with $\tau = \mu_n$, $\nu = n_n$, we find that the inner integral is equal to $\mathtt{dnorm}(y; \mu_n, \sigma/\sqrt{w_n})$, where $w_n = (n_n + 1)/n_n$. Thus, using $f(\sigma^2|\boldsymbol{x})$ in (H.14) and (H.3) to obtain the integral, we obtain

$$\begin{aligned}
f(y|\boldsymbol{x}) &= \int_0^\infty f(\sigma^2|\boldsymbol{x}) \, \mathtt{dnorm}(y; \mu_n, \sigma/\sqrt{w_n}) \, d\sigma^2 \\
&= C \int_0^\infty \left(\frac{1}{\sigma^2}\right)^{(r_n+3)/2} \exp\left(-\frac{1}{\sigma^2} \frac{[r_n\sigma_n^2 + w_n(y - \mu_n)^2]}{2} \right) d\sigma^2 \\
&= \frac{1}{\sigma_n/\sqrt{w_n}} \frac{\Gamma[(r_n+1)/2]}{\Gamma(r_n/2)} \frac{1}{\sqrt{\pi r_n}} \left[1 + \frac{1}{r_n}\left(\frac{y - \mu_n}{\sigma_n/\sqrt{w_n}} \right)^2 \right]^{-(r_n+1)/2}, \qquad \text{(H.28)}
\end{aligned}$$

where $C = (r_n\sigma_n^2/2)^{r_n/2} \sqrt{w_n}/\left[\Gamma(r_n/2)\sqrt{2\pi}\right]$. Equivalently, the posterior predictive pdf $f(y|\boldsymbol{x})$ implies that, conditional on $\boldsymbol{x}$,

$$\frac{Y - \mu_n}{\sigma_n/\sqrt{w_n}} \sim t(r_n).$$

That is, $f(y|\boldsymbol{x})$ is a location-scale distribution with location μ_n, scale $\sigma_n/\sqrt{w_n}$, and a Student's t-distribution with r_n degrees of freedom as the standardized distribution.

To simulate draws from the posterior predictive distribution of Y in (H.28), one can generate t^* from a $t(r_n)$ distribution. Then

$$y^* = \mu_n + t^*\sigma_n/\sqrt{w_n},$$

where μ_n and σ_n are defined in (H.13).

H.6 POSTERIOR PREDICTIVE DISTRIBUTIONS BASED ON JEFFREYS PRIOR DISTRIBUTIONS

H.6.1 Posterior Predictive Distribution for the Binomial Distribution Based on Jeffreys Prior Distribution

Section H.4.1 showed that the Jeffreys prior distribution for the binomial distribution is $f(\pi) = \mathtt{dbeta}(\pi; 1/2, 1/2)$, which is a conjugate prior distribution for the $\mathtt{dbinom}(x; n, \pi)$ distribution. Then using (H.6), the posterior distribution for π is $f(\pi|x) = \mathtt{dbeta}(\pi; x + 1/2, n - x + 1/2)$. Using (H.26), the posterior predictive distribution for a future observation $Y \sim \mathrm{BINOM}(m, \pi)$ has a beta-binomial distribution with pmf given by

$$f(y|x) = \mathtt{dbetabinom}(y; m, x + 1/2, n - x + 1/2),$$

where y is a nonnegative integer and $y \leq m$. This is the basis for the Jeffreys prediction interval method in Section 6.2.5.

H.6.2 Posterior Predictive Distribution for the Poisson Distribution Based on Jeffreys Prior Distribution

In Section H.4.2, we show that for the Poisson distribution, the posterior distribution based on the Jeffreys prior distribution is the conjugate distribution $f(\lambda|\boldsymbol{x}) = \texttt{dgamma}(\lambda; x + 1/2, n)$. Consequently, the posterior predictive distribution for a new observation $Y \sim \text{POIS}(m\lambda)$ has a negative binomial distribution with pmf given by

$$
\begin{aligned}
f(y|\boldsymbol{x}) &= \int_0^\infty \texttt{dpois}(y; m\lambda)\texttt{dgamma}(\lambda; x + 1/2, n)\, d\lambda \\
&= \frac{\Gamma(y + x + 1/2)}{\Gamma(x + 1/2)\,\Gamma(y + 1)} \left(\frac{n}{n + m}\right)^{x+1/2} \left(1 - \frac{n}{n + m}\right)^y \\
&= \texttt{dnbinom}[y; x + 1/2, n/(n + m)],
\end{aligned}
$$

where y is a nonnegative integer. This is the basis for the Jeffreys prediction interval method in Section 7.2.5.

H.6.3 Posterior Predictive Distribution for the Normal Distribution Based on the Modified Jeffreys Prior Distribution

Here we use the posterior distributions for $f(\mu|\sigma^2, \boldsymbol{x})$ and $f(\sigma^2|\boldsymbol{x})$ obtained in (H.22) and the identity $f(\mu, \sigma^2|\boldsymbol{x}) = f(\mu|\boldsymbol{x}, \sigma^2)\, f(\sigma^2|\boldsymbol{x})$. For a new observation from $Y \sim \text{NORM}(\mu, \sigma)$, which has a pdf given by $f(y|\mu, \sigma^2) = \texttt{dnorm}(y; \mu, \sigma)$, the posterior predictive distribution is

$$
\begin{aligned}
f(y|\boldsymbol{x}) &= \int_0^\infty \int_{-\infty}^\infty f(\sigma^2|\boldsymbol{x})\, f(\mu|\sigma^2, \boldsymbol{x})\, f(y|\mu, \sigma^2)\, du\, d\sigma^2 \\
&= \int_0^\infty f(\sigma^2|\boldsymbol{x}) \left[\int_{-\infty}^\infty f(y|\mu, \sigma^2)\, f(\mu|\sigma^2, \boldsymbol{x})\, du\right] d\sigma^2 \qquad \text{(H.29)} \\
&= \int_{-\infty}^\infty f(\sigma^2|\boldsymbol{x})\, \texttt{dnorm}(y; \bar{x}, \sigma/\sqrt{w})\, d\sigma^2,
\end{aligned}
$$

where (H.1) was used to evaluate the inner integral in (H.29) and $w = n/(n + 1)$. Finally, using the expression for $f(\sigma^2|\boldsymbol{x})$ in (H.20) gives

$$
f(y|\boldsymbol{x}) = C \int_0^\infty \frac{1}{\sigma^{n+2}} \exp\left(-\frac{B}{\sigma^2}\right) d\sigma^2, \qquad \text{(H.30)}
$$

where $B = [w(y - \bar{x})^2 + (n - 1)s^2]/2$ and

$$
C = \frac{1}{\Gamma[(n - 1)/2]} \left[\frac{(n - 1)s^2}{2}\right]^{(n-1)/2} \frac{\sqrt{w}}{\sqrt{2\pi}}.
$$

Using (H.3) to evaluate the integral in (H.30) gives

$$
f(y|\boldsymbol{x}) = \frac{1}{s/\sqrt{w}} \frac{\Gamma(n/2)}{\Gamma[(n - 1)/2]\,\sqrt{\pi(n - 1)}} \left[1 + \frac{1}{n - 1}\left(\frac{y - \bar{x}}{s/\sqrt{w}}\right)^2\right]^{-n/2}. \qquad \text{(H.31)}
$$

Equivalently, the posterior predictive pdf $f(y|\boldsymbol{x})$ in (H.31) implies that, conditional on $\boldsymbol{x}$,

$$\frac{Y - \bar{x}}{s/\sqrt{w}} \sim t(n - 1). \tag{H.32}$$

That is, $f(y|\boldsymbol{x})$ is a location-scale distribution with location $\bar{x}$, scale $s/\sqrt{w} = s\,(1 + 1/n)^{1/2}$, and a Student's t-distribution with $n - 1$ degrees of freedom as the standardized distribution.

From (H.32), a $100(1 - \alpha)\%$ Bayesian prediction interval for a future observation is

$$[\underset{\sim}{Y},\ \widetilde{Y}] = \bar{x} \mp t_{(1-\alpha/2;n-1)} s/\sqrt{w}$$

$$= \bar{x} \mp t_{(1-\alpha/2;n-1)} \left(1 + \frac{1}{n}\right)^{1/2} s.$$

This interval is the same as the classical non-Bayesian prediction interval in (4.7) for $m = 1$.

Appendix *I*

Probability of Successful Demonstration

I.1 DEMONSTRATION TESTS BASED ON A NORMAL DISTRIBUTION ASSUMPTION

This section provides technical details for computing the probability of successful demonstration shown in Figures 9.1a–9.1d in Chapter 9.

I.1.1 Probability of Successful Demonstration Based on a Normal Distribution One-Sided Confidence Bound on a Quantile

Confidence intervals for a quantile can be used to demonstrate with $100(1 - \alpha)\%$ confidence that $x_{p^\dagger} \leq x^\dagger$, where $x_{p^\dagger}$ is the $p^\dagger$ quantile of a $\mathrm{NORM}(\mu, \sigma)$ distribution and $p^\dagger$ and $x^\dagger$ are specified. We need to introduce $p^\dagger$ here to distinguish this *specified* value from an actual probability p. Suppose that the available data are a random sample $x_1, \ldots, x_n$ from the $\mathrm{NORM}(\mu, \sigma)$ distribution. The data are summarized by the sample mean $\bar{x}$ and the sample standard deviation s, defined in Section 3.1.2.

The demonstration test is successful with $100(1 - \alpha)\%$ confidence if the one-sided upper $100(1 - \alpha)\%$ confidence bound $\widetilde{x}_{p^\dagger}$ is less than or equal to $x^\dagger$. Thus, the probability of successful demonstration is

$$p_{\mathrm{dem}} = \Pr(\widetilde{x}_{p^\dagger} \leq x^\dagger),$$

where the $100(1 - \alpha)\%$ one-sided upper confidence bound on $x_{p^\dagger}$ can be computed from (4.2) as

$$\widetilde{x}_{(p^\dagger)} = \bar{x} - t_{(\alpha; n-1, \delta_{p^\dagger})} \frac{s}{\sqrt{n}},$$

Statistical Intervals: A Guide for Practitioners and Researchers, Second Edition.
William Q. Meeker, Gerald J. Hahn and Luis A. Escobar.
© 2017 John Wiley & Sons, Inc. Published 2017 by John Wiley & Sons, Inc.
Companion Website: www.wiley.com/go/meeker/intervals

where $t_{(\alpha;n-1,\delta_{p^\dagger})}$ is the α quantile of a noncentral t-distribution with $n-1$ degrees of freedom and noncentrality parameter $\delta_{p^\dagger} = -z_{(p^\dagger)}\sqrt{n}$ (see Section C.3.9 for more information about the noncentral t-distribution). The underlying theory for this interval is given in Section E.3.3.

Let $p = \Pr(X \le x^\dagger)$ which implies that $x^\dagger = x_p = \mu + \sigma z_{(p)}$, where $z_{(p)} = \Phi_{\text{norm}}^{-1}(p)$ is the p quantile of the $\text{NORM}(0,1)$ distribution. Consequently, using $\bar{X}$ and S for the random sample mean and random sample standard deviation, the probability of a successful demonstration is

$$p_{\text{dem}} = \Pr\left(\bar{X} - t_{(\alpha;n-1,\delta_{p^\dagger})}\frac{S}{\sqrt{n}} \le x^\dagger\right) = \Pr\left(\bar{X} - t_{(\alpha;n-1,\delta_{p^\dagger})}\frac{S}{\sqrt{n}} \le \mu + \sigma z_{(p)}\right)$$

$$= \Pr\left[\frac{\sqrt{n}(\bar{X}-\mu)/\sigma - z_{(p)}\sqrt{n}}{S/\sigma} \le t_{(\alpha;n-1,\delta_{p^\dagger})}\right] = \Pr\left[T \le t_{(\alpha;n-1,\delta_{p^\dagger})}\right],$$

where T has a noncentral t-distribution with $n-1$ degrees of freedom and noncentrality parameter $\delta_p = -z_{(p)}\sqrt{n}$ (i.e., $T \sim t_{(n-1,\delta_p)}$). Thus

$$p_{\text{dem}} = \text{pt}[\text{qt}(\alpha;n-1,\delta_{p^\dagger});n-1,\delta_p], \tag{I.1}$$

where pt and qt are, respectively, the cdf and quantile functions of the noncentral t-distribution given in Section C.3.9. Note that when $x_{p^\dagger} = x^\dagger = x_p$, $p^\dagger = p$ and $p_{\text{dem}} = \text{pt}[\text{qt}(\alpha;n-1,\delta_p);n-1,\delta_p] = \alpha$, as can be seen in Figures 9.1a–9.1d.

I.1.2 Probability of Successful Demonstration Based on a One-Sided Confidence Bound on a Normal Distribution Probability

Because $x_{p^\dagger} \le x^\dagger$ is equivalent to $p^\dagger \le p$, the demonstration test can be also formulated as a demonstration test to show that $p^\dagger$ is less than or equal to p. The demonstration test is successful with $100(1-\alpha)\%$ confidence if the one-sided lower $100(1-\alpha)\%$ confidence bound $\underset{\sim}{p}$ for p (Section 4.5) is greater than or equal to $p^\dagger$ (i.e., $\underset{\sim}{p} \ge p^\dagger$). Consequently, the probability of a successful demonstration for this procedure is also given by (I.1).

I.1.3 Sample Size to Achieve a p_{dem} for a Demonstration Based on a Normal Distribution One-Sided Confidence Bound on a Quantile

The relationship in (I.1) can be used to find the sample size n needed to conduct a test to demonstrate with $100(1-\alpha)\%$ confidence that $p \ge p^\dagger$ (or equivalently that $x_{p^\dagger} \le x^\dagger$). For this purpose, one specifies a desired probability of successful demonstration p_{dem} and solves the equation

$$p_{\text{dem}} = \text{pt}\left[\text{qt}(\alpha;n-1,\delta_{p^\dagger});n-1,\delta_p\right] \tag{I.2}$$

numerically for n. For example, suppose that we need the smallest sample size to demonstrate with $1-\alpha = 90\%$ confidence that $p \ge p^\dagger = 0.50$ such that $p_{\text{dem}} \ge 0.70$ when $p = 0.80$. Root-finding numerically the value n satisfying

$$\text{pt}[\text{qt}(0.10;n-1,\delta_{0.50});n-1,\delta_{0.80}] = 0.70,$$

where $\delta_\gamma = -z_\gamma\sqrt{n}$, gives $n = 5.6$. Rounding this up, the required sample size is $n = 6$. Another example is given in Section 9.2.4.

I.2   DISTRIBUTION-FREE DEMONSTRATION TESTS

This section provides technical details for computing the probability of successful demonstration shown in Figures 9.2a–9.2d in Chapter 9.

I.2.1   Probability of Successful Demonstration Based on a Distribution-Free One-Sided Confidence Bound on a Quantile

In a manner that is similar to Section I.1.1, distribution-free confidence intervals for a quantile can be used to demonstrate, with $100(1 - \alpha)\%$ confidence, that $y_{p^\dagger} \le y^\dagger$, where $y_{p^\dagger}$ is the $p^\dagger$ quantile of a distribution with unspecified form and $p^\dagger$ and $y^\dagger$ are specified. (We use y here instead of x to distinguish from the binomial random variable X which is also used in this presentation.)

The demonstration will be successful if $\widetilde{y}_{p^\dagger}$, a distribution-free upper confidence bound for the $p^\dagger$ quantile of a continuous distribution (with unspecified form), is less than or equal to the specified value $y^\dagger$. From the computational method in Section 5.2.3, $\widetilde{y}_{p^\dagger} = y_{(u)}$ where $u = \texttt{qbinom}(1 - \alpha; n, p^\dagger) + 1$, implying that there must be at least u conforming units or, equivalently, no more than

$$c = n - u = n - \texttt{qbinom}(1 - \alpha; n, p^\dagger) - 1 \tag{I.3}$$

nonconforming units in the sample of size n. Thus, the probability of successful demonstration is

$$p_{\mathrm{dem}} = \Pr(\widetilde{y}_{p^\dagger} \le y^\dagger) = \Pr(X \le c) = \texttt{pbinom}(c; n, 1 - p), \tag{I.4}$$

where $X \sim \mathrm{BINOM}(n, 1 - p)$ is the number of nonconforming units in the sample of size n and $1 - p$ is the actual proportion nonconforming (p is the actual proportion conforming).

I.2.2   Probability of Successful Demonstration Based on a One-Sided Confidence Bound on a Binomial Distribution Probability

The approach in Section I.2.1 is equivalent to having a successful demonstration when there are no more than c *nonconforming* units in the sample of size n from a population. In this case, the demonstration will be successful if $\underset{\sim}{\pi} \ge \pi^\dagger$, where $\underset{\sim}{\pi}$ is the one-sided lower confidence bound on the proportion *conforming*, obtained from the conservative method in Section 6.2.2. (We use π instead of p here to be consistent with our notation for the binomial probability parameter.) Thus, the probability of a successful demonstration for this procedure is also given by (I.4).

Note that c can take integer values in the range -1 to $n - 1$. When $c = -1$, the demonstration test is infeasible in the sense that the demonstration fails even when all the units in the sample are conforming. That is, $p_{\mathrm{dem}} = 0$, regardless the number of nonconforming units in the sample, because $\underset{\sim}{\pi} < \pi^\dagger$. This indicates that a larger n is needed to have a positive probability of a successful demonstration test.

I.2.3   Sample Size to Achieve a p_{dem} for a Demonstration Based on a One-Sided Confidence Bound on a Probability

Similar to (I.2), combining (I.3) and (I.4), we see that the sample size n needed to conduct a test to demonstrate with $100(1 - \alpha)\%$ confidence that $\pi \ge \pi^\dagger$, with a probability of successful demonstration larger than or equal to p_{dem}, is the smallest integer n satisfying the relationship

$$\texttt{pbinom}(n - \texttt{qbinom}(1 - \alpha; n, \pi^{\dagger}) - 1; n, 1 - \pi) \geq p_{\text{dem}}.$$

For example, suppose that we need the smallest sample size to demonstrate with $1 - \alpha = 90\%$ confidence that $\pi \geq \pi^{\dagger} = 0.50$ such that $p_{\text{dem}} \geq 0.70$ when $\pi = 0.80$. Solving numerically the equation

$$\texttt{pbinom}(n - \texttt{qbinom}(1 - 0.10, n, 0.50) - 1; n, 1 - 0.80) \geq 0.70,$$

we find that $n = 9$ is the required sample size and $c = 9 - \texttt{qbinom}(1 - 0.10; 9, 0.50) - 1 = 2$ is the maximum number of nonconforming units in the sample that will allow a successful demonstration test. Another example is given in Section 9.5.3.

Statistical Intervals: A Guide for Practitioners and Researchers, Second Edition.
William Q. Meeker, Gerald J. Hahn and Luis A. Escobar.
© 2017 John Wiley & Sons, Inc. Published 2017 by John Wiley & Sons, Inc.
Companion Website: www.wiley.com/go/meeker/intervals

Table J.12	Smallest sample size for the range formed by the smallest and largest observations to contain at least $100\beta\%$ of the sampled population
Table J.13	Smallest sample size for the largest (smallest) observation to exceed (be exceeded by) at least $100\beta\%$ of the sampled population
Table J.14a–J.14c	Largest number k of m future observations that will be contained in (bounded by) the distribution-free $100(1 - \alpha)\%$ two-sided prediction interval (one-sided prediction bound) obtained by removing $\nu - 2$ (or $\nu - 1$) extreme observations from the ends (end) of a previous sample of size n
Table J.15	Smallest sample size for the range formed by the smallest and largest observations to contain, with $100(1 - \alpha)\%$ confidence, all m future observations from the previously sampled population
Table J.16a	Smallest sample size for the largest (smallest) observation to exceed (be exceeded by), with $100(1 - \alpha)\%$ confidence, all m future observations from the previously sampled population
Table J.16b	Smallest sample size for the largest (smallest) observation to exceed (be exceeded by), with $100(1 - \alpha)\%$ confidence, at least $m - 1$ of m future observations from the previously sampled population
Table J.16c	Smallest sample size for the largest (smallest) observation to exceed (be exceeded by), with $100(1 - \alpha)\%$ confidence, at least $m - 2$ of m future observations from the previously sampled population
Table J.17a	Smallest sample size such that a two-sided $100(1 - \alpha)\%$ confidence interval for the mean of a normal distribution will be no wider than $\pm k\sigma$ (σ known)
Table J.17b	Correction for sample size so that, with $100(1 - \gamma)\%$ probability, the sample will provide a normal distribution $100(1 - \alpha)\%$ confidence interval that is at least as narrow as the desired length
Table J.18	Smallest sample size needed to estimate a normal distribution standard deviation with a given upper bound on percent error
Table J.19	Smallest sample size for normal distribution two-sided tolerance intervals such that the probability of enclosing at least the population proportion β is $1 - \alpha$ and the probability of enclosing at least β^* is no more than δ
Table J.20	Smallest sample size for normal distribution one-sided tolerance bounds such that the probability of enclosing at least the population proportion β is $1 - \alpha$ and the probability of enclosing at least β^* is no more than δ
Table J.21	Smallest sample size for distribution-free tolerance intervals and tolerance bounds such that the probability of enclosing at least the population proportion β is $1 - \alpha$ and the probability of enclosing at least β^* is no more than δ

Table J.1a Factors for calculating two-sided 95% statistical intervals for a normal distribution. The two-sided 95% statistical interval is $\bar{x} \pm c_{(0.95;n)}s$, where $c_{(0.95;n)}$ is the appropriate tabulated value and $\bar{x}$ and s are, respectively, the sample mean and standard deviation of a sample of size n. A similar table first appeared in Hahn (1970b). Adapted with permission of the American Society for Quality.

Number of given observations	Factors for confidence intervals for the mean μ	Factors for tolerance intervals to contain at least $100\beta\%$ of the distribution			Factors for simultaneous prediction intervals to contain all m future observations						Factors for prediction intervals to contain the mean of $n = m$ future observations
		β			m						
n		0.90	0.95	0.99	1	2	5	10	20	n	
4	1.59	5.37	6.34	8.22	3.56	4.41	5.56	6.41	7.21	5.29	2.25
5	1.24	4.29	5.08	6.60	3.04	3.70	4.58	5.23	5.85	4.58	1.76
6	1.05	3.73	4.42	5.76	2.78	3.33	4.08	4.63	5.16	4.22	1.48
7	0.92	3.39	4.02	5.24	2.62	3.11	3.77	4.26	4.74	4.01	1.31
8	0.84	3.16	3.75	4.89	2.51	2.97	3.57	4.02	4.46	3.88	1.18
9	0.77	2.99	3.55	4.63	2.43	2.86	3.43	3.85	4.25	3.78	1.09
10	0.72	2.86	3.39	4.44	2.37	2.79	3.32	3.72	4.10	3.72	1.01
12	0.64	2.67	3.17	4.16	2.29	2.68	3.17	3.53	3.89	3.63	0.90
15	0.55	2.49	2.96	3.89	2.22	2.57	3.03	3.36	3.69	3.56	0.78
20	0.47	2.32	2.76	3.62	2.14	2.48	2.90	3.21	3.50	3.50	0.66
25	0.41	2.22	2.64	3.46	2.10	2.43	2.83	3.12	3.40	3.49	0.58
30	0.37	2.15	2.55	3.35	2.08	2.39	2.78	3.06	3.33	3.48	0.53
40	0.32	2.06	2.45	3.22	2.05	2.35	2.73	2.99	3.25	3.49	0.45
60	0.26	1.96	2.34	3.07	2.02	2.31	2.67	2.93	3.17	3.53	0.37
∞	0.00	1.64	1.96	2.58	1.96	2.24	2.57	2.80	3.02	∞	0.00

Number of given observations	Factors for confidence intervals for the mean μ	Factors for tolerance intervals to contain at least $100\beta\%$ of the distribution β			Factors for simultaneous prediction intervals to contain all m future observations m						Factors for prediction intervals to contain the mean of $n = m$ future observations
n		0.90	0.95	0.99	1	2	5	10	20	n	
4	2.92	9.42	11.12	14.41	6.53	7.94	9.88	11.32	12.70	9.41	4.13
5	2.06	6.65	7.87	10.22	5.04	5.97	7.25	8.22	9.15	7.25	2.91
6	1.65	5.38	6.37	8.29	4.36	5.07	6.06	6.80	7.53	6.25	2.33
7	1.40	4.66	5.52	7.19	3.96	4.56	5.38	6.01	6.62	5.69	1.98
8	1.24	4.19	4.97	6.48	3.71	4.24	4.95	5.50	6.04	5.32	1.75
9	1.12	3.86	4.58	5.98	3.54	4.01	4.66	5.15	5.63	5.07	1.58
10	1.03	3.62	4.29	5.61	3.41	3.85	4.44	4.89	5.34	4.89	1.45
12	0.90	3.28	3.90	5.10	3.23	3.63	4.15	4.54	4.94	4.65	1.27
15	0.77	2.97	3.53	4.62	3.07	3.43	3.89	4.23	4.58	4.44	1.09
20	0.64	2.68	3.18	4.17	2.93	3.25	3.65	3.96	4.26	4.26	0.90
25	0.56	2.51	2.98	3.91	2.85	3.15	3.53	3.81	4.08	4.17	0.79
30	0.50	2.39	2.85	3.74	2.80	3.09	3.45	3.71	3.97	4.12	0.71
40	0.43	2.25	2.68	3.52	2.74	3.01	3.35	3.60	3.84	4.07	0.61
60	0.34	2.11	2.51	3.30	2.68	2.94	3.26	3.49	3.71	4.05	0.49
∞	0.00	1.64	1.96	2.58	2.58	2.81	3.09	3.29	3.48	∞	0.00

Table J.1b Factors for calculating two-sided 99% statistical intervals for a normal distribution. The two-sided 99% statistical interval is $\bar{x} \pm c_{(0.99;n)} s$, where $c_{(0.99;n)}$ is the appropriate tabulated value and $\bar{x}$ and s are, respectively, the sample mean and standard deviation of a sample of size n. A similar table first appeared in Hahn (1970b). Adapted with permission of the American Society for Quality.

| Number of given observations | Factors for confidence intervals for the standard deviation σ | | Factors for prediction intervals for the standard deviation of $m = n$ future observations | |
| | Factor for calculating | | Factor for calculating | |
n	Lower limit	Upper limit	Lower limit	Upper limit
4	0.57	3.73	0.25	3.93
5	0.60	2.87	0.32	3.10
6	0.62	2.45	0.37	2.67
7	0.64	2.20	0.41	2.41
8	0.66	2.04	0.45	2.23
9	0.68	1.92	0.47	2.11
10	0.69	1.83	0.50	2.01
12	0.71	1.70	0.54	1.86
15	0.73	1.58	0.58	1.73
20	0.76	1.46	0.63	1.59
25	0.78	1.39	0.66	1.51
30	0.80	1.34	0.69	1.45
40	0.82	1.28	0.73	1.38
60	0.85	1.22	0.77	1.29
∞	1.00	1.00	1.00	1.00

Table J.2a Factors for calculating two-sided 95% statistical intervals for a normal distribution. The two-sided 95% statistical interval is $[c_{L(0.95;n)}s, \ c_{U(0.95;n)}s]$, where $c_{L(0.95;n)}$ and $c_{U(0.95;n)}$ are the appropriate tabulated factors and s is the sample standard deviation of a sample of size n. A similar table first appeared in Hahn (1970b). Adapted with permission of the American Society for Quality.

| Number of given observations | Factors for confidence intervals for the standard deviation σ | | Factors for prediction intervals for the standard deviation of $m = n$ future observations | |
| | Factor for calculating | | Factor for calculating | |
n	Lower limit	Upper limit	Lower limit	Upper limit
4	0.48	6.47	0.15	6.89
5	0.52	4.40	0.21	4.81
6	0.55	3.48	0.26	3.87
7	0.57	2.98	0.30	3.33
8	0.59	2.66	0.34	2.98
9	0.60	2.44	0.37	2.74
10	0.62	2.28	0.39	2.56
12	0.64	2.06	0.43	2.31
15	0.67	1.85	0.48	2.07
20	0.70	1.67	0.54	1.85
25	0.73	1.56	0.58	1.72
30	0.74	1.49	0.61	1.64
40	0.77	1.40	0.66	1.52
60	0.81	1.30	0.71	1.40
∞	1.00	1.00	1.00	1.00

Table J.2b Factors for calculating two-sided 99% statistical intervals for a normal distribution. The two-sided 99% statistical interval is $[c_{L(0.99;n)}s, \ c_{U(0.99;n)}s]$, where $c_{L(0.99;n)}$ and $c_{U(0.99;n)}$ are the appropriate tabulated factors and s is the sample standard deviation of a sample of size n. A similar table first appeared in Hahn (1970b). Adapted with permission of the American Society for Quality.

Number of given observations n	Factors for confidence bounds for the mean μ	Factors for tolerance bounds to exceed (be exceeded by) at least $100\beta\%$ of the distribution			Factors for simultaneous prediction bounds to exceed (be exceeded by) all m future observations						Factors for prediction bounds to exceed (be exceeded by) the mean of $n = m$ future observations
		β			m						
		0.90	0.95	0.99	1	2	5	10	20	n	
4	1.18	4.16	5.14	7.04	2.63	3.40	4.47	5.28	6.06	4.21	1.66
5	0.95	3.41	4.20	5.74	2.34	2.95	3.79	4.42	5.03	3.79	1.35
6	0.82	3.01	3.71	5.06	2.18	2.72	3.43	3.97	4.49	3.58	1.16
7	0.73	2.76	3.40	4.64	2.08	2.57	3.22	3.70	4.17	3.45	1.04
8	0.67	2.58	3.19	4.35	2.01	2.47	3.07	3.52	3.95	3.37	0.95
9	0.62	2.45	3.03	4.14	1.96	2.40	2.97	3.38	3.79	3.32	0.88
10	0.58	2.35	2.91	3.98	1.92	2.35	2.89	3.28	3.67	3.28	0.82
12	0.52	2.21	2.74	3.75	1.87	2.27	2.78	3.14	3.50	3.24	0.73
15	0.45	2.07	2.57	3.52	1.82	2.20	2.67	3.01	3.34	3.21	0.64
20	0.39	1.93	2.40	3.30	1.77	2.13	2.57	2.89	3.19	3.19	0.55
25	0.34	1.84	2.29	3.16	1.74	2.09	2.52	2.82	3.11	3.20	0.48
30	0.31	1.78	2.22	3.06	1.73	2.07	2.48	2.78	3.05	3.21	0.44
40	0.27	1.70	2.13	2.94	1.71	2.04	2.44	2.72	2.99	3.24	0.38
60	0.22	1.61	2.02	2.81	1.68	2.01	2.40	2.67	2.92	3.30	0.31
∞	0.00	1.28	1.64	2.33	1.64	1.95	2.32	2.57	2.80	∞	0.00

Table J.3a Factors for calculating one-sided 95% statistical bounds for a normal distribution. The one-sided 95% statistical bound is $\bar{x} + c'_{(0.95;n)} s$ or $\bar{x} - c'_{(0.95;n)} s$, where $c'_{(0.95;n)}$ is the appropriate tabulated value and $\bar{x}$ and s are, respectively, the sample mean and standard deviation of a sample of size n. A similar table first appeared in Hahn (1970b). Adapted with permission of the American Society for Quality.

Number of given observations	Factors for confidence bounds for the mean μ	Factors for tolerance bounds to exceed (be exceeded by) at least $100\beta\%$ of the distribution			Factors for simultaneous prediction bounds to exceed (be exceeded by) all m future observations						Factors for prediction bounds to exceed (be exceeded by) the mean of $n = m$ future observations
		β			m						
n		0.90	0.95	0.99	1	2	5	10	20	n	
4	2.27	7.38	9.08	12.39	5.08	6.31	8.07	9.43	10.76	7.63	3.21
5	1.68	5.36	6.58	8.94	4.10	4.94	6.13	7.04	7.95	6.13	2.37
6	1.37	4.41	5.41	7.33	3.63	4.30	5.22	5.94	6.64	5.41	1.94
7	1.19	3.86	4.73	6.41	3.36	3.93	4.70	5.30	5.90	5.00	1.68
8	1.06	3.50	4.29	5.81	3.18	3.69	4.37	4.90	5.43	4.73	1.50
9	0.97	3.24	3.97	5.39	3.05	3.52	4.14	4.62	5.09	4.55	1.37
10	0.89	3.05	3.74	5.07	2.96	3.39	3.97	4.41	4.85	4.41	1.26
12	0.78	2.78	3.41	4.63	2.83	3.22	3.74	4.13	4.52	4.23	1.11
15	0.68	2.52	3.10	4.22	2.71	3.07	3.53	3.88	4.22	4.08	0.96
20	0.57	2.28	2.81	3.83	2.60	2.93	3.34	3.65	3.95	3.95	0.80
25	0.50	2.13	2.63	3.60	2.54	2.85	3.24	3.52	3.80	3.89	0.70
30	0.45	2.03	2.52	3.45	2.50	2.80	3.17	3.44	3.71	3.86	0.64
40	0.38	1.90	2.36	3.25	2.46	2.74	3.09	3.35	3.59	3.83	0.54
60	0.31	1.76	2.20	3.04	2.41	2.68	3.02	3.26	3.49	3.84	0.44
∞	0.00	1.28	1.64	2.33	2.33	2.57	2.88	3.09	3.29	∞	0.00

Table J.3b Factors for calculating one-sided 99% statistical bounds for a normal distribution. The one-sided 99% statistical bound is $\bar{x} + c'_{(0.99;n)} s$ or $\bar{x} - c'_{(0.99;n)} s$, where $c'_{(0.99;n)}$ is the appropriate tabulated value and $\bar{x}$ and s are, respectively, the sample mean and standard deviation of a sample of size n. A similar table first appeared in Hahn (1970b). Adapted with permission of the American Society for Quality.

Number of given observations	Factors for confidence bounds for the standard deviation σ		Factors for prediction bounds for the standard deviation of $m = n$ future observations	
	Factor for calculating		Factor for calculating	
n	Lower bound	Upper bound	Lower bound	Upper bound
4	0.62	2.92	0.33	3.05
5	0.65	2.37	0.40	2.53
6	0.67	2.09	0.44	2.25
7	0.69	1.92	0.48	2.07
8	0.71	1.80	0.51	1.95
9	0.72	1.71	0.54	1.85
10	0.73	1.65	0.56	1.78
12	0.75	1.55	0.60	1.68
15	0.77	1.46	0.63	1.58
20	0.79	1.37	0.68	1.47
25	0.81	1.32	0.71	1.41
30	0.83	1.28	0.73	1.36
40	0.85	1.23	0.77	1.31
60	0.87	1.18	0.81	1.24
∞	1.00	1.00	1.00	1.00

Table J.4a Factors for calculating one-sided 95% statistical bounds for a normal distribution. The one-sided 95% statistical bound is $c'_{L(0.95;n)} s$ or $c'_{U(0.95;n)} s$, where $c'_{L(0.95;n)}$ and $c'_{U(0.95;n)}$ are the appropriate tabulated factors for lower and upper bounds, respectively, and s is the sample standard deviation of a sample size n. A similar table first appeared in Hahn (1970b). Adapted with permission of the American Society for Quality.

Number of given observations	Factors for confidence bounds for the standard deviation σ		Factors for prediction bounds for the standard deviation of $m = n$ future observations	
	Factor for calculating		Factor for calculating	
n	Lower bound	Upper bound	Lower bound	Upper bound
4	0.51	5.11	0.18	5.43
5	0.55	3.67	0.25	4.00
6	0.58	3.00	0.30	3.31
7	0.60	2.62	0.34	2.91
8	0.62	2.38	0.38	2.64
9	0.63	2.20	0.41	2.46
10	0.64	2.08	0.43	2.31
12	0.67	1.90	0.47	2.11
15	0.69	1.73	0.52	1.92
20	0.72	1.58	0.57	1.74
25	0.75	1.49	0.61	1.63
30	0.76	1.43	0.64	1.56
40	0.79	1.35	0.68	1.46
60	0.82	1.27	0.74	1.36
∞	1.00	1.00	1.00	1.00

Table J.4b Factors for calculating one-sided 99% statistical bounds for a normal distribution. The one-sided 99% statistical bound is $c'_{L(0.99;n)}$ or $c'_{U(0.99;n)} s$, where $c'_{L(0.99;n)}$ and $c'_{U(0.99;n)}$ are the appropriate tabulated factors for lower and upper bounds, respectively, and s is the sample standard deviation of a sample size n. A similar table first appeared in Hahn (1970b). Adapted with permission of the American Society for Quality.

		β = 0.500					β = 0.700					β = 0.800					
n	1 − α:	0.50	0.80	0.90	0.95	0.99	0.50	0.80	0.90	0.95	0.99	0.50	0.80	0.90	0.95	0.99	n
2		1.243	3.369	6.808	13.652	68.316	1.865	5.023	10.142	20.331	101.732	2.275	6.110	12.333	24.722	123.699	2
3		0.942	1.700	2.492	3.585	8.122	1.430	2.562	3.747	5.382	12.181	1.755	3.134	4.577	6.572	14.867	3
4		0.852	1.335	1.766	2.288	4.028	1.300	2.026	2.673	3.456	6.073	1.600	2.486	3.276	4.233	7.431	4
5		0.808	1.173	1.473	1.812	2.824	1.236	1.788	2.239	2.750	4.274	1.523	2.198	2.750	3.375	5.240	5
6		0.782	1.081	1.314	1.566	2.270	1.198	1.651	2.003	2.384	3.446	1.477	2.034	2.464	2.930	4.231	6
7		0.764	1.021	1.213	1.415	1.954	1.172	1.562	1.853	2.159	2.973	1.446	1.925	2.282	2.657	3.656	7
8		0.752	0.979	1.143	1.313	1.750	1.153	1.499	1.749	2.006	2.668	1.423	1.849	2.156	2.472	3.284	8
9		0.742	0.947	1.092	1.239	1.608	1.139	1.451	1.672	1.896	2.455	1.407	1.791	2.062	2.337	3.024	9
10		0.735	0.923	1.053	1.183	1.503	1.128	1.414	1.613	1.811	2.297	1.393	1.746	1.990	2.234	2.831	10
11		0.729	0.903	1.021	1.139	1.422	1.119	1.385	1.566	1.744	2.175	1.382	1.710	1.932	2.152	2.682	11
12		0.724	0.886	0.996	1.103	1.357	1.112	1.360	1.527	1.690	2.078	1.374	1.680	1.885	2.086	2.563	12
13		0.720	0.873	0.974	1.073	1.305	1.106	1.339	1.495	1.645	1.999	1.366	1.654	1.846	2.031	2.466	13
14		0.717	0.861	0.956	1.048	1.261	1.100	1.322	1.467	1.608	1.933	1.360	1.633	1.812	1.985	2.386	14
15		0.714	0.851	0.941	1.027	1.224	1.096	1.307	1.444	1.575	1.877	1.355	1.614	1.783	1.945	2.317	15
16		0.711	0.842	0.927	1.008	1.193	1.092	1.293	1.423	1.547	1.829	1.350	1.598	1.758	1.911	2.259	16
17		0.709	0.835	0.915	0.992	1.165	1.089	1.282	1.405	1.522	1.788	1.346	1.584	1.736	1.881	2.207	17
18		0.707	0.828	0.905	0.978	1.141	1.086	1.271	1.389	1.501	1.751	1.342	1.571	1.717	1.854	2.163	18
19		0.705	0.822	0.895	0.965	1.120	1.083	1.262	1.375	1.481	1.719	1.339	1.559	1.699	1.830	2.123	19
20		0.704	0.816	0.887	0.953	1.101	1.081	1.253	1.362	1.464	1.690	1.336	1.549	1.683	1.809	2.087	20
21		0.702	0.811	0.879	0.943	1.084	1.079	1.246	1.350	1.448	1.664	1.333	1.540	1.669	1.789	2.055	21
22		0.701	0.806	0.872	0.934	1.068	1.077	1.239	1.340	1.434	1.640	1.331	1.531	1.656	1.772	2.026	22
23		0.700	0.802	0.866	0.925	1.054	1.075	1.232	1.330	1.420	1.619	1.329	1.523	1.644	1.755	2.000	23
24		0.699	0.798	0.860	0.917	1.042	1.073	1.226	1.321	1.408	1.599	1.327	1.516	1.633	1.741	1.976	24
25		0.698	0.795	0.855	0.910	1.030	1.072	1.221	1.313	1.397	1.581	1.325	1.509	1.623	1.727	1.954	25
26		0.697	0.791	0.850	0.903	1.019	1.070	1.216	1.305	1.387	1.565	1.323	1.503	1.613	1.714	1.934	26
27		0.696	0.788	0.845	0.897	1.009	1.069	1.211	1.298	1.378	1.550	1.322	1.497	1.604	1.703	1.915	27
28		0.695	0.785	0.841	0.891	1.000	1.068	1.207	1.291	1.369	1.535	1.320	1.492	1.596	1.692	1.898	28
29		0.694	0.783	0.837	0.886	0.991	1.067	1.202	1.285	1.360	1.522	1.319	1.487	1.589	1.682	1.882	29
30		0.694	0.780	0.833	0.881	0.983	1.066	1.199	1.279	1.353	1.510	1.318	1.482	1.581	1.672	1.866	30
35		0.691	0.770	0.817	0.859	0.950	1.061	1.182	1.255	1.320	1.459	1.312	1.462	1.551	1.632	1.803	35
40		0.689	0.761	0.805	0.843	0.925	1.058	1.170	1.236	1.296	1.420	1.308	1.446	1.528	1.602	1.756	40
50		0.686	0.750	0.787	0.820	0.889	1.054	1.152	1.209	1.260	1.365	1.303	1.424	1.495	1.558	1.688	50
60		0.684	0.741	0.775	0.804	0.864	1.051	1.139	1.190	1.235	1.327	1.299	1.408	1.471	1.527	1.641	60
120		0.679	0.718	0.740	0.759	0.797	1.044	1.104	1.137	1.166	1.225	1.290	1.365	1.406	1.442	1.514	120
240		0.677	0.704	0.719	0.731	0.756	1.040	1.082	1.104	1.124	1.162	1.286	1.337	1.365	1.390	1.437	240
480		0.676	0.694	0.705	0.713	0.730	1.038	1.067	1.083	1.096	1.122	1.284	1.320	1.339	1.355	1.387	480
∞		0.674	0.674	0.674	0.674	0.674	1.036	1.036	1.036	1.036	1.036	1.282	1.282	1.282	1.282	1.282	∞

Table J.5a Factors $g_{(1-\alpha;\beta,n)}$ for calculating normal distribution two-sided $100(1 - \alpha)\%$ tolerance intervals to control the center of the distribution. The factors in this table were computed with an algorithm provided by Robert E. Odeh.

n	β = 0.900					β = 0.950					β = 0.990					n
1 − α:	0.50	0.80	0.90	0.95	0.99	0.50	0.80	0.90	0.95	0.99	0.50	0.80	0.90	0.95	0.99	
2	2.869	7.688	15.512	31.092	155.569	3.376	9.032	18.221	36.519	182.720	4.348	11.613	23.423	46.944	234.877	2
3	2.229	3.967	5.788	8.306	18.782	2.634	4.679	6.823	9.789	22.131	3.415	6.051	8.819	12.647	28.586	3
4	2.039	3.159	4.157	5.368	9.416	2.416	3.736	4.913	6.341	11.118	3.144	4.850	6.372	8.221	14.405	4
5	1.945	2.801	3.499	4.291	6.655	2.308	3.318	4.142	5.077	7.870	3.010	4.318	5.387	6.598	10.220	5
6	1.888	2.595	3.141	3.733	5.383	2.243	3.078	3.723	4.422	6.373	2.930	4.013	4.850	5.758	8.292	6
7	1.850	2.460	2.913	3.390	4.658	2.199	2.920	3.456	4.020	5.520	2.876	3.813	4.508	5.241	7.191	7
8	1.823	2.364	2.754	3.156	4.189	2.167	2.808	3.270	3.746	4.968	2.836	3.670	4.271	4.889	6.479	8
9	1.802	2.292	2.637	2.986	3.860	2.143	2.723	3.132	3.546	4.581	2.806	3.562	4.094	4.633	5.980	9
10	1.785	2.235	2.546	2.856	3.617	2.124	2.657	3.026	3.393	4.294	2.783	3.478	3.958	4.437	5.610	10
11	1.772	2.189	2.473	2.754	3.429	2.109	2.604	2.941	3.273	4.073	2.764	3.410	3.849	4.282	5.324	11
12	1.761	2.152	2.414	2.670	3.279	2.096	2.560	2.871	3.175	3.896	2.748	3.353	3.759	4.156	5.096	12
13	1.752	2.120	2.364	2.601	3.156	2.085	2.522	2.812	3.093	3.751	2.735	3.306	3.684	4.051	4.909	13
14	1.744	2.093	2.322	2.542	3.054	2.076	2.490	2.762	3.024	3.631	2.723	3.265	3.620	3.962	4.753	14
15	1.737	2.069	2.285	2.492	2.967	2.068	2.463	2.720	2.965	3.529	2.714	3.229	3.565	3.885	4.621	15
16	1.731	2.049	2.254	2.449	2.893	2.061	2.439	2.682	2.913	3.441	2.705	3.198	3.517	3.819	4.507	16
17	1.726	2.030	2.226	2.410	2.828	2.055	2.417	2.649	2.868	3.364	2.698	3.171	3.474	3.761	4.408	17
18	1.721	2.014	2.201	2.376	2.771	2.050	2.398	2.620	2.828	3.297	2.691	3.146	3.436	3.709	4.321	18
19	1.717	2.000	2.178	2.346	2.720	2.045	2.381	2.593	2.793	3.237	2.685	3.124	3.402	3.663	4.244	19
20	1.714	1.987	2.158	2.319	2.675	2.041	2.365	2.570	2.760	3.184	2.680	3.104	3.372	3.621	4.175	20
21	1.710	1.975	2.140	2.294	2.635	2.037	2.351	2.548	2.731	3.136	2.675	3.086	3.344	3.583	4.113	21
22	1.707	1.964	2.123	2.272	2.598	2.034	2.338	2.528	2.705	3.092	2.670	3.070	3.318	3.549	4.056	22
23	1.705	1.954	2.108	2.251	2.564	2.030	2.327	2.510	2.681	3.053	2.666	3.054	3.295	3.518	4.005	23
24	1.702	1.944	2.094	2.232	2.534	2.027	2.316	2.494	2.658	3.017	2.662	3.040	3.274	3.489	3.958	24
25	1.700	1.936	2.081	2.215	2.506	2.025	2.306	2.479	2.638	2.984	2.659	3.027	3.254	3.462	3.915	25
26	1.698	1.928	2.069	2.199	2.480	2.022	2.296	2.464	2.619	2.953	2.656	3.015	3.235	3.437	3.875	26
27	1.696	1.921	2.058	2.184	2.456	2.020	2.288	2.451	2.601	2.925	2.653	3.004	3.218	3.415	3.838	27
28	1.694	1.914	2.048	2.170	2.434	2.018	2.279	2.439	2.585	2.898	2.650	2.993	3.202	3.393	3.804	28
29	1.692	1.907	2.038	2.157	2.413	2.016	2.272	2.427	2.569	2.874	2.648	2.983	3.187	3.373	3.772	29
30	1.691	1.901	2.029	2.145	2.394	2.014	2.265	2.417	2.555	2.851	2.645	2.974	3.173	3.355	3.742	30
35	1.684	1.876	1.991	2.094	2.314	2.006	2.234	2.371	2.495	2.756	2.636	2.935	3.114	3.276	3.618	35
40	1.679	1.856	1.961	2.055	2.253	2.001	2.211	2.336	2.448	2.684	2.628	2.905	3.069	3.216	3.524	40
50	1.672	1.827	1.918	1.999	2.166	1.992	2.177	2.285	2.382	2.580	2.618	2.861	3.003	3.129	3.390	50
60	1.668	1.807	1.888	1.960	2.106	1.987	2.154	2.250	2.335	2.509	2.611	2.830	2.956	3.068	3.297	60
120	1.656	1.752	1.805	1.851	1.943	1.974	2.087	2.151	2.206	2.315	2.594	2.743	2.826	2.899	3.043	120
240	1.651	1.716	1.753	1.783	1.844	1.967	2.045	2.088	2.125	2.197	2.585	2.688	2.744	2.793	2.887	240
480	1.648	1.694	1.718	1.739	1.780	1.963	2.018	2.048	2.073	2.121	2.580	2.652	2.691	2.724	2.787	480
∞	1.645	1.645	1.645	1.645	1.645	1.960	1.960	1.960	1.960	1.960	2.576	2.576	2.576	2.576	2.576	∞

Table J.5b Factors $g_{(1-\alpha;\beta,n)}$ for calculating normal distribution two-sided $100(1-\alpha)\%$ tolerance intervals to control the center of the distribution. The factors in this table were computed with an algorithm provided by Robert E. Odeh.

n	p=0.250					p=0.150					p=0.100					n
1−α:	0.600	0.800	0.900	0.950	0.990	0.600	0.800	0.900	0.950	0.990	0.600	0.800	0.900	0.950	0.990	
2	1.784	4.869	9.847	19.748	98.829	2.333	6.302	12.730	25.522	127.707	2.702	7.272	14.682	29.431	147.264	2
3	1.330	2.455	3.617	5.214	11.830	1.773	3.210	4.705	6.765	15.322	2.072	3.724	5.448	7.826	17.714	3
4	1.180	1.914	2.557	3.330	5.892	1.596	2.526	3.348	4.340	7.644	1.876	2.944	3.890	5.035	8.853	4
5	1.100	1.666	2.121	2.631	4.137	1.503	2.217	2.795	3.445	5.378	1.774	2.593	3.257	4.007	6.239	5
6	1.049	1.520	1.879	2.264	3.324	1.444	2.036	2.489	2.978	4.331	1.710	2.388	2.908	3.471	5.032	6
7	1.012	1.421	1.722	2.035	2.856	1.402	1.915	2.292	2.687	3.731	1.665	2.251	2.685	3.139	4.340	7
8	0.984	1.349	1.611	1.877	2.551	1.371	1.827	2.154	2.488	3.341	1.631	2.153	2.527	2.911	3.892	8
9	0.962	1.294	1.527	1.760	2.336	1.346	1.760	2.050	2.342	3.067	1.605	2.078	2.410	2.744	3.577	9
10	0.944	1.250	1.461	1.670	2.175	1.326	1.707	1.969	2.229	2.862	1.583	2.018	2.318	2.616	3.342	10
11	0.929	1.214	1.408	1.598	2.049	1.309	1.663	1.903	2.139	2.703	1.565	1.970	2.244	2.514	3.160	11
12	0.916	1.184	1.364	1.539	1.948	1.295	1.627	1.849	2.066	2.576	1.550	1.929	2.183	2.430	3.014	12
13	0.905	1.158	1.326	1.489	1.865	1.283	1.596	1.803	2.004	2.471	1.537	1.895	2.132	2.360	2.894	13
14	0.895	1.135	1.294	1.446	1.795	1.272	1.569	1.764	1.951	2.383	1.526	1.865	2.087	2.301	2.794	14
15	0.886	1.115	1.266	1.409	1.735	1.262	1.545	1.730	1.906	2.308	1.516	1.839	2.049	2.250	2.708	15
16	0.879	1.098	1.241	1.377	1.683	1.254	1.525	1.700	1.866	2.243	1.507	1.816	2.015	2.205	2.635	16
17	0.872	1.082	1.219	1.349	1.638	1.246	1.506	1.673	1.831	2.187	1.499	1.795	1.986	2.165	2.570	17
18	0.866	1.068	1.200	1.323	1.597	1.240	1.489	1.650	1.800	2.137	1.492	1.777	1.959	2.130	2.513	18
19	0.860	1.056	1.182	1.300	1.562	1.233	1.474	1.628	1.772	2.092	1.485	1.760	1.935	2.099	2.463	19
20	0.854	1.044	1.166	1.279	1.529	1.228	1.461	1.609	1.747	2.053	1.479	1.745	1.913	2.070	2.417	20
21	0.850	1.033	1.151	1.260	1.500	1.222	1.448	1.591	1.724	2.017	1.473	1.731	1.893	2.044	2.377	21
22	0.845	1.024	1.137	1.243	1.474	1.217	1.437	1.575	1.703	1.984	1.468	1.718	1.875	2.020	2.339	22
23	0.841	1.014	1.125	1.227	1.449	1.213	1.426	1.560	1.683	1.954	1.464	1.707	1.858	1.999	2.305	23
24	0.837	1.006	1.113	1.212	1.427	1.209	1.416	1.546	1.666	1.926	1.459	1.696	1.843	1.979	2.274	24
25	0.833	0.998	1.102	1.199	1.406	1.205	1.407	1.533	1.649	1.901	1.455	1.685	1.828	1.960	2.246	25
26	0.830	0.991	1.092	1.186	1.387	1.201	1.398	1.521	1.634	1.878	1.451	1.676	1.815	1.943	2.219	26
27	0.827	0.984	1.083	1.174	1.370	1.198	1.390	1.510	1.619	1.856	1.448	1.667	1.802	1.927	2.194	27
28	0.824	0.978	1.074	1.163	1.353	1.194	1.383	1.499	1.606	1.836	1.444	1.659	1.791	1.912	2.172	28
29	0.821	0.971	1.066	1.153	1.338	1.191	1.376	1.489	1.593	1.817	1.441	1.651	1.780	1.898	2.150	29
30	0.818	0.966	1.058	1.143	1.323	1.188	1.369	1.480	1.582	1.799	1.438	1.644	1.770	1.884	2.130	30
35	0.807	0.941	1.025	1.101	1.262	1.176	1.340	1.441	1.532	1.725	1.425	1.612	1.726	1.828	2.046	35
40	0.798	0.922	0.999	1.069	1.215	1.166	1.318	1.410	1.493	1.668	1.415	1.588	1.692	1.785	1.982	40
50	0.784	0.893	0.960	1.021	1.146	1.151	1.284	1.364	1.436	1.585	1.399	1.551	1.641	1.722	1.889	50
60	0.774	0.872	0.932	0.986	1.098	1.141	1.260	1.332	1.395	1.527	1.388	1.525	1.605	1.677	1.824	60
120	0.743	0.811	0.851	0.887	0.960	1.109	1.190	1.238	1.279	1.364	1.355	1.448	1.501	1.548	1.642	120
240	0.723	0.769	0.797	0.822	0.870	1.087	1.143	1.175	1.203	1.259	1.333	1.397	1.433	1.464	1.526	240
480	0.708	0.741	0.760	0.777	0.810	1.072	1.111	1.133	1.152	1.190	1.317	1.362	1.387	1.408	1.450	480
∞	0.674	0.674	0.674	0.674	0.674	1.036	1.036	1.036	1.036	1.036	1.282	1.282	1.282	1.282	1.282	∞

Table J.6a Factors $g''_{(1-\alpha,p,n)}$ for calculating normal distribution two-sided $100(1-\alpha)\%$ tolerance intervals to simultaneously control both tails of the distribution to $100p\%$ or less. The factors in this table were provided by Robert E. Odeh.

n	p = 0.050					p = 0.025					p = 0.005				
1 − α:	0.600	0.800	0.900	0.950	0.990	0.600	0.800	0.900	0.950	0.990	0.600	0.800	0.900	0.950	0.990
2	3.246	8.708	17.574	35.225	176.251	3.717	9.954	20.082	40.251	201.393	4.635	12.386	24.984	50.073	250.531
3	2.512	4.487	6.554	9.408	21.281	2.893	5.151	7.516	10.785	24.388	3.636	6.450	9.402	13.485	30.483
4	2.288	3.565	4.700	6.074	10.664	2.645	4.105	5.405	6.980	12.246	3.341	5.164	6.788	8.760	15.354
5	2.174	3.152	3.948	4.847	7.531	2.520	3.639	4.550	5.582	8.661	3.194	4.592	5.733	7.025	10.887
6	2.102	2.912	3.535	4.209	6.085	2.442	3.368	4.082	4.855	7.008	3.104	4.263	5.156	6.125	8.827
7	2.052	2.753	3.271	3.815	5.258	2.388	3.190	3.783	4.407	6.063	3.042	4.045	4.789	5.571	7.650
8	2.015	2.639	3.086	3.546	4.723	2.348	3.061	3.574	4.100	5.451	2.997	3.889	4.531	5.192	6.888
9	1.986	2.551	2.948	3.348	4.346	2.317	2.964	3.418	3.876	5.021	2.961	3.771	4.340	4.915	6.353
10	1.963	2.483	2.840	3.197	4.066	2.292	2.887	3.296	3.704	4.702	2.933	3.678	4.192	4.704	5.956
11	1.944	2.426	2.754	3.076	3.849	2.271	2.824	3.199	3.568	4.454	2.910	3.602	4.072	4.535	5.649
12	1.927	2.380	2.682	2.978	3.675	2.253	2.771	3.118	3.456	4.256	2.890	3.539	3.974	4.398	5.403
13	1.913	2.340	2.622	2.895	3.533	2.239	2.727	3.050	3.363	4.094	2.874	3.486	3.891	4.284	5.201
14	1.901	2.306	2.571	2.825	3.414	2.226	2.689	2.993	3.284	3.958	2.859	3.440	3.821	4.187	5.033
15	1.890	2.276	2.526	2.765	3.312	2.214	2.656	2.942	3.216	3.843	2.847	3.400	3.760	4.103	4.890
16	1.881	2.249	2.487	2.713	3.225	2.204	2.626	2.898	3.157	3.743	2.835	3.365	3.706	4.030	4.767
17	1.872	2.226	2.452	2.666	3.149	2.195	2.600	2.859	3.104	3.657	2.825	3.334	3.659	3.966	4.659
18	1.864	2.204	2.421	2.625	3.081	2.187	2.577	2.825	3.058	3.580	2.816	3.305	3.616	3.909	4.564
19	1.857	2.185	2.393	2.588	3.021	2.179	2.555	2.793	3.016	3.512	2.808	3.280	3.578	3.858	4.480
20	1.851	2.168	2.368	2.555	2.968	2.173	2.536	2.765	2.978	3.451	2.801	3.257	3.544	3.812	4.405
21	1.845	2.152	2.345	2.524	2.919	2.166	2.518	2.739	2.944	3.396	2.794	3.236	3.513	3.770	4.337
22	1.839	2.138	2.324	2.497	2.876	2.161	2.502	2.715	2.913	3.346	2.788	3.217	3.484	3.731	4.275
23	1.834	2.124	2.304	2.471	2.835	2.155	2.487	2.694	2.884	3.301	2.782	3.199	3.458	3.696	4.219
24	1.830	2.112	2.287	2.448	2.799	2.150	2.474	2.674	2.858	3.259	2.776	3.182	3.433	3.664	4.168
25	1.825	2.100	2.270	2.426	2.765	2.146	2.461	2.655	2.833	3.221	2.772	3.167	3.411	3.634	4.120
26	1.821	2.089	2.254	2.406	2.734	2.142	2.449	2.638	2.811	3.185	2.767	3.153	3.390	3.607	4.076
27	1.817	2.079	2.240	2.387	2.705	2.138	2.438	2.621	2.790	3.153	2.763	3.140	3.370	3.581	4.036
28	1.814	2.070	2.226	2.370	2.678	2.134	2.427	2.606	2.770	3.122	2.758	3.127	3.352	3.557	3.998
29	1.810	2.061	2.214	2.353	2.653	2.130	2.417	2.592	2.752	3.093	2.755	3.115	3.335	3.535	3.963
30	1.807	2.053	2.202	2.338	2.629	2.127	2.408	2.579	2.734	3.067	2.751	3.104	3.319	3.513	3.930
35	1.793	2.017	2.151	2.273	2.530	2.113	2.369	2.522	2.661	2.955	2.736	3.057	3.250	3.424	3.792
40	1.783	1.989	2.112	2.223	2.455	2.101	2.338	2.478	2.605	2.870	2.723	3.021	3.197	3.356	3.687
50	1.767	1.947	2.054	2.149	2.346	2.085	2.292	2.414	2.522	2.747	2.705	2.966	3.119	3.255	3.536
60	1.755	1.918	2.013	2.097	2.270	2.072	2.259	2.368	2.464	2.661	2.693	2.928	3.064	3.184	3.431
120	1.720	1.831	1.894	1.949	2.059	2.037	2.164	2.236	2.298	2.423	2.655	2.815	2.905	2.983	3.139
240	1.697	1.773	1.816	1.853	1.925	2.013	2.101	2.149	2.191	2.273	2.630	2.741	2.802	2.854	2.956
480	1.681	1.734	1.764	1.788	1.837	1.997	2.058	2.091	2.119	2.175	2.614	2.690	2.732	2.767	2.836
∞	1.645	1.645	1.645	1.645	1.645	1.960	1.960	1.960	1.960	1.960	2.576	2.576	2.576	2.576	2.576

Table J.6b Factors $g''_{(1-\alpha,p,n)}$ for calculating normal distribution two-sided $100(1-\alpha)$% tolerance intervals to simultaneously control both tails of the distribution to $100p$% or less. The factors in this table were provided by Robert E. Odeh.

n	p = 0.01 0.010	0.025	0.050	0.100	0.200	p = 0.05 0.010	0.025	0.050	0.100	0.200	p = 0.10 0.010	0.025	0.050	0.100	0.200	n
2	0.564	0.761	0.954	1.225	1.672	0.000	0.273	0.475	0.717	1.077	-0.707	-0.143	0.138	0.403	0.737	2
3	0.782	0.958	1.130	1.361	1.710	0.295	0.478	0.639	0.840	1.126	-0.072	0.159	0.334	0.535	0.799	3
4	0.924	1.088	1.246	1.455	1.760	0.443	0.601	0.743	0.922	1.172	0.123	0.298	0.444	0.617	0.847	4
5	1.027	1.182	1.331	1.525	1.801	0.543	0.687	0.818	0.982	1.209	0.238	0.389	0.519	0.675	0.883	5
6	1.108	1.256	1.396	1.578	1.834	0.618	0.752	0.875	1.028	1.238	0.319	0.455	0.575	0.719	0.911	6
7	1.173	1.315	1.449	1.622	1.862	0.678	0.804	0.920	1.065	1.261	0.381	0.507	0.619	0.755	0.933	7
8	1.227	1.364	1.493	1.658	1.885	0.727	0.847	0.958	1.096	1.281	0.431	0.550	0.655	0.783	0.952	8
9	1.273	1.406	1.530	1.688	1.904	0.768	0.884	0.990	1.122	1.298	0.472	0.585	0.686	0.808	0.968	9
10	1.314	1.442	1.563	1.715	1.922	0.804	0.915	1.017	1.144	1.313	0.508	0.615	0.712	0.828	0.981	10
11	1.349	1.474	1.591	1.738	1.937	0.835	0.943	1.041	1.163	1.325	0.538	0.642	0.734	0.847	0.993	11
12	1.381	1.502	1.616	1.758	1.950	0.862	0.967	1.062	1.180	1.337	0.565	0.665	0.754	0.863	1.004	12
13	1.409	1.528	1.638	1.776	1.962	0.887	0.989	1.081	1.196	1.347	0.589	0.685	0.772	0.877	1.013	13
14	1.434	1.551	1.658	1.793	1.973	0.909	1.008	1.098	1.210	1.356	0.610	0.704	0.788	0.890	1.022	14
15	1.458	1.572	1.677	1.808	1.983	0.929	1.026	1.114	1.222	1.364	0.629	0.721	0.802	0.901	1.029	15
16	1.479	1.591	1.694	1.822	1.992	0.948	1.042	1.128	1.234	1.372	0.647	0.736	0.815	0.912	1.036	16
17	1.499	1.608	1.709	1.834	2.000	0.965	1.057	1.141	1.244	1.379	0.663	0.750	0.827	0.921	1.043	17
18	1.517	1.625	1.724	1.846	2.008	0.980	1.071	1.153	1.254	1.385	0.678	0.763	0.839	0.930	1.049	18
19	1.534	1.640	1.737	1.857	2.015	0.995	1.084	1.164	1.263	1.391	0.692	0.775	0.849	0.939	1.054	19
20	1.550	1.654	1.749	1.867	2.022	1.008	1.095	1.175	1.271	1.397	0.705	0.786	0.858	0.946	1.059	20
21	1.565	1.667	1.761	1.876	2.028	1.021	1.107	1.184	1.279	1.402	0.716	0.796	0.867	0.953	1.064	21
22	1.579	1.680	1.772	1.885	2.034	1.033	1.117	1.193	1.286	1.407	0.728	0.806	0.876	0.960	1.068	22
23	1.592	1.691	1.782	1.893	2.039	1.044	1.127	1.202	1.293	1.412	0.738	0.815	0.884	0.966	1.073	23
24	1.605	1.702	1.791	1.901	2.045	1.054	1.136	1.210	1.300	1.416	0.748	0.823	0.891	0.972	1.076	24
25	1.616	1.713	1.801	1.908	2.049	1.064	1.145	1.217	1.306	1.420	0.757	0.831	0.898	0.978	1.080	25
26	1.628	1.723	1.809	1.915	2.054	1.074	1.153	1.225	1.311	1.424	0.766	0.839	0.904	0.983	1.084	26
27	1.638	1.732	1.817	1.922	2.058	1.083	1.161	1.231	1.317	1.427	0.774	0.846	0.911	0.988	1.087	27
28	1.648	1.741	1.825	1.928	2.063	1.091	1.168	1.238	1.322	1.431	0.782	0.853	0.917	0.993	1.090	28
29	1.658	1.749	1.833	1.934	2.067	1.099	1.175	1.244	1.327	1.434	0.790	0.860	0.922	0.997	1.093	29
30	1.667	1.757	1.840	1.940	2.070	1.107	1.182	1.250	1.332	1.437	0.797	0.866	0.928	1.002	1.096	30
35	1.708	1.793	1.871	1.965	2.087	1.141	1.212	1.276	1.352	1.451	0.828	0.893	0.951	1.020	1.108	35
40	1.741	1.823	1.896	1.986	2.101	1.169	1.236	1.297	1.369	1.462	0.854	0.916	0.970	1.036	1.119	40
50	1.793	1.869	1.936	2.018	2.122	1.212	1.274	1.329	1.396	1.480	0.894	0.950	1.000	1.059	1.134	50
60	1.833	1.903	1.966	2.042	2.138	1.245	1.303	1.354	1.415	1.493	0.924	0.976	1.022	1.077	1.146	60
120	1.963	2.016	2.063	2.119	2.189	1.352	1.395	1.433	1.478	1.535	1.020	1.059	1.093	1.134	1.184	120
240	2.061	2.100	2.135	2.176	2.227	1.431	1.463	1.492	1.525	1.565	1.092	1.121	1.146	1.175	1.211	240
480	2.134	2.163	2.189	2.218	2.255	1.491	1.514	1.535	1.558	1.588	1.145	1.166	1.184	1.205	1.231	480
∞	2.326	2.326	2.326	2.326	2.326	1.645	1.645	1.645	1.645	1.645	1.282	1.282	1.282	1.282	1.282	∞

Table J.7a Factors $g'_{(\gamma;p,n)}$ for calculating two-sided confidence intervals and one-sided confidence bounds for normal distribution quantiles or normal distribution one-sided tolerance bounds; see Section 4.6.3. The factors in this table were computed with an algorithm provided by Robert E. Odeh.

n	γ:	p = 0.20					p = 0.30					p = 0.40					n
		0.010	0.025	0.050	0.100	0.200	0.010	0.025	0.050	0.100	0.200	0.010	0.025	0.050	0.100	0.200	
2		-3.204	-1.229	-0.521	-0.084	0.288	-7.494	-2.968	-1.431	-0.610	-0.091	-13.821	-5.508	-2.718	-1.286	-0.493	2
3		-0.792	-0.380	-0.127	0.111	0.377	-1.662	-0.966	-0.580	-0.261	0.043	-2.728	-1.657	-1.090	-0.650	-0.276	3
4		-0.405	-0.158	0.021	0.209	0.432	-0.949	-0.592	-0.355	-0.130	0.111	-1.563	-1.061	-0.746	-0.463	-0.187	4
5		-0.227	-0.038	0.110	0.272	0.470	-0.670	-0.418	-0.237	-0.052	0.155	-1.145	-0.811	-0.583	-0.363	-0.133	5
6		-0.117	0.043	0.173	0.319	0.499	-0.513	-0.312	-0.160	0.001	0.186	-0.924	-0.668	-0.483	-0.297	-0.096	6
7		-0.040	0.103	0.220	0.355	0.522	-0.409	-0.237	-0.103	0.041	0.211	-0.784	-0.572	-0.413	-0.249	-0.068	7
8		0.020	0.150	0.258	0.384	0.540	-0.333	-0.181	-0.060	0.073	0.231	-0.685	-0.501	-0.361	-0.213	-0.046	8
9		0.067	0.188	0.290	0.408	0.556	-0.275	-0.136	-0.025	0.099	0.247	-0.611	-0.447	-0.319	-0.183	-0.028	9
10		0.107	0.220	0.316	0.428	0.569	-0.227	-0.100	0.004	0.121	0.261	-0.552	-0.403	-0.286	-0.159	-0.013	10
11		0.140	0.247	0.339	0.446	0.580	-0.188	-0.069	0.029	0.139	0.272	-0.504	-0.367	-0.258	-0.138	0.000	11
12		0.169	0.271	0.359	0.461	0.591	-0.155	-0.043	0.050	0.156	0.283	-0.464	-0.336	-0.233	-0.121	0.011	12
13		0.194	0.292	0.376	0.475	0.599	-0.126	-0.020	0.069	0.170	0.292	-0.430	-0.310	-0.213	-0.105	0.020	13
14		0.216	0.310	0.392	0.487	0.608	-0.101	0.001	0.086	0.182	0.300	-0.401	-0.287	-0.194	-0.092	0.029	14
15		0.236	0.327	0.406	0.498	0.615	-0.079	0.019	0.100	0.194	0.307	-0.375	-0.267	-0.178	-0.080	0.037	15
16		0.254	0.342	0.419	0.508	0.621	-0.059	0.035	0.114	0.204	0.314	-0.352	-0.248	-0.163	-0.069	0.044	16
17		0.271	0.356	0.430	0.518	0.627	-0.041	0.050	0.126	0.213	0.320	-0.332	-0.232	-0.150	-0.059	0.050	17
18		0.286	0.369	0.441	0.526	0.633	-0.024	0.063	0.137	0.222	0.325	-0.313	-0.217	-0.138	-0.050	0.056	18
19		0.299	0.380	0.451	0.534	0.638	-0.009	0.075	0.147	0.230	0.330	-0.296	-0.204	-0.127	-0.041	0.061	19
20		0.312	0.391	0.460	0.541	0.643	0.004	0.086	0.156	0.237	0.335	-0.281	-0.192	-0.117	-0.034	0.066	20
21		0.324	0.401	0.468	0.548	0.647	0.017	0.097	0.165	0.243	0.340	-0.267	-0.180	-0.108	-0.027	0.070	21
22		0.335	0.410	0.476	0.554	0.651	0.029	0.107	0.173	0.250	0.344	-0.254	-0.170	-0.099	-0.020	0.075	22
23		0.345	0.419	0.484	0.560	0.655	0.040	0.116	0.181	0.256	0.347	-0.242	-0.160	-0.091	-0.014	0.078	23
24		0.355	0.427	0.491	0.565	0.659	0.050	0.124	0.188	0.261	0.351	-0.230	-0.151	-0.084	-0.008	0.082	24
25		0.364	0.435	0.497	0.570	0.662	0.059	0.132	0.194	0.266	0.354	-0.220	-0.142	-0.077	-0.003	0.086	25
26		0.373	0.442	0.503	0.575	0.665	0.068	0.139	0.200	0.271	0.358	-0.210	-0.134	-0.070	0.002	0.089	26
27		0.381	0.449	0.509	0.580	0.669	0.077	0.147	0.206	0.276	0.361	-0.201	-0.127	-0.064	0.007	0.092	27
28		0.388	0.456	0.515	0.584	0.671	0.085	0.153	0.212	0.280	0.363	-0.192	-0.120	-0.058	0.011	0.095	28
29		0.396	0.462	0.520	0.588	0.674	0.092	0.159	0.217	0.284	0.366	-0.184	-0.113	-0.053	0.015	0.098	29
30		0.403	0.468	0.525	0.592	0.677	0.099	0.165	0.222	0.288	0.369	-0.176	-0.106	-0.048	0.020	0.100	30
35		0.433	0.494	0.547	0.610	0.688	0.130	0.191	0.244	0.305	0.380	-0.143	-0.079	-0.025	0.037	0.111	35
40		0.457	0.514	0.565	0.624	0.698	0.155	0.212	0.261	0.319	0.389	-0.116	-0.057	-0.007	0.051	0.121	40
50		0.495	0.547	0.592	0.645	0.712	0.193	0.244	0.288	0.340	0.403	-0.077	-0.024	0.021	0.072	0.134	50
60		0.523	0.571	0.612	0.661	0.723	0.221	0.268	0.308	0.355	0.413	-0.047	0.000	0.041	0.088	0.145	60
120		0.611	0.646	0.676	0.712	0.756	0.308	0.341	0.370	0.404	0.445	0.041	0.074	0.103	0.136	0.176	120
240		0.676	0.701	0.723	0.749	0.780	0.370	0.393	0.414	0.438	0.468	0.103	0.126	0.147	0.170	0.199	240
480		0.723	0.741	0.757	0.775	0.798	0.414	0.431	0.446	0.463	0.484	0.147	0.163	0.178	0.194	0.215	480
∞		0.842	0.842	0.842	0.842	0.842	0.524	0.524	0.524	0.524	0.524	0.253	0.253	0.253	0.253	0.253	∞

Table J.7b Factors $g'_{(\gamma;p,n)}$ for calculating two-sided confidence intervals and one-sided confidence bounds for normal distribution quantiles or normal distribution one-sided tolerance bounds; see Section 4.6.3. The factors in this table were computed with an algorithm provided by Robert E. Odeh.

n	γ:	p = 0.01					p = 0.05					p = 0.10					n
		0.800	0.900	0.950	0.975	0.990	0.800	0.900	0.950	0.975	0.990	0.800	0.900	0.950	0.975	0.990	
2		9.156	18.500	37.094	74.234	185.617	6.464	13.090	26.260	52.559	131.426	5.049	10.253	20.581	41.201	103.029	2
3		5.010	7.340	10.553	15.043	23.896	3.604	5.311	7.656	10.927	17.370	2.871	4.258	6.155	8.797	13.995	3
4		4.110	5.438	7.042	9.018	12.387	2.968	3.957	5.144	6.602	9.083	2.372	3.188	4.162	5.354	7.380	4
5		3.711	4.666	5.741	6.980	8.939	2.683	3.400	4.203	5.124	6.578	2.145	2.742	3.407	4.166	5.362	5
6		3.482	4.243	5.062	5.967	7.335	2.517	3.092	3.708	4.385	5.406	2.012	2.494	3.006	3.568	4.411	6
7		3.331	3.972	4.642	5.361	6.412	2.407	2.894	3.399	3.940	4.728	1.923	2.333	2.755	3.206	3.859	7
8		3.224	3.783	4.354	4.954	5.812	2.328	2.754	3.187	3.640	4.285	1.859	2.219	2.582	2.960	3.497	8
9		3.142	3.641	4.143	4.662	5.389	2.268	2.650	3.031	3.424	3.972	1.809	2.133	2.454	2.783	3.240	9
10		3.078	3.532	3.981	4.440	5.074	2.220	2.568	2.911	3.259	3.738	1.770	2.066	2.355	2.647	3.048	10
11		3.026	3.443	3.852	4.265	4.829	2.182	2.503	2.815	3.129	3.556	1.738	2.011	2.275	2.540	2.898	11
12		2.982	3.371	3.747	4.124	4.633	2.149	2.448	2.736	3.023	3.410	1.711	1.966	2.210	2.452	2.777	12
13		2.946	3.309	3.659	4.006	4.472	2.122	2.402	2.671	2.936	3.290	1.689	1.928	2.155	2.379	2.677	13
14		2.914	3.257	3.585	3.907	4.337	2.098	2.363	2.614	2.861	3.189	1.669	1.895	2.109	2.317	2.593	14
15		2.887	3.212	3.520	3.822	4.222	2.078	2.329	2.566	2.797	3.102	1.652	1.867	2.068	2.264	2.521	15
16		2.863	3.172	3.464	3.749	4.123	2.059	2.299	2.524	2.742	3.028	1.637	1.842	2.033	2.218	2.459	16
17		2.841	3.137	3.414	3.684	4.037	2.043	2.272	2.486	2.693	2.963	1.623	1.819	2.002	2.177	2.405	17
18		2.822	3.105	3.370	3.627	3.960	2.029	2.249	2.453	2.650	2.905	1.611	1.800	1.974	2.141	2.357	18
19		2.804	3.077	3.331	3.575	3.892	2.016	2.227	2.423	2.611	2.854	1.600	1.782	1.949	2.108	2.314	19
20		2.789	3.052	3.295	3.529	3.832	2.004	2.208	2.396	2.576	2.808	1.590	1.765	1.926	2.079	2.276	20
21		2.774	3.028	3.263	3.487	3.777	1.993	2.190	2.371	2.544	2.766	1.581	1.750	1.905	2.053	2.241	21
22		2.761	3.007	3.233	3.449	3.727	1.983	2.174	2.349	2.515	2.729	1.572	1.737	1.886	2.028	2.209	22
23		2.749	2.987	3.206	3.414	3.681	1.973	2.159	2.328	2.489	2.694	1.564	1.724	1.869	2.006	2.180	23
24		2.738	2.969	3.181	3.382	3.640	1.965	2.145	2.309	2.465	2.662	1.557	1.712	1.853	1.985	2.154	24
25		2.727	2.952	3.158	3.353	3.601	1.957	2.132	2.292	2.442	2.633	1.550	1.702	1.838	1.966	2.129	25
26		2.718	2.937	3.136	3.325	3.566	1.949	2.120	2.275	2.421	2.606	1.544	1.691	1.824	1.949	2.106	26
27		2.708	2.922	3.116	3.300	3.533	1.943	2.109	2.260	2.402	2.581	1.538	1.682	1.811	1.932	2.085	27
28		2.700	2.909	3.098	3.276	3.502	1.936	2.099	2.246	2.384	2.558	1.533	1.673	1.799	1.917	2.065	28
29		2.692	2.896	3.080	3.254	3.473	1.930	2.089	2.232	2.367	2.536	1.528	1.665	1.788	1.903	2.047	29
30		2.684	2.884	3.064	3.233	3.447	1.924	2.080	2.220	2.351	2.515	1.523	1.657	1.777	1.889	2.030	30
35		2.652	2.833	2.995	3.145	3.334	1.900	2.041	2.167	2.284	2.430	1.502	1.624	1.732	1.833	1.957	35
40		2.627	2.793	2.941	3.078	3.249	1.880	2.010	2.125	2.232	2.364	1.486	1.598	1.697	1.789	1.902	40
50		2.590	2.735	2.862	2.980	3.125	1.852	1.965	2.065	2.156	2.269	1.461	1.559	1.646	1.724	1.821	50
60		2.564	2.694	2.807	2.911	3.038	1.832	1.933	2.022	2.103	2.202	1.444	1.532	1.609	1.679	1.764	60
120		2.488	2.574	2.649	2.716	2.797	1.772	1.841	1.899	1.952	2.015	1.393	1.452	1.503	1.549	1.604	120
240		2.437	2.497	2.547	2.591	2.645	1.733	1.780	1.819	1.854	1.896	1.358	1.399	1.434	1.465	1.501	240
480		2.403	2.444	2.479	2.509	2.545	1.706	1.738	1.766	1.790	1.818	1.335	1.363	1.387	1.408	1.433	480
∞		2.326	2.326	2.326	2.326	2.326	1.645	1.645	1.645	1.645	1.645	1.282	1.282	1.282	1.282	1.282	∞

Table J.7c Factors $g'_{(\gamma;p,n)}$ for calculating two-sided confidence intervals and one-sided confidence bounds for normal distribution quantiles or normal distribution one-sided tolerance bounds; see Section 4.6.3. The factors in this table were computed with an algorithm provided by Robert E. Odeh.

n	p = 0.20					p = 0.30					p = 0.40					n
γ:	0.800	0.900	0.950	0.975	0.990	0.800	0.900	0.950	0.975	0.990	0.800	0.900	0.950	0.975	0.990	
2	3.417	6.987	14.051	28.140	70.376	2.357	4.881	9.843	19.726	49.344	1.577	3.343	6.778	13.602	34.038	2
3	2.016	3.039	4.424	6.343	10.111	1.441	2.228	3.277	4.722	7.547	0.991	1.602	2.399	3.484	5.593	3
4	1.675	2.295	3.026	3.915	5.417	1.199	1.693	2.265	2.954	4.112	0.819	1.219	1.672	2.209	3.102	4
5	1.514	1.976	2.483	3.058	3.958	1.080	1.456	1.861	2.315	3.020	0.729	1.042	1.370	1.732	2.287	5
6	1.417	1.795	2.191	2.621	3.262	1.006	1.318	1.638	1.982	2.490	0.672	0.935	1.199	1.478	1.884	6
7	1.352	1.676	2.005	2.353	2.854	0.955	1.225	1.495	1.775	2.176	0.631	0.862	1.087	1.317	1.642	7
8	1.304	1.590	1.875	2.170	2.584	0.917	1.158	1.393	1.633	1.967	0.600	0.808	1.006	1.205	1.478	8
9	1.266	1.525	1.779	2.036	2.391	0.888	1.107	1.317	1.528	1.816	0.576	0.766	0.945	1.121	1.358	9
10	1.237	1.474	1.703	1.933	2.246	0.864	1.066	1.257	1.446	1.701	0.556	0.733	0.896	1.055	1.267	10
11	1.212	1.433	1.643	1.851	2.131	0.844	1.032	1.208	1.381	1.610	0.540	0.705	0.856	1.002	1.194	11
12	1.192	1.398	1.593	1.784	2.039	0.827	1.004	1.168	1.327	1.537	0.526	0.681	0.823	0.958	1.134	12
13	1.174	1.368	1.551	1.728	1.963	0.813	0.980	1.133	1.282	1.475	0.513	0.661	0.794	0.921	1.084	13
14	1.159	1.343	1.514	1.681	1.898	0.800	0.959	1.104	1.243	1.424	0.502	0.643	0.770	0.889	1.041	14
15	1.145	1.321	1.483	1.639	1.843	0.789	0.940	1.078	1.210	1.379	0.493	0.628	0.748	0.861	1.005	15
16	1.133	1.301	1.455	1.603	1.795	0.779	0.924	1.056	1.180	1.340	0.484	0.614	0.729	0.836	0.972	16
17	1.123	1.284	1.431	1.572	1.753	0.770	0.910	1.035	1.154	1.306	0.477	0.601	0.712	0.814	0.944	17
18	1.113	1.268	1.409	1.543	1.716	0.762	0.896	1.017	1.131	1.275	0.470	0.590	0.696	0.795	0.918	18
19	1.104	1.254	1.389	1.518	1.682	0.755	0.885	1.001	1.110	1.248	0.463	0.580	0.682	0.777	0.895	19
20	1.096	1.241	1.371	1.495	1.652	0.748	0.874	0.986	1.091	1.223	0.458	0.570	0.669	0.761	0.875	20
21	1.089	1.229	1.355	1.474	1.625	0.742	0.864	0.972	1.073	1.200	0.452	0.562	0.658	0.746	0.856	21
22	1.082	1.218	1.340	1.455	1.600	0.736	0.855	0.960	1.057	1.180	0.447	0.554	0.647	0.732	0.838	22
23	1.076	1.208	1.326	1.437	1.577	0.731	0.846	0.948	1.043	1.161	0.443	0.546	0.637	0.720	0.822	23
24	1.070	1.199	1.313	1.421	1.556	0.726	0.838	0.937	1.029	1.144	0.438	0.540	0.628	0.708	0.808	24
25	1.065	1.190	1.302	1.406	1.537	0.722	0.831	0.927	1.017	1.128	0.434	0.533	0.619	0.698	0.794	25
26	1.060	1.182	1.291	1.392	1.519	0.717	0.824	0.918	1.005	1.113	0.430	0.527	0.611	0.687	0.781	26
27	1.055	1.174	1.280	1.379	1.502	0.713	0.812	0.910	0.994	1.099	0.427	0.522	0.604	0.678	0.769	27
28	1.051	1.167	1.271	1.367	1.486	0.710	0.812	0.901	0.984	1.086	0.423	0.516	0.596	0.669	0.758	28
29	1.047	1.160	1.262	1.355	1.472	0.706	0.806	0.894	0.974	1.073	0.420	0.511	0.590	0.661	0.748	29
30	1.043	1.154	1.253	1.344	1.458	0.703	0.801	0.886	0.965	1.062	0.417	0.506	0.583	0.653	0.738	30
35	1.026	1.127	1.217	1.299	1.400	0.688	0.778	0.856	0.927	1.014	0.404	0.486	0.556	0.619	0.696	35
40	1.013	1.106	1.188	1.263	1.356	0.677	0.760	0.831	0.896	0.976	0.394	0.470	0.535	0.593	0.663	40
50	0.993	1.075	1.146	1.211	1.291	0.659	0.732	0.795	0.852	0.920	0.379	0.446	0.503	0.554	0.615	50
60	0.978	1.052	1.116	1.174	1.245	0.647	0.713	0.769	0.820	0.881	0.367	0.428	0.480	0.525	0.580	60
120	0.936	0.986	1.029	1.068	1.113	0.609	0.655	0.693	0.727	0.767	0.333	0.375	0.411	0.442	0.478	120
240	0.907	0.942	0.971	0.997	1.028	0.584	0.615	0.642	0.665	0.692	0.309	0.339	0.363	0.385	0.410	240
480	0.887	0.912	0.932	0.950	0.971	0.566	0.588	0.606	0.622	0.641	0.293	0.313	0.331	0.346	0.363	480
∞	0.842	0.842	0.842	0.842	0.842	0.524	0.524	0.524	0.524	0.524	0.253	0.253	0.253	0.253	0.253	∞

Table J.7d Factors $g'_{(\gamma:p,n)}$ for calculating two-sided confidence intervals and one-sided confidence bounds for normal distribution quantiles or normal distribution one-sided tolerance bounds; see Section 4.6.3. The factors in this table were computed with an algorithm provided by Robert E. Odeh.

$1-\alpha$	n	m: 1	2	3	4	5	6	7	8	9	10	12	16	20	40	60	80	100
	4	2.631	3.329	3.742	4.032	4.255	4.435	4.585	4.713	4.826	4.925	5.095	5.357	5.555	6.143	6.468	6.692	6.861
	5	2.335	2.909	3.246	3.483	3.665	3.812	3.936	4.041	4.134	4.216	4.356	4.573	4.738	5.229	5.502	5.690	5.833
	6	2.177	2.685	2.982	3.190	3.350	3.480	3.589	3.682	3.764	3.836	3.960	4.153	4.299	4.736	4.980	5.148	5.276
	7	2.077	2.546	2.818	3.008	3.155	3.273	3.373	3.458	3.533	3.599	3.713	3.889	4.023	4.426	4.652	4.807	4.925
	8	2.010	2.452	2.706	2.884	3.021	3.132	3.225	3.305	3.375	3.437	3.543	3.708	3.834	4.213	4.425	4.571	4.683
	9	1.960	2.383	2.625	2.794	2.924	3.029	3.118	3.193	3.259	3.318	3.419	3.576	3.696	4.056	4.258	4.398	4.505
	10	1.923	2.331	2.564	2.726	2.851	2.951	3.036	3.108	3.172	3.228	3.325	3.476	3.591	3.936	4.131	4.265	4.368
0.90	15	1.819	2.188	2.395	2.539	2.648	2.737	2.811	2.875	2.930	2.980	3.065	3.197	3.297	3.601	3.772	3.891	3.982
	20	1.772	2.122	2.318	2.454	2.556	2.639	2.709	2.768	2.820	2.866	2.945	3.068	3.162	3.445	3.604	3.715	3.800
	25	1.745	2.085	2.275	2.405	2.504	2.583	2.650	2.707	2.757	2.801	2.877	2.994	3.084	3.354	3.506	3.612	3.693
	30	1.727	2.061	2.246	2.373	2.470	2.547	2.612	2.667	2.716	2.758	2.832	2.946	3.033	3.294	3.442	3.544	3.623
	60	1.685	2.003	2.178	2.298	2.388	2.460	2.520	2.572	2.617	2.656	2.724	2.829	2.909	3.149	3.283	3.377	3.448
	120	1.665	1.976	2.146	2.262	2.349	2.419	2.477	2.526	2.569	2.607	2.673	2.773	2.849	3.078	3.206	3.294	3.362
	∞	1.645	1.949	2.114	2.226	2.311	2.378	2.434	2.481	2.523	2.560	2.622	2.718	2.791	3.008	3.129	3.212	3.276
	4	3.558	4.412	4.923	5.285	5.564	5.789	5.978	6.140	6.282	6.407	6.622	6.954	7.206	7.954	8.370	8.655	8.872
	5	3.041	3.697	4.087	4.364	4.577	4.751	4.896	5.021	5.131	5.229	5.395	5.655	5.852	6.441	6.771	6.997	7.170
	6	2.777	3.333	3.662	3.896	4.076	4.223	4.346	4.452	4.545	4.628	4.770	4.991	5.159	5.665	5.949	6.145	6.295
	7	2.616	3.114	3.407	3.614	3.774	3.905	4.014	4.108	4.191	4.265	4.391	4.589	4.739	5.194	5.449	5.626	5.761
	8	2.508	2.968	3.236	3.426	3.573	3.692	3.792	3.879	3.954	4.022	4.138	4.319	4.457	4.876	5.112	5.275	5.400
	9	2.431	2.863	3.115	3.292	3.429	3.540	3.634	3.714	3.785	3.848	3.956	4.125	4.255	4.647	4.869	5.022	5.140
	10	2.373	2.785	3.024	3.192	3.321	3.426	3.515	3.591	3.658	3.717	3.820	3.980	4.102	4.474	4.685	4.831	4.942
0.95	15	2.215	2.574	2.778	2.921	3.031	3.120	3.194	3.258	3.314	3.365	3.451	3.585	3.689	4.002	4.180	4.305	4.400
	20	2.145	2.480	2.670	2.801	2.902	2.983	3.052	3.110	3.162	3.208	3.286	3.409	3.503	3.788	3.951	4.064	4.151
	25	2.105	2.427	2.608	2.734	2.829	2.907	2.971	3.027	3.075	3.119	3.193	3.309	3.397	3.666	3.820	3.926	4.008
	30	2.079	2.393	2.569	2.690	2.783	2.857	2.920	2.973	3.020	3.062	3.133	3.244	3.329	3.587	3.734	3.837	3.915
	60	2.018	2.312	2.475	2.587	2.672	2.740	2.797	2.846	2.888	2.926	2.991	3.091	3.167	3.398	3.529	3.619	3.689
	120	1.988	2.274	2.431	2.538	2.619	2.685	2.739	2.785	2.826	2.862	2.923	3.018	3.090	3.308	3.430	3.515	3.580
	∞	1.960	2.236	2.388	2.491	2.569	2.631	2.683	2.727	2.766	2.800	2.858	2.948	3.016	3.220	3.335	3.414	3.474
	4	6.530	7.942	8.800	9.411	9.884	10.268	10.590	10.867	11.110	11.325	11.693	12.264	12.698	13.991	14.711	15.207	15.582
	5	5.044	5.972	6.536	6.940	7.253	7.509	7.725	7.911	8.074	8.219	8.469	8.857	9.154	10.045	10.545	10.891	11.153
	6	4.355	5.071	5.503	5.814	6.055	6.253	6.420	6.564	6.690	6.803	6.998	7.302	7.535	8.239	8.637	8.912	9.122
	7	3.963	4.562	4.922	5.181	5.382	5.546	5.685	5.806	5.912	6.006	6.169	6.425	6.621	7.217	7.556	7.791	7.970
	8	3.712	4.238	4.552	4.778	4.953	5.097	5.218	5.323	5.416	5.499	5.641	5.865	6.038	6.564	6.863	7.072	7.231
	9	3.537	4.014	4.297	4.500	4.657	4.787	4.896	4.990	5.074	5.148	5.277	5.479	5.634	6.110	6.382	6.572	6.717
	10	3.408	3.850	4.111	4.297	4.442	4.560	4.660	4.747	4.824	4.892	5.010	5.196	5.339	5.778	6.029	6.205	6.340
0.99	15	3.074	3.426	3.631	3.776	3.888	3.979	4.056	4.123	4.182	4.234	4.325	4.468	4.578	4.916	5.112	5.249	5.354
	20	2.932	3.247	3.428	3.556	3.655	3.735	3.802	3.860	3.912	3.957	4.036	4.160	4.256	4.550	4.720	4.839	4.931
	25	2.852	3.148	3.317	3.435	3.526	3.600	3.663	3.716	3.763	3.806	3.878	3.992	4.079	4.348	4.503	4.612	4.696
	30	2.802	3.085	3.247	3.359	3.446	3.516	3.575	3.625	3.670	3.710	3.778	3.885	3.968	4.221	4.366	4.468	4.547
	60	2.684	2.939	3.082	3.182	3.258	3.319	3.371	3.415	3.454	3.488	3.547	3.639	3.710	3.925	4.048	4.134	4.200
	120	2.629	2.871	3.006	3.100	3.171	3.229	3.277	3.318	3.354	3.386	3.441	3.526	3.591	3.789	3.901	3.980	4.040
	∞	2.576	2.806	2.934	3.022	3.089	3.143	3.188	3.226	3.260	3.289	3.340	3.419	3.479	3.661	3.764	3.835	3.889

Table J.8 Factors $r_{(1-\alpha;m,n)}$ for calculating normal distribution two-sided $100(1-\alpha)\%$ prediction intervals for all m future observations using the results of a previous sample of n observations. The factors in this table were computed and provided to us by Robert E. Odeh.

$1-\alpha$	n	m: 1	2	3	4	5	6	7	8	9	10	12	16	20	40	60	80	100
	4	1.831	2.484	2.873	3.150	3.364	3.538	3.684	3.809	3.919	4.017	4.184	4.443	4.640	5.226	5.552	5.776	5.946
	5	1.680	2.240	2.567	2.798	2.975	3.120	3.241	3.345	3.436	3.517	3.656	3.871	4.035	4.526	4.799	4.988	5.131
	6	1.594	2.105	2.399	2.606	2.764	2.893	3.001	3.093	3.174	3.246	3.370	3.562	3.708	4.146	4.391	4.560	4.688
	7	1.539	2.020	2.294	2.485	2.631	2.750	2.849	2.935	3.010	3.076	3.190	3.367	3.502	3.906	4.133	4.290	4.409
	8	1.501	1.961	2.221	2.402	2.540	2.652	2.745	2.826	2.896	2.959	3.066	3.232	3.359	3.741	3.955	4.103	4.215
	9	1.472	1.917	2.167	2.341	2.473	2.580	2.669	2.746	2.813	2.873	2.975	3.134	3.255	3.619	3.823	3.965	4.072
0.90	10	1.451	1.884	2.126	2.294	2.422	2.525	2.611	2.685	2.750	2.807	2.906	3.059	3.175	3.526	3.723	3.859	3.963
	15	1.389	1.792	2.013	2.165	2.280	2.372	2.450	2.516	2.573	2.625	2.712	2.848	2.952	3.263	3.438	3.560	3.652
	20	1.361	1.749	1.961	2.105	2.215	2.302	2.375	2.437	2.492	2.540	2.623	2.750	2.848	3.140	3.304	3.418	3.505
	25	1.344	1.724	1.931	2.071	2.177	2.262	2.332	2.392	2.445	2.491	2.571	2.694	2.787	3.068	3.226	3.335	3.418
	30	1.333	1.708	1.911	2.049	2.152	2.235	2.304	2.363	2.414	2.460	2.537	2.657	2.748	3.021	3.174	3.280	3.361
	60	1.307	1.669	1.864	1.995	2.093	2.171	2.236	2.292	2.340	2.383	2.455	2.567	2.652	2.904	3.046	3.143	3.218
	120	1.294	1.651	1.841	1.969	2.064	2.141	2.204	2.257	2.304	2.345	2.415	2.523	2.605	2.847	2.982	3.075	3.146
	∞	1.282	1.632	1.818	1.943	2.036	2.111	2.172	2.224	2.269	2.309	2.376	2.480	2.559	2.791	2.919	3.008	3.075
	4	2.631	3.401	3.871	4.209	4.472	4.686	4.867	5.023	5.160	5.282	5.491	5.816	6.063	6.805	7.218	7.503	7.719
	5	2.335	2.952	3.320	3.583	3.788	3.955	4.095	4.216	4.323	4.418	4.581	4.835	5.029	5.612	5.940	6.166	6.337
	6	2.177	2.715	3.033	3.259	3.433	3.576	3.696	3.799	3.890	3.971	4.111	4.328	4.495	4.996	5.279	5.475	5.623
	7	2.077	2.570	2.857	3.060	3.217	3.345	3.452	3.545	3.627	3.699	3.824	4.019	4.168	4.620	4.875	5.051	5.186
	8	2.010	2.471	2.738	2.926	3.071	3.189	3.288	3.374	3.449	3.516	3.631	3.811	3.948	4.365	4.601	4.765	4.890
	9	1.960	2.400	2.652	2.830	2.966	3.077	3.170	3.250	3.321	3.384	3.492	3.660	3.790	4.182	4.404	4.558	4.675
0.95	10	1.923	2.346	2.587	2.757	2.887	2.992	3.081	3.157	3.224	3.284	3.387	3.547	3.670	4.042	4.254	4.400	4.512
	15	1.819	2.198	2.411	2.559	2.671	2.762	2.839	2.904	2.962	3.013	3.101	3.237	3.342	3.660	3.841	3.967	4.063
	20	1.772	2.132	2.331	2.469	2.574	2.659	2.730	2.790	2.843	2.891	2.972	3.098	3.194	3.486	3.652	3.767	3.856
	25	1.745	2.094	2.286	2.419	2.519	2.600	2.668	2.726	2.776	2.821	2.898	3.018	3.109	3.386	3.543	3.652	3.736
	30	1.727	2.069	2.257	2.386	2.484	2.562	2.628	2.684	2.733	2.776	2.851	2.966	3.055	3.321	3.472	3.577	3.657
	60	1.685	2.010	2.187	2.308	2.399	2.471	2.532	2.584	2.629	2.669	2.737	2.843	2.923	3.164	3.300	3.395	3.467
	120	1.665	1.982	2.154	2.270	2.358	2.428	2.486	2.536	2.579	2.618	2.683	2.784	2.860	3.089	3.217	3.306	3.374
	∞	1.645	1.955	2.121	2.234	2.319	2.386	2.442	2.490	2.531	2.568	2.630	2.726	2.799	3.016	3.137	3.220	3.283
	4	5.077	6.305	7.073	7.632	8.070	8.430	8.734	8.996	9.228	9.434	9.788	10.341	10.764	12.035	12.747	13.238	13.611
	5	4.105	4.943	5.459	5.833	6.126	6.367	6.571	6.748	6.904	7.043	7.282	7.658	7.946	8.819	9.312	9.653	9.913
	6	3.635	4.298	4.701	4.993	5.221	5.409	5.567	5.705	5.827	5.935	6.122	6.417	6.643	7.332	7.724	7.996	8.204
	7	3.360	3.926	4.267	4.513	4.705	4.862	4.996	5.111	5.213	5.305	5.462	5.710	5.902	6.486	6.819	7.051	7.229
	8	3.180	3.685	3.987	4.203	4.372	4.511	4.628	4.730	4.819	4.900	5.038	5.257	5.425	5.941	6.237	6.443	6.601
	9	3.053	3.517	3.792	3.988	4.141	4.267	4.373	4.465	4.546	4.619	4.744	4.942	5.094	5.562	5.831	6.019	6.163
0.99	10	2.959	3.393	3.648	3.830	3.972	4.087	4.185	4.270	4.345	4.412	4.528	4.710	4.851	5.284	5.532	5.707	5.840
	15	2.711	3.067	3.273	3.418	3.531	3.622	3.699	3.766	3.824	3.877	3.967	4.109	4.219	4.557	4.752	4.889	4.994
	20	2.602	2.927	3.112	3.242	3.342	3.423	3.492	3.551	3.603	3.649	3.728	3.853	3.950	4.246	4.416	4.536	4.628
	25	2.542	2.849	3.023	3.145	3.238	3.314	3.377	3.432	3.480	3.523	3.597	3.712	3.801	4.073	4.230	4.340	4.425
	30	2.503	2.799	2.966	3.083	3.172	3.244	3.305	3.357	3.402	3.443	3.513	3.622	3.707	3.964	4.111	4.215	4.295
	60	2.411	2.682	2.833	2.938	3.017	3.082	3.135	3.181	3.221	3.257	3.318	3.414	3.487	3.709	3.835	3.924	3.991
	120	2.368	2.627	2.771	2.870	2.946	3.006	3.056	3.100	3.137	3.171	3.228	3.317	3.386	3.591	3.707	3.789	3.851
	∞	2.326	2.575	2.712	2.806	2.877	2.934	2.981	3.022	3.057	3.089	3.143	3.226	3.289	3.479	3.587	3.661	3.718

Table J.9 Factors $r'_{(1-\alpha;m,m,n)}$ for calculating normal distribution one-sided $100(1-\alpha)\%$ prediction bounds for all m future observations using the results of a previous sample of n observations. The factors in this table were computed and provided to us by Robert E. Odeh.

	0.50		0.60		0.70		0.80		0.90		0.95		0.99	
n	ℓ	u	ℓ	u	ℓ	u	ℓ	u	ℓ	u	ℓ	u	ℓ	u
10	1	3	1	3	1	3	1	3	1	3	1	3	1	3
	0.5811		0.5811*		0.5811*		0.5811*		0.5811*		0.5811*		0.5811*	
15	1	3	1	3	1	4	1	4	1	4	1	4	1	4
	0.6100		0.6100		0.7386		0.7386*		0.7386*		0.7386*		0.7386*	
20	1	3	1	4	1	4	1	5	1	5	1	5	1	5
	0.5554		0.7455		0.7455		0.8352		0.8352*		0.8352*		0.8352*	
25	2	5	2	5	1	5	1	5	1	6	1	6	1	6
	0.6308		0.6308		0.8302		0.8302		0.8948*		0.8948*		0.8948*	
30	2	5	2	5	2	6	1	6	1	7	1	7	1	7
	0.6408		0.6408		0.7431		0.8844		0.9318		0.9318*		0.9318*	
35	3	6	2	5	2	6	2	7	1	7	1	8	1	8
	0.5621		0.6084		0.7460		0.8224		0.9198		0.9550		0.9550*	
40	3	6	3	7	2	6	2	7	1	8	1	9	1	9
	0.5709		0.6777		0.7133		0.8200		0.9433		0.9697		0.9697*	
45	4	7	3	7	3	8	2	8	1	8	1	9	1	10
	0.5126		0.6824		0.7653		0.8719		0.9156		0.9593		0.9792*	
50	4	7	4	8	3	8	3	9	2	9	1	10	1	11
	0.5199		0.6276		0.7661		0.8304		0.9083		0.9703		0.9855*	
60	5	9	4	8	4	9	3	9	3	11	2	11	1	13
	0.5874		0.6142		0.7210		0.8053		0.9127		0.9520		0.9925	
70	6	10	5	10	5	11	4	11	3	12	2	12	1	14
	0.5542		0.6826		0.7539		0.8415		0.9318		0.9504		0.9906	
80	7	11	6	11	6	12	5	12	4	13	3	14	2	16
	0.5262		0.6497		0.7226		0.8116		0.9109		0.9626		0.9925	
90	8	12	7	12	6	12	6	14	4	14	3	15	2	17
	0.5020		0.6210		0.7103		0.8334		0.9198		0.9621		0.9917	
100	8	13	8	14	7	14	7	15	6	16	4	16	3	19
	0.5958		0.6701		0.7590		0.8103		0.9025		0.9523		0.9935	
150	13	18	12	19	12	20	11	21	9	22	8	23	6	25
	0.5036		0.6599		0.7161		0.8219		0.9253		0.9603		0.9904	
200	18	24	17	25	16	25	15	26	14	28	12	29	10	32
	0.5134		0.6476		0.7120		0.8066		0.9000		0.9561		0.9911	
250	22	29	21	29	21	31	19	32	18	34	16	35	13	38
	0.5394		0.6017		0.7034		0.8307		0.9077		0.9558		0.9916	
300	27	35	26	35	25	36	24	38	22	40	20	41	17	44
	0.5538		0.6137		0.7107		0.8198		0.9164		0.9575		0.9907	
400	36	45	35	46	34	47	33	49	31	51	28	52	25	56
	0.5468		0.6409		0.7218		0.8158		0.9040		0.9540		0.9903	
500	46	56	45	57	44	58	42	60	39	62	37	64	33	68
	0.5411		0.6262		0.7009		0.8189		0.9142		0.9563		0.9910	
600	56	66	54	67	53	69	51	70	48	73	46	75	42	80
	0.5014		0.6237		0.7218		0.8044		0.9116		0.9520		0.9905	
700	65	76	64	78	62	79	60	81	57	84	54	86	50	91
	0.5116		0.6202		0.7160		0.8145		0.9115		0.9559		0.9902	
800	75	87	73	88	72	90	70	92	67	95	63	97	58	102
	0.5187		0.6233		0.7097		0.8042		0.9008		0.9546		0.9901	
900	84	97	83	99	81	100	79	103	76	106	72	108	67	114
	0.5298		0.6244		0.7090		0.8168		0.9042		0.9543		0.9910	
1000	94	107	92	108	91	111	88	113	85	117	81	119	76	125
	0.5067		0.6011		0.7070		0.8127		0.9082		0.9546		0.9902	

Table J.10a Order statistics ℓ and u and actual confidence levels for two-sided distribution-free confidence intervals for the 0.10 quantile for various sample sizes n and nominal $100(1 - \alpha)\%$ confidence levels. An * indicates that a symmetric confidence interval with the desired confidence cannot be achieved without a larger sample size.

	1 − α						
n	0.50	0.60	0.70	0.80	0.90	0.95	0.99
10	2 4	1 4	1 4	1 5	1 5	1 5	1 5
	0.5033	0.7718	0.7718	0.8598	0.8598*	0.8598*	0.8598*
15	2 5	2 5	2 6	1 5	1 6	1 7	1 7
	0.6686	0.6686	0.7718	0.8006	0.9038	0.9468*	0.9468*
20	3 6	3 7	3 7	2 7	1 7	1 8	1 9
	0.5981	0.7072	0.7072	0.8441	0.9018	0.9563	0.9785*
25	4 7	4 8	3 8	3 9	2 9	2 10	1 11
	0.5460	0.6569	0.7927	0.8550	0.9258	0.9553	0.9907
30	5 8	5 9	4 9	4 10	2 10	2 11	1 13
	0.5056	0.6161	0.7486	0.8162	0.9284	0.9639	0.9957
35	6 10	5 9	5 10	4 11	4 12	2 12	2 14
	0.5822	0.6015	0.7108	0.8648	0.9051	0.9617	0.9908
40	7 11	6 11	6 12	5 12	4 13	4 14	2 15
	0.5533	0.6779	0.7512	0.8366	0.9283	0.9521	0.9906
45	8 12	7 12	7 13	6 13	5 14	4 15	3 17
	0.5284	0.6491	0.7237	0.8103	0.9097	0.9621	0.9923
50	9 13	8 13	7 13	7 15	5 15	5 16	3 18
	0.5066	0.6235	0.7105	0.8359	0.9208	0.9507	0.9925
60	10 15	10 16	9 16	8 16	7 18	6 19	5 21
	0.5802	0.6562	0.7426	0.8024	0.9265	0.9658	0.9912
70	12 17	12 18	11 18	10 19	9 20	7 21	5 23
	0.5448	0.6202	0.7051	0.8229	0.9018	0.9617	0.9914
80	14 19	13 20	13 21	12 22	11 23	9 24	7 26
	0.5151	0.6725	0.7294	0.8334	0.9047	0.9652	0.9924
90	16 22	15 22	15 23	14 24	12 25	11 26	8 28
	0.5630	0.6439	0.7014	0.8078	0.9148	0.9532	0.9904
100	18 24	17 24	16 25	15 26	13 27	12 28	10 31
	0.5397	0.6186	0.7401	0.8321	0.9188	0.9533	0.9916
150	27 34	26 35	25 36	24 37	22 39	20 40	17 43
	0.5249	0.6418	0.7389	0.8162	0.9182	0.9579	0.9913
200	37 45	36 46	35 47	33 48	31 50	29 52	25 55
	0.5170	0.6198	0.7081	0.8157	0.9076	0.9585	0.9915
250	46 55	45 56	44 58	42 59	40 61	38 63	34 67
	0.5232	0.6156	0.7293	0.8215	0.9037	0.9524	0.9910
300	56 66	55 67	53 68	52 70	49 72	46 74	42 78
	0.5272	0.6112	0.7212	0.8047	0.9035	0.9563	0.9903
400	75 86	74 88	72 89	70 91	67 94	64 96	59 101
	0.5082	0.6167	0.7121	0.8109	0.9088	0.9542	0.9911
500	94 107	93 109	91 110	89 112	86 116	82 118	77 124
	0.5326	0.6275	0.7119	0.8017	0.9062	0.9556	0.9915
600	114 128	112 129	110 131	108 134	104 137	101 140	95 146
	0.5239	0.6144	0.7162	0.8148	0.9081	0.9536	0.9908
700	133 148	132 150	130 152	127 155	123 158	119 161	113 168
	0.5215	0.6039	0.7005	0.8135	0.9020	0.9526	0.9907
800	153 169	151 171	149 173	146 175	142 180	138 183	131 190
	0.5196	0.6224	0.7104	0.8002	0.9068	0.9534	0.9909
900	172 189	170 191	168 193	165 196	161 201	156 204	149 211
	0.5212	0.6184	0.7025	0.8036	0.9043	0.9544	0.9901
1000	192 210	190 212	187 214	184 217	180 222	175 225	167 233
	0.5225	0.6148	0.7142	0.8080	0.9030	0.9518	0.9908

Table J.10b   Order statistics ℓ and u and actual confidence levels for two-sided distribution-free confidence intervals for the 0.20 quantile for various sample sizes n and nominal $100(1 − \alpha)\%$ confidence levels. An * indicates that a symmetric confidence interval with the desired confidence cannot be achieved without a larger sample size.

	1 − α												
n	0.50		0.60		0.70		0.80		0.90		0.95		0.99	
10	4	7	4	7	3	7	3	8	3	9	2	9	1	10
	0.6563		0.6563		0.7734		0.8906		0.9346		0.9785		0.9980	
15	6	9	6	10	5	10	5	11	4	11	4	12	3	13
	0.5455		0.6982		0.7899		0.8815		0.9232		0.9648		0.9926	
20	9	13	9	13	8	13	7	13	7	15	6	15	5	17
	0.6167		0.6167		0.7368		0.8108		0.9216		0.9586		0.9928	
25	11	15	10	15	10	16	9	16	8	17	8	18	6	19
	0.5756		0.6731		0.7705		0.8314		0.9245		0.9567		0.9906	
30	14	18	13	18	13	19	11	19	11	20	10	21	8	23
	0.5269		0.6384		0.7190		0.8504		0.9013		0.9572		0.9948	
35	16	20	15	21	14	21	14	22	13	23	12	24	10	26
	0.5004		0.6895		0.7570		0.8245		0.9105		0.9590		0.9940	
40	18	23	18	24	17	24	16	25	15	26	14	27	12	29
	0.5704		0.6511		0.7318		0.8461		0.9193		0.9615		0.9936	
45	20	25	20	26	19	27	18	27	17	29	16	30	14	32
	0.5386		0.6287		0.7673		0.8161		0.9275		0.9643		0.9934	
50	23	28	22	29	22	30	20	30	20	32	19	33	16	35
	0.5201		0.6778		0.7376		0.8392		0.9081		0.9511		0.9934	
60	28	34	27	34	26	35	25	36	24	37	23	39	21	41
	0.5574		0.6337		0.7549		0.8450		0.9075		0.9604		0.9901	
70	33	39	32	40	31	40	30	41	29	43	27	44	24	46
	0.5233		0.6575		0.7180		0.8118		0.9041		0.9586		0.9914	
80	37	44	37	45	36	46	35	47	33	48	32	50	29	52
	0.5660		0.6258		0.7336		0.8179		0.9071		0.9552		0.9903	
90	42	49	41	50	41	51	39	52	38	54	36	55	33	58
	0.5392		0.6572		0.7055		0.8298		0.9071		0.9554		0.9920	
100	47	54	46	55	45	56	44	57	42	59	41	61	38	64
	0.5159		0.6318		0.7287		0.8067		0.9114		0.9540		0.9907	
150	71	80	70	81	69	82	68	84	65	86	63	88	60	92
	0.5375		0.6308		0.7115		0.8073		0.9139		0.9591		0.9910	
200	96	106	95	107	93	108	91	110	89	113	87	115	82	119
	0.5193		0.6026		0.7112		0.8210		0.9098		0.9520		0.9913	
250	120	131	119	133	117	134	115	136	112	139	110	141	105	146
	0.5133		0.6231		0.7177		0.8160		0.9125		0.9503		0.9906	
300	145	157	143	158	142	160	139	162	136	165	134	168	128	173
	0.5108		0.6135		0.7005		0.8159		0.9061		0.9502		0.9907	
400	194	208	192	209	190	211	188	214	184	217	181	221	175	227
	0.5155		0.6046		0.7063		0.8059		0.9012		0.9544		0.9907	
500	243	259	241	260	239	263	236	265	232	269	229	273	222	280
	0.5252		0.6045		0.7164		0.8054		0.9021		0.9508		0.9905	
600	292	309	290	311	288	314	285	317	280	321	276	325	269	333
	0.5123		0.6087		0.7111		0.8083		0.9059		0.9546		0.9910	
700	342	360	339	362	337	365	334	368	329	373	325	377	316	385
	0.5034		0.6153		0.7097		0.8009		0.9035		0.9505		0.9909	
800	391	411	389	413	386	416	382	419	377	424	373	429	364	437
	0.5202		0.6035		0.7109		0.8092		0.9035		0.9522		0.9902	
900	440	461	438	464	435	467	431	470	426	476	421	480	412	490
	0.5161		0.6136		0.7136		0.8064		0.9043		0.9508		0.9907	
1000	490	512	487	514	484	517	480	521	474	527	470	532	460	542
	0.5131		0.6068		0.7033		0.8052		0.9063		0.9500		0.9905	

Table J.10c Order statistics ℓ and u and actual confidence levels for two-sided distribution-free confidence intervals for the 0.50 quantile for various sample sizes n and nominal $100(1 - \alpha)\%$ confidence levels.

Each cell below shows the achieved coverage/confidence (decimal, with `*` where noted) and the value ν (in parentheses).

n	β = 0.750, 1−α = 0.90	0.95	0.99	β = 0.900, 1−α = 0.90	0.95	0.99	β = 0.950, 1−α = 0.90	0.95	0.99	β = 0.990, 1−α = 0.90	0.95	0.99
10	0.9437 (1)	0.9437* (1)	0.9437* (1)	0.6513* (1)	0.6513* (1)	0.6513* (1)	0.4013* (1)	0.4013* (1)	0.4013* (1)	0.0956* (1)	0.0956* (1)	0.0956* (1)
15	0.9198 (2)	0.9866 (1)	0.9866* (1)	0.7941* (1)	0.7941* (1)	0.7941* (1)	0.5367* (1)	0.5367* (1)	0.5367* (1)	0.1399* (1)	0.1399* (1)	0.1399* (1)
20	0.9087 (3)	0.9757 (2)	0.9968 (1)	0.8784* (1)	0.8784* (1)	0.8784* (1)	0.6415* (1)	0.6415* (1)	0.6415* (1)	0.1821* (1)	0.1821* (1)	0.1821* (1)
25	0.9038 (4)	0.9679 (3)	0.9930 (2)	0.9282 (1)	0.9282* (1)	0.9282* (1)	0.7226* (1)	0.7226* (1)	0.7226* (1)	0.2222* (1)	0.2222* (1)	0.2222* (1)
30	0.9021 (5)	0.9626 (4)	0.9980 (2)	0.9576 (1)	0.9576 (1)	0.9576* (1)	0.7854* (1)	0.7854* (1)	0.7854* (1)	0.2603* (1)	0.2603* (1)	0.2603* (1)
35	0.9024 (6)	0.9590 (5)	0.9967 (3)	0.9750 (1)	0.9750 (1)	0.9750* (1)	0.8339* (1)	0.8339* (1)	0.8339* (1)	0.2966* (1)	0.2966* (1)	0.2966* (1)
40	0.9038 (7)	0.9567 (6)	0.9953 (4)	0.9195 (2)	0.9852 (1)	0.9852* (1)	0.8715* (1)	0.8715* (1)	0.8715* (1)	0.3310* (1)	0.3310* (1)	0.3310* (1)
50	0.9084 (9)	0.9547 (8)	0.9930 (6)	0.9662 (2)	0.9662 (2)	0.9948 (1)	0.9231 (1)	0.9231* (1)	0.9231* (1)	0.3950* (1)	0.3950* (1)	0.3950* (1)
60	0.9141 (11)	0.9548 (10)	0.9912 (8)	0.9470 (3)	0.9862 (2)	0.9982 (2)	0.9539 (1)	0.9539 (1)	0.9539* (1)	0.4528* (1)	0.4528* (1)	0.4528* (1)
80	0.9260 (15)	0.9579 (14)	0.9953 (11)	0.9120 (5)	0.9647 (4)	0.9978 (2)	0.9139 (2)	0.9835 (2)	0.9835* (1)	0.5525* (1)	0.5525* (1)	0.5525* (1)
100	0.9005 (20)	0.9624 (18)	0.9946 (15)	0.9424 (6)	0.9763 (5)	0.9922 (4)	0.9629 (2)	0.9629 (2)	0.9941 (2)	0.6340* (1)	0.6340* (1)	0.6340* (1)
200	0.9196 (42)	0.9595 (40)	0.9927 (36)	0.9071 (15)	0.9680 (13)	0.9919 (11)	0.9377 (6)	0.9736 (5)	0.9910 (4)	0.8660* (1)	0.8660* (1)	0.8660* (1)
300	0.9210 (65)	0.9544 (63)	0.9916 (58)	0.9301 (23)	0.9542 (22)	0.9903 (19)	0.9350 (10)	0.9659 (9)	0.9934 (7)	0.9510 (1)	0.9510 (1)	0.9510* (1)
400	0.9092 (89)	0.9548 (86)	0.9922 (80)	0.9254 (32)	0.9643 (30)	0.9908 (27)	0.9010 (15)	0.9645 (13)	0.9906 (11)	0.9095 (2)	0.9820 (2)	0.9820* (1)
500	0.9027 (113)	0.9575 (109)	0.9910 (103)	0.9249 (41)	0.9607 (39)	0.9921 (35)	0.9135 (19)	0.9657 (17)	0.9945 (14)	0.9602 (2)	0.9602 (2)	0.9934 (2)
600	0.9153 (136)	0.9520 (133)	0.9905 (126)	0.9043 (51)	0.9591 (48)	0.9901 (44)	0.9247 (23)	0.9680 (21)	0.9938 (18)	0.9389 (3)	0.9830 (3)	0.9976 (2)
800	0.9120 (184)	0.9542 (180)	0.9909 (172)	0.9146 (69)	0.9591 (66)	0.9912 (61)	0.9199 (32)	0.9606 (30)	0.9935 (26)	0.9015 (5)	0.9583 (4)	0.9971 (3)
1000	0.9001 (233)	0.9509 (228)	0.9901 (219)	0.9081 (88)	0.9515 (85)	0.9901 (79)	0.9194 (41)	0.9566 (39)	0.9907 (35)	0.9339 (6)	0.9713 (5)	0.9973 (3)

Table J.11   Value ν needed to obtain a two-sided distribution-free tolerance interval (one-sided distribution-free tolerance bound) by removing $\nu - 2$ (or $\nu - 1$) observations from the ends (end) of a sample of size n. The interval will contain (exceed or be exceeded by) at least $100\beta\%$ of the sampled distribution with $100(1 - \alpha)\%$ confidence. An * indicates that a symmetric confidence interval or bound with the desired confidence level cannot be achieved without a larger sample size. See also Figures 5.3a and 5.3b. A similar table appeared in Somerville (1958). Adapted with permission of the Institute of Mathematical Statistics.

β				$1 - \alpha$			
	0.50	0.75	0.90	0.95	0.98	0.99	0.999
0.500	3	5	7	8	9	11	14
0.550	4	6	8	9	11	12	16
0.600	4	6	9	10	12	14	19
0.650	5	7	10	12	15	16	22
0.700	6	9	12	14	17	20	27
0.750	7	10	15	18	21	24	33
0.800	9	13	18	22	27	31	42
0.850	11	18	25	30	37	42	58
0.900	17	27	38	46	56	64	89
0.950	34	53	77	93	115	130	181
0.960	42	67	96	117	144	164	227
0.970	56	89	129	157	193	219	304
0.980	84	134	194	236	290	330	458
0.990	168	269	388	473	581	662	920
0.995	336	538	777	947	1165	1325	1843
0.999	1679	2692	3889	4742	5832	6636	9230

Table J.12 Smallest sample size for the range formed by the smallest and largest observations to contain, with $100(1 - \alpha)\%$ confidence, at least $100\beta\%$ of the sampled population. A similar table appeared in Dixon and Massey (1969). Adapted with permission of McGraw-Hill, Inc.

β				$1 - \alpha$			
	0.50	0.75	0.90	0.95	0.98	0.99	0.999
0.500	1	2	4	5	6	7	10
0.550	2	3	4	6	7	8	12
0.600	2	3	5	6	8	10	14
0.650	2	4	6	7	10	11	17
0.700	2	4	7	9	11	13	20
0.750	3	5	9	11	14	17	25
0.800	4	7	11	14	18	21	31
0.850	5	9	15	19	25	29	43
0.900	7	14	22	29	38	44	66
0.950	14	28	45	59	77	90	135
0.960	17	34	57	74	96	113	170
0.970	23	46	76	99	129	152	227
0.980	35	69	114	149	194	228	342
0.990	69	138	230	299	390	459	688
0.995	139	277	460	598	781	919	1379
0.999	693	1386	2302	2995	3911	4603	6905

Table J.13 Smallest sample size for the largest (smallest) observation to exceed (be exceeded by), with $100(1 - \alpha)\%$ confidence, at least $100\beta\%$ of the sampled population. A similar table appeared in Dixon and Massey (1969). Adapted with permission of McGraw-Hill, Inc.

Table J.14a Largest number k of m future observations that will be contained in (bounded by) the distribution-free $100(1-\alpha)\%$ two-sided prediction interval (one-sided prediction bound) obtained by removing $\nu - 2$ (or $\nu - 1$) extreme observations from the ends (end) of a previous sample of size n. This table is similar to tables that first appeared in Danziger and Davis (1964). Adapted with permission of the Institute of Mathematical Statistics.

		$\nu = 1$							$\nu = 2$							$\nu = 3$							$\nu = 4$						
$1-\alpha$	n \ m:	5	10	25	50	75	100	∞	5	10	25	50	75	100	∞	5	10	25	50	75	100	∞	5	10	25	50	75	100	∞
0.50	5	5	9	22	44	65	87	0.871	4	7	17	34	52	69	0.686	3	5	13	25	38	50	0.500	2	3	8	16	23	31	0.314
	10	5	9	23	47	70	93	0.933	4	9	21	42	63	84	0.838	4	8	19	37	56	74	0.741	3	7	16	32	48	65	0.645
	25	5	10	25	49	73	97	0.973	5	10	24	47	70	94	0.934	5	9	23	45	67	90	0.894	4	9	22	43	64	86	0.855
	50	5	10	25	50	74	99	0.986	5	10	24	49	73	97	0.967	5	10	24	48	71	95	0.947	5	10	23	47	70	93	0.927
	75	5	10	25	50	75	99	0.991	5	10	25	49	74	98	0.978	5	10	24	48	73	97	0.965	5	10	24	48	72	95	0.951
	100	5	10	25	50	75	99	0.993	5	10	25	49	74	98	0.983	5	10	25	49	73	97	0.973	5	10	24	48	72	96	0.963
0.75	5	4	7	19	38	57	76	0.758	2	5	13	27	41	54	0.546	1	3	9	18	27	36	0.359	1	2	4	9	14	19	0.194
	10	4	9	22	43	65	87	0.871	4	7	18	37	56	75	0.753	3	6	16	32	48	64	0.645	2	4	13	27	40	54	0.542
	25	5	9	24	47	71	94	0.946	4	9	22	44	67	89	0.896	4	8	21	42	63	84	0.849	4	8	20	40	60	80	0.804
	50	5	10	24	49	73	97	0.973	5	9	23	47	71	94	0.947	4	9	23	46	69	92	0.923	4	9	22	45	67	90	0.900
	75	5	10	25	49	74	98	0.982	5	10	24	48	72	96	0.965	5	9	23	47	71	94	0.948	5	9	23	46	70	93	0.933
	100	5	10	25	49	74	99	0.986	5	10	24	48	73	97	0.973	5	9	24	48	72	96	0.961	5	9	23	47	71	95	0.949
0.90	5	3	6	15	31	47	62	0.631	1	3	10	20	30	41	0.416	1	2	5	12	18	24	0.247	0	1	2	5	8	11	0.112
	10	4	7	19	39	59	79	0.794	3	6	16	32	49	65	0.663	2	5	13	27	40	54	0.550	1	4	10	21	33	44	0.448
	25	4	9	22	45	68	91	0.912	4	8	21	42	63	84	0.853	3	7	19	39	59	79	0.801	3	7	18	37	55	74	0.752
	50	5	9	23	47	71	95	0.955	5	9	22	45	69	92	0.924	4	8	22	44	66	89	0.897	4	8	21	43	64	86	0.871
	75	5	9	24	48	72	96	0.970	5	9	23	47	70	94	0.949	4	9	23	46	69	92	0.931	4	8	22	45	67	90	0.913
	100	5	10	24	48	73	97	0.977	5	9	23	47	71	95	0.962	4	9	23	47	70	94	0.948	4	9	23	46	69	92	0.934
0.95	5	2	5	13	27	40	54	0.549	1	3	8	16	25	33	0.343	0	1	4	9	13	18	0.189	0	0	1	3	5	7	0.076
	10	3	7	18	36	55	73	0.741	2	5	14	29	44	59	0.606	2	4	11	23	36	48	0.493	1	3	9	18	28	38	0.393
	25	4	8	21	43	66	88	0.887	3	7	20	40	61	81	0.824	3	7	18	37	56	75	0.769	3	6	17	34	52	70	0.718
	50	4	9	23	46	70	93	0.942	4	8	22	44	67	90	0.909	4	8	21	43	65	87	0.879	3	7	20	41	62	84	0.852
	75	4	9	23	47	71	95	0.961	4	9	23	46	69	93	0.938	4	8	22	45	68	90	0.918	4	8	21	44	66	88	0.900
	100	5	9	24	48	72	96	0.970	5	9	23	47	70	94	0.953	4	9	22	46	69	92	0.938	4	8	22	45	68	91	0.924
0.99	5	1	3	9	19	29	39	0.398	0	1	4	10	15	21	0.222	0	0	2	4	7	9	0.106	0	0	0	1	2	2	0.033
	10	2	5	14	30	46	62	0.631	1	4	11	23	35	48	0.496	1	2	8	18	27	37	0.388	0	2	6	13	20	28	0.297
	25	3	7	19	40	61	82	0.832	3	6	17	36	55	74	0.763	2	5	16	33	51	68	0.704	2	5	14	30	46	63	0.651
	50	4	8	21	44	67	90	0.912	3	7	20	42	64	85	0.874	3	7	19	40	61	82	0.842	3	6	18	38	59	79	0.813
	75	4	8	22	46	69	92	0.940	4	8	21	44	67	90	0.915	3	7	20	43	65	87	0.893	3	7	20	41	63	85	0.872
	100	4	9	23	46	70	94	0.955	4	8	22	45	68	92	0.935	3	8	21	44	67	90	0.919	3	8	21	43	65	88	0.903

1−α	n	ν=5 m=5	10	25	50	75	100	∞	ν=6 m=5	10	25	50	75	100	∞	ν=7 m=5	10	25	50	75	100	∞	ν=8 m=5	10	25	50	75	100	∞
0.50	5	1	1	3	6	10	13	0.129																					
	10	3	6	14	27	41	55	0.548	2	5	11	23	34	45	0.452	2	4	9	18	27	35	0.355	1	3	6	13	19	26	0.259
	25	4	8	21	41	61	82	0.816	4	8	20	39	58	78	0.776	4	7	19	37	55	74	0.737	3	7	18	35	52	70	0.697
	50	5	9	23	46	68	91	0.907	5	9	22	45	67	89	0.887	4	9	22	46	65	87	0.868	4	9	21	43	64	85	0.848
	75	5	10	24	47	71	94	0.938	5	9	23	46	70	93	0.925	5	9	23	46	69	91	0.911	5	9	23	45	67	90	0.898
	100	5	10	24	48	72	95	0.953	5	10	24	47	71	94	0.943	5	10	23	47	70	93	0.934	5	9	23	46	69	92	0.924
0.75	5	0	0	1	2	4	5	0.056																					
	10	2	4	11	22	33	44	0.445	1	3	8	17	26	35	0.351	1	2	6	13	19	26	0.261	0	1	4	8	13	17	0.176
	25	3	7	19	37	56	75	0.760	3	7	17	35	53	71	0.718	3	6	16	33	50	67	0.675	3	6	15	31	47	63	0.634
	50	4	8	22	43	65	87	0.877	4	8	21	42	64	85	0.855	4	8	20	41	62	83	0.833	4	8	20	40	60	80	0.812
	75	4	9	23	45	68	91	0.918	4	9	22	45	67	90	0.903	4	9	22	44	66	88	0.888	4	8	21	43	65	87	0.873
	100	5	9	23	46	70	93	0.938	4	9	23	46	69	92	0.927	4	9	22	45	68	91	0.916	4	9	22	45	67	90	0.904
0.90	5	0	0	0	1	1	2	0.021																					
	10	1	3	8	17	26	34	0.354	1	2	6	12	19	26	0.267	0	1	4	8	13	18	0.188	0	0	2	5	8	11	0.116
	25	3	6	17	34	52	69	0.705	2	6	15	32	48	65	0.660	2	5	14	30	45	60	0.617	2	5	13	27	42	56	0.574
	50	4	8	20	41	62	83	0.846	3	7	19	40	60	81	0.822	3	7	19	39	58	78	0.799	3	7	18	37	57	76	0.776
	75	4	8	21	44	66	88	0.896	4	8	21	43	65	87	0.880	4	8	20	42	63	85	0.864	4	8	20	41	62	83	0.848
	100	4	9	22	45	68	91	0.922	4	8	22	44	67	90	0.909	4	8	21	44	66	88	0.897	4	8	21	43	65	87	0.885
0.95	5	0	0	0	0	0	1	0.010																					
	10	1	2	6	14	21	29	0.304	0	1	4	10	15	21	0.222	0	1	3	6	10	14	0.150	0	0	1	3	6	8	0.087
	25	2	5	15	32	49	65	0.670	2	4	14	30	45	61	0.625	2	4	13	27	42	56	0.580	2	4	12	25	38	52	0.538
	50	3	7	19	40	60	81	0.826	3	6	18	38	58	78	0.801	3	6	18	37	56	76	0.777	3	6	17	36	54	73	0.753
	75	4	8	21	43	65	86	0.882	3	7	20	42	63	85	0.865	3	7	20	41	62	83	0.848	3	7	19	40	60	81	0.832
	100	4	8	21	44	67	89	0.911	4	8	21	43	66	88	0.898	4	8	21	43	65	87	0.885	4	8	20	42	63	85	0.873
0.99	5	0	0	0	0	0	0	0.002																					
	10	0	1	4	9	15	20	0.218	0	0	2	6	10	13	0.150	0	0	1	3	6	8	0.093	0	0	0	1	2	4	0.048
	25	1	4	13	28	43	58	0.602	1	4	12	25	39	53	0.556	1	3	10	23	36	48	0.512	1	3	9	21	33	44	0.469
	50	2	6	17	37	56	76	0.785	2	6	17	35	54	73	0.758	2	5	16	34	52	70	0.733	2	5	15	33	50	68	0.708
	75	3	7	19	40	61	83	0.853	3	7	19	39	60	81	0.834	2	6	18	38	58	79	0.817	2	6	17	37	57	77	0.799
	100	3	7	20	42	64	86	0.888	3	7	20	41	63	85	0.874	3	7	19	40	62	83	0.860	3	7	19	40	61	82	0.847

Table J.14b Largest number k of m future observations that will be contained in (bounded by) the distribution-free $100(1 − \alpha)\%$ two-sided prediction interval (one-sided prediction bound) obtained by removing $\nu − 2$ (or $\nu − 1$) extreme observations from the ends (end) of a previous sample of size n. This table is similar to tables that first appeared in Danziger and Davis (1964). Adapted with permission of the Institute of Mathematical Statistics.

Table J.14c Largest number k of m future observations that will be contained in (bounded by) the distribution-free $100(1 - \alpha)\%$ two-sided prediction interval (one-sided prediction bound) obtained by removing $\nu - 2$ (or $\nu - 1$) extreme observations from the ends (end) of a previous sample of size n. This table is similar to tables that first appeared in Danziger and Davis (1964). Adapted with permission of the Institute of Mathematical Statistics.

			ν = 9							ν = 10							ν = 11							ν = 12						
1 − α	n	m:	5	10	25	50	75	100	∞	5	10	25	50	75	100	∞	5	10	25	50	75	100	∞	5	10	25	50	75	100	∞
0.50	5																													
	10		1	2	4	8	12	16	0.162	0	1	2	3	5	7	0.067														
	25		3	7	17	33	49	66	0.658	3	6	16	31	46	62	0.618	3	6	15	29	43	58	0.579	3	5	14	27	40	54	0.539
	50		4	8	21	42	62	83	0.828	4	8	20	41	61	81	0.808	4	8	20	40	59	79	0.788	4	8	19	38	58	77	0.768
	75		5	9	22	44	66	89	0.885	5	9	22	44	65	87	0.872	4	9	22	43	65	86	0.858	4	9	21	42	64	85	0.845
	100		5	9	23	46	69	91	0.914	5	9	23	45	68	90	0.904	5	9	22	45	67	89	0.894	5	9	22	44	66	88	0.884
0.75	5																													
	10		0	1	2	4	7	9	0.096	0	0	0	1	2	2	0.028														
	25		3	5	14	29	44	59	0.593	2	5	13	27	41	54	0.552	2	5	12	25	38	50	0.512	2	4	11	23	35	46	0.473
	50		4	7	19	39	59	78	0.790	3	7	19	38	57	76	0.769	3	7	18	37	55	74	0.748	3	7	17	36	54	72	0.727
	75		4	8	21	42	64	85	0.859	4	8	21	42	63	84	0.844	4	8	20	41	61	82	0.830	4	8	20	40	60	81	0.816
	100		4	9	22	44	66	89	0.893	4	8	22	43	65	88	0.883	4	8	21	43	65	86	0.872	4	8	21	42	64	85	0.861
0.90	5																													
	10		0	0	1	2	3	5	0.055	0	0	0	0	0	1	0.010														
	25		2	4	12	25	38	52	0.533	2	4	11	23	35	48	0.492	1	3	10	21	32	44	0.452	1	3	9	19	29	40	0.413
	50		3	6	18	36	55	74	0.753	3	6	17	35	53	71	0.731	3	6	16	34	51	69	0.709	3	6	16	33	50	67	0.687
	75		3	7	20	40	61	82	0.833	3	7	19	39	60	80	0.817	3	7	19	39	58	78	0.802	3	7	18	38	57	77	0.787
	100		4	8	21	42	64	86	0.873	4	8	20	42	63	85	0.862	4	8	20	41	62	83	0.850	3	7	20	40	61	82	0.839
0.95	5																													
	10		0	0	0	1	2	3	0.037	0	0	0	0	0	0	0.005														
	25		1	4	11	23	35	48	0.496	1	3	10	21	32	44	0.456	1	3	9	19	29	40	0.417	1 [a]	2	8	17	26	36	0.379
	50		3	6	16	34	53	71	0.730	2	6	16	33	51	68	0.707	2	5	15	32	49	66	0.684	2	5	15	31	47	64	0.662
	75		3	7	19	39	59	79	0.816	3	7	18	38	58	78	0.800	3	6	18	37	57	76	0.784	3	6	17	36	55	74	0.769
	100		3	7	20	41	62	84	0.860	3	7	20	40	61	83	0.848	3	7	19	40	61	81	0.836	3	7	19	39	60	80	0.825
0.99	5																													
	10		0	0	0	0	0	1	0.016	0	0	0	0	0	0	0.001														
	25		1	2	8	19	29	40	0.429	0	2	7	17	27	36	0.390	0	2	6	15	24	33	0.352	0	1	6	13	21	29	0.316
	50		2	5	14	31	48	65	0.684	2	4	14	30	46	63	0.660	1	4	13	29	45	60	0.637	1	4	13	28	43	58	0.615
	75		2	6	17	36	55	75	0.782	2	6	16	35	54	73	0.766	2	5	16	34	53	71	0.749	2	5	15	33	52	70	0.733
	100		3	6	18	39	59	80	0.834	2	6	18	38	58	79	0.821	2	6	18	37	57	77	0.809	2	6	17	37	56	76	0.796

			$1 - \alpha$				
m	0.50	0.75	0.90	0.95	0.98	0.99	0.999
1	3	7	19	39	99	199	1999
2	6	14	38	78	198	398	3998
3	8	20	56	116	296	596	5996
4	11	27	75	155	395	795	7995
5	13	33	93	193	493	993	9993
6	15	40	112	232	592	1192	11992
7	18	46	130	270	690	1390	13990
8	20	53	149	309	789	1589	15989
9	23	59	167	347	887	1787	17987
10	25	66	186	386	986	1986	19986
11	28	72	204	424	1084	2184	21984
12	30	79	223	463	1183	2383	23983
13	32	85	241	501	1281	2581	25981
14	35	91	260	540	1380	2780	27980
15	37	98	278	578	1478	2978	29978
16	40	104	297	617	1577	3177	31977
17	42	111	315	655	1675	3375	33975
18	44	117	334	694	1774	3574	35974
19	47	124	352	732	1872	3772	37972
20	49	130	371	771	1971	3971	39971
21	52	137	389	809	2069	4169	41969
22	54	143	408	848	2168	4368	43968
23	57	150	426	886	2266	4566	45966
24	59	156	445	925	2365	4765	47965
25	61	163	463	963	2463	4963	49963
30	73	195	556	1156	2956	5956	59956
35	85	227	648	1348	3448	6948	69948
40	98	260	740	1541	3941	7941	79941
45	110	292	833	1733	4433	8933	89933
50	122	324	925	1926	4926	9926	99926
60	146	389	1110	2311	5911	11911	119911
70	170	453	1295	2696	6896	13896	139896
80	194	518	1480	3080	7881	15881	159881
90	218	583	1665	3465	8866	17866	179866
100	242	647	1850	3850	9851	19851	199851

Table J.15   Smallest sample size for the range formed by the smallest and largest observations to contain, with $100(1 - \alpha)\%$ confidence, all m future observations from the previously sampled population.

				$1 - \alpha$			
m	0.50	0.75	0.90	0.95	0.98	0.99	0.999
1	1	3	9	19	49	99	999
2	2	6	18	38	98	198	1998
3	3	9	27	57	147	297	2997
4	4	12	36	76	196	396	3996
5	5	15	45	95	245	495	4995
6	6	18	54	114	294	594	5994
7	7	21	63	133	343	693	6993
8	8	24	72	152	392	792	7992
9	9	27	81	171	441	891	8991
10	10	30	90	190	490	990	9990
11	11	33	99	209	539	1089	10989
12	12	36	108	228	588	1188	11988
13	13	39	117	247	637	1287	12987
14	14	42	126	266	686	1386	13986
15	15	45	135	285	735	1485	14985
16	16	48	144	304	784	1584	15984
17	17	51	153	323	833	1683	16983
18	18	54	162	342	882	1782	17982
19	19	57	171	361	931	1881	18981
20	20	60	180	380	980	1980	19980
21	21	63	189	399	1029	2079	20979
22	22	66	198	418	1078	2178	21978
23	23	69	207	437	1127	2277	22977
24	24	72	216	456	1176	2376	23976
25	25	75	225	475	1225	2475	24975
30	30	90	270	570	1470	2970	29970
35	35	105	315	665	1715	3465	34965
40	40	120	360	760	1960	3960	39960
45	45	135	405	855	2205	4455	44955
50	50	150	450	950	2450	4950	49950
60	60	180	540	1140	2940	5940	59940
70	70	210	630	1330	3430	6930	69930
80	80	240	720	1520	3920	7920	79920
90	90	270	810	1710	4410	8910	89910
100	100	300	900	1900	4900	9900	99900

Table J.16a Smallest sample size for the largest (smallest) observation to exceed (be exceeded by), with $100(1 - \alpha)\%$ confidence, all m future observations from the previously sampled population.

			$1 - \alpha$				
m	0.50	0.75	0.90	0.95	0.98	0.99	0.999
2	1	2	3	5	9	13	44
3	1	3	6	9	15	22	75
4	2	4	8	12	21	32	107
5	2	5	10	16	28	41	137
6	3	6	12	19	34	50	168
7	3	7	14	23	40	59	199
8	4	8	17	26	46	68	230
9	4	9	19	30	52	77	260
10	4	10	21	33	58	86	291
11	5	11	23	37	64	95	322
12	5	12	25	40	70	104	352
13	6	13	27	44	76	113	383
14	6	14	30	47	82	122	414
15	6	15	32	51	88	131	444
16	7	16	34	54	95	140	475
17	7	17	36	58	101	149	506
18	8	18	38	61	107	158	536
19	8	19	40	65	113	167	567
20	9	20	43	68	119	176	597
21	9	21	45	72	125	185	628
22	9	22	47	75	131	194	659
23	10	23	49	79	137	203	689
24	10	24	51	82	143	212	720
25	11	25	53	86	149	221	751
30	13	30	64	103	180	266	904
35	15	35	75	120	210	311	1057
40	17	40	86	138	240	356	1210
45	19	45	97	155	271	401	1363
50	21	50	108	172	301	446	1516
60	25	60	129	207	362	536	1822
70	29	70	151	242	422	626	2129
80	33	80	172	277	483	716	2435
90	38	90	194	311	544	806	2741
100	42	100	216	346	605	896	3047

Table J.16b Smallest sample size for the largest (smallest) observation to exceed (be exceeded by), with $100(1 - \alpha)\%$ confidence, at least $m - 1$ of m future observations from the previously sampled population.

				$1 - \alpha$			
m	0.50	0.75	0.90	0.95	0.98	0.99	0.999
3	1	1	2	3	5	7	17
4	1	2	4	5	8	11	26
5	1	3	5	7	11	15	36
6	2	3	6	9	14	18	45
7	2	4	7	11	16	22	54
8	2	5	8	12	19	26	63
9	3	5	10	14	22	29	72
10	3	6	11	16	25	33	81
11	3	6	12	18	27	37	90
12	3	7	13	19	30	40	99
13	4	8	14	21	33	44	108
14	4	8	15	23	35	48	117
15	4	9	17	24	38	51	126
16	4	9	18	26	41	55	135
17	5	10	19	28	43	59	144
18	5	10	20	30	46	62	153
19	5	11	21	31	49	66	162
20	5	12	22	33	51	70	171
21	6	12	24	35	54	73	180
22	6	13	25	36	57	77	189
23	6	13	26	38	59	81	198
24	6	14	27	40	62	84	207
25	7	15	28	42	65	88	216
30	8	18	34	50	78	106	261
35	9	20	40	59	92	124	306
40	11	23	45	67	105	142	351
45	12	26	51	76	119	161	396
50	13	29	57	84	132	179	441
60	16	35	69	102	159	215	531
70	18	41	80	119	186	252	621
80	21	47	92	136	213	288	711
90	24	53	103	153	239	325	801
100	26	59	115	170	266	361	891

Table J.16c Smallest sample size for the largest (smallest) observation to exceed (be exceeded by), with $100(1 - \alpha)\%$ confidence, at least $m - 2$ of m future observations from the previously sampled population.

k	$1 - \alpha$							
	0.50	0.75	0.80	0.90	0.95	0.98	0.99	0.999
0.01	4550	13234	16424	27056	38415	54119	66349	108109
0.02	1138	3309	4106	6764	9604	13530	16588	27028
0.03	506	1471	1825	3007	4269	6014	7373	12013
0.04	285	828	1027	1691	2401	3383	4147	6757
0.05	182	530	657	1083	1537	2165	2654	4325
0.06	127	368	457	752	1068	1504	1844	3004
0.07	93	271	336	553	784	1105	1355	2207
0.08	72	207	257	423	601	846	1037	1690
0.09	57	164	203	335	475	669	820	1335
0.10	46	133	165	271	385	542	664	1082
0.11	38	110	136	224	318	448	549	894
0.12	32	92	115	188	267	376	461	751
0.13	27	79	98	161	228	321	393	640
0.14	24	68	84	139	196	277	339	552
0.15	21	59	73	121	171	241	295	481
0.16	18	52	65	106	151	212	260	423
0.17	16	46	57	94	133	188	230	375
0.18	15	41	51	84	119	168	205	334
0.19	13	37	46	75	107	150	184	300
0.20	12	34	42	68	97	136	166	271
0.21	11	31	38	62	88	123	151	246
0.22	10	28	34	56	80	112	138	224
0.23	9	26	32	52	73	103	126	205
0.24	8	23	29	47	67	94	116	188
0.25	8	22	27	44	62	87	107	173
0.30	6	15	19	31	43	61	74	121
0.35	4	11	14	23	32	45	55	89
0.40	3	9	11	17	25	34	42	68
0.45	3	7	9	14	19	27	33	54
0.50	2	6	7	11	16	22	27	44
0.60	2	4	5	8	11	16	19	31
0.70	1	3	4	6	8	12	14	23
0.80	1	3	3	5	7	9	11	17
0.90	1	2	3	4	5	7	9	14
1.00	1	2	2	3	4	6	7	11
1.20	1	1	2	2	3	4	5	8
1.40	1	1	1	2	2	3	4	6
1.60	1	1	1	2	2	3	3	5
1.80	1	1	1	1	2	2	3	4
2.00	1	1	1	1	1	2	2	3

Table J.17a   Smallest sample size such that a two-sided $100(1 - \alpha)\%$ confidence interval for the mean of a normal distribution will be no wider than $\pm k\sigma$ (σ known). A similar table appeared in Dixon and Massey (1969). Adapted with permission of McGraw-Hill, Inc.

1 − α = 0.75

n	γ'	1 − γ: 0.50	0.60	0.70	0.80	0.90	0.95	0.99
2	0.366	3	4	4	5	6	6	7
3	0.402	4	5	5	6	7	8	9
4	0.420	5	6	7	7	9	10	11
5	0.430	6	7	8	9	10	11	13
6	0.437	7	8	9	10	11	12	15
8	0.447	9	10	11	12	14	15	18
10	0.453	11	12	13	15	17	18	21
15	0.463	16	17	19	21	23	25	28
20	0.468	21	23	24	26	29	31	35
25	0.471	26	28	30	32	35	37	41
30	0.474	31	33	35	37	41	43	48
40	0.478	41	43	46	48	52	55	61
50	0.480	51	54	56	59	63	67	73
60	0.482	61	64	67	70	75	78	85
70	0.483	71	74	77	81	86	90	97
80	0.484	81	84	88	91	97	101	109
90	0.485	91	94	98	102	108	112	121
100	0.486	101	105	108	113	119	124	133

1 − α = 0.90

n	γ'	1 − γ: 0.50	0.60	0.70	0.80	0.90	0.95	0.99
2	0.206	4	5	5	6	6	7	8
3	0.272	5	6	6	7	8	8	10
4	0.310	6	7	7	8	9	10	12
5	0.334	7	8	8	9	11	12	13
6	0.351	8	9	10	11	12	13	15
8	0.374	10	11	12	13	15	16	18
10	0.389	12	13	14	15	17	19	21
15	0.411	17	18	20	21	23	25	28
20	0.423	22	23	25	27	29	32	35
25	0.432	27	29	30	33	35	38	42
30	0.438	32	34	36	38	41	44	49
40	0.446	42	44	46	49	53	56	61
50	0.452	52	54	57	60	64	67	74
60	0.456	62	64	67	71	75	79	86
70	0.460	72	75	78	81	86	90	98
80	0.462	82	85	88	92	97	102	110
90	0.465	92	95	99	103	108	113	122
100	0.466	102	105	109	113	119	124	133

1 − α = 0.95

n	γ'	1 − γ: 0.50	0.60	0.70	0.80	0.90	0.95	0.99
2	0.123	5	5	5	6	7	7	8
3	0.187	6	6	7	7	8	9	10
4	0.232	6	7	7	9	10	11	12
5	0.263	7	8	9	10	11	12	14
6	0.286	8	9	10	11	12	13	15
8	0.317	10	11	12	14	15	16	19
10	0.337	12	13	15	16	18	19	22
15	0.369	17	19	20	22	24	26	29
20	0.387	22	24	26	27	30	32	36
25	0.399	27	29	31	33	36	38	43
30	0.408	32	34	36	39	42	44	49
40	0.421	42	45	47	50	53	56	62
50	0.429	52	55	57	60	65	68	74
60	0.436	62	65	68	71	76	80	86
70	0.441	72	75	78	82	87	91	98
80	0.444	82	85	89	93	98	102	110
90	0.448	92	96	99	103	109	114	122
100	0.450	102	106	110	114	120	125	134

1 − α = 0.99

n	γ'	1 − γ: 0.50	0.60	0.70	0.80	0.90	0.95	0.99
2	0.032	6	6	6	7	8	8	9
3	0.065	7	7	8	8	9	10	11
4	0.100	8	8	9	10	11	12	13
5	0.131	9	9	10	11	12	13	15
6	0.156	10	10	11	12	14	15	17
8	0.197	12	13	14	15	16	17	20
10	0.226	14	15	16	17	19	20	23
15	0.274	19	20	21	23	25	27	30
20	0.303	24	25	27	29	31	33	37
25	0.324	29	30	32	34	37	39	44
30	0.339	34	36	38	40	43	46	50
40	0.360	44	46	48	51	55	58	63
50	0.375	54	56	59	62	66	69	75
60	0.386	64	66	69	73	77	81	88
70	0.394	74	77	80	83	88	92	100
80	0.401	84	87	90	94	99	104	112
90	0.406	94	97	101	105	110	115	123
100	0.411	104	107	111	115	121	126	135

Table J.17b Correction for sample size so that, with $100(1 − \gamma)$% probability, the sample will provide a normal distribution $100(1 − \alpha)$% confidence interval that is at least as narrow as the desired length; see Section 8.2.3. This table was adapted from a similar table that appeared in Kupper and Hafner (1989). Adapted with the permission of the American Statistical Association.

				$1 - \alpha$			
$100p$	0.80	0.85	0.90	0.95	0.98	0.99	0.999
0.7	7310	11000	16800	27600	43000	55200	97300
0.8	5600	8450	12900	21200	32900	42300	74500
0.9	4430	6690	10200	16700	26000	33400	58800
1	3600	5420	8250	13600	21100	27000	47700
2	914	1370	2070	3390	5270	6750	11900
3	413	614	927	1510	2340	3000	5280
4	236	349	524	852	1320	1690	2960
5	154	226	338	547	844	1080	1890
6	109	159	236	381	586	749	1310
7	81	118	174	280	431	550	962
8	63	91	134	215	330	421	735
9	51	73	107	171	261	333	580
10	42	60	87	139	212	269	469
15	21	28	40	63	95	120	207
20	13	17	24	36	54	67	115
25	9	12	16	24	35	43	73
30	7	9	12	17	24	30	51
35	6	7	9	13	18	22	37
40	5	6	7	10	14	17	28
45	4	5	6	8	11	14	22
50	4	4	5	7	9	11	18
55	3	4	5	6	8	9	15
60	3	3	4	5	7	8	12
65	3	3	4	5	6	7	10
70	3	3	3	4	5	6	9
75	2	3	3	4	5	5	7
80	2	3	3	3	4	5	6
85	2	2	3	3	4	4	5
90	2	2	2	3	3	3	5
95	2	2	2	2	3	3	4
100	2	2	2	2	2	2	2

Table J.18 Smallest sample size needed to estimate a normal distribution standard deviation with a given upper bound on percent error equal to $100p$; see Section 8.2.3.

β = 0.50

	β* = 0.54				β* = 0.56				β* = 0.58				β* = 0.60			
δ:	0.20	0.10	0.05	0.01	0.20	0.10	0.05	0.01	0.20	0.10	0.05	0.01	0.20	0.10	0.05	0.01
0.80	174	272	370	594	80	124	167	265	47	71	96	151	31	47	62	98
0.90	279	401	517	776	128	182	235	351	75	105	135	200	50	69	88	130
0.95	383	526	654	942	177	239	298	429	103	139	172	245	68	91	113	160
0.99	628	801	962	1313	288	368	442	597	169	213	255	343	112	141	168	224

β = 0.75

	β* = 0.79				β* = 0.81				β* = 0.83				β* = 0.85			
δ:	0.20	0.10	0.05	0.01	0.20	0.10	0.05	0.01	0.20	0.10	0.05	0.01	0.20	0.10	0.05	0.01
0.80	194	304	413	664	85	132	179	285	48	73	98	155	30	46	61	95
0.90	311	447	578	869	137	195	251	376	76	108	138	205	48	68	86	126
0.95	429	586	735	1060	189	257	321	459	106	142	176	251	67	89	110	155
0.99	700	899	1081	1469	310	395	473	639	173	219	261	351	109	137	163	218

β = 0.90

	β* = 0.93				β* = 0.94				β* = 0.95				β* = 0.96			
δ:	0.20	0.10	0.05	0.01	0.20	0.10	0.05	0.01	0.20	0.10	0.05	0.01	0.20	0.10	0.05	0.01
0.80	154	240	326	522	81	125	169	270	48	74	99	157	31	47	62	97
0.90	246	353	456	685	130	185	238	355	77	109	140	207	49	69	88	129
0.95	340	463	581	836	179	243	303	435	107	144	179	254	68	91	112	158
0.99	555	711	855	1159	293	374	448	605	175	221	264	355	111	140	167	222

β = 0.95

	β* = 0.970				β* = 0.975				β* = 0.980				β* = 0.985			
δ:	0.20	0.10	0.05	0.01	0.20	0.10	0.05	0.01	0.20	0.10	0.05	0.01	0.20	0.10	0.05	0.01
0.80	139	216	294	471	81	125	169	269	50	77	104	164	33	49	66	103
0.90	223	319	412	617	130	185	238	354	81	114	146	217	52	73	92	136
0.95	307	419	524	755	179	243	303	434	111	150	187	266	72	96	118	167
0.99	502	641	772	1046	293	373	447	604	182	231	276	371	117	148	176	235

β = 0.99

	β* = 0.996				β* = 0.997				β* = 0.998				β* = 0.999			
δ:	0.20	0.10	0.05	0.01	0.20	0.10	0.05	0.01	0.20	0.10	0.05	0.01	0.20	0.10	0.05	0.01
0.80	117	182	247	396	73	112	152	241	45	69	92	145	26	39	51	80
0.90	188	269	347	519	117	166	213	318	72	101	130	192	41	57	72	106
0.95	259	353	442	634	161	218	272	390	99	133	166	235	57	75	93	130
0.99	424	542	651	881	264	336	402	543	162	206	245	329	92	116	138	183

Table J.19 Smallest sample size for normal distribution two-sided $100(1 - \alpha)$% tolerance intervals such that the probability of enclosing at least the population proportion β is $1 - \alpha$ and the probability of enclosing at least β^* is no more than δ. This table is similar to a table that appeared in Faulkenberry and Daly (1970). Adapted with permission of the American Statistical Association.

Table J.20 Smallest sample size for normal distribution one-sided $100(1-\alpha)\%$ tolerance bounds such that the probability of enclosing at least the population proportion β is $1-\alpha$ and the probability of enclosing at least β^* is no more than δ. This table is similar to a table that appeared in Faulkenberry and Daly (1970). Adapted with permission of the American Statistical Association.

$\beta = 0.50$

$1-\alpha$		$\beta^* = 0.56$				$\beta^* = 0.58$				$\beta^* = 0.60$				$\beta^* = 0.62$			
	δ:	0.20	0.10	0.05	0.01	0.20	0.10	0.05	0.01	0.20	0.10	0.05	0.01	0.20	0.10	0.05	0.01
0.80		125	199	272	441	70	111	153	247	45	71	97	157	31	49	67	108
0.90		199	290	377	572	112	162	211	321	72	104	135	204	50	72	93	141
0.95		273	378	477	694	154	212	267	389	98	135	170	248	68	94	118	171
0.99		444	574	695	953	249	323	390	534	160	206	249	340	111	143	172	235

$\beta = 0.75$

$1-\alpha$		$\beta^* = 0.79$				$\beta^* = 0.81$				$\beta^* = 0.83$				$\beta^* = 0.85$			
	δ:	0.20	0.10	0.05	0.01	0.20	0.10	0.05	0.01	0.20	0.10	0.05	0.01	0.20	0.10	0.05	0.01
0.80		208	328	448	723	90	141	192	308	49	76	103	166	30	47	63	101
0.90		333	482	625	944	144	207	268	404	79	113	145	217	49	69	89	133
0.95		459	630	793	1148	199	272	341	492	109	148	185	266	68	91	114	162
0.99		748	964	1162	1585	325	417	502	682	178	228	273	369	111	141	168	226

$\beta = 0.900$

$1-\alpha$		$\beta^* = 0.93$				$\beta^* = 0.94$				$\beta^* = 0.95$				$\beta^* = 0.96$			
	δ:	0.20	0.10	0.05	0.01	0.20	0.10	0.05	0.01	0.20	0.10	0.05	0.01	0.20	0.10	0.05	0.01
0.80		148	231	315	506	77	120	163	261	46	70	95	151	29	44	59	93
0.90		237	341	441	663	124	178	229	343	73	104	134	199	46	65	83	123
0.95		327	447	560	808	172	233	292	419	101	137	171	244	64	86	107	151
0.99		534	685	824	1120	281	359	431	583	166	212	253	341	105	133	159	212

$\beta = 0.95$

$1-\alpha$		$\beta^* = 0.970$				$\beta^* = 0.975$				$\beta^* = 0.980$				$\beta^* = 0.985$			
	δ:	0.20	0.10	0.05	0.01	0.20	0.10	0.05	0.01	0.20	0.10	0.05	0.01	0.20	0.10	0.05	0.01
0.80		131	205	279	448	76	118	160	256	47	73	98	156	30	46	62	97
0.90		211	303	391	587	123	175	225	337	76	108	138	205	49	68	87	129
0.95		291	397	498	716	169	230	287	412	105	142	177	252	67	90	112	158
0.99		476	610	733	994	277	354	424	573	172	219	262	352	111	140	166	222

$\beta = 0.99$

$1-\alpha$		$\beta^* = 0.996$				$\beta^* = 0.997$				$\beta^* = 0.998$				$\beta^* = 0.999$			
	δ:	0.20	0.10	0.05	0.01	0.20	0.10	0.05	0.01	0.20	0.10	0.05	0.01	0.20	0.10	0.05	0.01
0.80		111	173	235	376	69	106	144	229	42	65	87	138	24	36	48	76
0.90		178	255	329	493	111	157	202	302	68	96	123	182	39	54	68	100
0.95		246	335	419	603	153	207	258	370	94	127	157	224	54	71	88	124
0.99		402	515	618	837	250	319	382	515	154	195	233	313	88	110	131	174

β = 0.50

1 − α	β* = 0.56				β* = 0.58				β* = 0.60				β* = 0.62			
δ:	0.20	0.10	0.05	0.01	0.20	0.10	0.05	0.01	0.20	0.10	0.05	0.01	0.20	0.10	0.05	0.01
0.80	199	314	435	704	112	183	249	392	77	112	158	253	49	81	108	177
0.90	316	463	601	906	176	260	337	506	115	168	214	322	81	115	149	227
0.95	438	602	748	1098	245	340	423	614	158	213	268	392	106	147	188	275
0.99	701	903	1092	1497	391	510	615	843	252	325	391	535	174	224	274	370

β = 0.75

1 − α	β* = 0.79				β* = 0.81				β* = 0.83				β* = 0.85			
δ:	0.20	0.10	0.05	0.01	0.20	0.10	0.05	0.01	0.20	0.10	0.05	0.01	0.20	0.10	0.05	0.01
0.80	325	503	689	1107	140	216	291	474	75	118	161	258	49	75	101	157
0.90	508	730	955	1434	219	318	411	613	118	171	223	335	73	109	136	206
0.95	697	958	1205	1745	299	416	514	752	167	229	282	407	103	139	176	251
0.99	1128	1450	1758	2398	490	630	760	1030	271	344	415	560	164	211	253	344

β = 0.90

1 − α	β* = 0.93				β* = 0.94				β* = 0.95				β* = 0.96			
δ:	0.20	0.10	0.05	0.01	0.20	0.10	0.05	0.01	0.20	0.10	0.05	0.01	0.20	0.10	0.05	0.01
0.80	245	385	534	849	135	213	278	449	78	124	169	267	54	78	113	169
0.90	402	566	739	1102	210	301	390	588	128	187	233	346	91	116	152	222
0.95	549	748	934	1344	298	402	493	715	179	239	298	425	116	154	191	275
0.99	887	1134	1367	1860	482	612	727	989	287	362	435	589	197	236	287	374

β = 0.95

1 − α	β* = 0.970				β* = 0.975				β* = 0.980				β* = 0.985			
δ:	0.20	0.10	0.05	0.01	0.20	0.10	0.05	0.01	0.20	0.10	0.05	0.01	0.20	0.10	0.05	0.01
0.80	272	427	580	923	180	249	339	536	110	157	226	339	85	110	157	226
0.90	446	628	807	1202	282	377	492	718	184	258	306	446	132	158	209	306
0.95	601	832	1036	1481	361	506	624	877	234	311	386	554	153	208	260	361
0.99	997	1279	1511	2057	606	755	901	1233	398	504	580	779	259	344	398	529

β = 0.99

1 − α	β* = 0.996				β* = 0.997				β* = 0.998				β* = 0.999			
δ:	0.20	0.10	0.05	0.01	0.20	0.10	0.05	0.01	0.20	0.10	0.05	0.01	0.20	0.10	0.05	0.01
0.80	551	906	1137	1811	427	551	790	1137	299	427	551	790	132	299	299	427
0.90	926	1297	1538	2358	667	798	1051	1538	388	531	667	1051	388	388	531	667
0.95	1182	1693	2064	2902	913	1050	1312	1941	628	773	913	1312	473	473	628	773
0.99	2010	2539	3052	4047	1453	1736	2010	2669	1001	1157	1453	1736	662	838	1001	1157

Table J.21   Smallest sample size for distribution-free 100(1 − α)% tolerance intervals and tolerance bounds such that the probability of enclosing at least the population proportion β is 1 − α and the probability of enclosing at least β* is no more than δ.

References

Abernethy, R. B., J. E. Breneman, C. H. Medlin, and G. L. Reinman (1983). *Weibull Analysis Handbook*. Air Force Wright Aeronautical Laboratories Technical Report AFWAL-TR-83-2079. Available from the National Technical Information Service, Washington, DC. [300]

Adcock, C. (1997). Sample size determination: A review. *The Statistician 46*, 261–283. [161]

Adimari, G. (1998). An empirical likelihood statistic for quantiles. *Journal of Statistical Computation and Simulation 60*, 85–95. [97]

Agresti, A. and B. A. Coull (1998). Approximate is better than exact for interval estimation of binomial proportions. *The American Statistician 52*, 119–126. [125]

Aitchison, J. (1964). Bayesian tolerance regions. *Journal of the Royal Statistical Society, Series B 26*, 161–175 (discussion: 192–210). [323]

Albert, J. (2007). *Bayesian Computation with R*. Springer. [323]

Altman, D. G., D. Machin, T. N. Bryant, and M. J. E. Gardner (2000). *Statistics with Confidence: Confidence Intervals and Statistical Guidelines* (Second Edition). BMJ Books. [36]

Arnold, B. C., N. Balakrishnan, and H. N. Nagaraja (2008). *A First Course in Order Statistics*, Volume 54 of *Classics in Applied Mathematics*. Society for Industrial and Applied Mathematics. Unabridged republication of the 1992 original. [439, 482]

Ascher, H. and H. Feingold (1984). *Repairable Systems Reliability: Modeling, Inference, Misconceptions and Their Causes*. Marcel Dekker. [129]

Atkinson, A. C. (1985). *Plots, Transformations, and Regression: An Introduction to Graphical Methods of Diagnostic Regression Analysis*. Oxford University Press. [69]

Statistical Intervals: A Guide for Practitioners and Researchers, Second Edition.
William Q. Meeker, Gerald J. Hahn and Luis A. Escobar.
© 2017 John Wiley & Sons, Inc. Published 2017 by John Wiley & Sons, Inc.
Companion Website: www.wiley.com/go/meeker/intervals

Barbe, P. and P. Bertail (1995). *The Weighted Bootstrap*. Springer. [266]

Bates, D. M. and D. G. Watts (1988). *Nonlinear Regression Analysis and Its Applications*. John Wiley & Sons, Inc. [67]

Beran, R. (1990). Calibrating prediction regions. *Journal of the American Statistical Association 85*, 715–723. [421]

Beran, R. (2003). The impact of the bootstrap on statistical algorithms and theory. *Statistical Science 18*, 175–184. [265]

Beran, R. and P. Hall (1993). Interpolated nonparametric prediction intervals and confidence intervals. *Journal of the Royal Statistical Society, Series B 55*, 643–652. [97]

Berkson, J. (1966). Examination of randomness of α-particle emissions. In F. David (Editor), *Festschrift for J. Neyman, Research Papers in Statistics*. John Wiley & Sons, Inc. [214]

Beyer, W. H. (1968). *Handbook of Tables for Probability and Statistics* (Second Edition). Chemical Rubber Company. [161]

Bhattacharyya, G. K. (1985). The asymptotics of maximum likelihood and related estimators based on Type II censored data. *Journal of the American Statistical Association 80*, 398–404. [455]

Billingsley, P. (2012). *Probability and Measure* (Anniversary Edition). John Wiley & Sons, Inc. [458]

Bisgaard, S. and M. Kulahci (2011). *Time Series Analysis and Forecasting by Example*. John Wiley & Sons, Inc. [67, 201]

Bjørnstad, J. F. (1990). Predictive likelihood: A review. *Statistical Science 5*, 242–254. [243]

Blyth, C. R. (1986). Approximate binomial confidence limits. *Journal of the American Statistical Association 81*, 843–855. [125]

Blyth, C. R. and H. A. Still (1983). Binomial confidence intervals. *Journal of the American Statistical Association 78*, 108–116. [125]

Boag, J. W. (1949). Maximum likelihood estimates of the proportion of patients cured by cancer therapy. *Journal of the Royal Statistical Society, Series B 11*, 15–53. [389]

Boos, D. D. and L. A. Stefanski (2013). *Essential Statistical Inference*. Springer. [243, 454, 463]

Bowden, D. C. (1968). Tolerance intervals in regression (query). *Technometrics 10*, 207–209. [69]

Box, G. E. P. and D. R. Cox (1964). An analysis of transformations (with discussion). *Journal of the Royal Statistical Society, Series B 26*, 211–252. [67, 69]

Box, G. E. P., J. S. Hunter, and W. G. Hunter (2005). *Statistics for Experimenters: Design, Innovation, and Discovery* (Second Edition). John Wiley & Sons, Inc. [22]

Box, G. E. P., G. M. Jenkins, G. C. Reinsel, and G. M. Ljung (2015). *Time Series Analysis: Forecasting and Control* (Fifth Edition). John Wiley & Sons, Inc. [67, 192, 201]

Box, G. E. P. and G. C. Tiao (1973). *Bayesian Inference in Statistical Analysis*. Addison-Wesley. [322]

Boyles, R. A. (2008). The role of likelihood in interval estimation. *The American Statistician 62*, 22–26. [36]

Brown, L. D., T. T. Cai, and A. DasGupta (2001). Interval estimation for a binomial proportion. *Statistical Science 16*, 101–133. With comments and a rejoinder by the authors. [125]

Brown, L. D., T. T. Cai, and A. DasGupta (2002). Confidence intervals for a binomial proportion and asymptotic expansions. *Annals of Statistics 30*, 160–201. [125, 126]

Brown, L. D., T. T. Cai, and A. DasGupta (2003). Interval estimation in exponential families. *Statistica Sinica 13*, 19–49. [125, 147]

Browne, W. J. and D. Draper (2006). A comparison of Bayesian and likelihood-based methods for fitting multilevel models. *Bayesian Analysis 1*, 473–514. [365]

Brush, G. G. (1988). *How to Choose the Proper Sample Size*. ASQC Basic References in Quality Control: Statistical Techniques, Vol. 12. American Society for Quality. [161]

Buonaccorsi, J. P. (1987). A note on confidence intervals for proportions in finite populations. *The American Statistician 41*, 215–218. [126]

Buonaccorsi, J. P. (2010). *Measurement Error: Models, Methods, and Applications*. CRC Press. [67]

Butler, R. W. (1986). Predictive likelihood inference with applications. *Journal of the Royal Statistical Society, Series B 48*, 1–38. [243]

Butler, R. W. (1989). Approximate predictive pivots and densities. *Biometrika 76*, 489–501. [243]

Byrne, J. and P. Kabaila (2005). Comparison of Poisson confidence intervals. *Communications in Statistics: Theory and Methods 34*, 545–556. [147]

Cai, T. T. (2005). One-sided confidence intervals in discrete distributions. *Journal of Statistical Planning and Inference 131*, 63–88. [126]

Cai, T. T. and H. Wang (2009). Tolerance intervals for discrete distributions in exponential families. *Statistica Sinica 19*, 905. [126, 147]

Canavos, G. C. and L. A. Kautauelis (1984). The robustness of two-sided tolerance limits for normal distributions. *Journal of Quality Technology 16*, 144–149. [69]

Carlin, B. P. and T. A. Louis (2009). *Bayesian Methods for Data Analysis* (Third Edition). CRC Press. [322, 365]

Carroll, R. J. and D. Ruppert (1988). *Transformation and Weighting in Regression*. Monographs on Statistics and Applied Probability. Chapman & Hall. [67]

Carroll, R. J., D. Ruppert, L. A. Stefanski, and C. M. Crainiceanu (2006). *Measurement Error in Nonlinear Models: A Modern Perspective* (Second Edition). CRC Press. [67]

Casella, G. and R. L. Berger (2002). *Statistical Inference*. Duxbury Press. [243, 330, 474]

Casella, G. and C. P. Robert (1999). *Monte Carlo Statistical Methods*. Springer. [322]

Chatfield, C. (1995). Model uncertainty, data mining and statistical inference. *Journal of the Royal Statistical Society, Series A 158*, 419–444 (discussion: 444–466). [22]

Chatfield, C. (2002). Confessions of a pragmatic statistician. *The Statistician 51*, 1–20. [22]

Chatterjee, S. and A. Bose (2005). Generalized bootstrap for estimating equations. *Annals of Statistics 33*, 414–436. [266]

Chen, S. X. and P. Hall (1993). Smoothed empirical likelihood confidence intervals for quantiles. *Annals of Statistics 21*, 1166–1181. [97]

Chernick, M. R. (2008). *Bootstrap Methods: A Practitioner's Guide*. John Wiley & Sons, Inc. [265]

Chew, V. (1968). Simultaneous prediction intervals. *Technometrics 10*, 323–330. [69, 471]

Chiang, A. K. L. (2001). A simple general method for constructing confidence intervals for functions of variance components. *Technometrics 43*, 356–367. [294, 389]

Chiang, C.-T., L. F. James, and M.-C. Wang (2005). Random weighted bootstrap method for recurrent events with informative censoring. *Lifetime Data Analysis 11*, 489–509. [266]

Chung, J. H. and D. B. De Lury (1950). *Confidence Limits for the Hypergeometric Distribution*. University of Toronto Press. [126]

Clopper, C. J. and E. S. Pearson (1934). The use of confidence or fiducial limits illustrated in the case of the binomial. *Biometrika 26*, 404–413. [104, 125]

Cochran, W. G. (1977). *Sampling Techniques* (Third Edition). John Wiley & Sons, Inc. [22]

Coles, S. (2001). *An Introduction to Statistical Modeling of Extreme Values*. Springer. [269]

Congdon, P. (2007). *Bayesian Statistical Modelling*. John Wiley & Sons, Inc. [322, 365]

Cryer, J. D. and K.-S. Chan (2008). *Time Series Analysis with Applications in R* (Second Edition). Springer. [67]

Danziger, L. and S. A. Davis (1964). Tables of distribution-free tolerance limits. *Annals of Mathematical Statistics 35*, 1361–1365. [97, 550, 551, 552]

DasGupta, A. (2008). *Asymptotic Theory of Statistics and Probability*. Springer. [265]

David, H. A. and H. Nagaraja (2003). *Order Statistics* (Third Edition). John Wiley & Sons, Inc. [97, 498]

Davidian, M. and D. M. Giltinan (1995). *Nonlinear Models for Repeated Measurement Data*. CRC Press. [365]

Davis, C. B. and R. J. McNichols (1999). Simultaneous nonparametric prediction limits. *Technometrics 41*, 89–101. [98]

Davison, A. C. and D. V. Hinkley (1997). *Bootstrap Methods and Their Application*. Cambridge University Press. [265]

Deming, W. E. (1950). *Some Theory of Sampling*. John Wiley & Sons, Inc. [5]

Deming, W. E. (1953). On the distinction between enumerative and analytic surveys. *Journal of the American Statistical Association 48*, 244–255. [5]

Deming, W. E. (1975). On probability as a basis for action. *The American Statistician 29*, 146–152. [5, 6, 12]

Deming, W. E. (1986). *Out of the Crisis*. Massachusetts Institute of Technology, Center for Advanced Engineering Study. [5]

Desu, M. and D. Raghavarao (2012). *Sample Size Methodology*. Elsevier. [161]

Dirikolu, M. H., A. Aktaş, and B. Birgören (2002). Statistical analysis of fracture strength of composite materials using Weibull distribution. *Turkish Journal of Engineering and Environmental Sciences 26*, 45–48. [269]

Dixon, W. J. and F. J. Massey, Jr. (1969). *Introduction to Statistical Analysis* (Third Edition). McGraw-Hill. [161, 549, 557]

Draper, N. R. and H. Smith (1981). *Applied Regression Analysis* (Second Edition). John Wiley & Sons, Inc. [69]

Eberhardt, K. R., R. W. Mee, and C. P. Reeve (1989). Computing factors for exact two-sided tolerance limits for a normal distribution. *Communications in Statistics: Simulation and Computation 18*, 397–413. [67]

Edwards, A. W. F. (1985). *Likelihood.* Cambridge University Press. [243]

Efron, B. (1979). Bootstrap methods: Another look at the jackknife. *Annals of Statistics 7*, 1–26. [264]

Efron, B. (1987). Better bootstrap confidence intervals. *Journal of the American Statistical Association 82*, 171–185. [265]

Efron, B. (2003). Second thoughts on the bootstrap. *Statistical Science 18*, 135–140. [265]

Efron, B. and R. Tibshirani (1986). Bootstrap methods for standard errors, confidence intervals, and other measures of statistical accuracy. *Statistical Science 1*, 54–77. [265]

Efron, B. and R. Tibshirani (1993). *An Introduction to the Bootstrap.* Chapman & Hall. [256, 265]

Faulkenberry, G. D. and J. C. Daly (1970). Sample size for tolerance limits on a normal distribution. *Technometrics 12*, 813–821. [175, 560, 561]

Faulkenberry, G. D. and D. L. Weeks (1968). Sample size determination for tolerance limits. *Technometrics 10*, 343–348. [175]

Fertig, K. W. and N. R. Mann (1977). One-sided prediction intervals for at least p out of m future observations from a normal population. *Technometrics 19*, 167–177. [69, 483]

Fitzmaurice, G. M., N. M. Laird, and J. H. Ware (2012). *Applied Longitudinal Analysis.* John Wiley & Sons, Inc. [365]

Flegal, J. M., M. Haran, and G. L. Jones (2008). Markov chain Monte Carlo: Can we trust the third significant figure? *Statistical Science 23*, 250–260. [323]

Fligner, M. A. and D. A. Wolfe (1976). Some applications of sample analogues to the probability integral transformation and a coverage property. *The American Statistician 30*, 78–85. [97]

Fligner, M. A. and D. A. Wolfe (1979a). Methods for obtaining a distribution-free prediction interval for the median of a future sample. *Journal of Quality Technology 11*, 192–198. [98]

Fligner, M. A. and D. A. Wolfe (1979b). Nonparametric prediction intervals for a future sample median. *Journal of the American Statistical Association 74*, 453–456. [98]

Fraser, D. A. S. (2011). Is Bayes posterior just quick and dirty confidence? *Statistical Science 26*, 299–316. [36]

Fujino, Y. (1980). Approximate binomial confidence limits. *Biometrika 67*, 677–681. [125]

Fuller, W. A. (1987). *Measurement Error Models.* John Wiley & Sons, Inc. [67]

Garthwaite, P. H., J. B. Kadane, and A. O'Hagan (2005). Statistical methods for eliciting probability distributions. *Journal of the American Statistical Association 100*, 680–701. [323]

Garwood, F. (1936). Fiducial limits for the Poisson distribution. *Biometrika 28*, 437–442. [147]

Gelman, A. (2006a). Multilevel (hierarchical) modeling: What it can and cannot do. *Technometrics 48*, 432–435. [355, 365]

Gelman, A. (2006b). Prior distributions for variance parameters in hierarchical models (comment on article by Browne and Draper). *Bayesian Analysis 1*, 515–534. [351, 365, 376]

Gelman, A., J. B. Carlin, H. S. Stern, D. B. Dunson, A. Vehtari, and D. B. Rubin (2013). *Bayesian Data Analysis*. CRC Press. [322, 350, 365, 512, 513]

Gelman, A. and J. Hill (2006). *Data Analysis Using Regression and Multilevel/Hierarchical Models*. Cambridge University Press. [69, 365]

Gentle, J. E. (2009). *Computational Statistics*. Springer. [9]

Gentle, J. E. (2013). *Random Number Generation and Monte Carlo Methods*. Springer. [9]

Gibbons, J. D. (1997). *Nonparametric Methods for Quantitative Analysis* (Third Edition). American Sciences Press. [70, 97]

Gilat, D. and T. P. Hill (1996). Strongly-consistent, distribution-free confidence intervals for quantiles. *Statistics & Probability Letters 29*, 45–53. [97]

Gitlow, H., S. Gitlow, A. Oppenheim, and R. Oppenheim (1989). *Tools and Methods for the Improvement of Quality*. Irwin. [7, 22]

Goos, P. and B. Jones (2011). *Optimal Design of Experiments: A Case Study Approach*. John Wiley & Sons, Inc. [22]

Graybill, F. A. (1976). *Theory and Application of the Linear Model*. Duxbury Press. [68]

Greenwood, J. A. and M. M. Sandomire (1950). Sample size required for estimating the standard deviation as a percent of its true value. *Journal of the American Statistical Association 45*, 257–260. [156, 161]

Groves, R. M., F. J. Fowler Jr., M. P. Couper, J. M. Lepkowski, E. Singer, and R. Tourangeau (2009). *Survey Methodology* (Second Edition). John Wiley & Sons, Inc. [22]

Guenther, W. C. (1975). The inverse hypergeometric—a useful model. *Statistica Neerlandica 29*, 129–144. [97]

Guilbaud, O. (1983). Nonparametric prediction intervals for sample medians in the general case. *Journal of the American Statistical Association 78*, 937–941. [98]

Guttman, I. (1970). *Statistical Tolerance Regions, Classical and Bayesian*. Charles Griffin. [68, 323]

Hahn, G. J. (1969). Factors for calculating two-sided prediction intervals for samples from a normal distribution. *Journal of the American Statistical Association 64*, 878–888. [69, 484]

Hahn, G. J. (1970a). Additional factors for calculating prediction intervals for samples from a normal distribution. *Journal of the American Statistical Association 65*, 1668–1676. [69, 484]

Hahn, G. J. (1970b). Statistical intervals for a normal population. Part I: Tables, examples and applications. *Journal of Quality Technology 2*, 115–125. [22, 43, 45, 529, 530, 531, 532, 533, 534]

Hahn, G. J. (1971). How abnormal is normality? *Journal of Quality Technology 3*, 18–22. [19, 69]

Hahn, G. J. (1972a). Prediction intervals for a regression model. *Technometrics 14*, 203–213. [69]

Hahn, G. J. (1972b). Simultaneous prediction intervals to contain the standard deviations or ranges of future samples from a normal distribution. *Journal of the American Statistical Association 67*, 938–942. [69]

Hahn, G. J. (1975). A simultaneous prediction limit on the means of future samples from an exponential distribution. *Technometrics 17*, 341–345. [184]

Hahn, G. J. (1977). A prediction interval on the difference between two future sample means and its application to a claim of product superiority. *Technometrics 19*, 131–134. [70, 184]

Hahn, G. J. (1982). Removing measurement error in assessing conformance to specifications. *Journal of Quality Technology 14*, 117–121. [4]

Hahn, G. J. and R. Chandra (1981). Tolerance intervals for Poisson and binomial variables. *Journal of Quality Technology 13*, 100–110. [126, 147]

Hahn, G. J. and W. Q. Meeker (1993). Assumptions for statistical inference. *The American Statistician 47*, 1–11. [22]

Hahn, G. J. and W. Nelson (1973). A survey of prediction intervals and their applications. *Journal of Quality Technology 5*, 178–188. [69, 126]

Hahn, G. J. and S. S. Shapiro (1967). *Statistical Models in Engineering*. John Wiley & Sons, Inc. [19, 60, 453]

Hall, D. B. (1989). Analysis of surveillance data: A rationale for statistical tests with comments on confidence intervals and statistical models. *Statistics in Medicine 8*, 273–278. [36]

Hall, I. J., R. R. Prairie, and C. K. Motlagh (1975). Nonparametric prediction intervals. *Journal of Quality Technology 7*, 109–114. [97]

Hall, P. (1988). Theoretical comparison of bootstrap confidence intervals. *Annals of Statistics 16*, 927–985. With discussion and a reply by the author. [265, 287, 293]

Hall, P. (1992). *The Bootstrap and Edgeworth Expansion*. Springer. [263, 265]

Hall, P. and M. A. Martin (1989). A note on the accuracy of bootstrap percentile method confidence intervals for a quantile. *Statistics & Probability Letters 8*, 197–200. [265]

Hamada, M., V. Johnson, L. M. Moore, and J. Wendelberger (2004). Bayesian prediction intervals and their relationship to tolerance intervals. *Technometrics 46*, 452–459. [323]

Hamada, M. and S. Weerahandi (2000). Measurement system assessment via generalized inference. *Journal of Quality Technology 32*, 241. [294, 372, 373, 389]

Hannig, J. (2009). On generalized fiducial inference. *Statistica Sinica 19*, 491–54. [294]

Hannig, J. (2013). Generalized fiducial inference via discretization. *Statistica Sinica 23*, 489–514. [294]

Hannig, J., H. Iyer, R. C. S. Lai, and T. C. M. Lee (2016). Generalized fiducial inference: A review and new results. *Journal of the American Statistical Association*. doi: 10.1080/01621459.2016.1165102. [294]

Hannig, J., H. Iyer, and P. Patterson (2006). Fiducial generalized confidence intervals. *Journal of the American Statistical Association 101*, 254–269. [294, 390, 489, 493, 495]

Harlow, L. L., S. A. Mulaik, and J. H. E. Steiger (1997). *What If There Were No Significance Tests?* Psychology Press. [36]

Harville, D. A. (1997). *Matrix Algebra from a Statistician's Perspective*. Springer. [514]

Helsel, D. R. (2005). *Nondetects and Data Analysis*. John Wiley & Sons, Inc. [225]

Hesterberg, T. C. (2015). What teachers should know about the bootstrap: Resampling in the undergraduate statistics curriculum. *The American Statistician 69*, 371–386. [265]

Hillmer, S. C. (1996). A problem-solving approach to teaching business statistics. *The American Statistician 50*, 249–256. [22]

Ho, Y. H. and S. M. Lee (2005). Iterated smoothed bootstrap confidence intervals for population quantiles. *Annals of Statistics 33*, 437–462. [265]

Hoff, P. D. (2009). *A First Course in Bayesian Statistical Methods*. Springer. [322, 365, 518]

Hong, Y., W. Q. Meeker, and L. A. Escobar (2008a). Avoiding problems with normal approximation confidence intervals for probabilities. *Technometrics 50*, 64–68. [244]

Hong, Y., W. Q. Meeker, and L. A. Escobar (2008b). The relationship between confidence intervals for failure probabilities and life time quantiles. *IEEE Transactions on Reliability 57*, 260–266. [244]

Hong, Y., W. Q. Meeker, and J. D. McCalley (2009). Prediction of remaining life of power transformers based on left truncated and right censored lifetime data. *Annals of Applied Statistics 3*, 857–879. [266]

Houf, R. E. and D. B. Berman (1988). Statistical analysis of power module thermal test equipment performance. *IEEE Transactions on Components, Hybrids, and Manufacturing Technology 11*, 516–520. [371]

Hox, J. (2010). *Multilevel Analysis: Techniques and Applications*. Routledge. [365]

Hyndman, R. J. and Y. Fan (1996). Sample quantiles in statistical packages. *The American Statistician 50*, 361–365. [253]

Jackson, J. E. (1991). *A User's Guide to Principal Components*. John Wiley & Sons, Inc. [241]

Jeffreys, H. (1946). An invariant form for the prior probability in estimation problems. *Proceedings of the Royal Society of London, Series A 186*, 453–461. [516]

Jeng, S.-L. and W. Q. Meeker (2000). Comparisons of approximate confidence interval procedures for Type I censored data. *Technometrics 42*, 135–148. [244, 461]

Jílek, M. (1982). Sample size and tolerance limits. *Trabajos de Estadística y de Investigación Operativa 33*, 64–78. [175]

Jin, Z., Z. Ying, and L. Wei (2001). A simple resampling method by perturbing the minimand. *Biometrika 88*, 381–390. [266]

JMP (2016). *Using JMP®*. SAS Institute. [9]

Johnson, R. A. and D. W. Wichern (2002). *Applied Multivariate Statistical Analysis* (Fifth Edition). Prentice Hall. [240]

Junk, G. A., R. F. Spalding, and J. J. Richard (1980). Areal, vertical, and temporal differences in ground water chemistry: II. Organic constituents. *Journal of Environmental Quality 9*, 479–483. [225]

Kalbfleisch, J. (1985). *Probability and Statistical Inference*. Springer. [243]

Kass, R. E. and L. Wasserman (1996). The selection of prior distributions by formal rules (Corr: 1998V93 p412). *Journal of the American Statistical Association 91*, 1343–1370. [516]

Katz, L. (1953). Confidence intervals for the number showing a certain characteristic in a population when sampling is without replacement. *Journal of the American Statistical Association 48*, 256–261. [126]

Katz, L. (1975). Presentation of a confidence interval estimate as evidence in a legal proceeding. *The American Statistician 29*, 138–142. [36]

Klinger, D., Y. Nakada, and M. Menendez (1990). *AT&T Reliability Manual*. Van Nostrand Reinhold. [311]

Koehler, E., E. Brown, and S. J.-P. Haneuse (2009). On the assessment of Monte Carlo error in simulation-based statistical analyses. *The American Statistician 63*, 155–162. [323]

Konijn, H. S. (1973). *Statistical Theory of Sample Survey Design and Analysis*. Elsevier. [126]

Krantz, D. H. (1999). The null hypothesis testing controversy in psychology. *Journal of the American Statistical Association 94*, 1372–1381. [36]

Krishnamoorthy, K. and T. Mathew (2009). *Statistical Tolerance Regions: Theory, Applications, and Computation*. John Wiley & Sons, Inc. [69, 126, 147, 294, 481]

Krishnamoorthy, K. and J. Peng (2011). Improved closed-form prediction intervals for binomial and Poisson distributions. *Journal of Statistical Planning and Inference 141*, 1709–1718. [126, 147]

Krishnamoorthy, K., Y. Xia, and F. Xie (2011). A simple approximate procedure for constructing binomial and Poisson tolerance intervals. *Communications in Statistics: Theory and Methods 40*, 2243–2258. [126, 147]

Kupper, L. L. and K. B. Hafner (1989). How appropriate are popular sample size formulas? *The American Statistician 43*, 101–105. [161, 558]

Kutner, M. H., C. J. Nachtsheim, J. Neter, and W. Li (2005). *Applied Linear Statistical Models*. McGraw-Hill Irwin. [69, 192]

Lahiri, S. (2003). *Resampling Methods for Dependent Data*. Springer. [265]

Landon, J. and N. D. Singpurwalla (2008). Choosing a coverage probability for prediction intervals. *The American Statistician 62*, 120–124. [36]

Lawless, J. F. (2003). *Statistical Models and Methods for Lifetime Data* (Second Edition). John Wiley & Sons, Inc. [60, 67, 70, 208, 293, 484]

Leemis, L. M. and K. S. Trivedi (1996). A comparison of approximate interval estimators for the Bernoulli parameter. *The American Statistician 50*, 63–68. [125]

Lehmann, E. L. (1983). *Theory of Point Estimation*. John Wiley & Sons, Inc. [455]

Lehmann, E. L. (1986). *Testing Statistical Hypotheses* (Second Edition). John Wiley & Sons, Inc. [433, 437, 440, 443]

Lemon, G. H. (1977). Factors for one-sided tolerance limits for balanced one-way-ANOVA random-effects model. *Journal of the American Statistical Association 72*, 676–680. [70]

Lenth, R. V. (2001). Some practical guidelines for effective sample size determination. *The American Statistician 55*, 187–193. [161]

Levy, P. S. and S. Lemeshow (2008). *Sampling of Populations: Methods and Applications* (Fourth Edition). John Wiley & Sons, Inc. [22]

Li, J., D. R. Jeske, and J. Pettyjohn (2009). Approximate and generalized pivotal quantities for deriving confidence intervals for the offset between two clocks. *Statistical Methodology 6,* 97–107. [294]

Liao, C. T., T. Y. Lin, and H. K. Iyer (2005). One- and two-sided tolerance intervals for general balanced mixed models and unbalanced one-way random models. *Technometrics 47,* 323–335. [377, 389]

Lieberman, G. J. (1961). Prediction regions for several predictions from a single regression line. *Technometrics 3,* 21–27. [69]

Lieberman, G. J. and R. G. Miller, Jr. (1963). Simultaneous tolerance intervals in regression. *Biometrika 50,* 155–168. [69]

Lieblein, J. and M. Zelen (1956). Statistical investigation of the fatigue life of deep-groove ball bearings. *Journal of Reseach, National Bureau of Standards 57,* 273–316. [60, 62]

Limam, M. M. T. and D. R. Thomas (1988a). Simultaneous tolerance intervals for the linear regression model. *Journal of the American Statistical Association 83,* 801–804. [69]

Limam, M. M. T. and D. R. Thomas (1988b). Simultaneous tolerance intervals in the one-way model with covariates. *Communications in Statistics: Simulation and Computation 17,* 1007–1019. [70]

Lindley, D. V. (1997). The choice of sample size. *The Statistician 46,* 129–138 (discussion: 139–166). [161]

Little, R. J. (2014). Survey sampling: Past controversies, current orthodoxy, future paradigms. In X. Lin, C. Genest, D. L. Banks, G. Molenberghs, D. W. Scott, and J.-L. Wang (Editors), *Past, Present, and Future of Statistical Science,* Chapter 37. CRC Press. [22]

Liu, J., D. J. Nordman, and W. Q. Meeker (2016). The number of MCMC draws needed to compute Bayesian credible bounds. *The American Statistician 70,* 275–284. [323, 350]

Loh, W.-Y. (1987). Calibrating confidence coefficients. *Journal of the American Statistical Association 82,* 155–162. [421]

Loh, W.-Y. (1991). Bootstrap calibration for confidence interval construction and selection. *Statistica Sinica 1,* 477–491. [421]

Lohr, S. L. (2010). *Sampling: Design and Analysis* (Second Edition). Duxbury Press. [22]

Lumley, T. (2010). *Complex Surveys: A Guide to Analysis Using R.* John Wiley & Sons, Inc. [22]

Lunn, D., C. Jackson, N. Best, A. Thomas, and D. Spiegelhalter (2012). *The BUGS Book: A Practical Introduction to Bayesian Analysis.* CRC Press. [322, 323, 350, 365]

Lunn, D. J., A. Thomas, N. Best, and D. Spiegelhalter (2000). WinBUGS—a Bayesian modelling framework: Concepts, structure, and extensibility. *Statistics and Computing 10,* 325–337. [323]

Mace, A. E. (1964). *Sample Size Determination.* Reinhold Publishing. [161]

MacKay, R. J. and R. W. Oldford (2000). Scientific method, statistical method and the speed of light. *Statistical Science 15,* 254–278. [22]

Maller, R. A. and X. Zhou (1996). *Survival Analysis with Long-Term Survivors*. John Wiley & Sons, Inc. [389]

Maltz, M. and R. McCleary (1977). The mathematics of behavioral change: Recidivism and construct validity. *Evaluation Quarterly 1*, 421–438. [389]

Mathews, P. (2010). *Sample Size Calculations: Practical Methods for Engineers and Scientists*. Mathews Malnar and Bailey. [161]

Matney, T. and A. Sullivan (1982). Compatible stand and stock tables for thinned and unthinned loblolly pine stands. *Forest Science 28*, 161–171. [246]

McKane, S. W., L. A. Escobar, and W. Q. Meeker (2005). Sample size and number of failure requirements for demonstration tests with log-location-scale distributions and failure censoring. *Technometrics 47*, 182–190. [175]

Mee, R. W. (1984). β-expectation and β-content tolerance limits for balanced one-way ANOVA random model. *Technometrics 26*, 251–254. [70]

Mee, R. W. (1989). Normal distribution tolerance intervals for stratified random samples. *Technometrics 31*, 99–105. [70]

Mee, R. W., K. R. Eberhardt, and C. P. Reeve (1991). Calibration and simultaneous tolerance intervals for regression. *Technometrics 33*, 211–219. [69]

Mee, R. W. and D. B. Owen (1983). Improved factors for one-sided tolerance limits for balanced one-way ANOVA random model. *Journal of the American Statistical Association 78*, 901–905. [70]

Meeker, W. Q. (1987). Limited failure population life tests: Application to integrated circuit reliability. *Technometrics 29*, 51–65. [368, 389, 461]

Meeker, W. Q. and L. A. Escobar (1995). Teaching about approximate confidence regions based on maximum likelihood estimation. *The American Statistician 49*, 48–53. [243, 462]

Meeker, W. Q. and L. A. Escobar (1998). *Statistical Methods for Reliability Data*. John Wiley & Sons, Inc. [36, 60, 61, 67, 161, 208, 214, 215, 216, 235, 243, 244, 289, 290, 300, 361, 368, 430, 436]

Meeker, W. Q., L. A. Escobar, and Y. Hong (2009). Using accelerated life tests results to predict product field reliability. *Technometrics 51*, 146–161. [291]

Meeker, W. Q. and G. J. Hahn (1980). Prediction intervals for the ratios of normal distribution sample variances and exponential distribution means. *Technometrics 22*, 357–366. [70, 184]

Meeker, W. Q. and G. J. Hahn (1982). Sample sizes for prediction intervals. *Journal of Quality Technology 14*, 201–206. [180, 181, 184]

Mendenhall, W. (1968). *Introduction to Linear Models and the Design and Analysis of Experiments*. Duxbury Press. [70]

Meyer, M. A. and J. M. Booker (1991). *Eliciting and Analyzing Expert Judgment: A Practical Guide*. Society for Industrial and Applied Mathematics. [323]

Miller, Jr., R. G. (1981). *Simultaneous Statistical Inference* (Second Edition). Springer. [34, 69]

Minitab (2016). *Minitab 17*. Minitab Inc. [9]

Montgomery, D. C. (2009). *Design and Analysis of Experiments* (Seventh Edition). John Wiley & Sons, Inc. [22]

Montgomery, D. C., E. A. Peck, and G. G. Vining (2015). *Introduction to Linear Regression Analysis*. John Wiley & Sons, Inc. [69]

Morris, M. D. (2011). *Design of Experiments: An Introduction Based on Linear Models*. Texts in Statistical Science Series. CRC Press. [22]

Murphy, R. B. (1948). Non-parametric tolerance limits. *Annals of Mathematical Statistics 19*, 581–589. [79, 80, 97]

Nair, V. N. (1981). Confidence bands for survival functions with censored data. *Biometrika 68*, 99–103. [216, 244]

Nair, V. N. (1984). Confidence bands for survival functions with censored data: A comparative study. *Technometrics 26*, 265–275. [244]

Natrella, M. G. (1963). *Experimental Statistics*. National Bureau of Standards, Handbook 91. US Government Printing Office. [161]

Nelson, W. (1970). Confidence intervals for the ratio of two Poisson means and Poisson predictor intervals. *IEEE Transactions on Reliability R-19*, 42–49. [147]

Nelson, W. (1972a). Charts for confidence limits and test for failure rates. *Journal of Quality Technology 4*, 190–195. [147]

Nelson, W. (1972b). A short life test for comparing a sample with previous accelerated test results. *Technometrics 14*, 175–186. [69]

Nelson, W. (1982). *Applied Life Data Analysis*. John Wiley & Sons, Inc. [36, 60, 70, 147, 208]

Nelson, W. (1984). Fitting of fatigue curves with nonconstant standard deviation to data with runouts. *Journal of Testing and Evaluation 12*, 69–77. [67]

Nelson, W. (1990). *Accelerated Testing: Statistical Models, Test Plans, and Data Analyses*. John Wiley & Sons, Inc. [67]

Nelson, W. (2000). Weibull prediction of a future number of failures. *Quality and Reliability Engineering International 16*, 23–26. [243]

Newcombe, R. G. (1998). Two-sided confidence intervals for the single proportion: Comparison of seven methods. *Statistics in Medicine 17*, 857–872. [125]

Newton, M. A. and A. E. Raftery (1994). Approximate Bayesian inference with the weighted likelihood bootstrap. *Journal of the Royal Statistical Society, Series B 56*, 3–48. With discussion and a reply by the authors. [265]

Nordman, D. J. and W. Q. Meeker (2002). Weibull prediction intervals for a future number of failures. *Technometrics 44*, 15–23. [243]

Odeh, R. E. (1990). Two-sided prediction intervals to contain at least k out of m future observations from a normal distribution. *Technometrics 32*, 203–216. [69, 484]

Odeh, R. E. and M. Fox (1975). *Sample Size Choice*. Marcel Dekker. [161]

Odeh, R. E. and D. B. Owen (1980). *Tables for Normal Tolerance Limits, Sampling Plans, and Screening*. Marcel Dekker. [68, 70, 71, 161, 481]

Odeh, R. E. and D. B. Owen (1983). *Attribute Sampling Plans, Tables of Tests and Confidence Limits for Proportions*. Marcel Dekker. [125, 126, 161]

Odeh, R. E., D. B. Owen, Z. W. Birnbaum, and L. Fisher (1977). *Pocket Book of Statistical Tables*. Marcel Dekker. [161]

O'Hagan, A. (1998). Eliciting expert beliefs in substantial practical applications. *The Statistician 47*, 21–35. [323]

O'Hagan, A., C. E. Buck, A. Daneshkhah, J. R. Eiser, P. H. Garthwaite, D. J. Jenkinson, J. E. Oakley, and T. Rakow (2006). *Uncertain Judgements: Eliciting Experts' Probabilities*. John Wiley & Sons, Inc. [323]

O'Hagan, A. and J. E. Oakley (2004). Probability is perfect, but we can't elicit it perfectly. *Reliability Engineering & System Safety 85*, 239–248. [323]

Olwell, D. and A. Sorell (2001). Warranty calculations for missiles with only current-status data, using Bayesian methods. In *Proceedings of the Annual Reliability and Maintainability Symposium*, 133–138. IEEE. [290, 387]

Oppenlander, J. E., J. Schmee, and G. J. Hahn (1988). Some simple robust estimators of normal distribution tail percentiles and their properties. *Communications in Statistics: Theory and Methods 17*, 2279–2301. [208]

Ostrouchov, G. and W. Q. Meeker (1988). Accuracy of approximate confidence bounds computed from interval censored Weibull and lognormal data. *Journal of Statistical Computation and Simulation 29*, 43–76. [461]

Owen, A. B. (1988). Empirical likelihood ratio confidence intervals for a single functional. *Biometrika 75*, 237–249. [97]

Owen, D. B. (1964). Control of percentages in both tails of the normal distribution. *Technometrics 6*(4), 377–387. [481]

Owen, D. B. (1965). A special case of a bivariate non-central t-distribution. *Biometrika 52*(3/4), 437–446. [481]

Owen, D. B. (1966). Control of percentages in both tails of the normal distributionerratum. *Technometrics 8*, 570. [481]

Owen, D. B. and W. Frawley (1971). Factors for tolerance limits which control both tails of the normal distribution. *Journal of Quality Technology 3*(2), 69–79. [481]

Patterson, P., J. Hannig, and H. Iyer (2004). Fiducial generalized confidence intervals for proportion of conformance. Technical report, Technical Report 2004/11, Colorado State University, Fort Collins, CO. [390]

Pawitan, Y. (2001). *In All Likelihood: Statistical Modelling and Inference Using Likelihood*. Oxford University Press. [243]

Pinheiro, J. C. and D. M. Bates (2000). *Mixed-Effects Models in S and S-PLUS*. Springer. [365]

Plummer, M. (2003). JAGS: A program for analysis of Bayesian graphical models using Gibbs sampling. In K. Hornik, F. Leisch, and A. Zeileis) (Editors), *Proceedings of the 3rd International Workshop on Distributed Statistical Computing*. [323]

Polson, N. G., J. G. Scott, et al. (2012). On the half-Cauchy prior for a global scale parameter. *Bayesian Analysis 7*, 887–902. [365]

Poudel, K. P. and Q. V. Cao (2013). Evaluation of methods to predict Weibull parameters for characterizing diameter distributions. *Forest Science 59*, 243–252. [246]

R Core Team (2016). *R: A Language and Environment for Statistical Computing*. R Foundation for Statistical Computing. [9, 323, 423]

Raftery, A. E. and S. Lewis (1992). How many iterations in the Gibbs sampler? *Bayesian Statistics 4*, 763–773. [323]

Rao, C. R. (1973). *Linear Statistical Inference and Its Applications*. John Wiley & Sons, Inc. [455]

Raudenbush, S. W. and A. S. Bryk (2002). *Hierarchical Linear Models: Applications and Data Analysis Methods*. Sage. [365]

Ripley, B. D. (2009). *Stochastic Simulation*. John Wiley & Sons, Inc. [9]

Ritz, C. and J. C. Streibig (2008). *Nonlinear Regression with R*. Springer. [67]

Rizzo, M. L. (2007). *Statistical Computing with R*. CRC Press. [9, 323]

Robert, C. P. and G. Casella (2010). *Introducing Monte Carlo Methods with R*. Springer. [323]

Ross, G. J. S. (1970). The efficient use of function minimization in non-linear maximum-likelihood estimation. *Applied Statistics 19*, 205–221. [299]

Ross, G. J. S. (1990). *Nonlinear Estimation*. Springer. [299]

Ross, S. (2012). *A First Course in Probability* (Ninth Edition). Pearson. [125, 146, 215]

Ross, S. (2014). *Introduction to Probability Models* (Eleventh Edition). Academic Press. [125, 146]

Rubin, D. B. (1981). The Bayesian bootstrap. *Annals of Statistics 9*, 130–134. [265]

Ryan, T. P. (2013). *Sample Size Determination and Power*. John Wiley & Sons, Inc. [161]

Sahu, S. and T. Smith (2006). A Bayesian method of sample size determination with practical applications. *Journal of the Royal Statistical Society, Series A 169*, 235–253. [161]

Scheffé, H. (1959). *The Analysis of Variance*. John Wiley & Sons, Inc. [69]

Scheuer, E. M. (1990). Let's teach more about prediction. In *Proceedings of the ASA Section on Statistical Education*. American Statistical Association. [22]

Seber, G. A. and A. J. Lee (2012). *Linear Regression Analysis* (Second Edition). John Wiley & Sons, Inc. [69]

Seber, G. A. F. and C. J. Wild (1989). *Nonlinear Regression*. John Wiley & Sons, Inc. [67, 365]

Semiglazov, V. F., V. N. Sagaidack, V. M. Moiseyenko, and E. A. Mikhailov (1993). Study of the role of breast self-examination in the reduction of mortality from breast cancer: The Russian Federation/World Health Organization Study. *European Journal of Cancer Part A: General Topics 29*, 2039–2046. [17]

Severini, T. A. (2000). *Likelihood Methods in Statistics*. Oxford University Press. [243]

Shao, J. and D. Tu (1995). *The Jackknife and Bootstrap*. Springer. [265]

Singer, J. D. and J. B. Willett (2003). *Applied Longitudinal Data Analysis: Modeling Change and Event Occurrence*. Oxford University Press. [365]

Smith, A. F. and A. E. Gelfand (1992). Bayesian statistics without tears: A sampling–resampling perspective. *The American Statistician 46*, 84–88. [323]

Snedecor, G. W. and W. G. Cochran (1967). *Statistical Methods* (Sixth Edition). Iowa State University Press. [22]

Snijders, T. A. (2011). *Multilevel Analysis.* Springer. [365]

Somerville, P. N. (1958). Tables for obtaining non-parametric tolerance limits. *Annals of Mathematical Statistics 29,* 599–601. [97, 548]

Spiegelhalter, D., A. Thomas, N. Best, and D. Lunn (2014). *OpenBUGS User Manual. Version 3.2.3.* [323]

Stan Development Team (2015a). Stan: A C++ library for probability and sampling, version 2.8.0. [323]

Stan Development Team (2015b). *Stan Modeling Language Users Guide and Reference Manual, Version 2.8.0.* [323]

Stein, C. (1945). A two-sample test for a linear hypothesis whose power is independent of the variance. *Annals of Mathematical Statistics 16,* 243–258. [161]

Stuart, A. and K. J. Ord (1994). *Distribution Theory.* Kendall's Advanced Theory of Statistics, Volume 1. Edward Arnold. [453]

Thatcher, A. R. (1964). Relationships between Bayesian and confidence limits for predictions (with discussions). *Journal of the Royal Statistical Society, Series B 26,* 176–210. [126]

Thoman, D. R., L. J. Bain, and C. E. Antle (1970). Maximum likelihood estimation, exact confidence intervals for reliability, and tolerance limits in the Weibull distribution. *Technometrics 12,* 363–371. [293]

Thomas, D. L. and D. R. Thomas (1986). Confidence bands for percentiles in the linear regression model. *Journal of the American Statistical Association 81,* 705–708. [69]

Tillitt, D. E., D. M. Papoulias, J. J. Whyte, and C. A. Richter (2010). Atrazine reduces reproduction in fathead minnow (*Pimephales promelas*). *Aquatic Toxicology 99,* 149 – 159. [225]

Tomsky, J. L., K. Nakano, and M. Iwashika (1979). Confidence limits for the number of defectives in a lot. *Journal of Quality Technology 11,* 199–204. [126]

Trindade, D. C. (1991). Can burn-in screen wearout mechanisms? Reliability modeling of defective subpopulations—A case study. In *29th Annual Proceedings of the International Reliability Physics Symposium, Las Vegas, NV,* 260–263. IEEE. [389]

Tsui, K.-W. and S. Weerahandi (1989). Generalized p-values in significance testing of hypotheses in the presence of nuisance parameters. *Journal of the American Statistical Association 84*(406), 602–607. [294]

Turner, D. L. and D. C. Bowden (1977). Simultaneous confidence bands for percentile lines in the general linear model. *Journal of the American Statistical Association 72,* 886–889. [69]

Turner, D. L. and D. C. Bowden (1979). Sharp confidence bands for percentile lines and tolerance bands for simple linear models. *Journal of the American Statistical Association 74,* 885–888. [69]

Vander Wiel, S. A. and W. Q. Meeker (1990). Accuracy of approximate confidence bounds using censored Weibull regression data from accelerated life tests. *IEEE Transactions on Reliability 39,* 346–351. [461]

Vardeman, S. B. (1992). What about the other intervals? (Corr: 93V47 p238). *The American Statistician 46*, 193–197. [22]

Venables, W. N. and B. D. Ripley (2002). *Modern Applied Statistics with S*. Springer. [365]

Villa, E. R. and L. A. Escobar (2006). Using moment generating functions to derive mixture distributions. *The American Statistician 60*, 75–80. [510]

Wald, A. and J. Wolfowitz (1946). Tolerance limits for a normal distribution. *Annals of Mathematical Statistics 17*, 208–215. [481]

Wallis, W. A. (1951). Tolerance intervals for linear regression. In *Proceedings of the Second Berkeley Symposium on Mathematical Statistics and Probability*, 43–51. University of California Press. [69]

Wang, H. (2008). Coverage probability of prediction intervals for discrete random variables. *Computational Statistics & Data Analysis 53*, 17–26. [126, 147]

Wang, H. and F. Tsung (2009). Tolerance intervals with improved coverage probabilities for binomial and Poisson variables. *Technometrics 51*, 25–33. [126, 147]

Weaver, B. P., M. S. Hamada, S. B. Vardeman, and A. G. Wilson (2012). A Bayesian approach to the analysis of gauge R&R data. *Quality Engineering 24*, 486–500. [376]

Weerahandi, S. (1993). Generalized confidence intervals (Corr: 94V89 p726). *Journal of the American Statistical Association 88*, 899–905. [294]

Weerahandi, S. (1995). *Exact Statistical Methods for Data Analysis*. Springer. [294]

Weerahandi, S. (2004). *Generalized Inference in Repeated Measures: Exact Methods in MANOVA and Mixed Models*. John Wiley & Sons, Inc. [294]

Wei, W. W.-S. (2005). *Time Series Analysis: Univariate and Multivariate Methods* (Second Edition). Addison-Wesley. [67, 201]

Wild, C. J. and M. Pfannkuch (1999). Statistical thinking in empirical enquiry. *International Statistical Review 67*, 223–248 (discussion: 248–265). [22]

Wilks, S. S. (1941). Determination of sample sizes for setting tolerance limits. *Annals of Mathematical Statistics 12*, 91–96. [175]

Wilson, E. B. (1927). Probable inference, the law of succession, and statistical inference. *Journal of the American Statistical Association 22*, 209–212. [125]

Wu, C. J. and M. S. Hamada (2009). *Experiments: Planning, Analysis, and Optimization* (Second Edition). John Wiley & Sons, Inc. [22]

Xie, Y., Y. Hong, L. Escobar, and W. Meeker (2017). Simultaneous prediction intervals for the (log)-location-scale family of distributions. *Journal of Statistical Computation and Simulation*. DOI 10.1080/00949655.2016.1277426. [294, 321, 487]

Yuan, M., Y. Hong, L. A. Escobar, and W. Q. Meeker (2017). Two-sided tolerance intervals for the (log)-location-scale family of distributions. *Quality Technology & Quantitative Management*. DOI: 10.1080/16843703.2016.1226594. [294, 319, 481]

Zheng, X. and W.-Y. Loh (1995). Bootstrapping binomial confidence intervals. *Journal of Statistical Planning and Inference 43*, 355–380. [421]

Index

See Appendix A for a listing of notation. Page numbers for some of the terms listed in Appendix A can be found in this subject Index. Particular data sets and applications are listed under Examples, indicating the first occurrence of an application within a chapter. A listing of probability distributions is given under Probability distributions, distribution name. For pages giving technical information and applications of the more important probability distributions, look under the probability distribution name (e.g., Weibull, applications). When there are more than a few page references for an entry and if the first listed page(s) are not the most appropriate for general first purposes, the suggested page number is set in bold. Instead of an Author index, page numbers are given at the end of each references, pointing to the page(s) where the reference is cited.

Statistical Intervals: A Guide for Practitioners and Researchers, Second Edition.
William Q. Meeker, Gerald J. Hahn and Luis A. Escobar.
© 2017 John Wiley & Sons, Inc. Published 2017 by John Wiley & Sons, Inc.
Companion Website: www.wiley.com/go/meeker/intervals

WILEY SERIES IN PROBABILITY AND STATISTICS

Multivariate Density Estimation: Theory, Practice, and Visualization, 2nd Edition
by David W. Scott

Foundations of Linear and Generalized Linear Models
by Alan Agresti

Applied Econometric Time Series, 4th Edition
by Walter Enders

Correspondence Analysis: Theory, Practice and New Strategies
by Eric J. Beh, Rosaria Lombardo

The Fitness of Information: Quantitative Assessments of Critical Evidence
by Chaomei Chen

Nonparametric Hypothesis Testing: Rank and Permutation Methods with Applications in R
by Stefano Bonnini, Livio Corain, Marco Marozzi, Luigi Salmaso

Applied Bayesian Modelling, 2nd Edition
by Peter Congdon

Information and Exponential Families in Statistical Theory
by O. Barndorff-Nielsen

Markov Chains: Analytic and Monte Carlo Computations
by Carl Graham

Lower Previsions
by Matthias C. M. Troffaes, Gert de Cooman

Introduction to Imprecise Probabilities
by Thomas Augustin (Editor), Frank P. A. Coolen (Editor), Gert de Cooman (Editor), Matthias C. M. Troffaes (Editor)

Exploration and Analysis of DNA Microarray and Other High-Dimensional Data, 2nd Edition
by Dhammika Amaratunga, Javier Cabrera, Ziv Shkedy

Examples and Problems in Mathematical Statistics
by Shelemyahu Zacks

Applied Linear Regression, 4th Edition
by Sanford Weisberg

Understanding Uncertainty, Revised Edition
by Dennis V. Lindley

Fast Sequential Monte Carlo Methods for Counting and Optimization
by Reuven Y. Rubinstein, Ad Ridder, Radislav Vaisman

Multivariate Time Series Analysis: With R and Financial Applications
by Ruey S. Tsay

Nonparametric Statistical Methods, 3rd Edition
by Myles Hollander, Douglas A. Wolfe, Eric Chicken

Quantile Regression: Theory and Applications
by Cristina Davino, Marilena Furno, Domenico Vistocco

Design and Analysis of Clinical Trials: Concepts and Methodologies, 3rd Edition
by Shein-Chung Chow, Jen-Pei Liu

Loss Models: From Data to Decisions, 4th Edition
by Stuart A. Klugman, Harry H. Panjer, Gordon E. Willmot

Using the Weibull Distribution: Reliability, Modeling and Inference
by John I. McCool

Causality: Statistical Perspectives and Applications
by Carlo Berzuini (Editor), Philip Dawid (Editor), Luisa Bernardinell (Editor)

Methods of Multivariate Analysis, 3rd Edition
by Alvin C. Rencher, William F. Christensen

Modelling Under Risk and Uncertainty: An Introduction to Statistical, Phenomenological and
Computational Methods
by Etienne de Rocquigny

Introduction to Linear Regression Analysis, 5th Edition
by Douglas C. Montgomery, Elizabeth A. Peck, G. Geoffrey Vining

Bayesian Analysis of Stochastic Process Models
by David Insua, Fabrizio Ruggeri, Mike Wiper

Optimal Learning
by Warren B. Powell, Ilya O. Ryzhov

Geostatistics: Modeling Spatial Uncertainty, 2nd Edition
by Jean-Paul Chiles, Pierre Delfiner

Probability and Measure, Anniversary Edition
by Patrick Billingsley

Design and Analysis of Experiments, Volume 3, Special Designs and Applications
by Klaus Hinkelmann (Editor)

Empirical Model Building: Data, Models, and Reality, 2nd Edition
by James R. Thompson

The Analysis of Covariance and Alternatives: Statistical Methods for Experiments,
Quasi-Experiments, and Single-Case Studies, 2nd Edition
by Bradley Huitema

Loss Models: From Data to Decisions (One Year Online: Preparation for Actuarial Exam C/4),
3rd Edition
by Stuart A. Klugman, Harry H. Panjer, Gordon E. Willmot

Approximate Dynamic Programming: Solving the Curses of Dimensionality, 2nd Edition
by Warren B. Powell

Statistical Analysis of Profile Monitoring
by Rassoul Noorossana, Abbas Saghaei, Amirhossein Amiri

Applied Longitudinal Analysis, 2nd Edition
by Garrett M. Fitzmaurice, Nan M. Laird, James H. Ware

A Primer on Experiments with Mixtures
by John A. Cornell

Statistical Methods for Quality Improvement, 3rd Edition
by Thomas P. Ryan

Latent Variable Models and Factor Analysis: A Unified Approach, 3rd Edition
by David J. Bartholomew, Martin Knott, Irini Moustaki

Multiple Comparison Procedures
by Yosef Hochberg, Ajit C. Tamhane

Time Series Analysis and Forecasting by Example
by Soren Bisgaard, Murat Kulahci

An Elementary Introduction to Statistical Learning Theory
by Sanjeev Kulkarni, Gilbert Harman

Mixtures: Estimation and Applications
by Kerrie Mengersen, Christian Robert, Mike Titterington

Dirichlet and Related Distributions: Theory, Methods and Applications
by Kai Wang Ng, Guo-Liang Tian, Man-Lai Tang

Data Analysis: What Can Be Learned From the Past 50 Years
by Peter J. Huber

Statistical Methods in Diagnostic Medicine, 2nd Edition
by Xiao-Hua Zhou, Nancy A. Obuchowski, Donna K. McClish

Handbook of Monte Carlo Methods
by Dirk P. Kroese, Thomas Taimre, Zdravko I. Botev

Statistical Methods for Fuzzy Data
by Reinhard Viertl

Cluster Analysis, 5th Edition
by Brian S. Everitt, Sabine Landau, Morven Leese, Daniel Stahl

Multistate Systems Reliability Theory with Applications
by Bent Natvig

Biostatistical Methods: The Assessment of Relative Risks, 2nd Edition
by John M. Lachin

Spatial Statistics and Spatio-Temporal Data: Covariance Functions and Directional Properties
by Michael Sherman

Multilevel Statistical Models, 4th Edition
by Harvey Goldstein

Bias and Causation: Models and Judgment for Valid Comparisons
by Herbert I. Weisberg

Statistical Inference for Fractional Diffusion Processes
by B. L. S. Prakasa Rao

Applied Statistics: Analysis of Variance and Regression, 3rd Edition
by Ruth M. Mickey, Olive Jean Dunn, Virginia A. Clark

Analysis of Ordinal Categorical Data, 2nd Edition
by Alan Agresti

Permutation Tests for Complex Data: Theory, Applications and Software
by Fortunato Pesarin, Luigi Salmaso

Generalized Linear Models: with Applications in Engineering and the Sciences, 2nd Edition
by Raymond H. Myers, Douglas C. Montgomery, G. Geoffrey Vining, Timothy J. Robinson

Random Data: Analysis and Measurement Procedures, 4th Edition
by Julius S. Bendat, Allan G. Piersol

Multivariate Statistics : High-Dimensional and Large-Sample Approximations
by Yasunori Fujikoshi, Vladimir V. Ulyanov, Ryoichi Shimizu

Latent Class and Latent Transition Analysis: With Applications in the Social, Behavioral, and Health Sciences
by Linda M. Collins, Stephanie T. Lanza

Batch Effects and Noise in Microarray Experiments: Sources and Solutions
by Andreas Scherer

Bayesian Analysis for the Social Sciences
by Simon Jackman

Planning and Analysis of Observational Studies
by William G. Cochran

Biostatistical Methods: The Assessment of Relative Risks
by John M. Lachin

System Reliability Theory: Models and Statistical Methods
by Arnljot Høyland, Marvin Rausand

Statistical Methods in Diagnostic Medicine
by Xiao-Hua Zhou, Donna K. McClish, Nancy A. Obuchowski

Linear Statistical Models
by James H. Stapleton

Statistical Methods for Comparative Studies: Techniques for Bias Reduction
by Sharon Roe Anderson, Ariane Auquier, Walter W. Hauck, David Oakes, Walter Vandaele, Herbert I. Weisberg

Simulation and the Monte Carlo Method
by Reuven Y. Rubinstein

Empirical Model Building
by James R. Thompson

Bayesian Networks: An Introduction
by Timo Koski, John Noble

Loss Distributions
by Robert V. Hogg, Stuart A. Klugman

Multiple Time Series
by Edward James Hannan

Smoothing of Multivariate Data: Density Estimation and Visualization
by Jussi Klemela

Experiments: Planning, Analysis, and Optimization, 2nd Edition
by C. F. Jeff Wu, Michael S. Hamada

Linear Statistical Models, 2nd Edition
by James H. Stapleton

ExamPrep (Online) for Loss Models: From Data to Decisions, 3e
by Stuart A. Klugman, Harry H. Panjer, Gordon E. Willmot

The Collected Works of Shayle R. Searle
by Shayle R. Searle

The Collected Works of George A.F. Seber
by George A. F. Seber

Modern Regression Methods, Set, 2nd Edition
by Thomas P. Ryan

Robust Methods in Biostatistics
by Stephane Heritier, Eva Cantoni, Samuel Copt, Maria-Pia Victoria-Feser

Decision Theory: Principles and Approaches
by Giovanni Parmigiani, Lurdes Inoue

Statistical Tolerance Regions: Theory, Applications, and Computation
by Kalimuthu Krishnamoorthy, Thomas Mathew

Stage-Wise Adaptive Designs
by Shelemyahu Zacks

Statistical Control by Monitoring and Adjustment 2e & Statistics for Experimenters: Design, Innovation, and Discovery 2e Set
by George E. P. Box

Statistical Control by Monitoring and Adjustment, 2nd Edition
by George E. P. Box, Alberto Luceño, Maria del Carmen Paniagua-Quinones

Statistical Analysis of Designed Experiments: Theory and Applications
by Ajit C. Tamhane

Student Solutions Manual to Accompany Introduction to Time Series Analysis and Forecasting
by Douglas C. Montgomery, Cheryl L. Jennings, Murat Kulahci, Rachel T. Johnson (Photographer), James R. Broyles, Christopher J. Rigdon

Robust Statistics, 2nd Edition
by Peter J. Huber, Elvezio M. Ronchetti

Markov Processes and Applications: Algorithms, Networks, Genome and Finance
by Etienne Pardoux

Stochastic Processes for Insurance and Finance
by Tomasz Rolski, Hanspeter Schmidli, V. Schmidt, Jozef Teugels

Modern Regression Methods, 2nd Edition
by Thomas P. Ryan

Statistical Rules of Thumb, 2nd Edition
by Gerald van Belle

Statistical Meta-Analysis with Applications
by Joachim Hartung, Guido Knapp, Bimal K. Sinha

Fundamentals of Queueing Theory, 4th Edition
by Donald Gross, John F. Shortle, James M. Thompson, Carl M. Harris

Linear Models: The Theory and Application of Analysis of Variance
by Brenton R. Clarke

Generalized, Linear, and Mixed Models, 2nd Edition
by Charles E. McCulloch, Shayle R. Searle, John M. Neuhaus

Multivariable Model - Building: A Pragmatic Approach to Regression Anaylsis based on
Fractional Polynomials for Modelling Continuous Variables
by Patrick Royston, Willi Sauerbrei

Introduction to Time Series Analysis and Forecasting
by Douglas C. Montgomery, Cheryl L. Jennings, Murat Kulahci

Meta Analysis: A Guide to Calibrating and Combining Statistical Evidence
by Elena Kulinskaya, Stephan Morgenthaler, Robert G. Staudte

The EM Algorithm and Extensions, 2nd Edition
by Geoffrey McLachlan, Thriyambakam Krishnan

Applied Survival Analysis: Regression Modeling of Time to Event Data, 2nd Edition
by David W. Hosmer, Jr., Stanley Lemeshow, Susanne May

Design and Analysis of Experiments Set
by Klaus Hinkelmann, Oscar Kempthorne

Univariate Discrete Distributions, Set, 3rd Edition
by Norman L. Johnson, Adrienne W. Kemp, Samuel Kotz

Applied Multiway Data Analysis
by Pieter M. Kroonenberg

Simulation and the Monte Carlo Method, 2nd Edition Set
by Reuven Y. Rubinstein

Statistical Advances in the Biomedical Sciences: Clinical Trials, Epidemiology, Survival
Analysis, and Bioinformatics
by Atanu Biswas (Editor), Sujay Datta (Editor), Jason P. Fine (Editor), Mark R. Segal (Editor)

Modern Applied U-Statistics
by Jeanne Kowalski, Xin M. Tu

Simulation and the Monte Carlo Method, 2nd Edition
by Reuven Y. Rubinstein, Dirk P. Kroese

Models for Probability and Statistical Inference: Theory and Applications
by James H. Stapleton

Design and Analysis of Experiments, Volume 1, Introduction to Experimental Design,
2nd Edition
by Klaus Hinkelmann, Oscar Kempthorne

Student Solutions Manual to Accompany Simulation and the Monte Carlo Method,
2nd Edition
by Dirk P. Kroese, Thomas Taimre, Zdravko I. Botev, Reuven Y. Rubinstein

A Matrix Handbook for Statisticians
by George A. F. Seber

Bootstrap Methods: A Guide for Practitioners and Researchers, 2nd Edition
by Michael R. Chernick

Nonparametric Analysis of Univariate Heavy-Tailed Data: Research and Practice
by Natalia Markovich

Inference and Prediction in Large Dimensions
by Denis Bosq, Delphine Blanke

Periodically Correlated Random Sequences: Spectral Theory and Practice
by Harry L. Hurd, Abolghassem Miamee

Management of Data in Clinical Trials, 2nd Edition
by Eleanor McFadden

Nonparametric Statistics with Applications to Science and Engineering
by Paul H. Kvam, Brani Vidakovic

The Construction of Optimal Stated Choice Experiments: Theory and Methods
by Deborah J. Street, Leonie Burgess

Bayes Linear Statistics, Theory and Methods
by Michael Goldstein, David Wooff

Nonlinear Regression Analysis and Its Applications
by Douglas M. Bates, Donald G. Watts

Long-Memory Time Series: Theory and Methods
by Wilfredo Palma

Response Surfaces, Mixtures, and Ridge Analyses, 2nd Edition
by George E. P. Box, Norman R. Draper

An Introduction to Categorical Data Analysis, 2nd Edition
by Alan Agresti

Structural Equation Modeling: A Bayesian Approach
by Sik-Yum Lee

Simulation and Monte Carlo: With applications in finance and MCMC
by J. S. Dagpunar

Modern Experimental Design
by Thomas P. Ryan

Variations on Split Plot and Split Block Experiment Designs
by Walter T. Federer, Freedom King

Bayesian Statistical Modelling, 2nd Edition
by Peter Congdon

Regression Analysis by Example, 4th Edition
by Samprit Chatterjee, Ali S. Hadi

Measurement Error Models
by Wayne A. Fuller

The Theory of Response-Adaptive Randomization in Clinical Trials
by Feifang Hu, William F. Rosenberger

Reliability and Risk: A Bayesian Perspective
by Nozer D. Singpurwalla

Visual Statistics: Seeing Data with Dynamic Interactive Graphics
by Forrest W. Young, Pedro M. Valero-Mora, Michael Friendly

Operational Risk: Modeling Analytics
by Harry H. Panjer

Statistical Methods in Spatial Epidemiology, 2nd Edition
by Andrew B. Lawson

Applied MANOVA and Discriminant Analysis, 2nd Edition
by Carl J. Huberty, Stephen Olejnik

Nonparametric Regression Methods for Longitudinal Data Analysis: Mixed-Effects Modeling
Approaches
by Hulin Wu, Jin-Ting Zhang

Improving Almost Anything: Ideas and Essays, Revised Edition
by George E. P. Box

Robust Statistics: Theory and Methods
by Ricardo A. Maronna, Douglas R. Martin, Victor J. Yohai

Longitudinal Data Analysis
by Donald Hedeker, Robert D. Gibbons

Theory of Preliminary Test and Stein-Type Estimation with Applications
by A. K. Md. Ehsanes Saleh

Matrix Algebra Useful for Statistics
by Shayle R. Searle

Linear Models for Unbalanced Data
by Shayle R. Searle

Introductory Stochastic Analysis for Finance and Insurance
by X. Sheldon Lin, Society of Actuaries

Uncertainty Analysis with High Dimensional Dependence Modelling
by Dorota Kurowicka, Roger M. Cooke

Stochastic Simulation
by Brian D. Ripley

Variance Components
by Shayle R. Searle, George Casella, Charles E. McCulloch

Exploring Data Tables, Trends, and Shapes
by David C. Hoaglin (Editor), Frederick Mosteller (Editor), John W. Tukey (Editor)

Modes of Parametric Statistical Inference
by Seymour Geisser, Wesley O. Johnson

Latent Curve Models: A Structural Equation Perspective
by Kenneth A. Bollen, Patrick J. Curran

Introduction to Nonparametric Regression
by K. Takezawa

Bayesian Statistics and Marketing
by Peter E. Rossi, Greg M. Allenby, Rob McCulloch

Aspects of Multivariate Statistical Theory
by Robb J. Muirhead

Statistical Methods for Forecasting
by Bovas Abraham, Johannes Ledolter

Counting Processes and Survival Analysis
by Thomas R. Fleming, David P. Harrington

Analysis of Financial Time Series, 2nd Edition
by Ruey S. Tsay

Contemporary Bayesian Econometrics and Statistics
by John Geweke

Markov Processes: Characterization and Convergence
by Stewart N. Ethier, Thomas G. Kurtz

Fractal-Based Point Processes
by Steven Bradley Lowen, Malvin Carl Teich

Univariate Discrete Distributions, 3rd Edition
by Norman L. Johnson, Adrienne W. Kemp, Samuel Kotz

Clinical Trials: A Methodologic Perspective, 2nd Edition
by Steven Piantadosi

Statistics for Experimenters: Design, Innovation, and Discovery, 2nd Edition
by George E. P. Box, J. Stuart Hunter, William G. Hunter

Bayesian Models for Categorical Data
by Peter Congdon

Recent Advances in Quantitative Methods in Cancer and Human Health Risk Assessment
by Lutz Edler (Editor), Christos Kitsos (Editor)

Design and Analysis of Experiments, Volume 2, Advanced Experimental Design
by Klaus Hinkelmann, Oscar Kempthorne

Combinatorial Methods in Discrete Distributions
by Charalambos A. Charalambides

Robust Statistics: The Approach Based on Influence Functions
by Frank R. Hampel, Elvezio M. Ronchetti, Peter J. Rousseeuw, Werner A. Stahel

Finding Groups in Data: An Introduction to Cluster Analysis
by Leonard Kaufman, Peter J. Rousseeuw

Markov Decision Processes: Discrete Stochastic Dynamic Programming
by Martin L. Puterman

Robust Statistics
by Peter J. Huber

Image Processing and Jump Regression Analysis
by Peihua Qiu

Matrix Analysis for Statistics, 2nd Edition
by James R. Schott

Flowgraph Models for Multistate Time-to-Event Data
by Aparna V. Huzurbazar

Constrained Statistical Inference: Order, Inequality, and Shape Constraints
by Mervyn J. Silvapulle, Pranab Kumar Sen

Extreme Value and Related Models with Applications in Engineering and Science
by Enrique Castillo, Ali S. Hadi, N. Balakrishnan, Jose M. Sarabia

Preparing for the Worst: Incorporating Downside Risk in Stock Market Investments
by Hrishikesh (Rick) D. Vinod, Derrick Reagle

Accelerated Testing: Statistical Models, Test Plans, and Data Analysis
by Wayne B. Nelson

Statistics of Extremes: Theory and Applications
by Jan Beirlant, Yuri Goegebeur, Johan Segers, Jozef Teugels, Daniel De Waal (Contributions by), Chris Ferro (Contributions by)

Regression With Social Data: Modeling Continuous and Limited Response Variables
by Alfred DeMaris

Biostatistics: A Bayesian Introduction
by George G. Woodworth

Multivariate Observations
by George A. F. Seber

Measurement Errors in Surveys
by Paul P. Biemer (Editor), Robert M. Groves (Editor), Lars E. Lyberg (Editor), Nancy A. Mathiowetz (Editor), Seymour Sudman (Editor)

Spatial Statistics
by Brian D. Ripley

Regression Diagnostics: Identifying Influential Data and Sources of Collinearity
by David A. Belsley, Edwin Kuh, Roy E. Welsch

Applied Bayesian Modeling and Causal Inference from Incomplete-Data Perspectives
by Andrew Gelman (Editor), Xiao-Li Meng (Editor)

Analyzing Microarray Gene Expression Data
by Geoffrey McLachlan, Kim-Anh Do, Christophe Ambroise

Discriminant Analysis and Statistical Pattern Recognition
by Geoffrey McLachlan

Biostatistics: A Methodology For the Health Sciences, 2nd Edition
by Gerald van Belle, Lloyd D. Fisher, Patrick J. Heagerty, Thomas Lumley

Applied Spatial Statistics for Public Health Data
by Lance A. Waller, Carol A. Gotway

Generalized Least Squares
by Takeaki Kariya, Hiroshi Kurata

Financial Derivatives in Theory and Practice, Revised Edition
by Philip Hunt, Joanne Kennedy

Discrete Distributions: Applications in the Health Sciences
by Daniel Zelterman

Time Series: Applications to Finance
by Ngai Hang Chan

Planning, Construction, and Statistical Analysis of Comparative Experiments
by Francis G. Giesbrecht, Marcia L. Gumpertz

Random Graphs for Statistical Pattern Recognition
by David J. Marchette

Statistics for Research, 3rd Edition
by Shirley Dowdy, Stanley Wearden, Daniel Chilko

Applied Statistics: Analysis of Variance and Regression, 3rd Edition
by Ruth M. Mickey, Olive Jean Dunn, Virginia A. Clark

Applied Life Data Analysis
by Wayne B. Nelson

Environmental Statistics: Methods and Applications
by Vic Barnett

Numerical Issues in Statistical Computing for the Social Scientist
by Micah Altman, Jeff Gill, Michael P. McDonald

Weibull Models
by D. N. Prabhakar Murthy, Min Xie, Renyan Jiang

Numerical Methods in Finance: A MATLAB-Based Introduction
by Paolo Brandimarte

Robust Regression and Outlier Detection
by Peter J. Rousseeuw, Annick M. Leroy

Statistical Methods for Rates and Proportions, 3rd Edition
by Joseph L. Fleiss, Bruce Levin, Myunghee Cho Paik

Nonlinear Regression
by George A. F. Seber, C. J. Wild

The Theory of Measures and Integration
by Eric M. Vestrup

A User's Guide to Principal Components
by J. Edward Jackson

A History of Probability and Statistics and Their Applications before 1750
by Anders Hald

Statistical Size Distributions in Economics and Actuarial Sciences
by Christian Kleiber, Samuel Kotz

Order Statistics, 3rd Edition
by Herbert A. David, H. N. Nagaraja

An Introduction to Multivariate Statistical Analysis, 3rd Edition
by Theodore W. Anderson

Exploratory Data Mining and Data Cleaning
by Tamraparni Dasu, Theodore Johnson

Categorical Data Analysis, 2nd Edition
by Alan Agresti

Methods of Multivariate Analysis, 2nd Edition
by Alvin C. Rencher

Lévy Processes in Finance: Pricing Financial Derivatives
by Wim Schoutens

Applied Bayesian Modelling
by Peter Congdon

Statistical Design and Analysis of Experiments: With Applications to Engineering and Science, 2nd Edition
by Robert L. Mason, Richard F. Gunst, James L. Hess

Linear Regression Analysis, 2nd Edition
by George A. F. Seber, Alan J. Lee

Case Studies in Reliability and Maintenance
by Wallace R. Blischke (Editor), D. N. Prabhakar Murthy (Editor)

Subjective and Objective Bayesian Statistics: Principles, Models, and Applications, 2nd Edition
by S. James Press

Statistical Models and Methods for Lifetime Data, 2nd Edition
by Jerald F. Lawless

Advanced Calculus with Applications in Statistics, 2nd Edition
by André I. Khuri

Models for Investors in Real World Markets
by James R. Thompson, Edward E. Williams, M. Chapman Findlay, III

The Statistical Analysis of Failure Time Data, 2nd Edition
by John D. Kalbfleisch, Ross L. Prentice

Statistical Analysis with Missing Data, 2nd Edition
by Roderick J. A. Little, Donald B. Rubin

Elements of Applied Stochastic Processes, 3rd Edition
by U. Narayan Bhat, Gregory K. Miller

Regression Models for Time Series Analysis
by Benjamin Kedem, Konstantinos Fokianos

Randomization in Clinical Trials: Theory and Practice
by William F. Rosenberger, John M. Lachin

Methods of Multivariate Statistics
by Muni S. Srivastava

Statistical Methods for the Analysis of Biomedical Data, 2nd Edition
by Robert F. Woolson, William R. Clarke

Statistical Group Comparison
by Tim Futing Liao

Bayesian Methods for Nonlinear Classification and Regression
by David G. T. Denison, Christopher C. Holmes, Bani K. Mallick, Adrian F. M. Smith

Statistical Process Adjustment for Quality Control
by Enrique del Castillo

Comparison Methods for Stochastic Models and Risks
by Alfred Müller, Dietrich Stoyan

Experiments with Mixtures: Designs, Models, and the Analysis of Mixture Data, 3rd Edition
by John A. Cornell

Linear Statistical Inference and its Applications, 2nd Edition
by C. Radhakrishna Rao

Approximation Theorems of Mathematical Statistics
by Robert J. Serfling

Runs and Scans with Applications
by N. Balakrishnan, Markos V. Koutras

Matrix Algebra for Applied Economics
by Shayle R. Searle, Lois Schertz Willett

Biostatistical Methods in Epidemiology
by Stephen C. Newman

Limit Distributions for Sums of Independent Random Vectors: Heavy Tails in Theory and Practice
by Mark M. Meerschaert, Hans-Peter Scheffler

Probability and Finance: It's Only a Game!
by Glenn Shafer, Vladimir Vovk

Foundations of Time Series Analysis and Prediction Theory
by Mohsen Pourahmadi

The Subjectivity of Scientists and the Bayesian Approach
by S. James Press, Judith M. Tanur

Mathematics of Chance
by Jirí Andel

A Course in Time Series Analysis
by Daniel Peña, George C. Tiao, Ruey S. Tsay

Finite Mixture Models
by Geoffrey McLachlan, David Peel

Sensitivity Analysis: Gauging the Worth of Scientific Models
by A. Saltelli (Editor), K. Chan (Editor), E. M. Scott (Editor)

Spatial Tessellations: Concepts and Applications of Voronoi Diagrams, 2nd Edition
by Atsuyuki Okabe, Barry Boots, Kokichi Sugihara, Sung Nok Chiu

Continuous Multivariate Distributions, Volume 1, Models and Applications, 2nd Edition
by Samuel Kotz, N. Balakrishnan, Norman L. Johnson

Statistical Methods for the Reliability of Repairable Systems
by Steven E. Rigdon, Asit P. Basu

Bayesian Theory
by José M. Bernardo, Adrian F. M. Smith

Reliability: Modeling, Prediction, and Optimization
by Wallace R. Blischke, D. N. Prabhakar Murthy

Fourier Analysis of Time Series: An Introduction, 2nd Edition
by Peter Bloomfield

Directional Statistics
by Kanti V. Mardia, Peter E. Jupp

Simulation: A Modeler's Approach
by James R. Thompson

Shape and Shape Theory
by D. G. Kendall, D. Barden, T. K. Carne, H. Le

Applied Regression Including Computing and Graphics
by R. Dennis Cook, Sanford Weisberg

Convergence of Probability Measures, 2nd Edition
by Patrick Billingsley

Comparative Statistical Inference, 3rd Edition
by Vic Barnett

Statistical Modeling by Wavelets
by Brani Vidakovic

Statistical Analysis of Categorical Data
by Chris J. Lloyd

Fractional Factorial Plans
by Aloke Dey, Rahul Mukerjee

Nonparametric Statistical Methods, Solutions Manual, 2nd Edition
by Myles Hollander, Douglas A. Wolfe

Stochastic Processes for Insurance and Finance
by Tomasz Rolski, Hanspeter Schmidli, V. Schmidt, Jozef Teugels

Practical Nonparametric Statistics, 3rd Edition
by W. J. Conover

Stochastic Dynamic Programming and the Control of Queueing Systems
by Linn I. Sennott

Regression Graphics: Ideas for Studying Regressions Through Graphics
by R. Dennis Cook

Statistical Shape Analysis
by Ian L. Dryden, Kanti V. Mardia

Statistical Methods for Reliability Data
by William Q. Meeker, Luis A. Escobar

Records
by Barry C. Arnold, N. Balakrishnan, H. N. Nagaraja

Collected Works of Jaroslav Hájek: With Commentary
by M. Hušková, R. Beran, V. Dupac

Applied Regression Analysis, 3rd Edition
by Norman R. Draper, Harry Smith

Modern Simulation and Modeling
by Reuven Y. Rubinstein, Benjamin Melamed

Statistical Tests for Mixed Linear Models
by André I. Khuri, Thomas Mathew, Bimal K. Sinha

Multivariate Statistical Inference and Applications
by Alvin C. Rencher

Limit Theorems in Change-Point Analysis
by Miklós Csörgö, Lajos Horváth

Applied Survival Analysis
by Chap T. Le

The Theory of Canonical Moments with Applications in Statistics, Probability, and Analysis
by Holger Dette, William J. Studden

Geometrical Foundations of Asymptotic Inference
by Robert E. Kass, Paul W. Vos

Leading Personalities in Statistical Sciences: From the Seventeenth Century to the Present
by Norman L. Johnson (Editor), Samuel Kotz (Editor)

Sequential Estimation
by Malay Ghosh, Nitis Mukhopadhyay, Pranab Kumar Sen

A Weak Convergence Approach to the Theory of Large Deviations
by Paul Dupuis, Richard S. Ellis

Survey Measurement and Process Quality
by Lars E. Lyberg (Editor), Paul P. Biemer (Editor), Martin Collins (Editor), Edith D. De Leeuw (Editor), Cathryn Dippo (Editor), Norbert Schwarz (Editor), Dennis Trewin (Editor)

Discrete Multivariate Distributions
by Norman L. Johnson, Samuel Kotz, N. Balakrishnan

Construction and Assessment of Classification Rules
by David J. Hand

Methods for Statistical Data Analysis of Multivariate Observations, 2nd Edition
by R. Gnanadesikan

Survival Analysis with Long-Term Survivors
by Ross A. Maller, Xian Zhou

Aspects of Statistical Inference
by A. H. Welsh

Operational Subjective Statistical Methods: A Mathematical, Philosophical, and Historical
Introduction
by Frank Lad

Probability: A Survey of the Mathematical Theory, 2nd Edition
by John W. Lamperti

Time Series Analysis: Nonstationary and Noninvertible Distribution Theory
by Katsuto Tanaka

Adaptive Sampling
by Steven K. Thompson, George A. F. Seber

A Probabilistic Analysis of the Sacco and Vanzetti Evidence
by Joseph B. Kadane, David A. Schum

Assessment: Problems, Developments and Statistical Issues
by Harvey Goldstein (Editor), Toby Lewis (Editor)

A Guide to Chi-Squared Testing
by Priscilla E. Greenwood, Michael S. Nikulin

Epidemiological Research Methods
by Donald R. McNeil

Sequential Stochastic Optimization
by R. Cairoli, Robert C. Dalang

Bayesian Methods and Ethics in a Clinical Trial Design
by Joseph B. Kadane (Editor)

Introduction to Statistical Time Series, 2nd Edition
by Wayne A. Fuller

Design and Analysis of Experiments for Statistical Selection, Screening, and Multiple
Comparisons
by Robert E. Bechhofer, Thomas J. Santner, David M. Goldsman

Continuous Univariate Distributions, Volume 2, 2nd Edition
by Norman L. Johnson, Samuel Kotz, N. Balakrishnan

Business Survey Methods
by Brenda G. Cox (Editor), David A. Binder (Editor), B. Nanjamma Chinnappa (Editor),
Anders Christianson (Editor), Michael J. Colledge (Editor), Phillip S. Kott (Editor)

Continuous Univariate Distributions, Volume 1, 2nd Edition
by Norman L. Johnson, Samuel Kotz, N. Balakrishnan

Fractals, Random Shapes and Point Fields: Methods of Geometrical Statistics
by Dietrich Stoyan, Dr. Helga Stoyan

Statistical Factor Analysis and Related Methods: Theory and Applications
by Alexander T. Basilevsky

Bayesian Theory
by José M. Bernardo, Adrian F. M. Smith

Hilbert Space Methods in Probability and Statistical Inference
by Christopher G. Small, Don L. McLeish

Monotone Structure in Discrete-Event Systems
by Paul Glasserman, David D. Yao

Outliers in Statistical Data, 3rd Edition
by Vic Barnett, Toby Lewis

Numerical Methods for Stochastic Processes
by Nicolas Bouleau, Dominique Lépingle

Alternative Methods of Regression
by David Birkes, Dr. Yadolah Dodge

Resampling-Based Multiple Testing: Examples and Methods for p-Value Adjustment
by Peter H. Westfall, S. Stanley Young

Multivariate Density Estimation: Theory, Practice, and Visualization
by David W. Scott

Nonsampling Error in Surveys
by Judith T. Lessler, William D. Kalsbeek

Exploring the Limits of Bootstrap
by Raoul LePage (Editor), Lynne Billard (Editor)

Forecasting with Dynamic Regression Models
by Alan Pankratz

Fundamentals of Exploratory Analysis of Variance
by David C. Hoaglin (Editor), Frederick Mosteller (Editor), John W. Tukey (Editor)

Statistical Intervals: A Guide for Practitioners
by Gerald J. Hahn, William Q. Meeker

MACSYMA for Statisticians
by Barbara Heller

LISP-STAT: An Object-Oriented Environment for Statistical Computing and Dynamic
Graphics
by Luke Tierney

Statistical Methods in Engineering and Quality Assurance
by Peter W. M. John

Robust Estimation and Testing
by Robert G. Staudte, Simon J. Sheather

Nonlinear Multivariate Analysis
by Albert Gifi

Graphical Models in Applied Multivariate Statistics
by J. Whittaker

Influence Diagrams, Belief Nets and Decision Analysis
by Robert M. Oliver (Editor), James Q. Smith (Editor)

Introductory Statistics for Business and Economics, 4th Edition
by Thomas H. Wonnacott, Ronald J. Wonnacott

Computation for the Analysis of Designed Experiments
by Richard Heiberger

Structural Equations with Latent Variables
by Kenneth A. Bollen

Sensitivity Analysis in Linear Regression
by Samprit Chatterjee, Ali S. Hadi

Multivariate Statistical Simulation: A Guide to Selecting and Generating Continuous
Multivariate Distributions
by Mark E. Johnson

Nonlinear Statistical Models
by A. Ronald Gallant

The Theory and Practice of Econometrics, 2nd Edition
by George G. Judge, William E. Griffiths, R. Carter Hill, Helmut Lütkepohl, Tsoung-Chao Lee

Multivariate Analysis: Methods and Applications
by William R. Dillon, Matthew Goldstein

Introduction to Mathematical Statistics, 5th Edition
by Paul G. Hoel

Forecasting with Univariate Box - Jenkins Models: Concepts and Cases
by Alan Pankratz

Matrix Algebra Useful for Statistics
by Shayle R. Searle

Survival Models and Data Analysis
by Regina C. Elandt-Johnson, Norman L. Johnson

Sampling Techniques, 3rd Edition
by William G. Cochran

Statistical Concepts and Methods
by Gouri K. Bhattacharyya, Richard A. Johnson

Applications of Statistics to Industrial Experimentation
by Cuthbert Daniel

An Introduction to Probability Theory and Its Applications, Volume 2, 2nd Edition
by William Feller

An Introduction to Probability Theory and Its Applications, Volume 1, 3rd Edition
by William Feller

Printed and bound by CPI Group (UK) Ltd, Croydon, CR0 4YY

16/04/2025

14658371-0005